LABOUR MARKET
ECONOMICS
THIRD EDITION

Theory, Evidence, and Policy in Canada

MORLEY GUNDERSON
Professor of Economics and Director of the
Centre for Industrial Relations
University of Toronto

W. CRAIG RIDDELL
Professor of Economics
University of British Columbia

McGraw-Hill Ryerson
Toronto Montreal New York Auckland Bogotá Caracas Lisbon
London Madrid Mexico Milan New Delhi Paris San Juan
Singapore Sydney Tokyo

To our parents,
Ann and Magnus Gunderson
and
Ethel and William Riddell

LABOUR MARKET ECONOMICS
Theory, Evidence and Policy in Canada
Third Edition

4 5 6 7 8 9 0 BBM 2 1 0 9 8 7 6 5

Care has been taken to trace ownership of copyright material contained in
this text. The publisher will gladly accept any information that will enable
them to rectify any reference or credit in subsequent editions.

Sponsoring editor: Jennifer Mix
Supervising editor: Rosalyn Steiner
Copy editor: Lenore Gray
Cover design and text: Dianna Little
Cover art: Nina Berkson, Montreal-based illustrator. Represented in
Canada and the U.S.A. by Link, Toronto.

Canadian Cataloguing in Publication Data
   Gunderson, Morley, 1945-
      Labour market economics : theory, evidence and policy in Canada

   3rd ed.
   Includes bibliographical references and index.
   ISBN 0-07-551222-X

   1. Labor supply — Canada.   2. Labor economics —
   Canada.   I. Riddell, W. Craig (William Craig),
   1946-   . II. Title.

HD5728.G87 1993   331.12'0971   C92-095773-0

PRINTED IN CANADA

CONTENTS

Chapter 18   Human Capital Theory: Applications to Education and Training   450

Chapter 19   Wage Structures: By Occupation and Region   482

Chapter 20   Wage Structures: By Industry, Firm Size, and Individuals   503

Chapter 26   Wage Changes, Price Inflation, and Unemployment   681

PREFACE

In this third edition of *Labour Market Economics: Theory, Evidence and Policy in Canada*, we have tried to maintain the strengths of the earlier editions, while expanding and revising to incorporate new theoretical developments, new empirical evidence, and the constantly changing policy issues. As the subtitle implies, the emphasis is still on a *balance* between economic *theory*, empirical *evidence* and policy *relevance*. As a result, each chapter is organized around the following questions: Why are we interested in this phenomenon? What does economic theory imply about the issue? What is the empirical evidence; that is, what are the facts? What are the policy implications and trade-offs — specifically, what can we expect to happen from alternative policy scenarios?

Our purpose is to provide an integrated and modern treatment of labour market issues. All too often courses in labour market analysis have relied too heavily on description and institutional detail and too little on economic theory or econometric analysis. At the same time, application of economic theory to labour market behaviour without an understanding of the institutional setting is naïve and unwise. This book tries to bridge the gap between economic theory and labour market analysis by indicating that the field of labour economics is fertile ground for the application of economic theory and of econometric analysis to interesting and relevant policy issues.

As the title itself indicates, the emphasis of this book is on the *economic* aspects of the *market* for labour. For some persons, such an emphasis may appear misguided, given various peculiar characteristics of the labour market — a variety of factors with goals that often conflict; an abundance of sociological, legislative, and institutional constraints; and a complex price (wage) structure with moral overtones because of the human element and because wages are often called upon both to allocate labour efficiently and to curb poverty. Because of these and other peculiarities, some have argued that the labour market is fundamentally different from other markets and that, therefore, economics is largely irrelevant — or at most can play only a minor role — in analyzing labour market phenomena.

Although there are important differences between labour markets and many other markets — and they are what makes labour economics so interesting — a basic theme of this book is that these differences are ones of degree and not of kind. Rather than making economics irrelevant to the analysis of the labour market, these characteristics make economics even more relevant in understanding some of the basic underlying forces and in analyzing the impact, and the reasons for the emergence, of a variety of sociological, legislative, and institutional constraints. In essence, the complexity of the labour market makes a basic theoretical framework even more necessary in order to understand the basic underlying forces, and not miss the forest for the trees.

The book is meant to provide a consistent theoretical framework that can be applied not only to current policy issues but also to new policy issues as they change and emerge over time. Policies that are relevant today may not be relevant tomorrow, but a good theoretical foundation will always be relevant to analyzing changing issues. Although the emphasis throughout the book is on Canadian data, examples, institutions, problems and policies, these are discussed within the context of a more general theoretical framework that is applicable to the labour problems of most developed countries.

Many of the changes in this third edition reflect recent theoretical and empirical developments in the subject, as well as changing issues of policy importance and practical relevance. Examples of where significant changes were made in this third edition include:

- A new introductory chapter providing an overview, a discussion of the similarities and differences between the labour market and other markets, and a motivation with respect to the variety of current issues of policy importance and practical relevance;

- International issues pertaining to global competition, free trade as well as the growth of trade blocs such as the European Economic Community, with particular emphasis on Canada's changing competitive position with respect to the various dimensions of unit labour cost — wages, fringe benefits, productivity and the exchange rate;

- Human resource management issues pertaining to such factors as executive compensation, optimal compensation arrangements, employee cooperation, pension plans, and changing workplace practices, all of which have important implications for the work force as well as the performance of the firm;

- Changing demand side conditions emanating from such interrelated factors as international competition, mergers and acquisitions, privatization, deregulation and industrial restructuring from manufacturing to services and the "information economy";

- The aging work force, especially issues pertaining to skill shortages, promotion opportunities, and pensions and retirement;

- Equity issues in the labour market, including practical issues of design and implementation associated with pay equity (equal pay for work of equal value) as well as employment equity for women, visible minorities, disabled persons and Aboriginal peoples;

- The assimilation of immigrants into the Canadian labour market, and the impact of immigration on wages and employment;

- Wage polarization, growing wage inequality, and increasing economic returns to higher education;

- The impact of workers' compensation and disability payments;

- "Efficiency wages" and their implications for human resource management and industrial and trade policy;

- The changing pattern of unionization especially between Canada and the United States;

- Changing government policies with respect to reallocating from income maintenance through unemployment insurance to more active labour market policies pertaining to training;

- A new appendix on indifference curve analysis for students who have not previously been exposed to this concept, which is used extensively in the book;

- A new chapter providing an integrated treatment of the role of imperfect information, risk and uncertainty, and incentives on labour market behaviour;

- A new appendix on education as a signal of productivity, and discussion of new evidence on human capital versus signaling explanations of the relationship between education and earnings;

- Treatment of recent developments relating to unemployment, including the rise in unemployment during the 1980s, differences among Canada, the United States, and Europe in unemployment experience, the increase in long-term unemployment, and new research on "hysteresis" and persistence in unemployment;

- A new section on the implications for unemployment of sectoral adjustment and insider-outsider theories of wage determination;

- Exhibits throughout, illustrating interesting and important applications and policy issues;

- A streamlining and merging of some chapters, in particular those on wage structures as well as household production and fertility.

By trying to integrate labour market analysis with the mainstream of economic theory, the text reflects the prevailing neoclassical paradigm with its emphasis on microeconomic foundations involving optimization subject to constraints. However, alternative perspectives and paradigms are discussed — some would say not enough, given their potential, and others would say too much, given their actual impact.

Each chapter ends with a series of questions designed to recapitulate some of the material of the chapter, to integrate material across chapters, to suggest possible criticisms of the analyses, and to extend the analyses to some issues that have not been dealt with specifically. In some instances, the questions are ones that currently have not been answered satisfactorily by labour economists, and it may even be that they are unanswerable, given our current state of knowledge.

In addition, each chapter ends with a set of references and further readings, mostly from the economics and labour journals that are listed in the key at the beginning of the text. The references focus on material since 1980; references prior to that time are contained in the first and second editions. Obviously, these references will soon become dated; hence, there is no substitute for a judicious reading of the appropriate current journals to keep abreast of current developments in the field.

The text was written for students who have had an introductory course in economics and preferably — but not necessarily — a course in microeconomics. Formal mathematical analysis is sometimes utilized. To the greatest extent possible we have confined mathematical analysis to separate self-contained sections that can be omitted by those without the requisite training. The text is also suitable as a basis for graduate labour courses in business schools as well as economics departments.

Various people have aided in the preparation of this book. Our colleagues and fellow labour economists within Canada are too numerous to mention; however, we are particularly indebted to our colleagues at the University of British Columbia and the University of Toronto as well as to a number of anonymous referees who made comments and suggestions for improvements on the second edition. Outside of Canada, useful comments on the earlier editions, and material for this edition, were provided by David Bloom of Columbia University, Richard Freeman of Harvard and Mark Killingsworth of Rutgers. At McGraw-Hill Ryerson, our editor Jennifer Mix has provided useful guidance, suggestions and prodding, and Lenore Gray and Rosalyn Steiner have provided valued editorial assistance. We especially wish to thank our wives, Melanie and Rosemarie, for their support and encouragement.

KEY TO JOURNAL ABBREVIATIONS
USED IN REFERENCES AND
FURTHER READINGS

AER   American Economic Review
BPEA   Brookings Papers on Economic Activity
CJE   Canadian Journal of Economics
CPP   Canadian Public Policy
EER   European Economic Review
EI   Economic Inquiry
EJ   Economic Journal
IER   International Economic Review
ILRR   Industrial and Labour Relations Review
IR   Industrial Relations
IRRA   Industrial Relations Research Association Proceedings
JEL   Journal of Economic Literature
JEP   Journal of Economic Perspectives
JHR   Journal of Human Resources
JLR   Journal of Labor Research
JOLE   Journal of Labor Economics
JPE   Journal of Political Economy
JPubEc   Journal of Public Economics
MLR   Monthly Labor Review
QJE   Quarterly Journal of Economics
R.E. Stats.   Review of Economics and Statistics
R.E. Studies   Review of Economic Studies
RI/IR   Relations Industrielles/Industrial Relations
SEJ   Southern Economic Journal

Other journals have been included in the references, without an abbreviation for their title. These include, but are not restricted to:

Econometrica
Economica
Journal of Econometrics
Journal of Law and Economics
Rand Journal of Economics

The references cited in the text tend to focus on articles published since the mid-1980s. Earlier references can be found in the first and second editions of the text.

CHAPTER 1

Introduction to Labour Market Economics

Labour economics studies the outcomes of decisions we can all expect to make over our lifetime. As a subject of inquiry it is both practically relevant to our everyday lives and socially relevant to the broader issues of society. This makes labour economics interesting, practical, and socially relevant, as well as controversial.

DECISIONS BY INDIVIDUALS, FIRMS, AND GOVERNMENTS

Labour market decisions that affect our everyday well-being are made by each of the three main actors or participants in the labour market: individuals, firms, and governments.

For individuals, the decisions include when to enter the labour force; how much education, training, and job search to undertake; what occupation and industry to enter; how many hours to work; whether to move to a different region; when to accept a job; when to quit or look for another job; what wage to demand; whether to join a union or employee association; and when to retire. It may also involve decisions on how to allocate time between market work and household responsibilities. As we will see subsequently, it may even involve decisions we normally don't think of in the context of economic analysis, such as the decision to have children and when to have them!

Many aspects of labour market behaviour are positive experiences, such as obtaining a job, getting a wage increase or promotion, or receiving a generous pension. Other experiences are negative, such as being unemployed or permanently displaced from one's job, or experiencing discrimination or poverty.

1

Employers also have to make decisions on such aspects as how many workers to hire, what wages and fringe benefits to pay, what hours of work to require, when to lay off workers and perhaps ultimately close a plant, what to subcontract, and how to design an effective pension and retirement policy. These decisions are increasingly being made under pressures from global competition, free trade, industrial restructuring, deregulation, and privatization. As well, the decisions of employers must be made in the context of a dramatically changing work force with respect to such factors as age, gender, and ethnic diversity. The legislative environment within which employers operate also is constantly changing in such areas as human rights and anti-discrimination legislation; employment standards laws with respect to such areas as minimum wages, maternity leave, and hours of work and overtime; legislation on workers' compensation and occupational health and safety; legislation on pensions and on mandatory retirement; and labour relations laws that regulate the process of collective bargaining.

Governments (i.e., through legislators, policymakers, and the courts) also have difficult decisions to make so as to establish the environment in which employees and employers interact. In part this involves a balancing act between providing rights and protection to individuals, while not jeopardizing the competi- tiveness of employers. It also involves decisions as to what to provide publicly in such areas as training, information, unemployment insurance, workers' compensation, vocational rehabilitation, income maintenance, pensions, and even public sector jobs.

Labour economics deals with these decisions on the part of individuals, employers, and government — decisions that affect our everyday lives and that lead to consequences we can all expect to experience. While this makes labour economics interesting and relevant — and controversial — it also puts a premium on a solid framework from which to analyze these decisions and their consequences. The purpose of this book is to provide and apply such a framework.

Parts of the book are the equivalent of "eating spinach"; they provide the basic tools and theoretical framework that are crucial to understanding labour market behaviour. In many cases the immediate application of economic theory to labour market issues facilitates understanding the basic tools of economics itself; that is, the application to interesting and important issues in the labour market facilitates a deeper understanding of economic theory.

In labour economics, however, the ultimate usefulness of economic theory is in understanding labour market behaviour. At the risk of carrying the food analogy too far, "the proof of the pudding is in the eating": the usefulness of basic economic theory in the labour area must be demonstrated by its ability to help us understand labour market behaviour. This is one of the objectives of this book.

SUBJECT MATTER OF LABOUR MARKET ECONOMICS

Labour market economics involves analyzing the determinants of the various dimensions of labour supply and demand and their interaction in alternative market structures to determine wages, employment, and unemployment. Behind this simple description, however, lies a complex array of behaviours, decision-making, and dimensions that often interrelate labour economics with other disciplines and areas of study.

Labour supply, for example, involves a variety of dimensions. It includes population growth, which involves decisions pertaining to fertility and family formation, as well as to immigration and emigration, all of which are amenable to economic analysis. Labour supply also involves the dimension of labour force participation to determine what portion of the population will participate in labour market activities as opposed to other activities such as household work, education, retirement, or pure leisure. For those who participate in the labour market, there is the labour supply dimension of hours of work, including trends and cyclical patterns, as well as such phenomena as overtime, moonlighting, part-time work, worksharing, flexible work-time arrangements, and compressed work weeks. This ties labour economics to such areas as demography and personnel and human resource planning.

In addition to these *quantity* dimensions of labour supply, there are also *quality* dimensions that are amenable to economic analysis. Such quality dimensions include education, training, and health. These dimensions, along with labour mobility, are often analyzed as human capital investment decisions, emphasizing that they involve incurring costs today in exchange for benefits in the future. Quality dimensions of labour supply also include work effort and intensity — dimensions that are analyzed in the context of efficiency wage theory and optimal compensation systems. This ties labour economics to issues of personnel and human resource management as well as to the key policy issue of productivity, often involving education and training.

Labour supply analysis also involves determining the work incentive effects of income maintenance and tax-transfer schemes. Such schemes include demogrants (e.g., Old Age Security Pension), negative income taxes, wage subsidies, income taxes, unemployment insurance, welfare, disability payments, workers' compensation, and private and public pension plans. This ties labour economics to the interesting and controversial areas of poverty and income distribution as well as tax and social welfare policy.

The labour *demand* side of the picture focuses on how firms vary their demand for labour in response to changes in the wage rate and other elements of labour cost including fringe benefits and legislatively imposed costs. The impact of quasi-fixed costs that may be involved in hiring, in training, and even in terminating workers are also important. Since the demand for labour is a derived demand — derived from the demand for the firm's output — this side of the picture is influenced by such issues as free trade, global competition, industrial restructuring, privatization, public sector retrenchment, mergers and acquisitions, and technological change. It is increasingly important to analyze the demand side in the context of changes that are occurring in the international global environment.

The various dimensions of labour supply and demand are interesting and informative not only in their own right but also because their interaction determines other key labour market outcomes — wages, employment, and unemployment. These outcomes are influenced by the interaction of labour supply and demand in alternative market structures, including the degree of competition in the product market as well as the labour market. They are also influenced by unions and collective bargaining, as well as by legislative interventions including minimum wages and equal pay and equal opportunity laws. This highlights the interrelationship of labour economics with the areas of industrial relations and labour law.

The various wage structures that emerge include wage differentials by occupation (e.g., engineer versus secretary), by industry (e.g., public versus private sector), by region (e.g., Newfoundland versus Ontario), and by personal characteristics (e.g., males versus females). Economic analysis outlines the determinants of these wage structures, how and why they change over time, and how they are influenced by such factors as demand changes, nonwage aspects of employment, and noncompetitive factors including unions, legislation, occupational licensing, and labour market imperfections. For some wage structures, like male-female wage differentials, the economic analysis of discrimination is informative in indicating why wage differentials arise and persist, and how they will be affected by legislation and other policy initiatives.

Wage differentials between union and nonunion workers can be explained by basic economic analysis incorporating bargaining theory. In order to fully understand how unions affect wages, it is also necessary to understand why unions arise in the first place and how they affect the nonwage aspects of employment. Increasingly, labour economics has been applied to understanding why certain institutional arrangements exists in the labour market, including unions, seniority-based wage increases, and personnel practices such as mandatory retirement.

There is increasing recognition that wages can have strong effects on various incentives: to acquire education and training or to move, to stay with a job or quit or retire early, and even to work hard and more productively. These effects are analyzed in various areas of labour market economics: human capital theory, optimal compensation systems, efficiency wages, and private pension plan analysis. This highlights the interrelationship of labour economics to the areas of industrial relations and personnel and human resource management.

The area of unemployment is analyzed at both the microeconomic and macroeconomic level. At the micro level, the emphasis is on theories of job search, implicit contracts and efficiency wages, as well as the impact of unemployment insurance. At the macro level, the relationship between wage changes, price inflation, productivity, and unemployment is emphasized.

CURRENT POLICY ISSUES

The previous discussion of the subject matter of labour market economics highlighted a number of areas of current policy concern to both public policymakers (e.g., legislators, government policymakers, and the courts) as well private policymakers (e.g., employers through their personnel and human resource management decisions and unions through their collective bargaining actions). Labour market economics should be able to shed light on many of these issues, illustrating its social relevance as well as practical relevance in private decision-making.

In the labour supply area, such current policy questions abound. What is the work incentive effect of income maintenance and tax-transfer programs such as income taxes, unemployment insurance, welfare, wage subsidies, disability payments, and workers' compensation? How can such programs be designed to achieve their broader social objectives while minimizing any adverse incentive effects that could jeopardize their existence? Why has there been a dramatic increase in the labour force participation of women, especially married women? What accounts for the trend towards early retirement? Can employers use features of their private pension plans to encourage early retirement? What accounts for the long-run decline in the average work week that has occurred? Why do some workers require an overtime premium to work longer hours, while others are willing to moonlight at a second job at a rate that is less than the rate in their primary job? Why has there been an increase in alternative work-time arrangements, including part-time work, flex-time, and compressed work weeks? Under what conditions would workers voluntarily accept worksharing arrangements, for example, a four-day work week in return for no layoffs? What is the impact of increased commuting time and daycare costs on labour force participation and hours of work decisions? What factors influence how we allocate time over our life cycle and across different activities? What factors influence our decision to have children, as well as when to have them?

In the labour demand area, many policy questions have been brought to the fore because of new competitive pressures, many emanating from global competition. What is the impact of wages and other elements of labour cost on the competitiveness of Canadian firms? How is this competitiveness affected by legislated policies with respect to such factors as wages, employment, hours of work, termination, anti-discrimination, human rights, health and safety, workers' compensation, and labour relations legislation dealing with the formation of unions and the conduct of collective bargaining? What is the impact on the labour market of free trade, global competition, industrial restructuring, mergers and acquisitions, privatization, technological change, and the shift from manufacturing to services and high-technology industries? How will employers change their employment decisions in response to legislation on minimum wages, equal pay for work of equal value, severance pay, and advance notice of termination? Can new jobs be created by policies to reduce the use of overtime through higher overtime premiums or regulations on maximum hours of work?

With respect to wage determination and wage structures many policy issues also abound. What has been the influence of increased competitive pressures on wages and wage structures as well as the returns to education? What accounts for the increased wage polarization that has occurred? Do public sector workers receive higher pay than comparable private sector workers? How important are fringe benefits in the total compensation picture, and why do fringe benefits prevail in the first place? How quickly do immigrants assimilate into the Canadian labour market? Are workers in risky occupations compensated for that risk through compensating wage premiums? Do workers "pay" for pension benefits by receiving lower wages than they would receive in the absence of such pensions? Why do male-female wage differentials prevail and have they been changing over time, especially in response to legislative initiatives? What exactly is meant by the new policies of pay equity (equal pay for work of equal value) and employment equity, and how do they operate? How can employers design optimal compensation systems to induce productivity from their work force, including executives? Why does mandatory retirement exist, and what will happen to wage structures if it is banned? Why do wages seem to rise with seniority even if there is no productivity increase with seniority? Why do some employers seem to pay wages that are excessively high for their recruitment and retention needs?

With respect to unemployment, many old questions prevail and new ones are added. Why has unemployment drifted upwards, even in times of seeming prosperity and inflation? Is unemployment different now than it was in earlier periods? Is permanent job loss more of a problem? What has happened to the unemployment of youths as the baby-boomers age through the labour force? Does our aggregate unemployment rate reflect a large number of people experiencing short bouts of unemployment, or a small number of people experiencing long bouts of unemployment? Why don't workers and unions take wage cuts to ameliorate unemployment? Why has the relationship between unemployment and inflation worsened? What impact has unemployment insurance had on our unemployment rate?

These are merely illustrative of the myriad of current policy questions that are relevant for public policy as well as for private personnel and human resource practices and for collective bargaining. They indicate the social and practical relevance of labour economics, as well as the controversy that surrounds the field. The tools of labour market economics developed in this text will provide answers to many of these questions and shed light on others.

SIMILARITIES AND DIFFERENCES BETWEEN THE LABOUR MARKET AND OTHER MARKETS

A key issue in labour market economics is whether the labour market is so different from other markets that the basic tools of economics do not apply. The following discussion illustrates some of these peculiarities and their implications for labour markets.

Various Actors with a Variety of Goals

As discussed previously, the labour market has three main actors or stakeholders — labour, management, and government — each with different subgroups and objectives or agendas. For example, within labour there is union labour and nonunion labour, and within organized labour there are different types of unions (e.g., craft, industrial, professional associations) as well as possible differences between the leadership and the rank-and-file membership. Within management, the goals of stockholders may differ from the goals of chief executive officers which in turn may differ from the goals of middle management. With increased emphasis on employee participation, the distinction between middle management and workers may be blurred. Within the catch-all of government there are federal, provincial, and local levels. As well, there are often differences in the units who make the laws (e.g., legislators and regulators), who enforce the laws (e.g., courts, administrative agencies, and tribunals), and who design the laws and operate particular government programs (e.g., policy units and bureaucrats). As well, the distinction between making, designing, and administering the law and policies often becomes blurred in practice. For example, the jurisprudence and case law of the courts and administrative tribunals can shape the interpretation of the law such that it can be more important than establishing new laws.

Sociological, Institutional, and Legislative Constraints

More so than most markets, the labour market is riddled with sociological, institutional, legislative, and other constraints. In the sociological area, for example, family and community ties and roles can affect labour mobility, the role of women in the labour market and the household, as well as their career and educational choices. Social norms may also influence what is considered appropriate wage structures as well as who should do certain jobs. In the labour market area it is important to remember the maxim that "economics is about how we make choices, and sociology is about why we have no choices to make"!

Institutions are also important in labour markets. This is most obvious with respect to unions, especially in Canada where almost 40 percent of the work force is unionized. Large corporations may dominate a local labour market and hence set wages, and multinational corporations may follow the wage and workplace practices of their parent company as well as use the threat of plant relocation decisions to influence bargaining. Wage patterns may become institutionalized through pattern bargaining, albeit these may be breaking down under current competitive conditions. Earlier perspectives on labour economics in fact were characterized as "institutional," emphasizing the importance of institutions and institutional arrangements in understanding labour market behaviour.

Legislative constraints are also crucial in labour markets. Employment standards laws set minimal or "community" standards with respect to such things as minimum wages, hours of work and overtime, statutory holidays, parental leave, and severance pay and advance notice requirements. Human rights and antidiscrimination legislation are important, as are laws on health and safety and workers' compensation. Labour relations laws also establish the framework for the formation of unions and the conduct of collective bargaining. Separate statutes often exist for workers in the public sector, sometimes circumscribing the right to strike and providing alternative dispute resolution procedures including binding wage arbitration. The labour market may also be affected by other laws and regulations, such as those that affect immigration, free trade, unemployment insurance, pensions, training and education, and even environmental protection.

Market Imperfections

The labour market, more so than most other markets, is subject to market imperfections and other constraints. For example, imperfect information may make it difficult to decide which type of education to undertake or whether to make a geographic or occupational move. Asymmetric information (e.g., the employer has information not available to workers) may make it difficult to demand compensating wage premiums for workplace hazards, or to agree to wage concessions in return for employment security. Transactions costs may make it difficult to finance human capital investments in areas like education, training, or mobility. Uncertainty and risk may make it difficult to make occupational choice decisions that are affected by uncertain future demand and supply conditions.

Complex Price, Serving a Variety of Functions

The essential price that gets determined in the labour market — the price of labour services or the wage rate — is a complex price that is called upon to serve a variety of functions. It is a price that reflects a variety of factors: the returns to investments in education, training, or mobility; compensation for risk or undesirable working conditions; a "rent" or surplus for being a member of a union or professional association or perhaps being employed in a particular industry; the result of discrimination; and the result of supply and demand decisions on the part of other participants in the labour market. As well, the wage rate is only one component of total compensation, with fringe benefits making up an increasingly larger share.

The wage rate is also called upon to perform a variety of functions. It allocates labour to its efficient uses, that is, to particular occupations, industries, and regions. It encourages optimal investments in human capital like education, training, labour market information, job search, and mobility. It is used to enhance performance, work effort, and productivity. It is also called upon to provide an adequate standard of living and family income, and to alleviate poverty and the manifestations of discrimination. In the labour market, issues of efficiency and equity and fairness are intricately related.

In part for these reasons, there are often strong moral overtones to labour market issues and the resultant wage that emerges. This is compounded by the fact that even though the wage is the price of labour services, the fact remains that the labour services and the labourer are inseparable. Dignity, perceptions of self-worth, and social attitudes are tied to the wage. Hence, phrases like a decent wage, a living wage, the social wage, a just wage, and equal pay for work of equal value are common expressions in labour economics.

Clearly, the labour market has a number of characteristics that make it different from many other markets. This raises the issue of whether the basic tools of economics apply in a market with such a complex array of participants and goals, a variety of sociological, legal and institutional constraints, and market imperfections, and a complex price that is called upon to serve a variety of functions.

Our perception is that these are differences in degree and not in kind; other markets have some of these peculiarities as well as others of their own. Moreover, these differences make the basic tools of economics more relevant, not less relevant, to understanding labour market behaviour and the impact of these peculiarities of the labour market. Economic analysis deals with decision-making and trade-offs under risk, uncertainty, transactions costs, information problems, and market imperfections. It deals with the interplay among various market participants often with conflicting objectives. It deals with the impact of legal, institutional, and even sociological constraints — after all, the methodology of economics is optimizing subject to constraints.

Economics has even proven useful in understanding why many of these constraints and institutional arrangements arise in the first place. As illustrated later in the text, economic efficiency rationales can be given to explain the existence of a variety of phenomena that may appear to be "irrational" or inefficient: pure seniority-based wage increases; the simultaneous existence of moonlighting and overtime; wages that appear "excessive" relative to what is necessary to recruit and retain workers; strikes that appear "irrational"; an unwillingness to accept wage concessions in spite of high unemployment; the provision of job security in the public sector; payment through fringe benefits rather than cash compensation; and the existence of personnel practices like mandatory retirement. In some cases, these phenomena may be the result of mistakes or of bargaining power; but in other cases, they may be the efficient institutional arrangement to deal with other problems in labour markets. Economics can shed light on why these arrangements arise in the first place, as well as indicating their impact on the labour market. This often requires an interdisciplinary understanding that blends labour economics with other areas such as industrial relations, personnel and human resource management, labour law, sociology, psychology, and history.

ALTERNATIVE PERSPECTIVES

The importance of the different disciplines, as well as the peculiarity of labour markets and the importance of issues of equity and fairness, have given rise to a variety of perspectives on how best to analyze labour market behavior. As discussed in more detail in the text, these perspectives include neoclassical economics, segmented labour market views, and more radical perspectives.

The neoclassical view emphasizes the application of conventional, modern economic tools to the analysis of labour markets to understand both the impact of many of the peculiar features of labour markets as well as why these features arise in the first place. The methodology involves the optimizing behaviour of the parties subject to constraints. The emphasis is on delineating the trade-offs involved, and on how markets function and malfunction.

The segmented labour market perspective emphasizes that the labour market is segmented into noncompeting groups with competition being precluded because of forces of custom, tradition, and discrimination. The role of institutions (especially unionism) is emphasized, as is the importance of administered wage determination rules within the internal labour markets of firms. The radical perspective adds an emphasis on the importance of class conflict, political power, and social relations.

SUMMARY OBSERVATIONS AND FORMAT OF BOOK

The previous discussion highlighted how labour economics is both practically relevant to our everyday lives and socially relevant to broader social issues. It also indicated how labour economics interrelates to a variety of other disciplines. The discussion also emphasized that the peculiarities of labour markets make economics more relevant, not less relevant, to understanding not only the impact of these peculiarities and institutional features, but also why they arise in the first place. These issues make the study of labour markets relevant, controversial, and interesting.

In covering these issues, the text is divided into six parts. Part 1 deals with the various dimensions of labour supply. The underlying theory of labour supply is first outlined, emphasizing the income-leisure choice framework modified to incorporate household production functions. This is then applied to analyze the work incentive effects of income maintenance programs as well as the determinants of the labour force participation decision and the hours of work decision. The retirement decision is analyzed, emphasizing the role of private and public pension plans, and special topics are covered including the impact of commute time and daycare costs. The economic theory of fertility and family formation is covered as part of household production theory.

Part 2 deals with labour demand, emphasizing how employers change their demand for labour in response to changes in the wage rate and other determinants of demand. The conditions under which the adverse employment effect of a wage increase will be large or small are analyzed and applied to the potential adverse employment effect of minimum wage legislation. Also analyzed are the effects on labour demand and employment of recent trends such as global competition, free trade, industrial restructuring, privatization, and deregulation. Canada's labour cost position in the competitive international environment is outlined. The impact on labour demand and employment of quasi-fixed costs such as hiring and training costs and the cost of certain fringe benefits are also analyzed, especially in the context of the willingness of employers to engage in worksharing or in restrictions on overtime to create new jobs.

Part 3 deals with the interaction of labour demand and labour supply in alternative market structures to determine wages and employment. The competitive norm (firms that are competitive in both the product market and the labour market) is contrasted with the situation where the firm has market power in the product market (e.g., monopolist) and in the labour market (e.g., monopsonist). Alternative perspectives of the labour market are also portrayed, emphasizing the institutionalist view, the segmented labour market perspective, and the radical view.

Unions are analyzed in Part 4. Union growth and the different incidence of unions across industries and regions is discussed, as is a comparison of unionization across Canada and the United States. Union preferences are analyzed and used in the context of contract theory and bargaining theory. The impact of unions on wages, employment, nonwage aspects of employment, and productivity is also analyzed.

Part 5 deals with wages and wage structures, developing the theory of compensating wage differentials and human capital theory that deals with the costs and benefits of education, training, mobility, labour market information, and job search. Various wage structures are analyzed, by occupation, industry, and region. Issues such as the assimilation of immigrants into the labour market are dealt with. Particular attention is paid to public-private sector wage differentials as well as male-female wage differentials and the impact of policy initiatives including pay and employment equity. Optimal compensation systems are analyzed and applied to explain, for example, the existence of institutional rules and personnel practices like mandatory retirement.

Unemployment is dealt with in Part 6. The meaning and measurement of unemployment are analyzed, as are its causes and consequences and the macroeconomic relationship between aggregate wage changes, inflation, and unemployment.

Throughout the text, international comparisons are made where feasible. As indicated in Table 1.1, on an international basis Canada is second only to the United States in terms of real Gross Domestic Product per capita. Furthermore,

Table 1.1
REAL GROSS DOMESTIC PRODUCT PER CAPITA[1], VARIOUS COUNTRIES, 1960, 1970, 1980, 1990
(as % of U.S. GDP)

COUNTRY	1960	1970	1980	1990
United States	100.0	100.0	100.0	100.0
Canada	71.0	76.7	89.9	93.2
Norway	56.0	60.5	77.1	80.7
Japan	29.4	56.5	65.9	80.1
Sweden	66.1	75.4	74.9	74.8
Germany	60.7	66.6	72.5	74.0
France	53.7	64.7	71.3	73.2
Denmark	61.3	67.9	68.9	71.8
Belgium	50.1	59.9	68.2	69.8
United Kingdom	65.7	63.9	65.0	69.3
Italy	44.7	56.7	66.0	68.6
Netherlands	60.1	68.1	70.7	68.3
Austria	46.4	54.0	64.6	67.1
Korea	9.6	13.0	20.9	37.9

1. Adjusted for differences in the exchange rate on the basis of purchasing power parity, i.e., what it cost in terms of U.S. dollars to buy the same bundle of goods and services. The countries are ranked, in descending order from high to low, in terms of 1990 GDP per capita.

Source: Computed from data provided by the U.S. Department of Labor, Bureau of Labor Statistics, Office of Productivity and Technology, January 1992.

our position has improved in that regard since 1960. However, other countries have improved their relative position even more. Similar international comparison will be made throughout the text, especially with respect to our comparative labour market picture.

The emphasis in the text is on a balance of theory, empirical evidence, and application to important issues of social policy and private employment practices. This is developed in a context that emphasizes the increasingly competitive international environment in which the Canadian labour market is situated. The intent is to show the social and practical relevance of labour economics to a wide range of ever-changing and interesting important issues.

REFERENCES AND FURTHER READINGS

Dunlop, J. Policy decisions and research in economics and industrial relations. *ILLR* 30 (April 1977) 275–282.

Economic Council of Canada. *People and Jobs: A Study of the Canadian Labour Market*. Ottawa: Economic Council of Canada, 1976.

Economic Council of Canada. *In Short Supply: Jobs and Skills in the 1980s*. Ottawa: Economic Council of Canada, 1982.

Freeman, R. *Labor Markets in Action*. Cambridge: Harvard University Press, 1989.

Gunderson, M. Labour relations in Canada. *The Canadian Economy*. Toronto: Gage, 1977.

Gunderson, M. Labour economics and industrial relations. *The State of the Art in Industrial Relations*, G. Hebert, H. Jain and N. Meltz (eds). Queen's University Industrial Relations Centre & University of Toronto Centre for Industrial Relations, 1988.

Kerr, C. Labor markets: their character and consequences. *AER* 40 (May 1950) 278–291.

Lazear, E. Labor economics and the psychology of organizations. *JEP* 5 (Spring 1991) 89–110.

Mitchell, D. *Human Resource Management: An Economic Approach*. Boston: Wadsworth, 1989.

Riddell, W.C. (ed.). *Work and Pay: The Canadian Labour Market*. Toronto: University of Toronto Press, 1985.

Riddell, W.C. (ed.). *Adapting to Change: Labour Market Adjustment in Canada*. Toronto: University of Toronto Press, 1986.

Sapsford, D. And Z. Tzannatos (eds). *Current Issues in Labour Economics*. New York: St. Martin's Press, 1989.

Solow, R. *The Labor Market as a Social Institution*. Oxford: Blackwell, 1990.

Wise, L. *Labor Market Policies and Employment Patterns in the United States*. London: Westview Press, 1989.

PART 1

Labour Supply

Part 1 deals with the various dimensions of labour supply, especially the individual's labour force participation and hours of work decisions. Chapter 2 develops the basic income-leisure choice model which is the theoretical framework that underlies the individual's labour supply decision. This is expanded on in Chapter 3 to incorporate the household production function perspective, which is also applied to a decision not usually thought of in the economic realm: the decision to have children.

The income-leisure choice framework is then applied in Chapter 4 to analyze the work incentive effects of various income maintenance schemes. These include demogrants (e.g., Old Age Security Pensions), welfare payments, negative income tax proposals, wage subsidies to assist the working poor, unemployment insurance including unemployment insurance-assisted worksharing, and disability payments including workers' compensation.

In Chapter 5 the income-leisure choice framework is applied to the decision to participate in the labour market, with particular emphasis on the labour force participation of women. That chapter also deals with how the labour force and its associated concepts of employment and unemployment are defined. Particular attention is given to the hidden unemployment that may exist because "discouraged workers" may be categorized as outside of the labour force if they have given up looking for work.

Chapter 6 analyzes the labour force participation decisions of a group that is growing in importance: older workers and their retirement decisions. The determinants of these decisions are analyzed with particular attention paid to institutional rules, such as mandatory retirement, and to the features of both public and private pension schemes.

For those who participate in the labour force, their hours-of-work decision is analyzed in Chapter 7. The historical decline in working hours is documented and analyzed. The income-leisure choice framework is then used to explain such phenomena as moonlighting and overtime, and it is used to highlight the benefits from alternative work-time arrangements such as flex-time and compressed work weeks.

Chapter 8 is a more advanced level treatment of some of the newer and sometimes more difficult topics of labour supply. As such, it can be skipped or selectively read by those focusing on the more basic treatment. Specific topics include the impact of fixed costs, including commute time; the effect of daycare subsidies; problems of sample selection bias; nonlinearities in the wage line created by such factors as taxes and income maintenance programs; and the allocation of work-time over the life cycle.

The focus of Part 1 is on the various *quantity* dimensions of labour supply: population size (as affected by birth rate), the portion of the population that participates in the labour force, and their hours of work. Other quantity dimensions are discussed later in the book (e.g., how population size is affected by immigration).

Labour supply also has a *quality* dimension. The quality dimension pertaining to investment in human capital (e.g., education, training, mobility, health) is analyzed later, in Chapter 18 on Human Capital Theory. The quality dimension pertaining to such factors as motivation, morale, effort, and commitment are more appropriately dealt with in texts on human resource management, personnel, and organization behaviour. Nevertheless, some of these areas are also touched on in subsequent chapters, especially in Chapter 23 on Optimal Compensation Systems.

CHAPTER 2

Income-Leisure Choice Theory

Quantity dimensions of labour supply involve labour force participation and hours of work. Labour force participation involves the decisions of when to first enter the labour force, whether to remain in the labour force or interrupt one's working career, and when to retire. The quantity of labour supplied will also depend on the number of hours individuals in the labour force would prefer to work.

The basic theoretical framework underlying all of these dimensions of labour supply is the income- or work-leisure choice model. This chapter develops the work-leisure choice model and uses it to examine the effects of a wage change on preferred hours of work, and thus to derive the individual's supply curve of labour. A major extension of the basic model is discussed in the following chapter. In subsequent chapters the model is applied to analyze the work incentive effects of alternative income maintenance schemes, the labour force participation and retirement decisions, the various dimensions of hours of work, and the life cycle allocation of time in a multi-period context that distinguishes between short-run and long-run changes.

BASIC INCOME-LEISURE MODEL

The income-leisure choice theory is simply an application of standard micro-economic theory (indifference curve analysis) to the household's choice between income and nonlabour market activities including household work and education, as well as pure leisure activities. The phrase "leisure" is somewhat of a misnomer since it includes activities that obviously are not leisure activities. The term is used here, however, because it is the phraseology used by Robbins (1930) in his original work on the subject and it has been retained in the literature as a summary term for "non-labour market activities."

The individual is indifferent (has the same welfare) between the various combinations of income and leisure as given by the indifference curve U_0 in

Figure 2.1(a).[1] The slope of the indifference curve exhibits a diminishing marginal rate of substitution between income and leisure. For example, at point A, the individual has an abundance of income and hence is willing to give up considerable income to obtain more leisure; hence the steeply sloped indifference curve at point A. At point B, the individual has an abundance of leisure and hence is willing to give up considerable leisure (i.e., work more) to obtain more labour income; hence, the relatively flat indifference curve at point B. At intermediate points such as C, income and leisure are more substitutable because the individual is not saturated with either. Higher indifference curves such as U_1 represent higher levels of utility or welfare since they involve more of both income and leisure.

The individual will try to reach the highest indifference curve possible, constrained by the potential income constraint as given in Figure 2.1(b). This is a "potential" income constraint because it indicates varying amounts of income that can be obtained by giving up leisure and working at the market wage rate. The actual amount of income will depend on the chosen amount of work in addition to the amount of income received from other sources, referred to as "nonlabour income" and shown as the amount Y_N in Figure 2.1(b). Alternative phrases for the potential income constraint include wage, budget, wealth, income, and full-income constraint.

The slope of the potential income constraint line depends on the individual's labour market wage rate. Persons with a high market wage rate, such as W_1 where $W_1 > W_0$, will be able to earn more income by giving up leisure and working more; hence the slope of W_1 is steeper than the slope of W_0. Note that work is measured from right to left; for analytical purposes it is regarded as the residual between the maximum amount of leisure available and the amount of leisure chosen.

Perhaps the simplest way to understand how wage changes (as well as other factors such as taxes and transfers which will be examined later) affect the potential income constraint is to first mark the end point on the leisure axis. This end point is the maximum amount of leisure available; depending on the units in which leisure is measured, it could be 24 hours per day, 7 days per week, 52 weeks per year, or some combination such as 8736 hours per year. The potential income constraint is then derived by plotting the income people receive as they give up leisure and work additional units; that is, as they move leftward from the maximum leisure end point. For a constant wage rate, this income constraint would be a straight line, since a unit of work would yield the same additional income throughout the range of work time. The end point on the income axis

---

1. Indifference curves and budget constraints are used extensively in developing the theory of labour supply. Those unfamiliar with indifference curve analysis should review the appendix to this chapter before proceeding further. The derivation of the individual's budget constraint is described in the main body of the chapter.

Figure 2.1
INCOME-LEISURE INDIFFERENCE CURVES, WAGE CONSTRAINT,
AND EQUILIBRIUM

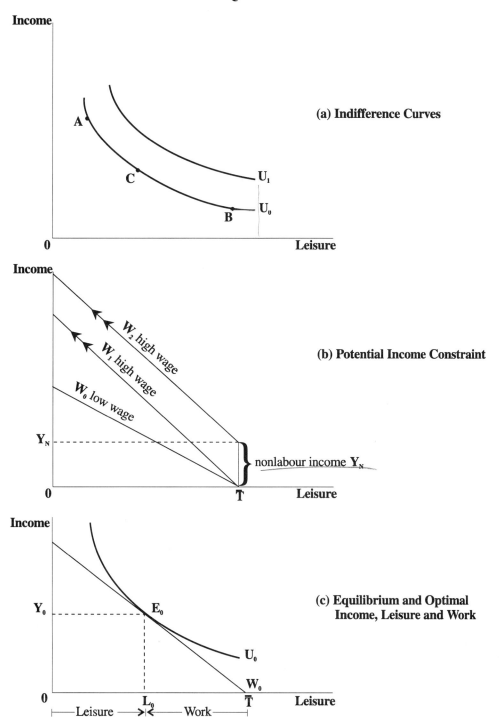

would be the maximum income the individual could attain by working all of the time; that is, by having zero leisure. For individuals with income from sources other than market work, the budget constraint begins vertically above the point $\overline{T}$ at which a maximum amount of leisure is attained with the vertical distance (OY_N in Figure 2.1(b)) being the amount of nonlabour income. This gives rise to a higher budget constraint, W_2, which involves the same high wage W_1 (hence W_1 and W_2 are parallel), albeit W_2 has the nonlabour income Y_N.

By putting the individual's potential income constraint and indifference curve together (Figure 2.1(c)), we can obtain the individual's optimal amount of income and leisure (and hence we can obtain the optimal work or labour supply). The utility maximizing individual will reach the highest indifference curve possible, constrained by the labour market earnings as given by the potential income constraint W_0. The highest feasible indifference curve in Figure 2.1(c) involves the point of tangency E_0 with optimal income Y_0, leisure L_0, and labour supply or work $\overline{T} - L_0$. One can easily verify that E_0 is the utility-maximizing equilibrium by seeing what would happen if the individual were at any point other than E_0.

INCOME AND SUBSTITUTION EFFECTS OF A WAGE CHANGE

What will happen to the equilibrium amount of work effort if wages are increased? On the one hand, the individual may work more because the returns for work are greater; that is, the opportunity cost of leisure or the income forgone by not working is higher and hence the person may substitute away from leisure. This is termed the substitution effect and, in the case of a wage increase, it leads to increased work. On the other hand, the higher wage rate means that for each quantity of work the person now has more income from which to buy more of all goods including leisure. Alternatively stated, with the higher wage rate, individuals can reach their target level of income sooner and hence they can afford not to have to work as much. This is termed the income effect, and in the case of a wage increase it leads to reduced work, assuming that leisure is a normal good. In the case of a wage change, therefore, the income and substitution effects work in the opposite direction; hence, it is ultimately an empirical proposition as to whether a wage increase would increase or decrease the supply of work effort.[2] (See Exhibit 2.1.)

---

2. The fact that when leisure is a normal good the income and substitution effects of a price (wage) change act in *opposing* directions implies that consumer behaviour with respect to leisure time is fundamentally different from that with respect to other goods for which the income and substitution effects work in the *same* direction, when the good is normal. This difference arises because an increase in the price of leisure makes leisure time relatively more expensive but *raises* real income while an increase in the price of another commodity makes that good relatively more expensive and *lowers* real income.

EXHIBIT 2.1   IMPORTANCE OF PAYING ATTENTION TO BOTH INCOME AND SUBSTITUTION EFFECTS

The capital of Brazil, Brasilia, was constructed as a planned new city. It was built in the undeveloped heartland of Brazil in the Amazon jungle area. This movement of the capital from the popular coastal area was done, in part, to encourage development in that otherwise isolated area.

Because of that isolation, it was difficult to recruit civil servants to work in the new capital. Wages were raised in order to get them to voluntarily move to the new capital. This worked; in essence, it induced a substitution effect whereby the higher wages elicited a voluntary increase in labour supply.

Unfortunately, the higher wages also induced an income effect that worked at cross purposes with the objective of encouraging the civil service to permanently move to the new capital and to become part of a new integrated community. That is, with their new higher income, many of the civil servants could afford to maintain another residence in Rio de Janiero on the coast. They could also afford to fly regularly out of Brasilia, leaving it to spend the greater wealth!

Over time, a more permanent, committed community developed. Nevertheless, it highlights the importance of paying attention to both the income and substitution effects of wage changes.

The income and substitution effects of a wage change are illustrated in Figure 2.2. After the wage increase from w_0 to w_1, the new equilibrium is E_2. Like all price changes, this increase in the price of leisure can be decomposed into an income effect and a substitution effect. The income effect can be illustrated by drawing a line parallel to the potential income constraint w_0 but tangent to the new indifference curve U_1. Since the lines are parallel, the relative price of leisure is the same at E_0 and E_1. Consequently the difference between E_0 and E_1 is purely a matter of having more potential income from which the individual can buy more of all goods including leisure. Hence if leisure is a normal good E_1 lies above and to the right of E_0 and consequently involves less work effort and more leisure. The difference between E_1 and E_2 is now a pure (or income-compensated) substitution effect, since household utility or satisfaction is held constant (as evidenced by E_1 and E_2 being on the same indifference curve U_1) and only the relative price of leisure is different. The relative price of leisure is higher on the steeper wage constraint w_1 as evidenced by the fact that giving up a unit of leisure (i.e., working more) yields more income. Consequently, the individual will substitute away from the more expensive leisure and work more: the pure (compensated) substitution effect of a wage increase will unambiguously increase work effort. In Figure 2.2, the substitution effect dominates the income effect; hence, in this illustration the wage increase resulted in a net increase in the amount of work. However, the income effect can dominate the substitution effect, in which case a wage increase will result in a decrease in the amount of work supplied.

DERIVING THE INDIVIDUAL SUPPLY CURVE OF LABOUR

As the early writings of Robbins (1930) indicate, the income-leisure choice framework can be used to derive the individual's labour supply schedule, which indicates the amount of labour that will be offered at various wage rates. As Figure 2.2 indicates, we already have two points on that supply schedule: at wage rate w_0 the amount of labour supply (work) is $\overline{T} - L_0$, and at wage rate w_1 the amount of labour supply is $\overline{T} - L_2$. In this particular case, labour supply increased as wages increased because the substitution effect outweighed the income effect. By varying the wage rate, we can obtain a schedule of corresponding equilibrium amounts of work and hence trace out the individual's labour supply curve.

Figure 2.2
INCOME AND SUBSTITUTION EFFECTS OF A WAGE INCREASE

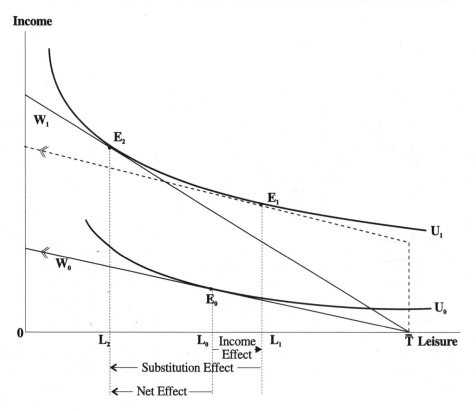

Derivation of an individual's labour supply schedule is further illustrated by the specific example of Figure 2.3 where leisure and work are measured in units of hours per day and wages in units of wages per hour. Income is the hourly wage multiplied by the hours of work. Figure 2.3(a) illustrates the equilibrium hours of work (measured from right to left) of 8 hours per day for a wage rate of $6/hr. The corresponding daily income is $48 per day. This gives us one point on the individual's labour supply schedule of Figure 2.3(c): at $6/hr the labour supply would be 8 hours. In Figure 2.3(b), wages are then raised to $10/hr and the individual is observed to increase work effort to 12 hours per day. This yields a second point on the labour supply schedule of Figure 2.3(c). In this particular case, the substitution effect of the wage increase outweighs the income effect so that hours of work have increased. In Figure 2.3(b), wages are next raised to $14/hr and the individual is observed to decrease work effort to 10 hours per day. This yields a third point on the labour supply schedule of Figure 2.3(c). At the higher wage rate, the income effect of the wage increase begins to dominate the substitution effect and the labour supply schedule becomes backward bending. In fact, this may be a reasonable approximation for many individuals: at low wage rates they have an abundance of unmet needs, so that higher wages will induce them to work more in order to fulfill these needs; at higher wages many of their needs are fulfilled, so that additional wage increases will be used to purchase leisure.

CRITICISMS OF THE WORK-LEISURE FRAMEWORK

The work-leisure choice framework has been criticized in a variety of areas. Many of these criticisms, however, are not really criticisms of the basic framework. Rather they are extensions of the model and reminders of its limitations.

There is the belief that many individuals do not really have a choice between work and leisure. This may especially be the case for poor persons at a subsistence level of income, or people who are permanently attached to the labour force or who are required to work a fixed period of time if they are to work. These criticisms should not negate the viability of the work-leisure framework; in fact, they can easily be incorporated into the basic model. Poor persons existing at a subsistence income have a wage constraint close to the origin and an indifference curve between income and leisure that is flat to the right of the existing equilibrium — since they are at subsistence, they are unable to give up any income even for large increases in leisure. To the left of the existing equilibrium, their indifference curve could have a low slope indicating that they are willing to give up considerable leisure to attain the higher income and consumption patterns of middle-class status.

People who are required to work a fixed period of time would not face a continuous wage constraint, but rather would be relegated to a specific point on that wage constraint, or to not working at all. This can be depicted in the

Figure 2.3
DERIVING THE INDIVIDUAL SUPPLY CURVE OF LABOUR

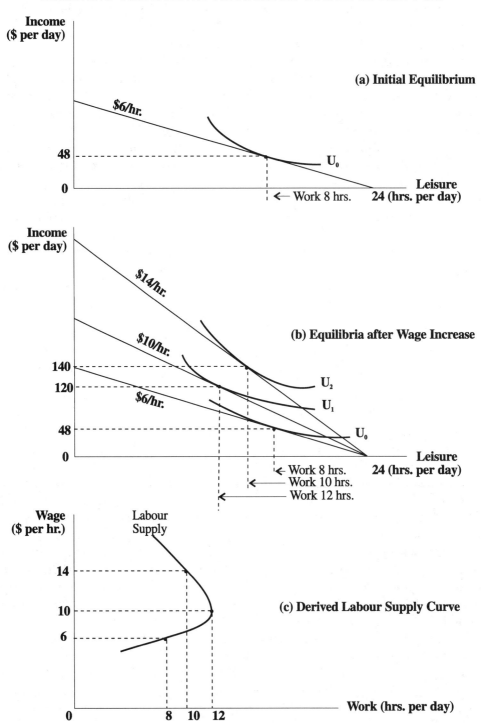

work-leisure choice framework and, as we will see later, gives rise to interesting implications with respect to overtime and moonlighting. The point is simply that rather than negating the viability of the work-leisure model, these observations highlight its usefulness in analyzing the impact of the different preferences and of the objective constraints facing individuals. In addition, the interpretation of the constraint need not involve the wage constraint of a single job. Workers may face a menu of wage-hours combinations across different jobs and firms.

The work-leisure choice framework has also been criticized on the grounds that leisure may not be a "normal" good. For low-income people, leisure may be an inferior good because they don't have the income to really enjoy the leisure. An increase in their income, other things being equal, may actually encourage work incentives since, for example, the more expensive vacation or the chance to send one of the children to a university now becomes a reality. Again, this does not negate the theory, since it does not deny the possibility that leisure is an inferior good. In fact, the basic theory of consumer demand merely states that an increase in income will lead to increased expenditure on all normal goods: this increased expenditure on leisure could well mean a lesser quantity of leisure and more quality leisure, in which case work incentives are increased.

APPENDIX: HOUSEHOLD CHOICE AND INDIFFERENCE CURVE ANALYSIS

The economic theory of consumer behaviour and labour supply has two main building blocks: preferences and constraints. Preferences summarize what the individual or household wishes to achieve. Constraints summarize what is feasible; that is, what the individual is able to achieve. The key assumption of the theory is that individuals choose from among the feasible outcomes that outcome which yields the highest level of satisfaction or well-being.

The individual's preferences can be expressed formally in terms of a utility function that shows the level of satisfaction or utility associated with any specific basket of commodities and leisure. The utility function can be written as $U(X_1, X_2, \ldots X_n, L)$ where X_i is the amount of good i consumed and L is the amount of "leisure" (nonwork time) consumed. For exposition of the basic theory of labour supply it is useful to illustrate the individual's preferences graphically. For this purpose the utility function can be expressed as $U(X,L)$ where X is a basket of commodities (or "composite commodity"). This function thus shows the level of utility or satisfaction associated with any particular combination of leisure (nonwork) time and basket of goods and services.

In order for the theory to make testable predictions about behaviour, it is necessary to make some assumptions about the general nature of the preferences of individuals. The following four fundamental assumptions form the basis of the theory of consumer-worker choice:

A1. *Complete ordering:* the individual can order all possible bundles (combinations of X and L). This assumption means that for any two combinations of X and L, say bundle A and bundle B, she can state that A is preferred to B, B is preferred to A, or she is indifferent between A and B. In other words, the individual is assumed to know what she derives satisfaction from and, therefore, to be able to say which bundle would yield higher utility or whether both would yield the same utility.

A2. *More of X or L is preferred:* both X and L are assumed to be "goods," not "bads," in the sense that the more of X consumed (holding constant the amount of L consumed), the higher the level of satisfaction. Similarly the more leisure time consumed (holding constant the consumption of goods and services), the higher her level of utility.

A3. *Transitivity of preferences:* if bundle A is preferred to B and bundle B is preferred to C, then bundle A must be preferred to bundle C. This assumption implies that preferences have a logical consistency similar to that which holds among objects of different weights: if object A weighs more than B and B weighs more than C, then object A weighs more than object C.

A4. *Diminishing marginal rate of substitution:* the individual becomes less willing to substitute X for L the lower the quantity of X consumed. More precisely, holding utility constant, the smaller the amount X consumed, the greater is the amount of L that would be needed to compensate the individual for one less unit of X. This important assumption is discussed further below.

These four assumptions imply that the individual's utility function has certain properties. Graphically these assumptions imply that individual preferences can be illustrated in the form of an *indifference map* as shown in Figure 2.A1. Each of the curves labelled U_0, U_1, U_2, and U_3 is called an *indifference curve*. An indifference curve shows combinations of X and L that yield the same utility or satisfaction to the consumer-worker. These curves are downward sloping and have a "convex" shape. They do not cross or intersect. Indifference curves above and to the right of other indifference curves show combinations of X and L that are preferred to those on curves lower and to the left (that is, all bundles on U_2 are preferred to any bundle on U_1).

Indifference curve analysis is a very useful graphic tool that will help your understanding of the basic theory of labour supply. If you have not previously been exposed to this valuable tool, you may be satisfied with the following statement (which is true, by the way!): assumptions A1 to A4 imply that individual preferences have the form shown in the indifference map in Figure 2.A1. If you don't like taking such claims "on faith" and want to understand why this statement is true, you should refer to a good text on intermediate microeconomics.

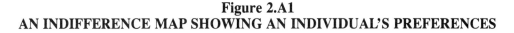

Figure 2.A1
AN INDIFFERENCE MAP SHOWING AN INDIVIDUAL'S PREFERENCES

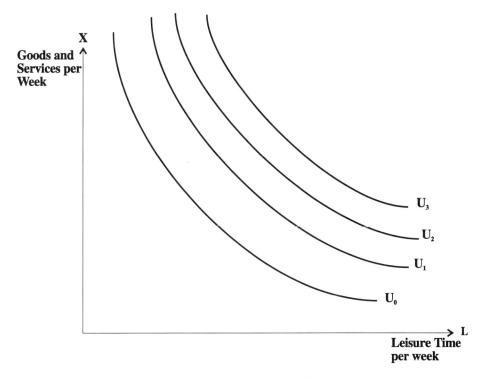

The "convex" shape of the indifference curves results from the assumption of diminishing marginal rate of substitution. This is illustrated in Figure 2.A2. At point A, the individual consumes a large amount of goods and services (24 units per week) but has limited leisure time (15 hours per week). Her utility is unchanged if she gives up six units of goods and services per week in exchange for one extra hour of leisure, thereby moving to point B on the indifference curve. Thus, beginning at the bundle A, she is willing to exchange six units of goods and services for one additional hour of leisure. However, beginning at bundle C (which consists of 13 units of goods and services and 18 hours of leisure), she is only willing to sacrifice one unit of goods and services for an additional hour of leisure (i.e., moving from bundle C to bundle D). At point E, which consists of 22 hours of leisure and only eight units of goods and services, she is only willing to give up a half unit of goods and services for one extra hour of leisure. Thus the fewer goods and services consumed, the less willing she becomes to give up additional units of X for extra hours of leisure.

This property of preferences can also be explained in terms of the amount of leisure time. Beginning at the bundle F, which consists of 23 hours of leisure and

Figure 2.A2
DIMINISHING MARGINAL RATE OF SUBSTITUTION AND "CONVEX" PREFERENCES

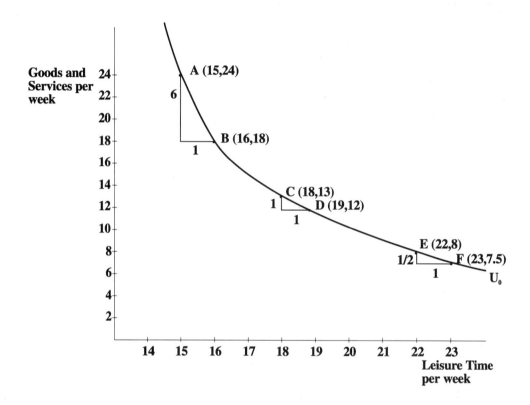

7.5 units of goods and services, the individual is willing to give up one hour of leisure for an additional half unit of goods and services. However, at a point such as D which provides 19 hours of leisure (compared to 26 at point F), one additional unit of goods and services is needed to compensate her for one less hour of leisure. At point B, where the amount of leisure is only 16 hours per week, six additional units of goods and services are required to compensate for sacrificing one additional hour of leisure. Thus the less of either good she consumes, the less willing she is to give up additional units of that good in exchange for more units of the other.

QUESTIONS

1. In Figure 2.1(c), why is E_0 a stable equilibrium? In other words, why would the individual not move along the potential income constraint line W_0 to some other feasible combination of income and leisure? Why would she not move to a feasible point below the wage constraint line? Why would she not move to a point above the indifference curve U_0?

2. What would happen to an individual's labour supply schedule if leisure is an inferior good?

3. Use the basic work-leisure choice framework to analyze the possible labour supply response of various groups to changes in their wage rate. The different groups could include the following: poor who are at minimum subsistence and who aspire to middle-class consumption patterns; wealthy who have acquired an abundance of material goods and who now aspire to be members of the idle rich; workers who have a fairly strong attachment to the labour force and who are reluctant to change their hours of work; workers who have a weak attachment to the labour force and who have viable alternatives to labour market work; "workaholics" who have strong preferences for labour market work.

4. Use a static, partial-equilibrium, income-leisure choice diagram to illustrate the effect of an increase in life expectancy (i.e., our endowment of time) on the following, assuming a constant wage rate that extends over one's lifetime: (a) leisure over one's lifetime, (b) income over one's lifetime, (c) labour supply over one's lifetime.

5. Illustrate the case where an individual responds differently to a wage increase and a wage decrease of the same magnitude. Specifically, have the person become "locked in" to a certain consumption pattern associated with the higher wage.

6. Depict the following situations in the basic work-leisure choice diagram: an individual works beyond his normal working hours, becomes less productive, and hence is paid less per hour; an individual works beyond her normal working hours, and an overtime premium is paid; the person takes a training program and increases his labour market productivity and wage; the government subsidizes the individual's wage by 50 percent; the government imposes a tax of 20 percent on labour market earnings; the government imposes a progressive income tax; the government gives everyone a basic demogrant of $1000 per year.

7. Suppose an individual is on the backward-bending portion of her labour supply curve. Is there a sufficiently high wage such that her hours of work will go to zero?

REFERENCES AND FURTHER READINGS

Abbott, M. and O. Ashenfelter. Labor supply, commodity demand, and the allocation of time. *R.E. Studies* 42 (October 1977) 389–411.

Ashenfelter, O. and J. Heckman. The estimation of income and substitution effects in a model of family labor supply. *Econometrica* 42 (January 1974) 73–85.

Barzel, Y. and R. McDonald. Assets, subsistence and the supply curve for labor. *AER* 63 (September 1973) 621–633.

Berg, E. Backward sloping labour supply functions in dual economies — the African case. *QJE* 75 (August 1961).

Darrough, M. A model of consumption and leisure in an intertemporal framework: a systematic treatment using Japanese date. *IER* 18 (October 1977) 677–696.

Dowell, R. Risk preference and the work-leisure trade-off. *EI* 23 (October 1985) 691–702.

Feldstein, M. Estimating the supply curve of working hours. *Oxford Economic Papers* 20 (August 1968) 74–80.

Finegan, T. Hours of work in the United States: a cross-sectional analysis. *JPE* 70 (October 1962) 452–470.

Hanock, G. The backward-bending supply of labor. *JPE* 73 (December 1965) 636–642.

Heckman, J. Life cycle consumption and labour supply. *AER* 64 (March 1974) 188–194.

Heckman, J. Shadow prices, market wages and labor supply. *Econometrica* 42 (July 1974) 679–694.

Juster, T. and F. Stafford. The allocation of time. *JEL* 29 (June 1991) 471–522.

Lewis, H.G. Economics of time and labor supply. *AER* 64 (May 1975) 29–36.

Malathy, R. Estimating substitution and income effects of female labour supply. *Journal of Quantitative Economics* 7 (January 1991) 43–63.

McCabe, P. Optimal leisure-effort choice with endogenously determined earnings. *JOLE* 1 (July 1983) 308–329.

Murray, M. A reinterpretation of the traditional income-leisure model with application to in-kind subsidy programs. *JPubEc* 14 (August 1980) 69–82.

Owen, J. *The Price of Leisure*. Montreal: McGill-Queen's University Press, 1970.

Owen, J. The demand for leisure. *JPE* 79 (January/February 1971) 56–76.

Robbins, L. On the elasticity of demand for income in terms of effort. *Economica* (June 1930) 123–129.

Ryder, H., F. Stafford and P. Stephan. Labor, leisure and training over the life cycle. *IER* 17 (October 1976) 651–674.

Sharir, S. The income leisure model: a diagrammatic extension. *Economic Record* (March 1975) 93–98.

Sharir, S. and Y. Weiss. The role of absolute and relative wage differentials in the choice of work activity. *JPE* 82 (November/December 1974) 1269–1275.

Smith, J. Assets, savings and labor supply. *EI* (October 1977) 551–573.

Wales, T. Estimation of a labour supply curve for self-employed business proprietors. *IER* 14 (February 1973) 69–80.

Wales, T. and A. Woodland. Estimation of household utility functions and labor supply response. *IER* 17 (June 1976) 397–410.

CHAPTER 3

Household Production Functions: Application to Fertility

Perhaps the most substantial modification of the basic work-leisure model is that offered by exponents of the "household production" view, sometimes termed the "economics of the family" or the "new home economics," as evidenced in the writings of Gary Becker (1965, 1976). This perspective emphasizes that households are producers as well as consumers and hence it is important to analyze how the household allocates its scarce resources, especially time, to affect the economic well-being of the household. The view highlights the misnomer of the term "leisure" as it has historically been used in the income-leisure choice framework. Rather, the household time component can consist of pure leisure, household work, and the use of household time in other activities that may involve both a consumption and a production component.

The household's allocation of its scarce time resource, then, is the focus of the household production function approach. This approach focuses on technological considerations of production rather than relying on differences in "tastes" as determinants of certain elements of behaviour. It extends the explanatory power of consumer demand analysis by expanding the concept of "price" to include time costs, hence making it easier to explain certain patterns of behaviour that are hard to explain in the conventional framework. It is a framework that has been applied, for example, to the decision to have children, as well as the demand for sleep, both of which are illustrated subsequently.

BASIC FRAMEWORK

The basic notion behind the household production perspective is that the household combines inputs of market goods, X, and household time, T, to produce commodities, which ultimately yield satisfaction or utility, U, to the household. Such commodities (to use Becker's 1965 terminology), or characteristics (to use Lancaster's 1966 terminology) include eating, sleeping, or viewing a play, all of which require combinations of household time and market goods to produce and hence ultimately consume such commodities.

More formally stated, household utility is a function of the commodities it consumes; that is, $U = f(Z_1, Z_2 \ldots Z_n) = f(Z)$ where $Z = (Z_1 \ldots Z_n)$ is a vector of the various commodities consumed by the household. Each of these commodities in turn is also produced by the household via the household production function $Z = g(X,T)$, which involves various combinations of goods $X = (X_1, X_2 \ldots X_m)$ and time as inputs in the production function. Indirectly then, household utility is a function of market goods and household time as they are combined in the household production function; that is, $U = f(Z) = f(g(X,T)) = h(X,T)$.

UTILITY ISOQUANTS

Figure 3.1(a) depicts such a household utility isoquant (or indifference curve U_0) indicating how a given constant level of utility can be produced by varying combinations of market goods, X, and household time, T. The notation X can represent either an individual good X_i or a basket of market goods $X = (X_1, X_2 \ldots X_m)$. In the latter case the Hicksian aggregation conditions which allow us to treat a basket of goods like a single good are assumed to hold; that is, the relative prices of the goods in the basket are held constant. In this context, X is usually referred to as a "composite commodity."

The slope of the utility isoquant exhibits a diminishing marginal rate of substitution between the inputs. That is, in the upper left segment, where an abundance of market goods and little household time are used to produce commodities that yield utility, a large increase in the use of market goods is required to offset any further reduction in the use of household time, in order to maintain household production and hence utility at the same level. For example, if a household were trying to reduce its meal preparation time from one hour down to one-half hour per day, then it would take a larger increase in the use of market goods (e.g., more elaborate appliances, pre-packaged food) in order to maintain utility, than if the household were trying to cut its meal preparation time from four to three and one-half hours. The latter reduction could be accomplished with a substitution of simple market goods like basic appliances and equipment.

The slope of the utility isoquant can be thought of as the "home wage" or the household's implicit valuation of its time since it shows how many market goods and services, X, can be saved by using an additional unit of household time, T. (This will be important later in developing the notion of a reservation wage or the minimum market wage at which an individual will enter the labour market.) An exogenous event that would increase one's valuation of time or "home wage" in the household (e.g., having children or having a sick relative that needs family care) would rotate the isoquant to the right. This is depicted by the utility isoquant U_0' in Figure 3.1(b) with the "home wage" (slope of the utility isoquant)

Figure 3.1
UTILITY ISOQUANTS AND HOUSEHOLD PRODUCTION

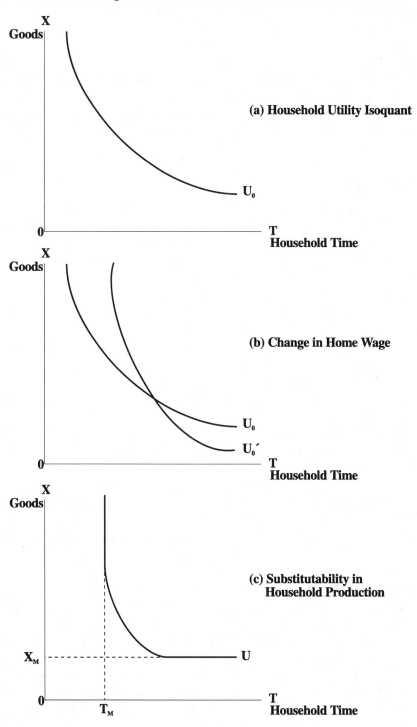

being higher for U_0' than for U_0. This is illustrated most clearly where the isoquants intersect since a small reduction in household time requires more inputs of goods and services to keep utility constant under U_0' than U_0.

The shape of the utility isoquant illustrates the substitutability between household time and market goods in the production of household utility. For example, Figure 3.1(c) illustrates the situation where neither market goods nor household time are good substitutes for each other below certain minimum threshold levels of goods, X_m, and time, T_m. However, they may be good substitutes in the intermediate range beyond those threshold levels. This could be the case if, for example, a minimum threshold amount of both household time and market goods is necessary to raise children (i.e., to attain a given amount of household satisfaction from raising a family). That is, it may be difficult to attain satisfaction from raising children if one cannot afford certain basic necessities or if one cannot afford to spend at least a certain amount of time with them. Alternatively stated, the threshold level, T_m, indicates that it is not always possible to completely "buy your way out" of raising children.

HOUSEHOLD EQUILIBRIUM AND INCOME AND SUBSTITUTION EFFECTS

Higher utility isoquants represent higher utility levels since more utility is associated with more market goods or household time or both. As under conventional consumer demand theory, households will try to maximize utility subject to their budget constraint. The slope of the budget constraint is determined by the market wage rate. A high labour *market* wage gives rise to a steep budget constraint, indicating that if the individual gives up household time for labour market work, more market goods, X, can be purchased. (Remember that the "home wage" is depicted in the utility isoquant with a high home wage giving rise to a steeper utility isoquant.)

Figure 3.2 indicates the equilibrium, E_0, amount of household time, T_0, and market goods and services, X_0 which yield the highest possible level of household satisfaction, U_0, given the individual's market wage, W_0, and home wage. (The home wage is depicted implicitly in the slope of the utility isoquant.) An increase in the market wage from W_0 to W_1 would rotate the budget constraint upwards yielding a new equilibrium E_2. As in conventional consumer demand theory, this price change can be decomposed into income and substitution effects. The income effect, from E_0 to E_1, implies that the household can afford to consume more of both goods- (X) and time- (T) intensive activities given the increased real income from the increased market wage. The pure or income-compensated substitution effect, from E_1 to E_2, implies that the household will utilize more goods-intensive and fewer time-intensive activities (i.e., substitute away from T and into X) to produce a given level of household satisfaction U_1.

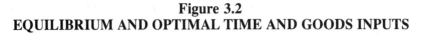

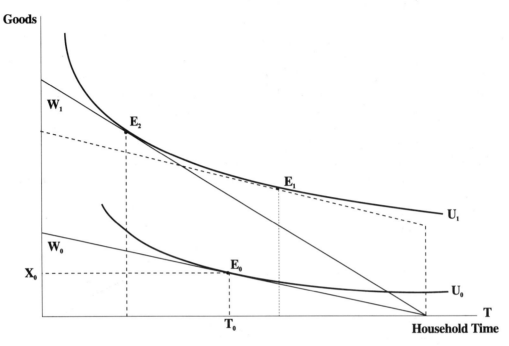

Figure 3.2
EQUILIBRIUM AND OPTIMAL TIME AND GOODS INPUTS

As with the basic income-leisure choice model, the income and substitution effects work in the opposite direction for the household time input T; therefore, E_2 could be to the right or left of E_0 depending upon whether the income or substitution effect dominates. (In Figure 3.1 the substitution effect dominates so E_2 is to the left of E_0.) However, the use of goods and services, X, as an input into the household production function unambiguously increases (i.e., E_2 is above E_0) because the income and substitution effects work in the same direction towards the purchase of goods.

In this regard, there are two major differences between conventional consumer demand theory and the household production function approach. First, the household production approach emphasizes that commodities consumed require both goods and time input. This focuses attention on the extent to which particular commodities are goods-intensive or time-intensive, and how the choice between goods-intensive and time-intensive commodities responds to exogenous changes, in particular changes in the market wage rate or the opportunity cost of time.

This aspect of household behaviour is illustrated in Figure 3.3, which shows the choice between a commodity which is relatively goods-intensive and one which is relatively time-intensive. The initial equilibrium, based on the market

Figure 3.3
HOUSEHOLD CHOICE BETWEEN GOODS- AND TIME-INTENSIVE COMMODITIES

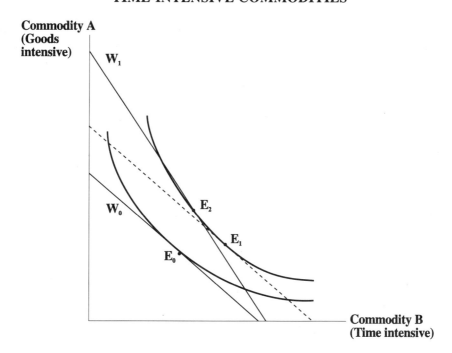

wage W_0, is at E_0. (The choice of hours of work cannot be shown without extending the diagram to three dimensions; however, optimal hours of work are determined as shown in Figure 3.2 and these determine the position of the budget constraint marked W_0 in Figure 3.3.) An increase in the market wage to W_1 enables the household to consume more of both commodities (and may affect the choice of hours of market work, as shown in Figure 3.2), but also increases the relative price of the time-intensive commodity resulting in the steeper budget constraint denoted by W_1 in Figure 3.3. The movement to the new equilibrium can be decomposed into the income effect (E_0 to E_1) which results in increased consumption of both commodities providing both are "normal," and the substitution effect (E_1 to E_2) which results in increased consumption of the goods-intensive commodity and decreased consumption of the time-intensive commodity. Thus the income and substitution effects work in the same direction for the goods-intensive commodity but in opposite directions for the time-intensive commodity.

The second difference between conventional consumer demand theory and the household production function approach is that the substitution effect in the latter approach has two reinforcing components. The substitution in household *production* implies that for a given amount of every commodity consumed, the household will substitute away from more costly time inputs and toward relatively less costly goods inputs, following a wage increase that increases the opportunity cost of time. In addition, the substitution in household *consumption* implies that the household will substitute away from time-intensive commodities and toward goods-intensive commodities, again reducing the use of the more costly time input if the market wage increases. Basically, the household production function approach reminds us that price changes can yield substitution effects in production as well as in the consumption of the commodities that yield ultimate satisfaction.

SOME IMPLICATIONS

The commodities that are the ultimate source of satisfaction generally require both time and goods in their production and consumption. This observation has significant implications. As real wages increase, the demand for goods-intensive commodities will increase at a relatively rapid rate due to the fact that the income and substitution effects reinforce each other. The consequences of this tendency are evident in many aspects of life: the growth of fast-food outlets and prepared foods; attempts to combine activities, sometimes with additional goods input (e.g., jogging with recorded music); even the choice of forms of exercise (popularity of jogging compared to walking). Our demand for sleep may even be affected by the fact that it is a very time-intensive activity! (See Exhibit 3.1). As real incomes rise in the future, households will be driven to more goods-intensive as opposed to time-intensive forms of consumption. Nonetheless, the tension between time available for market work and time available for household production and consumption may increase as the opportunity cost of time rises — a phenomenon captured nicely by the title of S.B. Linder's (1970) book *The*

EXHIBIT 3.1 THE ECONOMICS OF SLEEP

Sleep occupies more of our time than any other activity, with approximately one-third of our adult life devoted to sleeping. Casual observation suggests that many of our students undertake that same time allocation during class-time!

Presumably, we derive satisfaction from sleeping and it is an important input into our production of other activities. Hence, it should be amenable to economic analyses.

Biddle and Hameresh (1990), in fact, provide empirical evidence from a variety of sources indicating that the demand for sleep is negatively related to the price or cost of sleeping. That is, other things equal, people of higher potential earnings power sleep less because it costs them more (in terms of forgone income) to sleep longer.

Source: J. Biddle and D. Hamermesh. The demand for sleep. *Journal of Political Economy*, 98 (October 1990) 922–943.

Harried Leisure Class. (Incidentally, the book is well worth reading — if you can find the time!)

The household production function perspective has other implications that may not have been obvious from the conventional exposition of household commodity demand and labour supply behaviour. It reminds us, for example, that the consumption decisions of households depend not only upon their income and relative commodity prices, but also upon factor prices (especially the opportunity cost of their time) and factor productivity. Also, households with an efficient production function have a larger opportunity or choice set than those with less efficient production functions. That is, households are potentially better off (i.e., have a larger choice set) not only if they have high market wages or endowments of nonlabour income (the determinants of the budget constraint emphasized in the conventional income-leisure choice framework) but also if they are endowed with more efficient household production functions.

Emphasis on the household as a *producer* as well as a *consumer* also reminds us of the importance in household consumptions and time allocation decisions of various laws of production. The principle of comparative advantage emphasizes that it may be economically efficient for different members of the household to specialize in their respective areas of comparative advantage. This may mean, for example, that it is economically efficient for one party to specialize in labour market tasks and the other to specialize in household tasks, depending upon their relative earning power in the labour market and productivity in household tasks. Such specialization may be artificially induced, however, if female market wages are reduced because of discrimination. This can lead to a vicious circle whereby females specialize in household tasks because their market wages are lower due to discrimination, and this household specialization prevents them from acquiring the labour market experience that would give them a potential comparative advantage to specialize in labour market tasks.

Economies of scale may also be an important law of production that has relevance to the concept of the household production function. For example, the average *per person* cost of preparing a meal may be high for a single person but not for a family of six; hence, inducing the single person to enter co-operative arrangements or to eat more often in restaurants. Similarly, the extra cost of watching an additional child may be small once the fixed costs are incurred for watching one child; hence, the emergence of co-operative daycare arrangements. Also, within the same family, the extra cost of raising a third child may be less than that of the second child which in turn is less than that of the first child; hence, the rationale for per capita poverty measures that do not simply divide family income by the number of family members.

The emphasis on the household as a producer also reminds us of the importance of the laws of derived demand in analyzing how the demand for inputs of household time and market goods (as used in the household production function) would respond to changes in their prices. These laws of demand will be analyzed in more detail in the subsequent chapter on the demand for labour. One implication that emerges is that the reallocation of time from household to labour market tasks in response to an increase in the market wage will be small if there are few good substitute inputs for household labour, if there are few substitute commodities for the sorts of things produced in the household, or if the time cost is a small proportion of the total cost of household production.

The household production function approach also emphasizes the importance of the family as a decision-making unit and hence the importance of joint decision-making amongst family members with respect to various elements of behaviour. For example, the decisions to get married, participate in labour market activities, and have children are all intricately related for all family members. This also means, for example, that a change in the wage rate or price of time for one family member may affect not only their own allocation decisions, but also the decisions of other family members. Formally stated, cross-price effects can be important across family members.

The importance of allocating time over the life cycle is also highlighted in the household production function approach. For example, it may be economically rational to engage in time-intensive activities when the opportunity cost of time is low, perhaps because productivity is lower at certain stages in the life cycle. Thus, it may be efficient to work intensively and for long hours during the peak-earnings years of one's career and to engage in more leisure time when one's earnings are lower. For example, it may be efficient for students to take time off for time-intensive travel before they have established a career in the labour market, or for older workers to partake in early or partial retirement programs once they have passed the peak earnings years of their careers.

Lastly, from a methodological point of view, the household production function perspective emphasizes downplaying the importance of tastes and preferences in determining behaviour. Rather, the emphasis is on the objective constraints faced by households in their consumption and production decisions. Such constraints reflect a complicated array of income and prices including endowments of nonlabour wealth, time, and the household production function, as well as the market wages and home wages of various family members and the prices of goods and services that can be purchased directly or used as inputs into the household production function.

APPLICATION TO FERTILITY AND FAMILY FORMATION

Largely based on the pioneering work of Becker (1960) and Mincer (1963), economists have applied the basic framework of consumer demand theory and household production theory to the decision to have children. Montgomery and Trussell (1986) provide a current review of some of the economic literature and relate it to the demographic literature.

Although many find it obnoxious to think of children, even analogously, as if they are "consumer durables," few would deny that at least some families alter their childbearing decisions because they cannot yet afford to have children (perhaps until they have at least saved for the down payment on a house or have finished paying for their education) or because it is too costly to have a child (perhaps because it would mean an interruption in a wife's career in the labour market). As long as these and other economic factors affect the decision to have children for some families, then economic factors will have some predictive power in explaining variations in birth rates.

As suggested by consumer demand theory, the basic variables affecting the fertility decision are income, the price or cost of a child, the price of related goods, and tastes and preferences. Household production theory reminds us of the importance of time as a key input and of substitution effects in production as well as consumption. As we will see, however, non-economic factors also play a strong role.

Variables Affecting Fertility

Income

Economic theory predicts that, *other things being equal*, there will be a positive relationship between income and the desired number of children, assuming children are analogous to "normal goods." The problem is that in the real world

EXHIBIT 3.2   ARE CHILDREN NORMAL GOODS?

In economics, normal goods are ones that we spend more on as our income increases; that is, they have a positive income elasticity of demand. At first glance it appears that children are not like normal goods, in that higher income families and wealthier countries tend to be associated with small family sizes. Children, like potatoes, seem more like inferior goods which we purchase less of as our income rises!

The problem with the empirical evidence is that the *gross* negative relationship between income and family size is the result of two opposing forces. On the one hand, high-income families and countries have more wealth which should lead to larger family sizes, if children are like normal goods (i.e., a pure income effect). On the other hand, women in high-income families and countries tend to have high potential earnings power and this makes the (opportunity) cost of children higher. That is, women of high earnings power forgo more income if they take time out of the labour market to bear and raise children.

As indicated subsequently, the negative effect of the higher "price" of having a child tends to outweigh the positive income effect whereby higher earnings are used to "buy" more children. When the potential earnings of women is controlled (e.g., through econometric analysis) a positive relationship between income and family size prevails. Children are like normal goods!

other things are not equal. Specifically, factors such as contraceptive knowledge and the cost of having children tend to be related to the income variable, so that it becomes difficult to separate the pure effect of income alone. (See Exhibit 3.2.)

Price or Cost of Children

Economic theory also predicts that the demand for children is negatively related to the price or cost of having children. Cost estimates averaging about $4000 per year per child have been cited.[1] Although direct costs such as the additional food, clothing, and housework associated with raising a child are an important aspect of the cost of children, the main element in the cost of having a child is the income forgone by the spouse (in our society this tends to be the wife) who takes time away from labour market activity to bear and raise the child.

EXHIBIT 3.3   CHINA'S NEW POPULATION CONTROL POLICY

In 1979, China adopted a strict policy of one child per family to control its burgeoning population growth. The program has been strongly criticized for its harsh sanctions, including heavy pressure to have abortions. As well, the program is failing in the countryside where most of China's population resides.

A more recent policy is being tried based on economic incentives rather than sanctions — the "carrot" rather than the "stick." With assistance from the United Nations, loans are being provided for women in rural communities to become entrepreneurs in their rural communities. As stated in the *Wall Street Journal*, "The details are complex, but the idea is simple: turn unschooled peasant women into entrepreneurs, making them too busy, and too successful, to want to have broods of babies."

Source: *Wall Street Journal*, January 2, 1992.

An increase in the potential earnings of wives can have both income and substitution effects on the decision to have children. The consumption substitution effect says that wives with high potential earnings will have fewer children because the higher opportunity cost (forgone income) of having a child induces them to substitute away from the expensive alternative of having children and to engage in less expensive activities that do not impinge as much on their earnings capacity. The income effect, on the other hand, says that wives with a high earnings capacity can contribute more to family income and this will enable the

---

1. Whittington, Alm and Peters (1990) citing five U.S. studies in the early 1980s, with the range from $2300 to $7300 per child.

family to spend more on all commodities, implying greater expenditures on children if children are "normal goods." Thus the income and substitution effects of a change in the wife's earnings capacity have opposing effects on desired family size.

Household production theory also emphasizes that there is a substitution effect in household production as well as consumption. That is, women with high wages and hence a higher opportunity cost of time, may substitute away from time-intensive and towards goods-intensive ways of raising children. The substitution effects in both consumption and production reinforce each other to reduce the use of the more costly time input. However, the substitution effect in production may mute some of the decline in demand for children emanating from the substitution in consumption. That is, a wage increase will not lead to as much of a reduction in the demand for time-intensive children if the household can easily substitute market goods and services (e.g., daycare, disposable diapers, domestic help) for household time. As these substitutes become more available over time (as should be the case given that the demand for them is increasing) the demand for children may increase even if the market wages of women increase.

Price of Related Goods

A rise in the price of complementary goods (e.g., medical expenses, daycare, education) would tend to reduce our desired number of children. Conversely, a fall in their price — perhaps from public subsidies — could encourage larger family sizes. In most cases, the cost of any one of these related commodities is probably not large enough to have any appreciable effect on family size. However, dramatic changes in the private cost of some of these items — such as free university tuition or universally free daycare — could have an impact, as could any trend in the overall extent of state support for medical care, family allowances, maternity leave, education, or daycare.

Tastes and Preferences

Economists tend to regard tastes and preferences as exogenously given from outside the economic system. In the area of family formation, our tastes and preferences have dramatically changed, related to our ideas on religion, family planning, and the women's liberation movement in general. These factors have all changed over time in a fashion that would encourage smaller family sizes.

Some would argue that these changes were not really exogenous to the economic system, but rather were a result of some of the more fundamental economic changes. For example, such factors as improved job opportunities for females may have made women's liberation more necessary to ensure more equal employment opportunities. The dramatic increase in the number of women working outside the home may have changed attitudes towards women, their role in society, and the nature of the family. Cause and effect works both ways: tastes and preferences both shape, and are shaped, by the economic system.

Education also has an important bearing on tastes and preferences. Not only does it raise the forgone income from raising children, but increased education may also widen our horizon for other goods (travel, entertainment outside the home), enable family planning, and encourage self-fulfillment through means other than having children.

Technology

The term technology is often used by economists as a general rubric to describe the general technological and environmental factors that influence our economic decisions. In the area of fertility and family formation, birth control knowledge and contraceptive devices have had an important impact on the number of children, primarily by equilibrating the actual with the desired number of children. Medical advances with vasectomies and tubal ligation can have a similar impact. The reduction in infant mortality that has occurred over time should also reduce the number of births since, in times of high infant mortality, it was often necessary to have large families simply to have a few children survive to adult age.

Most of the technological advances discussed so far are ones that would encourage or enable smaller families. Some, such as reduced danger and discomforts during pregnancy, and advances in fertility drugs and operations, work in the opposite direction to encourage or enable more childbearing. These have been especially important for women who have postponed childbearing to later in their life. Other advances, such as processed food and disposable diapers, have made care for children easier. Household production theory has emphasized how the technology has emerged in part because of an increased demand for controlling fertility — not only its quantity, but also the spacing and timing of births.

Empirical Results

The empirical evidence on fertility behaviour is confusing to interpret because of the difficulty of holding other factors constant when observing actual behaviour. For example, it is difficult to isolate the effect of family income, independent of the effect of the wife's education, because highly educated wives tend to marry husbands with high incomes. If we observe fewer children in high-income families, is this because of their high income, or is it because wives in such families tend to be highly educated and hence their forgone income or cost of having children is high?

Statistical techniques such as multiple regression analysis have been used by econometricians to isolate the effect of a single variable, while holding the impact of all other factors constant. The results of such studies generally confirm the economic predictions.[2] Specifically, there tends to be a negative relationship between the number of children and the cost of having children as measured, for example, by the wife's potential labour market earnings. As well, there tends to be a weak positive relationship between income and family size, after holding constant the impact of other variables, notably the earnings potential of the wife.[3] This positive relationship is especially likely to prevail between income and *expenditures* on children, since higher income is used to "purchase" more "quality" (as measured by expenditures per child) as well as quantity (number of children).

When the earnings potential of the wife is *not* held constant then there tends to be a negative relationship between fertility and family size; that is, poor

---

EXHIBIT 3.4   CAN PUBLIC POLICY AFFECT FERTILITY?

Public policy can obviously affect fertility through such mechanism as regulations on abortions as well as public support for research and development on fertility and the application of new techniques in fertility clinics. As well, pro-natalist policies can explicitly be adopted as was recently the case in Quebec where grants are given for the birth of each child, with the magnitude of the grant increasing for subsequent children. This policy is designed explicitly to offset, in part at least, the dramatic decline in the birth rate that has been going on in that province.

Other policies can have unintended impacts on fertility, largely by reducing the cost associated with having a child. For example, Family Allowance benefits provide a monthly lump-sum payment for each child up to the age of 18. The Child Tax Credit provides a tax credit for each child. Maternity benefits under the Unemployment Insurance Act provide unemployment insurance benefits (60 percent of earnings for up to 15 weeks) and the recipient is entitled to return to her previous, or comparable, job.

Hyatt and Milne (1991) provide empirical evidence for Canada indicating that all three policies have a positive but quantitatively small impact on fertility. For example, in the 1980s, a 1 percent increase in the real value of maternity benefits would result in a 0.26 percent increase in fertility. Alternatively stated, if the 1986 replacement rate had been 100 percent of earnings instead of 60 percent, fertility would have been 1.96 children per family instead of the 1.67 recorded that year.

Source: D. Hyatt and W. Milne. Can public policy affect fertility? *CPP* 17 (1991a) 77–85.

---

2. Becker (1960), Mincer (1963), Cain and Weininger (1973), Butz and Ward (1979), Iglesias and Riboud (1988), and Hyatt and Milne (1991a, 1991b).

3. Based on Canadian data over the period 1948 to 1975, Hyatt and Milne (1991b) estimate the elasticity of demand for children to be 1.38 with respect to male income, and –1.55 with respect to female wages. These are fairly close to the U.S. estimates obtained by Butz and Ward (1979) for the same period. (continued next page)

countries and poor families tend to have larger family sizes. This is due, however, to the low earnings potential of women in such situations, with this negative effect of their potential earnings dominating the positive effect of income. This highlights that a viable population control policy, for less-developed countries especially, is to improve the job opportunities for women (see Exhibit 3.3 earlier). This will raise the (opportunity) cost of having children, and if the empirical estimates are applicable, this should outweigh any tendency to have more children because of the additional income generated by their employment.

The empirical evidence also indicates that government policies can affect fertility by altering the economic determinants factors that influence family size. Based on Canadian data, Hyatt and Milne (1991) indicate that family allowances, the Child Tax Credit, and maternity benefits under unemployment insurance all had a positive effect on family size (Exhibit 3.4). Based on U.S. Data, Whittingham, Alm and Peters (1990) indicated that the personal exemption for dependents had a positive effect on fertility. Leibowitz (1990) provides evidence

---

EXHIBIT 3.5   DID THE "FAMILY ALLOWANCE" OF THE POOR LAWS INCREASE BIRTH RATES?

The poor laws of early nineteenth century England essentially provided a system of "child allowances" whereby workers often received additional income from their parish if they had more children. Malthus argued against such laws on the grounds that they simply encouraged larger family sizes which in the long run would simply increase poverty.

There has been, and continues to be, considerable debate on the impact of the Poor Laws. Recent econometric evidence, however, indicates that after controlling for the effect of other variables affecting birth rates, the Poor Law system of family allowances did lead to higher birth rates. Malthus appears to have been correct at least in his argument that family allowances leads to higher birth rates.

Source: G. Boyer. Malthus was right after all: poor relief and birth rates in Southeastern England. *Journal of Political Economy* 97 (February 1989) 93–114.

This implies that a 1 percent increase in male income gives rise to a 1.38 percent increase in the number of children born, other things held constant. For example, if real average male income increases from $30,000 to $33,000 (a 10 percent increase) then the number of children born would increase by 13.8 percent (10 × 1.38%). Based on the average fertility rate of 1.67 in 1987, this implies an increase in fertility of 0.23 children (0.138 × 1.67 = 0.23) or an increase in average fertility from 1.67 to 1.90.

The negative wage elasticity of –1.55 implies that a 1 percent increase in the wages of women gives rise to 1.55 percent reduction in the fertility rate, other things held constant. For example, if real average female wages increase from $30,000 to $33,000 (as above) this would reduce fertility by 0.26 children (0.155 × 1.67 = 0.26) or a reduction in average fertility from 1.67 to 1.41.

indicating that families with free medical care for three to five years had 29 percent more births than did families where medical care was not free. Even in nineteenth century England, the "family allowance" of the poor laws apparently led to increased birth rates (see Exhibit 3.5).

QUESTIONS

1. Briefly outline the Becker-Lancaster approach to consumer demand theory ("household production function" approach: the "new microeconomics"; the new "home economics") and its applicability to labour market economics. Specifically, indicate some labour market implications of this theory that we would not have been able to derive from conventional consumer demand theory.

2. Utilize your knowledge of household production theory to indicate

    (a) the expected effect of a wage increase on labour supply;

    (b) the expected effect, on the demand for children, of an increase in the wage of a wife;

    (c) why the income-compensated substitution effect on labour supply may be larger for women than men;

    (d) the effect, on an individual's "home wage," of the disability of a spouse who requires home and not institutional care;

    (e) the effect, on an individual's "home wage," if the government were to pay the family for the disability in (d).

3. In recent years economists have applied economic theory, especially the Becker-Lancaster "new consumer demand theory" or "household production function" approach to the decision to have children.

    (a) Briefly outline the relevance of this approach to the family formation decision.

    (b) Indicate the expected effect of an increase in the wages of women on the family formation decision, being careful to distinguish between substitution effects in production and consumption.

    (c) Discuss the implications of the theory for the spacing (i.e., how far apart children are in age) and timing (i.e., when parents will have them in terms of the parents' life cycle) of children.

4. The household production function approach focuses on technological considerations of production rather than "tastes" to explain various aspects of behaviour. Discuss the virtues of this approach.

5. Based on your knowledge of the economic theory of fertility and family formation, indicate the expected impact on the birth rate of increases in each

of the following: women's wages, family allowances, the basic income tax exemption for each dependent, and the education of women.

6. Why has there been a long-run decline in fertility over time, even though income has risen?

7. What does the economic theory of fertility tell us about a viable population policy for less-developed economies experiencing a population explosion?

8. What does the economic theory of fertility tell us about the expected population problem in the year 2000?

9. What does the economics of fertility tell us about family size and urban versus rural location?

10. Cain (1971, p. 412) has stated that "there are a number of prices included in the overall price of a child, and each component price is potentially changeable by means of a wide variety of policy actions." Discuss this statement and indicate the policies that can have an impact on the decision to have children.

11. In the regression equation, $C = .10 Y_H - .20 E_W$, relating births per family, C, to the husband's income Y_H, and the wife's education E_W, evaluate the income elasticity of demand for children for the average family. Assume that the average husband's income is 9 when Y_H is measured in units of $1000, and that the average number of children per family is 3. Evaluate the income elasticity of demand for children for poorer families whose husband's average income is 5 when Y_H is measured in units of $1000 and whose average number of children is 2. From the information given in the regression equation, can you calculate the pure or compensated substitution effect for changes in the wife's income?

REFERENCES AND FURTHER READINGS

A.   Household Production

Becker, G. A theory of the allocation of time. *EJ* 75 (September 1965) 493–517.

Becker, G. *The Economic Approach to Human Behavior.* Chicago: University of Chicago Press, 1976.

Becker, G., K. Murphy and R. Tamura. Human capital, fertility and economic growth. *JPE* 98 (October 1990) S12–S37.

Ben-Porath, Y. The production of human capital and the life cycle of earnings. *JPE* 75 (March 1986) 352–365.

Biddle, J. and D. Hamermesh. Sleep and the allocation of time. *JPE* 98 (October 1990) 922–943.

Bovlier, B. and M. Rosenzweig. Schooling, search and spouse selection: testing economic theories of marriage and household behavior. *JPE* 92 (August 1984) 712–732.

Chiappori, P. Collective labor supply and welfare. *JPE* 100 (June 1992) 437–467.

Chiswick, C. The value of housewife's time. *JHR* 17 (Summer 1982) 413–425.

Chiswick, B. Labor supply and investment in child quality: a study of Jewish and non-Jewish women. *R.E. Stats.* 58 (November 1986) 700–703.

Corman, H. The demand for education for home production. *EI* 24 (April 1986) 213–230.

Graham, J. and C. Green. Estimating the parameters of a household production function with joint products. *R.E. Stats.* 66 (May 1984) 277–282.

Gronau, R. Leisure home production and work — the theory of the allocation of time revisited. *JPE* 85 (December 1977) 1099–1123.

Gronau, R. Home production — a forgotten industry. *R.E. Stats.* (May 1980) 408–415.

Gronau, R. Home production: a survey. *Handbook of Labor Economics.* Vol. 1. O. Ashenfelter and R. Layard (eds.). New York: Elsevier, 1986.

Grossbard-Shechtman, S. A theory of allocation of time in markets for labour and marriage. *EJ* 94 (December 1984) 868–882.

Grossbard-Schechtman, S. and S. Neuman. Women's labor supply and marital choice. *JPE* 96 (December 1988) 1298–1302.

Grossman, M. and T. Joyce. Unobservables, pregnancy resolutions, and birth weight production functions in New York City. *JPE* 98 (October 1990) 983–1007.

Hill, C. and F. Stafford. Parental care of children: time diary estimates of quantity predictability and variety. *JHR* 15 (Spring 1980) 219–239.

Juster, F. and F. Stafford (eds.). *Time, Goods, and Well Being.* The University of Michigan, Survey Research Center, 1983.

Kooreman, P. and A. Kapteyn. A disaggregated analysis of the allocation of time within the household. *JPE* 95 (April 1987) 223–249.

Lancaster, K. A new approach to consumer theory. *JPE* 74 (January 1966) 132–157.

Linder, S. *The Harried Leisure Class.* New York: Columbia University Press, 1970.

Lommerud, K. Marital division of labor with risk of divorce. *JOLE* 7 (January 1989) 113–127.

Manser, M. and M. Brown. Marriage and household decision-making. *IER* 21 (February 1980) 31–44.

Mincer, J. Market prices, opportunity costs and income effects. *Measurement in Economics: Studies in Mathematical Economics and Econometrics in Memory of Yehuda Grunfeld.* C. Chris et al. (eds.). Stanford, Calif.: Stanford University Press, 1963.

Pollak, R. and M. Wachter. The relevance of the household production function and its implication for the allocation of time. *JPE* 83 (April 1975) 255–277.

Ray, R. Measuring the costs of children: an alternative approach. *JPubEc* 22 (October 1983) 89–102.

Rosenzweig, M. and T. Schultz. Estimating the household production function: heterogeneity, the demand for health inputs, and their effect on birth weight. *JPE* 91 (October 1983) 723–746.

Williams, A. and N. Terleckyj. Household production and consumption. *JPE* 86 (February 1978) 168–170.

B.   Fertility and Family Formation

Anderson, K.H. The sensitivity of wage elasticities to selectivity bias and the assumption of normality: an example of fertility demand estimation. *JHR* 17 (Fall 1982) 594–605.

Barro, R. and G. Becker. Fertility choice in a model of economic growth. *Econometrica* 57 (March 1989) 481–501.

Becker, G. An economic analysis of fertility. *Demographic and Economic Change in Developed Countries.* Princeton: Princeton University Press, 1960, 209–231.

Becker, G. and H. Lewis. On the interaction between the quantity and quality of children. *JPE* 81 (March/April 1973) S279–S288.

Becker, G. and N. Tomes. Child endorsements and the quantity and quality of children. *JPE* 84 (August 1976) S143–S162.

Becker, G. and R. Barro. A reformulation of the economic theory of fertility. *QJE* 103 (February 1988) 1–26.

Behrman, J. and B. Wolf. A more general approach to fertility determination in a developing country. *Economica* 51 (August 1984) 319–340.

Ben-Porath, Y. Fertility responses to child mortality. *JPE* 84 (August 1976) S163–S178.

Bloom, D. and J. Trusell. What are the determinants of delayed childbearing and permanent childlessness in the United States? *Demography* 21 (November 1984) 591–611.

Boyer, G. Malthus was right after all: poor relief and birth rates in Southeastern England. *JPE* (February 1989) 93–114.

Butz, W. and M. Ward. The emergence of countercyclical U.S. fertility. *AER* 69 (June 1979) 318–328.

Cain, G. Issues in the economics of a population policy for the United States. *AER* 61 (May 1971) 408–417.

Cain, G. and M. Dooley. Estimation of a model of labor supply, fertility, and wages of married women. *JPE* 84 (August 1976) S179–S200.

Cain, G. and A. Weininger. Economic determination of fertility: results from cross-sectional aggregate data. *Demography* 10 (May 1973) 205–223.

Calhoun, C. Estimating the distribution of desired family size and excess fertility. *JHR* 24 (Fall 1989) 709–724.

Carliner, G., C. Robinson and N. Tomes. Female labour supply and fertility in Canada. *CJE* 13 (February 1980) 46–64.

Chiswick, B. Differences in education and earnings across racial and ethnic groups: tastes, discriminations, and investments in child quality. *QJE* 103 (August 1988) 571–598.

Cigno, A. Fertility and the tax-benefit system. *EJ* 96 (December 1986) 1035–1051.

Cigno, A. and J. Ermisch. A microeconomic analysis of the timing of births. *European Economic Review* 33 (April 1989) 737–760.

Coale, A. and S. Watkins, (eds.). *The Decline of Fertility in Europe.* Princeton: Princeton University Press, 1985.

Conger, D. and J. Campbell. Simultaneity in the birth rate equation: the effects of education, labor force participation, income and health. *Econometrica* 46 (May 1978) 631–641.

Cremer, H. and P. Pestieau. Bequests, filial attention and fertility. *Economica* 58 (August 1991) 359–376.

Deaton, A. and J. Muellbauer. On measuring child costs. *JPE* 94 (August 1986) 720–744.

Dooley, M.D. Labour supply and fertility of married women: an analysis with grouped and individual data from the 1970 U.S. census. *JHR* 17 (Fall 1982) 499–532.

Easterlin, R. Economic-demographic interactions and long swings in economic growth. *AER* 56 (December 1966) 1063–1105.

Eckstein, Z. and K. Wolpin. Endogenous fertility and optimal population size. *JPubEc* 27 (June 1985) 93–106.

Ermisch, J. Econometric analysis of birth rate dynamics in Britain. *JHR* 23 (Fall 1988) 563–576.

Falaris, E. An empirical study of the timing and spacing of children. *SEJ* 54 (October 1987) 287–300.

Fleisher, B. and G. Rhodes. Fertility, women's wage rates, and labor supply. *AER* 69 (March 1979) 14–24.

Gregory, P., J. Campbell and B. Cheng. A cost-inclusive simultaneous equation model of birth rates. *Econometrica* 40 (July 1972) 41–47.

Gregory, P., J. Campbell and B. Cheng. A simultaneous equation model of birth rates in the United States. *R.E. Stats.* 54 (November 1972) 374–380.

Hanushek, E. The trade-off between child quantity and quality. *JPE* 100 (February 1992) 84–117.

Hashimoto, M. Economics of postwar fertility in Japan. *JPE* 82 Part 2 (March/April 1974) S170–S194.

Heckman, J., J. Hotz and J. Walker. New evidence on the timing and spacing of births. *AER* 75 (May 1985) 179–184.

Heckman, J. and J. Walker. The relationship between wages and income and the timing and spacing of births. *Econometrica* 58 (November 1990) 1411–1441.

Hotz, U.J. and R. Miller. An empirical analysis of life cycle fertility and female labor supply. *Econometrica* 56 (January 1988) 91–118.

Hyatt D. and W. Milne. Can public policy affect fertility? *CPP* 17 (1991a) 77–85.

Hyatt D. and W. Milne. Countercyclical fertility in Canada: some empirical results. *Canadian Studies in Population* 18 (1991b) 1–16.

Iglesias, F. and M. Riboud. Intergenerational effects on fertility behaviour and earnings mobility in Spain. *R.E. Stats.* 70 (May 1988) 253–256.

Journal of Political Economy. New economic approaches to fertility. T. Paul Schultz (ed.) 81 Part 2 (March/April 1973).

Keeley, M. The economics of family formation. *Economic Enquiry* 15 (April 1977) 238–250.

Kramer, W. and K. Neusser. The emergence of countercyclical U.S. fertility: Note. *AER* 47 (March 1984) 201–202.

Lam, D. Marriage markets and assortive matings with household public goods. *JHR* 23 (Fall 1988) 462–487.

Leibenstein, H. An interpretation of the economic theory of fertility. *JEL* 12 (June 1974) 457–479.

Leibenstein, H. An economic theory of fertility decline. *QJE* 89 (February 1975) 1–31. Comment by W. Cullison and reply 91 (May 1977) 345–350.

Leibowitz, A. Home investments in children. *JPE* 82 Part 2 (March/April 1974) S111–S131.

Leibowitz, A. The response of births to changes in health care costs. *JHR* 25 (Fall 1990) 697–711.

Leung, S. A stochastic dynamic analysis of parental sex preferences and fertility. *QJE* 106 (November 1991) 1063–1088.

Mincer, J. Market prices, opportunity costs, and income effects. *Measurement in Economics*. Stanford University Press, 1963, 67–82.

Moffitt, R. The estimation of fertility equations on panel data. *JHR* 19 (Winter 1984) 22–34.

Moffitt, R. Optimal life-cycle profiles of fertility and labour supply. *Research in Population Economics* 5. T. Schultz and K. Wolpin (eds.) (1984) 29–50.

Montgomery, M. and J. Trussel. Models of marital status and child-bearing. *Handbook of Labor Economics*. Vol. I. O. Ashenfelter and R. Layard (eds.). New York: Elsevier Science Publishers, 1986.

Neher, P. Peasants, procreation and pensions. *AER* 61 (June 1971) 380–389. Comment by M. Ferber and reply 62 (June 1972) 451–452, and comment by W. Robinson and reply 62 (December 1972) 977–979.

Nerlove, M., A. Razin and E. Sadka. Some welfare theoretic implications for endogenous fertility. *IER* 27 (February 1986) 3–32.

Newman, J. and C. McCulloch. A hazard rate approach to the timing of births. *Econometrica* 52 (July 1984) 939–961.

Robinson, C. and N. Tomes. Family labour supply and fertility: a two-regime model. *CJE* 15 (November 1982) 706–734.

Rosenzweig, M. The demand for children in farm households. *JPE* 85 (February 1977) 123–146.

Rosenzweig, M. Birth spacing and sibling inequality: asymmetric information within the family. *IER* 27 (February 1986) 55–76.

Rosenzweig, M. and T. Schultz. Market opportunities, genetic endowments and intra-family resource distribution: child survival in rural India. *AER* 72 (September 1982) 803–815.

Rosenzweig, M. and T. Schultz. Consumer demand and household production: the relationship between fertility and child mortality. *AER* 73 (May 1983) 38–42.

Rosenzweig, M. and T. Schultz. The demand for and supply of births. *AER* 75 (December 1985) 992–1015.

Rosenzweig, M. and D. Seiver. Education and contraceptive choice: a conditional demand framework. *IER* 23 (February 1982) 171–198.

Rosenzweig, M. and K. Wolpin. Life-cycle labour supply and fertility: causal inferences from household models. *JPE* 88 (April 1980) 328–348.

Sah, R. The effects of child mortality changes in fertility choice and parental welfare. *JPE* 99 (June 1991) 582–606.

Sander, W. Farm women, work and fertility. *QJE* 101 (August 1986) 653–657.

Schultz, T. An economic model of family planning and fertility. *JPE* 77 (March/April 1969) 153–180.

Schultz, T. The influence of fertility on labour supply of married women: simultaneous equation estimates. *Research in Labour Economics*. Vol. 2. R. Ehrenberg (ed.). Greenwich, Conn.: JAI Press, 1978.

Schultz, T. Changing world prices, women's wages and the fertility transitions: Sweden, 1860–1910. *JPE* 93 (December 1985) 1126–1154.

Schultz, T. *The Economics of Population*. Reading, Mass.: Addison-Wesley, 1981.

Sprague, A. Post-war fertility and female labour force participation rates. *EJ* 98 (September 1988) 682–700.

Tomes, N. A model of fertility and children's schooling. *EI* 19 (April 1981) 209–234.

Ward, M. and W. Butz. Completed fertility and its timing. *JPE* 88 (October 1980) 917–940.

Tsakloglou, P. Estimation and comparison of two simple models of equivalence scales for the cost of children. *EJ* 101 (March 1991) 343–357.

Whittington, L., J. Alm and H. Peters. Fertility and the personal exemption. *AER* 80 (June 1990) 545–556.

Willis, R. What have we learned about the economics of the family? *AER* 77 (May 1987) 68–81.

Winegarden, C. Women's fertility, market work and marital status: a test of the new household economics with international data. *Economica* 51 (November 1984) 447–456.

Wolpin, K. An estimable dynamic stochastic model of fertility and child mortality. *JPE* 92 (October 1984) 852–874.

CHAPTER 4

Work Incentive Effects of Alternative Income Maintenance Schemes

As a response to the problem of income loss and poverty, various income maintenance schemes have been proposed to raise the income of certain groups and to supplement low wages. One concern with many of these proposals is that they may reduce work incentives — i.e., result in lower labour force participation or fewer hours worked among program participants. If an income maintenance program provides little incentive to work, this would raise the cost of the program to taxpayers and hence make the program politically less acceptable. In addition, it could prevent the recipients from acquiring the on-the-job training and labour market experience that may raise their income in the long run. Clearly, it is important to understand the exact way in which income maintenance schemes could alter work incentives. The income maintenance programs examined here are demogrants, negative income taxes, welfare, wage subsidies, unemployment insurance, unemployment-insurance-assisted worksharing, and disability payments and workers' compensation.

STATIC PARTIAL EQUILIBRIUM EFFECTS IN THEORY

The basic income-leisure choice framework provides a convenient starting point to analyze the work incentive effects of alternative income maintenance schemes. The analysis is restricted to the static, partial equilibrium effects of alternative income maintenance programs. That is, it does not consider dynamic changes that can occur over time, or general equilibrium effects that can occur as the impact of the program works its way through the whole economic system.

Income support programs affect behaviour by altering the individual's opportunities (their potential income constraint). To analyze the effect of specific programs on the income constraint, ask first what is the effect on nonlabour income, and second, what is the effect on the slope of the constraint — i.e., what happens as the individual gives up leisure and works more? Throughout this analysis, Y is defined as income after taxes and transfer payments, and E as labour market earnings, equal to wages times hours worked.

Demogrant

Perhaps the simplest income maintenance program to analyze is the demogrant. As the name implies, a demogrant means an income grant to a specific demographic group, such as female-headed families with children, or all persons aged 60 and over, or all family units irrespective of their wealth. Examples of demogrants in Canada include the Family Allowance program which provides monthly payments to families in respect of each child under the age of 18, and the Old Age Security program which provides monthly payments to individuals over the age of 65.

As illustrated by the dashed line in Figure 4.1(a), the demogrant would shift the potential income constraint vertically upwards by the amount of the grant. The slope of the new income constraint would be equal to the slope of the original constraint since the relative price of leisure has not changed. Thus there is no substitution effect involved with the demogrant; the new equilibrium, E_d, would be above and to the right of the original equilibrium; that is, work incentives would unambiguously be reduced. This occurs because the demogrant involves only a leisure-inducing pure income effect. The increase in actual take-home income is less than the amount of the demogrant because some of the demogrant was used to buy leisure, hence reducing earned income. This can readily be seen in Figure 4.1(a); if the individual did not alter working time, the outcome would be at the point E_1 vertically above the original equilibrium E_0.

Welfare

In Canada, welfare or social assistance programs are administered by the provinces but financed partly (approximately 50 percent) by the federal government under the Canada Assistance Plan. Benefits vary by province and according to other factors such as family type (single parent, couple) and employability. Welfare benefits also depend on the needs of the family and the family's assets and other sources of income. As illustrated in Table 4.1, the total income (including welfare payments) of single parents with one child, tends to be about 70 percent of the poverty line level of income with considerable variation by provinces.

For those who are eligible for welfare their new potential income constraint is given by the dashed line in Figure 4.1(b). If they do not work, they are given the welfare payment. Hence, at the point of maximum leisure (zero work), their income constraint shifts vertically upwards by the amount of the welfare

Figure 4.1
WORK INCENTIVE EFFECTS OF INCOME MAINTENANCE PLANS

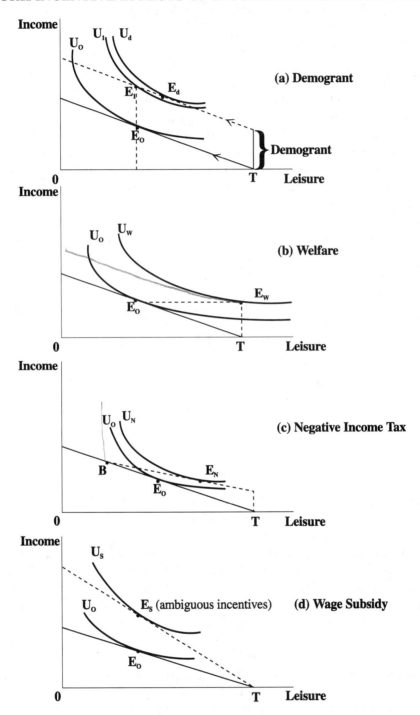

Table 4.1
INCOME[a] OF WELFARE RECIPIENTS BY PROVINCE, 1986
(Single Parent, One Child, Age 2)

PROVINCE	ANNUAL INCOME[a]	INCOME AS % OF POVERTY LINE[b]
	$	
Newfoundland	9,559	71.7
Prince Edward Island	9,739	84.3
Nova Scotia	9,074	68.0
New Brunswick	7,911	59.3
Quebec	9,101	64.8
Ontario	10,249	72.9
Manitoba	8,925	63.5
Saskatchewan	9,804	73.5
Alberta	9,860	70.2
British Columbia	8,861	63.0
Average (unweighted)	9,308	69.1

Notes:
a. Includes family allowances, child tax credits, and provincial tax credits as well as welfare.
b. Statistics Canada's low income cut-offs.

Source: National Council of Welfare. *Welfare in Canada: The Tangled Safety Net.* Ottawa: National Council of Welfare, 1987.

payment. Under *many* welfare programs, as individuals work and receive labour market earnings they are required to forgo welfare payments by the exact amount of their labour market earnings. In this sense, there is a 100 percent tax on earnings. In fact, the implicit tax may even be greater than 100 percent, if, for example, they also lose medical or housing subsidies. (See Fortin, 1985, for calculations of the implicit tax rates facing welfare recipients in Quebec and Ontario.) Their potential income constraint is thus horizontal at the amount of the welfare payment: as they work and earn income, they forgo a comparable amount in welfare and hence their income does not increase. Every dollar earned results in a dollar reduction in welfare. Of course, once they reach their original labour market wage constraint, then their take-home pay will be indicated by their original wage constraint: at this point their welfare payments have been reduced to zero so they cannot be "taxed" any further by being required to give up welfare.

If the welfare payment is sufficiently high, the individual would have a strong incentive to move to the corner solution at E_w where he would not work at all.

e is no incentive to work more (move to the left of E_w), because of the 100 percent implicit tax on income from work that arises because the individual has to give up an equivalent amount of welfare for every dollar earned. Even though the person's take-home pay, Y, is lower at E_w than E_0, he chooses E_w because it involves considerably more leisure. Clearly welfare has extreme potential adverse effects on work incentives. Of course, for many people on welfare, work is not a viable alternative if they are perhaps disabled or unemployable. Yet for others, work would be a viable alternative if there would not be this 100 percent implicit tax on earned income.

This analysis suggests a variety of ways of reducing the number of people on welfare. Traditionally we think of making eligibility requirements more stringent or reducing the magnitude of welfare for those who are eligible. These changes would, of course, work. In Figure 4.1(b), for example, if the welfare payment were lowered to an amount lower than the height of U_0 at the point of maximum leisure, there would be no incentive to go on welfare since the individual would be maximizing utility at E_0. Although successful in reducing the number of people on welfare, these changes may have undesirable side effects, not the least of which are denying welfare to those in need and providing inadequate income support to those who are unemployable.

One alternative to these policies would be to increase the market wage rate of those on welfare and thereby encourage them to voluntarily leave welfare and earn income.[1] In Figure 4.1(b), an increase in the market wage rate would pivot the wage constraint upwards from T. At some higher wage rate, the individual clearly would be induced to move to a higher indifference curve that would involve more work effort than under welfare (i.e., a new equilibrium to the left of E_w). The increased market wage could come about through training, job information, mobility, a government wage subsidy, or institutional pressures such as minimum wages or unionization. Obviously these policies may be costly, or in the case of minimum wages and unionization may involve a loss of jobs. However, they could have the benefit of voluntarily reducing the number of people on welfare and hence increasing work incentives.

Another way of improving work incentives would be to reduce the 100 percent implicit tax on welfare. In some welfare programs, this is accomplished by requiring welfare recipients to give up only a portion of their welfare if they earn income by working. For example, if recipients are required to give up 50 cents in welfare for every dollar earned in the labour market, they would have some incentive to work because the implicit tax rate would be 50 percent. In Figure 4.1(b), this could be shown by a wage constraint starting at E_w, with a negative slope of 50 percent of the slope of the labour market wage constraint, reflecting the fact that the recipient takes home 50 percent of every dollar earned by

---

1. This is emphasized in the recent Ontario Social Assistance Review Committee (1988) recommendation for welfare reform.

working. The negative income tax, discussed below, is a general scheme designed to ensure that individuals receiving income support face an implicit tax on income from work that is significantly less than 100 percent.

An alternative solution to reducing the number of welfare recipients would be to alter the preferences of welfare recipients away from being on welfare and towards labour market activity. In Figure 4.1(b), this would imply changing the shape of the indifference curves. If, for example, at all points to the right of E_0, the indifference curve U_0 were flat, then the individual would not have opted for the welfare equilibrium E_w. The flat indifference curve would indicate a reluctance to accept any cut in income even to get substantial increases in leisure. Traditionally, preferences have been altered by attaching a social stigma to being on welfare. Alternatively, preferences could be altered towards income-earning activities, perhaps by making potential recipients feel more a part of the non-welfare society or perhaps by attempting to break the intergenerational cycle of welfare.

Recently, welfare reform has also emphasized the distinction between "employables" and "nonemployables" and has often set a requirement that the employables work or be registered in a training program as a condition of eligibility for the receipt of welfare. Such programs — termed workfare — have often been directed at single-parent families, given their potential to engage in paid employment. One of the problems, of course, is that this can entail expensive daycare requirements that may be added to the welfare expenses. Nevertheless, it may encourage work incentives as well as the longer run employability of recipients by providing them with work experience or training.

Negative Income Tax

Negative income tax or guaranteed annual income plans involve an income guarantee, and an implicit tax rate of less than 100 percent applied to labour market earnings. Income after taxes and transfers would be $Y = G + (1 - t)E$, where G is the basic guarantee, t is the implicit tax rate, and Y and E as defined earlier are take-home pay and labour market earnings respectively. Most negative income tax plans differ in so far as they involve different values of the basic guarantee and the tax rate. The term *negative* income tax is used because recipients will receive more from the guarantee than they will pay out in taxes, even though they do face a positive implicit tax rate.

Although a comprehensive guaranteed annual income program has never been implemented in Canada, some programs which apply to particular groups — for example, the Child Tax Credit which supplements the income of families with children and the Guaranteed Income Supplement which supplements the income of individuals over 65 — have the negative income tax design. That is, there is a basic guarantee received by those with income below a certain level; those with higher income receive less income supplementation according to the program's implicit tax rate; and those whose income exceeds the program's breakeven point receive no benefits.

A negative income tax plan with a constant rate is illustrated by the dashed line in Figure 4.1(c), presented earlier. As with the demogrant, at the point of maximum leisure the basic income guarantee shifts the potential income constraint upwards by the amount of the guarantee: even if the individual does not work, she has positive income equal to the amount of the guarantee. Unlike welfare, as the individual works, income assistance is not reduced by the full amount of labour market earnings. However, income support does decline as income from work increases; thus labour market earnings are subject to a positive implicit tax rate. Take-home pay does not rise as fast as labour market earnings; hence, the income constraint under the negative income tax plan is less steeply sloped than the original labour market income constraint. At the point B, often referred to as the breakeven point, income received from the negative income tax program has declined to zero. Thus to the left of this point the original income constraint applies.

Assuming leisure is a normal good, the new equilibrium for recipients of the negative income tax plan will unambiguously lie to the right of the original equilibrium: work incentives are unambiguously reduced relative to a situation in which there is no other form of income support. This occurs because the income and substitution effects both work in the same direction to reduce work effort. The tax increase on earned income reduces the relative price of leisure, inducing a substitution into leisure and hence reducing work effort. The tax increase also has an income effect (working in the opposite direction); however, for *recipients* this is outweighed by the guarantee, so that on net their new potential income constraint is always above the original constraint — that is why they are defined as recipients. Because the potential income of recipients is increased, they will buy more of all normal goods including leisure. Thus the income effect works in the same direction as the substitution effect to reduce work effort.

The income-leisure choice framework predicts that work incentives will be unambiguously reduced as a result of a negative income tax plan. But this does not negate the viability of such a program. The adverse work incentive effects may be small, or the increased "leisure" may be used productively as in job search, mobility, education, or increased household activities. In addition, the reduction in labour supply may have other desirable side effects, such as raising the wages of low-wage labour since it is now a relatively scarcer factor of production. Perhaps most important, the adverse incentive effects were predicted when a negative income tax was imposed in a world without other taxes and transfers. In most circumstances a negative income tax scheme is intended to replace some welfare programs which have even larger adverse incentive effects. In this sense work incentives may increase relative to incentives under welfare (as will be discussed in more detail in subsequent sections). While the basic conclusions from the work-leisure model should not be ignored, they must be kept in proper perspective. Empirical information is needed on the magnitude of any adverse work incentive effects and on the form in which "leisure" is taken. In addition, the effects on work incentives will depend on the programs that the negative income tax scheme is intended to replace.

Wage Subsidy

Since one of the problems with the negative income tax and welfare is that the tax on earnings may discourage work effort, some have suggested that rather than tax additional earnings, the government should subsidize wages in an attempt to encourage work. Although there are a variety of forms of a wage subsidy — most often associated with the proposals of Kesselman (1969, 1971, 1973) — the simplest wage subsidy schemes have the common result that the recipient's per-hour wage rate is supplemented by a government subsidy.

The static, partial equilibrium effect of the wage subsidy is illustrated by the dashed line in Figure 4.1(d), given earlier. For the recipients, it is exactly like a wage increase, hence the potential income constraint rotates upwards from the point of maximum leisure T. If the person does not work (i.e., is at T), his income is still zero even though his wage is subsidized. However, as the person works more, his take-home pay rises more under the wage subsidy than would be the case if he were only receiving his market wage.

Just as an increase in wages has both an income and a substitution effect working in opposite directions in so far as they affect work incentives, so will the wage subsidy have an ambiguous effect on work incentives. The higher wage means higher potential income from which the individual will buy more of all normal goods, including leisure; hence, work incentives are reduced via the income effect. This income effect will be at work even though the individual has to work to receive the income: the increased leisure could come in the form of reduced hours or longer vacations or periodic withdrawals from the labour force or reduced work from another family member. The higher wage also means that the price (opportunity cost) of leisure has now increased; hence, work incentives are increased via this substitution effect. On net, theory does not indicate which effect dominates; hence the work incentive effects of a wage subsidy are ultimately an empirical proposition.

Although the work incentive effects of a wage subsidy are theoretically indeterminant, Kesselman (1969) and Garfinkel (1973) show that, other things equal (the recipients' welfare, their post-transfer income, or the size of the subsidy), the adverse work incentive effects of the wage subsidy are not as great as those of the negative income tax. This is illustrated in Figure 4.2 for the case where the recipients' welfare is held constant at U_0 so that they are indifferent between the two plans. The equilibrium E_S under the wage subsidy *must* lie to the left of the equilibrium E_N under the negative income tax. That is so because the potential

Figure 4.2
WORK INCENTIVE EFFECTS OF WAGE SUBSIDY VS. NEGATIVE INCOME TAX

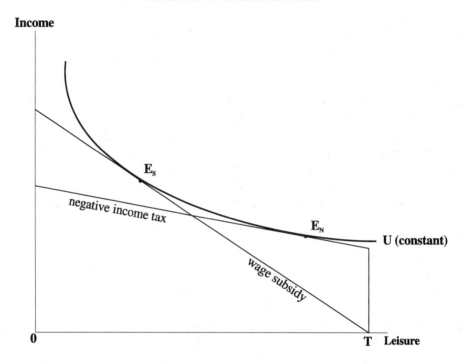

income constraint under the wage subsidy is steeper than the negative income tax constraint since it involves a net subsidy to the wage of recipients. Since the wage subsidy constraint must be tangent to U_0 in order for the recipients to be indifferent between the two plans, this tangency must be to the left of E_N because only to the left of E_N is the indifference curve more steeply sloped than at E_N (when we have less leisure, more income is required to give up a unit of leisure and remain indifferent).

Although the work incentive effects are greater under a wage subsidy than a negative income tax, this does not necessarily make the wage subsidy a better plan. One distinct disadvantage of the wage subsidy is that it does nothing to maintain the income of those who are unable to work. Although it may help the working poor, it does nothing to help those who legitimately cannot work. For the latter group, a negative income tax plan would at least provide a guaranteed minimum income. It would also be possible to have both plans and allow individuals to choose between them.

Because funds for income assistance are limited, an important aspect of an income maintenance scheme is the extent to which the program helps those most in need of assistance. Poverty and need are usually defined in terms of family income. For this reason, the negative income tax, which is based upon family income, is more likely to target payments to those in greatest need than a wage subsidy, which is based on individual wage rates. Many low-wage earners — especially youth and spouses of the primary income earner — are members of families with average or above-average income, resulting in a weak association between low individual wages and low family income. Simpson and Hum (1986) compare negative income tax and wage subsidy programs with equal budgets using microdata on Canadian families in 1981. They conclude that the negative income tax is clearly superior to wage subsidies in terms of its ability to divert payments to those families in greatest need of assistance.

Unemployment Insurance

Unemployment insurance is the largest income security program in the country, covering approximately 90 percent of the labour force. The replacement rate is 60 percent of lost earnings (subject to a maximum) with the benefit duration depending upon the rate of unemployment in the region. In order to qualify for benefits, individuals must have worked 10 to 20 weeks, depending upon the unemployment rate of the region. A two-week waiting period exists before benefits are paid, and this can go up to six weeks for persons who quit or are fired or refuse suitable work.

The income-leisure choice framework can be used to analyze the static, partial equilibrium effect that unemployment insurance will have on work incentives. The theoretical impact of these features of the UI system can be captured by a stylized example of an unemployment insurance scheme that gives recipients 60 percent of their weekly pay for a maximum period of one-half of a year (26 weeks) and that requires a minimum of 12 weeks of employment in order to qualify for benefits.

This hypothetical scheme is illustrated in Figure 4.3, using a one-year time horizon. The solid line AF is the potential income constraint in the absence of a UI program and the dashed line ABCDF is the constraint with unemployment insurance. If the individual works less than 12 weeks during the year, he does not qualify for UI benefits and his income constraint is the segment DF. If the individual works 12 weeks, he is eligible for benefits for an additional 26 weeks; thus his income equals 12 weeks of labour market earnings plus 26 times 60 percent of his normal weekly earnings; i.e., annual income $Y = 12E + (0.60) 26E$ where E is weekly earnings. This gives the point C in Figure 4.3, the vertical distance CD equalling $(0.60) 26E$. As the person works additional weeks (moves from right to left from the point C), he can earn his weekly earnings as well as his unemployment insurance for a maximum period of 26 weeks. Thus his unemployment insurance constraint line is parallel to his original wage constraint between 40 and 26 weeks of leisure. If he works beyond 26 weeks in the year he takes home his weekly earnings, but forgoes the chance to collect the 60 percent

Figure 4.3
INCOME CONSTRAINT FOR A SIMPLIFIED
UNEMPLOYMENT INSURANCE SCHEME

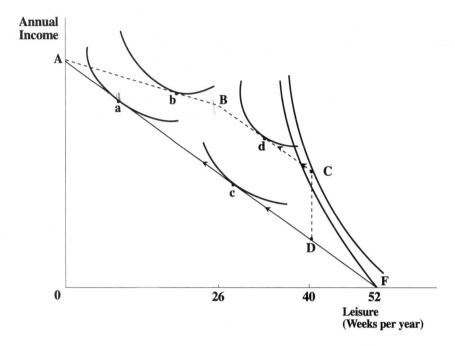

of his weekly earnings through unemployment insurance. This occurs because he cannot legally work and collect unemployment insurance at the same time. Thus between 26 and 0 weeks of leisure, the unemployment insurance constraint has a slope of 40 percent, reflecting the fact that work beyond 26 weeks only increases potential income by 40 percent since one forgoes the unemployment insurance benefits of 60 percent of weekly earnings.

The nature of the work incentive effects of unemployment insurance depend on where the individual would be located in the absence of the UI system. Three different possibilities are shown in Figure 4.3. For individuals working more than 26 weeks (e.g., at the point a on AF), the income and substitution effects work in the same direction to *potentially* decrease weeks worked (moving to the point B on AB). For individuals such as those originally at the point c on AF, the unemployment insurance system constitutes a pure income effect, again *potentially* decreasing work incentives (moving to the point d on BC). These adverse work incentive effects are potential because the vast majority of people will not collect unemployment insurance. Many employees would not leave their job because they may not be able to return once they have exhausted unemployment insurance. However, for those with a guaranteed or reasonably certain job — be it seasonal, or with family friends, or in household activity, or perhaps one of many low-paid dead end jobs that can't attract other workers — there may be an incentive to collect unemployment insurance. In addition, the incentive would certainly be there for those who have lost their job, perhaps because of a recessionary phase in the business cycle, and hence who have no labour market alternatives.

The third possibility illustrated in Figure 4.3 applies to individuals who would be outside the labour force in the absence of a UI program. These individuals are originally located at the point F, reflecting the high value of their time devoted to nonmarket activities. Many such individuals would choose to remain outside the labour force and neither work nor collect unemployment insurance. Others, however, would enter the labour force and work a sufficient number of weeks in order to qualify for unemployment insurance benefits, locating at the point C in Figure 4.3. In this case the UI program has *increased* work incentives by making it attractive for these individuals to devote less time to nonmarket activities and to enter the labour force for relatively brief periods. Similar conclusions apply to those who would work fewer than 12 weeks in the absence of a UI system (i.e., those who would be located in the segment DF).

The previous discussion of the incentive effects of unemployment insurance — as with other income maintenance programs — assumes that if individuals want to increase their labour supply, the jobs are available for them to do so. However, in regions or periods of high unemployment this may be an unrealistic assumption. Individuals may be constrained by the lack of jobs on the demand side. Phipps (1990, 1991) shows that both theoretically and empirically, this can have an important effect in altering the work incentive effects of income maintenance programs like unemployment insurance. Specifically, reforms to UI to increase work incentives may be thwarted, in part at least, by the fact that increasing the *incentive* to work (a supply side phenomenon) does not guarantee that more people will work, if the jobs are not available (a demand side phenomenon).

In summary, work incentives may be either increased or decreased by unemployment insurance. Some of those individuals who would, in the absence of a UI system, either be outside the labour force or work fewer weeks than are required to qualify for UI benefits will increase their weeks worked per year in order to receive unemployment insurance. In contrast, some of those individuals who would, in the absence of a UI system, work more than the minimum number of weeks required to qualify for UI will reduce their weeks worked per year and increase the number of weeks during which unemployment insurance benefits are received. Both these consequences of unemployment insurance may increase the unemployment rate, the former because it draws into the labour force individuals with marginal employment attachments who devote much of their time to nonmarket activities, and the latter because it reduces employment and increases unemployment of those with strong labour force attachment. The impact of the unemployment insurance system on unemployment is discussed further in the chapter on unemployment.

Unemployment-Insurance-Assisted Worksharing

One of the more recent innovative income maintenance schemes pertains to modifications to the unemployment insurance system to encourage worksharing. As defined in Meltz, Reid and Swartz (1981, p. 3), "Worksharing is an attempt to combat unemployment by reducing the number of hours each employee works rather than laying employees off when there is a reduction in the demand for labour." In the context of unemployment-insurance (UI)-assisted worksharing, the rationale was to encourage the sharing of unemployment amongst a broader work force rather than having it concentrated in the hands of a few. As pointed out by Reid (1982, 1985) this could be not only more equitable but also more efficient if the larger work force willingly accepted unemployment insurance payments for a small reduction in its work week. Whether UI-assisted worksharing is a socially desirable method of combating unemployment depends on a number of factors such as whether the reduction in demand is temporary or permanent in nature, how the total UI costs associated with the reduction in labour demand are affected, and how the UI program is financed. These issues are discussed further in the chapter on unemployment. Rather than having the unemployment concentrated, for example, in 20 percent of the work force who may be laid off (and receiving unemployment insurance), under UI-assisted worksharing the whole work force of the firm may willingly accept a 20 percent (e.g., one day per week) reduction in their work in return for receiving unemployment insurance for the one day that they are each engaging in worksharing. This, of course, would require modification of the conventional unemployment insurance rules which are based on continuous unemployment, not on being "unemployed" for one day per week.

In 1977 on an experimental basis, and in 1982 on an economy-wide basis, Canada modified its unemployment insurance scheme to allow firms to apply for unemployment insurance for its work force that willing engaged in such worksharing. The scheme (discussed in Reid 1982, 1985) basically provides unemployment insurance benefits for giving up an average of one day of work per week so that other workers in the firm would not have to be laid off.

Such an unemployment-insurance-assisted worksharing scheme is illustrated in Figure 4.4, assuming UI benefits (i.e., the replacement rate) to be 60 percent of earnings. The initial equilibrium at E_0 shows a hypothetical situation for a representative worker who works five days per week and takes home a weekly income of Y_0. An unemployment-insurance-assisted worksharing scheme that would allow each worker to choose any amount of work reduction would be illustrated by the dashed line budget constraint above the original budget

Figure 4.4
UNEMPLOYMENT-INSURANCE-ASSISTED WORKSHARING

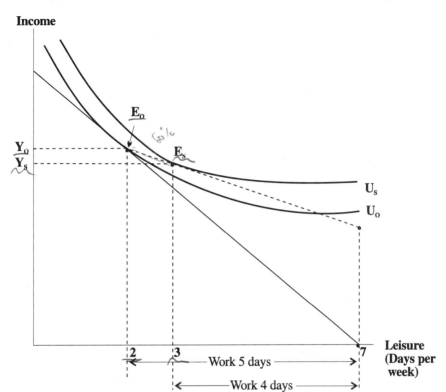

constraint. For any work reduction (i.e., movement to the right of E_0) income would not fall by the loss of a day's pay for every day of work reduction, as would be depicted by the original budget constraint. Rather, it would fall by only 40 percent since 60 percent would be supported by the receipt of unemployment insurance. Figure 4.4 depicts the case where the parties agree to a one-day reduction in the work week, and shows that this corresponds to the typical worker's preferred reduction (i.e., E_s is an equilibrium).

In reality, the negotiated reduction need not always correspond to the preferences of the work force, in which case E_s need not be tangent to the new budget constraint; however, it is likely to involve a higher level of utility (i.e., $U_S > U_0$) because the worker gets an extra day of leisure by giving up only 40 percent of a day's pay, the remainder coming from unemployment insurance. Clearly, the unemployment-insurance-assisted worksharing can provide an incentive for workers to voluntarily reduce their working time and enter into worksharing arrangements.

Disability Payments and Workers' Compensation

Income maintenance programs for the disabled also exist in a variety of forms such as workers' compensation, disability pension entitlements from public and private pensions, long-term disability insurance (with premiums usually paid from joint contributions from employers and employees), and court awards for personal injury. In most cases, of course, the work incentive effects of such income maintenance programs are not of interest as payment is made precisely because the disability legitimately prevents a person from working and the person may have little discretion over how much to work. In some cases, however, there may be more discretion over whether to work and if so, how much to work or when to return to work. In such cases, the work incentive effects of disability income maintenance programs become a legitimate policy concern.

Prior to examining the impact of disability compensation, it is informative to examine the effect of the disability or injury itself.

Effect of Disability on Budget Constraint

Depending upon its nature, the injury may have a number of different effects on the individual's budget constraint (i.e., ability to earn income). A partially disabling injury could reduce the amount of time the individual can spend at the job, but not affect the person's performance and hence wage at the job. This is illustrated in Figure 4.5(a). Without any disabling injury, the worker's choice set is Y H_f, given the market wage rate as illustrated by the slope of the budget constraint. In this particular case the worker chooses E_0 with corresponding utility of U_0 and hours of work (measured right to left) of H_0. The partially disabling injury, for example, could reduce the amount of time the worker is able to work from H_0 to H_p. The worker's choice set is now limited to E_p H_f with a corresponding drop in utility to U_p. If the injury were completely disabling so that the person could not work at all, then the person would be constrained to H_f, with no labour market income and utility reduced to U_f. This again highlights the misnomer associated with the phrase "leisure" — it is a catch-all for all activities outside of the labour market, ranging from pure leisure to being unable to work because of disability.

In addition to affecting one's ability to earn income, a disabling injury may also require medical expenses. This affects the budget constraint, reducing income by the amount of medical expenses at each level of hours of work, analogous to a reduction in nonlabour income. If the disabling injury led to such medical expenses but did not affect one's wage or ability to work in the labour market, then this would lead to a lower budget constraint like the dashed line Y_mM in Figure 4.5(b). The individual's wage is not affected (i.e., Y_mM is parallel to Y H_f), and the individual can work as much as before (i.e., Y_m M is the full choice set); however, there are some fixed medical costs equal to H_f M which also equals Y Y_m. If the individual does not work (i.e., the injury is a full disability

Figure 4.5
EFFECT OF DISABILITY

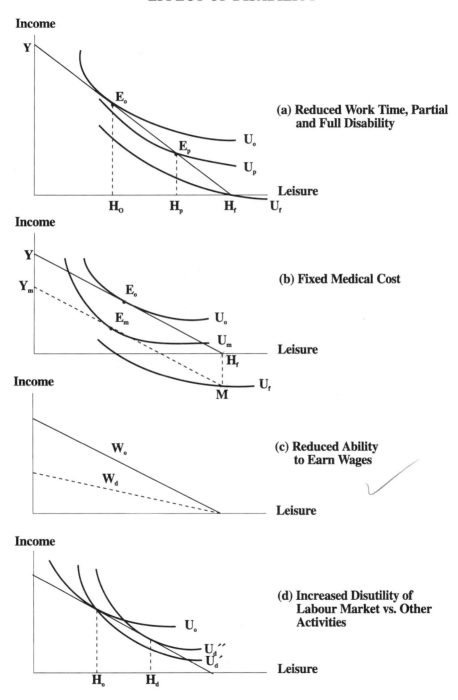

(a) Reduced Work Time, Partial
   and Full Disability

(b) Fixed Medical Cost

(c) Reduced Ability
   to Earn Wages

(d) Increased Disutility of
   Labour Market vs. Other
   Activities

so that the individual is compelled to move to H_f), then the medical costs are debts, as illustrated by the point M being below the axis (i.e., income is negative).

Because such fixed costs associated with medical expenses shift the budget constraint inwards without altering its slope, they will have a pure income effect, unambiguously reducing leisure and increasing hours of work, assuming leisure to be a normal commodity. The reduction in wealth associated with the injury cost means the household will buy less of all normal commodities, including leisure. Alternatively stated, the additional expenses mean that the individual would have to work more to prevent expenditure on goods from falling by the full amount of the medical expenses. This is illustrated in Figure 4.5(b) by the equilibrium E_m with its associated level of utility U_m, where E_m is to the left of E_0 and U_m is less than U_0.

If, in addition to these fixed medical costs, the injury also prevented the person from doing any labour market work, then the segment Y_m M of the budget constraint would not be available. In effect, the individual's "choice" set would be point M, with the corresponding reduced utility level U_f.

A disability that reduced one's ability to earn wages in the labour market would rotate the budget constraint downwards as in Figure 4.5(c). This would give rise to conventional income and substitution effects and hence have an indeterminate effect on hours of work. That is, the returns to work are reduced and hence one would work less (substitution effect). But one's wealth is also decreased, hence reducing one's ability to purchase commodities including leisure, thereby inducing more work (income effect).

A disability may also have a differential effect on the disutility associated with work or the ability to enjoy nonlabour market activities. For example, if the disability gave rise to more pain and suffering associated with labour market activities than with household work, it would rotate the household utility isoquant as in Figure 4.5(d) so that the indifference curves U_d' and U_d'' are steeper in the disabled state, U_d, than in the non-disabled state, U_0. That is, to remain at the same level of utility, the disabled individual would require more income in return for a given increase in hours of work. For a given market wage rate (as given by the slope of the budget constraint) this would induce a substitution from labour market to nonlabour market activities, reducing work time from H_0 to H_d (work being measured from right to left on the diagram).

Obviously a disability may have a variety of the above effects. They were treated separately here only for expositional purposes. It is a relatively straightforward matter to portray them in various realistic combinations. For example, the disability may reduce the time one is able to work, reduce the wage, impose medical costs, and increase the disutility associated with labour market work as opposed to other activities. In such circumstances, the budget constraint and utility isoquants will change to reflect each of these effects and the incentive to work will be affected accordingly.

Effect of Compensation

To compensate for the previously discussed type of disability, various forms of compensation exist including workers' compensation, disability pension entitlements from public and private pensions, long-term disability insurance, and court awards. The effect of such compensations upon well-being or utility and work incentives depends upon the form of compensation.

Workers' compensation, for instance, is designed to support the recipient's income so it is not reduced by as much as it would be in the absence of the compensation. For example, it may compensate for two-thirds of the loss in one's former potential income, in which case the worker's compensation budget constraint appears as the dashed line $E_0 Y_d H_f$ in Figure 4.6(a). That is, as the worker is compelled to reduce work activity (move to the right of E_0), income does not fall by the full drop in take-home pay (as it would along $E_0 H_f$), but rather it falls by the proportion that it is not supported by workers' compensation. If workers' compensation were two-thirds of one's income loss then the fall in income would be one-third.

The policy challenge is to design a compensation scheme that compensates those who are legitimately disabled and hence in need of compensation and yet provides an incentive for them to return to work where feasible. The potential for an adverse work incentive effect for those who could legitimately return to work is exhibited by the fact that the workers' compensation budget constraint is like the previously discussed budget constraint under a negative income tax. That is, the additional income enables one to afford to work less (income effect); further, the reduced opportunity cost of leisure time that comes about because one forgoes disability payments as one increases hours of work (substitution effect) also discourages work activity. Obviously, for those who have no discretion over their ability to engage in more work activity, the incentive effects are not at issue. For some, however, they may discourage a return to work activity that they are capable of doing.

The availability of the full budget constraint $E_0 Y_d H_f$ presupposes that such compensation is available for any combination of partial disabilities that would reduce the person's work between H_0 and H_f. That is, depending upon preferences, workers could locate anywhere along the segment $E_0 Y_d$. If the compensation were available only if one could prove full disability, H_f, then the worker's options would be only the points Y_d and E_0. Assuming Y_d were above the indifference curve, U_0, as shown in Figure 4.6(a), then the worker would choose the point Y_d.

Figure 4.6(b) illustrates such a situation where the worker receives workers' compensation for the full disability and has no incentive to return to work, assuming of course that such an option were feasible. The compensation ($E_c H_f$) is such that the worker is better off by not working, and receiving the compensation (point E_c) than by returning to work (point E_0). That is, utility under

Figure 4.6
EFFECT OF COMPENSATION

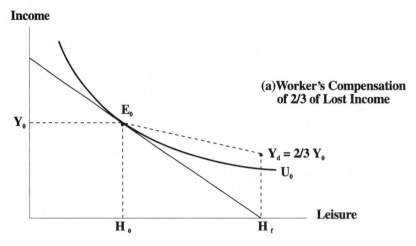

Income

Y_0

E_0

(a) Worker's Compensation
of 2/3 of Lost Income

$Y_d = 2/3 \ Y_0$

U_0

Leisure

H_0 H_f

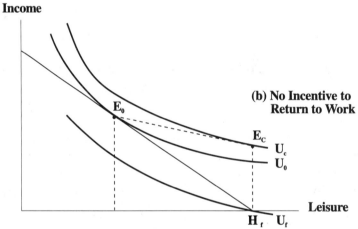

Income

E_0

(b) No Incentive to
Return to Work

E_C

U_c

U_0

H_f U_f

Leisure

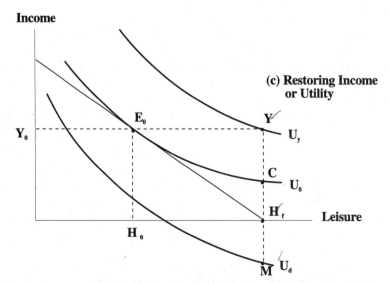

Income

Y_0

E_0

(c) Restoring Income
or Utility

Y

U_y

C

U_0

H_f

Leisure

H_0

M U_d

compensation (U_c) is greater than utility under work (U_0). Even though income by working would be greater than income under compensation, it is not sufficient to offset the disutility associated with the additional work.

However, not providing any compensation runs the risk that individuals who legitimately cannot return to work have their utility level reduced to U_f. Clearly, lowering the compensation level so that U_c is below U_0 would induce a return to work from those who can legitimately engage in such work; however, it runs the risk of penalizing those who legitimately cannot return to work. Also, when one considers the considerable medical costs and pain and suffering that are often associated with disability, it becomes less likely that compensating someone for a portion of their lost earnings would make them better off.

Different amounts of compensation are also involved depending upon the position to which the recipient is to be returned. Figure 4.6(c) illustrates the situation where a permanently disabling injury prevents the individual from working at all (i.e., forces the individual to locate at H_f) and gives rise to medical costs of $H_f M$. These events reduce the individual's well-being from U_0 to U_d. A court award to compensate for the individual's loss of earnings, $Y H_f$, plus medical costs, $H_f M$, would lead to a total award of YM. This would clearly leave the individual better off than before ($U_y > U_0$) because the individual is not experiencing the disutility associated with working. If, however, the disability results in pain and suffering then the individual's ability to enjoy goods and leisure consumption may be reduced and the individual could be worse off than before, even with the compensation award of YM (and possibly even with some additional compensation for pain and suffering).

If the court wanted to restore an individual to his former level of well-being or utility, U_0, then an award of CM would be appropriate. The problem is to assess the value the individual may attach to not having to work. This is given in the diagram by YC, which is the income reduction the individual would have willingly accepted to reduce work activity from H_0 to H_f (equal to $E_0 Y$) and to remain indifferent to being at E_0 with utility U_0. The problem is that YC is not observed; all that is observed is the person's income $Y H_f$.

This clearly illustrates the dilemma any court or administrative tribunal would have in arriving at compensation to "restore the person in whole." Different amounts are involved if this is interpreted to be the previous level of income as opposed to welfare or well-being.

ILLUSTRATIVE EVIDENCE OF INCENTIVE EFFECTS

Evidence on incentive effects of income maintenance programs comes from both nonexperimental and experimental sources. Nonexperimental studies relate measures of labour supply to various parameters related to the income maintenance schemes, notably wages and nonlabour income. In addition, results from experiments in the U.S. are available on the incentive effects of various negative income tax plans.[2]

Results from the nonexperimental literature are of dubious use in analyzing the work incentive effects of low-income families, not only because of their wide variation in results, but also because they generally apply to middle- and high-income families. In spite of their wide variation in results, they generally confirm the static predictions of the work-leisure model. (These results are outlined in more detail subsequently in Chapter 8.) Other things being equal, an increase in nonlabour income (such as would come about from a demogrant, a negative income tax guarantee, welfare, or unemployment insurance) reduces labour supply, confirming the existence of a pure income effect. Other things being equal (including family income), a decrease in wages (such as would come about from the tax rate in a negative income tax or welfare) also decreases labour supply, confirming the existence of a pure or compensated substitution effect. For most groups in the economy, the income effect appears to dominate the substitution effect, so that wage increases have been associated with a reduction in labour supply. For women, especially married women, most studies conclude that the substitution effect dominates the income effect, so that wage increases have been associated with an increase in labour supply. However, as discussed subsequently, some recent research has questioned this conclusion and has found that, for women who work full time, wage increases are associated with a decrease in labour supply.

This suggests that for most income maintenance programs, work incentives will be altered as predicted by the static theory. Demogrants involve only a pure income effect which would reduce incentives similar to increases in nonlabour income. Negative income taxes, welfare, and unemployment insurance have income and pure substitution effects working in the same direction to reduce incentives. Wages subsidies with their income and substitution effects working in the opposite direction would probably reduce work incentives for most groups except perhaps females, especially married females and those who work part-time.

Moffitt (1992) provides a comprehensive review of the effect of U.S. welfare programs on various aspects of work incentives. He finds that the incentive effects are generally in the direction predicted by economic theory. Specifically, higher guarantees reduce the incentive to work as do higher tax-back rates. There is some evidence of the intergenerational transmission of welfare dependence in that children of parents on welfare are more likely themselves to be on welfare, even after controlling for other determinants of welfare. There is also some evidence that, even after controlling for other determinants of being on

---

2. These results are discussed in more detail in Cain (1974), Peckman and Timpane (1975), Danziger, Haveman and Plotnick (1981), Moffitt (1981, 1992), and Basilevsky and Hum (1984).

welfare, the longer one is on welfare, the less likely one is to leave welfare. This "negative duration dependence" may occur because skills deteriorate or employers regard welfare as a negative signal, or because those who stay on welfare have conventionally unobserved characteristics (termed unobserved heterogeneity) that are not controlled for in the statistical analysis and that make them less likely to leave welfare. "Workfare" requirements that require welfare recipients to undertake training seem to have a positive effect on their earnings. Although the studies are subject to more controversy, there is also some evidence that welfare can affect migration decisions, inducing potential recipients to move from low welfare areas to high welfare areas.

The available evidence from the experimental literature comes from one Canadian and four U.S. negative income tax experiments: New Jersey and Pennsylvania (1968–72); rural areas of North Carolina and Iowa (1970–72); Seattle and Denver (1970–78); and Gary, Indiana (1971–74). The Canadian experiment was carried out in Manitoba during the period 1975–79; however, analysis of the results of this program was only recently carried out (see Hum and Simpson, 1992). Different families were given varying amounts of basic guaranteed income and their labour market earnings were subjected to various tax rates. Their work behaviour was then compared to that of a control group of similar families that were not under a negative income tax plan.

Moffitt (1981, p. 24) summarizes the effect on work incentives of the four U.S. experiments, based on an average tax rate of 0.50 and a guarantee level about equal to the poverty line. The conclusions are: (1) overall hours of work are unequivocally reduced by the presence of the negative income tax; (2) the disincentive effects vary considerably by demographic group with the reduction in hours of work ranging from 1 to 8 percent for husbands, 0 to 55 percent for wives, and 12 to 28 percent for female heads of families; (3) the work reduction for men occurred more in the form of reduced employment to zero hours of work for a small number of the men rather than marginal reductions in the hours of work of those who remained employed; (4) some marginal reductions in hours of work did occur through reduced overtime; (5) the reduced employment often occurred in the form of a lengthening of time spent between jobs for men and increased school attendance for youths; however, these responses did not appear to lead to subsequent wage gains; (6) for most groups, disincentive effects generally occurred in response to both the guarantee and the implicit tax rate, although the groups seemed more responsive to changes in the guarantee than the tax rate; (7) the longer the program the greater the disincentive effects, suggesting that the disincentive effects may be larger for a permanent as opposed to temporary program.

Hum and Simpson (1992) found somewhat smaller impacts on hours of work in the Canadian income experiment. Hours worked declined between 0.8 and 1.6 percent for men, between 2.4 and 3 percent for married women, and between 3.8 and 5.3 percent among single women. Evidence on the incentive

effects of unemployment insurance — including past and proposed reforms —
is discussed in the subsequent chapter on unemployment.

INCOME MAINTENANCE: CURRENT POLICY ISSUES

Recent proposals for changes to the Canadian income security system illustrate
several of the issues raised in this chapter. In its 1985 Report, the Royal Commission on the Economic Union and Development Prospects for Canada (the
Macdonald Commission) recommended fundamental reform of existing income
maintenance programs. The centrepiece of its proposal was the Universal Income Security Program (UISP), a negative income tax program that would
supplement the income of families with low incomes. Also recommended were
substantial changes to the existing unemployment insurance (UI) system and the
introduction of a Transitional Adjustment Assistance Program (TAAP) designed
to assist individuals to adjust to changing economic circumstances through relocation and retraining. Subsequently, in its 1986 Report, the Commission of
Inquiry on Unemployment Insurance (the Forget Commission) also recommended substantial changes to the unemployment insurance system and proposed a negative income tax scheme for supplementing low family incomes.
Proposals similar to those of the Macdonald and Forget Commissions were also
made by the Newfoundland Royal Commission on Employment and Unemployment (the Hause Commission) in its 1986 Report. Although the details of the
proposals made in these Commission Reports differ in important ways, the
general nature of the proposed reforms — in particular the adoption of a universal negative income tax scheme — are similar.

A common theme of these three Commission Reports is the need for substantial reform of existing income maintenance programs. Several factors contribute
to the need for reform. The adverse incentives associated with the existing
welfare or social assistance programs are an important consideration. As discussed previously, most welfare programs involve an implicit 100 percent tax on
income from work, as welfare benefits are reduced dollar-for-dollar in response
to any employment income. The tax can exceed 100 percent if those who earn
income not only forgo their welfare benefits but also additional benefits such as
subsidized housing or dental care. Relative to welfare, the negative income tax
scheme provides stronger work incentives as individuals who are able to do some
market work earn higher after-tax incomes. In the short run this employment
income reduces the amount of income recipients receive from the state (although not dollar-for-dollar, so that the implicit tax rate is less than 100 percent)
and thus reduces the burden on other taxpayers. In the long run the need for
income assistance may be eliminated as individuals gain work experience and
skills.

A second reason for reform is the existence of a significant number of the
"working poor" — individuals who are employed but whose family incomes fall

below the poverty line. In Canada these individuals receive little or no assistance from the state. A negative income tax scheme would supplement the income of these families, thus reducing the amount of poverty in Canada. Some reduction in the hours of work of these individuals may occur; as noted previously, relative to a situation of no income assistance, the negative income tax produces income and substitution effects which tend to reduce hours of work. The magnitude of these adverse work effects will depend on the program parameters — specifically, the income guarantee and the implicit tax rate.

The Macdonald Commission's proposed UISP would provide a low basic income guarantee (in 1984 dollars, $2750 per adult and $750 per child under 18, with an extra $2000 for single parents and an extra $1075 for the single elderly) and a low rate (20 percent). This combination of design parameters minimizes adverse work incentives for those who are able to work yet provides substantial additional income for the working poor. However, for those who are unable to work, the income guarantee (for example, $7000 per year for a family of four) is not sufficient to meet basic family needs. For this reason, the Macdonald Commission proposed a "two-tier" system, with unemployables receiving the basic income guarantee from the federal government and additional assistance from the relevant provincial government. Those able to work would not be eligible for these provincial "top-ups."

This type of two-tier negative income tax scheme, in which a distinction is made between employables and unemployables, has significant advantages over a single-tier scheme. In particular, it is extremely difficult to design a single-tier negative income tax system which (a) has a minimum income guarantee that is sufficient to meet the basic needs of those who are unable or not expected to work, (b) has a low implicit tax rate so that incentives to work are maintained, and (c) is feasible in terms of total cost.

A third aspect of reform is the unemployment insurance program. Although the primary purpose of this program is to provide insurance against the risk of income loss associated with unemployment, UI has evolved into an income maintenance program in certain regions (especially the Maritimes and eastern Quebec) and industries (especially fishing, forestry, and construction). However, UI is not well suited as an income maintenance program for several reasons. Income maintenance objectives focus on family income and need, whereas UI benefits are based on the individual worker's income and are not related to assets or need. In addition, income support programs attempt to reconcile family income and need over periods of six months to one year whereas UI is based on a much shorter accounting period (weekly). Unemployment insurance could be altered further to achieve income distribution goals — for example, by basing UI benefits on family income and need (e.g., number of dependents) and testing for other income and assets — but such changes would reduce the program's insurance role. The proposals of the Macdonald and Forget Commissions are designed to focus UI strictly on the insurance objective and to leave income distribution objectives to programs explicitly designed for that purpose — such as the Macdonald Commission's UISP. In addition, the proposals are designed to reduce some of the adverse work incentive effects that have been found to be associated with the existing unemployment insurance program. (These are discussed further in the chapter on unemployment.)

In summary, recent analyses of Canada's income security programs have concluded that the work incentive effects of these programs — in particular, welfare and unemployment insurance — are a matter of policy concern. At the same time, significant gaps exist in these programs, especially with respect to the working poor. In order to remedy these deficiencies, public inquiries have recommended replacing existing income support programs (such as the Canada Assistance Plan, Child Tax Credit, Family Allowances, and Guaranteed Income Supplement) by a universal two-tier negative income tax scheme.

Recent reforms of the unemployment insurance program have occurred under the federal government's Labour Force Development Strategy. The reforms involved reallocating funds from the unemployment insurance fund to the training and mobility components of the Canadian Job Strategy (CJS) program. The CJS is the main program of the federal government for dealing with labour adjustment issues. It involves six main components, often targeted to different groups:

(1) *Job Development* — training and work experience for the long-term unemployed;

(2) *Job Entry* — training and work experience for youths and women re-entering the labour force;

(3) *Skill Shortages* — training and relocation assistance for employers with skill shortages;

(4) *Skill Investment* — training and worksharing for workers affected by technological and other market changes;

(5) *Community Futures* — small business development and training and relocation;

(6) *Innovations* — support for innovative solutions to labour market problems.

This reallocation from unemployment insurance to training and mobility assistance under the CJS has occurred in response to the criticism from a variety of high-profile sources[3] that Canada allocates too much of its adjustment assistance resources in the direction of income maintenance and not enough in the direction of positive adjustment assistance programs like retraining or mobility. Of the seven major industrialized nations (Canada, the United States, the United

3. For example, the Advisory Council on Adjustment (1989), Employment and Immigration Canada (1989) and the Economic Council of Canada (1991).

Kingdom, Germany, France, Japan, and Sweden) Canada ranks highest in terms of its public expenditures, such as unemployment insurance, that go towards income maintenance, and lowest in terms of the percent that goes to more active labour market policies such as training.[4]

While income maintenance programs may provide important income support, they may discourage or slow down the reallocation of labour towards the expanding job opportunities. More active labour market policies, for example, retraining and mobility, in contrast, should facilitate market adjustments in the direction of the expanding job opportunities. In the often-cited phraseology of the Advisory Committee on Adjustment (1989), passive income maintenance programs like unemployment insurance emphasize the "safety net" approach to adjustment assistance. In contrast, more active adjustment assistance programs, such as retraining and mobility, emphasize the "trampoline effect" of helping workers bounce back into the labour market. The concern is that emphasis on the safety net may actually encourage people to stay in declining sectors and regions, and that this may slow down the adjustment process in the direction of basic market forces towards the expanding sectors.

Income maintenance programs (e.g., unemployment insurance) subsidize adjustment when it involves unemployment but not when it involves other forms of adjustment. This may discourage marginal adjustments on the part of workers and employers. Marginal adjustments on the part of workers include mobility, reductions in hours of work or even wages, and decisions to enter certain occupations or industries. Marginal adjustments on the part of employers include attrition, reduced new hiring, changes in hours worked, and the internal redeployment of workers. For example, since adjustment through unemployment is being subsidized, it may be in the interest of both employers and employees to utilize layoffs as an adjustment procedure rather than to internally redeploy redundant workers, perhaps in retraining efforts, so as to prepare them for the changing job opportunities. By discouraging continuous adjustment at the margin, passive income maintenance programs may lead to more disruptive inframarginal adjustments (e.g., plant closings, mass layoffs) if they ultimately occur.

While passive income maintenance programs such as unemployment insurance have these potential disadvantages, this must be put in perspective. They can provide important income support for people who are unemployed, and thereby fulfill their insurance role. They can encourage some labour market work to build up eligibility for UI — in that vein, they are a work-conditioned income maintenance scheme since eligibility and benefits depend upon one's history of work. They may even be a viable form of worksharing for scarce jobs. For some recipients, the income support may enable them to be able to afford to search for a new job, or to not have to simply take the first job that comes along

---

4. Economic Council of Canada, 1991, p. 131.

— a job that may not be conducive to their long-run employment prospects. There is also concern that proposals to reallocate resources from passive income maintenance programs (e.g., UI) to more active ones (e.g., training) are really just veiled attempts to reduce support in general. As well, adjustment assistance in areas such as retraining will carry its own set of problems, not the least of which is the possibility of simply displacing private training efforts that would go on in the absence of public support. Clearly, as in most policy areas, difficult trade-offs are involved.

QUESTIONS

1. Use the work-leisure choice diagram to illustrate that, under normal circumstances, if an individual is given a demogrant, take-home pay will not increase by as much as the demogrant. Why is this so?

2. Illustrate the work incentive effects of a demogrant on a disabled person who is unable to work.

3. The negative income tax involves a positive guarantee which has a pure income effect reducing work effort. It also involves a tax increase on the earned income of recipients, and this has both an income and substitution effect working in the opposite direction in their effect on work incentives. Consequently, we are unable to predict unambiguously the static, partial equilibrium effects that a negative income tax plan would have on work incentives. Indicate why this last statement is wrong.

4. Use the work-leisure choice diagram to illustrate the following cases where an individual is: unwilling to go on welfare because the welfare payment is too low; indifferent between welfare and work; induced to move off welfare because of an increase in his wage; induced to move off welfare because of a change in preferences between income and leisure; induced to move off welfare by a reduction of the 100 percent implicit tax on earnings.

5. Use the work-leisure framework to illustrate that if we hold the post-subsidy income of the recipients constant or the size of the subsidy constant, a negative income tax will involve more adverse work incentives than would a wage subsidy.

6. In an unemployment insurance diagram like Figure 4.3, depict the case where the individual must work eight weeks before becoming eligible to collect unemployment insurance. Depict the case of a two-third rather than 60 percent replacement rate.

7. In a diagram like Figure 4.3 depict the workers' compensation system in Ontario that recently went to a wage-loss system of compensating for 90 percent of the pre-injury wage.

REFERENCES AND FURTHER READINGS

(References to the incentive effects of unemployment insurance are contained in the subsequent chapter on unemployment.)

Advisory Council on Adjustment. *Adjusting to Win*. Ottawa: Supply and Services, 1989.

Ashenfelter, O. and M. Plant. Nonparametric estimates of the labor-supply effects of negative income programs. *JOLE* 8 (January 1990) S396–S415.

Barr, N. and R. Hall. The probability of dependence on public assistance. *Economica* 48 (May 1981) 109–124.

Basilevsky, A. and D. Hum. *Experimental Social Programs and Analytical Methods: An Evaluation of the U.S. Income Maintenance Projects*. New York: Academic Press, 1984.

Besley, T. and S. Coate. Workfare versus welfare: incentive arguments for work requirements in poverty alleviation programs. *AER* 82 (March 1992) 249–261.

Betson, D. and J. Bishop. Reform of the tax system to stimulate labor supply: efficiency and distributional effects. *IRRA* (September 1980) 307–315.

Blau, D. and P. Robins. Labor supply response to welfare programs: a dynamic analysis. *JOLE* 4 (January 1986) 82–104.

Berkowitz, M. and J. Burton Jr. *Permanent Disability Benefits in Workers' Compensation*. Kalamazoo, Mich.: W.E. Upjohn Institute, 1987.

Blomquist, N. The effect of income taxation on the labour supply of married men in Sweden. *JPubEc* 22 (November 1983) 169–198.

Blomquist, N. Nonlinear taxes and labor supply. *European Economic Review* 32 (July 1988) 1213–1226.

Blomquist, N. and U. Hansson-Brusewitz. The effect of taxes on male and female labour supply in Sweden. *JHR* 25 (Summer 1990) 317–357.

Bosworth, B. and G. Burtless. Effects of tax reform on labor supply, investment and saving. *JEP* 6 (Winter 1992) 3–25.

Bound, J. The health and earnings of rejected disability insurance applicants. *AER* 79 (June 1989) 482–503.

Browning, E. Alternative programs for income redistribution: the NIT and the NWT. *AER* 63 (March 1973) 38–49.

Burkhauser, R. and R. Vaeman. *Disability and Work: The Economics of American Policy*. Baltimore: The Johns' Hopkins Press, 1982.

Burtless, G. and D. Greenberg. Inferences concerning labor supply behavior based on limited-duration experiments. *AER* 72 (June 1982) 488–497.

Burtless, G. and D. Greenberg. Measuring the impact of NIT experiment on work effort. *ILRR* 36 (July 1983) 592–605.

Burtless, G. and J. Hausman. The effect of taxation on labor supply: evaluating the Gary negative income tax experiment. *JPE* 86 (December 1978) 1103–1130.

Cain, G. (ed.) Symposium articles — the graduated work incentive experiment. *JHR* 9 (Spring 1974).

Cain, G. and H. Watts. *Income Maintenance and Labor Supply.* Chicago: Rand McNally, 1973.

Cogan, J. Labor supply and negative income taxation: new evidence from the New Jersey-Pennsylvania experiment. *EI* 21 (October 1983) 465–484.

Cowell, F. Taxation and labour supply with risky activities. *Economica* 48 (November 1981) 365–380.

Cowell, F. Tax evasion with labour income. *JPubEc* 26 (February 1985) 19–34.

Coyte, P. Optimal tax-subsidy schemes in a screening model of the labor market with information externalities. *Public Finance* 45 (No. 1, 1990) 37–58.

Danziger, S., R. Haveman and R. Plotnick. How income transfers affect work, saving and income distribution. *JEL* 19 (September 1981) 975–1028.

Daymont, T. and P. Andrisani. Some causes and consequences of disability benefit receipt. *IRRA* (December 1983) 177–188.

Dionne, G. and P. St.-Michel. Workers' compensation and moral hazard. *R.E. Stats.* 73 (May 1991) 236–244.

Economic Council of Canada. *Employment in the Service Economy.* Ottawa: Supply and Services, 1991.

Enberg, N., P. Gottschalk and D. Wolf. A random-effects target model of work-welfare transitions. *Journal of Econometrics* 43 (Jan./Feb. 1990) 63–75.

Employment and Immigration Canada. *The Canadian Jobs Strategy: A Review.* Ottawa: Employment and Immigration Canada, 1988.

Fallick, B. Unemployment insurance and the rate of re-employment of displaced workers. *R.E. Stats.* 73 (May 1991) 228–235.

Fenn, P. Sickness duration, residual disability, and income replacement: an empirical analysis. *EJ* 91 (March 1981) 158–173.

Fenn, P. and I. Vlachonikolis. Male labour force participation following illness or injury. *Economica* 53 (August 1986) 379–392.

Flood, L. Effects of taxes on nonmarket work: the Swedish case. *JPubEc* 36 (July 1988) 259–267.

Fortin, B. Income security in Canada. In *Income Distribution and Economic Security in Canada.* F. Vaillancourt (ed.). Toronto: University of Toronto Press, 1985, 153–186.

Fracker, T. and R. Moffit. The effect of food stamps on labor supply. *JPubEc* 35 (February 1988) 25–56.

Garfinkel, I. A skeptical note on the optimality of wage subsidy programs. *AER* 63 (June 1973) 447–453.

Garfinkel, I. and S. Masters. *Estimating Labor Supply Effects of Income Maintenance Alternatives.* New York: Academic Press, 1978.

Graham, J. and A. Beller. The effect of child support payments on the labor supply of female heads. *JHR* 24 (Fall 1989) 644–688.

Green, C. Negative taxes and the monetary incentives to work: the static theory. *JHR* 3 (Summer 1968) 280–288.

Greenberg, D. and H. Halsey. Systematic misreporting and effects of income maintenance experiments on work effort: evidence from the Seattle-Denver experiment. *JOLE* 1 (October 1983) 380–407.

Grossman, J. The work disincentive effect of extended unemployment compensation. *R.E. Stats.* 71 (February 1989) 159–163.

Gwartney, J. and R. Stroup. Labor supply and tax rates: a correction of the record. *AER* 73 (June 1983) 446–451. Also, comments and replies 76 (March 1986) 277–285.

Hansson, I. and C. Stuart. Taxation, government spending, and labor supply: a diagrammatic exposition. *EI* 21 (October 1983) 584–587.

Haveman, R. and B. Wolfe. Disability transfers and early retirement: a causal relationship? *JPubEc* 24 (June 1984) 47–66.

Haveman, R., P. de Jong and B. Wolfe. Disability transfers and the work decisions of older men. *QJE* 106 (August 1991) 939–950.

Heckman, J. Effects of child-care programs on women's work effort. *JPE* 82 (March/April 1974) S136–S163.

Hemming, R. Income tax progressivity and labor supply. *JPubEc* 14 (August 1980) 95–100.

Hosek, J. Determinant of family participation in the AFDC-unemployed fathers program. *R.E. Stats.* 62 (August 1980) 466–470.

Hum, D. and W. Simpson. *Income Maintenance, Work Effort, and the Canadian M-income Experiment.* Ottawa: Economic Council of Canada, 1992.

Hurd, M. and J. Pencavel. A utility-based analysis of the wage subsidy program. *JPubEc* 15 (April 1981) 185–202.

Hutches, R. Entry and exit transitions in a government transfer program. *JHR* 16 (Spring 1981) 217–237.

Ippolito, R. Income tax policy and lifetime labor supply. *JPubEc* 26 (April 1985) 327–348.

Jacobson, S. The effects of pay incentives on teacher absenteeism. *JHR* 24 (Spring 1989) 280–286.

Jenkins, S. Lone mothers' employment and full-time work probabilities. *EJ* 102 (March 1992) 310–320.

Johnson, T. and J. Pencavel. Forecasting the effects of a negative income tax program. *ILRR* 35 (January 1982) 211–234.

Johnson, W. The effect of a negative income tax on risk-taking in the labor market. *EI* 18 (July 1980) 395–407.

Johnson, W. and J. Ondrich. The duration of post-injury absences from work. *R.E. Stats.* 72 (November 1990) 578–586.

Journal of Human Resources. Special issue on Taxation and Labor Supply 25 (Summer 1990).

Keeley, M. *Labor Supply and Public Policy.* New York: Academic Press, 1981.

Keeley, M., P. Robbins, R. Spiegelman and R. West. The estimation of labor supply models using experimental data. *AER* 68 (December 1975) 873–887.

Keeley, M., P. Robbins, R. Spiegelman and R. West. The labor supply effects and costs of alternative income tax programs. *JHR* 13 (Winter 1978) 3–36.

Kenym, P. and P. Dawkins. A time series analysis of labour absence in Australia. *R.E. Stats.* 71 (May 1989) 232–239.

Kesselman, J. Labour supply effects on income, income-work, and wage subsidies. *JHR* 4 (Summer 1969) 275–292.

Kesselman, J. Conditional subsidies in income maintenance. *WEJ* 9 (March 1971) 1–20.

Kesselman, J. A comprehensive approach to income maintenance: SWIFT. *JPubEc* 2 (February 1973) 59–88.

Kesselman, J. Egalitarianism of earnings and income taxes. *JPubEc* 5 (April/May 1976) 285–302.

Kesselman, J. and I. Garfinkel. Professor Friedman meet Lady Rhys-Williams: NIT vs. CIT. *JPubEc* 10 (October 1978) 179–216.

Killingsworth, M. Must a negative income tax reduce labor supply? *JHR* 11 (Summer 1976) 354–365.

Krueger, A. Incentive effects of workers' compensation insurance. *JPubEc* 41 (February 1990) 73–99.

Leonesio, M. Predicting the effects of in-kind transfers on labor supply. *SEJ* 54 (April 1988) 901–912.

Leonesio, M. In-kind transfers and work incentives. *JOLE* 6 (October 1988) 515–529.

Leuthold, J. The effect of taxation on the hours worked by married women. *ILRR* 31 (July 1978) 520–526.

Levy, F. The labor supply of female household heads, or AFDC work incentives don't work too well. *JHR* 14 (Winter 1979) 76–97.

Lindbeck, A. Tax effects versus budget effects on labor supply. *EI* 20 (October 1982) 473–489.

Marvel, H. An economic analysis of the operation of social security disability insurance. *JHR* 17 (Summer 1982) 393–412.

Masters, S. and I. Garfinkel. *Estimating the Labor Supply Effects of Income-Maintenance Alternatives.* New York: Academic Press, 1978.

McElwain, A. and J. Swofford. The social security payroll tax and the life-cycle work pattern. *JHR* 21 (Spring 1986) 279–287.

Meltz, N., F. Reid and G. Swartz. *Sharing the Work.* Toronto: University of Toronto Press, 1981.

Micklewright, J. On earnings-related unemployment benefits and their relation to earnings. *EJ* 95 (March 1985) 133–145.

Moffit, R. The negative income tax: would it discourage work? *MLR* 104 (April 1981) 23–27.

Moffit, R. Work incentives of the AFDC system. *AER* 76 (May 1986) 219–223.

Moffit, R. Incentive effects of the U.S. Welfare system: a review. *JEL* 30 (March 1992) 1–61.

Moffit, R. and K. Kehrer. The effect of tax and transfer programs on labor supply: the evidence from the income maintenance experiments. In R. Ehrenberg (ed.) *Research in Labor Economics*. Greenwich, Conn.: JAI Press, 1981.

Moffitt, R. and A. Ransarajan. The work incentives of AFDC tax rates. *JHR* 26 (Winter 1991) 165–179.

Munro, A. In-kind transfers, cash grants and the supply of labor. *European Economic Review* 33 (October 1989) 1597–1604.

Nagi, S. and L. Hadley. Disability behavior, income change and motivation to work. *ILRR* 25 (January 1972) 223–235.

Narendranathan, W., S. Nickell, and J. Stern. Unemployment benefits revisited. *EJ* 95 (June 1985) 307–329.

Ontario, Social Assistance Review Committee. *Transitions.* Toronto: Ministry of Community and Social Services, 1988.

Parsons, D. The decline in male labor force participation. *JPE* 88 (February 1980) 117–134. Also comments by R. Haveman and B. Wolf, and reply, 92 (June 1984) 532–549.

Parsons, D. The male labour force participation decision: health, reported health, and economic incentives. *Economica* 49 (February 1982) 81–92.

Phipps, S. Quantity-constrained household responses to unemployed insurance reform. *EJ* 100 (March 1990) 124–140.

Phipps, S. Behavioral response to UI reform in constrained and unconstrained models of labour supply. *CJE* 24 (February 1991) 34–54.

Phipps, S. Equity and efficiency consequences of unemployment insurance reform in Canada. *Economica* 58 (May 1991) 199–214.

Plotnick, R. Turnover in the AFDC population. *JHR* 18 (Winter 1983) 65–81.

Prescott, D., R. Swidinsky and C. Wilton. Labour supply estimates for low-income female heads of households using M income data. *CJE* 19 (February 1986) 134–141.

Rea, S. Incentive effects of alternative negative income tax plans. *JPubEc* 3 (August 1974) 237–249.

Rea, S. Investment in human capital under a negative income tax. *CJE* 10 (November 1977) 607–620.

Reid, F. UI-assisted worksharing as an alternative to layoffs: the Canadian experience. *ILRR* 35 (April 1982) 319–329.

Reid, F. Reductions in worktime: an assessment of employment sharing to reduce unemployment. *Work and Pay.* W.C. Riddell (ed.). Toronto: University of Toronto Press, 1985.

Rice, P. Juvenile unemployed, relative wages and social security in Great Britain. *EJ* 382 (June 1986) 352–374.

Rosen, H. Tax illusion and the labour supply of married women. *R.E. Stats.* 58 (May 1976) 167–172.

Rosen, H. Taxes in a labor supply model with joint wage-hours determination. *Econometrica* 44 (May 1976) 485–507.

Ruser, J. Workers' compensation and occupational injuries and illness. *JOLE* 9 (October 1991) 325–350.

Scheffler, R. and G. Iden. The effect of disability on labor supply. *ILRR* 28 (October 1974) 122–132.

Slade, F. Older men, disability insurance and the incentive to work. *IR* 23 (Spring 1984) 260–269.

Solon, G. Work incentive effects of taxing unemployment benefits. *Econometrica* 53 (March 1985) 295–306.

Steinberg, D. Induced work participation and the returns to experience for welfare women. *Journal of Econometrics* 41 (July 1989) 321–340.

Stelcner, M. and J. Breslaw. Income taxes and the labor supply of married women in Quebec. *SEJ* 51 (April 1985) 1053–1072.

Swisher, I. The disabled and the decline in men's labor force participation. *MLR* 96 (November 1973) 53.

Triest, R. The effect of income tax on labor supply when deductions are endogenous. *R.E. Stats.* 74 (February 1992) 91–99.

Watson, H. Tax evasion and labor markets. *JPubEc* 27 (July 1985) 231–246.

Weaver, C. (ed.). *Disability and Work: Incentives, Rights and Opportunities*. Washington, D.C.: AEI Press, 1991.

Winkler, A. The incentive effects of medicaid on women's labor supply. *JHR* 26 (Spring 1991) 308–337.

Woodbury, S. and R. Spiegelman. Bonuses to workers to reduce unemployment: randomized trials in Illinois. *AER* 77 (September 1987) 513–530.

CHAPTER 5

Labour Force Participation

The labour force participation decision is basically a decision to participate in labour market activities as opposed to other activities such as household work, education, or retirement. As such it influences the size and composition of our labour force and it has an impact on household activities, education, and retirement programs.

The policy implications of these changes can be dramatic. Changes in the size and composition of our labour force affect our growth and unemployment rates, as well as the occupational and sex composition of the labour force. The latter, in turn, affect such factors as relative wages, demands for unionization, day care, and equal pay and equal employment opportunity legislation. Changes in household activities can involve family formation and mobility. Retirement programs can be affected in so far as new labour force participants will add contributions to pension funds, while those who retire (i.e., do not participate in the labour force) will be a drain on the funds.

Clearly, various public policy decision are both affected by, and have an impact on, the labour force participation decision. Consequently, it is useful to be able to forecast changes in labour force participation and to predict the impact of policy changes on this decision. This requires an understanding of the determinants of the labour force participation decision. The basic theoretical framework for analyzing this decision is the income-leisure choice model, developed formally in Chapter 2. The collection of international studies in Layard and Mincer (1985) illustrates the usefulness of this basic economic framework in understanding labour force participation patterns in a variety of countries with very different economic and other conditions. These studies also illustrate the interrelatedness of decisions on participation, education, and family formation.

In this chapter, the income-leisure choice framework is applied to the labour force participation decision. Before presenting the theory and empirical evidence, however, it is necessary to discuss a variety of definitions associated with the labour force concept.

DEFINING THE LABOUR FORCE

As illustrated in Figure 5.1, the labour force consists of those persons in the eligible population who participate in labour market activities, as either employed or unemployed. The eligible population is that portion of the population that is surveyed[1] as potential labour force participants (i.e., civilian noninstitutional population, 15 years and over, excluding the Yukon, Northwest Territories, and reservations). Persons from that potential population of labour force participants (POP) are categorized as either in the labour force (LF) or not in the labour force (NLF). Those in the labour force are either employed[2] (E), or unemployed (U), with the latter being not employed but actively seeking work. Those not in the labour force are usually students, retired people, persons in the household, or some "discouraged workers" who have simply given up looking for work. According to labour force definitions, the latter are not categorized as unemployed because they are not actively seeking work.

The labour force participation rate (LFPR) is the portion of the eligible population that participates in the labour force (LFPR = LF/POP). The unemployment rate (UR) is the proportion of the labour force that is unemployed (UR = U/LF).

The Canadian Labour Force Survey, conducted by Statistics Canada, is currently based on a monthly sample of approximately 62,000 households. The results are published monthly, and the survey is described in more detail, in *The Labour Force* (Cat. No. 71–001). Between 1945 and 1952, the survey was quarterly. In 1976 it was conducted monthly and expanded substantially.

While the Labour Force Survey provides the most often used estimates for our labour force, employment and unemployment figures, other sources are available. In particular, the Canadian Census is conducted every five years, the most recent being in 1991 referring to activity during 1990. The census is more comprehensive (not being based on a sample from a larger population), and consequently includes richer details on such factors as unemployment by industry and occupation. However, its use is limited because it is conducted only every five years, there is a considerable time lag before the results are published, and its reliability on labour force questions may be questioned because, unlike the Labour Force Survey, it does not focus only on labour force activity.

LABOUR FORCE PARTICIPATION THEORY

The income-leisure choice model provides the underlying theoretical framework for both the labour force participation decision and the hours-of-work

1. The survey is published monthly as Statistics Canada, *The Labour Force*, Catalogue No. 71-001.
2. People are categorized as employed if they are normally employed but they happen to not be at work at the time of the survey because, for example, they are ill, or on strike, or on a short-term layoff subject to recall.

Figure 5.1
LABOUR FORCE CONCEPTS
(numbers refer to 1990)

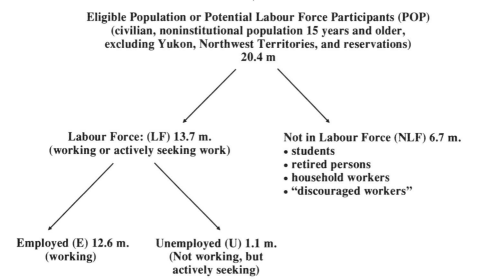

Population of Canada

Eligible Population or Potential Labour Force Participants (POP)
(civilian, noninstitutional population 15 years and older,
excluding Yukon, Northwest Territories, and reservations)
20.4 m

Labour Force: (LF) 13.7 m.
(working or actively seeking work)

Not in Labour Force (NLF) 6.7 m.
• **students**
• **retired persons**
• **household workers**
• **"discouraged workers"**

Employed (E) 12.6 m.
(working)

Unemployed (U) 1.1 m.
(Not working, but
actively seeking)

Some Definitions
Labour force participation rate: LFPR = LF/POP = 13,681,000/20,430,000 = 67.0%
Unemployment rate: UR = U/LF = 1,109,000/13,681,000 = 8.1%

Source: Statistics Canada, *Labour Force Annual Averages, 1990*, Catalogue No 71-220, Annual.

decision of those who participate. The main difference between the two decisions is that the participation decision involves a discrete choice between two labour force states (in or out of the labour force) while the hours decision involves a choice over a continuous variable.

The labour force participation decision is illustrated in Figure 5.2. Panel (a) shows an individual who will not participate given the individual's preferences concerning income and leisure, the amount of nonlabour income received per period (Y_N) and the market wage W_0 (which determines the slope of the potential income constraint). The highest possible utility is attained at the point on the

Figure 5.2
THE LABOUR FORCE PARTICIPATION DECISION

(a) Equilibrium of a Nonparticipant

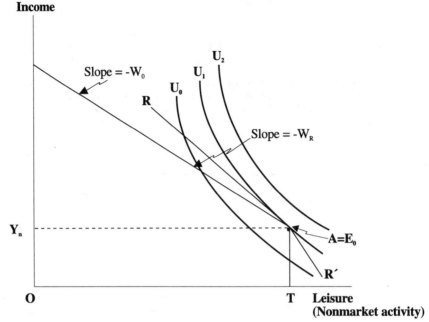

(b) Equilibrium of a Participant

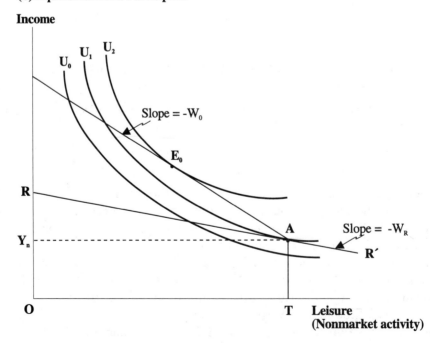

budget constraint corresponding to maximum "leisure" (zero hours of market work). This outcome is referred to as a "corner solution" because the household equilibrium occurs at one of the two extreme points on the potential income constraint. Figure 5.1(b) shows an individual who will participate given the individual's preferences and constraints. In this case the household equilibrium occurs in between the two extreme points on the potential income constraint, and is referred to as an "interior solution." With an interior solution the household equilibrium is characterized by a tangency between the budget constraint and the highest attainable indifference curve. At a corner solution, the tangency condition usually does not hold; the slopes of the highest attainable indifference curve and budget constraint typically differ, the indifference curve being more steeply sloped than the budget constraint.

A useful way of viewing the participation decision is to compare the individual's marginal rate of substitution (MRS) between income and leisure at the point of maximum leisure (point A in Figures 5.2(a) and 5.2(b)) with the market wage. The former measures the individual's *willingness* to exchange nonmarket time for income beginning from an initial position in which all time is devoted to nonmarket activities, while the market wage measures the individual's *ability* to exchange leisure time for income. When the MRS at zero hours of work (slope of indifference curve at A) exceeds the wage rate (slope of budget constraint), as at point A in Figure 5.2(a), the individual's implicit value of nonmarket time is high relative to the explicit market value of that time and the individual will therefore not participate. When the MRS at zero hours of work is less than the wage rate, as at the point A in Figure 5.2(b), the individual's implicit value of leisure time is less than the explicit market value of that time and the individual will participate in the labour market. In this case the individual increases hours of work until the marginal rate of substitution between income and leisure equals the wage rate, thereby exhausting all "gains from trade" associated with exchanging nonmarket for market time.

In analyzing labour force participation behaviour, economists often use the concept of a "reservation wage" which is defined as the wage rate at which an individual would be indifferent between participating and not participating in the labour force; that is, the wage at which an individual would be indifferent between work in the labour market as opposed to engaging in nonlabour market activities such as household work, retirement, or leisure activities, all of which require time. If the market wage rate were equal to the reservation wage w_R, the individual's potential income constraint would be given by the line RR' in Figures 5.2(a) and 5.2(b). That is, the reservation wage is the slope of the individual's indifference curve at zero hours of work. If the market wage is less than the reservation wage, as in Figure 5.2(a), the individual will not participate in the labour force; if the market wage exceeds the reservation wage, as in Figure 5.2(b), the individual will participate in the labour force since the return from engaging in labour market activity exceeds that individual's valuation of time in nonlabour market activities. (The concept of the reservation wage is developed more formally in Chapter 8).

The effects of changes in the wage rate and nonlabour income on labour force participation are illustrated in Figure 5.3. For a participant an increase in the wage rate has both income and substitution effects; because these are opposite in sign, hours of work may either increase or decrease. However, even if the income effect dominates the substitution effect is that hours of work decline, an increase in the wage rate can never cause a participant to withdraw from the labour force. This prediction of income-leisure choice theory is illustrated in Figure 5.3(a). As the wage increases, the household can achieve higher levels of utility and would never choose to not participate in the labour force which yields the utility level U_0.

For a nonparticipant, an increase in the wage rate may result in the individual entering the labour force or it may leave labour force participation unchanged, depending on whether the now higher market wage exceeds or is less than the individual's reservation wage. As illustrated in Figure 5.3(b), the increase in the wage from W_0 to W_1 leaves the individual's preferred outcome at E_0; that is, out of the labour force. However, the further wage increase to W_2 results in the individual entering the labour force (point E_2). This prediction can be explained in terms of income and substitution effects. An increase in the wage rate raises the opportunity cost of leisure time and tends to cause the individual to substitute market work for nonmarket activities. However, because the individual works zero hours initially, the income effect of an increase in the wage rate is zero. Thus for nonparticipants a wage increase has only a substitution effect which tends to increase hours of work.

These predictions can also be expressed in terms of the reservation wage. An increase in the market wage will not alter the labour force status of participants because the market wage already exceed their reservation wage. However, an increase in the market wage may cause some nonparticipants to enter the labour force if the increase in the wage is sufficiently large that the now higher market wage exceeds the reservation wage.

Together these two predictions of the income-leisure choice model imply that an increase in the wage rate can never reduce and may increase labour force participation.

The effects of an increase in nonlabour income on labour force participation are opposite to those of an increase in the wage rate. If leisure is a normal good, an increase in nonlabour income will never cause a nonparticipant to enter the labour force and may cause some participants to withdraw from the labour force. These predictions follow from the fact that an increase in nonlabour income has a pure income effect. Thus if leisure is a normal good, the amount of time devoted to nonmarket activities must either rise (if the individual was a labour force participant in the initial equilibrium) or not decline (if the individual was already at the maximum leisure point initially).

Figure 5.3 (a)
THE EFFECT OF A WAGE INCREASE ON A PARTICIPANT

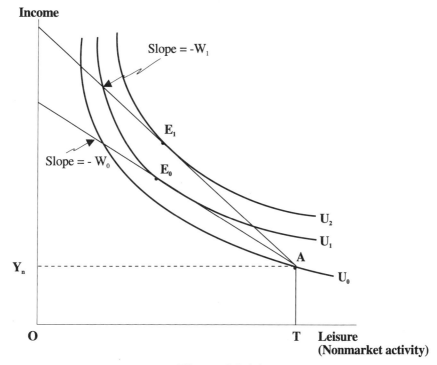

Figure 5.3 (b)
THE EFFECT OF A WAGE INCREASE ON A NONPARTICIPANT

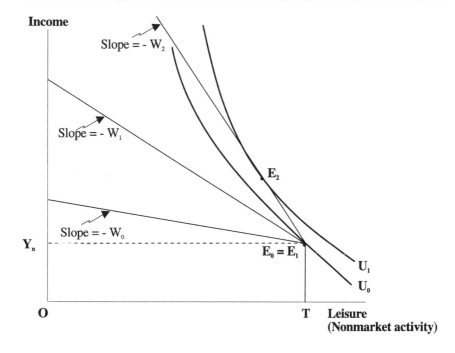

These predictions of the income-leisure choice model can alternatively be stated using the concept of a reservation wage. If leisure is a normal good, an increase in nonlabour income will increase the individual's reservation wage. (Proof of this statement is left as an exercise; see question 1 at the end of this chapter.) Thus some participants will withdraw from the labour force, their reservation wage having risen above the market wage, while nonparticipants will remain out of the labour force, their reservation wage having been above the market wage even prior to the increase in nonlabour income.

The determinants of an individual's labour force participation decision can be conveniently categorized as to whether the variable affects the individual's reservation wage, market wage or both. Other things equal, a variable that raises the individual's reservation wage would decrease the probability of participation in labour market activities. Such variables could be observable in the sense that data are often available on characteristics such as the presence of children, the availability of nonlabour market income, and the added importance of household work as on a farm. Or the variables could be unobservable in the sense that data are not directly available on the characteristic, such as a preference for household work.

While some such variables may affect primarily an individual's reservation wage and hence have an unambiguous effect on their labour force participation decision, others may affect both their market wage and their reservation wage and hence have an indeterminate effect on their labour force participation decision. An increase in a person's age, for example, may be associated with a higher market wage that makes participation in labour market activities more likely. However, it may also raise their reservation wage if the disutility associated with work increases relative to leisure time, and this may induce them to retire from the labour force.

An alternative way to analyze the participation decision is in terms of the allocation of time between market work and nonmarket activities over the individual's lifetime. That is, the income-leisure choice model can be used to examine the individual's preferences and opportunities between lifetime income (measured in present value) and years devoted to market and nonmarket activities. The slope of the budget constraint is the annual wage and the nonlabour wealth is the present value of the individual's lifetime nonlabour income. In this framework, a nonparticipant (such as in Figure 5.2(a)) is someone who will not engage in some market work throughout his or her lifetime. For an individual who will engage in some market work activity, as shown in Figure 5.2(b), the equilibrium E_0 shows the division of the individual's lifetime into nonmarket activity (out of the labour force) and market activity (in the labour force).

The effects of changes in the annual wage and nonlabour market wealth on labour force participation can be analyzed using this lifetime income-leisure choice framework. An increase in the annual wage will have both an income effect (which tends to reduce the number of years worked over the lifetime, assuming that leisure is a normal good) and a substitution effect (which tends to increase the number of years worked). Thus an increase in the wage may either increase or decrease the amount of labour force participation over the lifetime, depending on the magnitudes of the income and substitution effects. An increase in nonlabour market wealth will have a pure income effect, thus reducing the amount of lifetime labour force participation (assuming leisure time is a normal good).

The lifetime income-leisure choice model shows how individuals allocate their lifetimes between market work and nonmarket activities. However, the model does not show which periods the individual will be in the labour force and which out of the labour force. The current period income-leisure choice model can be used to indicate how this choice is determined. During periods in which the value of time spent in nonlabour market activities (and therefore the individual's reservation wage) is high relative to the current market wage, the individual will be out of the labour force. For most individuals these periods occur early and late in their lifetimes when the alternative nonmarket activities, education and retirement, have a high value. Reductions in the amount of labour force participation over the lifetime (resulting, for example, from an increase in the market wage) thus generally take the form of later entry into and/or earlier retirement from the labour force.

EMPIRICAL EVIDENCE, MARRIED FEMALES

The basic theory of labour force participation is often tested with data on married women since they have considerable flexibility to respond to the determinants of labour force participation. In addition, their responses have been dramatic in recent years, and this has policy implications for such things as sex discrimination, family formation, and the demand for daycare facilities.

Empirical evidence in both Canada and the United States tends to confirm the predictions based on the economic theory of labour force participation. For Canada, some illustrative figures are given in Table 5.1.

Labour force participation rates are higher for women with high education, reflecting their high opportunity cost of not working, as well as, perhaps, preferences for labour market as opposed to household work. After the age of 20, labour force participation rates decline continuously with age. The impact of the income effect is illustrated by the decline in their participation as family income increases, albeit this decline in participation does not occur until the husband's income is over $30,000. With higher family income, the household can afford to buy more of everything, including the opportunity to have some members not participate in labour market activities.

These results are also confirmed when the determinants of the labour force participation of married women are analyzed via multiple regression analysis, which enables one to estimate the separate, partial effect of one variable while

Table 5.1
LABOUR FORCE PARTICIPATION RATES OF WOMEN, CANADA 1986

ALL WOMEN, TOTAL	55.9
Education	
Less than grade 9	24.8
Grade 9–13 without certificate or diploma	46.4
Grade 9–13 with certificate or diploma	66.3
Trades certificate or diploma	69.7
Some university without certificate or diploma	68.7
University with certificate or diploma	74.9
University degree	81.3
Age	
Age 15–19	45.6
Age 20–24	80.9
Age 25–34	73.7
Age 35–44	72.2
Age 45–54	62.7
Age 55–64	36.2
Age 65 and over	4.7
WOMEN WITH HUSBAND\PARTNER PRESENT, TOTAL	**57.3**
Husband's Income	
Under $10,000	46.9
$10,000–19,999	53.9
$20,000–29,999	63.0
$30,000–39,999	62.0
$40,000–49,999	60.3
$50,000–59,999	57.5
$60,000 and over	54.9

Source: Statistics Canada, *Women and the Labour Force*, 1986 Census Focus on Canada Series, Catalogue 98-125.

holding other factors constant.[3] The probability that a woman will participate in the labour force is consistently found to be negatively related to the number of children (especially young children) and the husband's income, and positively related to the woman's potential market wage and education attainment.

MAJOR LONG-RUN TRENDS

The basic theory of labour force participation also helps explain the major long-run trends that have been observed with respect to labour force participation in Canada. The trends, as shown in Table 5.2, include a fairly stable and high participation rate for middle-aged males, a drop in the participation rates of older persons, a drop in the participation rate of younger persons until the 1960s, and an increase in the participation rate of females.

The high and fairly stable participation rate of middle-aged males indicates their continuous strong attachment to labour market activities. Until the 1960s, 98 to 99 percent of males 25–34 participated in labour market activities: the corresponding figure for men 35–64 was 95 to 97 percent. Only during the 1960s did a slight fall begin to occur, until by 1986 the participation rates were 95 percent for males 25–34 and 88 percent for males 35–64. Clearly, these are still high figures, yet they do indicate that in recent years labour force participation is not the sole activity for these persons. Continuing education and early retirement have extended even into the age groups that were once thought to be exclusively involved in labour market activity.

The continuous drop in the participation rate of older males over 65 — from 60 percent in 1921 to 15 percent in 1986 — indicates the increasing tendency to retire from labour market activities. To a large extent this may reflect a pure wealth or income effect; that is, as we become wealthier over time we buy more of everything, including leisure in the form of early retirement. It may also reflect increased social insurance benefits and the availability of private and public pension funds. The decrease in self-employment, where retirement comes later and where pensions have not always been arranged, may also have contributed to the growth of early retirement. The early retirement decision is particularly remarkable when one considers that health factors and life expectancy increases have enabled people to participate longer in labour force activities. The fact that they do not participate longer indicates the strength of the economic factors inducing retirement, namely increased income and pension coverage. Because of their obvious importance, and the growing policy concern in this area of retirement, these factors are examined in more detail in Chapter 6.

The drop in the participation rates of younger persons 14–19, at least until the 1960s, reflects their increased educational activities, as institutionalized in the higher legal school leaving age, the increase in the length of the school year and the decline in the importance of farm work. Whether this resulted from an increased awareness of the value of education, or simply greater "consumption" of educational activities as society became wealthier, remains an open question:

◄ 3. Canadian studies in this area include Spencer (1973), Skoulas (1974), Gunderson (1976), Nakamura, Nakamura and Cullen (1979), Nakamura and Nakamura (1981, 1983), Robinson and Tomes (1982, 1985), and Smith and Stelcner (1988). Nakamura and Nakamura (1985) provided a survey of these and related studies.

Table 5.2
LABOUR FORCE PARTICIPATION RATES BY AGE AND SEX, CANADA, CENSUS YEARS 1901–1986

YEAR	SEX	14–19[a]	20–24	25–34	35–64	65+	ALL AGES
1901	Male	n.a.	n.a.	n.a.	n.a.	n.a.	88
	Female	n.a.	n.a.	n.a.	n.a.	n.a.	16
1911	Male	n.a.	n.a.	n.a.	n.a.	n.a.	91
	Female	n.a.	n.a.	n.a.	n.a.	n.a.	19
1921	Male	68	94	98	97	60	90
	Female	30	40	20	12	7	20
1931	Male	57	94	99	97	57	87
	Female	27	47	24	13	6	22
1941	Male	55	93	99	96	48	86
	Female	27	47	28	15	6	23
1951	Male	54	94	98	95	40	84
	Female	34	49	25	20	5	24
1961	Male	41	94	98	95	31	81
	Female	32	51	29	30	6	29
1971	Male	47	87	93	89	24	76
	Female	37	63	45	42	8	40
1981	Male	49	91	95	89	17	78
	Female	45	77	66	53	6	52
1986	Male	48	90	95	88	15	78
	Female	46	81	74	59	5	56

Note: [a]For 1971 and 1981, the youngest age group is 15–19.

Source: M. Gunderson, "Work Patterns," in *Opportunity for Choice: A Goal for Women in Canada*, edited by G. Cook, Ottawa: Statistics Canada, 1976, p. 97 for 1901–1971, reproduced by permission of the Minister of Supply and Services Canada. Figures for 1981 were computed from the 1981 Census Cat. 92-915, Vol. 1, Table 1, pp. 1–1. Figures for 1986 were computed from the 1986 Census report 93–111, *The Nation: Labour Force Activity*, Table 3.

presumably both the investment and consumption aspects have contributed. The rural to urban shift that has gone on over time also would reduce participation, since in agricultural sectors younger persons tend to work on the farms. The rise in participation rates of younger persons since the 1960s likely reflects a rise in part-time employment which allows some to combine work and educational activities.

Perhaps the most dramatic change that has occurred is the increased labour force participation of females. The trend has been continuous — from 16 percent in 1901 to 56 percent in 1986. However, the most dramatic changes have occurred during the postwar period, especially the 1960s and 1970s. Although not shown in the data of Table 5.2, the increase has been most dramatic for married women, increasing, for example, from 10 percent in 1951 to 58 percent in 1986, which is now higher than that of unmarried women.

This rapid increase in the labour force participation of married women has led to the two-earner family now being the norm rather than the exception. As illustrated in Gunderson and Muszynski (1990, p. 15), between 1961 and 1985, the proportion of two-earner families went from 14 percent to 52 percent, while the proportion of families with the husband as the single earner went from 65 percent to only 12 percent. "Other" family types went from 16 to 23 percent, and single parent families went from 6 to 13 percent of all families. Thus the salient long-run trends include a moderate decline in male labour force participation and a substantial increase in the participation rate of females. Following Mincer (1962), economists have explained the contrasting time series behaviour of male and female labour supply in terms of the magnitudes of the income and substitution effects of real wage increases for male and female workers. These important differences can perhaps best be explained using the lifetime income-leisure choice model. Most males have traditionally been in the labour force for much of their lives and work a substantial number of hours per year. Thus the income effect of an increase in the real wage is substantial and is expected to be larger than the substitution effect. Increases in real wages thus lead to a reduction in lifetime labour supply for males. This reduction is divided among hours worked per week, weeks worked per year and years worked over the lifetime according to the individual's preferences and opportunities. Females, in contrast, have traditionally been outside the labour force for much of their lives or, if employed, have worked fewer hours than men. Thus the income effect of an increase in the real wage is small, and is likely to be dominated by the substitution effect. As a result, increases in real wages lead to an increase in lifetime labour supply of females, some of which takes the form of increased labour force participation.

Nakamura and Nakamura (1985), survey a number of empirical studies of the labour supply behaviour of Canadian women. These studies confirm that increases in the market wage have induced women to enter the labour force. For married women, increases in the husband's earnings act like an income effect, increasing the wife's reservation wage and reducing labour force participation. More will be said about these relationships in the section on the evidence on the elasticity of labour supply in Chapter 8.

The dramatic increase in the labour force participation of women also reflects a variety of additional factors, many of them a result as much as a cause of their increased participation. Increased female education that is more labour market oriented makes the opportunity cost of not working much higher. Having fewer children, and having them spaced more closely together, reduces the reservation wage and enables labour force participation. Household work has become easier, due to technological changes and substitutes for household production in such things as child care, cleaning, food preparation, and even entertainment, although there still does not appear to be a sharing of household tasks between husband and wife when both engage in labour market activity[4]. The decline in the participation of other groups, namely the young and the old, has helped create a vacuum that the large influx of women has filled: in essence, their increased participation has tended to offset the decreased participation of other groups. The rapid growth of white-collar and part-time jobs has also been conducive to the growth of female employment, as has been the increased social acceptance of women working outside the home. The women's liberation movement itself has helped change these attitudes, both on the part of society at large, and on the part of women themselves. Legislative changes — especially equal pay and equal employment opportunity legislation, as well as unemployment insurance — may have helped make labour force participation more attractive for women although, as we will see in later chapters, their impact may be ambiguous.

Clearly these various factors have had an impact by encouraging the labour force participation of women, and this impact has been sufficiently strong to offset the tendency — observed in most other groups — to participate less as we become wealthier over time. What is particularly noticeable concerning several of these factors that have influenced the labour force participation of women is that they are both a cause and an effect of their increased participation. This also highlights the importance of the simultaneous and interrelated nature of many household decisions, such as the decision to marry, have children, and participate in the labour force, as well as the choice of occupation and region in which people work. The participation decision, like other labour market and household decisions, cannot be viewed in isolation.

---

4. As indicated in Gunderson and Muszynsky (1990, p. 26) women who work in the labour market tend to average 32.7 hours of labour market work and 16 hours of household work per 5-day work week. Men tend to average 37.5 hours of labour market work, but only 6.6 of household work.

ADDED AND DISCOURAGED WORKER EFFECTS

Labour force participation is also responsive to changes in unemployment in the economy. Specifically, in periods of high unemployment people may become discouraged from looking for work and drop out of the labour force returning to household activities or to school or perhaps even entering early retirement. This is termed the "discouraged worker" effect. On the other hand, in periods of high unemployment, some may enter the labour force to supplement family income that may have deteriorated during the period of high unemployment. This is termed the "added worker" effect.

Changes in unemployment are a proxy for transitory changes in expected wages and other income, and thus the discouraged and added worker effects can be interpreted as short-run substitution and income effects respectively. That is, if unemployment is high, then opportunities in the labour market are lowered temporarily: the price of leisure (opportunity cost or forgone income from not working) is reduced temporarily and hence people substitute leisure for labour market activities. On the other hand, the high unemployment means that it is more likely that family income is lowered temporarily, and additional members may have to participate in labour market activities so as to restore that income.

Since the added (income) and discouraged (substitution) worker effects operate in the opposite direction, it is necessary to appeal to the empirical evidence to see which dominates. Again, most of the empirical tests have been based on data of married women since they are more likely to respond to changes in unemployment.

In the U.S., the empirical evidence clearly indicates the dominance of the discouraged worker effect for most married women. That is, in periods of high unemployment, women become discouraged from entering the labour force to look for work, and this dominates any tendency to add themselves to the labour force to maintain their family income. In Canada, however, the empirical evidence is mixed.

HIDDEN UNEMPLOYMENT

The discouraged worker effect also gives rise to the problem of hidden unemployment — a topic that we will return to in a later chapter on unemployment. During a recession when the unemployment rate is high, there may also be a large number of discouraged workers who do not look for work because they believe that no work is available. They are considered as outside of the labour force because they are not working or actively looking for work. However, because they would look for work were it not for the high unemployment rate, they are often considered as the hidden unemployed; that is, people who the unemployment figures tend to miss because they are considered as outside of the labour force rather than as unemployed persons in the labour force.

Table 5.3
LABOUR FORCE PARTICIPATION RATES[a] BY SEX,
VARIOUS COUNTRIES, 1990
(Civilian Labour Force Basis)

COUNTRY	MALE	FEMALE	BOTH SEXES
Sweden	72.9	63.7	68.2
Canada	75.9	58.4	67.0
United States	76.1	57.5	66.4
Australia	76.5	53.1	64.7
United Kingdom	76.0	53.2	64.1
Japan	77.0	49.1	62.6
Netherlands	70.9[b]	42.6[b]	57.3
France	65.9	45.9	55.4
Germany	68.2	43.1	54.9
Italy	63.5	32.5	47.3

a. Ranked in descending order, from highest to lowest labour force participation rate for both sexes. Some figures are preliminary estimates.
b. 1989 estimate.

Source: U.S. Department of Labor, Bureau of Labor Statistics, November 1991.

In this sense, in times of high unemployment our unemployment figures may understate the true unemployment by missing those discouraged workers who constitute the hidden unemployed. Recent developments in our Labour Force Survey, in fact, have been designed to ascertain this hidden or disguised unemployment, and to find out why they are not looking for jobs.

The notion of the discouraged worker is especially important as groups like married women and teenagers become a larger portion of our potential labour force. Such persons often have a loose attachment to the labour force, moving in and out in response to changing economic conditions. They often, but not always, have other family income upon which to rely. For these reasons they are often labelled secondary or marginal workers — unfortunate misnomers if they are used to belittle their employment problems or the contribution to the family income of such workers. This is especially the case if they are contrasted with the terms primary workers or breadwinners, terms that are often used to describe the employment position of males.

However, because of their flexibility with respect to labour market activities, married women and teenagers do constitute a large portion of discouraged workers who would enter the labour force and look for work if employment prospects increased. Hence, the recent emphasis in labour force surveys to find out exactly what these people are doing, and why they are not looking for work.

Clearly the decision to include such persons either in the category of unemployed, or as outside of the labour force, is a difficult one. What is more important, however, than labelling them as either unemployed or outside of the labour force, is to know their approximate magnitude and the degree to which they seriously would look for work if employment prospects improved. Only then can we attach meaning to the phrase "hidden unemployed."

SOME INTERNATIONAL COMPARISONS

Table 5.3 provides some international comparisons of labour force participation rates for 10 industrialized countries. Canada has high participation rates (69 percent) by international standards, second only to Sweden. There is considerable variation with Sweden and the English-speaking countries of Canada, the United States, Australia, and the United Kingdom having rates in the 60 percent range, while the European countries tend to have rates more in the 50 percent range. In all countries, the male participation rate is higher than the female rate, albeit the gap is largest in the European countries, which also have the lower male rates.

QUESTIONS

1. Using the income-leisure choice model, show that an increase in nonlabour income will increase the individual's reservation wage if leisure is a normal good.

2. Given the basic determinants of labour force participation, do you feel that the main trends in labour force participation will continue into the future?

3. Indicate the expected impact on the labour force participation of married women of changes in each of the following factors, other things held constant:

   (a) an increase in the education of women;

   (b) a more equal sharing of household responsibilities between husband and wife;

   (c) a reduction in the average number of children;

   (d) an increased tendency to have children spaced more closely together;

   (e) an increase in the earnings of husbands;

   (f) daycare paid out of general tax revenues;

   (g) allowing daycare expenses to be tax deductible;

   (h) paying housewives a fixed sum out of general tax revenues for household work.

4. Assume that the following regression equation has been estimated where P_W is the labour force participation rate of married women (measured in percent with the average $\overline{P} = 35.0$), Y_H is husband's income (measured in thousands of dollars with average $\overline{Y}_H = 10$), Y_w is wife's expected income (measured in thousands of dollars with average $\overline{Y}_W = 6$) and u_H is the male unemployment rate (measured in percent with average $\overline{u}_H = 6.0$):

$$P_W = -7Y_H + 18Y_W - 0.5u_H$$

(a) What is the expected effect of an increase of $1,000 in the income of husbands on the participation rate of their wives?

(b) What is the expected effect of an increase of $1,000 in the income of the wives themselves?

(c) Decompose the latter impact into its separate income and substitution effects.

(d) Given the magnitude of the latter two effects, what would be the impact on female participation of an equal pay policy that increased the expected earnings of females by $1,000 while at the same time decreasing the expected earnings of their husbands by $1,000?

(e) Calculate the pure income and the gross or uncompensated wage elasticities of participation, evaluation at the means.

(f) Does this equation shed any light on why the labour force participation of married women has increased over time, even though their nonlabour income has also increased?

REFERENCES AND FURTHER READINGS

Blank, R. The impact of state economic differentials on household welfare and labor force behavior. *JPubEc* 28 (October 1985) 25–58.

Blau, F. and A. Grossberg. Real wage and employment uncertainty and the labor force participation decisions of married women. *EI* 29 (October 1991) 678–695.

Blundell, R., J. Ham and C. Meghir. Unemployment and female labor supply. *EJ* 97 (Supplement, 1987) 44–64.

Boothby, O. The continuity of married women's labour force participation in Canada. *CJE* 17 (August 1984) 471–480.

Bowen, W. and T. Finegan. *The Economics of Labor Force Participation*. Princeton, N.J.: Princeton University Press, 1969.

Breton, A. *Marriage, Population and the Labour Force Participation of Women*. Ottawa: Economic Council of Canada, 1984.

Cain, G. *Married Women in the Labor Force*. Chicago: University of Chicago Press, 1966.

Carliner, G. Female labour force participation rates for nine ethnic groups. *JHR* 16 (Spring 1981) 286–293.

Cave, G. Job rationing, unemployment and discouraged workers. *JOLE* 1 (July 1983) 286–307.

Clark, K. and L. Summers. Labour force participation: timing and persistence. *R.E. Studies* 49 (No. 5, 1982) 835–844.

Cohen, M., R. Lerman and S. Rea. Area employment conditions and labor-force participation: a microstudy. *JPE* 79 (September/October 1971) 1151–1157.

Cohen, M., S. Rea and R. Lerman. *A Micro Model of Labor Supply.* BLS Staff Paper. Washington: U.S. Government Printing Office, 1970.

Connelly, R. The effect of child care costs on married women's labor force participation. *R.E. Stats.* 74 (February 1992) 83–90.

Denton, F. A simulation model of month-to-month labour force movement in Canada. *IER* 14 (June 1973) 293–311.

Dernburg, T. and K. Strand. Hidden unemployment 1953–1962. *AER* 56 (March 1966) 71–95.

Bonner, A. and F. Lazar. Employment expectations and labour force participation in Canada. *RI/IR* 29 (No. 2, 1974) 320–330.

Eckstein, Z. and K. Wolpin. Dynamic labor force participation of married women and work experience. *R.E. Studies* 56 (July 1989) 375–390.

Evan, W. Career interruptions following childbirth. *JOLE* 5 (April 1987) 255–277.

Finegan, T. Discouraged worker effect. *ILRR* 35 (October 1981) 88–102.

Fleisher, B. The economics of labor force participation: a review article. *JHR* 6 (Spring 1971) 139–148.

Fleisher, B. and G. Rhodes. Unemployment and the labor force participation of married women: a simultaneous model. *R.E. Stats.* 68 (November 1976) 398–406.

Flinn, C. and J. Heckman. Are unemployment and out of the labor force behaviorally distinct labor force states? *JOLE* 1 (January 1983) 28–42.

Gallaway, L., R. Vedder and R. Lawson. Why people work: an examination of inter-state differences in labor force participation. *JLR* 12 (Winter 1991) 47–59.

Goldin, C. The role of World War II in the rise of women's employment. *AER* 81 (September 1991) 741–756.

Goldin, C. Life-cycle labor force participation of married women: historical evidence and implications. *JOLE* 7 (January 1989) 20–47.

Gonul, F. Dynamic labor force participation decision of males in the presence of layoffs and uncertain job offers. *JHR* 24 (Spring 1989) 195–220.

Gregory, P. Fertility and labor force participation in the Soviet Union and Eastern Europe. *R.E. Stats.* 64 (February 1982) 18–31.

Grossband, S., A. Shoshana and S. Neuman. Women's labor supply and marital choice. *JPE* 96 (December 1988) 1294–1302.

Gunderson, M. Work Patterns. *Opportunity for Choice: A Goal for Women in Canada,* G. Cook (ed.) Ottawa: Statistics Canada, 1976, 94–103.

Gunderson, M. Logit estimates of labour force participation based on census cross tabulations. *CJE* 10 (August 1977) 453–462.

Gunderson, M. Probit and Logit estimates of labor force participation. *IR* 19 (Spring 1980) 216–220.

Gunderson, M. and L. Muszynski. *Women and Labour Market Poverty.* Ottawa: Advisory Council on the Status of Women, 1990.

Heckman, J. Shadow prices, market wages and labor supply. *Econometrica* 42 (July 1974) 679–694.

Heckman, J. Income, labor supply, and urban residence. *AER* 70 (September 1980) 805–811.

Heckman, J. and R. Willis. A beta-logistic model for the analysis of segmented labor force participation of married women. *JPE* 85 (February 1977) 27–58. Comment by J. Mincer and H. Ofek and reply 87 (February 1979) 197–211.

Hill, M. Female labor force participation in developing and developed countries. *R.E. Stats.* 65 (August 1983) 459–568.

Hill, M. Female labor supply in Japan. *JHR* 24 (Winter 1989) 143–161.

Johnson, W. and J. Skinner. Labor supply and marital separation. *AER* 76 (June 1986) 455–469.

Johnson, W. and J. Skinner. Accounting for changes in the labor supply of recently divorced women. *JHR* 23 (Fall 1988) 417–436.

Khandker, S. Labor market participation of married women in Bangladesh. *R.E. Stats.* 69 (August 1987) 536–541.

King, A. Industrial structure, the flexibility of working hours, and women's labor force participation. *R.E. Stats.* 60 (August 1978) 399–407.

Klermann, J. and A. Leibowitz. Childcare and women's return to work after childbirth. *AER* 80 (May 1990) 284–288.

Kuch, P. and S. Sharir. Added and discouraged worker effects in Canada. *CJE* 11 (February 1978) 112–120.

Layard, R., M. Barton and A. Zabalza. Married women's participation and hours. *Economica* 47 (February 1980) 51–72.

Layard, R., and J. Mincer (eds.). Trends in women's work, education and family building. *JOLE* 3, Part 2 (January 1985).

Leuthold, J. Income-splitting and women's labor-force participation. *ILRR* 38 (October 1984) 98–105.

Levy, F. The labour supply of female household heads. *JHR* 14 (Winter 1979) 76–97.

Long, J. and E. Jones. Labor force entry and exit by married women. *R.E. Stats.* 62 (February 1980) 1–6.

Lundberg, S. The added worker effect. *JOLE* 3 (January 1985) 11–37.

Maloney, T. Employment constraints and the labor supply of married women. *JHR* 22 (Winter 1987) 51–61.

Maloney, T. Unobserved variables and the elusive added worker effect. *Economica* 58 (May 1991) 173–188.

McCaffery, S. Vacancies, discouraged workers and labor market dynamics. *SEJ* 47 (July 1980) 21–29.

McKenna, C. Labour market participation in matching equilibrium. *Economica* 54 (August 1987) 325–334.

Michael, R. Consequences of the rise in female labor force participation. *JOLE* 3 (January 1985) S117–S146.

Mincer, J. Labor force participation of married women. *Aspects of Labor Economics.* Princeton, N.J.: Princeton University Press, 1962.

Nakamura, A. and M. Nakamura. A comparison of the labour force behavior of married women in the United States and Canada, with special attention to the impact of income taxes. *Econometrica* 49 (March 1981) 451–489.

Nakamura, A. and M. Nakamura. Part-time and full-time work behavior of married women: a model with a double truncated dependent variable. *CJE* 16 (May 1983) 229–257.

Nakamura, A. and M. Nakamura. A survey of research on the work behavior of Canadian women. *Work and Pay: The Canadian Labour Market*, W.C. Riddell (ed.). Toronto: University of Toronto Press, 1985.

Nakamura, A., M. Nakamura and D. Cullen. Job opportunities, the offered wage, and the labor supply of married women. *AER* 69 (December 1979) 787–805.

O'Brien, A. and C. Hawley. The labor force participation of married women under conditions of constraints on borrowing. *JHR* 21 (Spring 1986) 267–278.

Ong, P. Immigrant wives' labor force participation. *IR* 26 (Fall 1987) 296–303.

Owen, S. and H. Joshi. Does elastic retract: the effect of recession on women's labour force participation. *British Journal of Economics* 25 (March 1987) 125–143.

Parsons, D. The decline in male labour force participation. *JPE* 88 (February 1980) 117–134.

Peters, E. Marriage and divorce. *AER* 76 (June 1986) 437–454.

Proulx, Pierre-Paul. La variabilité cyclique des taux de participation à la main d'ouvre au Canada. *CJE* 2 (May 1969) 268–277. Comments by L. Officer and P. Anderson, and R. Swidinsky, 3 (February 1970) 145–151.

Quinlan, D. and J. Shackelford. Labour force participation rates of women and the rise of the two-earner family. *AER* 70 (May 1980) 209–212.

Ransom, M. The labor supply of married men. *JOLE* 5 (January 1987) 63–75.

Rea, S. Unemployment and the supply of labor. *JHR* 9 (Spring 1974) 279–289.

Robinson, C. and N. Tomes. Family labour supply and fertility: a two-regime model. *CJE* 15 (November 1982) 706–734.

Robinson, C. and N. Tomes. More on the labour supply of Canadian women. *CJE* 18 (February 1985) 156–163.

Rosenzweig, M. Neoclassical theory and the optimizing peasant: an econometric analysis of market family labour supply in a developing country. *QJE* 94 (February 1980) 31–55.

Sander, W. Women, work and divorce. *AER* 75 (June 1985) 519–523.

Sjedule, T. and K. Newton. *Discouraged and Additional Workers Revisited.* Ottawa: Economic Council of Canada, 1980.

Smith, J. (ed.). *Female Labor Supply.* Princeton, N.J.: Princeton University Press, 1980.

Smith, J. and M. Stelcner. Labour supply of married women in Canada, 1980. *CJE* 21 (November 1988) 857–870.

Spencer, B. Determinants of the labour force participation of married women: a micro-study of Toronto households. *CJE* 6 (May 1973) 222–238.

Sprague, A. Post-war fertility of female labour force participation rates. *EJ* 98 (September 1988) 682–700.

Stern, S. Measuring the effect of disability on labor force participation. *JHR* 24 (Summer 1989) 361–395.

Strand, K. and T. Dernberg. Cyclical variation and labor force participation. *R.E. Stats.* 46 (November 1964) 378–391.

Swan, N. The responses of labour supply to demand in Canadian regions. *CJE* 7 (August 1974) 418–433.

Swidinsky, R. Unemployment and labour force participation: the Canadian experience. *RI/IR* 28 (No. 1, 1973) 56–75.

Wachter, M. Intermediate swings in labor-force participation. *BPEA* (No. 2, 1977) 545–576.

Wales, T. Estimation of labor supply curve for self-employed business proprietors. *IER* 14 (February 1973) 69–80.

Walsh, W. A time series analysis of female participation rates disaggregates by marital status. *RI/IR* 37 (1982) 367–382.

CHAPTER 6

Retirement Decision and Pensions

The retirement decision is essentially a decision by older persons not to participate in the labour force. Hence it is amenable to analysis utilizing labour force participation theory, and its underlying income-leisure choice model. The retirement decision is treated separately simply because it is an area of increasing policy concern and, as the references indicate, it has developed its own empirical literature. This chapter applies the basic income-leisure framework to the retirement decision. Subsequently, Chapter 23 deals with the reasons for the existence of mandatory retirement rules that lead to a "bunching" of retirement dates, often around age 65.

The notion of retirement has many meanings, ranging from outright leaving of the labour force, to a reduction of hours worked, to simply moving into a less onerous job. The process itself may also be gradual, beginning with a reduction in time worked (perhaps associated with a job change) and ending in full retirement. Throughout this chapter, we will generally talk of retirement as leaving the labour force; however, the importance of various forms of quasi-retirement and the often gradual nature of the retirement process should be kept in mind.

The policy importance of the retirement decision stems from the fact that it can have an impact on so many elements of social policy. For the individuals themselves, and their families, the retirement decision has implications ranging from their financial status to their psychological state. For the economy as a whole, the retirement decision also has macroeconomic implications with respect to such factors as private savings, unemployment, and the size of the labour force, all of which have implications for the level of national income.

In addition, there is concern over the solvency of public pension funds if large numbers retire and few are in the labour force to pay into the fund. This problem may be especially acute around the turn of the century, when the post–World War II baby-boom population reaches potential retirement age and, depending on fertility factors and patterns of female labour force participation, when there may be few other participants in the labour force paying into the funds.

The policy importance of the retirement decision is further heightened by the fact that it is an area where policy changes can affect the retirement decision. This is especially the case with respect to such factors as the mandatory retirement age and the nature and availability of pension funds. However, to know the expected impact of changes in these factors, we must know the theoretical determinants of the retirement decision, and the empirical evidence on the retirement response.

THEORETICAL DETERMINANTS OF RETIREMENT

Mandatory Retirement Age

The term mandatory retirement provisions refers to both compulsory retirement provisions and automatic retirement provisions. Under automatic retirement, people *have* to retire at a specific age and cannot be retained by the company. Under compulsory retirement provisions, however, the company can compel the worker to retire at a specific age, but it can also retain the services of a worker, usually on a year-to-year basis.

The term mandatory retirement is somewhat of a misnomer, since there is no magic age embodied in *legislation* that says a person *must* retire by a specific age. The mandatory retirement age may be part of an employer's personnel policy, or it may be negotiated in a collective agreement. In addition, there is an age at which public pensions become available, although they do not *prevent* people from continuing to work. As well, aspects of labour legislation may not apply to workers beyond a specific age. Thus the so-called mandatory retirement age — of which age 65 appears to be the magic number in North America — is really a result of personnel policy and is influenced, but not determined, by government programs. It is neither fixed nor immutable.

This is illustrated by the fact that Europe and the United States appear to be moving in the opposite direction with respect to changes in the mandatory retirement age. Presumably to help alleviate problems of youth unemployment, in Europe the tendency is to encourage a lowering of the retirement age. In the U.S., on the other hand, the trend seems to be in the opposite direction. Recent legislation has removed any mandatory retirement age in the federal public service and has forbidden a mandatory retirement age below age 70 in most other sectors. Certainly, workers can retire before age 70, and employers may try to induce them to do so. Nevertheless, they cannot be forced to retire against their wishes.

In Canada (as of 1992), mandatory retirement is banned in Quebec, Manitoba, and the federal jurisdiction. Recently, the Supreme Court of Canada ruled that mandatory retirement did not constitute age discrimination, which is prohibited under Section 15(1) of the Charter of Rights. Essentially, this decision leaves it up to each individual jurisdiction as to whether it will ban mandatory retirement.

Approximately half of the Canadian work force is employed in jobs that are ultimately subject to a mandatory retirement provision (Gunderson and Pesando, 1988). Most of these persons have an occupational pension plan in which the mandatory retirement provision is a condition for receiving the pension. Mandatory retirement provisions are most prevalent and pensions most prominent in situations where workers are covered by a collective agreement. These are important observations since they remind us that mandatory retirement provisions tend to exist in situations where workers have a reasonable degree of bargaining power and the income protection of a pension plan. They are not simply imposed by management in a unilateral fashion. This will be an important consideration when the rationale for mandatory retirement is discussed in more detail in the later chapter on optimal compensation systems.

The fact that the mandatory retirement age is not immutable suggests that it can change in response to other basic forces that affect the retirement decision and that account for the existence of mandatory retirement itself. These factors are dealt with in Chapter 23 which looks at the rationale for mandatory retirement in the context of an optimal compensation system.

The remainder of this chapter deals with the basic forces that affect the retirement decision, either directly as people are induced to retire, or indirectly as pressure is exerted to change the mandatory retirement age or the age at which various retirement benefits become available.

Wealth and Earnings

Economic theory, in particular the income-leisure choice theory discussed earlier, indicates that the demand for leisure — as indicated, for example, by the decision to retire early — is positively related to one's wealth, and is related to expected earnings in an indeterminate fashion. The wealth effect is positive, reflecting a pure income effect: with more wealth we buy more of all normal goods, including leisure in the form of retirement. The impact of expected earnings is indeterminate, reflecting the opposing influences of income and substitution effects. An increase in expected earnings increases the income forgone if one retires and therefore raises the (opportunity) cost of retirement: this has a pure substitution effect reducing the demand for retirement leisure. On the other hand, an increase in expected earnings also means an increase in expected wealth and, just like wealth from nonlabour sources, this would increase the demand for retirement leisure. Since the income and substitution effects work in opposite directions, then the impact on retirement of an increase in expected earnings is ultimately an empirical proposition. Thus the increase in our earnings that has gone on over time, and that presumably will continue, should have an indeterminate effect on the retirement decision.

Health and the Nature of Work and the Family

As people age and approach retirement, their health and the health of their spouses can obviously influence the retirement decision. This can be the case especially if their accumulated wealth or pension income enables them to afford to retire because of health problems.

The changing nature of work may also influence the retirement decision. For example, the trend toward white-collar and professional jobs and away from more physically demanding blue-collar jobs may make it easier and more appealing for people to work longer. Emerging labour shortages may provide an incentive for employers to encourage older workers to return to the labour force. On the other hand, the permanent job loss of older workers due to plant closings and mass layoffs may make earlier retirement more attractive for these workers, especially if their industry-specific skills make it difficult for them to find alternative employment.

The changing nature of the family can also affect the retirement decision, albeit in complicated ways. The decline of the extended family may make retirement less attractive. The dominance of the two-income family may lead to one spouse not retiring from the labour market until the other also retires, given the joint nature of the retirement decision. However, the increased family income associated with both parties working may enable them both to be able to afford to retire early. The "deinstitutionalization" of health care with its emphasis on home care as opposed to institutional care may force the retirement of one spouse in order to care for the other.

In the income-leisure choice framework many of these factors affect the shape of the individual's indifference curve (preferences for labour market work versus preferences for leisure or household work in the form of retirement). Others can affect the budget constraint or wage line, which depicts the market returns to labour market work. For example, health problems that make labour market work more difficult will increase the slope of the indifference curve (requiring more income per unit of work to maintain the same level of utility). This, in turn, will induce a shift toward retirement leisure. Alternatively stated, it increases the individual's reservation wage and therefore makes retirement more likely if the reservation wage exceeds the market wage.

A permanent job loss will lead to a reduction in the market wage if alternative jobs are lower paid. This downward rotation of the wage line has opposing income and substitution effects, as previously discussed. The returns to work are reduced (substitution effect) and this may induce retirement as the market wage now may fall below the reservation wage. In contrast, income or wealth has fallen (income effect) and this may compel the individual to postpone retirement.

UNIVERSAL OLD AGE SECURITY PENSION

In addition to being affected by the mandatory retirement age, wealth and earnings, health, and the changing nature of work and the family, retirement decisions can be affected by various features of public and private pension plans. In fact, as the references indicate, a growing area of research in labour economics has involved indicating how these various features of pension plans can have intended and unintended effects on retirement decisions. This highlights that pensions are not simply forms of saving for retirement but are active policy and human resource management instruments that can be used to influence the retirement decision.

As indicated in Exhibit 6.1, there are a variety of sources of pension income in Canada. The universal "old age" pension is given to all persons over the age of 65. In addition, persons who have worked in the labour market likely will have built up eligibility for the Canada/Quebec Pension Plan (C/QPP). As well, many will have an employer-sponsored occupational pension plan. In fact, many receive pension income from all three sources.

All three tiers of the pension income support system also have features that affect the retirement decision. In fact, increased attention is being placed on utilizing these features of pension plans to influence the retirement decision, for purposes of public policy and human resource management.

The universal Old Age Security pension (OAS) is basically a demogrant, an unconditional grant, given to all persons over the age of 65. In the income-leisure choice framework the budget constraint is shifted outward in a parallel fashion, by the amount of the pension. The shift is parallel to the original budget constraint because the pension is given irrespective of the recipient's work behaviour. There is no change in the returns to work — recipients receive their full market wage as well as pension. (This ignores, of course, the income taxes or other methods that may be used to finance the program.)

As a demogrant, the old age pension has the work incentive effects of a demogrant (discussed previously in Chapter 4 and illustrated in Figure 4.1a). That is, there is a pure income effect, enabling the recipient to buy more of all normal commodities, including more leisure in the form of earlier retirement. Alternatively stated, the pension enables people to be able to afford to retire earlier.

The Guaranteed Income Supplement (GIS) may also be paid to persons over 65 and in need as determined by a means test. There is a 50 percent implicit "tax-back" feature to the GIS in that the payments are reduced by 50 cents for every dollar earned by persons who continue to work and earn income. As such, the GIS shifts the budget constraint upward by the amount of the supplement, and then reduces its slope by 50 percent (i.e., rotates it downward and to the left) because of the implicit tax-back feature. On net, the budget constraint has both shifted outward and reduced its slope. The outward shift or income effect

EXHIBIT 6.1   CANADA'S THREE-TIER PUBLIC AND PRIVATE PENSION SYSTEM

1. Universal Old Age Security Pension

- financed by general tax revenue

- demogrant or flat amount paid to all persons over age 65, irrespective of need or past contributions or work history

- may be supplemented by a means-tested Guaranteed Income Supplement, based on need

2. Social Insurance Pension: Canada/Quebec Pension Plan

- financed by compulsory employer and employee contributions through a payroll tax for all workers age 18-65, whether paid or self-employed

- benefits related to contributions based on payroll tax applied to past earnings, but not necessarily self-financing since funds for current pensioners comes from payments from current work force

- Canada/Quebec Pension Plan

- covered 89 percent of the labour force in 1987, with average contributions of $607

3. Employer Sponsored Occupational Pension Plans

- financed by employer, sometimes with employee contributions

- benefits depend upon type of plan: flat benefit plans pay a flat benefit per year of service; earnings based plans pay a percentage of earnings per year of service; and defined contribution plans pay on the basis of returns earned by contributions to the fund

- covered 37 percent of the labour force and 45 percent of paid workers in 1987, with average contributions of $2,816

- coverage differed by sex, being 41 percent of the male and 32 percent of the female labour force and 50 percent of the male and 38 percent of the female paid workers

Also, private savings, privately arranged pensions and savings through such mechanisms as Registered Retirement Pension Plans (RRSPs). In 1987, 27 percent of the labour force made RRSP contributions, averaging $2,590.

implies more income or wealth, which in turn implies earlier retirement. The reduction in the slope or substitution effect implies that the returns to labour market work are reduced (i.e., the opportunity cost of retirement is lower) which also increases the likelihood of retirement. That is, both the income and substitution effects of the GIS increase the probability of retiring from the labour force.

SOCIAL INSURANCE PENSIONS: CANADA/QUEBEC PENSION PLAN (C/QPP)

Social insurance pensions refer to public pension schemes that are financed by compulsory employer and employee contributions through a payroll tax, and that pay pensions based on past earnings to those who qualify by virtue of their age and work experience. Some social insurance pensions (e.g., Social Security in the United States, but not the C/QPP in Canada) also have a *retirement test* or work-income test whereby the pension gets reduced if the recipient works and continues to earn income. This is an implicit tax on earnings because it involves forgoing pension payments as one earns additional income. In addition, there is usually an explicit payroll tax on earnings used to finance the social insurance fund.

The retirement test tends to be prevalent in the vast majority of countries with social insurance. Developing economies usually require complete withdrawal from the labour force in order to receive the pension (i.e., implicit 100 percent tax on earnings) mainly because such countries cannot afford to pay pensions to people who also work. In contrast, some countries with the longest history of social insurance schemes have no retirement test, allowing the person to retain full earnings and pension. The motivation for this may be different: in France it appears to be because pension benefits are low and earnings are needed to supplement income; in Germany it appears to be because of a desire to encourage the work ethic.

In the Canadian Pension Plan the retirement test was eliminated in 1975. In the U.S., recipients who work are allowed to keep their full pension as long as their earnings do not exceed a specific threshold amount ($9,360 in 1990); thereafter, pension benefits are reduced by $.33 for every $1.00 earned (i.e., 33 percent implicit tax).

These features of the Canadian and American social insurance pension schemes can have a substantial impact on the retirement decision. The pension itself, like all fixed benefit payments, has a pure income effect inducing retirement. In addition, for those who work, the implicit tax of the pension reduction associated with the retirement test, and the explicit payroll tax used to finance the scheme, both lower the returns to work and hence make retirement more financially attractive. That is, both taxes involve a substitution effect towards retirement because the opportunity cost of leisure in the form of retirement is lowered by the amount of tax on forgone earnings. To be sure, the tax on earnings also involves an income effect working in the opposite direction; that is, our reduced after-tax income means we can buy less of everything including leisure in the form of retirement. However, this income effect is outweighed by the income effect of the pension itself, since for all potential recipients their income is *at least* as high when social insurance is available. Thus, as with the negative income tax plans analyzed in Chapter 4, both the substitution effect and the (net) income effect of the features of social insurance serve to unambiguously induce retirement.

These features of social insurance and their impact on retirement can be modelled somewhat more formally, along the lines of the income-leisure framework developed in Chapter 2. With Y defined as income after taxes and transfers, W the wage rate, T the maximum amount of leisure (L) available, B the pension received upon retiring, p the explicit payroll tax used to finance social insurance, and t the implicit tax involved in the pension reduction through the retirement test, the budget constraint with a social insurance pension is:

$$Y_p = B + (1 - p - t) W(T - L)$$

This can be compared to the constraint without social insurance, which is:

$$Y_o = W(T - L)$$

The new budget constraint is illustrated in Figure 6.1 for various possible social insurance schemes. Figure 6.1(a) illustrates the case when there is no retirement test (e.g., Canada/Quebec Pension Plans); that is, the recipient is given a pension and is allowed to work without forgoing retirement. (For simplicity, the payroll tax for financing the pension has been ignored. With the payroll tax as a fixed percentage of earnings, the budget constraint would have rotated downwards from the point B.) The new budget constraint is Y = B + W(T - L) when p = 0 and t = 0. Such a pension scheme simply has a pure income effect, encouraging potential recipients to retire. Although not shown in the diagram, the new indifference curve may be tangent to the new budget constraint, at a point such as E_p above and to the right of the original point E_0. The new equilibrium may also involve a corner solution at B, in which case the recipient retired completely, or it may be at a point like E_p, signifying partial retirement, or, if leisure is not a normal good, it may be at a point vertically above E_0, in which case the recipient continued to work the same as before and received the full pension.

Figure 6.1(b) illustrates the new budget constraint under a full retirement test whereby the pension recipient is required to give up $1.00 of pension for every $1.00 earned. This implicit 100 percent tax rate makes the budget constraint similar to the one for welfare discussed in Chapter 4, and the adverse work incentive effects are similar. In particular, there will be a strong incentive to retire completely (move to point B).

The typical case with a partial retirement test is illustrated in Figure 6.1(c). The first arm of the budget constraint, TB, indicates the pension benefits payable upon complete retirement. (Income is Y_b = TB when leisure is OT.) The second arm, BC, illustrates that up to a specific threshold amount of labour market earnings, $Y_c - Y_b$, for working T - L_c, the recipient may keep the full pension benefits of Y_b (= TB), so that her total income with pension and labour market earnings could be up to Y_c. As indicated previously, this threshold amount of earnings, Y_c-Y_b, was $9,360 in the U.S. in 1990. The new arm of the

Figure 6.1
BUDGET CONSTRAINTS UNDER SOCIAL INSURANCE PENSIONS
(Assuming payroll tax p = 0)

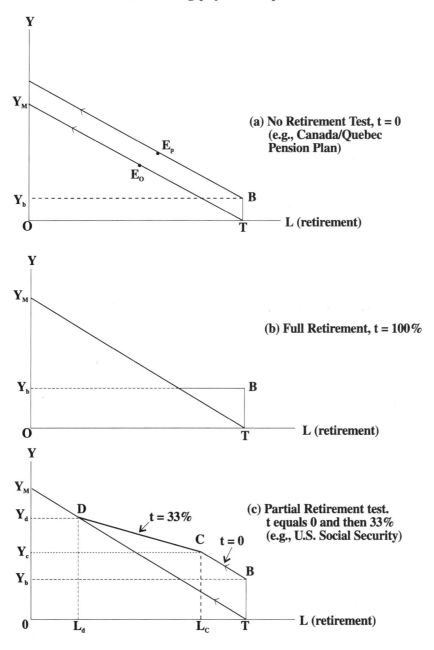

(a) No Retirement Test, t = 0
(e.g., Canada/Quebec
Pension Plan)

(b) Full Retirement, t = 100%

(c) Partial Retirement test.
t equals 0 and then 33%
(e.g., U.S. Social Security)

budget constraint is parallel to the original constraint of TY_m because the implicit tax is zero; that is, recipients who work keep their full labour market earnings.

The third arm, CD, illustrates the implicit tax that is involved when the recipient is required to give back a portion of the pension for the additional labour market earnings of $Y_d - Y_c$ that result from the additional work of $L_c - L_d$. An implicit tax of 33 percent results in the slope of CD being two-thirds of the slope of the original constraint TY_m; that is, for every dollar earned, the recipient forgoes \$.33 in pension. When the recipient's labour earnings exceed $Y_d - Y_b$, at point D, then the person would no longer receive any pension and any additional work activity of $L_d - 0$ would result in additional income as shown by the fourth arm of the budget constraint DY_m.

The work incentive effects of the new budget constraint, $TBCDY_m$, are such as to unambiguously induce retirement. Basically two things have happened. The budget constraint has shifted outward (upward to the right) from the original constraint of TY_m and this has a wealth or income effect encouraging the purchase of more leisure in the form of retirement. In addition, the slope of the new budget constraint is always equal to or less than the original constraint; that is, the opportunity cost of leisure is reduced and this would encourage the substitution of leisure for other commodities. Both the income and substitution effect of the partial retirement work in the same direction to encourage early retirement.

EMPLOYER SPONSORED OCCUPATIONAL PENSION PLANS

The third tier of pension income support consists of employer-sponsored or occupational pension plans, for those who work in establishments with such plans. As indicated in Exhibit 6.2, such plans can be classified on the basis upon which benefits are calculated. Earnings-based plans are most common (especially final-earnings plans), covering about three-quarters of all workers. Flat benefit plans, which predominate in the union sector, cover about 20 percent of plan members. Defined contribution plans cover only about 6 percent of plan members. However, because these tend to exist in small establishments with few employees, about half of all occupational pension plans are defined contribution plans. As well, there is a growing trend toward such plans, in part because of the extensive regulations being placed upon the defined benefit plans and the uncertainty and conflict over who "owns" the surplus assets that have often been generated in the defined benefit plans.

The defined benefit plans have a number of features that can influence the retirement decision of workers. In fact, such features can be used to encourage early retirement or to discourage postponed retirement past the normal retirement age established by the plan. This aspect is likely to grow in importance as

EXHIBIT 6.2 EMPLOYER SPONSORED OCCUPATIONAL PENSION PLANS, 1990
(% of members in parenthesis)

Defined Contribution (8%)

- Pension benefits equal the accumulated value of contributions made by, or on behalf of, employees
- Usually in smaller establishments
- 57% of plans covering 8% of members

Defined Benefit (92%)

Flat Benefit (20%)

- Fixed benefit for each year of service, e.g., $20/month per year of service with a maximum of 35 years of service (i.e., $700 month)
- Predominantly in union sector
- 7% of plans covering 20% of members

Earnings Based (72%)

- Pension based upon length of service and percent of earnings of final years or career average
- Typically 2% of earnings for each year of service, up to a maximum of 35 years of service

Final Earnings

- 21% of plans covering 60% of members

Career Average

- 13% of plans covering 11% of members

% of members

Defined Contribution 8%	Flat Benefit 20%	Career Average 11%	Final Earnings 61%

Earnings based 72%

Defined Benefit 92%

Source: *Pension Plans in Canada, 1989* Ottawa: Supply and Services, 1990. In 1990, 37% of the labour force and 45% of employed paid workers were in occupational pension plans.

Notes: Includes about 1% of composite and other plans.

the aging work force means that more people are in the age groups likely to be affected by early retirement, and as firms view early retirement as a viable way to "downsize" their work force. If the early retirement payments are sufficiently generous, this may also be in the mutual interest of employees, opening new promotion opportunities for younger workers, and enabling some of the retired employees to start new careers. As well, if mandatory retirement is banned or becomes less common, voluntary early retirement programs may become more prominent.

The potential work incentive effects of occupational pension plans can be illustrated by various features of a final-earnings plan. Such a plan, for example, could involve the payment of 2 percent of the final 3 years of earnings for each year of service (maximum of 35), which would imply a "replacement rate" of 70 percent (0.02 × 35) of final earnings. For a worker earning $30,000 per year in their final years, this would imply an annual pension of $21,000.

The features that most influence the retirement decision, in addition to the "normal" retirement age for pension eligibility, are: (1) the "backloading" or accumulation of pension benefit accruals, (2) early and special retirement provisions, and (3) postponed retirement provisions.

As indicated in Exhibit 6.3, eligibility for *early retirement* typically occurs at age 55 provided the worker has at least 10 years of service. It usually involves a subsidy in that the pension is actuarially reduced, but by an amount that does not compensate for the fact that it is received earlier and for a longer period.[1] If there is no actuarial reduction in the pension (termed *special retirement*) then a larger subsidy is typically involved. Special retirement is less common, usually occurring at ages 60–62 and requiring at least 20 years of service. *Postponed retirement* provisions can also exist if workers are allowed to continue working past the normal retirement age, usually 65. Postponed retirement usually involves a penalty because of incomplete actuarial adjustments and/or further benefit accruals are not allowed.

The potential retirement-inducing effect of these features of occupational pension plans can be illustrated through their impact on the expected pension

---

1. For example, the reduction formulae could be 5 percent per year for each year that early retirement precedes normal retirement. A worker who retired 10 years early would have an annual pension that is reduced by 50 percent based on a reduction formula of 5 percent for each of the 10 years of early retirement. In the previous example, the worker who received $21,000 per year after the normal retirement age of 65 would have an annual pension of $11,500 commencing at the early retirement age of 55. Even though the annual pension is substantially reduced, a subsidy may be involved because the employee receives the pension earlier (i.e., age 55) and for a longer period of time (i.e., age 55 until death). The actual magnitude of the subsidy, and whether it exists, depends upon such factors as inflation and wage growth, as well as actuarial reductions implied by the benefit reduction formula.

EXHIBIT 6.3   EARLY AND POSTPONED RETIREMENT FEATURES OF FINAL-EARNINGS PLANS

Early Retirement Provisions

- typically at age 55 with at least 10 years service

- in most earnings-based plans

- may be unsubsidized in that annual pension benefits are reduced by an "actuarially fair" amount to exactly compensate for the fact that it is received early (i.e., at age 55 instead of the normal retirement age of 65) and for a longer period of time

- or may be subsidized in that the early retirement pension is actuarially reduced by an amount that does not exactly compensate for the fact that it is received early and for a longer period of time.

Special Retirement Provisions

- typically at age 60–62 with at least 20 years of service

- exists for about 30 percent of workers in earnings-based plans in private sector

- extensively subsidized in that the special retirement pension is not actuarially reduced to compensate for the fact that it is received earlier and for a larger period of time

Postponed Retirement Provisions

- typically involve pension "penalties" for postponing retirement past the age of normal retirement (e.g., 65)

- penalties usually occur because of some combination of not allowing further accrual of pension benefits and/or not actuarially increasing previously accrued pension benefits to offset the fact that they are received later and for a shorter period

benefits (obligations of employers) that workers accrue as they work additional years with their employer.[2]

These pension benefit accruals, expressed as a percent of the workers wages, are illustrated in Figure 6.2 for a representative final-earnings plan.

The accruals are substantial, averaging around 18 percent of wages for workers in their 50s and 60s. This means that employer wage costs for such workers are augmented by about 18 percent to cover the pension obligations that are being accrued by such workers. It also means that for every additional year that

---

2. These are calculated as the annual present value of the stream of pension income to which a member is legally entitled at the end of the year, less the corresponding value of the previous year, adjusted for the interest factor (Wise and Kotlikoff 1985, Pesando and Gunderson 1989).

Figure 6.2
PENSION BENEFIT ACCRUALS AS % OF WAGES IN REPRESENTATIVE FINAL-EARNINGS PLAN

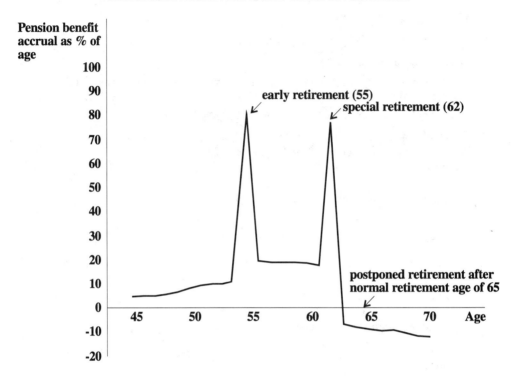

Source: Based on data given in Pesando and Gunderson (1988) for a representative final-earnings plan for an employee who enters the plan at age 30. Early retirement is at age 55 and 10 years of service with benefits being reduced by 5 percent for each year that early retirement precedes normal retirement at 65. Special retirement is at 62 and 20 years of service with (by definition) no reduction in benefits. If retirement is postponed past the normal retirement age of 65, there is a continued accrual of pension benefits but no actur-ial increase of previously accrued benefits. Inflation is assumed to be 5%, with other assumptions outlined in the original article.

such employees work, they are accumulating these substantial pension commit-ments that they will receive on retirement. The benefits are "backloaded" in the sense that they tend to get larger as the worker ages and accumulates service credits as well as seniority-based wage increases upon which pension benefits are calculated. This makes their total compensation profile (wages plus pension benefit accruals) steeper than otherwise would be the case. This in turn aug-ments any pure seniority effect whereby wages themselves rise with seniority.

In addition to giving rise to substantial benefit accruals that are backloaded, occupational pension plans also can give rise to sharp "spikes" or jumps in benefit accruals at the ages of early retirement (usually 55) and special retirement (usually around age 62 if such a feature exists). As well, negative accruals typically occur after any special retirement spike or after the age of normal retirement when postponed retirement provisions apply.

These features of occupational pension plans can have important incentive effects on the turnover decisions of younger workers and the retirement decisions of older workers. Specifically, the backloading of pension benefit accruals means that younger workers have an incentive to stay with the firm to receive these benefits, and older workers have an incentive not to retire too early and forgo those accruals. This is especially the case just before ages when early or special retirement features would apply. For example, in Figure 6.2, a worker who retired just before age 55 would forgo the wages that would be received at age 55 as well as the pension benefit accrual, which could amount to about 80 percent of wages. The same would be true of a worker who retired just before the special retirement age of 62. Such workers would have an incentive not to retire just before these early or special retirement dates, but to wait at least until these dates. As well, they have an incentive to retire after the special or normal retirement age because of the negative accruals that occur after those dates. The negative accruals were around 10 percent in the representative plan of Figure 6.2, but they can easily be in the neighbourhood of 40 percent depending upon other provisions of the plan (Pesando and Gunderson, 1988, p. 257).

In essence, the features of occupational pension plans can have important incentive effects on the retirement decision, discouraging early retirement that would lead to a loss of these backloaded accruals, especially just prior to the substantial accruals that can occur at early and special retirement dates.

Early retirement can be encouraged just after an early retirement date as the benefit accruals have peaked and dropped off thereafter, albeit if a subsequent special retirement date also exists then there may be a reluctance to retire just prior to that date and forgo the option of accruing the special retirement benefits.[3] Retirement is strongly encouraged at the normal retirement date even if postponed retirement is an option. These incentive effects suggest a bunching of retirement just after the dates of early, special, and normal retirement, a few retirements just prior to the dates of early and special retirement. As discussed subsequently, these empirical predictions are verified in the data on retirement decisions under occupational pension plans.

---

3. This highlights that the turnover and retirement incentives should be formally modelled through "option values" whereby working an additional year yields not only the pension accrual of that year, but also the option to work additional years and consequently to qualify for subsequent retirement provisions. Conversely, retiring means forgoing the income and benefit accruals of that year and subsequent years as well as forgoing the option to work additional years and consequently to qualify for subsequent retirement provisions. See Lazear and Moore (1988).

These incentive effects are based on "modelling" only some of the institutional features of occupational pension plans such as early and postponed retirement provisions. Other features such as vesting, bridging supplements, spouse benefits, golden handshakes, and retirement windows are discussed in articles in the references.[4] Clearly, pensions are an integral feature of the compensation system, as well as a key element in the human resource planning strategy of the firm. They can be used to affect retirement decisions and hence the size and age composition of the firm's work force, as well as to provide income support after retirement. The negative benefit accruals that tend to exist if retirement is postponed after the age of normal retirement may even serve as a substitute for mandatory retirement should mandatory retirement be banned.[5] The sizable pension obligations also highlight the cost to the firm of an aging work force where both wages and pension benefits tend to be higher.[6] Pension benefits may also have implications for the magnitude of the male-female wage gap — or more accurately, the gap in total compensation to include fringe benefits such as pensions (see Exhibit 6.4).

EMPIRICAL EVIDENCE

The empirical evidence on the determinants of the retirement decision is reviewed in Campbell and Campbell (1976), Boskin (1977), Mitchell and Fields (1982), Lazear (1986), and Hurd (1990). Basically, there have been two techniques — survey interviews and labour force participation studies — that have been used to estimate the relative importance of the determinants of the retirement decision, and these two techniques tend to give conflicting answers with respect to the importance of health factors as opposed to pension schemes.

Survey interviews asking people why they have retired, or would retire, usually find that ill health is the prime motivating factor — and that social insurance

---

4. Vesting refers to employees having the rights to the employer contribution as well as their own contribution. Surviving spouse benefits refer to the extent to which benefits are paid to a surviving spouse. Bridging supplements refer to additional payments that may be made between the date of early retirement and the date when the member reaches age 65 and thereby qualifies for social insurance pensions like the Canada/Quebec pension plan. Golden handshakes refer to special retirement packages given usually to senior executives. Retirement windows refer to special periods of time (often a year or less) whereby specific groups of workers can retire early and receive a special retirement bonus.

5. See Burkhauser and Quinn (1983) and Lazear (1983). Gunderson and Pesando (1988, 1991), however, indicate that the postponed retirement provisions that provide a strong penalty (negative benefit accruals) for delayed retirement are far from universal and they themselves can be banned by legislation.

6. The same general picture also emerges from the flat benefit plans that tend to predominate in the unionized sector. Periodic enrichments to the flat benefit formula, usually at the time of the renewal of the collective agreement, lead to a backloading of pension benefit accruals. Spikes in the accruals occur at dates of early and special retirement if they exist. Pesando and Gunderson (1991) simulate these effects for representative flat benefit plans in Canada.

EXHIBIT 6.4   DO PENSIONS EXACERBATE OR OFFSET MALE-FEMALE WAGE DIFFERENCES?

Women tend to have longer life expectancy than men. For example, at the age of 65, remaining life expectancy is slightly over 18 years for women and 14 years for men, for a difference of about 4 years at the age of normal retirement. In employer-sponsored occupational pension plans, therefore, employers can expect to pay out greater pension benefits to women than men, at least for this reason. Some have suggested that this may explain much of the male-female wage gap. That is, wages to women are lower in part to offset their higher pension cost associated with their greater longevity (Moore, 1987).

Empirical evidence presented in Pesando, Gunderson, and McLaren (1991), however, indicates that the greater pension benefits that go to women for this reason are offset by other characteristics of employer pension plans that result in smaller pension benefits going to women.

Specifically, within a given occupational pension plan: (1) The impact of the greater longevity of women is offset somewhat by the fact that employers do not have to pay out as much in surviving spouse benefits. In contrast, even though the years of pension payout is shorter for men because of their shorter longevity, the existence of surviving spouse benefits means that employers still incur pension obligations to the surviving spouse. Because women live longer than men, and because men tend to marry women who are younger than themselves, these surviving spouse benefits can continue on for a longer period if a man dies. (2) Because of childraising and household responsibilities, women tend not to accumulate the seniority and seniority-based wage increases that men accumulate. As such, women are less likely to receive the substantial "backloading" of pension benefit accruals or the sharp "spikes" that are associated with subsidized early and special retirement provisions. (3) Because of their lower wages women receive lower pension benefits then men, especially because women are likely to belong to an earnings-based plan where pension benefits are a percent of earnings. In essence, lower wages (including those attributable to wage discrimination and occupational segregation) get compounded into lower pension benefits.

These various factors refer to pension benefits within a given employer-sponsored occupational pension plan. Not only do women tend to receive smaller pension benefits within a given plan, but also they are less likely to be covered by a plan. In 1989, for example, 41 percent of the male labour force and 32 percent of the female labour force were covered by such plans. The corresponding coverage for employed paid workers was 50 percent for males and 38 percent for females.

For these reasons, pensions likely exacerbate rather than offset the male-female wage differential.

pensions have not induced early retirement.[7] The interview technique may yield biased results, however, in that respondents may feel that retiring for reasons of ill health is a more socially acceptable response than retiring because of social insurance pensions. In addition, in retrospective surveys taken after retirement, the respondent's health may have declined since retirement, and this may induce them to respond that ill health was a motivating factor.

In contrast to the interview surveys, econometric and statistical studies of the retirement decision generally find that retirement is affected by the retirement-inducing features of social insurance pensions,[8] albeit the precise magnitude of that effect is subject to debate. As well, some of the studies indicate that ill health induces early retirement especially if pension income is available.[9]

A number of empirical studies have also documented the potential and actual incentive effects of employer-sponsored occupational pension plans on the retirement decision.[10] Specifically, the various features of such plans (especially early and postponed retirement provisions) can reduce turnover prior to periods when large pension benefit accruals would occur, and they can induce retirement when the accruals are negative.

SOME POLICY ISSUES

The fact that social insurance pensions have a substantial impact on the retirement decision has important policy implications. This is especially so since many features of pension schemes are policy parameters, subject to change by policymakers. This is the case with respect to such features as the retirement test (implicit tax on earnings as pensions are reduced if one continues to work), the

---

7. This is the conclusion reached in the study by Steiner and Dorfman (1959) and in numerous studies by the U.S. Social Security Administration as reviewed in Campbell and Campbell (1976, pp. 372–3).

8. Feldstein (1974) based on international data, and Campbell and Campbell (1976), Boskin (1977), Quinn (1977), Boskin and Hurd (1978), Hall and Johnson (1978), Blinder, Gordon and Wise (1980), Burkhauser (1980), Hurd and Boskin (1984), Burtless and Moffitt (1984), Hauseman and Wise (1985), Burtless (1986), and Gustman and Steinmeier (1986) based on U.S. Data.

   Boskin (1977), for example, found that the income effect from expected social security pension benefits was seven times more powerful in inducing retirement than the income effect from other assets. This he attributed to a variety of factors: social security benefits are certain, since they are guaranteed for the remainder of one's life and often indexed against inflation; income from assets can be bequested, and therefore the elderly may not want to use up the income by retiring early; and people may be reluctant to borrow against the inputed income from assets like a house, because they may completely use up their assets.

9. Breslaw and Stelcner (1987), Quinn (1977), Hall and Johnson (1978) and Gustman and Steinmeier (1986).

10. Burkhauser (1979), Fields and Mitchell (1982), Hogarth (1988), Kotlikoff and Wise (1985, 1987, 1989), Lazear (1983), Lazear and Moore (1988) and Stock and Wise (1990) based on U.S. data; Pesando and Gunderson (1988, 1991) and Pesando, Gunderson and Shum (1991) based on Canadian data.

explicit payroll tax used to finance social insurance, the replacement ratio (amount of social insurance benefits relative to pre-retirement earnings), coverage and eligibility for social insurance, pensions for early retirement, and the indexing of pensions for inflation.

Experience outside North America also suggests that features of social insurance can be altered selectively to achieve very specific human resource objectives. Communist countries have often reduced the retirement test for older workers willing to work in specific occupations, industries, and geographic areas experiencing labour shortages. Many European countries and Japan have utilized social insurance to encourage early retirement in agriculture so as to promote efficiency by reducing the number of agricultural workers. Specifically, farmers may be required to sell their land or cease full-time farming in order to receive a pension.

Employers can also utilize features of employer-sponsored occupational pension plans as part of strategic human resource planning. The backloading of pension benefit accruals may deter unwanted turnover. The sharp "spikes" in accruals that occur at ages of early retirement may induce early retirement just after those spikes, especially if benefit accruals become negative. Negative accruals or penalties to postponed retirement may also encourage early retirement. These features are likely to become more prominent as the work force ages and employers want to use voluntary retirement as a way to open new promotion opportunities or to "downsize" their work force. This is especially the case if mandatory retirement is banned or becomes less prominent.

Clearly, retirement policies of both governments and employers can have a substantial impact on the retirement decision, and this in turn has implications for a variety of factors including the size, composition and deployment of our labour force, the extent of poverty, the magnitude of private savings, company personnel policies, and the solvency of social insurance funds. Because of this importance, the retirement decision merits careful analysis if it is to be altered by various policy instruments. Hopefully labour market analysis can contribute to this area of growing policy concern.

QUESTIONS

1. Outline the main areas of public policy that are affected by the retirement decision. Discuss the impact on these areas of a policy of making the mandatory retirement age illegal.

2. Outline the main features of public pension schemes that can be considered as policy parameters, subject to change by policymakers.

3. Discuss the extent to which the mandatory retirement age is both an exogenous determinant of the retirement decision, and an endogenous result of the retirement decision.

4. Discuss the expected impact on the retirement decision of an increase in each of the following factors:

(a) nonlabour wealth;

(b) labour market earnings;

(c) earnings of spouse;

(d) replacement ratio in social insurance pensions;

(e) retirement test;

(f) payroll tax to finance social insurance.

5. Use a static, partial equilibrium, income-leisure choice diagram to illustrate the effect of an increase in life expectancy (at the same wage rate) on the optimal accrual of lifetime income, leisure, and labour supply.

6. In an income-leisure diagram, depict the budget control for the Canadian Guaranteed Income Supplement.

7. Based on Figure 6.1(a) draw the indifference curves, before and after the pension, for persons who:

(a) retire completely;

(b) retire partially;

(c) work the same.

Compare their income in each case.

8. Based on Figure 6.1(b), draw the indifference curves for a person who is indifferent between retiring completely and continuing to work. Why would a person ever be indifferent between these two alternatives assuming that retirement is regarded positively? Based on the various factors given in the diagram, indicate how people may be induced to stay in the labour force rather than retire.

9. Based on Figure 6.1(c), draw the indifference curve for a person who:

(a) retires completely;

(b) works part-time and only earns income up to the maximum amount before it becomes retirement tested;

(c) works and earns income up to the maximum amount before it becomes retirement tested at a 100% implicit tax;

(d) works and earns so much income that no pension is forthcoming.

Compare the pension cost under each alternative.

10. Assume that an individual earned an average of $30,000 during his last 5 years prior to retirement. Depict the budget constraint for a social insurance pension with a replacement rate of 50%, and a retirement test allowing one

to earn up to $4,000 without having to forgo any pension, but which requires a pension reduction of $.50 for every dollar for income between $4,000 and $8,000, and requires one to completely forgo pension income if earnings exceed $8,000.

11. Discuss the advantage and disadvantage of the full retirement test versus no retirement test.

12. Do you think that the trend towards early retirement will continue into the future?

13. If the retirement test is an implicit tax on earnings and if a tax increase has an ambiguous effect on work incentives, why is it that the features of social insurance pensions can be said to unambiguously encourage retirement?

14. You have been asked to estimate the impact on retirement of a policy of raising the replacement rate from 40% to 60%, and of lowering the implicit tax of the retirement test from 50% to 30%. You have access to comprehensive microeconomic data where the individual is the unit of observation. Specify an appropriate regression equation for estimating the retirement response and indicate how you would simulate the impact of the change in the two policy parameters.

15. Based on a social insurance pension scheme where income, Yp, is

$$Yp = B + (1 + t)E$$

and B is the pension, t the tax-back rate of the retirement test, and e is earnings, solve for the following:

(a) the breakeven level of earnings, Eb, where the recipient no longer receives any pension income. (Hint: Income without the pension scheme is Yo = E and, therefore, the breakeven level if earnings, Eb, occurs when Yo = Yp).

(b) the breakeven level when there is also a threshold level of income, Yt, before the tax-back rate of the retirement test is applied, so that

$$Yp = B + Yt + (1 - t)E.$$

(c) the dollar value of the breakeven level of earnings in parts (a) and (b) when B = $6,000, Yt = $10,000 and t = 0.5.

16. In an income-retirement choice diagram, indicate the effect on an individual's utility function and/or wage line, of each of the following:

(a) retirement of ones's spouse

(b) illness of ones's spouse

(c) improvements in the health of the elderly

(d) permanent displacement from a high-wage job in the steel industry to a low-wage job in the service sector

(e) improvements in the retirement leisure industry (e.g., condominiums, recreation, travel) such that retirement is now relatively more attractive.

17. Discuss the various features of occupational pension plans that employers can strategically use to affect the voluntary early retirement decision. Discuss how you would change these features if you were an employer who wanted to achieve the following multiple objectives:

(a) downsizing of your work force,

(b) opening up of new promotion opportunities for your mid-career work force,

(c) reducing the turnover of your mid-career work force,

(d) maintaining your reputation as a good employer, by downsizing through voluntary methods, and by providing reasonable pension income to your retired employees.

REFERENCES AND FURTHER READINGS

Allen, S., R. Clark and A. McDermed. Why do pensions reduce mobility? *IRRA* 40 (December 1987) 204–219.

Anderson, K. and R. Burkhauser. The retirement health nexus: a new measure of an old puzzle. *JHR* 20 (Summer 1985) 315–330.

Anderson, K., R. Burkhauser and J. Quinn. Do retirement dreams come true? *ILRR* 39 (July 1986) 518–526.

Bazzoli, G. The early retirement decision: new empirical evidence on the influence of health. *JHR* 20 (Spring 1985) 214–234.

Bell, D. and W. Marclay. Trends in retirement eligibility and pension benefits. *MLR* 110 (April 1987) 18–25.

Berkovek, J. and S. Stern. Job exit behaviour of older men. *Econometrica* 59 (January 1991) 189–210.

Bloom, D. and R. Freeman. The fall in private pension coverage in the United States. *AER* 82 (May 1992) 539–545.

Bodie, Z., J. Shoven and D. Wise (eds.). *Issues in Pension Economics.* Chicago: University of Chicago Press, 1987.

Boskin, M. Social security and retirement decisions. *EI* 15 (January 1977) 1–25.

Boskin, M. and M. Hurd. Effect of social security on early retirement. *JPubEc* 10 (December 1978) 361–377.

Breslaw, J. and M. Stelcner. The effect of health on the labor force behaviour of elderly men in Canada. *JHR* 22 (Fall 1987) 490–517.

Burbidge, J. and L. Robb. Pensions and retirement behaviour. *CJE* 13 (August 1980) 421–437.

Burkhauser, R. The pension acceptance decision of older workers. *JHR* 14 (Winter 1979) 63–75.

Burkhauser, R. The early acceptance of social security: an asset maximization approach. *ILRR* 33 (July 1980) 484–492.

Burkhauser, R. and J. Turner. A time series analysis of social security and its effects on the market work of men at younger ages. *JPE* 86 (August 1978) 701–716.

Burkhauser, R. and J. Turner. Labour market experience of the almost old and the implications for income support. *AER* 72 (May 1982) 304–308.

Burkhauser, R. and J. Quinn. Is mandatory retirement overrated? *JHR* 18 (Summer 1983) 337–358.

Burtless, G. Social security, unanticipated benefit increases and the time of retirement. *R.E. Studies* 53 (October 1986) 781–805.

Burtless, G. and J. Hausman. "Double-dipping": the combined effects of Social Security and civil service pensions on employee retirement. *JPubEc* 18 (July 1982) 139–160.

Burtless, G. and R. Moffit. The effects of Social Security on the labor supply of the aged. *Retirement and Economic Behavior*, H. Aaron and G. Burtless (eds.). Washington: Brookings, 1984.

Burtless, G. and R. Moffit. The joint choice of retirement age and postretirement hours of work. *JOLE* 3 (April 1985) 209–236.

Butler, J., K. Anderson and R. Burkhauser. Work and health after retirement. *R.E. Stats.* 71 (February 1989) 46–53.

Campbell, C. and R. Campbell. Conflicting views on the effect of old-age and survivors insurance on retirement. *EI* 14 (September 1976) 369–387. Comment by V. Reno, A. Fox and L. Mallan and reply 15 (October 1977) 619–623.

Chang, F. Uncertain lifetimes, retirement and economic welfare. *Economica* 58 (May 1991) 215–232.

Chirikos, T. and G. Nestal. Occupational differences in the ability of men to delay retirement. *JHR* 26 (Winter 1991) 1–26.

Clark, R. *Japanese Retirement Systems.* Homewood, Illinois: Dow Jones-Irwin, 1990.

Clark, R. and A. McDermed. *The Choice of Pension Plans in a Changing Regulatory Environment.* Washington, D.C.: AEI Press, 1990.

Clark, R. and R. Anker. Labour force participation rates of older persons: an international comparison. *International Labour Review* 129 (No. 2, 1990) 255–271.

Cornwell, C., S. Dorsey and N. Mehrzad. Opportunistic behaviour by firms in implicit pension contracts. *JHR* 26 (Fall 1991) 704–725.

Crawford, V. and D. Lilien. Social security and the retirement system. *QJE* 96 (August 1981) 505–530.

Deaton, R. *The Political Economy of Pensions: Power, Politics and Social Change in Canada.* Vancouver: University of British Columbia Press, 1989.

Doeringer, P. (ed.) *Bridges to Retirement.* Ithaca, N.Y.: ILR Press, 1990.

Dorsey, S. A model and empirical estimates of worker pension coverage in the U.S. *SEJ* 49 (October 1982) 506–520. Comment by D. Cymrot and reply by S. Dorsey 51 (April 1985) 1245–1248, 1249–1251.

Dorsey, S. The economic function of private pensions. *JOLE* 5, Part 2 (October 1987) S171–S189.

Dowel, R. and K. McLaren. An intertemporal analysis of the independence between risk preference, retirement, and work rate decisions. *JPE* 94 (June 1986) 667–682.

Driffill, E. Life-cycles with terminal retirement. *IER* 21 (February 1980) 45–62.

Duggan, J. The labor-force participation of older workers. *ILRR* 37 (April 1984) 416–430.

Ethier, M. Survey of pension issues. *Income Distribution and Economic Security in Canada*, F. Vaillancourt (ed.). Toronto: University of Toronto Press, 1985, 215–49.

Evan, W. and D. Macpherson. The gender gap in pensions and wages. *R.E. Stats.* 72 (May 1990) 259–265.

Fields, G. and O. Mitchell. Economic determinants of the optimal retirement age: an empirical investigation. *JHR* 19 (Spring 1984) 245–262.

Fields, G. and O. Mitchell. *Retirement Pensions and Social Security*. Cambridge, Mass.: MIT Press, 1984.

Fields, G. and O. Mitchell. The effects of Social Security reforms on retirement ages and retirement incomes. *JPubEc* 25 (November 1984) 143–159.

Filer, R. and R. Petri. A job-characteristics theory of retirement. *R.E. Stats.* 70 (February 1988) 123–128.

Ghilarducci, T. Pensions and the uses of ignorance by unions and firms. *JLR* 11 (Spring 1990) 203–216.

Gordon, R and A. Blinder. Market wages, reservation wages, and retirement decisions. *JPubEc* 14 (October 1980) 277–308.

Gunderson, M. and J. Pesando. The case for allowing mandatory retirement. *CPP* 14 (March 1988) 32–39.

Gustman, A. and T. Steinmeier. The effects of pension and retirement policies in retirement in higher education. *AER* 81 (May 1991) 111–115.

Gustman, A. and T. Steinmeier. The stampede toward defined contribution plans: fact or fiction. *IR* 31 (Spring 1991) 361–369.

Gustman, A. and T. Steinmeier. Partial retirement and the analysis of retirement behavior. *ILRR* 37 (April 1984) 403–415.

Gustman, A. and T. Steinmeier. The 1983 Social Security reforms and labor supply adjustments of older individuals in the long run. *JOLE* 3 (April 1985) 237–253.

Gustman, A. and T. Steinmeier. A disaggregated structural analysis of retirement by race, difficulty of work and health. *R.E. Stats.* 58 (August 1986) 509–512.

Gustman, A. and T. Steinmeier. A structural retirement model. *Econometrica* (May 1986) 555–584.

Hannah, L. *Inventing Retirement: the Development of Occupational Pensions in Britain*. Cambridge: Cambridge University Press, 1986.

Hall, A. and T. Johnson. The determinants of planned retirement age. *ILRR* 33 (January 1980) 241–24.

Hamermesh, D. Social insurance and consumption: an empirical inquiry. *AER* 72 (March 1982) 101–113.

Hamermesh, D. Life-cycle effects on consumption and retirement. *JOLE* 2 (July 1984) 353–370.

Hanoch, G. and M. Honig. Retirement, wages, and labor supply of the elderly. *JOLE* 1 (April 1983) 131–151.

Hausman, G. and D. Wise. Social security, health status and retirement. *Pensions, Labor and Individual Choice*, D. Wise (ed.). Chicago: University of Chicago Press, 1985, 159–190.

Haveman, R., B. Wolfe and J. Warlick. Labor market behaviour of older men. *JPubEc* 36 (July 1988) 153–176.

Hogarth, J. Accepting an early retirement bonus. *JHR* 23 (Winter 1988) 21–33.

Honig, M. and C. Reimers. Is it worth eliminating the retirement test? *AER* 79 (May 1989) 103–107.

Honig, M. and G. Hanoch. Partial retirement as a separate mode of retirement behavior. *JHR* 20 (Winter 1985) 21–46.

Hu, S. Pensions in labor contracts. *IER* 29 (August 1988) 477–492.

Hu, S.C. Social security, the supply of labor and capital accumulation. *AER* 69 (June 1979) 274–283.

Hurd, M. Research on the elderly. *JEL* 28 (June 1990) 565–637.

Hurd, M. and M. Boskin. The effect of social security on retirement in the early 1970s. *QJE* 46 (November 1984) 767–790.

Ippolito, R. The implicit pension contract. *JHR* 22 (Summer 1987) 441–467.

Ippolito, R. Toward explaining early retirement after 1970. *ILRR* 43 (July 1990) 556–569.

Kahn, J. Social security, liquidity and early retirement. *JPubEc* 35 (February 1988) 97–117.

Kotlifkoff, L. and D. Wise. *The Wage Carrot and the Pension Stick*. Kalamazoo, Michigan: Upjohn Institute, 1989.

Kotlifkoff, L. and D. Wise. Labor compensation and the structure of private pension plans. *Pensions, Labor and Individual Choices*, D. Wise (ed.). Chicago: University of Chicago Press, 1985, 55–83.

Kotlifkoff, L. and D. Wise. The incentive effects of private pension plans. *Issues in Pension Economics*, Z. Bodie, J. Shoven and D. Wise (eds.). Chicago: University of Chicago Press, 1987.

Krashinsky, M. The case for eliminating mandatory retirement. *CPP* 14 (March 1988) 40–50.

Lazear, E. Pensions as severance pay. *Financial Aspects of the United States Pension System*, Z. Bodie and J. Shoven (eds.) Chicago: University of Chicago Press, 1983.

Lazear, E. and R. Moore. Pensions and turnover. *Pensions in the U.S. Economy*, Z. Bodie, J. Shoven and D. Wise (eds.). Chicago: University of Chicago Press, 1988.

Lazear, E. Pensions and deferred benefits as strategic compensation. *IR* 29 (Spring 1990) 263–280.

Lovejoy, L. The comparative value of pensions in the public and private sectors. *MLR* 111 (December 1988) 18–26.

Luzadis, R. and O. Mitchell. Explaining pension dynamics. *JHR* 26 (Fall 1991) 679–703.

McCarthy, T. The effect of social security on married women's labour force participation. *National Tax Journal* 43 (March 1990) 95–110.

Meng, R. Collective bargaining, firm size and pensions in Canada. *Economic Letters* 31 (November 1989) 99–103.

Mirkin, B. Early retirement as a labor force policy: an international overview. *MLR* 110 (March 1987) 19–33.

Mitchell, O. Worker knowledge of pension provisions. *JOLE* 6 (No. 1, 1988) 21–39.

Mitchell, O. and R. Luzadis. Changes in pension incentive through time. *ILRR* 42 (October 1988) 100–108.

Mitchell, O. and G. Fields. The economics of retirement behavior. *JOLE* 2 (January 1984) 84–105.

Mitchell, O. and G. Fields. The effects of pensions and earnings on retirement: a review essay. *Research in Labor Economics*, R. Ehrenberg (ed.). New Haven: JAI Press, 1982.

Montgomery, E., K. Shaw and M. Benedict. Pensions and wages. *IER* 33 (February 1992) 111–128.

Nader, J. Rational decision rules for early retirement inducements contained in corporate pension plans. *Journal of Risk and Insurance* 58 (March 1991) 101–108.

Pesando, J. and M. Gunderson. Retirement incentives contained in occupational pension plans and their implications for the mandatory retirement debate. *CJE* 21 (May 1988) 244–264.

Pesando, J., M. Gunderson and J. McLaren. Pension benefits and male-female wage differentials. *CJE* 24 (August 1991) 536–550.

Pesando, J., M. Gunderson and P. Shun. Incentive and redistributive effects of private sector union pension plans in Canada. *IR* 30 (Winter 1992) 179–194.

Pesando, J. and M. Gunderson. Does pension wealth peak at the age of early retirement? *IR* 30 (Winter 1991) 79–95.

Pesando, J. and S. Rea. *Public and Private Pensions in Canada*. Ontario Economic Council Study. Toronto: University of Toronto Press, 1977.

Quinn, J., R. Burkhauser and D. Myers. *Passing the Torch: The Influence of Economic Incentives on Work and Retirement*. Kalamazoo, Mich.: W.E. Upjohn Institute, 1990.

Quinn, J. Microeconomic determinants of early retirement: a cross section view of white married men. *JHR* 12 (Summer 1977) 329–346.

Quinn, J. Job characteristics and early retirement. *IR* 17 (October 1978) 315–328.

Rees, A. and S. Smith. *Faculty Retirement in the Arts and Sciences.* Princeton, N.J.: Princeton University Press, 1991.

Reid, F. Economic aspects of mandatory retirement: Canadian experience. *RI/IR* 43 (No. 1, 1988) 101–113.

Ricardo-Campbell, R. and E. Lazear (eds.) *Issues in Contemporary Retirement.* Stanford: Hoover Institution Press, 1988.

Ruhm, C. Bridge jobs and partial retirement. *JOLE* 8 (October 1990) 482–501.

Schell, B., R. Lebrasseur and R. Renaud. Predictors of acceptance of early retirement offers for workers. *RI/IR* 44 (No. 2, 1989) 376–392.

Schiller, B. And R. Weiss. The impact of private pensions on firm attachment. *R.E. Stats.* 61 (August 1979) 369–380.

Slade, F. Retirement status and state dependence. *JOLE* 5 (January 1987) 90–105.

Smith, S. Ending mandatory retirement in the Arts and Sciences. *AER* 81 (May 1991) 106–110.

Stern, S. Promotion and optimal retirement. *JOLE* 5 (October 1987) S107–S123.

Stock, J. And D. Wise. Pensions, the option value of work, and retirement. *Econometrica* 58 (September 1990) 1151–1180.

Stock, J. And D. Wise. The pension inducement to retire: an option value analysis. *Issues on the Economics of Aging*, D. Wise (ed.). Chicago: University of Chicago Press, 1990, 205–224.

Turner, J. And L. Dailey (eds.). *Pension Policy: An International Perspective.* Washington, D.C.: U.S. Dept. Of Labor, 1991.

Warlick, J. Participation of the aged in SSI. *JHR* 17 (Spring 1982) 236–260.

Wise, D. (ed.). *The Economics of Aging.* Chicago: University of Chicago Press, 1989.

Wolfe, J. Perceived longevity and early retirement. *R.E. Stats.* 65 (November 1983) 544–551.

Wolfe, J. A model of declining health and retirement. *JPE* 93 (December 1985) 1258–1267.

Woodland, A. Determinants of the labour force status of the aged. *EJ* 63 (June 1987) 97–114.

Zabalza, A. and D. Piachaud. Social security and the elderly: a simulation of policy changes. *JPubEc* 16 (October 1981) 145–170.

Zabalza, A., C. Pissarides and M. Barton. Social security and the choice between full-time work, part-time work and retirement. *JPubEc* 14 (October 1980) 245–276.

CHAPTER 7

Hours of Work

The hours-of-work aspect of labour supply has a variety of dimensions including hours per day, days per week, and weeks per year. Changes in any or all of these dimensions can alter the hours-of-work aspect of the labour supply decision. Phenomena such as the eight-hour day, the shorter work week, and increased vacation time are institutional embodiments of a reduction in hours of work. Similarly, moonlighting, overtime, flexible working time, and compressed work weeks are institutional arrangements that alter the typical pattern of hours of work.

In the short run, hours of work appear to be relatively fixed with little scope for individual variation. The eight-hour day, five-day work week, and fixed vacation period are fairly standard for most wage and salary earners. However, as we will analyze later in this chapter, the increased importance of flexible working hours is altering these arrangements. In addition, occupational choice provides an element of flexibility as people choose jobs partly on the basis of the hours of work required.

Greater flexibility is possible in the long run when firms and workers have sufficient time to adjust to new preferences and constraints they may face. The adjustment process may be subtle, as for example when workers choose jobs partly on the basis of the hours of work required. Or it may be overt, as for example when workers bargain through their unions for longer vacations or a reduced work week or workday.

The policy importance of the hours-of-work decision is illustrated in a variety of ways. Changes in the hours-of-work decision can affect not only the quantity but also the quality of our overall labour supply (and hence national output), as well as absenteeism, turnover, employment opportunities, and the demand for related activities, notably those involving leisure time and flexible working hours. Changes in hours of work, in turn, can be affected by changes in the age and sex structure of the labour force, the prominence of two-earner families, as well as government policies and laws and, of course, the basic economic determinants — wealth and expected earnings. Only by analyzing these determinants can we forecast the future time pattern of hours of work and predict the impact of alternative policies and institutional arrangements.

The focus of this chapter is on the supply side of the market; that is, on the determinants of the hours of work that individuals decide to offer or supply to the labour market. The demand side, emphasizing the firm's hours-of-work decision, is discussed in Chapters 9 and 10.

THEORETICAL DETERMINANTS OF HOURS OF WORK

Institutional Factors

At first glance it appears that hours of work are largely influenced by institutional factors. Unions have fought long and successfully for reduced working hours in a variety of forms. Labour standards legislation has specified maximum hours that can be worked and has required overtime premiums in certain circumstances. In addition, as we will analyze later, legislation can affect the number of hours for which it is optimal to employ certain types of labour. The age and sex structure of the work force can also affect hours worked because of the varying preferences for leisure and work activities amongst various elements of the work force.

However, rather than exerting independent influences on the choice of hours of work, many of these factors can be thought of as being the institutional embodiments through which the preferences of the work force are registered. Presumably unions have pushed for reduced hours because this is what the rank and file wanted, given their changing circumstances. Similarly, legislation governing hours of work would change in response to the pressures of the parties influencing the legislative process. What then are the basic forces that, in part at least, have worked through these institutional channels to alter the hours-of-work decision?

Economic Factors

As with the labour force participation decision, the basic economic determinants of the hours-of-work decision are one's nonlabour income and hourly wage — the respective income and price variables of economic theory. The income-leisure choice framework predicts that an increase in our nonlabour income will have a pure income effect increasing the demand for leisure and hence reducing the hours of work. An increase in the wage will have both an income and a substitution effect, each working in the opposite direction. On the one hand, the increased wage means that the price (opportunity cost, forgone income) of leisure has gone up, and hence the individual will "buy" less leisure and work more hours. On the other hand, the increased wage means that potential wealth is higher, and as with nonlabour income, this means an increased demand for leisure and a reduction in hours of work. To ascertain the net impact of these respective substitution and income effects of a wage change, we must appeal to the empirical evidence.

EMPIRICAL EVIDENCE

The pronounced and continuous decline over time in hours of work is illustrated in Table 7.1, which traces the standard work week in Canadian manufacturing. Between 1901 and 1981 the standard work week declined from almost 60 hours to less than 40 hours. The decline slowed down in the depression years of the 1930s, and the war years of the 1940s, and it appears to be slower in the postwar period. However, as the last column illustrates, when vacations and holidays are

Table 7.1
STANDARD WEEKLY HOURS IN MANUFACTURING,
CANADA, 1901–1981

YEAR	STANDARD WEEKLY HOURS[a]	HOURS NET OF VACATIONS & HOLIDAYS[b]
1901	58.6	n.a.
1911	56.5	n.a.
1921	50.3	n.a.
1931	49.6	n.a.
1941	49.0	n.a.
1951	42.6	40.7
1961	40.4	38.1
1971	39.8	36.7
1981	39.2	34.8

Notes:
a. Standard hours are usually determined by collective agreements or company policies, and they are the hours beyond which overtime rates are paid. The data apply to nonoffice workers.
b. Standard hours minus the average hours per week spent on holidays and vacations.

Source: Figures for 1901–1971 for standard weekly hours are from S. Ostry and M. Zaidi, *Labour Economics in Canada*, 3rd ed. Toronto: Macmillan, 1979, pp. 80, 81. Figures for 1951–1971 for hours net of vacations and holidays are from Labour Canada (1974, p. 6). Both sources used as their primary data the Survey of Working Conditions conducted annually by the Canada Department of Labour and published as *Wage Rates, Salaries and Hours of Labour*. This survey is also the source for the 1981 figure. Unfortunately, the survey was discontinued in the early 1980s.

considered the decline in average working hours is more noticeable. In essence, in recent years the work force has reduced its working hours more in the form of increased vacations and holidays rather than a reduction in hours worked per week. The decline in net weekly hours in the postwar period and in standard hours prior to World War II give a long-run trend reduction of about two hours per decade.

This long-run decline in hours of work suggests the dominance of the income effect over the substitution effect with respect to the hours-of-work decision. That is, as real wages have increased over time, the resulting increased income has been used to buy more leisure in the form of reduced hours of work, and this has dominated any tendency to buy less leisure because it is more expensive as wages increase. The income effect of the wage increase appears to have dominated the substitution effect, with the result that hours of work have similarly decreased over time as wages have increased.

MOONLIGHTING, OVERTIME, AND FLEXIBLE WORKING HOURS

Any analysis of the hours-of-work decision must confront the following basic question: why is it that some people moonlight at a second job at a wage less than their market wage on their first job, while others require an overtime premium to work more? This apparent anomaly occurs because people who moonlight are underemployed at the going wage on their main job while people who require an overtime premium are already overemployed at the going wage on their main job. Underemployment and overemployment, in turn, occur because different workers have different preferences and they tend to be confronted with an institutionally fixed work schedule. The fixed hours of work, in turn, can arise because of such factors as legislation, union pressure, or company personnel policy. The need for interaction among employees (or between employees and the firm's physical capital) may cause the employer to set standardized hours of work, rather than allowing each employee to work his or her preferred number of hours. Again, economic theory will enable us to analyze the labour market impact of these important institutional constraints.

The fixed hours-of-work phenomenon is illustrated in Figure 7.1(a). In this case, the worker is faced with two constraints. The first is the usual budget constraint L_tY_t as determined by the person's hourly wage. This restricts the worker's maximum choice set to the triangular area L_tY_tO. The second constraint of the fixed workday of L_tL_c hours (recall work is measured from right to left) restricts the worker's maximum choice set to the area L_cCY_cO: the worker can take no more leisure than OL_c (work no more than L_tI_{tc}), and no more income than OY_c even if he worked more than L_tL_c. In effect, this reduces the worker's realistic choice set to the point C since C will always be preferred to other points within L_cCY_cO.

Figure 7.1
FIXED HOURS CONSTRAINT, UNDEREMPLOYMENT,
AND MOONLIGHTING

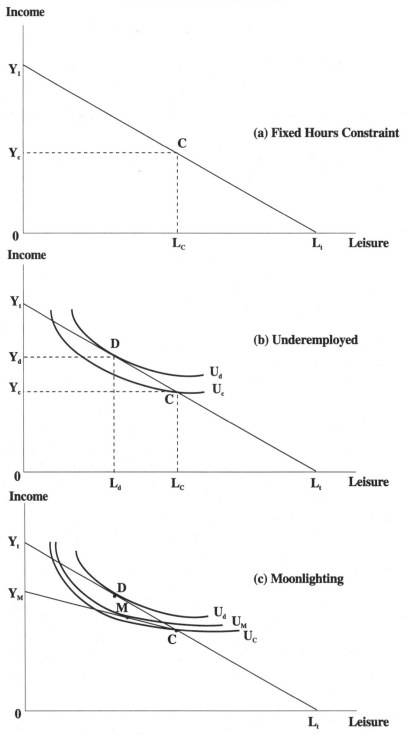

Moonlighting and Underemployment

Some individuals, however, may have preferences such that they would prefer to work more hours at the going wage rate. Figure 7.1(b) illustrates the case for an individual whose preferences (indifference curve U_d) are such that he would prefer to be at D, working L_tL_d hours at the going wage and taking home an income of Y_d. However, because of the hours of work constraint, the worker must be at C, obviously being less well off since $U_c < U_d$. In fact, the difference between U_d and U_c is a measure of how much the worker would be willing to give up in order to have the fixed hours constraint relaxed.

A variety of implications follow from this analysis. The worker is *underemployed* because he would like to work more at the going wage rate. Because of the additional constraint of the fixed working hours, the worker is also less well off ($U_c < U_d$) and may be seeking a different job that would enable him to achieve his desired equilibrium at D. In addition, the worker may be willing to *moonlight* and do additional work at a wage rate that is lower than the wage rate of the first job.

This moonlighting equilibrium is illustrated in Figure 7.1(c) by the budget constraint CY_m. (To simplify the diagram the details of Figure 7.1(b) have been omitted). This new budget constraint rotates downward from CY_t because the moonlighting wage, which is less than the regular wage as given by the slope of CY_t, applies only to hours of work beyond L_tL_c. In spite of the lower moonlighting wage, the worker is willing to work more hours (move from C to M) because of the greater utility associated with the move ($U_m > U_c$). That is, workers who are underemployed at the going wage rate would be willing to moonlight at a lower wage rate on their secondary job.

Overtime and Overemployment

Other individuals, however, may have preferences such that they would prefer to work fewer hours at the going wage rate. Figure 7.2(a) illustrates the situation where the worker would prefer (maximize utility) to be a D, working L_tL_d hours for an income of Y_d. However, because of the institutionally fixed work week, he is compelled to be at C, being less well off ($U_c < U_d$) even though he takes home more income ($Y_c > Y_d$).

Such a worker is *overemployed* at the going wage rate and consequently would support policies to reduce the institutionally fixed work week. In addition, she also may be seeking a different job that would enable her to work fewer hours at the going wage rate, and she may even exhibit absenteeism and tardiness as ways of moving towards her desired work week. Because such a worker is already overemployed at the going wage rate, she would not willingly work more hours at the wage rate; however, she may be induced to do so by an *overtime* premium.

Figure 7.2
OVEREMPLOYMENT AND OVERTIME

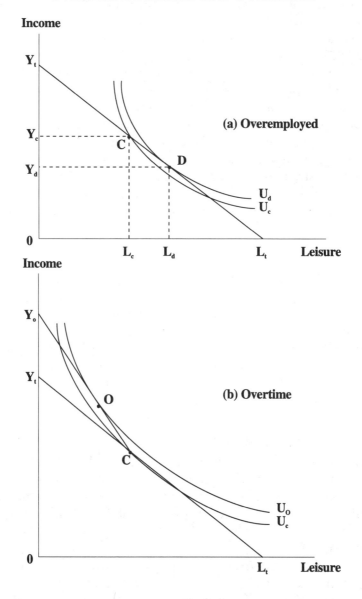

This overtime premium is illustrated in Figure 7.2(b) by the budget constraint CY_o that rotates upwards from CY_t because the overtime premium, which is greater than the normal wage which determines the slope of L_tY_t, applies only to overtime hours of work beyond L_tL_c. If, for example, the overtime premium is time-and-one-half, then the overtime budget constraint has a fifty percent greater slope than the regular straight-time budget constraint. As long as the worker is willing to give up some leisure for additional income, then ther overtime premium that will induce him to work more, for example, to mᴜ. point O on Figure 7.2(b) (on the indifference curve U_o). Because the worker is overemployed at the going wage rate, the overtime premium is necessary to get him to work more hours.

The person works longer hours even though he was overemployed at the going wage rate because the overtime premium results in a large substitution effect, by making the price (opportunity cost, income forgone) of leisure higher only for the overtime hours. The overtime premium has a small income effect because the budget constraint rotates upward *only* for the overtime hours; consequently, it does not have an income effect for the normal straight-time hours. Recall that the substitution effect was illustrated by a changed *slope* in the budget constraint, while the income effect was illustrated by a *parallel shift* of the constraint. Since the overtime premium changes the slope for the overtime hours, it is primarily a work-inducing substitution effect, with little leisure-inducing income effect.

Overtime Premium Versus Straight-time Equivalent

The importance of the relative absence of the income effect in the overtime premium can be illustrated by a comparison of the overtime premium with the straight-time equivalent. One might logically ask the question: if workers are constantly working overtime, why not institutionalize that into a longer workday and pay them the straight-time equivalent of their normal wage plus their overtime wage?

This alternative is illustrated in Figure 7.3. The overtime situation is illustrated by the budget constraint L_tCY_o with L_tC being the normal wage paid during the regular work period and CY_o being the overtime premium paid for overtime hours. The normal wage is assumed not to change as a result of the overtime premium. The regular work period would be L_tL_c hours and overtime hours would be L_cL_o. (To simplify the diagram, these points are not shown; however, as in Figure 7.2(a), they are simply the points on the horizontal leisure axis vertically below their corresponding equilibrium points.) The straight-time hourly equivalent for L_tL_o hours of work is given by the budget constraint L_tO, the slope off which is a weighted average of the slopes of the regular wage L_tC and the overtime premium CO. The straight-time hourly equivalent is derived by simply taking the earnings associated with the overtime plus regular time hours of work, L_tL_o, and determining the straight-time wage, L_tO, that would yield the same earnings.

A worker who is paid the straight-time equivalent, however, would not voluntarily remain at O, but rather would move to the point S which involves less work. This is so because the wage line L_tSO has a larger leisure-inducing income effect, whereas the overtime premium COY_o is dominated by the work-inducing

Figure 7.3
OVERTIME PREMIUM VERSUS STRAIGHT-TIME EQUIVALENT

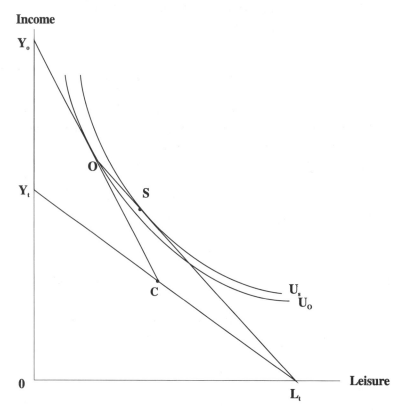

substitution effect (rotation of the budget constraint). In essence, since the overtime premium is paid only for hours *beyond* the regular workday, the person has to work more to get the additional income.

Overtime premiums, therefore, may be a rational way for employers to get their existing work force to voluntarily work more hours, even if they are over-employed at their regular workday. Employers, in turn, may want to work their existing work force longer hours, rather than hiring additional workers, so as to spread their fixed hiring costs over longer hours. The importance of these fixed hiring costs will be analyzed in more detail when labour demand is discussed.

Workers need not be overemployed at their going wage rate for the overtime premium to work. One can easily portray the situation for a worker in equilibrium at the regular wage rate who would unambiguously work more when offered an overtime wage premium, but who may work less if offered a straight-time hourly wage increase. Firms that want to work their existing work force longer hours may prefer a wage package that involves a low regular wage and a high overtime premium, to a package that costs them the same but that involves a straight-time wage increase.

Again, what at first glance appear to be costly and irrational actions on the part of firms — in this case the coexistence of overtime and moonlighting rates and a preference for overtime premiums over straight-time equivalent earnings — may well be rational actions when viewed in the larger picture where the parties are optimizing with respect to legal institutional constraints and when the varying preferences of individual workers are considered. Rather than rendering economic theory irrelevant in the force of such constraints, they highlight the usefulness of economics in analyzing the impact of the constraints and in explaining why, in fact, they may arise as an endogenous institutional response to the peculiarities of the labour market.

Allowing Choice in Working Hours

As illustrated previously, the composition of the work force has been changing dramatically in recent years. This is evidenced by such phenomena as the increased labour force participation of women, the dominance of the two-earner family, and the aging of the work force. Given such changes it is not surprising that these different groups would have different tastes and preferences for alternative work-time arrangements in the labour market. As well, they will face different household constraints.

For example, two-earner families may prefer part-time employment for one party, reduced hours of work for the other party, and flexible working-time arrangements, as well as the right to refuse overtime work for both parties. This would enable them to better combine labour market work with their household activities. In contrast, the one-earner family may prefer a long work week (e.g., with regular overtime) for the single earner so as to earn a similar family income to that of the dual-earner family. In essence, the growing diversity of the work force has given rise to a growing diversity of preferences for alternative work-time arrangements. Preferences for such arrangements are no longer dictated by the former stereotypical male "breadwinner" in a single-earner family.

This diversity of preferences is illustrated in the results of the Statistics Canada Survey of Work Reductions, conducted as a supplement to the June 1985 Labour Force Survey (Benimadhu, 1987). According to that survey only about one-third of the work force was content with its work-time arrangements. The two-thirds who were discontent were about evenly divided between wanting more work for more pay and wanting less work for a corresponding reduction in pay. The strongest preference for work-time reductions was from women in the childrearing years (25 to 34) and women with children under the age of five.

The basic income-leisure choice framework can be used to illustrate that there are gains to be had by employers providing alternative work-time arrange-

Figure 7.4
GAINS FROM ALTERNATIVE WORK SCHEDULES

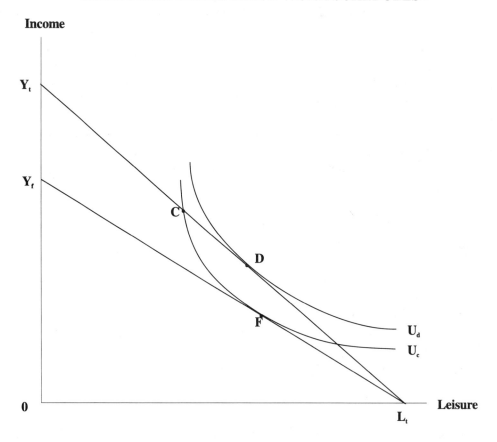

ments to meet the divergent tastes and preferences of an increasingly heterogeneous work force. These benefits are illustrated in Figure 7.4. Point C illustrates where workers are constrained to operate given the all-or-nothing choice of working L_tL_c hours (points on the leisure axis are not marked to simplify the diagram) at the going wage (slope of L_tY_t). However, many workers have different preferences. Some, for example, may prefer to be at point D. Because they are overemployed at the going wage rate, their discontent ($U_c < U_d$) may be exhibited in the form of costly absenteeism, high turnover, and perhaps reduced morale and productivity.

Obviously firms that allowed such workers to work their desired hours of work could save on these costs. Alternatively, such firms could lower their wage rates and still retain their work force. This is illustrated by the wage line L_tY_f, which could be lowered until the point of tangency, F, with the original utility curve U_c. Workers are equally well off at C and F (same level of utility U_c) even though F implies a lower wage rate, simply because they are at an equilibrium with respect to their hours of work. In essence, they are willing to give up wages in return for a work schedule that meets their preferences.

Competition for such jobs would ensure that firms could offer lower wages in return for more flexible work schedules. In this sense, the gains from flexible work schedules could be recouped by the firm to cover the other costs that may be associated with such schedules. Firms that offer more flexible hours need not lower wages, but may take the benefits in the form of reduced absenteeism, lower turnover, and improved worker morale. Various combinations of reduced wages (downward rotated wage line) and improved worker morale (higher indifference curve), of course, are possible.

While there are benefits from allowing workers to choose their preferred hours of work, there are costs, especially if such flexibility is given to different workers in the same establishment. Individual differences in hours worked can give rise to problems of monitoring, supervision, communication, scheduling, and coordination. One compromise is flexitime whereby the workers are required to work a fixed number of hours (e.g., eight per day), a certain "core" hours (e.g., 10:00 am to 3:00 pm), but the beginning and ending times are flexible. This helps meet the divergent preferences of different workers (e.g., late risers versus early risers); it enables some with childraising responsibilities to be home when children return from school; and it facilitates avoiding rush-hour commuting problems.

Compressed work weeks are another alternative working-time arrangement. Common compressed schedules include four 10-hour days or even three 12-hour days. These are often attractive to employees because of the longer "weekends" that are involved. They also enable the amortizing or spreading of fixed daily commute cost over a longer workday. For example, a two-hour commute over three 12-hour days is six hours of commute time per week, compared to ten hours based on a five-day week. As well, the 12-hour day may avoid rush-hour commutes. Such compressed work weeks have been used in some areas (e.g., nursing) as a recruiting device to help attract and retain personnel.

Clearly, there may be costs associated with such alternative work-time arrangements and these have to be weighed against the potential benefits. The point made here, and illustrated in the income-leisure framework, is that there are potential benefits to meeting the divergent tastes and preferences of workers. As well, these benefits are likely to be growing as a result of the increasing diversity of the work force.

QUESTIONS

1. Indicate the various ways in which workers may alter their hours-of-work decision and discuss the factors that may influence why a particular way is chosen. Discuss, for example, why increased vacation time may be chosen over a shorter workday.

2. Indicate the ways in which some of the recent decline in average hours worked may be explained by changes in the age and sex structure of our work force.

3. What factors may influence our decision to alter labour supply through changes in hours of work rather than labour force participation? Is it possible that some groups may choose to increase one dimension of labour supply while simultaneously decreasing another?

4. Utilize economic theory to suggest the relevant variables and their functional form that are appropriate for estimating the determinants of hours of work.

5. Discuss the econometric problems involved in estimating an hours-of-work equation based on microeconomic data where the individual is the unit of observation. If the full sample were to include those who worked and those who did not work, how might you construct a variable to reflect the expected wage for those who did not work?

6. Based on the diagrams of Figure 7.1, illustrate how an underemployed worker would respond to:

    (a) an offer to work as many more hours as the worker would like at the going wage;

    (b) payment of an overtime premium for hours of work beyond C;

    (c) an offer to work an additional fixed number of hours, as determined by the employee at the going wage.

7. Based on the diagrams of Figure 7.2, illustrate how an overemployed worker would respond to:

    (a) an offer to work as many more hours as the worker would like at the going wage;

    (b) payment of a moonlighting rate for hours of work beyond C.

8. Based on Figure 7.2(b), precisely illustrate the following overtime rates for hours worked beyond L_tL_c:

    (a) time-and-one-half,

    (b) double-time,

    (c) time-and-one-half for the first two hours of overtime, and double-time thereafter.

REFERENCES AND FURTHER READINGS

Altonji, J. and C. Parson. Labor supply, hours constraints, and job mobility. *JHR* 27 (Spring 1992) 256–278.

Altonji, J. and C. Parson. Labor supply preferences, hours constraints, and hours-wage trade-offs. *JOLE* 6 (April 1988) 254–276.

Benimadhu, P. *Hours of Work: Trends and Attitudes in Canada.* Ottawa: Conference Board, 1987.

Best, F. and J. Mattesich. Short-time compensation systems in California and Europe. *MLR* 103 (No. 7, 1980) 13–22.

Biddle, J. and G. Zarkin. Choice among wage-hours packages: an empirical investigation of male labor supply. *JOLE* 7 (October 1989) 415–437.

Blank, R. Simultaneously modeling the supply of weeks and hours of work among female household heads. *JOLE* 6 (April 1988) 177–204.

Booth, A. and F. Schiantarelli. The employment effects of a shorter working week. *Economica* 54 (May 1987) 237–248.

Borjas, G. The relationship between wages and weekly hours of work: the role of division bias. *JHR* 15 (Summer 1980) 409–423.

Bosworth, D., P. Dawkins, and A. Westaway. Explaining the incidence of shift-working in Great Britain. *EJ* 91 (March 1981) 145–157.

Brunello, G. The employment effects of shorter working hours: an application to Japanese data. *Economica* 56 (November 1989) 473–486.

Coyte, P. The supply of individual hours and labor force participation under uncertainty. *EI* 24 (January 1986) 155–171.

Dunn, L. An empirical study of labor market equilibrium under working hours constraints. *R.E. Stats.* 72 (May 1990) 250–258.

Earle, J. and J. Pencavel. Hours of work and trade unionism. *JOLE* 8 (January 1990) S150–S174.

Goldin, C. Maximum hours legislation and female employment. *JPE* (February 1988) 189–205.

Hanock, G. The backward-bending supply of labor. *JPE* 73 (December 1965).

Hansen, C. and T. Sargent. Straight time and overtime in equilibrium. *Journal of Monetary Economics* 21 (March 1988) 281–308.

Hart, R. *Working Time and Employment.* London: Allen and Unwin, 1987.

Hinrichs, K., W. Roche and C. Sirianni. *Working Time in Transition.* Philadelphia: Temple University Press, 1991.

Hoel, M. Employment and allocation effects of reducing the length of the workday. *Economica* 53 (February 1986) 75–86.

Hughes, J. The reduction in the working week: a critical look at Target 35. *BJIR* 18 (November 1980) 287–296.

Jakubson, G. The sensitivity of labor-supply parameter estimates to unobserved individual effects. *JOLE* 6 (July 1988) 302–329.

Jamal, M. and R.L. Crawford. Moonlighters: a product of deprivation or aspiration? *RI/IR* 36 (No. 2, 1981) 325–334.

Johnson, T. and J. Pencavel. Dynamic hours of work functions for husbands, wives and single females. *Econometrica* 52 (March 1984) 363–390.

Kahn, S. and K. Lang. The effect of hours constraints on labor supply estimates. *R.E. Stats.* 73 (November 1991) 605–611.

Kenym, P. and P. Dawkins. A time series analysis of labor absence in Australia. *R.E. Stats.* 71 (May 1989) 232–239.

Krishna, P. The economics of moonlighting: a double selection model. *R.E. Stats.* 72 (May 1990) 361–367 and errata (November 1990) 712.

Landes, E. The effect of state maximum-hours legislation on the employment of women in 1920. *JPE* 80 (June 1980) 476–494.

Larson, D.A. Labor supply adjustment over the business cycle. *ILRR* 34 (July 1981) 591–595.

Lewis, H. Hours of work and hours of leisure. *IRRA* (December 1956) 196–206.

Link, C. and R. Seattle. Wage incentives and married professional nurses: a case of backward-bending supply. *EI* 19 (January 1981) 132–143.

Long, J. and E. Jones. Part-week work by married women. *SEJ* 46 (January 1980) 716–725.

Maynes, J. Alternative working hours: who needs them? *The Journal of Industrial Relations* 22 (December 1980) 480–483.

McCoy, R. and M. Morand. *Short-time Compensation*. New York: Pergamon Press, 1986.

Meltz, N. and F. Reid. Reducing the impact of unemployment through worksharing: some industrial relations considerations. *The Journal of Industrial Relations* 25 (June 1983) 153–161.

Moffit, R. The estimation of a joint wage-hours labor supply model. *JOLE* 2 (October 1984) 550–564.

Moffitt, R. The Tobit model, hours of work and institutional constraints. *R.E. Stats.* 64 (August 1982) 510–515.

Mroz, T. The sensitivity of an empirical model of married women's hours of work to economic and statistical assumptions. *Econometrica* 55 (July 1987) 765–799.

Newton, K. and N. Leckie. Determinants of weekly work hours in Canada. *RI/IR* 34 (No. 2, 1979) 257–271.

Northrup, H.R. and T.D. Greis. The decline in average annual hours worked in the United States, 1947–1979. *JHR* 4 (Spring 1983) 95–114.

Olsen, E.O. The effort level, work time and profit maximization. *SEJ* 42 (April 1976) 644–652.

Ontario Task Force on Hours of Work and Overtime. *Working Times*. Toronto: Queen's Printer, 1987.

Owen, J. *The Price of Leisure*. Montreal: McGill-Queen's University Press, 1975.

Owen, J. Workweek and leisure: an analysis of trends, 1945–1975. *MLR* 99 (August 1976) 3–8.

Owen, J. Flexitime: some problems and solutions. *ILRR* 30 (January 1977) 152–160.

Owen, J. *Working Hours: An Economic Analysis*. Lexington, Mass.: Lexington Books, 1979.

Owen, J. *Reduced Working Hours*. Baltimore: John Hopkins Press, 1989.

Perlman, R. Observations on overtime and moonlighting. *SEJ* 33 (October 1966) 237–244.

Reid, F. Reductions in work time: an assessment of employment sharing to reduce unemployment. *Work and Pay: The Canadian Labour Market*, W.C. Riddell (ed.). Toronto: University of Toronto Press, 1985, 141–170.

Reid, F. *Hours of Work and Overtime in Ontario.* Background Report to the Ontario Task Force on Hours of Work and Overtime. Toronto: Queen's Printer, 1987.

Salkever, D. Effect of children's health on maternal hours of work: a preliminary analysis. *SEJ* 47 (July 1980) 156–166.

Smith, S. The growing diversity of work schedules. *MLR* 109 (November 1986) 7–13.

Szyszczak, E. *Partial Unemployment: The Regulation of Short-Time Working in Britain.* London: Mansell, 1990.

Wallace, John (Commissioner). *Part-time Work in Canada.* Ottawa: Labour Canada, 1983.

West, R.G. Marx's hypotheses on the length of the working day. *JPE* 91 (April 1983) 266–281.

CHAPTER 8

Additional Topics on Labour Supply

As pointed out in Killingsworth (1983) the area of labour within labour economics underwent dramatic changes in the 1970s and 1980s. In large part this has reflected the application of standard but sophisticated techniques of microeconomics and econometrics to analyze the various dimensions of labour supply. In fact, many of the procedures in econometrics developed as applications in the labour supply area. For example, they developed to handle econometric problems associated with the discrete nature of the labour force participation decision and the advent of large microeconomic data sets (where the individual person is the unit of observation) and panel or longitudinal data sets (which follow the same individual over a period of time). These developments in labour supply include fixed costs, discontinuities, sample selection bias, nonlinearities in budget constraints, and dynamic life cycle models.

FIXED COSTS

In the absence of any fixed costs associated with engaging in labour market activities, the reservation wage is the slope of the individual's indifference curve at zero hours of work. This slope is given by W_r, the slope of the line RR′ which is tangent to the indifference curve U_0 at the point E corresponding to zero hours of work in Figure 8.1(a). (Because the line RR′ is downward sloping, the reservation wage is the absolute value of the slope.)

If the individual's market wage was below the reservation wage W_r then the person would not engage in labour market activities. The individual would maximize utility by moving to the "corner solution" at point E in Figure 8.1(a).

If the market wage exceeds the reservation wage W_r then the individual will enter the labour market since the returns from engaging in labour market activity exceed the individual's valuation of their time in nonlabour market activities (their reservation wage, W_r). Their amount of labour supplied will be

Figure 8.1
LABOUR SUPPLY AND FIXED COSTS

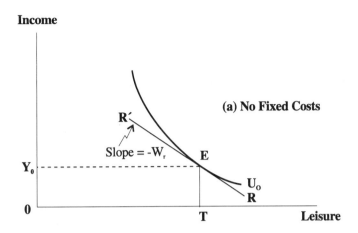

(a) No Fixed Costs

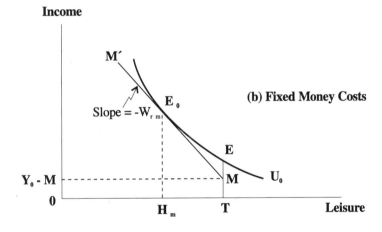

(b) Fixed Money Costs

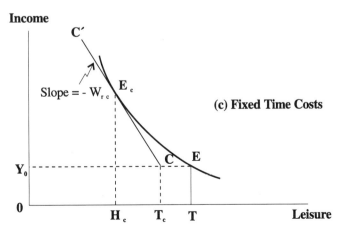

(c) Fixed Time Costs

determined in the usual fashion by the tangency of their highest attainable
indifference curve and budget constraint.

Fixed Money Costs

As illustrated by such works as Cogan (1981), Hausman (1980), and Smith
(1980), there may be fixed money costs associated with entering the labour
market and these "start-up" costs may be independent of the number of hours
worked. Such costs for a prospective employee could include daycare, buying a
car to drive to work, buying or renting accommodation closer to work, or the
purchase of special clothing. (Fixed costs for *employers* can also be important
and will be analyzed subsequently as they affect the demand for labour.) Such
fixed costs for workers are incurred only if the worker engages in labour market
activity; they can be avoided by not participating in the labour force.

Such fixed money costs are analogous to a vertical drop in the individual's
budget constraint immediately upon entering the labour market. This is depicted by the line EM in Figure 8.1(b). That is, immediately upon engaging in
labour market activities the individual incurs the fixed costs m = EM which are
analogous to a drop in income of EM. (The point M is slightly to the left of the
vertical line ET to indicate that the costs are incurred only if one engages in
labour market activity; that is, moves to the left of T.) These fixed costs are
avoided if one does not engage in labour market activity but remains at E.

In the presence of such fixed money costs, the individual's reservation wage is
depicted by the slope of line MM′ in Figure 8.1(b). This is the minimum wage
above which one would enter the labour market. It is the wage at which one
would be indifferent between engaging in labour market activities at E_0 and
engaging in nonlabour market activities at E. As in the previous case, depicted in
Figure 8.1(a), market wages below the reservation wage would not induce labour
force participation because the individual would be better off by engaging in
nonlabour market activities at E, attaining utility level U_0. Market wages above
the reservation wage would induce labour force participation and the amount of
work would be determined by the tangency of the new higher indifference curve
and the higher budget constraint (not shown in the diagram).

Clearly the individual's reservation wage in the presence of fixed money costs
is greater than the person's reservation wage in the absence of these fixed costs.
This is illustrated in Figure 8.1(b) by the fact that the slope of the line MM′ is
greater than the slope of the individual's indifference curve at the point E, the
latter slope being the person's reservation wage in the absence of fixed costs.
(This can be formally proved by the fact that the line MM′ is tangent to U_0 at the
point E_0 which is to the left of E of the same indifference curve, and the
indifference curve becomes steeper as one moves to the left of E because of the
hypothesis of diminishing marginal rate of substitution.) Simply stated, fixed
costs associated with entering the labour market increase the reservation wage

by making labour market work less attractive than nonlabour market activities such as household work, retirement, or leisure.

Fixed Time or Commute Costs

Fixed costs can also occur in the form of commuting time associated with getting to work. If it takes one hour to get to work and another to return home from work, then two hours per day are used up in commute time. This cost is fixed in the sense that it is independent of the hours of work; it can be avoided only by not participating in the labour market.

Such fixed costs are akin to taking a fixed amount of time out of each day before time is allocated to earning income in the labour market. This is illustrated in Figure 8.1(c) by the distance $TT_c = EC$. Labour market income is not earned until one moves to the left of C, and works in the labour market (assuming, of course, that the commute time is not directly compensated by the employer).

In the presence of such fixed commute costs, the individual's reservation wage is given by the slope of the line CC' in Figure 8.1(c). It is the minimum wage above which one would enter the labour market; it is the wage at which one would be indifferent between working in the labour market at E_c, and engaging in nonlabour market activities at E. Market wages below W_{rc} would not induce labour force participation because the individual would attain a higher level of utility, U_0, by engaging in nonlabour market activities at E. Market wages above W_{rc} would induce labour force participation and the amount of work would be determined by the tangency of the potential income constraint and the highest attainable indifference curve (not shown in the diagram).

As with fixed money costs, the individual's reservation wage, W_{rc}, is greater in the presence of fixed commute costs than in the absence of such fixed costs. The latter reservation wage in the absence of fixed commute costs is given by W_r, the slope of the individual's indifference curve at zero hours of work (point E in Figure 8.1(c)). Intuitively, the reservation wage is higher in the presence of fixed commute costs because the disutility of engaging in labour market activity is increased when commute costs have to be incurred. Thus, an increase in commute costs will increase reservation wage and decrease the probability of participating in the labour force.

HOURS OF WORK AND DISCONTINUITIES IN LABOUR SUPPLY

The previous discussion focused on the effect of fixed money and commute costs on the individual's reservation wage and subsequently on their labour force participation decision. Fixed costs were seen to raise reservation wages and hence discourage labour force participation. Fixed costs can also have an impact on the *hours of work* dimension of labour supply. This impact is more complicated than the labour force participation dimension.

The left panel of Figure 8.2(a) indicates the reservation wage as the slope of the indifference curve at zero hours of work (i.e., at the point E). The right panel indicates a conventional labour supply schedule as derived from the income-leisure choice framework as discussed previously. In the absence of any fixed costs, the reservation wage in the labour supply diagram would be the height of the labour supply schedule at zero hours of work. It is the minimum wage above which one would enter the labour market and supply positive hours of work. For market wages above the reservation wage, positive hours of work would be supplied as depicted by the labour supply schedule. (At higher wages this schedule may become backward-bending, as the income effect becomes larger and begins to exceed the substitution effect.)

Figure 8.2(b) depicts the situation in the presence of fixed money costs associated with labour market work. The left panel shows an initial equilibrium at E_0 with hours of work H_0 when there are no fixed costs. In this situation, the market wage, as shown by the slope of the budget constraint EE', exceeds the reservation wage and hence the person works H_0 hours.

The existence of fixed money costs of m = EM associated with entering the labour market is akin to a reduction in nonlabour income to Y_0 - m. This shifts the budget constraint down in a parallel fashion since market wages have not changed. If leisure is a normal good, this pure income effect, from the reduced income, increases hours of work from H_0 to H_m. Tracing this through for various possible wage levels would yield the hours-of-work supply schedule, S_m, in the right-hand panel of Figure 8.2(b). This hours-of-work schedule is to the right of the original schedule, S_0, when there are no fixed costs because one has to work more hours to make up for the income loss associated with the fixed costs.

In addition to increasing hours of work for those who are participating in labour market activity, fixed costs also create a discontinuity in the labour supply schedule. This can be seen in Figure 8.2(b). The distance TH_m in that diagram indicates the number of hours of work below which it would not be worthwhile to enter the labour market given the reservation wage W_{rm} created by the fixed costs EM. It is not possible to amortize the fixed costs of entering the labour force over so few hours; hence, there is a discontinuity in labour supply over that region. Geometrically, it is not possible for there to be a tangency between the budget constraint MM' and the indifference curve U_m in the region E_mE. This is so because the slope of MM' is always greater than the slope of the indifference curve to the right of E_m.

This discontinuity over the range of TH_m hours translates into the distance OH_m in the labour supply diagram of Figure 8.2(b). (The distance OH_m in the right panel of Figure 8.2(b) equals the distance TH_m in the left panel.) The discontinuity OH_m in the labour supply diagram illustrates the range of hours for

Figure 8.2
LABOUR SUPPLY DISCONTINUITIES

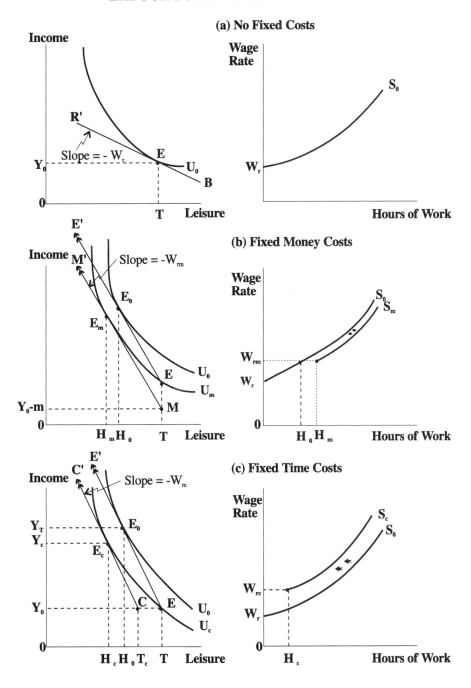

which it is not worthwhile entering the labour force given the fixed costs associated with entry. Such part-time work does not enable one to amortize the fixed costs of entry over so few hours. This illustrates how fixed costs discourage part-time work.

In summary, fixed money costs of entering the labour force increase reservation wages and this in turn decreases labour force participation. However, for those who do participate, hours of work will be increased. The fixed money costs also create a discontinuity in labour supply, making part-time work less likely. Intuitively, the fixed money costs make entering the labour force less attractive, especially for a few hours; however, once in the labour force, the individual will work more hours to compensate for the loss of income from the fixed money costs. Clearly, fixed costs have a complicated effect on the various dimensions of labour supply.

One would expect a similar pattern for an increase in fixed *time* or commute costs associated with labour market work. However, this case is complicated by the fact that commute time is time not available for labour market or other activities. The left panel of Figure 8.2(c) indicates the situation where the initial equilibrium E_0 with no fixed costs implies hours of work of H_0. Fixed commute costs of $T_cT = EC$ reduce the time available to allocate to labour market work or other activities like leisure. As with fixed money costs this involves an inward shift of the budget constraint in a parallel fashion since wages have not changed. This pure income effect leads to reduced purchases of all normal goods including leisure. Total hours spent in labour market work *and* commute time increase from H_0 to H_c. However, hours worked in the labour market are only T_cH_c since TT_c is commute time. Hours worked actually decrease since T_cH_c is less than TH_0. This is so because H_0H_c is less than TT_c, since the reduction in the individual budget constraint associated with the increased commute time TT_c is allocated between reduced leisure H_0H_c and a reduction in labour market income Y_TY_C. That is, $H_0H_c < TT_c$, and subtracting the common distance H_0T_c from each side yields $H_0H_c - H_0 T_c < TT_c - H_0T_c$ or $T_cH_c < TH_0$.

In essence, the increase in fixed commute costs associated with going to work will increase the *total* time spent in labour market and commuting activity for those who participate in the labour market. However, the increased commute time does not lead to an equivalent increase in labour market work since some of the wealth reduction associated with the increased commute time is absorbed by a reduction in the demand for other goods and services. Hence, hours of work actually decrease even though total work time plus commute time increases. This is illustrated by the leftward shift in the labour supply schedule in Figure 8.2(c). This is in contrast to fixed money costs which increase hours of work for those who participate in labour market activities.

The fixed commute costs also create a discontinuity in labour supply. In Figure 8.2(c) the discontinuity is illustrated by the segment T_cH_c which shows the hours for which it would not be worthwhile to enter the labour market since they would be too few to enable one to amortize the fixed commute costs. Such costs discourage part-time work.

This corresponding discontinuity is illustrated by the distance OH_c in the labour supply diagram of Figure 8.2(c). It is the range of hours over which it is not worthwhile to enter the labour force given the fixed commute time.

In summary, fixed commute costs like fixed money costs associated with entering the labour market will reduce the probability of participating in the labour force. For those who do participate, total hours spent commuting *and* at work will increase as commute time increases; however, hours spent at work by itself will actually decrease. This is so because the reduced wealth associated with the additional commute time will be allocated to reduced income as well as reduced leisure. Hence the total time allocated to work plus commuting will not increase by as much as the commute time increases and hours of work will thereby fall. In addition, there will be discontinuity in labour supply as part-time work becomes less attractive since it does not enable one to amortize the fixed commute time costs over sufficiently long hours.

EFFECT OF DAYCARE SUBSIDY

The importance off fixed costs can be illustrated by analyzing the impact of daycare subsidies[1] on both the labour force participation and the hours-of-work decision. To simplify the exposition it will be assumed that daycare costs are fixed upon entering the labour force and the subsidy eliminates these costs. Also, the impact of taxes needed to finance the daycare subsidies is ignored.

By removing the fixed costs associated with daycare, such subsidies would reduce reservation wages. This, in turn, would induce labour force participation amongst those who otherwise found the fixed daycare costs to be too inhibiting a barrier to labour force participation. Even part-time work may be attractive since there would no longer be a need to amortize the fixed daycare costs over longer working hours.

For those who are already participating in the labour force the daycare subsidy would not affect their participation decision; that is, it would never cause them to leave the labour force. This is so because if they were already participating then their reservation wage was already below their market wage, hence, lowering their reservation wage would only strengthen their decision to participate in the labour market.

The daycare subsidy, however, would induce those who already were participating in the labour force to work fewer hours. This is so because the subsidy is

---

1. The inhibiting effect of child-care costs on the labour supply of women is docu number of studies: Blan and Robbins (1988), Gustafsson and Stafford (1992 (1974), Leibowitz, Klerman and Waite (1992), Michalopoulos, Robbins an((1992), and Risen (1992).

like a demogrant, given to those who have children in daycare. This causes them to buy more of all normal commodities including leisure; hence, the reduction in hours worked. Alternatively stated, by eliminating the fixed cost of daycare, the subsidy reverses the effect of a fixed money cost as illustrated in Figure 8.2(b). That is, the subsidy would shift the labour supply schedule back from S_m to S_0, reducing hours of work and eliminating the discontinuity that made it unattractive to work part-time.

In summary, subsidies to daycare can have a complicated effect on the various dimensions of labour supply, encouraging labour force participation and part-time work, but reducing the hours of work for those already participating (assuming the subsidies are not on a per-hour basis). The net effect on total hours of work is therefore ambiguous. Obviously, making the subsidy available only to those who are not already participating in the labour force would ensure that total hours of work would increase. However, this can create inequities (and some interesting politics!) because similar groups who are already participating in the labour force would not receive the subsidy.

CORNER SOLUTIONS AND CENSORING

In the previous discussion, individuals whose market wage was below their reservation wage were simply lumped together and treated as nonparticipants in the labour force. In terms of the previous Figure 8.1(a), they were assumed to be at the "corner solution," or point E, since their marginal rate of substitution between income and leisure at E (i.e., slope of indifference curve or reservation wage) was greater than their market wage. All that we knew of these individuals was that the point of tangency between their indifference curve and budget line was not in the range of positive work hours.

Such individuals can be thought of as desiring to work negative hours of labour market work; however, all we can observe is that they do zero hours of labour market work. The notion of negative hours of labour market work can be thought of as an aversion to engaging in labour market work or a desire to engage in household work or other nonlabour market activities. When we observe that individuals are not participating in the labour market we only know that their market wage is less than their reservation wage. We do not know if they are marginally indifferent between participating or not (i.e., their point of tangency is near E) or if they are extremely unlikely to enter the labour market unless their market wage increased dramatically (i.e., their point of tangency is far to the right of E).

This notion of desired negative hours of work is illustrated in Figure 8.3 which is simply an extension of the previous Figure 8.1(a) designed to show the hypothetical tangency of the budget constraint and an indifference curve, when that tangency is in the region of negative desired hours of work (i.e., when the wage is below the individual's reservation wage). For this hypothetical individual, the

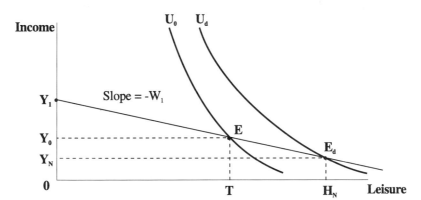

Figure 8.3
CORNER SOLUTION AND NEGATIVE DESIRED HOURS

desired equilibrium would be at E_d involving negative desired hours of work, H_n, and income of Y_n; i.e., if it were possible to buy additional leisure time at the market wage W_1, the individual would give up the income Y_0Y_n for the additional time TH_n. This situation could be a combination of having a low market wage and/or preferences against labour market work. Clearly, a higher market wage line emanating upward and to the left of E could induce the person to enter the labour force (i.e., yield a tangency with an indifference curve in the range of positive hours to the left of E). Alternatively, different preferences (i.e., different-shaped indifference curves) could yield a tangency to the budget line EY_1 to the left of E. Given the wage rate W_1 and the individual's preferences, however, their desired hours of work is negative at TH_n and to attain this they would be willing to give up income of Y_0Y_n.

All that we usually observe for such individuals, however, is that they are not participating in the labour market since data on labour supply basically indicates whether an individual is participating in the labour force and, if so, for how many hours of work. That is, we observe only points E (i.e., nonparticipants whose actual hours of work is zero) and tangencies to the left of E (i.e., hours of work for those who participate). All persons who have hypothetical tangencies to the right of E are simply aggregated together at the corner solution E as nonparticipants with zero hours of work.

In this situation where we do not have data on the individual's desired hours of work (the dependent variable in studies of labour supply) but we have information on their characteristics or explanatory variables (market wage and other relevant variables), the sample is said to be censored. If we do not have information on their behaviour that is under examination (i.e., dependent variable) nor on the characteristics believed to determine that behaviour (i.e., explanatory variables) the sample is said to be truncated with respect to those observations.

This would be the case, for example, if all high-wage individuals are excluded from eligibility in an income maintenance program in which case we would not have information on their characteristics or behaviour or numbers.

SAMPLE SELECTION BIAS

A number of implications can flow from the fact that we do not observe desired hours of work for those who are not participating in the labour force. One such implication is that estimated relationships between hours of work and explanatory variables believed to determine hours of work may be biased if they are estimated only from the observed data on those who are working positive hours. This select sample of workers may be different from a random draw of the population in terms of unobservable characteristics such as preferences for labour market work versus household work. In such circumstances inferences as to the expected behaviour of a random draw from the population, based upon the observed behaviour of those who are working positive hours, can be misleading unless special statistical techniques are used.

For example, it might be found that a one percent increase in wages led to a two percent increase in the expected hours of work of those who were already in the labour market. But this need not imply that the same relationship would prevail for a random draw from the population who may or may not be a labour force participant. The sample of those who are working may be systematically different in terms of unobservable characteristics than those who are not working in the labour market; that is in fact why one group may be sorted into labour market activities and another group into other activities. Inferences based on the estimated relationship from the subsample of workers may be subject to what is termed sample selection bias.

Diagrammatic Illustration

The bias can be illustrated diagrammatically. In Figure 8.4, hours of work, H, are measured on the vertical axis, and the wage, W, an individual can expect to receive is measured on the horizontal axis. At this stage we assume that information on each individual's expected wage is available; this assumption will later be relaxed. Assume that the population consists of nine individuals, with expected wages and hours of work represented by the Xs. Also, assume that the nine individuals can be grouped into three wage levels, low, medium, and high. Their variation in hours of work *within* each wage level (i.e., deviations from the true line) is attributable to unobservable characteristics that cannot be measured and hence directly controlled for in statistical analysis. For example, the three persons (Xs) above the true line are people with unusually high preferences for labour market work, those below the line have unusually low preferences for such work, and those on the line have average preferences for labour market work. On average, the relationship between hours of work and expected wages

Figure 8.4
SAMPLE SELECTION BIAS

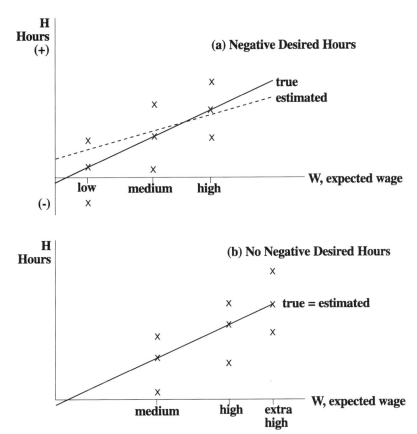

turns out to be positive as exhibited by the true line. This is akin to an upward-sloping labour supply schedule when wages are exhibited on the vertical axis and hours on the horizontal axis, and corresponds to the situation where the substitution effect dominates the income effect so that wage increases lead to more hours of work. Figure 8.4 reverses the axes because it is more common to portray the dependent variable on the vertical axis and the independent or explanatory variable on the horizontal axis.

The true line, based upon all nine observations, indicates what would happen, on average, to desired hours of work if wages increased. Desired hours of work is the equilibrium concept reflecting the optimizing decision of the individual. Desired hours can be negative as illustrated by the low-wage individual who falls below the horizontal axis.

However, in estimating the observed or actual (as opposed to desired) relationship between wages and hours of work, the only data on hours of work that is available is for the eight observations with positive hours of work. We do not observe the desired hours of work for that person who wishes negative hours. The estimated relationship is therefore based on the sample of the eight observations with positive actual hours of work.

In our particular example, the estimated line is "tilted clockwise" relative to the true line because the estimated line ignores the one low-wage observation with negative preferences toward labour market work. The estimated relationship is based upon a sample of persons who *disproportionately* have positive preferences for labour market work at each expected wage level; that is, of the eight persons in the sample from which the line is estimated, three have unusually positive preferences for labour market work (i.e., are above the true line), three have average preferences (are on the true line), but only two have negative preferences (are below the true line). The third observation with negative preferences (below the horizontal axis) is excluded from the sample altogether. Of crucial importance, that observation is excluded from the sample precisely because of the negative preference for labour market work. If that person had average or above-average preferences then he or she would be in the sample which is used to derive the estimated line.

The sample of eight from which the estimated relationship is derived is therefore systematically different from the total population of nine, in that the sample disproportionately consists of persons with positive preferences for labour market work. That is in fact why many of them are in the labour market working positive hours. This sample is said to be subject to a sample selection bias in the sense that it is not a random sample drawn from the population of potential workers and is therefore not representative of that population. It is a sample based upon the selection rule that to be included in the sample, the person must work a positive number of hours. The sample selection bias comes about because those persons with positive hours of work disproportionately have greater preferences for work than do those who are not working in the labour market. Relative to those who are not working, those who work in the labour market systematically differ in terms of the unobservable variable — preferences for labour market work.

This difference is important because the unobservable preference is correlated with the decision to work which itself is systematically related to wages; that is, within the sample of eight workers the low-wage workers are disproportionately represented by those with above-average preferences for labour market work. The one person with negative preferences and a low wage is excluded from the sample altogether. If the unobservable preferences were not correlated with the decision to work there would not be a difference between the estimated and true relationships.

For example, if the population consisted of the nine individuals in the bottom panel, Figure 8.4(b), then the estimated line would have the same slope whether it is based upon the nine observations, or if the three medium-wage or three high-wage observations were excluded. This is so because the unobservable preferences (deviations of the observations about the true line) are not correlated with the decision to work. That is, the remaining sample of six in either case would have two persons with above-average preferences for labour market work and two with below-average preferences (the same proportions as in the omitted group); hence, a sample selection bias would not occur since the selected sample would not differ from the population in terms of the unobservable preferences. This is so because the unobserved preferences are not correlated with the decision to work, which was used as a sample selection criterion.

Prime-Age Males and Married Women

This highlights the fact that sample selection bias need not always be a problem when select samples are used to make inferences about population behaviour. For example, if prime-age males tend to be participating in the labour force and work a standard work week then their situation may correspond to the bottom panel Figure 8.4(b). Few of them are at the margin of entering or leaving the labour force; that is, few are at zero or negative desired hours of work. In such circumstances, their hours-of-work response to a wage change may be estimated from a sample of workers with positive hours. Such a sample may give an accurate portrayal of the expected response of a random draw of all prime-age males (just as the estimated line of Figure 8.4(b) is an accurate portrayal of the true line).

Conversely, the situation for married women may be more like the situation portrayed in the top panel, Figure 8.4(a). Many married women, who may expect to receive a low wage if they enter the labour market, may be marginally indifferent between engaging in labour market work and household work. In such circumstances those who had a preference for labour market work disproportionately would be engaging in labour market tasks. Thus the estimated relationship between hours of work and expected wages based on the sample of those who work positive hours would be based on a biased sample of persons who are disproportionately labour market oriented in terms of unobservable characteristics like preferences. The estimated relationship may look more like the estimated line in Figure 8.4(a) in which case it would underestimate the true additional hours of work that would be associated with a wage increase for a randomly selected married woman.

Explanation of Bias in Terms of Response

This example can also be used to illustrate why the estimated relationship gives a smaller hours-of-work response than the true relationship in this case. We earlier indicated that the estimated line was "tilted clockwise" relative to the true line because it disproportionately reflected (was "pulled up" by) the positive preference for labour market work at the low-wage end (the observation with the negative preference for labour market work at that end having been excluded from the sample of participants). An alternative way of understanding the smaller hours-of-work response to a wage change based on the estimated relationship is to examine how hours of work *change* when wages *change*.

For illustrative purposes, assume that wages increase so that in Figure 8.4(a) low-wage persons become medium-wage, the medium become high and the high become extra-high. This means that the three observations in the low-wage category displace the three in the medium-wage category, the three in the medium displace the three in the high, and the three in the high move up to extra-high. On average, this would be reflected by movements along the true response line and the one low-wage individual with negative desired hours would now be a medium-wage person with small positive desired hours, observed as actual hours. The true response of that individual (in terms of the equilibrium concept of desired hours of work) is given by the true line. However, in terms of actual hours of work (which is all that is observed) it will appear that the individual went form *zero* hours to small positive hours; the change from *negative desired* hours to zero is not observed. Thus the estimated response (which does not include the change from negative desired to zero actual hours) underestimates the true response of desired hours of work.

The true hours-of-work response of married women, for example, may be larger than that observed when the relationship is estimated on the sample of those who work in the labour market since those who work only a few hours may have made a large response from negative desired hours to their desired positive hours. All we observe is that they appear to have gone from zero hours to a few actual, observed hours; in reality, their true response is larger, being from (unobserved) negative desired to positive (observed) actual hours. Thus, the true labour supply response of married women to a wage change may be larger than their response estimated from the sample of labour force participants only.

Appropriate Measure

This also highlights that the appropriate measure of labour supply response depends upon the question one wants answered. If one wanted to know the expected increase in hours of work associated with a wage increase *for those who are already participating* in the labour market, then the true response is appropriate. It would reflect the response off those observations who are already participating in the labour force (i.e., above the horizontal axis in Figure 8.4(a)). Similarly, if one wanted to know the response of *desired* hours, to include changes in the desired hours of those who are not working, the *true* response is appropriate. However, if one wanted to know the increase in *actual* hours that would be associated with a wage increase, whether those additional hours come from those who are already participating *or from those who will enter the labour force* in response to the wage change, then the *estimated* relationship (based upon nonparticipants being given zero hours of work) is appropriate even though it gives a smaller response than the true relationship.

To ascertain the actual increase in hours, whether from existing participants or from new entrants, the estimated relationship is appropriate because it reflects the fact that the actual (observed) increase in hours is less than the desired (unobserved) increase for those who also enter the labour force in response to the wage increase. For example, the actual increase in hours for the individual below the horizontal axis of Figure 8.4(a), in response to a wage increase from the low to medium category, would only be the observed hours of work above the horizontal axis. The increase in desired hours up to the horizontal axis would not be recorded as part of the hours response to the wage increase. This is appropriate, however, if one is interested in the actual observed response for potential participants as well as for those who are already participating in the labour force.

Clearly, the appropriate measure of labour supply responsiveness depends upon the questions one wants answered. If one is interested in the change in desired hours or in actual hours for those already participating, then the true relationship is correct. However, if one is interested in the change in actual observed hours for both participants and those who may enter, then the estimated relationship is appropriate.

Correcting for Selection Bias

When sample selection bias is likely to be important because a substantial number of observations are likely to be on the margin of being in or out of the select sample (e.g., married women on the margin of participating in labour market or household work), the obvious question is how to estimate the true relationship given that we only have data on actual hours of work and not desired hours. An explanation of this procedure (e.g., Heckman 1979) requires an understanding of econometrics beyond the prerequisites of this text. However, the procedure basically involves estimating the probability of being in the select sample (participating in the labour force in this case). Information from these predicted probabilities is then used to construct an artificial variable (called the inverse of Mill's ratio). This term is then used to correct the estimated relationship that is based on the select sample, to get a relationship corrected for selectivity bias. With the addition of the selectivity bias correction term one can convert the estimated line of Figure 8.4(a) to the true line.

Other Examples

The previous discussion illustrated the sample selection bias that can occur when one tries to make inferences about the hours-of-work behaviour of a random individual based upon estimates made from a select sample of individuals who differ from the population inn terms of unobservable characteristics. The bias occurs because these observable characteristics (preferences for labour market work in the previous example) are important determinants of being in the select sample; that is, people may sort themselves into the select sample in part based upon these unobservable characteristics.

Other examples of selection bias abound in the empirical literature of labour economics; in fact it has been a virtual growth area in recent years. In some instances the selection bias occurs because the individuals sort themselves into the select sample on the basis of unobservable characteristics, in other instances they may be selected by employers, and in other instances by the researcher. The important point is that the selection is not random; that is, the nature of the sample used by the researcher is itself endogenously determined by underlying forces, and the analysis of the data must take into account this endogeneity. While many of the examples are referred to elsewhere in the text, some are recapitulated here to illustrate their common features.

In the previous analysis, *expected* wages were discussed as an explanatory determinant of hours of work. This independent variable was depicted on the horizontal axis of Figure 8.4. Obviously this variable is not observed for persons who do not participate in labour market activities. One solution is to use an observed variable like education to serve as a proxy for the unobserved expected wage; the hypothesis being that, on average, more highly educated people would have a higher expected wage. Obviously this is an imperfect proxy, however, since education can reflect other factors.

Another solution is to estimate the wage that nonparticipants could expect to earn in the labour market if they did participate. This could be done by estimating the relationship between the wages of participants and their wage-determining characteristics and using this information to simulate the wage that nonparticipants could expect to earn if they were to participate in the labour market. However, labour force participants may be systematically different from nonparticipants not only in terms of observable characteristics (which can be controlled for in statistical analysis) but also in terms of unobservable characteristics (which are not directly controlled for). For example, participants may be more highly motivated or have a comparative advantage in labour market tasks; that may be part of the reason that they are sorted into such tasks either by themselves or by employers. If that is the case then their actual wage may be higher than the expected wage that a nonparticipant of similar *observable* characteristics could expect to receive upon participating in the labour market. That is, labour force participants may be systematically different than nonparticipants in terms of unobservable characteristics; hence, the earnings of participants may give a biased estimate of what nonparticipants with the same observable characteristics could expect to earn if they entered the labour market.

Formally, this sample selection bias can be corrected by the same three-step procedure outlined earlier to correct the bias in hours-of-work estimates. First, estimate the probability of being in the labour force, using the full sample of data from participants and nonparticipants. Second, use this information to construct a new variable — formally called the inverse of Mill's ratio. Third, include this new variable in estimating the determinants of wages, based on the select sample of participants. The resulting estimates will be corrected for the sample selection bias that occurs because wages are estimated on the selected sample of participants only. These corrected estimates of the determinants of wages can then be used to get estimates of what nonparticipants could expect to earn if they participated in the labour market.

Other similar examples abound. Migrants may be different from nonmigrants in terms of unobservable variables like motivation; hence, it may be inaccurate to use the earnings of the select sample of migrants to simulate what comparable nonmigrants (in terms of observable characteristics) could earn if they were to migrate. Union workers may be systematically different from nonunion workers in terms of unobservable characteristics, especially if employers hire from a queue of applicants for the high-wage union jobs or they alter the job requirements. Hence, it may be inaccurate to use the earnings of the select sample of union workers to estimate what comparable nonunion workers (comparable in terms of observable characteristics) would earn if they were unionized. Never-married females may be different from never-married males in terms of unobservable characteristics like career orientation. Hence it is likely to be inappropriate to use the earnings of the sample of never-married males (who may be less career oriented) to estimate what never-married females could expect to earn if they were paid like comparable males, where comparable is again defined in terms of observable but not unobservable characteristics.

The issue of sample selection bias can also be important in determining the true impact of various legislative activities on labour market behaviour. For example, the impact of an equal pay or equal employment opportunity law may be ascertained by comparing the wages or occupational distribution of males and females in jurisdictions with and without such legislation, after controlling for other observable variables that may affect wages and the occupational distribution. However, jurisdictions with and without such legislative initiatives may also differ in terms of unobservable factors such as discriminatory attitudes that can affect the wages and occupational distribution of females. In fact, such unobservable differences can also account for the differences in legislative initiatives. In such circumstances it may be the different attitudes, and not the independent effect of the legislative initiatives, that account for the difference in the labour market behaviour of women. That is, it could be erroneous to infer the effect of legislative initiatives to the population at large, when the effect of such legislative initiatives is based on estimates from the select sample of jurisdictions with the policy initiative in place. The select sample may be different from the population at large in terms of unobservable characteristics that influence behaviour.

As with the other examples, it would be possible to get estimates of the impact of the policy initiatives by correcting for the selectivity bias. This again requires estimating the determinants of the legislative initiative based on the full sample of jurisdictions with and without the initiative. This in turn is used to derive the correction factor which is included as a determinant of the legislative impact when that impact is based on estimates from the select sample of those jurisdictions with the legislation in place. This should provide an unbiased estimate of the expected effect of the legislation if it is adopted by any jurisdiction.

Clearly the sample selection procedure has a wide variety of applications and it can be very important from a theoretical, empirical, and policy perspective. As well, it rekindles a challenge for labour economists; that challenge involves incorporating an understanding of why certain institutional features of labour markets arise and persist. Conventionally in labour economics we regard institutional features such as unions and legislation as exogenously given or determined outside of the labour market. Their impact on various elements of labour market behaviour is then examined. However, the sample selection literature reminds us that this may give a biased estimate of their impact if the existence of these institutional features is determined in part by unobservable factors. In such circumstances we must regard these institutional features as endogenous, that is to explain their existence in terms of other predetermined or exogenous explanatory variables. This information is useful not only to explain why such institutional arrangements arise and change, but also to yield an unbiased estimate of the true impact of these institutional arrangements on other elements of labour market behaviour.

Alternatively stated, the sample selection bias literature reminds us that to fully understand the impact of certain institutional features of labour markets it is necessary to understand why these institutional arrangements arise in the first place. This challenge provides the opportunity to formally incorporate the analysis of the determinants of institutional features of labour markets into the larger picture of labour market analysis. Only then can we properly understand why these institutional features exist and their true impact.

NONLINEARITIES IN BUDGET CONSTRAINTS

In the conventional income-leisure choice framework, the budget constraint was linear (i.e., a straight line). This corresponds to the assumption that wages were constant or independent of hours of work or income. That is, working the first hour yielded the same additional income as the second hour and so forth. The picture was altered, however, with the introduction of such factors as income maintenance programs, overtime premiums, moonlighting, and fixed costs. These changes were shown to have potentially substantial impacts on various dimensions of labour supply. They can also have impacts on the appropriate methodology for analyzing labour supply and on the estimation of income and substitution effects (Hausman, 1980). Before illustrating these issues it is useful to introduce some microeconomic terminology.

A budget constraint is said to be nonconvex if a line segment joining two points in the budget choice set can fall outside of the set. Figure 8.5 illustrates four nonconvex budget constraints. Clearly in all cases, the dashed line joining two points within the choice set falls outside of the set of feasible choices. This is in contrast to budget constraints that are convex such as a linear budget constraint with a constant wage and no fixed costs, or a nonlinear budget constraint that reflects wages net of a progressive income tax,[2] or one that reflects a lower moonlighting rate in one's second job. In the latter cases, a straight line drawn between any two points in the feasible choice set (i.e., within the budget constraint) will always be within the constraint.

When the budget constraint is nonconvex, the previously discussed "local reservation wage" approach to analyzing labour force participation decisions is inappropriate. That decision rule indicated that the individual would participate in the labour force if the market wage exceeded the reservation wage at zero hours of work; otherwise the individual would not participate. In the presence of nonconvexities in the budget constraint, however, that decision rule could be erroneous in that it may not be consistent with utility maximization.

This is illustrated in Figure 8.6 which portrays the nonconvex budget constraint associated with a welfare program requiring the recipient to give up a dollar of welfare for every dollar earned in the labour market. This 100 percent implicit tax implies a *net* market wage of zero (the horizontal portion of the budget constraint) until all the welfare payments are exhausted, in which case the individual returns to the normal market wage line. Clearly the individual's local reservation wage, W_r, (at zero hours of work) is greater than the net market wage, W_0. Hence, by following the reservation wage decision rule the individual would not participate in the labour force, and would receive utility U_0. However, in this particular case the individual could attain a higher level of utility at U_p, by participating in the labour force and working H_p hours (work measured from right to left).

In this case the local reservation wage rule is inconsistent with global utility maximization because, by following the reservation wage rule, the individual is optimizing only with respect to the choices around zero hours of work. In that local region it does not appear to be worth entering the labour market because

---

2. If the individual faces a fixed gross hourly wage rate (i.e., independent of hours worked), the net or after-tax wage rate will decline as hours of work and therefore total income increases providing the tax system is progressive. This will result in a net-of-tax potential income constraint that is convex. The opposite case, which results when the income tax system is regressive, is shown in Figure 8.5(a).

Figure 8.5
NONCONVEX BUDGET CONSTRAINTS

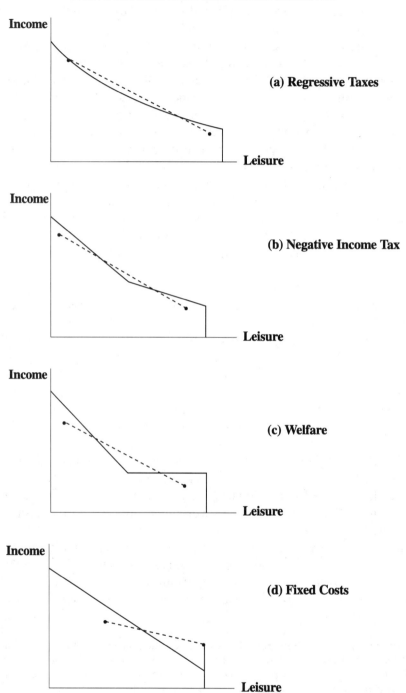

(a) Regressive Taxes

(b) Negative Income Tax

(c) Welfare

(d) Fixed Costs

Figure 8.6
LOCAL RESERVATION WAGE VS. UTILITY COMPARISONS

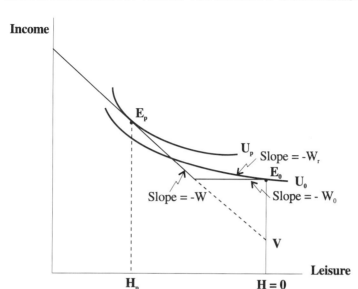

the disutility associated with an hour of work (the slope of the indifference curve or W_r) is greater than the return to work (the net wage of zero). However, by examining the wider set of choices around H_p hours of work it clearly pays to enter the labour market. After welfare is exhausted, the returns to work become sufficiently high to compensate for the disutility of work. This need not always be the case. The utility isoquant U_0 could become sufficiently steep that it always remains above the budget constraint so that E_0 is the global utility maximizing equilibrium. However, the nonconvexity of the budget constraint raises the *possibility* that the local reservation wage rule is inconsistent with utility maximization.

In the presence of nonconvexities in the budget constraint it is clearly important for the individual to examine the full range of choices that are available before decisions are made. This entails the equivalent of calculating the level of utility associated with the various combinations of wages and nonlabour income that are available and choosing the highest level of utility.[3]

---

3. These are termed indirect utility functions (when utility is expressed as a function of income and prices) as opposed to direct utility functions (where utility is expressed as a function of components like income and leisure that give rise to utility). When the direct utility function is maximized, the optimal amount of goods and leisure chosen will be functions of their prices and income. Therefore, these prices and income could be substituted in for the optimal goods and leisure, yielding utility (indirectly) to be a function of income and prices.

For example, in Figure 8.6 the individual would hypothetically estimate the level of utility, U_0, associated with the net market wage, W_0, and nonlabour market income of E_0 associated with the welfare payment if the individual does not work. Then the individual could estimate the level of utility, U_p, associated with the market wage, W, and the nonwage income, V, received from sources other than welfare.[4] Once these hypothetical calculations are made, the individual would compare the utility levels U_0 and U_p and choose the highest level of utility, in this particular case U_p.

In general, if the budget constraint is not convex, a local optimum such as E_0 need not be the global optimum. In such circumstances it is necessary to compare possible optimal points throughout the feasible set to ensure a global optimum. Informally stated, it is necessary for individuals to assess all of their options before decisions are made. By focusing only on small changes about their existing state, individuals may miss their wider range of opportunities. It follows that researchers analyzing labour supply behaviour must take account of the fact that local optima are not necessarily global maxima in the presence of nonconvex budget constraints.

DYNAMIC LIFE CYCLE MODELS

The previous discussion was couched in terms of static comparisons at a given point in time. Different equilibria were compared (a procedure termed comparative statics) assuming changes in income or relative prices that emanated exogenously from outside of the system, for example, from a change in taxes or transfer payments.

In a dynamic system, where changes are occurring over time and some may be anticipated and others not anticipated, the income and substitution effects that emanate from such income and price changes are more complicated. MaCurdy (1981) deals with many of these complications, and some of the simpler ones can be illustrated with a diagram based upon that source.

Figure 8.7 illustrates two hypothetical age-wage profiles, indicating how wages may at first increase over the life cycle and then decline later. The higher profile depicts the situation where an individual gets a higher wage at each and every age. The small "blip" in the higher profile illustrates the situation where an individual receives a temporary one-time only, unanticipated wage increase at age t. There are three possible sources of wage changes occurring in this diagram and each have different income and substitution effects and hence would

---

4. This nonwage income V (termed "virtual" income) is the level of nonwage income that the person would have to have, in combination with the known constant market wage, W, to yield a given level of utility, in this case the level of utility, U_p, associated with participating in the labour market. Alternatively stated, virtual income is the hypothetical nonwage income associated with extending the market wage line to zero hours of work.

Figure 8.7
DYNAMIC LIFE CYCLE WAGE CHANGES

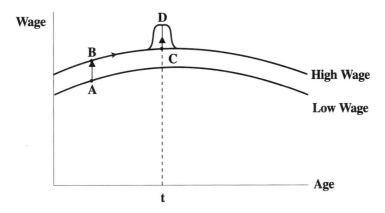

A→B   Permanent Unanticipated Wage Increase (smallest labour supply response)
B→C   Evolutionary Anticipated Wage Increase (largest labour supply response)
C→D   Transitory, Unanticipated Wage Increase (medium labour supply response)

lead to different labour supply responses. For illustrative purposes the three wage changes from A to B, B to C, and C to D are assumed to be of the same magnitude; as illustrated by the same vertical distance.

The difference between A and B reflects two persons whose wages differ by a permanent amount at each and every age in their life cycle. That difference would be associated with conventional income and substitution effects working in the opposite direction. That is, the high-wage individual at B would have a higher opportunity cost of leisure and hence work more (substitution effect). However, that individual would also have a higher expected lifetime income and hence be able to afford more leisure and work less (income effect). Since the income and substitution effects work in the opposite direction, the labour supply of the high-wage individual may be greater or less than the labour supply of the lower-wage individual over their lifetime.

The wage increase from B to C along an individual's given wage profile is an *evolutionary* wage change associated with aging over the life cycle. To the extent that such wage changes are fully anticipated and known with certainty, they will not create a leisure-inducing income effect as they occur. Rather, the income effect is spread over the life cycle, independent of when the wage change occurs. The evolutionary wage change, however, does have a substitution effect making leisure more expensive (and hence inducing work) in the years of peak earnings in one's career. Because there is no leisure-inducing income effect, the labour supply response to anticipated evolutionary, life-cycle wage changes (termed the intertemporal substitution response) will be larger than the labour supply response, at a given age (i.e., not adding up over the life cycle), to permanent unanticipated differences like the wage profile shift from A to B, the latter labour supply response being muted by the leisure-inducing income effect.

The wage difference between C and D represents a transitory, one-time wage difference between two individuals whose wage profiles are otherwise similar (both high wage profiles) except at age t. It may be associated, for example, with a period of overtime pay that is not expected to be repeated. This wage difference will have the usual work-inducing substitution effect at year t since the income forgone by not working is very high. There will also be a small leisure-inducing income effect spread over the life cycle associated with the fact that the individual with the higher wage at D will have a slightly higher lifetime wealth because of the transitory wage difference. Because the wage difference is transitory and not permanent, however, the labour supply response to the transitory wage difference at age t will be larger than the labour supply response to a permanent wage difference as between A and B, at age t, because the latter has a stronger more permanent leisure-inducing income effect. Basically, individuals with the higher transitory wage at D will be able to afford to work slightly less over their life cycle; however, they will likely work more at the particular age t to take advantage of the high temporary wage.

Clearly the three different sources of wage differences outlined above have different income effects associated with whether the wage differences are permanent or temporary and anticipated or not anticipated. The different income effects, in turn, imply different labour supply responses to the wage differences. The anticipated evolutionary wage change from B to C has a pure substitution effect and is not offset by any leisure-inducing income effect. Hence it will have the largest positive labour supply response to a wage change. The transitory wage difference between C and D mainly has a work-inducing substitution effect; however, there is a mild offsetting leisure-inducing income effect spread over the life cycle. The permanent wage difference between A and B, at a given age, in contrast has a larger offsetting leisure-inducing income effect and hence will have the smallest labour supply response to a wage increase (in fact, that response may be negative). The labour supply responsiveness to a wage change may thereby be ordered as the largest emanating from an evolutionary wage change (pure substitution effect), a medium response emanating from a transitory wage change (mainly substitution but slight offsetting income effect), and the smallest and possibly negative response emanating from a permanent wage change (reflecting offsetting substitution and income effects).

Clearly, in a dynamic life cycle context the income and substitution effects emanating from a wage change can differ depending upon whether the wage change is permanent or temporary, and anticipated or unanticipated. This in turn means that the labour supply response will differ upon the source of the wage increase in a dynamic life-cycle setting.

EVIDENCE ON THE ELASTICITY OF LABOUR SUPPLY

A large number of econometric studies have estimated the "shape" of the labour supply schedule — that is, the responsiveness of labour supply to changes in the wage rate. As discussed previously, there are a number of components to that responsiveness. The total or gross or *uncompensated* elasticity of labour supply is the percentage change in labour supply that results from a 1 percent increase in the wage rate. Its sign is theoretically indeterminate because it reflects both the expected negative income effect of the wage change, as well as the expected positive substitution effect. The *income* elasticity of labour supply is the percentage change in wages that results from a 1 percent increase in nonlabour income. It is expected to be negative reflecting the leisure-inducing effect of the wage increase. The *compensated* elasticity of labour supply is the percentage increase in labour supply that results from a 1 percent increase in the wage rate, after compensating the individual for the increase in income associated with the wage increase (i.e., after subtracting the income effect of the wage increase). The compensated wage elasticity of labour supply is expected to be positive since it reflects the pure substitution effect (movement along an isoquant) as the wage rate increases the price of "leisure," inducing a substitution away from more expensive leisure and into labour market activities. For this reason, the income-compensated wage elasticity is also often called the pure substitution elasticity.

Knowledge of the separate components of the labour supply elasticity may be important for policy purposes. For example, the uncompensated wage elasticity would be used to predict the labour supply response to a wage subsidy (since it has both income and substitution effects), but the compensated wage elasticity would be used if it was assumed that the higher income would be taxed back. The income elasticity would be relevant to predict the labour supply response to a lump-sum demogrant.

As indicated, a large number of econometric studies have estimated these various elasticities that constitute the labour supply response to a wage change. The results differ substantially depending upon such factors as the econometric technique and data. Especially for women, the results can differ depending upon whether the labour supply response refers to the participation decision, the hours of work decision conditional upon participation, or a combination of both. For women, this can be important because there tends to be more flexibility in their participation decision and hence it is important to take account of potential sample selection biases, since the subsample of labour force participants may be a select sample in terms of unobservable characteristics that can influence wages. As well, their labour supply decision may be more affected by discontinuities associated, for example, with fixed costs of entering the labour market, such as daycare costs. The differences in female labour supply responses often reflect differences in the extent to which these factors are taken into account in the estimation procedure.

In spite of the substantial variation that exists in the results of the different studies, a number of generalizations can be made. Table 8.1 provides an illustrative representation of those results based on a number of reviews that have been done in the literature (as cited in the source to the table). For example, Hanson and Stuart (1985) review approximately 50 labour supply studies for men and women. They calculate total elasticities for both sexes by weighing the male and female elasticities by their respective share of earnings. They calculate that for both sexes the overall elasticities were: uncompensated 0.10; compensated 0.25; and income elasticity -0.15. However, based on the 13 newer studies, which used more sophisticated econometric techniques to account for many of the previously discussed issues, the elasticities were: uncompensated 0.44; compensated 0.52; and income elasticity -0.08. (These numbers illustrate how the uncompensated or gross elasticity is simply the sum of the compensated and income elasticities.) The numbers used in the first row of Table 21.2 for both sexes are simply based on a "rounded approximation" of those elasticities from the 50 studies and the 13 newer ones. For that reason they are meant to be illustrative and representative rather than strict averages of the results of the different studies. The separate figures for males and females are also meant to be only illustrative.

The following generalizations are illustrated in Table 21.2:

1. The overall labour supply schedule for both sexes is likely to be slightly upward sloping; that is, a wage increase does lead to a slight increase in the amount of labour supplied to the labour market. The representative elasticity of 0.25 indicates that a 1 percent increase in real wages would lead to a one-quarter of 1 percent increase in labour supply. This small uncompensated total elasticity is a result of the positive pure compensated (substitution) elasticity slightly outweighing the negative income elasticity.

Table 8.1
COMPENSATED AND INCOME ELASTICITY OF LABOUR SUPPLY

SEX	UNCOMPENSATED (GROSS, TOTAL) WAGE ELASTICITY OF SUPPLY	COMPENSATED (SUBSTITUTION) WAGE ELASTICITY OF SUPPLY	INCOME ELASTICITY OF SUPPLY
Both sexes	0.25	0.40	−0.15
Males	−0.10	0.10	−0.20
Females	0.80	0.90	−0.10

Source: As discussed in the text, these are "representative illustrative" numbers based on different reviews of over 50 econometric studies of labour supply. The reviews include Hansson and Stuart (1985), Killingsworth (1983), Killingsworth and Heckman (1986), and Pencavel (1986).

2. For males, however, the labour supply schedule is likely to be slightly back-ward bending; that is, real wage increases are associated with a reduction in the amount of labour supplied to the labour market. This overall effect is very small and could well be zero (i.e., vertical or perfectly inelastic labour supply) or even slightly forward sloping. The small overall negative elasticity is a result of weak positive substitution elasticity being outweighed by a weak but slightly larger negative income elasticity.

3. For women, the labour supply schedule is more strongly forward sloping; that is, an increase in real wages is associated with a more substantial increase in the amount of labour supplied to the labour market. This is the result of a strong positive substitution elasticity outweighing the weak negative income elasticity.

4. The strong positive total elasticity for females is sufficiently strong to out-weigh the weak negative total elasticity for males, so that the aggregate supply schedule for both sexes is likely to be forward sloping, as discussed.

5. The substantial variation in the magnitudes of these effects across studies suggests that these representative numbers be used with caution. As well, there is some evidence[5] that the female labour supply response is closer to the male response than portrayed in many studies — a result that may not be surprising if female labour market behaviour is becoming more like male behaviour over time.

QUESTIONS

1. Illustrate the effect of both fixed money and time costs on the individual's reservation wage, participation, and hours-of-work decision.

2. Why might an increase in fixed time costs have a different effect than fixed commute costs on hours worked in the labour market?

3. Discuss the possible effects of fixed money and fixed time costs on the decision to work part time.

4. Analyze the effect, on female labour supply, of the following policy initiatives:

   (a) increasing the income tax deduction for child-care expenses;

   (b) changing the deduction to a credit;

   (c) direct subsidies to daycare centres.

5. Discuss the extent to which estimates of female wages, based upon data on females who are participating in the labour force, may give biased estimates

---

5. Nakamura, Nakamura, and Cullen (1979) and Robinson and Tomes (1985) for Canadian women, Nakamura and Nakamura (1981) for both Canadian and U.S. Women.

of the expected wage that a nonparticipant may receive if she enters the labour market. How would you handle the problem?

6. Discuss the extent to which there may be a sample selection bias in using the wages of migrants to infer the expected wage of nonmigrants if they migrated, or in using the wages of union members to infer the expected wage of nonunion workers if they were to change union status. Given other examples of where sample selection bias may be prominent.

7. What is meant by a nonconvex budget set? Indicate why the "local reservation wage" rule for analyzing labour force participation decisions is inappropriate when the income-leisure choice budget set is nonconvex. Indicate why an indirect utility function approach is appropriate under these and, in fact, under any circumstances. When would the two approaches give the same answer?

8. Discuss the labour supply response, in a dynamic life-cycle context, to the following wage changes: a permanent unanticipated wage increase; an evolutionary anticipated wage increase; a transitory, unanticipated wage increase.

REFERENCES AND FURTHER READINGS

Abbott, M. and O. Ashenfelter. Labour supply, commodity demand, and the allocation of time. *R.E. Studies* 43 (October 1976) 389–411 and correction 46 (October 1979) 567–596.

Abowd, J. and D. Card. Intertemporal labor supply and long-term employment contracts. *AER* 77 (March 1987) 50–68.

Abowd, J. and D. Card. On the covariance structure of earnings and hours changes. *Econometrica* 57 (March 1989) 411–445.

Alogoskoufis, G. On intertemporal substitution and aggregate labor supply. *JPE* 95 (October 1987) 938–960.

Altonji, J. and J. Ham. Intertemporal substitution, erogeneity and surprises: estimating life cycle models for Canada. *CJE* 23 (February 1990) 1–43.

Ball, L. Intertemporal substitution and constraints on labor supply. *EI* 28 (October 1990) 706–724.

Biddle, J. Intertemporal substitution and hours restrictions. *R.E. Stats.* 70 (May 1988) 347–351.

Blau, D. and P. Robins. Child-care costs and family labor supply. *R.E. Stats.* 70 (August 1988) 374–381.

Blundell, R. and I. Walker. Modelling the joint determination of household labour supplies and commodity demands. *EJ* 92 (June 1982) 351–364.

Blundell, R. and I. Walker. A life-cycle consistent empirical model of family labor supply using cross-section data. *R.E. Stud.* 53 (August 1986) 539–558.

Blundell, R. and I. Walker (eds.). *Unemployment, Search and Labor Supply.* Cambridge: Cambridge University Press, 1986.

Blundell, R., A. Duncan and C. Meghir. Taxation in empirical models of labour supply. *EJ* 102 (March 1992) 265–278.

Bover, O. Estimating intertemporal labour supply elasticities using structural models. *EJ* 99 (December 1989) 1026–1039.

Bover, O. Relaxing intertemporal substitutability. *JOLE* 9 (January 1991) 85–100.

Bover, O. Relaxing intertemporal separability: a rational habits model of labor supply estimated from panel data. *JOLE* 9 (January 1991) 85–100.

Browning, M., A. Deaton and M. Irish. A profitable approach to labor supply and commodity demands over the life cycle. *Econometrica* 53 (May 1985) 503–543.

Chiappori, P. Rational household labor supply. *Econometrica* 56 (January 1988) 63–90.

Cogan, J. Fixed costs and labour supply. *Econometrica* 49 (July 1981) 945–963.

Cotterman, R. The role of assets in labor supply functions. *EI* 19 (July 1981) 495–505.

Coyte, P. The supply of individual hours and labour force participation under uncertainty. *EI* 24 (January 1986) 155–172.

Deaton, A. and I. Muellbauer. Functional forms for labor supply and commodity demands with and without quantity restrictions. *Econometrica* 49 (November 1981) 1521–1532.

Dickinson, J. Parallel preference structures in labor supply and commodity demand: an adaptation of the Gorman polar form. *Econometrica* 48 (November 1980) 1711–1725.

Eaton, J. and R. Quandt. A model of rationing and labour supply. *Economica* 50 (August 1983) 221–234.

Eaton, J. and H. Rosen. Labor supply, uncertainty and efficient taxation. *JPubEc* 14 (December 1980) 365–374.

Elliot, J. and K. Sherony. Employer search activities and short-run aggregate labor supply. *SEJ* 52 (January 1986) 693–705.

Gustafsson, S. and F. Stafford. Childcare subsidies and labor supply in Sweden. *JHR* 27 (Winter 1992) 204–230.

Ham, J. Estimation of a labour supply model with censoring due to unemployment and underemployment. *R.E. Studies* 49 (July 1982) 335–354.

Ham, J. Testing whether unemployment represents intertemporal labor supply behaviour. *R.E. Studies* 53 (August 1986) 559–578.

Hansson, I. and C. Stuart. Tax revenue, and the marginal cost of public funds in Sweden. *JPubEc* 27 (August 1985) 333–353.

Haurin, D. Women's labor market reactions to family disruptions. *R.E. Stats.* 71 (February 1989) 54–61.

Hausman, J. The effect of wages, taxes, and fixed costs on women's labour force participation. *JPubEc* 14 (December 1980) 161–194.

Heckman, J. Shadow prices, market wages and labor supply. *Econometrica* 47 (January 1979) 153–162.

Heckman, J. and T. MaCurdy. A lifecycle model of female labour supply. *R.E. Studies* 47 (January 1980) 47–74.

Heckman, J. Effects of child-care programs on women's work effort. *JPE* 82 (No. 2, 1974) S136–S163.

Hotz, V., F. Kyland and G. Sedlack. Intertemporal preferences and labor supply. *Econometrica* 56 (March 1988) 335–360.

Hutchens, R. Layoffs and labor supply. *IER* 24 (February 1983) 37–56.

Ilmakunnas, S. and S. Pudney. A model of female labour supply in the presence of hours restrictions. *JPubEc* 41 (March 1990) 183–210.

Jakubson, G. The sensitivity of labor-supply parameter estimates to unobserved individual effects. *JOLE* 6 (July 1988) 302–239.

Joerding, W. Lifetime consumption labor supply and fertility: a complete demand system. *EI* 20 (April 1982) 255–276.

Johnson, W. and J. Skinner. Labor supply and marital separation. *AER* 76 (June 1986) 455–469.

Jones, S. Reservation wages and the cost of unemployment. *Economica* 56 (May 1989) 225–246.

Kahn, S. and K. Lang. The effect of hours constraints on labor supply estimates. *R.E. Stats.* 73 (November 1991) 605–611.

Kapteyn, A., P. Kooreman and A. van Soest. Quantity rationing and concavity in a flexible household labor supply model. *R.E. Stats.* 72 (February 1990) 55–62.

Killingsworth, M. *Labor Supply*. Cambridge: Cambridge University Press, 1983.

Killingsworth, M. and J. Heckman. Female labor supply: a survey. *Handbook of Labor Economics*, Vol. 1. O. Ashenfelter and R. Layard (eds.). New York: Elsevier, 1986.

Kohlase, J. Labor supply and housing demand for one and two-earner households. *R.E. Stats.* 68 (February 1986) 48–57.

Kooreman, P. and A. Kapteyn. Estimation of rational and unrational household labour supply functions using flexible functional forms. *EJ* 96 (June 1986) 398–412.

Lancaster, T. and A. Chesher. An econometric analysis of reservation wages. *Econometrica* 51 (November 1984) 1661–1676.

Leibowitz, A., J. Klerman and L. Waite. Employment of new mothers and child care choices. *JHR* 27 (Winter 1992) 112–133.

Link, C. and R. Settle. A simultaneous-equation model of labor supply, fertility and earnings of married women: the case of registered nurses. *SEJ* 47 (April 1981) 977–989.

Lundberg, S. Tied-wage-hours offers and the endogeneity of wages. *R.E. Stats.* 87 (August 1985) 405–410.

Lundberg, S. Labor supply of husbands and wives. *R.E. Stats.* 70 (May 1988) 224–235.

MaCurdy, T. An empirical model of labor supply in a life-cycle setting. *JPE* 89 (December 1981) 1059–1085.

MaCurdy, T. A simple scheme for estimating an intertemporal model of labor supply and consumption in the presence of taxes and uncertainty. *IER* 24 (June 1983) 265–290.

Michalopoulos, C., P. Robbins and I. Garfinkel. A structural model of labor supply and child care demand. *JHR* 27 (Winter 1992) 166–203.

Nakamura, A. and M. Nakamura. A comparison of the labour force behaviour of married women in the United States and Canada, with special attention to the impact of income taxes. *Econometrica* 49 (March 1981) 451–489.

Nakamura, A., M. Nakamura and D. Cullen. Job opportunities, the offered wage, and the labour supply of married women. *AER* 69 (December 1979) 785–805.

Pencavel, J. Labor supply of men: a survey. *Handbook of Labor Economics.* Vol. 1. O. Ashenfelter and R. Layard (eds.). New York: Elsevier, 1986.

Ransom, M. An empirical model of discrete and continuous choice in family labor supply. *R.E. Stats.* 59 (August 1987) 465–472.

Ribor, D. Child care and the labor supply of married women. *JHR* 27 (Winter 1992) 134–165.

Rivi Kanbur, S. Labour supply under uncertainty with piecewise linear tax regimes. *Economica* 50 (November 1983) 379–394.

Robinson, C. and N. Tomes. More on the labour supply of Canadian women. *CJE* 18 (February 1985) 156–163.

Rogerson, R. and P. Rupert. New estimates of intertemporal substitution. *Journal of Monetary Economics* 27 (April 1991) 255–269.

Sham, K. Life-cycle labor supply with human capital accumulation. *IER* 30 (May 1989) 431–456.

Smith, J. (ed.) *Female Labor Supply.* Princeton: Princeton University Press, 1980.

Snow, A. and R. Warren. Human capital investment and labor supply under uncertainty. *IER* 31 (February 1990) 195–206.

Snow, A. and R. Warren, Jr. Price level uncertainty, saving, and labor supply. *EI* 24 (January 1986) 97–106.

Tummers, M. and I. Woittiez. A simultaneous wage and labor supply model with hours restrictions. *JHR* 26 (Summer 1991) 393–423.

Vijverberg, W. Selectivity and distributional assumptions in static labor supply models. *SEJ* 57 (January 1991) 823–840.

Wales, T. Labour supply and commuting time. *Journal of Econometrics* 8 (October 1978) 215–226.

Wales, T. and A.D. Woodland. Estimation of household utility functions and labor supply response. *IER* 17 (June 1976) 397–410.

Wales, T. and A.D. Woodland. Estimation of the allocation of time for work, leisure, and housework. *Econometrica* 45 (January 1977) 115–132.

Wales, T. and A.D. Woodland. Labour supply and progressive taxes. *R.E. Studies* 46 (January 1979) 83–95.

Wales, T. and A. Woodland. Sample selectivity and the estimation of labor supply functions. *IER* 21 (June 1980) 437–468.

Yatchew, A. Labor supply in the presence of taxes. *R.E. Stats.* 67 (February 1985) 27–33.

Zabalza, A. The CES utility function, non-linear budget constraints and labour supply: results on female participation and hours. *EJ* 93 (June 1983) 312–330.

PART 2

Labour Demand

Part 2 deals with the demand side of the labour market, focusing on the factors that determine the employers' derived demand for labour under different market conditions.

Chapter 9 analyzes the demand for labour in the competitive norm — when the employer is a perfectly competitive employer in both the labour market and in the product market, and when there are no unions or employee associations that would give labour some monopoly power. The firm's demand for labour is derived in both the short run and the long run, and the determinants of the elasticity of demand for labour (i.e., responsiveness of employment to wage changes) are then analyzed. The concept is then illustrated by applying it to the policy issue of the employment effect of minimum wage legislation. Empirical evidence is presented on the magnitude of the elasticity of demand for labour and on the employment effect of minimum wage legislation.

Chapter 9 also contains an analysis of changes in labour demand that are occurring in response to such interrelated factors as global competition, free trade, industrial restructuring, technological change, privatization, and subcontracting. Canada's changing competitive position on an international basis is also documented with respect to the various dimensions of unit labour cost — wages, fringe benefits, and productivity — adjusted for the exchange rate. These are of particular importance given growing international competition, as well as free trade arrangements such as the Canada–U.S. Free Trade Agreement and the North American Free Trade Agreement to also include Mexico.

Chapter 10 deals with labour demand when there are quasi-fixed costs, employment costs that do not vary with the hours worked by the firm's work force. These can include hiring and training costs, some fringe benefits, and even expected termination costs. The general effects of these quasi-fixed costs on labour demand are analyzed and used to explain the existence of such phenomena as overtime, temporary help agencies, subcontracting, "labour hoarding" over the business cycle, and the reluctance of firms to engage in worksharing or to hire new employees. Special attention is also paid to the extent to which worksharing or restrictions on the use of overtime can be used to create new jobs.

CHAPTER 9

Demand for Labour in Competitive Labour Markets[1]

The general principles that determine the demand for any factor of production apply also to the demand for labour. In contrast to goods and services, factors of production are demanded not for final use or consumption but as inputs into the production of final goods and services. Thus the demand for a factor is necessarily linked to the demand for the goods and services that factor is used to produce. The demand for land suitable for growing wheat is linked to the demand for bread and cereal, just as the demand for construction workers is related to the demand for new buildings, roads, and bridges. For this reason, the demand for factors is called a "derived demand."

In discussing firms' decisions regarding the employment of factors of production, economists usually distinguish between short-run and long-run decisions. The short run is defined as a period during which one or more factors of production — referred to as fixed factors — cannot be varied, while the long run is defined as a period during which the firm can adjust all of its inputs. During both the short and long runs, the state of technical knowledge is assumed to be fixed; the very long run refers to the period during which changes in technical knowledge can occur. The amount of calendar time corresponding to each of these periods will differ from one industry to another and according to other factors. These periods are useful as a conceptual device for analyzing firm

---

1. This chapter does not deal directly with the extensive macroeconomic literature on how the demand for labour varies over the business cycle in response to changes in output, real wages, the price of other inputs, and termination policies. This literature has been useful in understanding productivity and employment and hours-of-work changes over the business cycle. Nickell (1986) reviews much of this literature and Hamermesh (1976) uses this literature to provide estimates of the elasticity of demand for labour based on time series data.

decision-making rather than for predicting the amount of time taken to adjust to change.

By the demand for labour we mean the quantity of labour services the firm would choose to employ at each wage. This desired quantity will depend on both the firm's objectives and its constraints. Usually we will assume that the firm's objective is to maximize profits. Also examined is the determination of labour demand under the weaker assumption of cost minimization. The firm is constrained by the demand conditions in its product markets, the supply conditions in its factor markets, and its production function which shows the maximum output attainable for various combinations of inputs, given the existing state of technical knowledge. In the short run the firm is further constrained by having one or more factors of production whose quantities are fixed.

The theory of labour demand examines the quantity of labour services the firm desires to employ given a market-determined wage rate or given the labour supply function the firm faces. In some circumstances, however, the employer and employees may negotiate explicit, or reach implicit, contracts involving both wages and employment. When this is the case, the wage-employment outcomes may not be on the labour demand curve because the contracts will reflect the preferences of both the employer and the employees with respect to both wages and employment. Explicit wage-employment contracts are discussed in Part 4 on unions; implicit wage-employment contracts are discussed in Parts 5 and 6.

CATEGORIZING THE STRUCTURE OF PRODUCT AND LABOUR MARKETS

Because the demand for labour is derived from the output produced by the firm, the way in which the firm behaves in the product market can have an impact on the demand for labour, and hence ultimately on wage and employment decisions. In general, the firm's product market behaviour depends on the structure of the industry to which the firm belongs. In decreasing order of the degree of competition, the four main market structures are: (1) perfect competition, (2) monopolistic competition, (3) oligopoly, and (4) monopoly.

Whereas the structure of the product market affects the firm's derived demand for labour, the structure of the labour market affects the labour supply curve that the firm faces. The supply curve of labour to an individual firm shows the amount of a specific type of labour (e.g., a particular occupational category) that the firm could employ at various wage rates. Analogous to the four product market structures there are four labour market structures. In decreasing order of the degree of competition the firm faces in hiring labour, these factor market structures are: (1) perfect competition, (2) monopsonistic competition, (3) oligopsony, and (4) monopsony. (The "opsony" ending is traditionally used to denote departures from perfect competition in factor markets.)

The product and labour market categorizations are independent in that there is no necessary relationship between the structure of the product market in which the firm sells its output and the labour market in which it buys labour services. The extent to which the firm is competitive in the product market (and hence the nature of its derived demand for labour curve) need not be related to the extent to which the firm is a competitive buyer of labour (and hence the nature of the supply curve of labour that it faces). The two *may* be related, for example if it is a large firm and hence dominates both the labour and product markets, but they need not be related. Hence for *each* product market structure, the firm can behave in at least four different ways as a purchaser of labour. Thus there are at least sixteen different combinations of product and labour market structure that can bear on the wage and employment decision at the level of the firm.

In general, however, the essence of the wage and employment decision by the firm can be captured by an examination of the polar cases of perfect competition (in product and/or factor markets) monopoly in the product market, and monopsony in the labour market. We will begin by examining the demand for labour in the short run under conditions of perfect competition in the labour market.

DEMAND FOR LABOUR IN THE SHORT RUN

The general principles determining labour demand can be explained for the case of a firm that produces a single output (Q) using two inputs, capital (K), and labour (N). In the short run the amount of capital is fixed at $K = K_o$, while the quantity of labour services can be varied by changing either the number of employees or hours worked by each employee (or both). For the moment we do not distinguish between variations in the number of employees and hours worked; however, this aspect of labour demand is discussed subsequently. Also discussed later is the possibility that labour may be a "quasi-fixed factor" in the short run.

The demand for labour in the short run can be derived by examining the firm's short-run output and employment decisions. Two decision rules follow from the assumption of profit maximization. First, because the costs associated with the fixed factor must be paid whether or not the firm produces (and whatever amount the firm produces), the firm will operate as long as it can cover its variable costs (i.e., if total revenue exceeds total variable costs). Fixed costs are sunk costs, and their magnitude should not affect what is the currently most profitable thing to do.

The second decision rule implied by profit maximization is that, if the firm produces at all (i.e., is able to cover its variable costs), it should produce the quantity Q^* at which marginal revenue (MR) equals marginal cost (MC). That is, the firm should increase output until the additional cost associated with the last unit produced equals the additional revenue associated with that unit. Because marginal revenue is a constant or decreasing function of output and marginal cost rises with output (due to diminishing marginal productivity), expanding output beyond the point at which MR = MC will lower profits because the addition to total revenue will be less than the increase in total cost. Producing a lower output would also reduce profits because, at output levels below Q^*, marginal revenue exceeds marginal cost; thus increasing output would add more to total revenue than to total cost, thereby raising profits.

The profit-maximizing decision rules can be stated in terms of the employment of inputs rather than in terms of the quantity of output to produce. Because concepts such as total revenue and marginal revenue are defined in terms of units of output, the terminology is modified for inputs. The total revenue associated with the amount of an input employed is called the total revenue product (TRP) of that input; similarly, the change in total revenue associated with a change in the amount of the input employed is called the marginal revenue product (MRP). Thus the profit-maximizing decision rules for the employment of the variable input can be stated as follows: (1) the firm should produce providing the total revenue product of the variable input exceeds the total costs associated with that input; otherwise the firm should shut down operations, and (2) if the firm produces at all, it should expand employment of the variable input to the point at which its marginal revenue product equals its marginal cost.

The short-run employment decision of a firm operating in a perfectly competitive labour market is shown in Figure 9.1. In a competitive factor market the firm is a price-taker; i.e., the firm can hire more or less of the factor without affecting the market price. Thus in a competitive labour market the marginal (and average) cost of labour is the market wage rate. The firm will therefore employ labour until its marginal revenue product equals the wage rate, which implies that the firm's short-run labour demand curve is its marginal revenue product of labour curve. For example, if the wage rate is w_0 the firm would employ N_0^* labour services. However, the firm will shut down operations in the short run if the total variable cost exceeds the total revenue product of labour, which will be the case if the average cost of labour (the wage rate) exceeds the average revenue product of labour. Thus at wage rates higher than w_1 in Figure 9.1 (the point at which the wage rate equals the average product of labour) the firm would choose to shut down operations. It follows that the firm's short-run labour demand curve is its marginal revenue product of labour curve below the point at which the average and marginal product curves intersect (i.e., below the point at which the ARP_N reaches a maximum).

The short-run labour demand curve is downward-sloping because of diminishing returns. Although the average and marginal products may initially rise as more labour is employed, both eventually decline as more units of the variable factor are combined with a given amount of the fixed factor. Because the firm

Figure 9.1
THE FIRM'S SHORT-RUN DEMAND FOR LABOUR

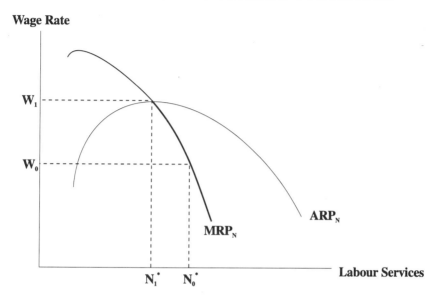

employs labour in the range in which its marginal revenue product is declining, a reduction in the wage rate is needed to entice the firm to employ more labour. Similarly, an increase in the wage rate will cause the firm to employ less labour, thus raising its marginal revenue product and restoring equality between the marginal revenue product and marginal cost.

At this point, it may be worth highlighting a common misconception concerning the reason for the downward-sloping demand for labour. The demand is for a given homogeneous type of labour; consequently, it does *not* slope downwards because, as the firm uses more labour, it uses poorer quality labour and hence pays a lower wage. It may well be true that when firms expand their work force they often have to use poorer quality labour. Nevertheless for analytical purposes it is useful to assume a given, homogeneous type of labour so that the impact on labour demand of changes in the wage rate for that type of labour can be analyzed. The change in the productivity of labour that occurs does so because of changes in the amount of the variable factor combined with a given amount of the fixed factor, not because the firm is delving more into the reserve of less qualified labour.

WAGES, THE MARGINAL PRODUCTIVITY OF LABOUR, AND THE PRODUCT MARKET

The demand schedule of a profit-maximizing firm which is a wage taker in the labour market is the locus of points for which the marginal revenue product of labour equals the wage rate. The marginal revenue product of labour (MRP_N) equals the marginal revenue of output (MR_Q) times the marginal physical product of labour (MPP_N). Thus there is a relationship between wages and the marginal productivity of labour.

The marginal revenue of output depends on the market structure in the product market. The two polar cases of perfect competition and monopoly are discussed here. A perfectly competitive firm is a price-taker in the product market. Because the firm can sell additional (or fewer) units of output without affecting the market price, the marginal revenue of output equals the product price; i.e., for a competitive firm $MRP_N = MR_Q \cdot MPP_N = p \cdot MPP_N$. Because it is the product of the market price and the marginal product of labour, this term is referred to as the value of the marginal product of labour; i.e., $p \cdot MPP_N = VMP_N$. A firm that is competitive in both the product and labour markets will thus employ labour services until the value of the marginal product of labour just equals the wage; i.e., the demand for labour obeys the equation

$$p \cdot MPP_N = VMP_N = w \tag{9.1}$$

Equation 9.1 follows from the MR = MC rule for profit maximization; the VMP_N is the increase in total revenue associated with a unit increase in labour input while the wage w is the accompanying increase in total cost. Alternatively stated, the labour demand curve of a firm that is competitive in both the product and labour markets is the locus of points for which the real wage w/p equals the marginal physical product of labour.

The polar case of non-competitive behaviour in the product market is that of monopoly. In this situation the firm is so large relative to the size of the product market that it can influence the price at which it sells its product: it is a price-setter, not a price-taker. In the extreme case of monopoly, the monopolist comprises the whole industry: there are no other firms in the industry. Thus, the industry demand for the product is the demand schedule for the product of the monopolist.

As is well known from standard microeconomic theory, the relevant decision-making schedule for the profit-maximizing monopolist is not the demand schedule for its product, but rather its marginal revenue schedule. In order to sell an additional unit of output, the monopolist has to lower the price of its product. Assuming that it cannot differentiate its homogeneous product to consumers, the monopolist will also have to lower the price on all units of its output, not just on the additional units that it wishes to sell. Consequently, its marginal revenue — the additional revenue generated by selling an additional unit of output — will fall faster than its price, reflecting the fact that the price decline applies to intramarginal units of output. The marginal revenue schedule for the monopolist will therefore lie below and to the left of its demand schedule. By equating marginal revenue with marginal cost so as to maximize profits, the monopolist will produce less output and charge a higher price (as given by the demand schedule, since this is the price that consumers will pay) than if it were a competitive firm on the product market.

This aspect of the product market has implications for the derived demand for labour. The monopolist's demand for labour curve is the locus of points for which

$$MR_Q \cdot MPP_N = MRP_N = w \qquad (9.2)$$

The differences between Equation 9.1 and 9.2 highlight the fact that when the monopolist hires more labour to produce more output, not only does the marginal physical product of labour fall (as is the case with the competitive firm), but also the marginal revenue from an additional unit of output, MR_Q, falls. This latter effect occurs because the monopolist, unlike the competitor, can sell more output only by lowering the product price and this in turn lowers revenue. Because both MPP_N and MR_Q fall when N increases in Equation 9.2, then the monopolist's demand for labour falls faster than it would if it behaved as a competitive firm in the product market, in which case only MPP_N would fall, as given in Equation 9.1.

DEMAND FOR LABOUR IN THE LONG RUN

In the long run the firm can vary all of its inputs. For expositional purposes we will continue to assume two inputs, labour (N) and capital (K) and one output (Q); however, the general principles apply to firms which employ many inputs and produce multiple outputs. As in the previous section, the demand for labour is derived by varying the wage rate that the firm faces for a given homogeneous type of labour, and tracing out the profit-maximizing (or cost-minimizing) quantity of labour that will be employed by the firm.

For conceptual purposes, the firm's production and employment decisions are usually exmined in two stages. First, the minimum cost method of producing any level of output is examined. Second, given that each output will be produced at minimum cost, the profit-maximizing level of output is chosen.

Isoquants, Isocosts, and Equilibrium

The first stage (cost minimization) is depicted geometrically in Figure 9.2. The top part of the diagram, Figure 9.2(a), gives the firm's isoquant Q_o, which shows the various combinations of labour N and capital K, that can produce a given level of output Q_o, via the firm's production function that can be written in general form as $Q = Q(K,N)$. The slope of the isoquant exhibits a diminishing marginal rate of substitution between the inputs; that is, in the upper left segment when an abundance of capital is used, a considerable increase in the use of capital is required for a small reduction in the use of labour, if output is to be maintained. In the lower right segment when an abundance of labour is used,

Figure 9.2
ISOQUANTS, ISOCOST, AND EQUILIBRIUM

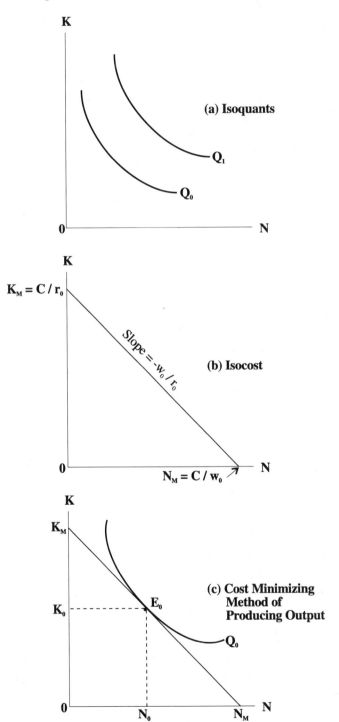

labour is a poor substitute for capital, and considerable labour savings are possible for small increases in the use of capital. In the middle segment of the isoquant, labour and capital are both good substitutes. Successively higher isoquants or levels of output, such as Q_1, can be produced by successively larger amounts of both inputs.

Figure 9.2(b) illustrates the firm's budget constraint or isocost line $K_m N_m$, depicting the various combinations of capital and labour the firm can employ, given their market price and the expenditures of the firm. Algebraically, the isocost line is $C = rK + wN$ where r is the price of capital and w the price of labour, or wage rate.[2] The position and shape of the isocost line can be determined by solving for the two intercepts or end points and the slope of the line between them. From the isocost equation, for fixed prices r_o and w_o, these end points are $N_m = C/w_o$ when $K = 0$, and $K_m = C/r_o$ when $N = 0$. The slope is simply minus the rise divided by the run or $-[C/r_o \div C/w_o] = -w_o/r_o$, that is, the price of labour relative to the price of capital. This is a straight line as long as w and r are constant for these given types of labour and capital, which is the case when the firm is a price-taker in both input markets.

A profit-maximizing firm will maximize the output that can be produced for a given level of cost or, alternatively, it will minimize the cost for a given level of output. Figure 9.2(c) combines the isoquants and isocost to illustrate the maximum output (isoquant) that the firm can attain given its resource expenditure constraint (isocost). This is clearly E_o where the isocost is tangent to the isoquant. The determination of the profit-maximizing level of output (the output at which marginal revenue equals marginal cost) is not shown in the diagram. What is shown is how to produce any output, including the profit-maximizing output, at minimum cost.

At E_o, the cost-minimizing amounts of labour and capital, respectively used to produce Q_o units of output, are N_o and K_o. If Q_o is the profit-maximizing output, this gives us one point on the demand curve for labour, as depicted later in Figure 9.3; that is, at the wage rate w_o the firm employs N_o units of labour.

Deriving the Firm's Labour Demand Schedule

The complete labour demand schedule, for the long run when the firm can vary both capital and labour, can be obtained simply by varying the wage rate and tracing out the new, equilibrium, profit-maximizing amounts of labour that

---

2. If the firm rents its capital equipment, the cost of capital is the rental price. If the firm owns its capital equipment, the implicit or opportunity cost of capital depends on the cost of the machinery and equipment, the interest rate at which funds can be borrowed or lent and the rate at which the machinery depreciates. In the analysis that follows, we will assume the firm is a price-taker in the market for capital.

Figure 9.3
DERIVING THE LABOUR DEMAND SCHEDULE

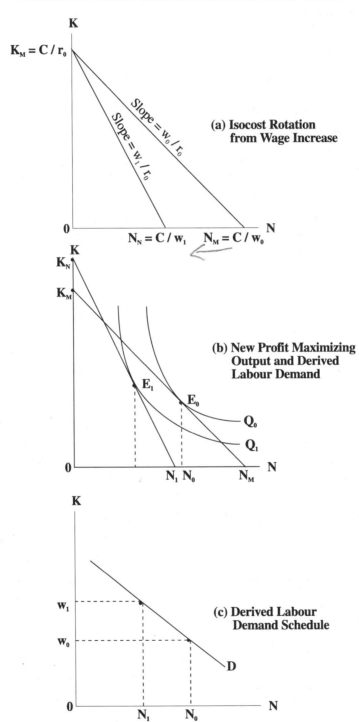

(a) Isocost Rotation from Wage Increase

(b) New Profit Maximizing Output and Derived Labour Demand

(c) Derived Labour Demand Schedule

would be employed. This is illustrated in Figure 9.3. For example, if total cost is held constant an increase in the wage rate from w_o to w_1 would rotate the isocost line downwards as in Figure 9.3(a). This is so because if the firm spent all of its expenditures on capital it would be at the same maximum point $K_m = C/r_o$. However, if it spent all of its budget on the higher priced labour then the maximum amount of labour it could employ would be reduced to a point like $N_n = C/w_1$ corresponding to the higher price of labour. Since $w_1 > w_o$ then $N_n < N_m$. Also, the slope of the new budget constraint would be greater than the original; that is, $w_1/r_o > w_o/r_o$.

As depicted in Figure 9.3(b), given the higher wage rate w_1, the firm will maximize profits by moving to a lower level of output Q_1, operating at E_1 and employing N_1 units of labour. This yields a second point on the firm's demand curve for labour depicted in Figure 9.3(c); that is, a lower level of employment N_1 corresponds to the higher wage rate w_1.

To complete the explanation of the firm's response to a wage increase, Figure 9.4 shows how the profit-maximizing output levels (Q_o and Q_1 in Figure 9.3(b)) are determined. The wage increase shifts up the firm's marginal and average cost curves. In a perfectly competitive industry, each firm reduces output which raises the market price of the product. In the new equilibrium the output of each firm and total output are lower than in the original equilibrium. The monopolist responds to the increase in costs by raising the product price and reducing output. The analysis of the intermediate market structures (monopolistic competition and oligopoly) is similar; in general an increase in costs, ceteris paribus, leads to an increase in the product price and a reduction in output. Note, however, that although the firm moves to a lower isoquant (Q_o to Q_1 in Figure 9.3(b)), its total costs may increase (as shown in Figure 9.3(b) where $K_n > K_m$) or decrease.[3] Producing a lower output tends to reduce total costs, but the higher wage tends to raise total costs. The net effect therefore depends on the magnitude of these offsetting forces. In this respect the analysis of the response of the firm to an increase in an input price is not exactly analogous to the response of the consumer to a commodity price increase. In consumer theory, income (or total expenditure on goods) is assumed to be exogenously determined; thus a change in the price of one good rotates the budget constraint as in Figure 9.3(a). In the theory of the firm, total cost (or total expenditure on inputs) is endogenous rather than exogenous and depends on the profit-maximizing levels of output and inputs given the output and input prices. As these change, total costs may change, as depicted previously in Figure 9.3(b).

---

3. As indicated previously, if total costs remained constant even though there was a wage increase, then the isocost line would simply rotate downwards from K_m (i.e., $K_n = K_m$). This is so because if the firm allocated all of its (constant) budget to capital, it would utilize the same amount of capital since only the price of labour has increased. However, if total costs increase and it is hiring no labour, then it must be utilizing more capital (i.e., $K_n > K_m$). Conversely, if total costs decreased then $K_n < K_m$.

Figure 9.4
THE EFFECT OF A COST INCREASE ON OUTPUT

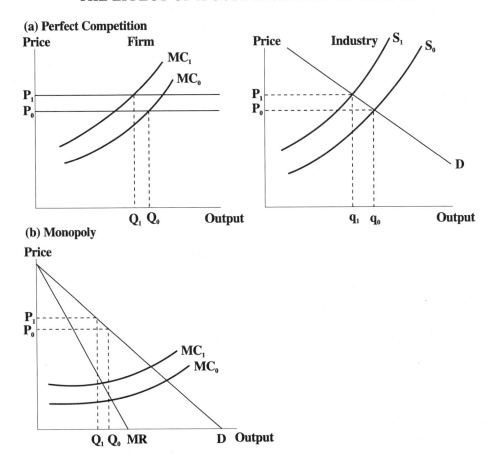

Clearly one could trace out the full demand schedule by varying the wage rate and observing the profit-maximizing amounts of labour that would be employed. Economic theory predicts that the demand schedule is downward-sloping (that is, higher wages are associated with a reduced demand for labour) both because the firm would substitute cheaper inputs for the more expensive labour (substitution effect), and because it would reduce its scale of operations because of the wage and hence cost increase (scale effect). For conceptual purposes it is useful to examine these two effects.

Separating Scale and Substitution Effects of a Wage Change

Figure 9.5 illustrates how the wage increase from w_0 to w_1 can be separated into its component parts — a substitution effect and a scale or output effect. The negative substitution effect occurs as the firm substitutes cheaper capital for the more expensive labour, therefore reducing the quantity of labour demanded. When labour is a normal input, the negative scale effect occurs as a wage increase leads to a higher marginal cost of production which, in turn, reduces the firm's optimal output and hence derived demand for labour. Thus the scale and substitution effects reinforce each other to lead to an unambiguously inverse relationship between wages and the firm's demand for labour.[4]

The scale effect can be isolated by hypothetically compensating the firm for its lost output so that it would operate on the original isoquant Q_0. This is illustrated by the hypothetical isocost line that is parallel to the new isocost w_1/r_0, but tangent to the original isoquant Q_0 at E_s. The only difference between E_s and E_1 is the scale of operation of the firm (Q_0 as opposed to Q_1), since the relative price of labour and capital are the same; that is, both are w_1/r_0. Therefore, N_s - N_1 can be thought of as the scale effect — the reduction in employment that comes about to the extent that the firm reduces its output in response to the wage increase. Except in the unusual circumstance in which labour is an inferior input, the scale effect works through the following scenario: wage increases imply cost increases, which lead to a reduced optimal scale of output, which in turn leads to a reduced demand for labour.

The difference between E_0 and E_s, on the other hand, is simply due to the different slopes of the isocost lines. They both represent the same output, Q_0. Consequently, the difference between E_0 and E_s reflects the pure substitution of capital for labour to produce the same output, Q_0. Therefore, N_0 - N_s can be thought of as a pure or compensated substitution effect, representing the substitution of relatively cheaper inputs for the inputs whose relative price has risen. Both scale and substitution effects work in the same direction to yield an unambiguously downward-sloping demand schedule for labour.

For capital, the substitution and scale effects of an increase in the wage rate work in opposite directions. The substitution effect (E_0 to E_s) increases the demand for capital as the firm uses more capital and less labour to produce a given level of output. However, the reduction in output from Q_0 to Q_1 causes a reduction in demand for all normal inputs; thus the scale effect (E_s to E_1) reduces the demand for capital. The overall or net effect on capital thus depends on which of the two effects is larger. In general, an increase in the wage may cause the firm to employ more or less capital.

THE RELATIONSHIP BETWEEN THE SHORT- AND LONG-RUN LABOUR DEMAND

The division of the firm's response to a wage change into the substitution and scale effects is also useful in understanding the difference between the short- and long-run labour demand. In the short run the amount of capital is fixed; thus

---

4. Nagatani (1978) proves that this is true even when labour is an inferior factor of production.

Figure 9.5
SUBSTITUTION AND SCALE EFFECTS OF A WAGE CHANGE

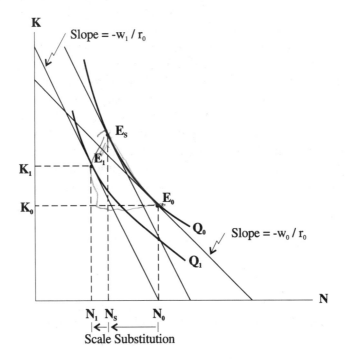

there is no substitution effect. The downward-sloping nature of the short-run labour demand curve is a consequence of a scale effect and diminishing marginal productivity of labour. In the long run the firm has the additional flexibility associated with varying its capital stock. Thus the response to a wage change will be larger in the long run than in the short run, ceteris paribus. This relationship is illustrated in Figure 9.6. The initial long-run equilibrium is at E_0. The short-run response to the wage increase from w_0 to w_1 involves moving to a new equilibrium at E_1. Here the firm is employing less labour (N_1 versus N_0) but the same amount of capital (K_0). In the long run the firm also adjusts its capital stock from K_0 to K_1, shifting the short-run labour demand curve to the left. The new long-run equilibrium is at E_2. The long-run labour demand curve consists of the locus of points such as E_0 and E_2 at which the firm has optimally adjusted employment of both labour and capital given the wage rate and other exogenous variables.

LABOUR DEMAND UNDER COST MINIMIZATION

Economic analysis generally assumes that firms in the private sector seek to maximize profits. However, organizations in the public and quasi-public sectors

— such as federal, provincial, and municipal public administration, Crown corporations, and educational and health institutions — generally have other goals. These goals will determine the quantity of output or services provided and the amount of labour and other productive inputs employed. Although the factors that determine the output of these organizations may vary, such organizations may seek to produce that output efficiently; i.e., at minimum cost.

The distinction between the scale and substitution effects is useful in analyzing the demand for labour in these circumstances. An organization whose output of goods and/or services is exogenously determined but which seeks to produce that output at minimum cost will respond to changes in wages by substituting between labour and other inputs. That is, with output fixed (determined by other factors) there will be a pure substitution effect in response to changes in the wage rate. Labour demand will therefore be unambiguously downward sloping but more inelastic than that of a profit-maximizing firm because of the absence of an output effect.

Although a cost-minimizing organization will respond to a wage increase by substituting capital for labour, its total expenditure on inputs (total costs) will nonetheless rise. This aspect is illustrated in Figure 9.5. The equilibrium E_s involves greater total costs than the original equilibrium E_0. The increase in total costs is, however, less than would be the case if the organization did not substitute capital for labour in response to the wage increase (i.e., if the firm remained at the point E_0 in Figure 9.5).

Figure 9.6
THE DEMAND FOR LABOUR IN THE SHORT RUN AND LONG RUN

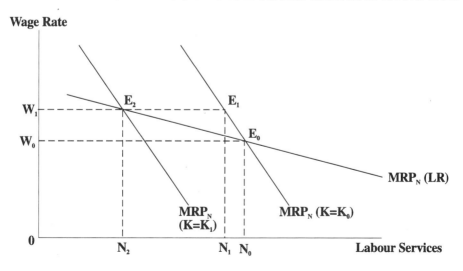

MATHEMATICAL EXPOSITION

The analysis of labour demand can also be carried out using mathematics. Profits π are defined as total revenues $R(Q)$ minus total costs, or

$$\pi(Q) = R(Q) - wN - rK \tag{9.3}$$

and the production function

$$Q = Q(N,K) \text{ where } \frac{\partial Q}{\partial N} > 0 \text{ and } \frac{\partial Q}{\partial K} > 0 \tag{9.4}$$

The firm is assumed to be a price-taker in the input markets; hence w and r are treated as fixed or exogenous variables. In the short run the capital input is also fixed at $K = K_o$.

In the short run the firm will choose labour in order to

$$\underset{\{N\}}{\text{Max}} \ R(Q(N,K_o)) - wN - rK_o$$

The necessary condition for a maximum is

$$\frac{dR}{dQ} \frac{\partial Q}{\partial N} (N,K_o) - w = 0 \tag{9.5}$$

$$\text{or } MR_Q \cdot MPP_N = w \tag{9.6}$$

This is the condition discussed previously; the firm should employ labour until the marginal revenue product of labour MRP_N equals the wage rate. The marginal revenue product is the marginal revenue of output MR_Q times the marginal physical product of labour MPP_N.

For firms that operate in a perfectly competitive output market, $R(Q) = p \cdot Q$ where p is fixed so $dR(Q)/dQ = p$; i.e., the marginal revenue of output is simply the product price. Thus Equation 9.6 becomes

$$p \cdot MPP_N = VMP_N = w \tag{9.7}$$

Unlike the competitive firm which treats the product price as fixed, the monopolist recognizes that the product price and output are negatively related; that is,

$$R(Q) = p(Q) \cdot Q \text{ where } dp/dQ < 0 \tag{9.8}$$

The monopolist thus employs labour until

$$MRP_N = [p(Q) + Q \frac{dp}{dQ}] \frac{\partial Q}{\partial N} = w \tag{9.9}$$

The term in square brackets is the marginal revenue of output MR_Q; $MR_Q <$ $p(Q)$ because $dp/dQ < 0$. If price p is fixed ($dp/dQ = 0$), Equation 9.9 simplifies to Equation 9.7.

In the long run the firm chooses N and K to

Max $R(Q(N,K)) - wN - rK$
$\{N,K\}$

The necessary conditions for a maximum are

$$\frac{dR}{dQ}\frac{\partial Q}{\partial N} - w = 0 \qquad\qquad (9.10)$$

$$\frac{dR}{dQ}\frac{\partial Q}{\partial K} - r = 0 \qquad\qquad (9.11)$$

Thus each input is employed to the point at which its marginal revenue product equals its price.

ELASTICITY OF DEMAND FOR LABOUR

The previous analysis indicated that the demand for labour is a negative function of the wage rate. Consequently, in this static, partial-equilibrium framework, an exogenous increase in wages, other things held constant, would lead to a reduction in the quantity of labour demanded. The exogenous increase in wages, for example, could emanate from a union wage demand, a wage parity scheme, or wage-fixing legislation, such as minimum wages, equal pay, fair-wage legislation, or extension legislation. Although there may be offsetting factors (to be discussed later), it is important to realize that economic theory predicts that there will be an adverse employment effect from these wage increases. The magnitude of the adverse employment effect depends on the elasticity of the derived demand for labour. As illustrated in Figure 9.7, if the demand for labour is inelastic (Figure 9.7(a)), then the adverse employment effect is small; if the demand is elastic (Figure 9.7(b)), then the adverse employment effect is large.

From a policy perspective it is important to know the expected magnitude of these adverse employment effects because they may offset other possible benefits of the exogenous wage increase. Consequently, it is important to know the determinants of the elasticity of demand for labour. As originally outlined by Marshall and formalized by Hicks (1963, pp. 241–246 and 374–384), the basic determinants of the elasticity of demand for labour are the availability of substitute inputs, the elasticity of supply of substitute inputs, the elasticity of demand for output, and the ratio of labour cost to total cost. These factors are related to the magnitude of the substitution and scale effects discussed previously. Each of these factors will be discussed in turn, in the context of an inelastic demand for labour (Figure 9.7(a)) which implies a wage increase being associated with a small adverse employment effect.

Availability of Substitute Inputs

The derived demand for labour will be inelastic, and hence the adverse employment effect of an exogenous wage increase will be small, if alternative inputs cannot be substituted easily for the higher-price labour. This would be depicted by an isoquant that is more L-shaped as opposed to a negatively sloped straight line; that is, the marginal rate of technical substitution between other inputs and labour is small. This factor relates to the magnitudes of the substitution effect.

The inability to substitute alternative inputs could be technologically determined, as for example if the particular type of labour is essential to the production process, or it could be institutionally determined, as for example if the union prevents such substitution as contracting-out or the use of nonunion labour or the introduction of new technology. Time also is a factor, since in the long run the possibility of substituting cheaper inputs is more feasible.

Examples of workers for whom substitute inputs may not *readily* be available could include construction tradespeople, teachers, and professionals with specialized skills. Even in these cases, however, substitutions are technically possible in the long run, especially when one considers alternative production processes and delivery systems (e.g., prefabricated construction, larger class size with more audio-visual aids, and the use of paraprofessionals).

In declining industries, it may be difficult to substitute new capital for higher priced labour because of the difficulty of attracting new capital to the industry.

Figure 9.7
INELASTIC AND ELASTIC DEMAND FOR LABOUR

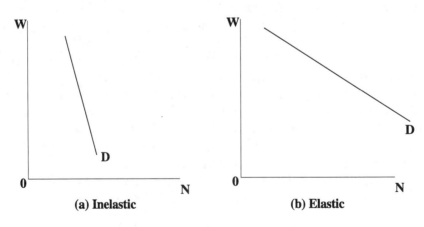

(a) Inelastic   (b) Elastic

As well, if the capital is industry specific it may have little alternative use, and hence there is little threat of firms moving their capital if complementary labour becomes too expensive. In such circumstances, the demand for labour may not only shift inwards (reflecting the declining derived demand for labour), but it may also become more inelastic (reflecting the lack of substitute capital). This can have opposing effects on wages in declining industries, with wages falling because of falling demand, but rising if unions push for higher wage increases, aware that there will not be much of an adverse employment effect given the more inelastic demand for labour. In such circumstances it may be perfectly rationale for workers to push for higher wage increases in spite of the declining nature of the industry. Even if profits are declining, unions can still bargain for a larger share of declining rents.

Elasticity of Supply of Substitute Inputs

The substitution of alternative inputs can also be affected by changes in the price of these inputs. If the substitutes are relatively *inelastic* in supply, so that an increase in the demand for the inputs will lead to an increase in their price, then this price increase may choke off some of the increased usage of these substitutes. In general, this is probably not an important factor, however, because it is unlikely that the firm or industry where the wage increase occurs is such a large purchaser of alternative inputs that it affects their price. Certainly in the long run, the supply of substitute inputs is likely to be more elastic.

Elasticity of Demand for Output

Since the demand for labour is derived from the output produced by the firm, then the elasticity of the demand for labour will depend on the price elasticity of the demand for the output or services produced by the firm. The elasticity of product demand determines the magnitude of the scale effect. If the demand for the output produced by the firm is inelastic, so that a price increase (engendered by a wage increase) will not lead to a large reduction in the quantity demanded, then the derived demand for labour will also be inelastic. In such circumstances, the wage increase can be passed on to consumers in the form of higher product prices without there being much of a reduction in the demand for those products and hence in the derived demand for labour.

This could be the case, for example, in construction, especially non-residential construction where there are few alternatives to building in a particular location. It could also be the case in tariff-protected industries, or in public services where few alternatives are available, or in most sectors in the short run. On the other hand, in fiercely competitive sectors, such as the garment trades or coal mining, where alternative products are available, then the demand for the product is probably quite price-elastic. In these circumstances, the derived demand for labour would be elastic and a wage increase would lead to a large reduction in the quantity of labour demanded.

Ratio of Labour Cost to Total Cost

The extent to which labour cost is an important component of total cost can also affect the responsiveness of employment to wage changes. Specifically, the demand for labour will be inelastic, and hence the adverse employment effect of a wage increase small, if labour cost is a small portion of total cost.[5] In such circumstances the scale effect would be small; that is, the firm would not have to reduce its output much because the cost increase emanating from the wage increase would be small. In addition, there is the possibility that if wage costs are a small portion of total cost then any wage increase more easily could be absorbed by the firm or passed on to consumers.

For obvious reasons, this factor is often referred to as the "importance of being unimportant." Examples of wage cost for a particular group of workers being a small portion of total cost could include construction craftworkers, airline pilots, and employed professionals (e.g., engineers and architects) on many projects. In such circumstances their wage increases simply may not matter much to the employer, and consequently their wage demands may not be tempered much by the threat of reduced employment. On the other hand, the wages of government workers and miners may constitute a large portion of the total cost in their respective trades. The resultant elastic demand for labour may thereby temper their wage demands.

Empirical Evidence

Clearly, knowledge of the elasticity of demand for particular types of labour is important for policy purposes so as to predict the adverse employment effect that may emanate from such factors as minimum wage and equal pay laws, or wage increases associated with unionization, occupational licensing, or arbitrated wage settlements. Estimates of the elasticity of the demand for the particular type of labour being affected would be useful to predict the employment effect of an exogenous wage increase. However, even without precise numerical estimates, judicious statements still can be made based on the

---

5. Hicks (1963, pp. 245–246) proves formally that this is true as long as the elasticity of demand for the final product is greater than the elasticity of substitution between the inputs; that is, as long as consumers can substitute away from the higher-priced product more easily than producers can substitute away from higher-priced labour. This factor thus depends on the magnitude of both the substitution and scale effects.

importance of the various factors that determine the elasticity of the demand for labour.

Hammermesh (1976, 1986) reviewed extensive empirical literature in the U.S. that can be used to calculate estimates of the elasticity of demand for labour. He concluded that, based largely on private sector data in the U.S. this elasticity ranged from a low of -.09 to a high of -.62, with a medium estimate of -.32, over a time period of one year. That is, a one percent increase in wages would lead to approximately a one-third of one percent reduction in employment after a one-year time period. This adverse employment effect was roughly equally divided between the substitution and scale effect; that is, the substitution and scale elasticities were approximately equal.

Canadian studies have obtained broadly similar estimates. For example, Woodland (1975) estimates labour demand functions for ten broadly defined Canadian industries (agriculture, manufacturing, forestry and mining, and so on) for the 1946–69 period. One labour input and two capital inputs (structures and equipment) are assumed in the empirical analysis. In each of the ten industries, the estimated labour demand curve is downward sloping. Estimated elasticities of labour demand are between zero and 0.5 in most industries. Eight of the ten industries exhibited statistically significant substitution among inputs. Woodland concludes that changes in relative input prices appear to play a significant role in determining the demand for factors of production.

MINIMUM-WAGE LEGISLATION

Minimum-wage laws provide a useful illustration of the practical relevance of our theoretical knowledge of the determinants of the derived demand for labour. In Canada, labour matters usually are under provincial jurisdiction; hence, each province has its own minimum-wage law. Federal labour laws cover approximately 10 percent of Canadian workers. Industries of an interprovincial or international nature — for example, interprovincial transportation and telephone communication, air transport, broadcasting, shipping, and banks — are under federal jurisdiction. The influence of federal laws may be larger to the extent that they serve as a model for comparable provincial legislation.

The rationale behind minimum-wage laws has not always been clear and explicit. Curbing poverty among the working poor, preventing "exploitation" of the unorganized nonunion sector, preventing "unfair" low-wage competition, and even discouraging the development of low-wage sectors have all been suggested as possible rationales. In the early days of union organizing, it is alleged, minimum-wage laws were also instituted to curb unionization — in that, if the wage of unorganized labour could be raised through government legislation, there would be less need for unions. As is so often the case with legislation, its actual impact may be different from its intended impact, or at least it may have unintended side effects. Economic theory may be of some help in shedding light on this issue.

Expected Impact

Economic theory predicts that a minimum wage in a competitive labour market will have an adverse employment effect; that is, employment will be reduced relative to what it would have been in the absence of the minimum wage. This is illustrated in Figure 9.8 where w_c and N_c are the equilibrium wage and level of employment, respectively, in a particular competitive labour market. After the imposition of the minimum wage, w_m, employers will reduce their demand for labour to N_m. Thus $(N_c - N_m)$ is the adverse employment effect associated with the minimum wage. In addition to the $(N_c - N_m)$ workers who would not be employed because of the minimum wage, an additional $(N_s - N_c)$ workers would be willing to work in this sector because the minimum wage is higher than the previous prevailing wage. Thus the queue of applicants for the reduced number of jobs is $(N_s - N_m)$, with $(N_c - N_m)$ representing workers laid off because of the minimum wage and $(N_s - N_c)$ representing new potential recruits who are attracted to the minimum-wage jobs.

The adverse employment effect $(N_c - N_m)$ occurs as firms substitute other relatively cheaper inputs for the higher-priced labour, and as they reduce output in response to the higher costs. These are the familiar substitution and scale effects, respectively, that give rise to the downward-sloping demand schedule for labour. (In the short run there would also be a disemployment effect, but the reduction in employment would be less than the long-run effect incorporating the scale and substitution effects.) The adverse employment effect need not mean an increase in measured unemployment: some who lose jobs in the sectors covered by the minimum-wage law may go to the uncovered sectors or drop out of the labour force.[6]

The magnitude of this adverse employment effect depends on the elasticity of the derived demand for labour. If the demand for labour were relatively inelastic then the adverse employment effect would be small. Unfortunately, in those sectors most affected by minimum-wage legislation — low-wage industries like textiles, sales, service, and tourism — the demand for labour is likely to be fairly elastic, especially in the long run, reflecting the availability of substitute inputs and products as well as the fact that labour costs are often a substantial proportion of the total cost. In addition, since many of these sectors are often thought of as "fiercely competitive," it is unlikely that the minimum-wage cost increases could be absorbed by the industry through a reduction in monopoly profits.

---

6. This indicates that minimum-wage laws may affect not only wages and employment in the sectors to which they apply but also in uncovered sectors through their impact on labour supply to these sectors. This aspect of minimum wages is discussed further in a subsequent chapter.

Figure 9.8
MINIMUM WAGE

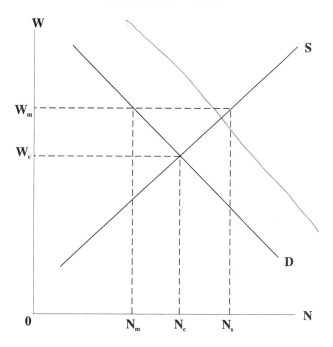

Possible Offsetting Factors

In a dynamic context, other things are changing at the same time as the minimum wage, and these changes may offset, in part at least, the adverse employment effect. For example, the demand for labour could increase perhaps because of an exogenous increase in the demand for output or because of an increase in the price of substitute inputs. While this may mitigate some, or perhaps even all, of the unemployment associated with the minimum wage, it is still true that employment would be even higher were it not for the minimum wage. Hence there is still an adverse employment effect relative to the situation with no minimum wage. This can easily be illustrated in Figure 9.8 by shifting the demand schedule upwards and to the right and tracing out the new level of employment with and without the minimum wage.

It is possible that some of the adverse employment effect may be offset by what could be labelled a "shock effect." Because of the cost pressure associated with the higher minimum wage, employers may be induced into utilizing other cost-saving devices that they should have introduced even without the minimum wage. The minimum wage simply serves as the catalyst for the introduction of cost-saving efficiencies. Similarly labour itself may be induced into becoming more productive, perhaps because of the higher wage or perhaps because of the queue of applicants vying for their jobs. The existence of these shock effects, however, requires that there be some slack in the system; that is, that firms were not maximizing profits or minimizing costs in the first place, because otherwise they would have instituted these possibilities even without the minimum wage.

There are possible situations when employers may be able to absorb the wage cost increases without reducing employment. If the firm were an oligopolist and would lose a substantial share of the market by raising its product price, it might try to absorb the wage cost increase. However, it is likely that the low-wage industries that are most affected by the minimum wage are likely to be competitive rather than oligopolistic.

Similarly, employers who dominate the local labour market may not have to pay a competitive wage rate. In such circumstances the minimum wage may simply compel the employer to pay what would be a competitive wage. (This situation of monopsony and its implications for minimum-wage legislation is developed more formally in Chapter 12.) Once again, however, it is unlikely that the industries that are most affected by minimum wages are those dominated by one or a few employers. Rather, the employers most affected tend to be small in size and competing in a labour market that has large pools of unskilled labour.

Thus while there is the theoretical possibility of these factors offsetting, in part at least, the adverse employment effect of minimum-wage laws, their practical relevance is in question. It is unlikely that they could be of great importance in the long run, especially in those low-wage sectors most affected by minimum wages. This leaves us with the inescapable conclusion that, in all probability, minimum-wage laws reduce employment opportunities. In addition, the impact will fall disproportionately on unskilled workers who are often most in need of work experience. This is the case, for example, with respect to younger workers and women who could utilize employment, even at low wages, as a means to acquire the on-the-job training and labour market experience necessary to move on to higher-paying jobs. For this reason, some would argue that minimum-wage laws actually harm the very people they are allegedly designed to help.

This does not necessarily mean that minimum-wage laws are undesirable, although that is likely to be the conclusion of many economists. The benefits of the wage increases to some have to be weighed against the costs of reduced employment opportunities to others. In addition, some would argue that even if minimum wages do reduce employment opportunities, it is in jobs that should not exist in the first place. It may be better to have these jobs eradicated and the unemployed workers trained for higher-wage jobs.

Empirical Evidence on Actual Impact

As may be expected, it is difficult to ascertain the actual impact of minimum-wage legislation, largely because so many other factors are also changing over time, and it may take a long time for all of the adjustments to occur. In addition, the adjustment processes may be subtle and difficult to document empirically. For example, a clothing manufacturer, when faced with a minimum-wage increase, may contract-out to households specific tasks, such as sewing on buttons or collars, or a restaurant may increase its usage of pre-packaged foods. How does one accurately compare the employment reduction in the clothing establishment or restaurant with the employment creation in the household or the food processing sector, to arrive at a net employment effect of the minimum wage?

In spite of the obvious difficulties, there have been numerous attempts to measure the employment impact of minimum-wage laws, mainly in the United States. After a thorough survey of the empirical evidence up until the late 1960s, Kaufman and Foran (1968) conclude that the overwhelming bulk of evidence supports the predictions of economic theory. Additional empirical evidence since that time seems to continue the support. This is especially the case with the large number of more recent studies — for example Moore (1971), Kosters and Welch (1972), Adie (1973), Katz (1973), Welch (1976), Mincer (1976), and Ragan (1977) — that have documented a significant adverse employment effect on teenagers. Holzer, Katz, and Kruger (1991) also indicate that workers tend to queue for minimum-wage jobs more so than for other jobs, suggesting that employers are unable to offset the minimum wage increase by altering the nonwage aspects of the job (see Exhibit 9.1).

Brown, Gilroy, and Kohen (1982) provide a survey of the U.S. evidence on the impact of the minimum wage on employment and unemployment. The most frequently studied group is teenage workers (16–19 years of age). Time-series studies typically find that a 10 percent increase in the minimum wage reduces teenage employment by 1 to 3 percent when other factors are held constant. Cross-section studies provide a wider range of estimated impacts, but most are in the range of a 0 to 3 percent decline in employment for a 10 percent increase in the minimum wage. Less research has been carried out on the impact on young adult workers (20–24 years of age); nonetheless these studies also find a negative impact, albeit smaller than that for teenagers. The impact on adult workers is uncertain. Higher minimum wages for younger workers may increase employment of those adults who are substitutes in production for youth. For other adults, the minimum wage will have the adverse employment impact as depicted in Figure 9.8. The U.S. empirical evidence does not enable one to conclude which of these effects is stronger.

Canadian econometric studies[7] also find significant minimum-wage impacts, especially for youths. Swidinsky (1980), for example, estimates employment elasticities of -0.10 and -0.27 for male and female teenagers respectively; that is,

---

7. For example, Cousineau (1979), Fortin and Phaneuf (1979), Maki (1979), Swidinsky (1980), and Schaafsma and Walsh (1983). Reviews of Canadian studies are contained in West and McKee (1980) and Mercier (1987).

EXHIBIT 9.1   DO FIRMS OFFSET MINIMUM WAGE INCREASES BY ALTERING NONWAGE ASPECTS OF THE JOBS?

Because it fixes wages above the competitive norm, minimum-wage legislation, like any form of wage fixing, should lead to a queue of applicants for such jobs (i.e., supply exceeds demand at the minimum wage as indicated in Figure 9.8). Given the potential queue of applicants, firms should be able to offset at least some of the wage increase by reducing the nonwage component of total compensation. This could include reducing fringe benefits, upgrading hiring standards, making the job more onerous, and letting working conditions deteriorate. If they completely offset the minimum wage, then there would be no queue of applicants. In practice, firms do not appear to be able to fully offset the minimum wage, since workers still queue for minimum-wage jobs more so than for other jobs.

Source: H.J. Holzer, L.F. Katz, and A.B. Krueger. Job queues and wages. *Quarterly Journal of Economics* 106 (August 1991) 739–768.

---

an increase in the minimum wage of 10 percent results in a decline in employment of 1.0 and 2.7 percent for male and female teenagers. Schaafsma and Walsh (1983) estimate larger impacts on teenagers. They also conclude that the minimum wage has a significant adverse employment impact on adult workers.

An alternative empirical procedure is to directly survey firms about the effect of minimum-wage revisions on their work force. The surveys usually concentrate on firms that employ low-wage workers, and typically find no adverse employment consequences of changes in the minimum wage. (See West and McKee, 1980, ch. 4, for a discussion of these studies.) The difficulty with this approach is that it does not distinguish between employment changes due to revisions to the minimum wage and those due to other factors. For example, if labour demand is increasing over time, economic theory indicates that increases in the minimum wage may not result in actual reductions in employment, even though the increase in employment will be lower than it would have been in the absence of the minimum-wage changes. Because the responses to minimum-wage revisions are varied, subtle (involving, for example, changes in the firm's product price and in the capital-labour mix), and take place gradually over time, it is important to control for other factors influencing employment. For this reason the survey approach is of little use in assessing the impact of minimum wages.

CHANGING DEMAND CONDITIONS AND GLOBAL COMPETITION

Since the demand for labour is derived from the output of firms, then labour demand (and consequently wages and employment) will be affected by changes in the product market conditions that are affecting employers. Such changes have been substantial in recent years, emanating from such interrelated factors as global competition, free trade, industrial restructuring, technological change, privatization, and subcontracting.

In recent years the competitive advantage of countries has depended less on traditional factors such as raw resources, capital intensive facilities, financial resources, technology, and even innovation. This is so in part because such resources are now more readily available in an increasingly global marketplace, and new innovations can be rapidly adapted and diffused. The emphasis is increasingly on competitive strategies that emphasize productivity and quality, often attained through a skilled, motivated, and committed work force. This puts increased emphasis on human capital investment decisions in areas like education, training, information and mobility, as well as human resource practices of employers in areas such as job design, employee involvement, and compensation practices.

The recent Free Trade Agreement (FTA) between Canada and the United States is another product market factor believed to affect labour markets. The *expected* effects on labour markets were predicted to be positive: job gains exceeding job losses, increases in real wages, and reductions in unemployment. However, controversy remains about its actual effects after the agreement, with opponents attributing mass layoffs and plant closings to the FTA. In all likelihood, it is difficult if not impossible to determine the independent impact of the FTA, in part because it is phased-in over a ten-year period, but more importantly because it is only one of the many interrelated forces affecting the labour market at this time (Betcherman and Gunderson, 1991). The other forces include those discussed in this section — global competition, industrial structuring, technological change, privatization, and subcontracting.

While there is not a consensus on the overall impact of the FTA, there is more agreement that it will create *adjustment consequences* in the labour market. Those consequences will be both on the downside (plant closings, mass layoffs) as well as on the upside (skill shortages, retraining). Even if the net job creation is positive, there will still be job losses as well as a need to adapt workers to the new jobs. As well, some industries will be more affected than others. These adjustment consequences will also prevail under a possible North American Free Trade Agreement (NAFTA) between Canada, the United States, and Mexico.

Industrial restructuring has also been prominent, associated mainly with the decline of blue-collar, often unionized, well-paid jobs in heavy manufacturing. As outlined in more detail later in the chapter on wage structures, these middle-wage jobs have been displaced by jobs at polar ends of the occupational distribution. At the high end are the professional, managerial, and administrative jobs; at the low end are the low-wage jobs in services and retailing. This phenomenon has been described as the "declining middle" and it has led to increased "wage polarization." The large factories of the heavy-manufacturing "smokestack" industries have often given way to more flexible, modular factories linked together by advanced communications and "just-in-time" delivery systems that require little inventory. Flexibility and adaptability are crucial, and the need for these elements has led to increases in subcontracting, permanent part-time work forces, and contingent work forces that are utilized as demand changes. The industrial restructuring has often been accompanied through mergers and acquisitions, often leading to work force reductions ("downsizing") and pressures for concession bargaining.

Technological change has continued its advance, especially in areas of office automation association with computers. Robotics has been introduced on the factory floor and the assembly line, and computer-assisted technology has been utilized in virtually every aspect of the production cycle.

In the public sector, privatization has been prominent for functions ranging from building maintenance to road maintenance. Crown corporations such as Air Canada have also been privatized.

These various changes emanating from the demand side of the labour market have significant implications for labour markets, as well as for human resource practices and industrial relations policies in general. They imply greater demands for a flexible and adaptable work force, just like the flexibility that is required from the product market side from which the demand for labour is derived. The work force must be willing and able to do a broader array of job assignments and to integrate quality control as an integral part of the job. The product market changes place greater emphasis on a work force that is trained in the general skills required for continuous upgrading and retraining and is also able to do a variety of tasks ("multiskilling") associated with the broader job assignments and the emphasis on quality control. As part of the individual worker's responsibility for quality control, there is less emphasis on direct supervision and more emphasis on employee involvement in decision-making.

The demand side changes are also pressuring for compensation that is more flexible and responsive to the changing economic conditions and to the particular economic circumstances of each individual enterprise. This in turn puts pressure towards the breakdown of historical pattern bargaining that often prevailed, whereby establishments frequently followed the wage settlements established by certain key pattern-setting firms. This may no longer be viable given the different demand conditions facing different firms.

The changes emanating from the demand side of the labour market have also led to numerous changes in workplace practices. Jobs have often been redesigned to provide for both *job enlargement* (the horizontal addition of a variety of tasks of similar complexity) and *job enrichment* (the vertical addition of tasks of different levels of complexity).

Whether these changes are part of a fundamental transformation or a continuous evolution of the workplace and the labour market is a debatable issue. What is certain is that the demand side forces are having a substantial impact. The impact is particularly pronounced because the forces tend to work in the same direction, thereby compounding their individual effects. When coupled with many of the supply side pressures highlighted previously (aging work force, continued labour force participation of women, dominance of the two-earner family), they emphasize the dynamism that will continue to characterize the Canadian labour market.

INTERNATIONAL COMPARISONS OF LABOUR COSTS

The changing demand conditions emanating from growing international competition highlight the importance of being competitive on an international basis. This will be increasingly important if world tariffs and nontariff barriers to trade continue to be reduced under arrangements such as the General Agreement on Tariffs and Trade (GATT). Competitive pressures will also be enhanced from the growing number of trading blocs such as the European Economic Community (EEC), the Closer Economic Relations agreement between Australia and New Zealand, the Canada–U.S. Free Trade Agreement (FTA), a possible North American Free Trade Agreement (NAFTA) to also include Mexico, and possible extensions to newly emerging free trade areas in South America such as Mercosur (Argentina, Brazil, Paraguay, and Uruguay) as well as the Andean Pact (Venezuela, Columbia, Ecuador, Peru, and Bolivia). Trade pressures are constantly being felt from the continued success of Japan as well as the Asian Newly Industrialized Countries (NICs) of Hong Kong, Korea, Singapore, and Taiwan. The transition of Soviet bloc and Eastern European economies to market-oriented economies will also open markets as well as provide pools of relatively untapped labour resources.

The greater openness to international trade puts more emphasis on labour costs as a determinant of international competitiveness. It also means that plant location decisions will be more important since plants can more easily locate in areas of lower labour cost and then export throughout the world. Such plant location decisions and international competitiveness will depend upon a variety of factors including the legislative and regulatory environment of a country, as well as its political stability. Labour costs will also be an important consideration.

The following tables provide some international comparisons of various aspects of labour costs. They highlight Canada's changing position in an increasingly integrated world economy. Since it is labour cost relative to productivity that influences competitiveness, the tables provide information on various dimensions: compensation costs (including some fringe benefits); productivity; and unit labour costs which reflect both compensation costs and productivity (i.e., labour cost per unit of output).

Table 9.1 compares compensation costs in Canada to a variety of countries throughout the world. The costs are converted to Canadian dollars on the basis of the exchange rate, and hence they reflect fluctuations in the exchange rate. They are specified on an hourly basis so they are adjusted for differences in hours worked. The costs for each year are indexed to the Canadian costs, which are set equal to 100.0 so that they are all expressed as percents of the Canadian costs. For example, in 1990, compensation costs in the highest-cost country,

Table 9.1
COMPENSATION COSTS[1], VARIOUS COUNTRIES[2], 1975, 1980, 1985, 1990
(As percent of Canadian compensation costs set equal to 100)

COUNTRY	1975	1980	1985	1990
Norway	117.1	139.5	98.2	136.5
Germany	109.7	147.3	88.6	134.4
Sweden	124.0	149.5	89.4	130.6
Switzerland	105.1	132.5	89.4	130.0
Finland	79.6	98.4	75.6	129.6
Belgium	110.7	156.6	82.6	118.2
Netherlands	113.6	144.1	81.0	113.8
Denmark	108.5	129.4	75.3	111.4
Austria	74.9	102.4	67.3	106.2
Italy	80.3	95.6	68.6	102.4
Canada	**100.0**	**100.0**	**100.0**	**100.0**
France	78.1	106.8	69.6	95.1
United States	109.8	117.6	120.5	92.2
Australia	95.7	101.2	75.4	80.9
Japan	52.7	67.0	59.6	78.9
United Kingdom	57.3	88.8	57.3	77.6
Ireland	52.3	71.1	54.8	73.4
Spain	44.7	71.2	44.3	72.4
Israel	38.9	45.3	37.6	53.4
New Zealand	17.3	63.7	41.4	52.0
Greece	29.2	44.6	33.9	42.1
				(continued)

Notes:
1. Defined as (1) all payments made directly to the worker, before payroll deductions of any kind, and (2) employer expenditures for legally required insurance programs and contractual and private benefit plans. In addition, for some countries, compensation is adjusted for other taxes on payrolls or employment (or reduced to reflect subsidies), even if they are not for the direct benefit of workers, because such taxes are regarded as labour costs. All figures are on an hours-worked basis and are exchange rate adjusted and therefore reflect exchange rates as well as wages and fringe benefits.
2. Listed in descending order, from high to low, in terms of 1990 compensation costs.

Table 9.1 (continued)

COUNTRY	1975	1980	1985	1990
Taiwan	6.9	11.9	13.9	24.7
Korea	5.7	11.6	11.6	23.8
Singapore	14.5	17.8	22.9	23.6
Hong Kong	13.1	18.0	16.0	20.0
Brazil	15.0	16.5	10.4	16.5
Mexico	N/A	N/A	14.8	11.2
Trade-weighted Measures[3]				
EEC[4]	87.2	118.1	72.9	105.7
OECD[5]	81.6	98.0	77.5	96.6
Asian NIEs[6]	8.6	13.7	14.7	23.4
All countries[7]	70.8	85.3	68.1	85.6

3. Averages for groups of countries, weighted by the amount of trade (exports and imports) with the United States as of 1986.
4. European Economic Community: Belgium, Denmark, France, Germany, Greece, Ireland, Italy, Luxembourg, the Netherlands, Portugal, Spain, and the United Kingdom.
5. Organization for Economic Cooperation and Development countries: Canada, Australia, Japan, New Zealand, and all European countries.
6. Asian Newly Industrialized economies: Hong Kong, Korea, Singapore, and Taiwan.
7. Brazil, Mexico, and Israel have been excluded in the original data source because their rapid rates of inflation distort the trade-weighted average percentage changes.

Source: U.S. Department of Labor, Bureau of Labor Statistics, Office of Productivity and Technology, 1992.

Norway, were 136.5 percent of Canadian costs (i.e., 36.5 percent above Canadian costs), while the costs in the lowest-cost country of Mexico were 11.2 percent of those in Canada.

The figures of Table 9.1 indicate that Canada is reasonably competitive on an international basis in terms of compensation costs. We are significantly below high-wage countries like Norway, Germany, Sweden, and Switzerland, but significantly above low-wage countries like Mexico and the Newly Industrialized Economies of Hong Kong, Korea, Singapore, and Taiwan. Of potential concern is that our compensation costs are higher than those of our major trading partners — the U.S. And Japan. In 1985, U.S. compensation costs were 20.5 percent above those in Canada; by 1990, they were 92 percent of those in Canada, much of this reflecting the appreciating Canadian dollar over that period. This is of particular concern, given the recent Free Trade Agreement with the U.S.

Compensation costs in other countries, however, have tended to increase even more than in Canada. For example, in 1975 such costs in the highest-wage countries tended to be 10 to 20 percent above those in Canada; by 1990, they were more than 30 percent above those in Canada. Compensation costs in the Newly Industrialized Economies of Asia have also increased steadily, as their industrialization process has matured and world demand for their goods has increased.

The continued success of higher-wage countries such as Germany and increasingly Japan (the latter were wage costs have increased dramatically in recent years) shows the viability of being able to compete on the basis of productivity and quality, as opposed to low wages. The extremely low wages in many of the Asian countries and in countries such as Mexico and Brazil highlight that it is not possible for Canada to compete on the basis of labour costs alone. Rather, it is important to be able to develop "market niches" on the basis of high productivity and high value-added production and quality service.

Table 9.2 provides some international comparisons that show the changes in compensation costs relative to productivity and the resulting changes in unit labour cost (compensation per unit of output). The figures are average annual rates of change in the different aspects of labour costs. As indicated in the last panel, unit labour costs increased more slowly in Canada (3.4 percent) than in the United States (4.9 percent) in the early 1980s (1979–1985). This was a result of our higher productivity growth of 2.4 percent per year versus 1.9 percent in the U.S. (first panel) and slower growth in hourly compensation, averaging 5.9 percent versus 6.9 percent in the U.S. (second panel). However, since then our productivity growth has lagged substantially behind that of the U.S. (first panel), while our compensation growth has been much higher (second panel) so that unit labour cost have increased substantially relative to the U.S. (third panel). This illustrates the importance of enhancing our productivity as the key to maintaining high wages *and* international competitiveness. As indicated, this is of particular concern given the recently negotiated Free Trade Agreement with the U.S.

While our unit labour costs have increased relative to those in the United States, they have not increased as much as those in most European and Scandinavian countries, at least since the early 1980s. This is illustrated most clearly in Table 9.3, which summarizes the change in unit labour costs and its productivity and compensation components from 1982 to 1990. Clearly, the United States has the slowest growth in unit labour cost, largely because its (by far) lowest growth in hourly compensation has been restrained relative to its average growth in productivity. Canada, is second, however, in its slow growth of unit labour costs. While its hourly compensation increases have been substantially larger than its productivity growth, this imbalance has not been as great as in most other countries, except for the United States.

Table 9.2
PRODUCTIVITY[1], HOURLY COMPENSATION[2], AND UNIT LABOUR COSTS[3] IN MANUFACTURING, VARIOUS COUNTRIES, 1979–1990

COUNTRY	1979–85	1987	1988	1989	1990
Productivity (output per hour) Average annual rates of change					
Canada	2.4	0.9	0.5	0.5	1.3
U.S.	1.9	4.1	4.0	0.5	2.5
Japan	3.9	8.3	6.0	5.2	3.6
Belgium	6.0	4.4	5.5	4.0	4.1
Denmark	2.1	−0.5	2.1	2.7	2.3
France	3.2	2.0	6.7	4.8	1.1
Germany	2.0	−1.7	3.5	4.6	4.5
Italy	5.0	2.8	2.6	3.1	3.2
Netherlands	4.2	−0.3	4.0	2.9	1.9
Norway	2.9	5.1	−0.5	3.7	2.3
Sweden	2.9	1.2	1.3	0.3	0.5
U.K.	4.5	6.1	5.5	4.8	0.9
Hourly Compensation (U.S. dollar basis) Average annual rates of change					
Canada	5.9	7.9	12.1	10.6	8.6
U.S.	6.9	2.2	3.9	4.0	5.3
Japan	3.1	19.2	15.0	−1.6	0.1
Belgium	−4.2	22.1	3.6	−1.7	25.7
Denmark	−3.8	29.3	4.2	−3.3	24.3
France	−0.3	19.6	3.2	−2.7	21.7
Germany	−2.1	26.4	6.3	−2.1	24.4
Italy	1.6	21.5	3.8	3.9	27.0
Netherlands	−3.6	23.9	4.7	−5.9	20.3
Norway	0.7	25.4	9.6	−1.7	17.1
Sweden	−2.5	20.1	10.1	3.7	19.0
U.K.	2.2	19.3	14.6	0.7	21.6

(continued)

Notes:
1. Output per hour.
2. Compensation is defined as (1) all payments made directly to the worker, before payroll deductions of any kind, and (2) employer expenditures for legally required insurance programs and contracted and private benefit plans. In addition, for some countries, compensation is adjusted for other taxes on payroll or employment (or reduced to reflect subsidies), even if they are not for the direct benefit of workers, because such taxes are regarded as turnover costs.
3. Hourly compensation per unit of output (i.e., adjusted for productivity).

Table 9.2   (continued)

COUNTRY	1979–85	1987	1988	1989	1990
	Unit Labour Cost (U.S. dollar basis) Average annual rates of change				
Canada	3.4	6.9	11.5	10.0	7.2
U.S.	4.9	–1.8	–0.1	3.5	2.7
Japan	–0.8	10.1	8.5	–6.4	–3.4
Belgium	–9.6	17.0	–1.8	–5.5	20.7
Denmark	–5.8	29.9	2.1	–5.9	21.5
France	–3.4	17.3	–3.3	–7.2	20.5
Germany	–4.0	28.5	2.7	–6.4	19.1
Italy	–3.3	18.1	1.2	0.8	23.1
Netherlands	–7.5	24.3	0.7	–8.5	18.0
Norway	–2.1	19.3	10.2	–5.2	14.4
Sweden	–5.3	18.6	8.7	3.4	18.4
U.K.	–2.2	12.4	8.7	–3.9	20.6

Source: U.S. Department of Labor, Bureau of Labor Statistics, Office of Productivity and Technology, February 1992.

In summary, Canada is reasonably competitive internationally on the basis of compensation costs. As well, our unit labour costs have been restrained relative to most European and Scandinavian countries because our compensation increases have not outpaced our relatively poor productivity performance as much as occurred in those countries. There are, however, two areas of concern. First, our competitive position relative to our major trading partner — the United States — has deteriorated significantly in recent years largely because our compensation has outpaced our stagnant productivity growth so that our unit labour costs have risen substantially relative to the U.S. Much of this reflects the appreciation of the Canadian dollar, relative to the U.S. dollar, but much also reflects our stagnant productivity growth. Second, even though our compensation increases have been more restrained than most European and Scandinavian countries, our productivity performance has been poor by international standards. In essence, increased productivity seems to be the key to maintaining international competitiveness while providing reasonable compensation increases.

This picture which prevails in the late 1980s and early 1990s is based on changes that have occurred mainly over the 1980s. This picture will likely change substantially over the next few years given the various components of

Table 9.3
CHANGE IN PRODUCTIVITY[1], HOURLY COMPENSATION[2],
AND UNIT LABOUR COSTS[3] IN MANUFACTURING 1982–1990,
VARIOUS COUNTRIES

COUNTRY	HOURLY PRODUCTIVITY	HOURLY COMPENSATION	UNIT LABOUR COST
Canada	21.8	56.7	28.7
U.S.	25.7	34.3	6.9
Japan	38.0	131.1	67.5
Belgium	40.5	102.8	44.4
Denmark	5.5	105.3	94.6
France	27.7	89.1	48.0
Germany	26.2	121.8	75.8
Italy	38.8	137.3	71.0
Netherlands	29.4	82.7	41.2
Norway	27.1	100.2	57.6
Sweden	18.2	103.9	72.5
U.K.	45.0	86.6	28.7

Notes:
1. Output per hour.
2. Compensation is defined as (1) all payments made directly to the worker, before payroll deductions of any kind, and (2) employer expenditures for legally required insurance programs and contractual and private benefit plans. In addition, for some countries, compensation is adjusted for other taxes on payrolls or employment (or reduced to reflect subsidies), even if they are not for the direct benefit of workers, because such taxes are regarded as labour costs.
3. Hourly compensation per unit of output (i.e., adjusted for productivity).

Source: U.S. Department of Labor, Bureau of Labor Statistics, Office of Productivity and Technology, February 1992.

productivity, wages, fringe benefits, and exchange rates that can influence a countries competitiveness. What is certain is that these factors will take on increased importance in an increasingly interdependent world.

QUESTIONS

1. The marginal productivity theory of labour demand has been criticized for being unrealistic in its assumptions, and for ignoring a wealth of information available to us from other sources. Evaluate.

2. Why is the point E_o in Figure 9.2(c) a stable equilibrium? Depict the firm's expansion path. Depict the isoquant for a firm whose technology is such that

it must utilize labour and capital in fixed proportions to produce its output. Depict the scale and substitution effect of a wage decrease in such a circumstance. Depict the firm's expansion path.

3. What is meant by an inferior factor of production? How would the firm's demand for labour be altered if labour were an inferior factor of production?

4. "The firm's demand for labour is a negative function of the wage because, as the firm uses more labour, it has to utilize poorer quality labour, and hence pays a lower wage." Discuss.

5. Discuss the firm's short-run demand for labour when its capital is fixed.

6. Derive the firm's demand schedule for labour if it were a monopolist that could influence the price at which it sells its output. That is, relax the assumption that product prices are fixed and trace through the implications.

7. Discuss the various forces from the general economic environment that have been influencing the demand for labour in Canada. Discuss the expected wage and employment impact of these forces.

8. Discuss Canada's changing international competitive position with respect to unit labour cost. Discuss the relative importance of the various dimensions of unit labour cost.

9. "Given our wage costs relative to those in the Asian Newly Industrialized Economies and in Mexico, it does not make sense to try to compete on the basis of restraining labour costs. Rather, we should concentrate on other ways of competing, including the development of market niches that involve a high wage, high value-added strategy." Discuss.

10. "Our real concern on a international basis is that our unit labour cost have recently increased relative to the United States, and this has occurred just at the time when we need to be most competitive, given the recent Free Trade Agreement with that country." Discuss.

REFERENCES AND FURTHER READINGS

A. Labour Demand

Archibald, G.C. Testing marginal productivity theory. *R.E. Stats.* 27 (June 1960) 210–213.

Bartel, A. and F. Lichtenberg. The comparative advantage of educated workers in implementing new technology. *R.E. Stats.* 69 (February 1987) 1–11.

Bassett, L. and T. Borcherding. The firm, the industry, and the long-run demand for factors of production. *CJE* 3 (February 1970) 140–144.

Beach, C. and F. Balfour. Estimated payroll tax incidence and aggregate demand for labour in the United Kingdom. *Economica* 50 (February 1983) 35–48.

Bean, C. and P. Turnbull. Employment in the British coal industry: a test of the labour demand model. *EJ* 98 (December 1988) 1092–1104.

Benjamin, D. Household composition, labor markets, and labor demand: testing for separation in agricultural household models. *Econometrica* 60 (March 1992) 287–322.

Betcherman, G. and M. Gunderson. Canada–U.S. free trade and labour relations. *Labour Law Journal* 41 (August 1990) 454–560.

Borgas, G. The substitutability of black, hispanic, and white labor. *EI* 21 (January 1983) 93–106.

Borjas, G. The sensitivity of labor demand functions to choice of dependent variable. *R.E. Stats.* 68 (February 1986) 58–66.

Burgess, S. and S. Nickell. Labour turnover in UK manufacturing. *Economica* 57 (August 1990) 295–318.

Burgess, S. and J. Dolado. Intertemporal rules with variable speed of adjustment. *EJ* 99 (1989) 347–365.

Clark, K. and R. Freeman. How elastic is the demand for labor? *R.E. Stats.* 62 (November 1980) 509–520.

Field, E. Free and slave labor in the Antebellum South: perfect substitutes or different inputs. *R.E. Stats.* 70 (November 1988) 654–659.

Forbes, K. and E. Zampelli. Growth, technology and the demand for scientists and engineers. *IR* 30 (Spring 1991) 254–301.

Freeman, R. An empirical analysis of the fixed coefficient "manpower requirements" model, 1960–1970. *JHR* 16 (Spring 1980) 176–199.

Freeman, R. and J. Medoff. Substitutions between production, labor and other inputs in unionized and nonunionized manufacturing. *R.E. Stats.* 64 (May 1982) 220–233.

Fuss, M. and M. Denny. The effect of factor prices and technological change on the occupational demand for labour: evidence from Canadian telecommunications. *JHR* 18 (Spring 1983) 161–176.

Gottschalk, P. A comparison of marginal productivity and earnings by occupation. *ILRR* 31 (April 1978) 368–378.

Gould, J.R. On the interpretation of inferior goods and factors. *Economica* 48 (November 1981) 397–406.

Grant, J. and D. Hamermesh. Labor market competition among youths, white women and others. *R.E. Stats.* 63 (August 1981) 354–360.

Grossman, J. The substitutability of natives and immigrants in production. *R.E. Stats.* 64 (November 1982) 596–603.

Hamermesh, D. Econometric studies of labour demand and their application to policy analysis. *JHR* 11 (Fall 1976) 507–525.

Hamermesh, D. The demand for labor in the long run. *Handbook of Labor Economics*, Vol. 1. O. Ashenfelter and R. Layard (eds.). New York: Elsevier, 1986.

Hamermesh, D. and J. Grant. Econometric studies of labour — labour substitution and their implications for policy. *JHR* 14 (Fall 1979) 518–542.

Hamermesh, D. Plant closing, labor demand and the value of the firm. *R.E. Stats.* 70 (November 1988) 580–586.

Hamermesh, D. Labor demand and the structure of adjustment cost. *AER* 79 (1989) 674–689.

Heckman, J. and G. Sedlacek. Heterogeneity, aggregation and market wage functions. *JPE* 93 (December 1985) 1077–1125.

Henderson, J. and R. Henderson. Perversity in factor demand curves. *Economica* 53 (November 1986) 515–518.

Hicks, J. *Theory of Wages*, 2nd ed. London: Macmillan, 1963; 241–246, 374.

Killingsworth, M. Substitution and output effects on labour demand: theory and policy implications. *JHR* 20 (Winter 1985) 142–152.

Meza, D. de. Perverse long-run and short-run factor demand curves. *Economica* 48 (August 1981) 299–304.

Michl, T. The productivity slowdown and the elasticity of demand for labor. *R.E. Stats.* 68 (August 1986) 532–536.

Monouchehr, M. and F. Rassekh. The tendency towards factor price equalization among OECD countries. *R.E. Stats.* 71 (November 1989) 636–642.

Montgomery, M. On the determinants of employer demand for part-time workers. *R.E. Stats.* 70 (February 1988) 112–117.

Nagatani, K. Substitution and scale effects in factor demands. *CJE* 11 (August 1978) 521–526.

Nickell, S. Dynamic models of labour demand. *Handbook of Labor Economics*, Vol. 1. O. Ashenfelter and R. Layard (eds.). New York: Elsevier, 1986.

Sargent, T. Estimation of dynamic labor demand schedules under rational expectations. *JPE* 86 (December 1978) 1009–1044.

Schaafsma, J. On estimating the time structure of capital labor substitution in the manufacturing sector: a model applied to 1949–1972 Canadian data. *SEJ* 44 (April 1978) 740–751.

Shapiro, M. The dynamic demand for capital and labor. *QJE* 101 (1986) 513–542.

Skolnik, M. An empirical analysis of the substitution between engineers and technicians in Canada. *RI/IR* 25 (April 1970) 284–399.

Swan, N. Difference in the response of the demand for labour to variations in output among Canadian regions. *CJE* 5 (August 1972) 373–385.

Symons, J. and R. Layard. Neoclassical demand for labour functions in six major economies. *EJ* 94 (December 1984) 788–799.

Tuckman, H. and C. Chang. Own-price and cross elasticities of demand for college faculty. *SEJ* 52 (January 1986) 735–744.

Woodland, A. Substitution of structures, equipment and labor in Canadian production. *IER* 16 (February 1975) 171–187.

B. Minimum Wage Legislation

Adams, F. The macroeconomic impacts of increasing the minimum wage. *Journal of Policy Modelling* 11 (No. 2, 1989) 179–189.

Alison, W. Effects of the minimum wage on the employment status of youths. *JHR* 26 (1991) 27–46.

Alpert, W. *The Minimum Wage in the Restaurant Industry.* New York: Praeger, 1986.

Ashenfelter, O. and R. Smith. Compliance with the minimum wage law. *JPE* 87 (April 1979) 333–351.

Behrman, J., R. Sickles and P. Taubman. The impact of minimum wages on the distribution of earnings for major race–sex groups: a dynamic analysis. *AER* 83 (September 1983) 766–778.

Beranek, W. The illegal alien work force, demand for unskilled labor, and the minimum wage. *JLR* 3 (Winter 1982) 89–100.

Betsey, C. and B. Dunson. Federal minimum wage laws and the employment of minority youth. *AER* 71 (May 1981) 379–384.

Bloch, F. Political support for minimum wage legislation. *JLR* 1 (Fall 1980) 245–254.

Brown, C., C. Gilroy and A. Kohen. The effect of the minimum wage on employment and unemployment. *JEL* 20 (June 1982) 487–528.

Brown, C., C. Gilroy and A. Kohen. Time-series evidence of the effect of the minimum wage on youth employment and unemployment. *JHR* 18 (Winter 1983) 3–31.

Brown, C. Minimum wage laws. *JEP* 2 (Summer 1988) 135–145.

Chang, Y. and I. Ehrlich. On the economics of compliance with the minimum wage law. *JPE* 93 (February 1985) 84–91.

Cousineau, J.-M. Impact du salaire minimum sur le chomage des jeunes et des femmes au Quebec. *RI/IR* 34 (No. 3, 1979) 403–417.

Cox, J. and R. Oaxaca. The political economy of minimum wage legislation. *EI* 20 (October 1982) 533–555.

Cox, J. and R. Oaxaca. Minimum wage effects with output stabilization. *EI* 24 (July 1986) 443–454.

Drazen, A. Optimal minimum wage legislation. *EJ* 383 (September 1986) 774–784.

Eccles, M. and R.B. Freeman. What! Another minimum wage study? *AER* 72 (May 1982) 226–232.

Ferris, J. Information and search: an alternative approach to the theory of minimum wage. *EI* 20 (October 1982) 490–510.

Flug, K. and O. Galor. Minimum wage in a general equilibrium model of international trade and human capital. *IER* 27 (February 1986) 149–164.

Fortin, P. Minimum wage in Quebec: price-employment and redistributive effects. *RI/IR* 34 (No. 4, 1979) 672–673.

Grenier, G. On compliance with the minimum wage law. *JPE* 90 (February 1982) 184–187.

Grossman, J. The impact of the minimum wage on other wages. *JHR* 18 (Summer 1983) 359–378.

Hamermesh, D. Minimum wages and the demand for labor. *EI* 20 (July 1982) 365–380.

Hashimoto, M. Minimum wage effects on training on the job. *AER* 72 (December 1982) 1070–1087.

Holzer, H. Job queues and wages. *QJE* 106 (August 1991) 739–768.

Imam, M. and J. Walley. Incidence analysis of a sector-specific minimum wage in a two-sector Harris-Todaro model. *QJE* 100 (February 1985) 207–224.

Johnson, W. and E. Browning. The distributional and efficiency effects of increasing the minimum wage: a simulation. *AER* 73 (March 1983) 204–211.

Jones, S. Minimum wage legislation in a dual labour market. *European Economic Review* 31 (August 1987) 1229–1246.

Kaufman, R. Nepotism and the minimum wage. *JLR* 4 (Winter 1983) 81–89.

Kaufman, R. The effect of statutory minimum rates of pay on employment in Great Britain. *EJ* 99 (December 1989) 1054–1053.

Keyes, W.A. The minimum wage and the Davis Bacon Act: employment effects on minorities and youth. *JLR* 3 (Fall 1982) 399–413.

Lang, K. Pareto improving minimum wage laws. *EI* (January 1987) 145–158.

Linneman, P. The economics impacts of minimum wage laws: a new look at an old question. *JPE* 90 (June 1982) 443–469.

Maki, D. The effect of changes in minimum wage rates on provincial unemployment rates, 1970–1977. *IR* 34 (1979) 418–429.

McKee, M. and E. West. Minimum wage effects on part-time employment. *EI* 22 (July 1984) 421–428.

McKenzie, R. The labour market effects of minimum wage laws: a new perspective. *JLR* 1 (Fall 1980) 255–264.

Mercier, J. Les effets du salaire minimum sur l'emploi des jeunes au Quebec. *RI/IR* 40 (No. 3, 1985) 431–456.

Mercier, J. Effets du salaire minimum sur l'emploi. *RI/IR* 42 (No. 4, 1987) 806–829.

Metzger, M.R. and R.S. Goldfarb. Do Davis-Bacon minimum wages raise product quality? *JLR* 4 (Summer 1983) 265–272.

Meyer, R. and D. Wise. The effects of minimum wage on the employment and earnings of youth. *JOLE* 1 (January 1983) 66–100.

Migné, J. Salaire minimum ou quand le diable se fait moine. *RI/IR* 32 (No. 3, 1977) 310–320.

Mincer, J. Unemployment effects of minimum wages. *JPE* 84 Part 2 (August 1976) S87–S104.

Nakosteen, R. and M. Zimmer. Minimum wages and labor market prospects of women. *SEJ* 56 (October 1989) 302–314.

Schaafsma, J. and W. Walsh. Employment and labour supply effects of the minimum wage: some pooled time-series estimates from Canadian provincial data. *CJE* 16 (February 1983) 86–97.

Sellekaerts, B. and S. Welch. An econometric analysis of minimum wage noncompliance. *IR* 23 (Spring 1984) 244–259.

Solon, G. The minimum wage and teenage unemployment: a reanalysis with attention to serial correlation and seasonality. *JHR* 20 (Spring 1985) 292–297.

Starr, G. Minimum wage fixing. *ILR* 20 (September/October 1981) 545–562.

Swidinsky, R. Minimum wages and teenage unemployment. *CJE* 13 (February 1980) 158–170.

Swidinsky, R. and D. Wilton. Minimum wages, wage inflation, and the relative wage structure. *JHR* 17 (Spring 1982) 163–177.

Tauchen, G. Some evidence on cross-sector effects of the minimum wage. *JPE* 89 (June 1981) 529–547.

Uri, N. and J. Mixon Jr. An economic analysis of the determinants of minimum wage voting behaviour. *Journal of Law and Economics* 23 (April 1980) 167–177.

Welch, F. and J. Cunningham. Effects of minimum wages on the level and age composition of youth employment. *R.E. Stats.* 60 (No. 1, 1978) 140–145.

Wellington, A. Effects of the minimum wage on the employment status of youths. *JHR* 26 (Winter 1991) 27–46.

Wessels, W. The effect of minimum wages in the presence of fringe benefits: an expanded model. *EI* 18 (April 1980) 239–313.

West, E.G. and M. McKee. *Minimum Wages: The New Issues in Theory, Evidence, Policy and Politics.* Ottawa: Economic Council of Canada and Institute for Research on Public Policy, 1980.

West, E.G. Job signalling and welfare improving minimum wage laws. *EI* 26 (July 1988) 525–532.

Yaniv, G. Enforcement and monopsonistic compliance with the minimum wage law. *SEJ* 55 (October 1988) 505–509.

Zaidi, M. *A Study of the Effects of the $1.25 Minimum Wage Under the Canadian Labor (Standards) Code.* Study No. 16, Task Force on Labour Relations. Ottawa: Privy Council Office, 1970.

CHAPTER 10

Labour Demand, Nonwage Benefits, and Quasi-Fixed Costs

The previous chapter analyzed the demand for labour when labour is a variable input into the production process. In these circumstances, a change in the amount of labour used in production (the number of employee hours) leads to a proportionate change in labour costs. Furthermore, the change in labour costs depends only on the overall magnitude of the change in the labour input and not on the composition of the change in terms of the number of employees versus hours worked per employee. However, some labour costs may be quasi-fixed in the sense that they are independent of the number of hours worked by the firm's work force. These quasi-fixed costs may arise because of the costs of hiring and training new workers, the costs of dismissing employees, and from nonwage benefits such as holiday pay, contributions to pension plans, and unemployment insurance.

Quasi-fixed costs have important implications for a variety of labour market phenomena such as work schedules, part-time work, overtime work, hiring and layoff behavior and unemployment.[1] This chapter examines firm's employment decisions when labour is a quasi-fixed factor of production.

NONWAGE BENEFITS AND TOTAL COMPENSATION

The analysis of labour income and labour demand is complicated by the fact that wages and salaries are only one aspect of total compensation, and therefore only one aspect of the cost of labour to the employer. Figure 10.1 illustrates the main components of total compensation and the most commonly used terminology for these components. Pay for time not worked (holiday and vacation pay, paternal

---

1. Seminal contributions in this area were made by Oi (1962) and Rosen (1966). Recent contributions include Hart (1984, 1988) and Hamermesh (1989).

Figure 10.1
COMPONENTS OF TOTAL COMPENSATION

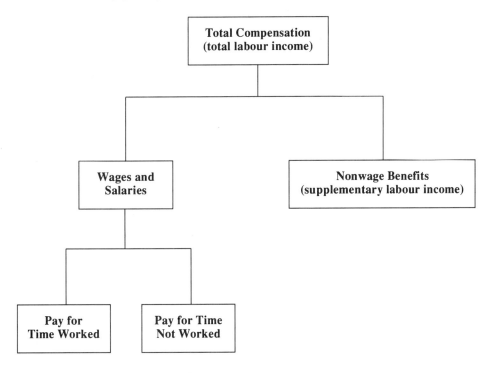

leave) is generally included in wage and salary income as reported by Statistics Canada. However, for some analytical purposes it may be desirable to treat this as a separate compensation category. In particular, for some employees, payment for time not worked is a quasi-fixed cost in the sense that the magnitude of holiday and vacation pay to which the employee is entitled is independent of hours worked.

As illustrated in Table 10.1, nonwage benefits (indirect labour costs) are a large and increasing component of total compensation. Between 1953 and 1989 nonwage employee benefits as a percentage of direct labour costs (gross payroll) increased from 15.1 to 33.5 percent. This approximate doubling in percentage terms tended to occur across most of the categories of nonwage benefits, so that the ranking of the components remained the same, with pay for time not worked, and pension and welfare plans being the largest components, although the latter grew the slowest. Legally required payments for workers' compensation, unemployment insurance, and CPP/QPP grew most rapidly.

The existence and growth of nonwage benefits are interesting phenomena to explain, especially given the basic economic proposition that unconditional cash transfers (e.g., wages) are generally more efficient than transfers-in-kind (e.g.,

Table 10.1
NONWAGE BENEFITS IN LARGE FIRMS:[1] COMPONENTS
AND CHANGES OVER TIME

NONWAGE BENEFIT CATEGORIES	GROSS PAYROLL			
	1953	**1957**	**1984**	**1989**
Pay For Time Not Worked	5.9	6.7	14.9	13.9
Pension and Welfare Plans	5.5	6.4	9.4	9.9
Payments Required by Law[2]	1.5	2.0	4.4	5.3
Bonuses, Profit-Sharing, Other	2.2	1.3	3.7	4.4
Total Nonwage Benefits	15.1	16.4	32.4	33.5

Notes:
1. Figures are for all industries. Separate figures for manufacturing and nonmanufacturing are not dramatically different.
2. Includes workers' compensation, unemployment insurance, and Canada/Quebec pension plans.

Source: KPMG Peat, Marwick, Stevenson, and Kellogg, *Employee Benefit Costs in Canada*, Toronto, 1989.

nonwage benefits). This proposition follows from the fact that recipients who receive cash could always buy the transfers-in-kind if they wanted them, but they are better off because they could buy other things if they wanted them more. Applied to employee benefits, why wouldn't employees and employers prefer wage compensation since it would enable employees to buy whatever benefits they want rather than being given some they may not value?

There are obviously a variety of possible reasons for the existence and growth of nonwage benefits. They are sometimes not taxed and, as taxes have increased over time, this increasingly makes nonwage benefits a preferred form of compensation. There may be economies of scale and administrative economies for group purchases through the employer of such things as pension and insurance plans. However, many of these economies still could be had through completely voluntary group plans. There is the possibility that workers think they are receiving nonwage benefits free in the sense that, if they did not have these benefits, their wages would not rise by a corresponding amount. This possibility, of course, can be true in times of wage controls to the extent that nonwage benefits are exempt or more difficult to control. More likely, workers may simply prefer nonwage benefits because it is an easy way of purchasing benefits that they value. In some cases, like increased vacations and holidays with pay, it is a way of institutionalizing the increased purchase of more leisure that accompanies their growing wealth.

Employers may have accepted the increased role of employee nonwage benefits for reasons that are beneficial to them. Paid vacations and holidays, for example, enable the planning of the production process more than if the workers' desire for increased leisure came in the form of lateness, absenteeism, work slowdowns, or turnover. Similarly, workers' compensation, unemployment insurance and health insurance can reduce the need for employers to have contingency plans for workers in industrial accidents or subject to layoffs or health problems. In addition, some employee benefits may alter the behaviour of workers in a fashion that is preferred by the employer. Subsidized cafeterias can reduce long lunch hours and keep employees in the plant, the provision of transportation can ensure that workers get to work on time, company health services can reduce absenteeism and provide a check on a worker's health, and pensions and other seniority-related benefits may reduce unwanted turnover. Employers may also prefer nonwage benefits because they reduce turnover and provide work incentives. Employee benefits such as company pension plans and seniority-based vacation pay represent deferred compensation in that these benefits are received later in the employee's career. Workers who leave the firm early in their careers do not receive these benefits. With some of their compensation in the form of deferred benefits, existing employees will be less likely to quit in order to seek or accept employment elsewhere. In addition, deferred compensation raises the cost of being fired for poor work, absenteeism, and so on, thus strengthening work incentives. These effects of deferred compensation on turnover and work incentives are discussed further in Chapter 23.

Deferred compensation can also act as a sorting or screening device, enabling the employer to hire those workers who are most likely to remain with the firm. Employers for whom turnover is very costly — perhaps because of the costs of hiring and training new workers — can screen out those potential employees who do not expect to remain with the firm for an extended period by making deferred compensation a significant proportion of total compensation, a compensation package that is most attractive to potential employees who expect to remain with the firm for a long period of time. Such screening mechanisms are especially valuable when the employer is unable to determine *ex ante* which job applicants are likely to leave the firm after a brief period and which are likely to remain with the firm.

Governments may also prefer nonwage benefits (and hence grant favourable tax treatment) because they may reduce pressures for government expenditures elsewhere. Employer-sponsored pensions may reduce the need for public pensions; contributions to the Canada (or Quebec) Pension Plan may reduce the need for public care for the aged, at least for the aged who have worked; and increased contributions for unemployment insurance directly reduce the amount the government has to pay to this fund. In addition, many nonwage benefits are part and parcel of the whole package of increased social security,

EXHIBIT 10.1   DID THE COST OF LABOUR FALL DURING THE 1980s?

Many Canadian workers experienced a decline in real wages during the 1980s. According to the theory of labour demand, this decline in real wages (brought about by nominal wages rising less rapidly than prices) would increase employment, other things being equal. However, as emphasized in this chapter, the cost of labour to the employer depends on both the wage rate and nonwage labour costs. As shown below, nonwage labour costs have been rising faster than prices. As a consequence, real total compensation per employee declined during the 1977–88 period, but not to as large an extent as real wages and salaries. This evidence indicates that data on wages and salaries alone would overstate the magnitude of the decline in real labour income and in the real cost of labour. These data also provide further evidence that nonwage labour costs represent an increasingly large fraction of total compensation.

Real Labour Income per Employee in Constant 1988 Dollars

	1977	1988	Percent change
Wages and salaries	27,916	26,999	−3.3
Supplementary labour income (nonwage benefits)	2,537	2,970	+17.1
Total compensation (total labour income)	30,453	29,969	−1.6

Source: H. Pold and F. Wong. The price of labour. *Perspectives on Labour and Income* (Autumn 1990) 42–49.

and hence the reasons for their growth are caught up with the reasons for the growth of the whole welfare state.

QUASI-FIXED LABOUR COSTS

General Definition

Employers can change their labour input by changing the number of employees, the hours per employee, or both. The way in which they adjust will depend upon the relative costs of the different options.

Variable costs of labour vary with the *total* hours of labour employed, whether the increase in total hours comes from an increase in the number of employees or from an increase in the hours per employee. For example, if the only cost of labour is the hourly wage rate, then costs will increase by the same percentage whether the labour input is expanded by increasing employment or hours per employee by a given percent. In such circumstances the firm's costs would increase by the same amount whether the labour input was expanded by increased employees or hours per employee.

Wage costs are therefore a variable labour cost, as are nonwage benefits that are proportional to wages. Other labour costs, however, are quasi-fixed costs in the sense that they are incurred per employee and are independent or largely independent of the hours of work per employee. They may be *pure* quasi-fixed costs in the sense that they are fixed per employee and are completely independent of the hours of work of each employee. Or they may be *mixed* variable and quasi-fixed costs in that they increase more rapidly for a given proportionate increase in the number of employees as opposed to hours per employee. In either circumstance, the firm would no longer be indifferent between whether the labour input (employees times hours per employee) expanded by increasing the number of employees or the hours per employee. Increasing the number of employees would be more costly given the quasi-fixed costs.

A distinction should also be made between *recurring* and *nonrecurring* or "one-time" quasi-fixed costs. Payroll taxes to finance public pensions and unemployment insurance are examples of recurring nonwage labour costs; these taxes are remitted regularly (e.g., monthly) to the relevant government agencies. Examples of nonrecurring costs include hiring and orienting new employees and dismissing employees whose services are no longer required.

Examples of Quasi-Fixed Costs

At the hiring stage, firms incur quasi-fixed costs associated with the recruiting, hiring, and training of new workers. These costs are independent of the hours the employees subsequently work. Under such circumstances there is an obvious incentive to try to amortize these fixed costs over longer hours and a longer worklife of these employees rather than to hire new employees and incur their additional fixed costs.

Expected termination costs themselves have a fixed cost component in the sense that at the *hiring stage* the firm may anticipate having to incur such costs should they subsequently lay off or dismiss the worker. Such costs become expected quasi-fixed costs at the hiring stage even though they may not be incurred until a subsequent period. Expectations of such costs are incurred when a new employee is hired but they do not increase, at least by the same proportion as wage costs, when an existing employee works additional hours.

As indicated in Meltz, Reid, and Swartz (1981) and Reid (1985, p. 163) employers' contributions (payroll taxes) to certain income maintenance programs may also have an element of fixed costs, when such contributions have a ceiling. Employer contributions to workers' compensation and public pensions, for example, have an annual income ceiling, and unemployment insurance a weekly earnings ceiling below which contributions are a certain percent of earnings and beyond which they are zero. Thus, once this ceiling is reached, it pays employers to work these people additional hours since no further payroll taxes are incurred, rather than to hire new employees and incur such payroll taxes. These payroll taxes are variable costs up until the ceiling, and thereafter they become fixed costs, creating the incentive to work existing employees more hours so as to amortize these costs.

Similarly, employment standards legislation often requires employees to have been employed for a certain minimum period before they are eligible for such benefits as paid holidays and advance notice of termination. Once the employee reaches this minimum period then these benefits become fixed costs for the employer; they are independent of the additional hours worked by the employee. As with other fixed costs, they encourage the employer to work existing employees longer hours and they discourage the hiring of new employees for whom such costs will be incurred.

Employers' contributions to life insurance and medical and dental plans can also have a quasi-fixed cost component to the extent that they are fixed per employee. In such circumstances costs are increased by hiring a new employee but not when an existing employee works more hours.

Not all nonwage benefits are ones that have a quasi-fixed component. Some, like employer-sponsored occupational pension plans, may have employer contributions based on a percent of earnings. In such circumstances the employers' contributions are proportional to earnings and are the same whether the labour input increases through an increase in the number of employees or hours per employee. The same situation prevails for vacation pay that is a percent of earnings.

General Effect of Quasi-Fixed Costs

As indicated, one general effect of such quasi-fixed costs is to increase the marginal cost of hiring an additional worker (sometimes termed the extensive margin of adjustment) relative to the marginal cost of working an existing worker longer hours (sometimes termed the intensive margin of adjustment). The quasi-fixed costs distort the labour expansion decision of the firm away from employment and towards more hours so as to minimize the quasi-fixed costs. Alternatively stated, the firm will want to amortize its fixed costs over the existing work force as much as possible rather than to hire new workers and incur the quasi-fixed costs. This is especially so for skilled workers where such quasi-fixed costs may be especially large, mainly because of training costs.

Just as on the supply side of the market (recall the discussion in Chapter 8), fixed money costs of work reduced the probability that an individual would enter the labour market but would increase the hours of work conditional upon having entered, so too quasi-fixed costs of employment have a differential effect on the demand side. Employers would be reluctant to hire more workers (and incur the fixed costs), but once the workers are hired, the employers would like them to work longer hours so as to amortize the fixed costs.

Nonrecurring fixed costs of employment alter the firm's hiring and layoff decisions, making these longer-term in nature. If the firm must incur some hiring and training costs when a new employee is hired, the firm will not hire the employee unless the addition to revenue is at least as great as the addition to costs, including the hiring and training costs. However, as long as the employee is expected to remain with the firm for several periods, it is not essential that the additional revenue equal or exceed the additional costs *each period*. Rather, the firm will continue hiring additional workers until the present value of additional future revenues equals the present value of additional costs. Thus rather than hiring until the current wage equals the current value of marginal product (the employment rule discussed previously, summarized in Equation 9.1), in the presence of nonrecurring fixed costs of employment the profit-maximizing employment rule becomes

$$H + T + \sum_{t=0}^{N} \frac{W_t}{(1+r)^t} = \sum_{t=0}^{N} \frac{VMP_t}{(1+r)^t} \qquad (10.1)$$

where H + T are hiring and training costs (assumed to be incurred in the first period for simplicity),

   W_t is the wage rate in period t,

   VMP_t is the expected value of the marginal product in period t,

   r is the firm's discount or interest rate, and

   N is the expected length of employment.

The left-hand side of Equation 10.1 is the present value of the costs of hiring an additional worker (i.e., the present value of the marginal costs) and the right-hand side is the present value of the marginal revenue.

Several implications of Equation 10.1 are worth noting. The hiring decision depends on the firm's expectations about the future, in particular the expected duration of employment, expected wage rates, and expected product market conditions (which determine the value of the marginal product of labour in future periods). Another implication is that, because of the hiring and training costs on the left-hand side of Equation 10.1,

$$\sum_{t=0}^{N} \frac{VMP_t}{(1+r)^t} > \sum_{t=0}^{N} \frac{W_t}{(1+r)^t} \qquad (10.2)$$

Thus when hiring an additional worker the firm must anticipate that the value of the marginal product will exceed the wage in most, if not all, future periods. As will be seen below, this prediction of the theory has implications for the firm's layoff behaviour in response to unanticipated changes in demand.

Some Phenomena Explained by Fixed Costs

The existence and growth of such fixed costs may help to explain, in part at least, the existence and growth of a number of labour market phenomena. The desire of employers to work their existing work-force overtime, in spite of overtime premium wages rates and possible fatigue effects, may be explained in part by the fact that fixed costs make it relatively cheaper to work the existing work force longer hours rather than hiring new workers. Ehrenberg and Schumann (1982, p. 14), for example, discuss a number of studies, based on United States data, indicating that employers' use of overtime hours are positively related to the ratio of their quasi-fixed employment cost to their overtime premium.

Quasi-fixed costs of employment may also explain, in part, the popularity of temporary help agencies and contracting-out. Employers avoid the quasi-fixed costs of hiring new employees when they contract-out or engage the services of temporary help agencies. As with the use of overtime hours, this can be a particularly important way of meeting new demand that may be temporary.

The layoff behaviour of firms will also be affected by quasi-fixed costs of employment. Once an employee has been hired and trained, the associated nonrecurring costs become sunk costs. Because these costs cannot be recovered, they will not influence the firm's decision-making with respect to its existing work force (for whom the hiring and training expenses have been incurred), even though they will influence decisions regarding potential employees (for whom the expenses have not been incurred).

At the time an employee is hired, the firm anticipates that the present value of future marginal products will exceed the present value of future wage costs, as shown in Equation 10.2. This excess of the value of the worker's contribution to the firm over the cost of that worker creates a "buffer" which reduces the risk that the worker will be laid off. In particular, even if there is an unanticipated decline in product demand, reducing the value of the marginal product of labour, the firm may nonetheless choose to not lay off any employees because the present value of the revenue associated with an employee, although lower than anticipated, remains higher than the present value of the costs. Because of the unanticipated decline in demand, the firm's return on its investment in hiring and training costs will be lower than expected; however, the return may still be positive, indicating that it is worthwhile retaining the employee. This will be the case especially if the decline in demand is believed to be temporary in nature, or if workers who are temporarily laid off may not be available for recall when demand returns to normal levels. In such circumstances the company may rationally engage in "labour hoarding" of such workers in a cyclical downturn and they may pay deferred compensation or seniority-related wages or benefits to discourage their experienced workers from leaving.

This implication of nonrecurring fixed employment costs is illustrated in Figure 10.2. Panel (a) shows the case in which there are no fixed costs of employment. At the wage W_o, employment will be N_o. A decline in labour demand from VMP to VMP^1 will result in a drop in employment from N_o to N_o^1. Panel (b) shows the case in which the employer incurs nonrecurring hiring and training costs H + T. At the wage W_o, employment will be N_o rather than N_o° as would occur in the absence of the hiring and training costs. For the N_o employees hired, the value of their marginal product is VMP_o; because of the fixed costs the employer does not expand employment to the point at which the VMP equals the wage. The "buffer" $VMP_o - W_o$ implies that employment will not necessarily fall in response to a decline in labour demand. For example, at the wage W_o, employment would remain at N_o even if the employees' value of the marginal product fell to VMP^1.

Fixed costs of employment can also explain the observation that layoff rates are much higher among low-skilled than high-skilled workers. Considerable

Figure 10.2a
NON-RECURRING FIXED EMPLOYMENT COSTS
AND CHANGES IN LABOUR DEMAND

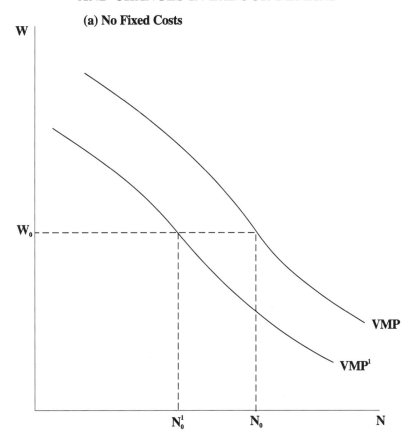

Figure 10.2b
NON-RECURRING FIXED EMPLOYMENT COSTS
AND CHANGES IN LABOUR DEMAND

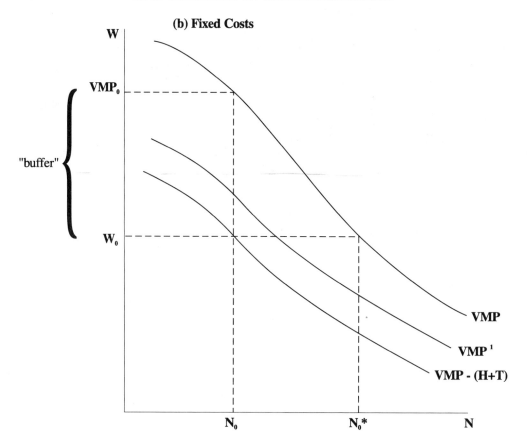

evidence exists that hiring and training costs are significantly larger for high-skilled workers. Thus they have a larger "buffer" (excess of value of marginal product over wage rate) protecting them from unanticipated declines in demand compared to unskilled and semiskilled workers. This prediction is illustrated by comparing panels (a) and (b) of Figure 10.2, the former showing the situation facing unskilled workers and the latter that for skilled workers.

Quasi-fixed costs may also foster the segmentation of labour markets into a protected sector (where employment stability is relatively guaranteed so as to amortize the quasi-fixed costs) and a sector that cannot compete for these jobs because firms would have to incur additional fixed costs upon hiring such workers. In essence, once an employee has been hired and trained, the firm is not indifferent between an existing employee and an otherwise identical potential employee.

EXHIBIT 10.2 NONWAGE BENEFITS AND PART-TIME WORK

Part-time employment, defined as working less than 30 hours per week, has grown rapidly in Canada during the past several decades, both in absolute terms and relative to full-time employment. For example, part-time employment constituted less than 4 percent of total employment in 1953, but has constituted 15 percent of total employment since the early 1980s.

What factors lie behind the growth of part-time and other forms of "flexible" or "nonstandard" employment, such as short-term and contract work, work within the temporary-help industry, and certain types of self-employment? The service sector accounts for much part-time employment, and the reasons for the growth of part-time work are interrelated with those accounting for the increase in service sector employment (Economic Council of Canada, 1991). Most part-time employees are women or young workers. Thus, supply side factors associated with the changing composition of the labour force — particularly the increased participation of women and youths — may also have played a role. However, evidence suggests that demand side factors also contributed to the rise in part-time employment. In particular, wages of part-time workers are lower than those of comparable full-time workers. The unadjusted earnings differential between part-time and full-time workers is 25 to 30 percent (Simpson, 1986; Economic Council of Canada, 1991, Chapter 5); adjusting for differences in productivity-related characteristics reduces this differential to about 10 percent (Simpson, 1986). Perhaps more important, hourly nonwage labour costs are generally substantially lower for part-time workers, due to their lower likelihood of being covered by unemployment insurance, workers' compensation, the Canada and Quebec Pension Plans, their much lower levels of employer-sponsored benefits, and their exclusion from some labour standards legislation (e.g., regulations requiring payment for statutory holidays). Because such non-wage labour costs have been growing rapidly, we would expect an increased tendency for employers to substitute part-time for full-time workers.

Evidence that such substitution has occurred in the U.K. is provided in an econometric study by Patricia Rice (1990). Changes in the structure of National Insurance payroll taxes increased significantly the average hourly cost of full-time youth workers, while reducing that of part-time adult workers. In most industries part-time adults and full-time youths were found to be substitutes in production. The consequence of these tax changes was a rise in part-time adult employment and a sharp decline in youth employment.

An important policy issue is whether employers should be required to provide the same benefits (on a prorated basis) to part-time workers as provided to full-time employers, as recommended by the Commission of Inquiry into Part-time Work (Labour Canada, 1983). Such a requirement would reduce the substantial gap in total compensation that now exists between those with "standard" full-time jobs and those with "nonstandard" jobs. However, this requirement would also encourage employers to substitute away from part-time employment. Women and youths would be the groups most likely to be affected by this policy. Whether these groups would be better or worse off would depend on whether the positive effects of higher nonwage benefits exceeded the negative consequences associated with reduced employment opportunities.

Sources: Economic Council of Canada (1991); Labour Canada (1983); Simpson (1986); Rice (1990).

To the extent that fixed costs inhibit the hiring of new workers (at the extensive margin), employers may try to meet their labour demand needs by expanding not only at the intensive margin of more hours, but also at the intensive margin of a greater *intensity* of work effort. That is, employers may try to work their existing workforce not only longer hours but also harder, rather than hiring new workers. Just as labour supply had quality and quantity dimensions so does labour demand, and these various dimensions may be affected by relative prices including fixed costs.

Quasi-fixed costs may also explain some of the employers' resistance to engage in worksharing practices. For example, having all of the companies' employees work a four-day as opposed to five-day work week, and hiring twenty percent more employees at the same wage, would increase labour costs if there were quasi-fixed costs associated with hiring the new employees. Similarly, having two workers engage in job sharing (e.g., by each working half of the week) would increase the quasi-fixed costs because of the additional employee. Such costs may also make it more expensive to engage part-time employees unless there are compensating offsets in the wage component or in some of the quasi-fixed costs like nonwage benefits.

WORKSHARING AND JOB CREATION

With the recent high levels of unemployment, increasing attention has been paid to worksharing as a way to create jobs, or at least to share the scarce number of jobs that seem to be available. This emphasis has been buttressed by workers' preferences for flexible worktime arrangements, in part because of the growth of the two-earner family and the changing age and sex composition of the work force.

Reduced work time could come in a variety of forms including delayed entry into the labour force, early retirement, a shorter work week, reduced overtime, or increased vacation time. The key question is: would any of these reductions in the worktime of some individuals lead to new jobs for other individuals?

Overtime Restrictions

In recent years the question has become extremely important with respect to reductions in the work week especially through restricting the use of overtime. The paradox of some workers working overtime, seemingly on a regular basis, while other workers in the same community and often in the same plant are on layoff, has led to increasing pressure to restrict the use of overtime. This restriction could come in a variety of forms: reducing the standard work week beyond which an overtime premium must be paid, increasing the overtime premium, establishing or reducing the statutory maximum hours that are allowed, making it more difficult to get special permission to exceed the statutory limit when exemptions are allowed, or granting workers the right to refuse overtime. Changes in any or all of these features of overtime legislation could be used to restrict the use of overtime.

Unfortunately we have very little empirical evidence on the extent to which changes in any of these features of overtime legislation would change the demand for overtime hours. Ehrenberg and Schumann (1982) provide empirical evidence based on United States data indicating that an increase in the overtime premium from time-and-one-half to double-time would reduce the use of overtime hours. However, the relationship was not overwhelmingly conclusive, being statistically insignificant for the manufacturing sector.

Overtime Restriction and Job Creation

Even if overtime hours can be reduced by policy changes like raising the overtime premium, there remains the question of the extent to which reductions in the use of overtime would translate into new jobs. On this issue there is also considerable controversy. Ehrenberg and Schumann (1982, p. 15) estimate that the reduction in the use of overtime that would result from a double-time premium could translate into an employment increase in the neighbourhood of three to four percent, assuming that the overtime hours reduction would be converted directly to new full-time positions. They indicate that this is an upper bound, however, because it ignores a number of the following adjustments that are likely to occur in response to increases in the overtime premium.

For example, firms may not comply fully with the legislation and some workers whose overtime hours were restricted may moonlight. More importantly, although firms may substitute new employees for the more expensive overtime hours, they would also reduce their *overall* demand for labour because it is now more expensive and because the demand for the firm's output will be reduced as it passes some of the cost increase on to consumers in the form of higher product prices. These are, respectively, the familiar substitution and scale effects that occur in response to an increase in the price of a factor of production.

As well, there may be productivity changes that affect the number of new jobs created if hours of work are reduced. If fatigue effects make overtime hours less productive than the hours of work of new workers or rehires, then the number of new jobs created will be diminished because the overtime hours can be replaced by fewer regular hours. Reid (1985, p. 163), for example, cites European evidence suggesting that the productivity improvements associated with a reduced work week may reduce the employment-creating potential of a reduced work week by one-half. That is, productivity would be sufficiently higher under the shorter work week so that the same output could be maintained with half as many hours in the new jobs to replace the reduced hours of the shorter work week.

Whether this seemingly large productivity offset would be permanent and apply to reductions in overtime hours (as opposed to the standard work week of everyone) remains an open question. Overtime hours may be more susceptible to fatigue effects but workers who work the overtime may already be the most productive of all workers; that is in fact why they are asked to work overtime. If that is the case, then reducing the overtime hours of the most productive workers may have the potential for creating a larger number of jobs for less productive workers, albeit at an obvious cost — a cost that could reduce the job-creating potential because of the substitution and scale effects discussed previously.

The job-creating potential of a reduction of overtime hours also may be offset by the fact that persons who are unemployed or on layoff (who are likely to be less skilled) may not be good substitutes for persons working overtime (who are likely to be more skilled). Unfortunately, the empirical literature that provides evidence on the substitutability between factors of production does not provide information on that point. That literature provides evidence on the substitutability between employment and hours of work of the labour force in general. And it provides some evidence on the substitutability between different groups of labour (e.g., skilled and unskilled, or females and youths). However, it does not do both simultaneously. Formally stated, we do not have estimates of the elasticity of substitution between overtime *hours* of skilled workers and the *employment* of less-skilled workers. In the absence of such information it is difficult to know by how much the job creation potential of reduced overtime hours would be offset by the fact that additional hours of new recruits or even those on layoff may not be good substitutes for reduced hours of those who work overtime.

The job-creating potential of reduced hours of work is further complicated by the fact that in the long run there are not really a fixed number of jobs in the economy. In fact, the notion that there is a fixed number of jobs is known as the "lump-of-labour fallacy." It is a fallacy because it ignores the fact that people who hold jobs have other effects on the economy (known as general equilibrium effects). For example, they may increase aggregate demand through their increased disposable income and they may reduce aggregate wages augmenting the supply of labour. Both of these effects would enhance employment by these groups being in the labour market.

Thus, while an individual who takes a job or works long hours at a particular establishment may be taking a job from someone else who could work at that establishment, this need not be the case for the economy as a whole. What is true for a given establishment may be a fallacy for the economy as a whole. Perhaps this is best illustrated by the example of the increased labour force participation of females. Few analysts today would say that every job taken by a female means one less job for a male. Even though that may be true for a particular job in a particular establishment (just as a job taken by a male means that particular job is not occupied by a female) it is not true for the economy as a whole. Females working in the labour market also create demands for other goods and services, especially those formerly produced in the household. This in turn creates jobs elsewhere.

In essence, a job held by one person does not mean that another person does not have a job. This applies to jobs held by females, by older workers who do not retire, or by persons who work long hours or overtime. Conversely, restricting their work in particular establishments (e.g., by mandatory retirement or by restricting overtime) may open up some new jobs in that establishment but this may not lead to more jobs in the economy as a whole when one considers the general equilibrium effects as previously discussed.

For all of these reasons, the job-creating potential of various forms of work-sharing is regarded with some scepticism. This is compounded by the fact that some regard worksharing basically as unemployment sharing, with the fear that any reduced measured unemployment that occurs may take the pressure off governments to follow appropriate policies to reduce unemployment. Nevertheless, as Reid (1985) points out, in times of cyclically high unemployment, worksharing may be a viable temporary method to share scarce jobs until the economy returns to full employment. In addition, it would be important to remove any impediments that may prevent the parties from entering into otherwise mutually beneficial voluntary worksharing arrangements.

Reducing Barriers to Employment Sharing

Reid (1985, p. 163) suggests a number of changes in public policy that could reduce the unintended barriers to employment sharing that are created by these policies. As indicated previously, the ceilings on payroll taxes for workers' compensation, public pensions, or unemployment insurance make it more attractive for employers to work their existing employees longer hours once the ceiling on the payroll tax has been reached, since no further taxes then have to be paid. Having the premiums based on hourly earnings with a ceiling based on hourly earnings would remove the bias against hiring new workers. Similarly, prorating employer contributions to medical, dental, and life insurance plans according to hours worked, rather than having fixed contributions per employee, would reduce their existing bias against employment sharing.

QUESTIONS

1. Explain the factors that affect the firm's choice between the number of employees and hours per employee.

2. Firms often respond to decreases in demand by laying off some of their work force. However, some groups of workers are more likely to be laid off than others. Provide an explanation for this phenomenon.

3. Explain how payroll taxes such as those used to help finance unemployment insurance, the Canada and Quebec Pension Plans, and workers' compensation may affect the firm's choice between workers and hours. How could these taxes be structured in order not to affect this choice?

4. Would it ever be rational for a firm to retain an employee whose current marginal revenue product is less than their current wage? Explain.

5. "In the absence of hiring, training, and other fixed costs of employment, the firm does not need to forecast the future when making employment decisions. However, in the presence of these fixed costs the firm's employment decision necessarily involves planning for and forecasting of the future." Discuss.

6. Many European countries have legislation regarding severance pay which makes termination of employees costly to employers. What impact would introducing such legislation have on employer behaviour relating to dismissals and new hires? What groups in society are likely to benefit from introducing this type of legislation? What groups are likely to lose?

REFERENCES AND FURTHER READINGS

Abraham, K. and S. Houseman. Job security and work force adjustment: how different are U.S. and Japanese practices? *Journal of Japanese and International Economies* 3 (1989) 500–521.

Archibald, G.C. and P.T. Chinloy. Job security versus income security. *Labour Market Adjustments in the Pacific Basin*, P.T. Chinloy and E. Stromsdorfer (eds.). Boston: Kluwer-Nijhoff Publishing, 1987.

Bell, D. Labour utilization and statutory non-wage cost. *Economica* 49 (August 1982) 335–344.

Bentolila, S. and G. Bertola. Firing costs and labour demand: how bad is eurosclerosis. *R.E. Studies* 57 (July 1990) 381–402.

Bertola, G. Job security, employment and wages. *European Economic Review* 34 (June 1990) 851–886.

Booth, A. and F. Schiantarelli. The employment effects of a shorter working week. *Economica* 54 (May 1987) 237–248.

Brunello, G. The employment effects of shorter working hours: an application to Japanese data. *Economica* 56 (November 1989) 473–486.

Burgess, M. and J. Dolado. Intertemporal rules with variable speed of adjustment. *EJ* 99 (1989) 347–365.

Commission of Inquiry into Part-time Work. *Part-time Work in Canada*. Ottawa: Labour Canada, 1983.

Dunn, L.F. An empirical study of labour market equilibrium under working hours constraints. *R.E. Stats.* 2 (May 1990) 250–258.

Dye, R.A. Self-selection via fringe benefits. *JOLE* 2 (July 1984) 388–411.

Economic Council of Canada. *Employment in the Service Economy.* Ottawa: Economic Council of Canada, 1991.

Ehrenberg, R. Absenteeism and the overtime decision. *AER* 60 (June 1970) 352–357.

Ehrenberg, R. Heterogeneous labour, the internal labour market and the dynamics of the employment-hours decision. *Journal of Economic Theory* 3 (March 1971) 85–104.

Ehrenberg, R. *Fringe Benefits and Overtime Behavior: Theoretical and Economic Analysis.* Lexington, Mass.: D.C. Heath, 1972.

Ehrenberg, R.G. and P.L. Schumann. *Longer Hours or More Jobs? An Investigation of Amending Hours Legislation to Create Employment.* Ithaca: New York State School of Industrial and Labor Relations, Cornell University, 1982.

Ehrenberg, R.G., Rosenberg, P. and J. Li. Part-time employment in the United States. *Employment, Unemployment and Labour Utilization*, Robert A. Hart (ed.). Boston: Unwin Hyman, 1988.

Emerson, M. Regulation or deregulation of the labour market. Policy regimes for the recruitment and dismissal of employees in the industrialized countries. *European Economic Review* 32 (April 1988) 775–818.

Fitzroy, F. and R. Hart. Hours, layoffs and unemployment insurance funding: theory and practice in an international perspective. *EJ* 95 (September 1985) 700–713.

Garbarino, J. Fringe benefits and overtime as barriers to expanding employment. *ILRR* 17 (April 1964) 426–442.

Hamermesh, D. Factor market dynamics and the incidence of taxes and subsidies. *QJE* 95 (December 1980) 751–764.

Hamermesh, D. Labour demand and the structure of adjustment costs. *AER.* 79 (September 1989) 674–689.

Hart, R.A. The employment and hours effects of a marginal employment subsidy. *Scottish Journal of Political Economy* 36 (November 1989) 385–395.

Hart, R. *The Economics of Non-Wage Labor Costs.* London: Allen and Unwin, 1984.

Hart, R. *Working Time and Employment.* London: Allen and Unwin, 1986.

Hoel, M. Employment and allocation effects of reducing the length of the workday. *Economica* 53 (February 1986) 75–85.

Kato, T. Bumping, layoffs and work-sharing. *EI* 24 (October 1986) 657–668.

Kerachsky, S. et. al. Work sharing programs: an evaluation of their use. *MLR* 109 (May 1986) 31–32.

Krueger, A.B. The evolution of unjust-dismissal legislation in the United States. *ILRR* 44 (July 1991) 644–660.

Laudadio, L. and M. Percy. Some evidence in the impact of non-wage labour costs on overtime work. *RI/IR* 28 (March 1973) 397–403.

Lazear, E.P. Job security provisions and employment. *QJE* 105 (August 1990) 699–726.

Leckie, N. and C. Caron. On non-wage labour income. *Perspectives on Labour and Income*. (Winter 1991) 47–55.

Mann, B. Capital heterogeneity, capital utilization and the demand for shiftworkers. *CJE* 17 (August 1984) 450–470.

Meltz, N., F. Reid and G. Swartz. *Sharing the Work: An Analysis of the Issues in Worksharing and Jobsharing*. Toronto: University of Toronto Press, 1981.

Morrison, C.J. Quasi-fixed inputs in U.S. and Japanese manufacturing: a generalized Leontief restricted cost function approach. *R.E. Stats.* 70 (May 1988) 275–287.

Nickell, S. Fixed costs, employment and labour demand over the cycle. *Economica* 45 (November 1978) 329–345.

Nickell, S. Unemployment and the structure of labour costs. *Carnegie-Rochester Conference Series on Public Policy* 11 (1979) 187–222.

Nickell, S. An investigation of the determinants of manufacturing employment in the United Kingdom. *R.E. Studies* 51 (1984) 529–557.

Oi, W. Labour as a quasi-fixed factor. *JPE* 70 (December 1962) 538–555.

Pold, H. and F. Wong. The price of labour. *Perspectives on Labour and Income* (Autumn 1990) 42–49.

Reid, F. Combatting unemployment through work time reductions. *CPP* 12 (June 1986) 275–285.

Reid, F. UI-assisted worksharing as an alternative to layoffs: the Canadian experience. *ILRR* 35 (April 1982) 319–329.

Reid, F. Reductions in work time: an assessment of employment sharing to reduce unemployment. *Work and Pay: The Canadian Labour Market*, W.C. Riddell (ed.) Toronto: University of Toronto Press, 1985.

Rice, P.G. Relative labour costs and the growth of part-time employment in British manufacturing industries. *EJ* 100 (December 1990) 1138–1146.

Rice, P.G. Juvenile unemployment, relative wages and social security in Great Britain. *EJ* 96 (1986) 352–374.

Rosen, S. Short-run employment variation in class I railroads in the U.S., 1937–1964. *Econometrica* 36 (October 1968) 511–529.

Shapiro, M. The dynamic demand for capital and labor. *QJE* 10 (1986) 513–542.

Simpson, W. Analysis of part-time pay in Canada. *CJE* 19 (November 1986) 798–807.

Symposium. Europeanizing American issues: New World jobs and Old World policy. *JLR* 10 (Winter 1989).

PART 3

Equilibrium in Nonunion Labour Markets

Now that the determinants of labour demand and labour supply are understood, we can analyze their interaction in determining the equilibrium price and quantity of labour — wages and employment respectively. Part 3 examines the interaction of labour demand and supply and the resulting labour market equilibrium under various circumstances.

Chapter 11 deals with equilibrium in competitive labour markets. As discussed previously in the development of labour demand, these are markets in which the employer is small relative to the size of the labour market. The employer is a "wage-taker" and employs labour at the prevailing market wage, which is determined by the interaction of demand and supply at the aggregate market level. Because the wage rate is given to the firm, it need only decide on the quantity of labour to employ at the prevailing market wage. However, the employer may possess market power in the product market; for example, the firm may be a monopolist or oligopolist in the product market.

Chapter 12 deals with equilibrium in situations in which the employer has some market power in the labour market. The employer is a "wage-setter" rather than a wage-taker. The polar case of monopsony occurs when the firm is the sole employer of that type of labour, analogous to monopoly in the product market where the firm is the sole supplier of that particular product. When the firm is a monopsonist in the labour market, it must choose both the wage rate and the amount of labour to employ.

The analysis in Chapters 11 and 12 abstracts from a number of phenomena, such as imperfect information, risk and uncertainty, and the impact of incentives on worker productivity. However, numerous aspects of labour market behaviour cannot be adequately understood without taking account of these phenomena. Indeed, much of what makes the labour market differ from other markets comes about because of the influence of incentives, risk, and imperfect information. Chapter 13 provides an introductory analysis of the role of imperfect information, risk and uncertainty, and incentives on the behaviour of employers and employees and on labour market equilibrium. The specific topics examined in that chapter include efficiency wages, which are associated with worker incentives; implicit contracts, which are associated with risk and uncertainty; and signalling and search, which are associated with imperfect information. This analysis provides the foundation for the study of dual labour markets (also examined in Chapter 13), as well as several other aspects of labour market behaviour examined in later parts of the book.

CHAPTER 11

Wages and Employment in Competitive Labour Markets

In this chapter we analyze the interaction of labour supply and demand when the firm is competitive in the labour market in which it purchases labour inputs. The firm may be operating in a competitive or noncompetitive (monopoly, oligopoly, monopolistic competition) product market. In the next chapter we analyze the situation when the firm is not competitive (i.e., possesses some market power) in the labour market. Throughout the analysis we are assuming that workers are selling their services on an individual basis; in Part 4 we will analyze the situation of collective bargaining via unionization.

In dealing with the interaction of supply and demand in various market structures, it is important to be specific about the level of aggregation that is being analyzed. In this section we begin at the level of aggregation of the individual firm, and consequently focus on the firm as the decision-making unit. Subsequently we deal with higher levels of aggregation such as the occupation, industry, region, and economy as a whole — the levels of aggregation at which the market wage is determined in competitive labour markets.

Throughout this chapter we maintain several important assumptions: (1) both employers and employees are fully informed about market opportunities; (2) there is absence of risk and uncertainty, or the ability to purchase insurance against all risks; and (3) workers are homogeneous with known productivity which is determined by the technology of production and is independent of the wage paid to the worker. Each of these assumptions is relaxed subsequently, beginning in Chapter 13. However, dealing with these issues here would unnecessarily complicate the topics addressed in this chapter.

COMPETITIVE PRODUCT AND LABOUR MARKETS

We first examine the case in which the firm is a competitive seller of its output on the product market and a competitive buyer of labour in the labour market.

In essence, the firm is so small relative to both markets that it can sell all of the
output it wants at the going price of the product, and it can buy all of the labour
it wants at the going wage rate. The firm is both a price- and wage-taker: it
cannot influence either the product price or the wage rate.

This situation is depicted in Figure 11.1(a) and (b) for two competitive firms.
In both cases, their supply schedules for a given homogeneous type of labour are
perfectly elastic (horizontal) at the market wage rate, w_c. The firms are wage-
takers not wage-setters, and consequently can employ all of the labour they want
at this market wage rate.

The market wage rate for this specific, homogeneous type of labour is determined by the interaction of supply and demand in the aggregate labour market
as depicted in Figure 11.1(c). This aggregate labour market could be a regional
labour market for a particular occupational category of labour. For example, it
could be the Halifax labour market for junior accounting clerks. By assuming
that the firms are in the same region and are hiring the same homogeneous
occupational type of labour, we are able to minimize the intervening influence
of these factors on the wage determination process, and to thereby focus on the
issue of wage and employment determination at the disaggregate or microeconomic level of the firm. In the subsequent chapters on wage structures, these
assumptions are relaxed sequentially and the resultant wage structures analyzed.

The demand schedules for labour in the two firms are simply the schedules of
the value of the marginal products of labour, defined as the marginal physical
product of labour times the price at which the firms can sell their products.
(These were derived formally in Chapter 9.) Since the firms are assumed to be
competitive sellers of their products, their product prices are fixed at p_1° and
p_2°. Only if the firms are selling the same output would their product prices have

Figure 11.1
COMPETITIVE PRODUCT AND LABOUR MARKETS

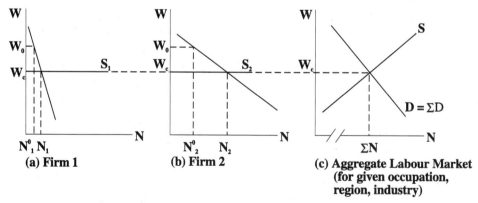

(a) Firm 1   (b) Firm 2   (c) Aggregate Labour Market
                                            (for given occupation,
                                            region, industry)

to be the same; otherwise p_1^* need not equal p_2^*. The magnitude and the elasticity of the demand for labour also are depicted as being different simply to emphasize that the market wage is the same irrespective of these factors. The demand schedules simply determine the level of employment in each firm: in this case N_1 and N_2 units of labour, respectively in firms one and two.

The market demand curve, as depicted in Figure 11.1(c), is simply the summation of the demand curves of the individual firms, such as those shown in Figures 11.1(a) and 11.1(b). Conceptually the market demand curve can be obtained as follows. For any specific wage rate, such as w_o in Figure 11.1, determine the quantity of labour that each firm in the market would wish to employ. For the two firms depicted in Figures 11.1(a) and 11.1(b), these quantities are N_1^o and N_2^o respectively. Adding these quantities gives the total market demand at that wage rate. Repeating this process for all wage rates traces out the market labour demand curve.

In summary, when the firm is a competitive buyer of labour, it faces a perfectly elastic supply of labour at the market wage. When the firm is a competitive seller of its output on the product market, it regards the price at which it sells its output on the product market as fixed, and its derived demand for labour schedule is the value of the marginal product of labour, defined as the marginal physical product of labour times the fixed price at which the firm sells its output. Because the labour supply schedule to the firm is perfectly elastic at the market wage rate, the intersection of the firm's labour supply and demand schedules determines the employment level of the firm for that particular type of labour — wages are determined elsewhere, specifically in the aggregate labour market for that particular type of labour.

SHORT RUN VERSUS LONG RUN

The previous analysis was based on the long-run assumption that the firm could get all of the labour it needed at the market-determined wage rate. That is, in order to expand its work force it need only hire additional workers at the going wage rate: there is no need to increase wages to attract additional workers.

In the short run, however, even a firm that is competitive in the labour market may have to raise its wages in order to attract additional workers. This situation, in the literature, is often referred to as "dynamic monopsony" (Baily 1975). In such circumstances the firm's short-run labour supply schedule could be upward-sloping, as depicted by the schedule S_s in Figure 11.2. In the short run, in order to expand its work force so as to meet an increase in the demand for labour from D to D', the firm may have to pay higher wages, perhaps by paying an overtime premium to its existing work force or by paying higher wages to attract local workers within the community. The resultant expansion of the work force can be depicted as a movement up the short-run supply curve in response to the

Figure 11.2
SHORT-RUN AND LONG-RUN LABOUR SUPPLY
SCHEDULES TO FIRM

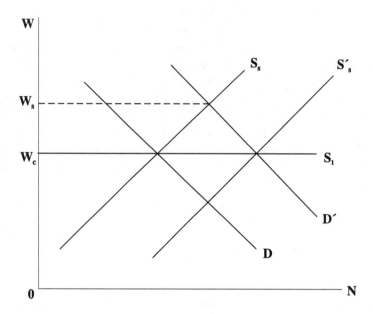

new higher wage of w_s, occasioned by the increase in the demand for labour from D to D′.

In the longer run, however, a supply influx of other workers will be forthcoming because the firm is paying an above-market wage (i.e., $w_s > w_c$) for that particular type of labour. The supply influx may not be instantaneous, but may occur in the long run because it may come from other firms or perhaps from outside of the labour force, and such adjustments take time.

The new supply influx in response to the higher wage could be depicted by the S_s′ supply schedule of labour. The supply influx would depress the temporarily high, short-run wage of w_s back to its long-run level of w_c. Thus the long-run supply of labour schedule to the firm, S_1, can be thought of as a locus of long-run equilibrium points, traced out by various shifting short-run supply schedules of the firm, as the firm tries to expand its work force.

In essence, temporary wage increases above the competitive norm are consistent with the firm being a competitive buyer of labour in the long run. In fact, short-run wage increases can be a market signal for the supply response that ensures that market forces operate in the longer run.

MONOPOLY

When an industry is competitive in the product market, the industry demand for labour is obtained by aggregating the labour demand of each of the firms in the industry, as illustrated in Figure 11.1. In the case of monopoly in the product market, the firm *is* the industry, and therefore there is no need to distinguish between the firm's labour demand and that of the industry as a whole.

As discussed in Chapter 9, the structure of the product market has implications for the derived demand for labour. As given by Equation 9.7 in Chapter 9, the labour demand schedule for the competitive firm on the product market is given by:

$$w^* = MPP_N \cdot p_Q^* = VMP_N \text{ [competitor]} \tag{9.7}$$

This was derived from the more general equation

$$w^* = MPP_N \cdot MR_Q = MRP_N \text{ [monopolist]} \tag{9.6}$$

which applies to the monopolist since its output price is not fixed. Rather, for the monopolist, marginal revenue is the relevant factor.

The difference between Equations 9.7 and 9.6 highlights the fact that when the monopolist hires more labour to produce more output, not only does the marginal physical product of labour fall (as is the case with the competitor), but also the marginal revenue from an additional unit of output, MR_Q, falls. This latter effect occurs because the monopolist, unlike the competitive firm, can sell more output only by lowering the product price and this, in turn, lowers marginal revenue. Because both MPP_N and MR_Q fall when N increases in Equation 9.6, then the monopolist's demand for labour falls faster than it would if it behaved as a competitive firm in the product market, in which case only MPP_N would fall, as given in Equation 9.7.

The difference between the labour demand schedule for a monopolist and the schedule that would prevail if the industry were competitive in the product market is illustrated in Figure 11.3. This comparison is most meaningful if it involves two situations that are identical except for the difference in market structure. To carry out the comparison, begin with a large number of price-taking firms. Aggregating the labour demand of each of the firms in this competitive industry gives the industry demand curve $D_c = \Sigma VMP_N$ in Figure 11.3. Now suppose these firms form a cartel and set the product price to maximize total industry profit, as would a monopolist which owned all of the firms in the industry. The labour demand schedule for the monopolist is $D_m = MPP_N \cdot MR_Q = MRP_N$. Since $MR_Q < P_Q$ then D_m lies below D_c.

The monopolist (or a cartel of firms acting like a single monopolist) raises the industry price relative to the competitive equilibrium level which reduces output and employment. Thus, at any particular wage (e.g., w^* in Figure 11.3),

Figure 11.3
MONOPOLIST VS. COMPETITIVE DEMAND FOR LABOUR

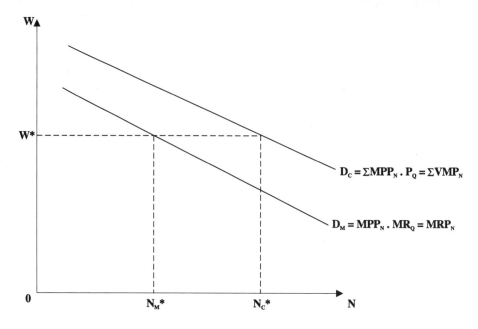

employment will be lower if the industry is monopolized than in an otherwise identical competitive industry.

PRODUCT MARKET STRUCTURE AND DEPARTURES FROM MARKET WAGES

In theory, the previous analysis has little implication for the wages that monopolists will pay relative to the wages paid by firms that are competitive in the product market. This is so because the firm's behaviour in the product market need not be related to the way the firm behaves in the labour market. Specifically, if the firms are competitive in the labour market then they all face a perfectly elastic supply schedule for labour in the labour market; that is, they are wage-takers, not wage-setters. In such circumstances firms facing different product market structures pay the same market wage, as illustrated in Figure 11.3.

There are reasons to believe, however, that some firms may pay wages somewhat above or below the equilibrium market wage even though the labour market is approximately competitive in nature. Some of these factors are related to product market structure; that is, they serve to imply that monopolists will pay more or less then the market wage. Other factors are unrelated to product market structure; these reasons for departures from market wages are discussed in a subsequent section of this chapter.

In terms of aspects related to the product market, perhaps the main factor suggesting that the monopolist would pay greater-than-market wages is that the monopolist earns monopoly profits, and labour — especially if organized — may be able to appropriate some of these profits. Alternatively stated, the monopolist may be under less pressure to be cost-conscious and hence may more readily yield to wage demands. Being sensitive to their public image, monopolists may also pay higher wages as a way of buying the image of being a good employer. This may especially be the case if the monopolist is regulated and subject to the review of a regulatory agency.

By virtue of simply being a large firm (even though it may be a small employer of certain types of labour), monopolists may also pay high wages. This could be the case, for example, if large firms must use formal methods to evaluate job applicants and hence they pay a high wage so as to have a queue of applicants from which to formally evaluate. In addition, large firms may have to pay a wage premium to compensate for the rigid work schedules and mass production techniques associated with larger size or to substitute for costly monitoring. Finally, by virtue of their size, large firms may have to follow administratively determined wage policies: unable to pay each worker its marginal product, such firms may pay wages approximating the productivity of the more productive workers within the group. In this way they minimize the risk of losing their more productive workers at a cost of having to pay slightly higher than average wages.

The previously discussed factors suggest why monopolists may pay high wages if they are competitive buyers of labour. Each of these factors, however, could also be associated with lower-than-average wages being paid by monopolists. For example, although monopoly profits could enable the monopolist to pay high wages, they could also enable them to resist wage demands, perhaps by holding out longer against strike activity. In terms of preserving its public image, especially if regulated, monopolists may want to be model employers by paying high wages, but they may also want to appear as cost-conscious to the consuming public and to regulatory agencies. By virtue of their size, monopolists may also be able to pay lower-than-average wages, for example, if their size provided employment security or ample opportunities for advancement, or if turnover was less costly to them because of the availability of standby work forces.

Clearly the various factors that affect the wage determination process do not yield unambiguous predictions on whether a firm will pay above- or below-average wages. Consequently, as is so often the case in economics, we must appeal to the empirical evidence to ascertain the net impact of the variety of factors. This evidence will be examined in a subsequent chapter on wage structures where factors such as the degree of concentration in the product market, the threat of unionization, firm size and the costs of turnover, are viewed as determinants of wages.

MONOPOLISTIC COMPETITION AND OLIGOPOLY

In between the polar cases of competition and monopoly in the product market are a variety of intermediate cases. Firms can be monopolistically competitive, a situation characterized by many firms which are small relative to the total market, but with products that are differentiated in some way, giving the firm some discretion in its price-setting. In such circumstances the demand for the firm's product is not perfectly elastic, as in the competitive case, but rather has some degree of inelasticity reflecting the fact that if the firm raises its price it will not lose all of its market, and if it lowers its price it will not gain all of the market. Under monopolistic competition, as is the case in perfect competition, there are no barriers to entry by new firms. This "free entry" property implies that firms cannot earn above-normal or monopoly profits in the long run.

Oligopoly industries are characterized by few firms that produce sufficiently similar products that the actions of one firm will affect the other firms. Consequently the firms will react to the actions of the other firms, and will take into account the possible reactions of their rivals in making their own decisions. There are many ways in which the firms can react; consequently, there is a large number of possible ways to categorize oligopoly situations. Oligopoly industries are generally characterized by some barriers to entry by new firms, so that above-normal profits may be earned by oligopolists in the long run.

The general conclusions reached in the previous analysis of perfect competition and monopoly continue to apply to the intermediate cases of monopolistic competition and oligopoly. In particular, market power in the product market is consistent with the firm being competitive in the labour market (or labour markets) in which it operates. Thus firms that exercise some discretion, or even exert considerable control, over the product price may be wage-takers in the labour market. In these circumstances, they would pay the market wage for the types of labour they employ, and could increase or decrease their employment without affecting the prevailing market wage.

However, several of the factors that may cause monopolists to depart from the market wage may also apply to oligopolists. Like monopolists, firms in an oligopoly typically earn above-normal profits that are not competed away by new entrants. Workers may succeed in capturing some of these "economic rents" — especially if the firm's management is satisfied with less than maximum possible profit, perhaps because of a separation of ownership and control. In these circumstances, workers benefit from a less cost-conscious management and shareholders earn a lower return than they would if the firm paid the market wage, albeit still a greater return than is available in competitive industries. Similarly, oligopolists, like monopolists, are often large firms and they may pay above-market wages for reasons relating to their size.

Under monopolistic competition, free entry implies that firms should not earn above-normal profits in the long run (though they may in the short run, as would occur if there were an unexpected increase in demand for the product). Such firms are also generally small in size, such as retail outlets that are differentiated by location and possibly also by the merchandise carried. Thus, the two characteristics that might cause otherwise wage-taking firms to pay above-market wages — economic rents and large size — are absent in the case of monopolistic competition, as they are in the case of perfect competition. For these reasons we would expect firms that are monopolistically competitive in the product market but perfectly competitive in the labour market to pay the prevailing market wage, over which changes in their employment levels will exert no influence.

Finally, it should be noted that this discussion applies to situations in which the product and labour markets are in equilibrium. Product market structure may also have implications for disequilibrium dynamics — the direction and speed of response to changes in market conditions. For example, because they have some discretion in price-setting, monopolistically competitive firms may respond differently to pressures for wage increases than would perfectly competitive firms, which are not able to pass on wage increases to customers in the form of price increases. Similarly, the response of an oligopolistic firm to a wage increase may depend on how its rivals are expected to react to an increase in the product price.

CONCLUDING OBSERVATIONS ON NONCOMPETITIVE PRODUCT MARKETS

Clearly the way in which a firm behaves in the product market can affect the way it behaves in the labour market. Since the firm's demand for labour is derived from the demand for the product or service produced by the firm, then whether the firm is a price-setter or a price-taker can influence its employment and *possibly* wage determination.

The term *possibly* is used with respect to the wage determination process because as long as the firm is competitive in the labour market — and it can be competitive in the labour market and not competitive in the product market — then it would have to pay at least the going wage for labour. This fact is important because it suggests that there may be an upward bias towards higher wages in firms that are not competitive in the product market. This upward bias occurs because noncompetitive forces in the product market usually have an indeterminate impact on the wages they would pay to their employees. However, the forces of competition in the labour market would ensure that noncompetitive firms in the product market do not pay below the competitive wage. These same forces, however, may be ignored by noncompetitive firms that pay greater-than-market wages, perhaps out of monopoly or oligopolistic profits.

For example, if a monopolist did not pay the market wage, it might not be able to recruit any work force. Consequently, the forces of competition in the labour market ensure a floor on wages paid by the monopolist. However, if the monopolist paid a greater-than-market wage out of monopoly profits, the same forces of competition need not ensure a ceiling; that is, the excess supplies of applicants could be ignored by the monopolist, who may be under less pressure to be cost conscious. Because competitive pressures in the labour market ensure a floor, but not necessarily a ceiling on wages paid in noncompetitive sectors of the product market, there may be an upward bias to wages paid in these sectors.

QUESTIONS

1. Monopoly and monopsony go hand in hand. Discuss.

2. "Given the variety of ways of classifying the firm's behaviour in the product and labour markets, and the various possible types of collusive action on the part of labour, we really have no theory of labour market behaviour. Rather, we have a series of explanations corresponding to the variety of ways employers and employees can act in the labour market. This is more description than good theory." Discuss.

3. Based on the demand schedules as given in Figure 11.1, what would happen to the respective labour demand schedules for firms 1 and 2 if they both became monopolists in their respective product markets? What would happen to the wages they pay? What would happen to the labour demand schedule in the aggregate labour market, and hence the wage and employment in that market?

4. For Figure 11.1, make up a hypothetical example for firms 1 and 2 with their differing labour demand schedules and indicate the reasons for the shape of their particular schedules.

5. What would happen to the analysis of this chapter if firm 2 in Figure 11.1 began to dominate the aggregate labour market and became a wage-setter rather than a wage-taker?

6. Strictly speaking, the analysis depicted in the diagram of Figure 11.1 is incorrect since the supply-of-labour schedule would not be the firm's relevant decision-making schedule, if the firm could affect the wage rate. Discuss.

7. It is deceptive to say that monopolistic firms do not behave as competitive firms. Both are simply maximizing profits subject to the constraints they face. However, the monopolist simply faces different constraints and this leads to different outcomes. Discuss.

8. Since the monopolist, by definition, is the only producer of its output, consumers cannot substitute other outputs, and consequently the demand for the product produced by the monopolist is price-inelastic. This in turn makes the monopolist's derived demand for labour inelastic, which in turn

means that the monopolist will pay a greater-than-competitive wage. Discuss.

9. It is not possible to say on theoretical grounds whether monopolists will pay higher or lower wages than competitive firms. Discuss.

10. In the field of industrial organization, considerable attention has been devoted to analyzing the impact of various constraints, imposed by regulatory agencies, on the price and output decision of regulated firms, especially in the transportation, communications, and utilities industries. What are the impact of these various constraints for the wage and employment decision of such regulated firms? Check your predictions by analyzing the impact of deregulation on wages and employment.

11. We really have no theory of noncompetitive pricing. Rather, we have a series of *ad hoc* stories corresponding to all of the different possible cases between the competitive and monopolistic norms. Consequently it is more fruitful to examine the implications of these two polar norms and to treat the noncompetitive cases as gravitating towards one or other of the norms. Discuss.

12. Set up a hypothetical product demand schedule for an oligopolist whose price decrease would be met by other oligopolists, but whose price increase would not be matched, leading to a reduction in market share. Derive the marginal revenue schedule. Explain why the large drop in marginal revenue occurs where it does and discuss the implication of this drop for the firm's labour demand schedule.

13. Competition in the labour market ensures a competitive floor on wages but not necessarily a competitive ceiling. Competition in the product market is necessary to ensure a competitive ceiling. Consequently there is an upward bias to the wages paid by firms that are not competitive in the product market. Discuss.

REFERENCES AND FURTHER READINGS

(This section does not contain references to the literature on the role of non-competitive product market conditions on the determination of aggregate money wages. This is contained in the references in Chapter 26. Related references on wage structures by industry, region, occupation, and firm size are contained in Chapters 19 and 20.)

Alcian, A. and R. Kessel. Competition, monopoly and the pursuit of pecuniary gain. *Aspects of Labor Economics*. Princeton University Press, 1962.

Baily, M.N. Dynamic monopsony and structural change. *AER* 65 (June 1975) 388–349.

Brown, C. and J.L. Medoff. The employer size wage effect. *JPE* (October 1989) 1027–1059.

Brown, W., J. Hayles, B. Hughes and L. Rowe. Product and labour markets in wage determination: some Australian evidence. *BJIR* 22 (July 1984) 169–176.

Dalton, J. and E. Ford. Concentration and professional earnings in manufacturing. *ILRR* 31 (April 1978) 379–384.

Dalton, J. and E. Ford. Concentration and labor earnings in manufacturing and utilities. *ILRR* 31 (October 1977) 45–60.

Hanushek, E. Alternative models of earnings determination and labour market structures. *JHR* 16 (Spring 1981) 238–259.

Hendricks, W. Unionism, oligopoly and rigid wages. *R.E. Stats.* 63 (May 1981) 198–205.

Jones, J. and L. Laudadio. Wage differentials and market imperfection: some cross section results in Canadian manufacturing industries. *RI/IR* (No. 3, 1975) 408–421.

Kowka, J. Monopoly, plant and union effects on worker wages. *ILRR* 36 (January 1983) 251–257.

Landon, J. The effect of product market concentration on wage levels. *ILRR* 23 (January 1970) 237–247.

Lawrence, C. and R. Lawrence. Manufacturing wage dispersion: an end game interpretation. *BPEA* (No. 1, 1985) 47–106.

Levinson, H. Unionism, concentration and wage changes. *ILRR* 20 (January 1967) 198–205.

Lichtenberg, F.R. and D. Siegel. The effect of ownership changes on the employment and wages of central office and other personnel. *The Journal of Law and Economics* 33 (October 1990) 383–408.

Mishel, L. Product markets, establishment size and wage determination. *IRRA.* (December 1982) 447–454.

Mitchell, D. Shifting norms in wage determination. *BPEA* (No. 2, 1985) 575–599.

Moore, T. The beneficiaries of trucking regulation. *The Journal of Law and Economics* 21 (October 1978) 327–344.

Pusel, T. Profitability, concentration and the interindustry variation in wages. *R.E. Stats.* 62 (May 1980) 248–253.

Richards, D. Wages and the dominant firm. *JLR* 4 (Spring 1983) 177–182.

Silvestre, J. There may be unemployment when the labour market is competitive and the output market is not. *EJ* 100 (September 1990) 899–913.

Wachtel, H. and C. Betsey. Employment and low wages. *R.E. Stats.* 54 (May 1972) 121–129.

Weiss, L. Concentration and labor earnings. *AER* 56 (March 1966) 96–117.

Yandle, B. The wages of regulation. *JLR* 5 (Fall 1984) 435–439.

CHAPTER 12

Monopsony in the Labour Market

In Chapter 11 we examined the wage and employment decision when the firm is both a competitive seller of its output in the product market and a competitive buyer of labour in the labour market. We also relaxed one of those assumptions — that of a competitive seller of its output — and examined the labour market implications when the firm is a noncompetitive seller of its output, but still a competitive buyer of labour in the labour market. In this chapter the assumption of being a competitive buyer of labour is relaxed. So as to trace out the implications of this single change, termed *monopsony*, the firm is still assumed to be a competitive seller of its product. The results of relaxing both assumptions simultaneously — that of competition in the labour market and the product market — follow in a straightforward fashion from the results of each separate case.

SIMPLE MONOPSONY

The situation in which a firm is sufficiently large relative to the size of the local labour market that it influences the wage at which it hires labour is referred to as monopsony. The monopsonist is a wage-setter, not a wage-taker. In order to attract additional units of labour, the monopsonist has to raise wages; conversely, if it lowers the wage rate it will not lose all of its work force.

Consequently the monopsonist faces an upward-sloping labour supply schedule rather than a perfectly elastic labour supply schedule at the going wage, as was the case when the firm was a competitive buyer of labour. This labour supply schedule shows the average cost of labour for the monopsonist because it indicates the wage that must be paid for each different size of the firm's work force. Since this same wage must be paid for each homogeneous unit of labour, then the wage paid at the margin becomes the actual wage paid to all off the workers, and this same average wage is paid to all.

The firm's labour supply or average-cost-of-labour schedule is not its relevant decision-making schedule. Rather, the relevant schedule is its *marginal* cost of

labour which lies above its average cost. This is so because when the firm has to
raise wages to attract additional units of labour, in the interest of maintaining
internal equity in the wage structure it also has to pay that higher wage to its
existing work force (intramarginal workers). Thus, the marginal cost of adding an
additional worker equals the new wage plus the addition to wage costs imposed
by the fact that this new higher wage must be paid to the existing work force.
Consequently, the marginal cost of adding an additional worker is greater than
the average cost, which is simply the wage.

This situation can be depicted by a simple hypothetical example. Suppose the
monopsonist employed only one worker at a wage of one dollar per hour. Its
average cost of labour would be the wage rate of one dollar. This would also be
its marginal cost; that is, the extra cost of hiring this worker. If the monopsonist
wanted to expand its work force, however, it would have to pay a higher wage of,
for example, $1.20 to attract an additional worker. Its average cost of labour is
the new wage of $1.20 (i.e., (1.20 + 1.20)/2); however, the marginal cost of
adding the new worker is the new wage of $1.20 *plus* the additional $0.20 per
hour it has to pay the first worker in order to maintain internal equity in the wage
structure. Thus the marginal cost of the second worker is $1.40 per hour, which
is greater than the average cost or wage of $1.20. If a third worker costs $1.40,
then the average cost or wage of the three workers would be $1.40, while the
marginal cost of adding the third worker would be $1.80, composed of $1.40 for
the third worker plus the additional $0.20 for each of the other two workers.
(This example is further extended in the first three columns of Table 12.1, which
is presented later to show the impact of minimum wages on a monopsony
situation.)

The monopsony situation is illustrated diagrammatically in Figure 12.1. The
essence of monopsony is that the firm faces an upward-sloping supply schedule
for labour and hence has a marginal cost-of-labour schedule that lies above the
supply or average cost schedule. The firm maximizes profits by hiring labour
until the marginal cost of an additional unit of labour just equals the marginal
revenue generated by the additional unit of labour. Marginal revenue is given by
the VMP schedule in this assumed case of the firm being a competitive seller of
its product. This equality of marginal cost and VMP occurs at the employment
level N_m.

The VMP curve for the monopsonist is not its demand curve for labour in the
sense of showing the various quantities of labour that will be demanded at
various wage rates. This is so because the monopsonist does not pay a wage equal
to the VMP of labour. For example, in Figure 12.1 at the wage w_m, the quantity
of labour demanded is N_m, not $w_m V_m$, as would be the case if VMP were a
demand schedule equating w with VMP. In essence, the demand for labour is
determined by the interaction of the MC and VMP schedules and this depends
on the shape of the supply schedule (and hence the MC schedule) as well as the

Figure 12.1
MONOPSONY

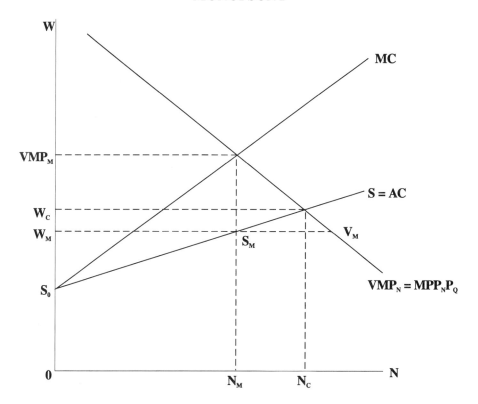

VMP schedule. This line of reasoning is analogous to that underlying the fact that a product market monopolist has no supply curve for its product.

IMPLICATIONS OF MONOPSONY

The level of employment, N_m, associated with monopsony is lower than the level, N_c, that would prevail if the monopsonist behaved as a competitive buyer of labour, equating the supply of labour with the demand. The monopsonist restricts its employment somewhat because hiring additional labour is costly since the higher wages have to be paid to the intramarginal units of labour.

For N_m units of labour, the monopsonist pays a wage of w_m as given by the supply schedule of labour. The supply schedule shows the amount of labour that will be forthcoming at each wage and, for a wage of w_m, then N_m units of labour will be forthcoming. This wage is less than the wage, w_c, that the monopsonist would pay if it employed the competitive amount of labour N_c. The monopsonist wage is also less than the value of the marginal product of N_m units of labour.

That is, the monopsonist pays a wage rate that is less than the value of the output produced by the additional unit of labour (although the value of that output does equal the marginal cost of producing it). This monopsony profit or difference between wages and the value of marginal product of labour has been termed a measure of the monopsonistic exploitation of labour, equal to $VMP_m - w_m$ per worker or $(VMP_m - w_m) N_m$ for the monopsonist's work force. This monopsony profit accrues to the firm because its wage bill, $w_m N_m$, is less than the market value of the marginal output contributed by the firm's labour force, $VMP_m N_m$.

Because the value of the marginal product of labour for the monopsonist is greater than the value of the marginal product if that firm hired competitively ($VMP_m > VMP_c$, the latter of which equals w_c) then welfare to society would be increased by transferring labour from competitive to monopsonistic labour markets. This seemingly paradoxical result — an increase in welfare by expanding noncompetitive markets relative to competitive markets — occurs simply because the value of the marginal product of labour is higher in the monopsonistic market. In effect, transferring labour is akin to breaking down the barriers giving rise to monopsony: it is allocating labour to its most productive use.

Although monopsony leads to a wage less than the value of the marginal product of labour, it is also true that intramarginal workers are receiving a seller's surplus; that is, the wage they are paid, w_m, is greater than their reservation wage as indicated by the supply schedule. The existing work force up until N_m units of labour is willing to work for the monopsonist for a wage that is less than the wage, w_m, they are paid. The wage that they are willing to work for is illustrated by the height of the labour supply schedule, which reflects their preferences for this firm and their opportunities elsewhere. However, to the extent that they are all paid the same wage, w_m, then the existing work force receives a seller's surplus or economic rent equal to the triangle $S_o S_m w_m$.

This is why the term monopsonist exploitation of labour should be used with care. It is true that the monopsonist pays a wage less than the value of the marginal product of labour. However, it is also true that the monopsonist pays a wage greater than the reservation wage (opportunity cost, supply price) that intramarginal employees could get elsewhere for their labour.

A final implication of monopsony is that there will be equilibrium vacancies (equal to $V_m - S_m$ in Figure 12.1) at the wage paid by the monopsonist. In other words the monopsonist would report vacancies at the wage it pays, but it will not raise wages to attract additional labour to fill these vacancies. In this sense the vacancies are an equilibrium, since there are no automatic forces that will reduce the vacancies. The monopsonist would like to hire additional labour *at the going wage* since the value of the marginal product exceeds the wage cost of an additional unit. However, because the higher wage would have to be paid to intramarginal units of labour, the value of the marginal product just equals the marginal cost, and that is why there are no forces to reduce the vacancies. The monopsonist is maximizing profits by having vacancies. It does not reduce the vacancies because it would have to raise wages to do so and the marginal cost of doing so is more than the marginal revenue.

CHARACTERISTICS OF MONOPSONISTS

As long as there is some inelasticity to the supply schedule of labour faced by a firm, then that firm has elements of monopsony power. To a certain extent most firms may have an element of monopsony power in the short run, in the sense that they could lower their wages somewhat without losing all of their work force. However, it is unlikely that they would exercise this power in the long run because it would lead to costly problems of recruitment, turnover, and morale. Facing an irreversible decline in labour demand, however, firms may well allow their wages to deteriorate as a way of reducing their work force.

In the long run, monopsony clearly will be less prevalent. It would occur when the firm is so large relative to the size of the local labour market that it influences wages. This could be the case, for example, in the classic one-industry towns in isolated regions. Such firms need not be large in an absolute sense: they are simply large relative to the size of the small local labour market, and this makes them the dominant employer. Monopsony may also be associated with workers who have particular preferences to remain employed with the monopsonist. In such circumstances, wages could be lowered and they would stay with the monopsonist because their skills or preferences are not transferable.

Thus a mining company could be a monopsonist, even if it were reasonably small, if it were located in an isolated region with no other firms competing for the types of labour it employed. The same company could even be a monopsonist for one type of labour, for example, miners, while having to compete for other types of labour, for example, clerical workers. Conversely, an even larger mining company might be a competitive employer of miners if it were situated in a less isolated labour market and had to compete with other firms. It is size relative to the local labour market that matters, not absolute size.

Monopsony may also be associated with workers who have specialized skills (specific human capital) that are useful mainly in a specific firm, or with workers who have particular preferences to remain employed with a specific firm. Because their skills or preferences are not completely transferable, such workers are tied to a single employer which therefore possesses a degree of monopsony power. Contractual arrangements in professional sports, for example, often effectively tie the professional athlete to a specific employer (team), giving the employer a degree of monopsony power. At a minimum the employer need pay a salary only slightly higher than the player's next-best-alternative salary, which would be considerably less. In practice, a much higher salary is usually paid in order to extract maximum performance. When players are free agents, as for

EXHIBIT 12.1   MONOPSONY IN THE LABOUR MARKET FOR NURSES

The labour market for nurses is often cited as being characterized by monopsony, in part because of a belief that hospitals — the main employer of nurses — possess considerable market power in wage-setting. In addition, the labour market for nurses has frequently displayed persistent shortages, an outcome that is consistent with monopsony. (Recall that a monopsonist will wish to employ more labour at the monopsony wage, and therefore will report unfilled job vacancies. However, the monopsonist will not wish to raise the wage in order to attract more applicants.)

However, sceptics point out that the market for nurses is often national or international in nature, and that nurses are geographically very mobile. Thus, even though a particular hospital may have considerable market power in its own town or city, it must compete in a larger labour market for nurses and may be close to a wage-taker in that larger market.

Thus, on *a priori* grounds there are arguments both for and against the hypothesis that the labour market for nurses is characterized by monopsony. As is frequently the case in labour economics, the debate can only be resolved by empirical analysis. Indeed, whatever beliefs are held on the basis of *a priori* reasoning, empirical analysis is needed to determine whether the extent of monopsony is quantitatively significant, or whether the amount of monopsony power is sufficiently small that the market is essentially competitive in nature.

The issue is of considerable practical importance. If monopsony is quantitatively important, unfilled vacancies can be expected to persist and market forces cannot be relied upon to eliminate the "shortage." Furthermore, there may be a role for public policy to alter the allocation of resources, as there is in the case of monopoly in the product market. In addition, proponents of pay equity legislation often claim individuals in female-dominated occupations such as nursing are paid less than those in male-dominated occupations which require similar skills and responsibility. Although lower earnings in the female-dominated occupations is generally attributed to discrimination by employers, monopsony wage-setting would provide an alternative explanation.

A number of empirical studies (e.g., Hurd, 1973; Link and Landon, 1975) tested for evidence of monopsony power by examining the cross-sectional relationship between wages of nurses and employer (hospital) concentration. Most, but not all, such studies find that higher employer concentration is associated with lower wages, evidence which is consistent with monopsony. However, such studies do not provide direct information on the quantitative significance of monopsony. Sullivan (1989) addressed this issue more directly using techniques similar to those used in industrial organization to measure the extent of monopoly power in product markets. His analysis suggests that U.S. hospitals have significant monopsony power in the short run, and that even over a longer time horizon may exercise considerable market power. In particular, the (inverse) elasticity of labour supply to hospitals is estimated to be 0.79 over a one-year horizon and 0.26 over a three-year horizon. These results indicate that cost-minimizing hospitals will find it in their interest to set wages of nurses below the value of their marginal product.

Source: D. Sullivan. Monopsony power in the market for nurses. *Journal of Law and Economics* 32 (October 1989) 5135–5178.

example when different leagues compete, the resultant salary explosions attest to the fact that the competitive salary is much higher than the monopsonist salary.

The professional sports example illustrates the important fact that, although monopsonists pay less than they would if they had to compete for labour, they need not pay low wages. In fact, unique specialized skills are often associated with high salaries, albeit they might even be higher if there were more competition for their rare services. Monopsony does not imply low wages: it only implies wages that are lower than they would be if there were competition for the particular skills.

PERFECT MONOPSONISTIC WAGE DIFFERENTIATION

The previous discussion focused on what could be labelled *simple monopsony* — a situation where the monopsonist did not differentiate its work force but rather paid the same wages to all workers of the same skill, both marginal and intramarginal workers. In terms of Figure 12.1, all workers were paid the wage, w_m, even though intramarginal workers would have been willing to work for a lower wage as depicted by their supply price. The resultant seller's surplus or economic rent (wage greater than the next-best-alternative) is appropriated by the intramarginal workers.

This highlights another implication of monopsony. The monopsonist may try to appropriate this seller's surplus by differentiating its otherwise homogeneous work force so as to pay each worker only their reservation wage. If the monopsonist were able to do this for each and every worker, the result could be labelled *perfect monopsonistic wage differentiation*. In this case the supply schedule would be its average cost of labour *and* its marginal cost, because it would not have to pay the higher wage to the intramarginal workers. In such circumstances the discriminating monopsonist would hire up to the point N_c in Figure 12.1.

In fact there is an incentive for the monopsonist to try to expand its work force by means that would make the cost of expansion peculiar only to the additional workers. In this fashion the monopsonist could avoid the rising marginal cost associated with having to pay higher wages to intramarginal workers. Thus monopsonists may try to conceal the higher wages paid to attract additional labour so as not to have to pay their existing work force the higher wage. Or they may try to use nonwage mechanisms, the costs of which are specific to only the new workers. Moving allowances, advertising, or other more expensive job search procedures, and paying workers for paper qualifications that are largely irrelevant for the job, are all ways in which monopsonist may try to expand their work force without raising wages for all workers.

IMPERFECT MONOPSONISTIC WAGE DIFFERENTIATION: APPLICATION TO SEX DISCRIMINATION

There may also be situations where monopsonists are unable to differentiate *each* worker so as to pay each only their reservation wage, but are able to differentiate between broad groups of otherwise homogeneous workers. This situation could be termed *imperfect monopsonistic wage differentiation*. In such circumstances the monopsonist only has to pay the same wage *within* groups: different wages can persist *between* groups. Such may be the case, for example, with respect to male and female workers, the example that will be pursued in this section.

The conditions necessary for imperfect monopsonistic discrimination are first that the markets for the different types of labour can be separated (otherwise the low-wage workers would simply move to the high-wage market), and second that there are different supply elasticities in each market (otherwise there are no monopsony rents for the firm to exploit). The importance of these two factors as preconditions for monopsony will become clearer as the analysis progresses.

The situation of imperfect monopsony is depicted in Figure 12.2 for a hypothetical example of a male and female work force for the firm. For illustrative purposes, males and females are assumed to be equally productive so that there is one VMP schedule for the combined male and female work force. However, the employer is able to segment or differentiate the male and female work force from each other. In other words, while there is internal competition *within* each work force, there is not competition *between* the male and female work forces. The lower supply schedule for females relative to males reflects their lower reservation wage, which, in turn, may reflect poor employment opportunities elsewhere as well as a low subjective evaluation of their own labour market worth.

Although the labour supply schedules in Figure 12.2 appear parallel, the wage elasticities for different employment levels may be different, whether the curves are parallel or not. Specifically, wage elasticity is defined as $\theta = \partial N / \partial w \cdot w/N$, and this depends not only on the slope of the supply schedule, $\partial N / \partial w$, but also on where the firm is on the supply schedule, w/N. For females, as will be demonstrated later, w is lower than males and N is higher; consequently w/N is smaller and female labour supply therefore more inelastic.

This lower elasticity for females could prevail for a variety of reasons. Females may have fewer employment opportunities elsewhere and they may be tied to the area of their husband's employment. Household responsibilities and commuting problems may tie them to a single particular local firm. Household and institutional discrimination, especially in educational institutions, may mean that they have been unable to accumulate continuous work experience or labour-market-oriented skills that would lead to competition for their services. In such

Figure 12.2
IMPERFECT MONOPSONISTIC WAGE DIFFERENTIATION
BETWEEN MALES AND FEMALES

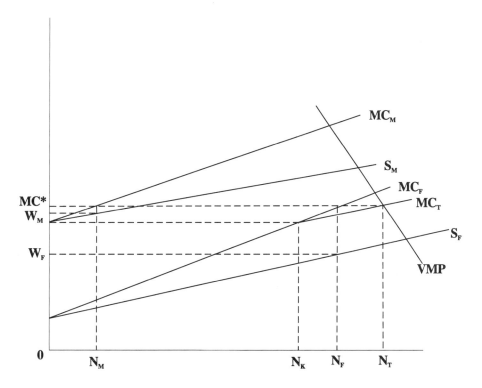

circumstances, female labour supply to an individual firm may be more inelastic than male labour supply.

Since the supply schedules for both males and females are upward sloping, the employer will have a derived marginal cost schedule which lies above each supply schedule. This reflects the fact that the employer has to pay higher wages to expand each work force, and because of competition *within* each group it has to pay these higher wages to the intramarginal workers within the group. The total marginal cost schedule, MC_T, is obtained by *horizontally* summing the male and female marginal cost schedules, because this indicates the total amount of labour that can be obtained for a given marginal cost. In other words as the firm expands its homogeneous work force, it would do so first by hiring only females because their marginal cost is below that of males. However, at some point, N_K in Figure 12.2, the marginal cost of female labour would just equal that of male labour and the firm would expand by hiring a mixture of males and females, always keeping its marginal cost as low as possible. At this point, N_K, the total marginal cost schedule is kinked downwards, reflecting the fact that it is a weighted average of the male and female marginal cost schedules.

The profit-maximizing employer will hire labour until the marginal cost (MC_T) just equals marginal revenue (VMP_N). This occurs at the employment level N_T in Figure 12.2. This total work force will consist of N_M males and N_F females because only here will the marginal cost of females just equal the marginal cost of males, and both will equal the common equilibrium marginal cost, MC^* (i.e., $MC_F = MC_M = MC^*$).

Male wages, w_M, are given by the height of the male supply schedule for N_M units of male labour, and female wages, W_f, are given by the height of the female supply schedule for N_F units of female labour. Male wages exceed female wages for the same assumed homogeneous type of labour and this is an equilibrium condition since the firm is maximizing profits by equating marginal cost with marginal revenue. Even though female wages are always lower than male wages over the relevant range of the supply schedules in this diagram, the firm would not hire exclusively females, because the marginal cost of doing so is not always below that of males. And it is the marginal cost of labour that is relevant for profit maximization.

Imperfect monopsonistic wage differentiation of this form can explain what otherwise appears to be a contradiction in the theory of discrimination: the persistence of a male–female wage differential for equally productive workers, and the persistence of a mixed work force. These phenomena appear to be contradictions of competitive theory because, under competition, firms would hire only low-wage females and this would increase the demand for females relative to males until the male–female wage differential disappeared. If some firms could resist competitive pressures and satisfy their desire to discriminate by paying higher wages to males, then males would gravitate to these employers and females would gravitate to nondiscriminating employers. Discrimination would thereby lead to all-male or all-female work forces in different firms (and presumably the lower-cost female firms expanding relative to the higher-cost male firms). That male–female wage differentials and mixed work forces seem to persist, therefore, appears at odds with competitive pressures. They are consistent, however, with imperfect monopsonistic wage differentiation.

Even here, however, there should be competitive pressures at work to break down the segmentation that enables males and females to be considered as two separate pools of labour, with females having a lower reservation wage. In essence, competition from other employers, facilitated by improved labour market information, should serve to raise the wage opportunities for females with other firms, and this would raise the female labour supply schedule of Figure 12.2, mitigating against monopsonistic wage discrimination.

MATHEMATICAL EXPOSITION

The results of the previous section can be demonstrated more rigorously by the use of mathematics. The profit function for the monopsonist is:

$$\pi = pQ(M,F) - w_m(M)M - w_f(F)F \tag{12.1}$$

The firm is still assumed to be a competitor in the product market, so that p is fixed. However, reflecting the fact that monopsonists are wage-setters and have to raise wages to attract additional labour, male and female wages are positively related to their respective employment levels. The profit-maximizing firm employs additional labour until $\partial\pi/\partial M = \partial\pi/\partial F = 0$ which, for males, implies

$$p\partial Q/\partial M = w_m + M \cdot \partial w_m/\partial M \tag{12.2}$$

— that is, the value of the marginal product of male labour (left-hand side) equals its marginal cost (right-hand side). Marginal cost in turn is the male wage rate plus the adjustment in that wage rate for the M intramarginal units of male labour.

The wage elasticity of supply of male labour to the firm is defined as

$$\theta_m = \partial M/\partial w_m \cdot w_m/M \tag{12.3}$$

Inverting, multiplying by w_m and adding w_m to each side yields

$$w_m(1 + 1/\theta_m) = w_m + M \cdot \partial w_m/\partial M \tag{12.4}$$

— which is the right-hand side of Equation 12.2. Substituting Equation 12.4 into 12.2 yields:

$$p\partial Q/\partial M = w_m(1 + 1/\theta_m) \tag{12.5}$$

The same derivation for females yields

$$p\partial Q/\partial F = w_f(1 + 1/\theta_f) \tag{12.6}$$

Under profit maximization, the values of the marginal product of males and females are equal; that is, $p\partial Q/\partial M = p\partial Q/\partial F$. Therefore, from Equations 12.5 and 12.6

$$w_m(1 + 1/\theta_m) = w_f(1 + 1/\theta_f) \tag{12.7}$$

For $w_m > w_f$, it must be the case that $(1 + 1/\theta_m) < (1 + 1/\theta_f)$. Subtracting one from each side yields $1/\theta_m < 1/\theta_f$, and multiplying by θ_f yields $\theta_f/\theta_m < 1$, which implies $\theta_f < \theta_m$. That is, for $w_m > w_f$ it must be the case that $\theta_m > \theta_f$: at their equilibrium levels of employment, male labour supply to the firm must be more elastic than female labour supply.

Equation 12.7 also illustrates why the labour supply elasticities have to be different for the male–female wage differential to persist. If $\theta_f = \theta_m$ then $w_m = w_f$.

MINIMUM WAGES AND MONOPSONY

Monopsony also has interesting implications for the employment impact of an exogenous wage increase, for example, from minimum wages, unionization, equal pay laws, or other forms of wage fixing. Specifically, over a specified range, an exogeneous wage increase may actually *increase* employment in a monopson-

Figure 12.3
MONOPSONY AND A MINIMUM WAGE

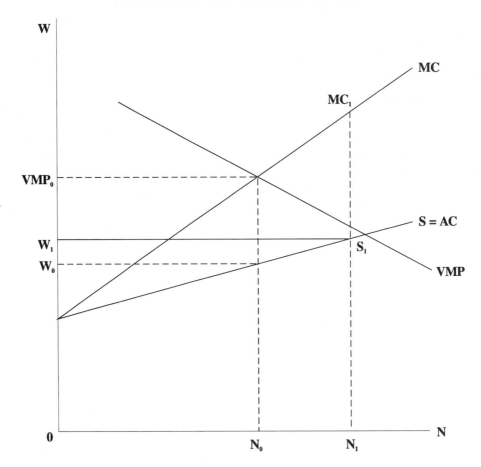

istic firm. This seemingly paradoxical result can be illustrated by the example of a minimum wage increase faced by a monopsonist. The proposition first will be demonstrated rigorously, and then explained heuristically and by way of an example.

Formal Exposition

Figure 12.3 illustrates a monopsonistic labour market with w_0 being the wage paid by the monopsonist for N_0 units of labour with a value of marginal product of VMP_0. With the imposition of an exogenous wage increase, perhaps from minimum wages, to w_1, the new labour supply schedule to the firm becomes horizontal at the minimum wage. This is so because the firm cannot hire at wages below the minimum wage. Thus the firm's labour supply schedule becomes w_1S_1S: to the right of S_1 it becomes the old supply schedule because firms can pay higher than the minimum wage.

The relevant marginal cost schedule, therefore, becomes $w_1S_1MC_1MC$. It is the same as the labour supply schedule for the horizontal portion w_1S_1 because both marginal and intramarginal workers receive the minimum wage and consequently marginal cost equals average cost equals the minimum wage. Similarly, to the right of S_1 the marginal cost schedule becomes the old marginal cost schedule MC_1MC reflecting the relevance of the portion of the old supply schedule S_1S. The jump in marginal cost between S_1 and MC_1 occurs because when the firm expands its work force beyond N_1 units of labour it has to raise wages to do so, and the wage increase applies to all intramarginal workers who previously were receiving the minimum wage.

After the imposition of the new minimum wage, the monopsonist would equate MC with MR (i.e., VMP) and hence employ N_1 units of labour at the minimum wage w_1. Wages have increased from w_o to w_1 and, paradoxically, employment has increased from N_o to N_1. In fact, for wage increases up to VMP_o, employment would increase, reaching a maximum increase for wages at the intersection of the supply and VMP schedules.

Heuristic Explanation

Why would wage increases ever lead to employment increases on the part of the firm? The answer to this seeming paradox lies in the fact that the minimum wage negates or makes redundant a certain portion of the firm's supply and marginal cost of labour schedule. Faced with the new constraint of the minimum wage, the firm no longer need concern itself, at least up to a point, with the fact that if it wants more labour it will have to raise wages and pay these higher wages to all workers. In other words it is no longer inhibited in its employment expansion by the rising marginal cost of such expansion. Consequently it will expand employment because although the wage it pays is higher, the marginal cost of expansion is lower since it already pays the minimum wage to its intramarginal workers.

This does not mean that the monopsonist would welcome a minimum-wage increase because then it wouldn't have to worry about the rising marginal cost of labour. The minimum wage obviously reduces profits to the monopsonist, otherwise it would have self-imposed a minimum wage without the need for legislation. The monopsonist is simply responding to a different set of constraints: the minimum-wage constraint simply leads to more employment, at least over a limited range of wage rates.

Hypothetical Example

The response of a monopsonist to a minimum wage increase is illustrated in the hypothetical example of Table 12.1. The symbols refer to those in Figure 12.3. The firm's labour supply schedule is given in column (2) with hypothetical

Table 12.1
HYPOTHETICAL EXAMPLE OF MONOPSONIST
RESPONDING TO MINIMUM WAGE

UNITS OF LABOUR	NO MINIMUM WAGE		MINIMUM WAGE		VALUE OF MARG. PROD.
	Wages	Marg. Cost	Wages	Marg. Cost	
N (1)	S = AC (2)	MC (3)	w_1S_1S = AC (4)	$w_1S_1MC_1MC$ (5)	VMP (6)
1	1.00	1.00	2.00	2.00	3.00
2	1.20	1.40	2.00	2.00	2.50
3	1.40	1.80	2.00	2.00	2.30
4	1.60	2.20	2.00	2.00	2.20
5	1.80	2.60	2.00	2.00	2.00
6	2.00	3.00	2.00	2.00	1.80
7	2.20	3.40	2.20	3.40	1.60

increments of $0.20 per hour necessary to attract an additional worker. The resultant marginal cost schedule is given in column (3), rising faster than the average wage cost because of the necessity to pay intramarginal workers the extra wage. In the absence of the minimum wage the monopsonist would equate MC with VMP and hence employ four workers.

With the imposition of a minimum wage equal to $2.00 per hour, the firm's labour supply schedule becomes as in column (4). For the seventh worker the minimum wage is redundant, since that worker would not work for less than $2.20 per hour, as indicated earlier in column (1). With the minimum wage the firm's new marginal cost of labour schedule becomes as in column (5). It is constant at the minimum wage for up until six workers because the firm simply pays the minimum wage to all workers. There is a large jump in MC associated with the seventh worker, however, because to acquire the seventh worker the firm has to pay not only the $2.20 for that worker but also an additional $0.20 for each of the previous six workers for a total of $3.40 (i.e., 2.20 + 6(.20)).

Given the new marginal cost schedule associated with the minimum wage, the firm will equate MC with VMP by employing five workers. The minimum wage increase results in an increase in employment from four to five workers.

Monopsony profits before the minimum wage were (VMP - w)N = (2.20 - 1.60)4 = 2.40 dollars per hour. After the minimum wage they are reduced to zero in this particular case because the minimum wage was set exactly equal to the VMP of the fifth unit of labour. (In Figure 12.3, for example, there still would be some monopsony profits even after the minimum wage.)

Practical Importance

While in theory minimum wage increases can lead to employment increases in situations of monopsony, in practice the importance of this factor depends on the extent to which monopsony is associated with workers who are paid below minimum wages. Earlier it was pointed out that the monopsonist may well pay high wages even though these wages are not as high as they would be if there were more competition for these workers. Examples of low wage labour markets that would be affected by minimum wages include the garment and textile trades and the service sector, such as hotels, restaurants, and theatres. Employers in these sectors tend to be small and the pool of low-wage labour from which they draw large. Consequently, the conditions giving rise to monopsony are not prevalent. There may be isolated instances of one-industry towns where the minimum wage would have an impact on otherwise monopsonistic wage determination. However, it is unlikely that these conditions are sufficiently prominent to negate the usual adverse employment effect that economic theory predicts will emanate from minimum-wage increases.

EVIDENCE OF MONOPSONY

Although the empirical evidence is by no means conclusive, there does appear to be evidence of monopsony in at least some particular labour markets. Scully (1974) finds evidence of monopsony in professional baseball, especially among the "star" players. The dramatic increase in player salaries following the introduction in 1977 of the free agent system (which significantly increased competition among teams for players, who were no longer tied to teams) also indicates that monopsony power was important in this labour market (Hill and Spellman, 1983). However, in their study of salary determination in the National Hockey League, Jones and Walsh (1988) find only modest evidence of monopsony effects on player salaries. Thus, the evidence relating to the importance of monopsony in markets for professional athletes is not entirely conclusive. The professional sports example is also interesting in that it highlights the point that monopsony power need not be associated with low salaries — just salaries that are lower than they would be in the presence of competition in the labour market.

Empirical studies carried out in the U.S., the U.K., and Canada have also found evidence of monopsony in the labour markets for teachers (e.g., Landon and Baird, 1971; Dahlby, 1981; Currie, 1991) and nurses (see Exhibit 12.1). In Canada, these two labour markets are now highly unionized; thus the employer-union bargaining models developed in Part 4 of this book may be more appropriate than the simple monopsony model of this chapter. In her study of the labour market for Ontario school teachers, Currie (1991) noted that both a bargaining model and a demand-supply framework are consistent with the data. The estimates associated with the demand-supply framework indicate that labour demand is very inelastic and that labour supply is slightly upward sloping; i.e., that school boards possess a small amount of monopsony power in wage-setting. Further evidence for newspaper printing employees and construction workers is found in Landon (1970) and Landon and Pierce (1971). The extent to which these results can be generalized, even within the occupations where some monopsony was found, remains an open question. In addition, the extent to which it is monopsony power, rather than other factors, that is associated with lower wages could be open to debate.

It is unlikely that monopsony can be an extremely important factor in the long run. Improved communications, labour market information, and labour mobility make the isolated labour market syndrome, necessary for monopsony, unlikely at least for large numbers of workers. As these factors improve over time, monopsony should diminish.

In contrast, monopsony may be quite common in the short run. Most firms can lower their offered wage and still recruit new employees, albeit perhaps at a slower pace than at a higher offered wage. Similarly, firms which offer a higher wage are likely to experience both more applicants and a higher acceptance rate of job offers. In these circumstances, firms face an upward-sloping labour supply curve in the short run, even though they may face a perfectly elastic labour supply curve over a longer horizon. This situation of "dynamic monopsony" is especially likely to occur in an environment of imperfect information in which workers are searching for jobs, and employers are searching for employees. These and other implications of imperfect information are examined further in the following chapter.

QUESTIONS

1. Combine the results of Chapter 11 on monopoly in the product market with Chapter 12 on monopsony in the labour market and illustrate the wage and employment determination process. Is it necessary that monopoly be accompanied by monopsony? Is it possible that the two conditions could go together?

2. If the monopsonist is a wage-setter and not a wage-taker than it can set wages in whatever fashion it wants. Discuss.

3. Monopsonists are at a disadvantage relative to competitive buyers of labour because, when the monopsonist wants to expand its work force, it has to raise wages while the competitive buyer can get all the labour it wants at the going wage. Discuss.

4. The monopsonist has an upward-sloping labour supply schedule because when it expands its work force it has to use better quality labour and hence has to pay a higher wage. Discuss.

5. The monopsonist's demand for labour depends on its elasticity of labour supply. Illustrate this proposition in Figure 12.1 by drawing both a more elastic and a more inelastic supply schedule through the point S_m.

6. Explain why the monopsonist has no unique demand schedule for labour.

7. Is it possible for a multimillion dollar professional sports player to be subject to monopsonistic exploitation of labour? Is it possible for workers who are subject to monopsonistic exploitation of labour to be receiving an economic rent on the sale of their labour market services? Could workers who are receiving an economic rent for their services ever be considered disadvantaged workers?

8. Monopsonists are not necessarily evil: they are simply acting like any other firms when they maximize profits subject to constraints. It is just that the constraints they face are different and this enables them to make a monopsony profit per worker equal to the difference between the wage they pay and the value of the marginal product of labour. The solution to this problem is therefore to alter the constraints faced by monopsonists. Discuss.

9. If the economic rent of $S_oS_mw_m$ in Figure 12.1 were taxed away from the workers, what would happen to the equilibrium employment of the monopsonist? Specifically, would these workers go elsewhere if only their earnings in this particular employment were taxed? Similarly, if the monopsonist appropriates this surplus, what would happen to its employment?

10. Discuss various ways in which monopsonists may try to differentiate their work force so as to appropriate any economic rent or seller's surplus of their workers.

11. Based on your knowledge of professional sports contracts, discuss ways in which monopsony behaviour may be exhibited.

12. One of the conditions necessary for monopsonistic wage discrimination against females as a group is that their labour supply to the firm must be more inelastic than that of males, at their respective wages and employment. Would you expect this to be the case, and why?

13. Based on the diagram of Figure 12.2, the slopes of the male and female labour supply schedules appear to be the same. How is it then that female labour supply must be more inelastic than male labour supply for the wages of females to be less than the wages of males in this situation of imperfect monopsonistic wage discrimination?

14. Based on the respective supply schedules of males and females in Figure 12.2, if other firms began to compete with this monopsonist for labour, what type of labour would they probably hire and what would this do to the respective supply schedules of this monopsonist? Specifically, would it be able to keep differentiating its pools of labour in the face of such competition?

15. The equilibrium of Figure 12.2 is unstable in the long run because, if the monopsonist is paying males a higher wage than equally productive females, it would obviously try to replace its costly male work force with a less costly female work force. Discuss.

16. Indicate how imperfect monopsonistic wage discrimination between males and females can explain the persistence of a male–female wage differential for equally productive workers and the persistence of a mixed work force within a firm. Could monopsony explain the persistence of these phenomena throughout the aggregate economy?

17. In the case of imperfect monopsonistic wage differentiation (as illustrated in Figure 12.2), would the monopsonist ever hire an all female work force? Why are the male and female marginal cost schedules not vertically summed to get a total marginal cost schedule? Would this monopsonist ever report vacancies of male and female labour at the going wage and, if so, exactly what is the measure of the vacancies for each?

18. The fact that a minimum wage, over a specific range, may actually increase employment for a monopsonist illustrates the proposition that a monopsonist does not have a demand curve for labour in the usual sense. For if it did, then a wage increase would have to be accompanied by a reduction in employment. Discuss.

19. Explain heuristically why a minimum-wage increase, over a certain range, would lead to a monopsonist actually increasing employment. Given this possibility, could the monopsony argument be relied upon to negate the critics of minimum-wage legislation who argue that minimum wages will have an adverse employment effect and hence harm some of the very people they were designed to help? Could minimum wages ever be applied selectively to monopsony situations? Could wage-fixing via unionization be applied more selectively?

20. Based on Figure 12.2, trace through the impact of equal pay for equal work legislation, assuming that the firm could not lower male wages.

21. Based on Figure 12.3, discuss the favourable impact on resource allocation of setting a minimum wage at the intersection of the S and VMP schedules. Heuristically, in what sense are resources allocated more efficiently at that point than at the monopsonist's equilibrium? If the monopsonist followed a policy of perfect wage differentiation in the absence of the minimum wage, what would be the implications for resource allocation? Compare the income distribution consequences of the two alternatives of minimum wages and perfect wage differentiation.

22. Based on the data of Table 12.1, how many units of labour would the monopsonist employ if it acted as a perfectly discriminating monopsonist? What are the consequences for efficient resource allocation?

23. Trace through the implications of the firm acting as an oligopsonist with a kinked supply curve; that is, where wage increases would be met by other oligopsonists but wage decreases would not, so that substantial losses of employment would result.

24. According to Link and Landon (1976, p. 151), "clear evidence of a relation between depressed wage levels and monopsony or oligopsony can be isolated empirically in labour markets such as nursing where highly specialized training is required, where the cross-elasticity of supply between occupations is small, where substantial inter-area variations in monopsony power are present, and where limited geographic mobility exists in the labour market." Discuss how each of these factors could lead to monopsony or oligopsony power being exercised.

25. Cohen (1972, p. 43) estimated the following equation to test for the existence of monopsony in the nursing market:

$$w_N/w_C = 3.69 - 0.30CR - 0.02NS$$

where w_n is the starting wage of registered nurses (in dollars per hour), w_c is an index of competitive wages in the area, CR is a concentration ratio giving the percentage of beds in the area provided by the four largest hospitals, and NS is a dummy variable coded one if the hospital has a nursing school, zero otherwise.

    Use this equation to predict the starting salaries of registered nurses in the following three cases, assuming throughout that the competitive wage index is unity (approximately the sample mean):

    (a) competitive case when CR = .20 and NS = 0;

    (b) typical case when CR = .60 and NS = 0;

    (c) oligopsony case when, for example, CR = .80 and NS = 1;

    (d) monopsony when, for example, CR = 1.00 and NS = 1.

REFERENCES AND FURTHER READINGS

Archibald, G. The factor gap and the level of wages. *Economic Record* 30 (November 1954) 187–199.

Baily, M. Dynamic monopsony and structural change. *AER* 65 (June 1975) 338–349.

Bradfield, M. Long-run equilibrium under pure monopsony. *CJE* 23 (August 1990) 700–704.

Chalkley, M. Monopsony wage determination and multiple unemployment equilibria in a non-linear search model. *R.E. Studies* 58 (January 1991) 181–193.

Cohen, H. Monopsony and discriminating monopsony in the nursing market. *Applied Economics* 4 (March 1972) 39–48.

Currie, J. Employment determination in a unionized public-sector labour market: the case of Ontario's school teachers. *JOLE* 9 (January 1991) 45–66.

Dahlby, B.G. Monopsony and the shortage of school teachers in England and Wales. *Applied Economics* 13 (September 1981) 303–319.

De-Meza, D. and J.R. Gould. Free access versus private property in a resource: income distributions compared. *JPE* 95 (December 1987) 1317–1325.

Ehrenberg, R. Heterogeneous labor, minimum hiring standards, and job vacancies in public employment. *JPE* 81 (November/December 1973) 1442–1450.

Hall, W. Unionization, monopsony power, and police salaries. *IR* 16 (February 1977) 94–100.

Hill, J.R. and W. Spellman. Professional baseball: the reserve clause and salary structure. *IR* 22 (Winter 1983) 1–19.

Hurd, R. Equilibrium vacancies in a labor market dominated by non-profit firms: the shortage of nurses. *R.E. Stats.* 55 (May 1973) 234–240.

Ioannides, Y. and C. Pissarides. Monopsony and the lifetime relation between wages and productivity. *JOLE* 3 (January 1985) 91–100.

Jones, J.C.H. and W.D. Walsh. Salary determination in the National Hockey League: the effects of skills, franchise characteristics, and discrimination. *ILRR* 41 (July 1988) 592–604.

Landon, J. and R. Baird. Monopsony in the market for public school teachers. *AER* 61 (December 1971) 966–971. Comment by R. Thornton and reply, *ILRR* 28 (July 1975) 574–578.

Landon, J. and W. Peirce. Discrimination, monopsony, and union power in the building trades: a cross section analysis. *IRRA* 24 (December 1971) 254–261.

Link, C. and J. Landon. Monopsony and union power in the market for nurses. *SEJ* 41 (April 1975) 649–659.

Link, C. and J. Landon. Market structure, nonpecuniary factors and professional salaries: registered nurses. *Journal of Economics and Business* 28 (Winter 1976) 151–155.

Magnac, T. Segmented or competitive labor markets. *Econometrica* 59 (January 1991) 165–187.

Mai, C. and J. Shih. Employment effect under monopsony wage discrimination. *SEJ* 49 (July 1982) 242–245.

Maurice, S. Monopsony and the effects of an externally imposed minimum wage. *SEJ* 41 (October 1974) 283–287.

Meltz, N.M. *Sorry no care available due to nursing shortage.* Toronto, Ontario: Nursing Association of Canada, 1988.

Nelson, P. The elasticity of labor supply to the individual firm. *Econometrica* 41 (September 1973) 853–866.

Raimond, H. Free agents' impact on the labour market for baseball players. *JLR* 4 (Spring 1983) 183–193.

Robinson, J. *The Economics of Imperfect Competition.* London: Macmillan, 1933, Reprinted 1965, 211–228.

Rottenberg, S. The baseball players' labor market. *JPE* (June 1956) 242–258.

Scully, G. Pay and performance in major league baseball. *AER* 64 (December 1974) 915–930.

Sullivan, D. Monopsony power in the market for nurses. *Journal of Law and Economics* 32 (October 1989) S135–S178.

Yett, D. Causes and consequences of salary differentials in nursing, *EI* 7 (March 1970) 78–99.

Yett, D. The chronic shortage of nurses: a public policy dilemma. *Empirical Studies in Health Economics*, H. Klarman (ed.) Baltimore: Johns Hopkins Press, 1970.

CHAPTER 13

Imperfect Information, Incentives, and Risk: Implications for Labour Market Equilibrium

The analysis in the previous two chapters abstracted from a number of phenomena, such as imperfect information, risk and uncertainty, and the effects of incentives on employee behaviour. Although such abstraction was necessary in order to focus on the role of market structure, there are numerous aspects of labour market behaviour that cannot be adequately understood in a world of perfect information, or in one in which employers and employees are assumed to not respond to incentives and/or uncertainty about the future. Indeed, many of the reasons why the labour market differs from other markets arise because of the role of incentives, risk, and imperfect information. This chapter provides an introductory analysis of these important topics, and their influence on labour market equilibrium. The specific topics examined in this chapter include efficiency wages, which are associated with worker incentives; implicit contracts, which are associated with risk and uncertainty; and search and matching, which are associated with imperfect information. The analysis of efficiency wages also provides a foundation for the study of dual labour markets, and these are also examined in this chapter.

SOME IMPLICATIONS OF COMPETITIVE LABOUR MARKET EQUILIBRIUM

In the absence of imperfect information, incentives, and risk and uncertainty, the basic theory of a competitive labour market makes several strong predictions. These are reviewed here to set the stage and motivate the analysis which follows.

One implication is that wages and employment will adjust to equate labour demand and labour supply. Furthermore, in markets with homogeneous workers (individuals who are the same in terms of productive characteristics) and homogeneous jobs (jobs which are equally desirable from the workers' point of view), wages will be equalized across workers. Firms would not pay more than the going market wage because they can employ as much labour as desired at the going wage rate. Workers would not accept less than the market wage because there are equally satisfactory jobs available at the going wage rate. This implication of the basic competitive equilibrium is illustrated in Figure 13.1 for the case of two industries or regions employing the same type of homogeneous labour. Panel (a) shows the situation in which the labour market is not in equilibrium (even though demand equals supply in each individual labour market) because employees in sector A are receiving a lower wage than employees in sector B. In the presence of full information, and in the absence of mobility costs (costs of changing from one sector to another), workers would move from sector A to sector B, thus increasing supply in B and reducing labour supply in sector A. As a consequence, wages would fall in sector B and rise in sector A. This process would continue until wages are equalized across the two sectors, as illustrated in panel (b) of Figure 13.1.

Mobility costs could account for some persistence of earnings differences across sectors, at least in the short run, especially if the sectors are geographically separated. However, these differences are unlikely to persist in the longer run. As older workers retire and young workers enter the labour market, new entrants will tend to choose the higher paying sector over the lower paying sector, thus bringing about wage equality across the homogeneous work force.

Note also that employers have no reason to pay their workers a wage higher than the going market wage because doing so would not affect the productivity of their work force. According to the basic theory of labour demand developed in Chapters 9 and 10, the productivity (marginal and average) of labour is technologically determined according to the amount of labour, capital, and other inputs employed in the production process. In particular, the productivity of labour is assumed in the basic theory to be independent of the remuneration received by the worker.

Another implication of the basic competitive equilibrium is the absence of "involuntary unemployment." This is also illustrated in panel (b) of Figure 13.1. In labour market equilibrium, there are no individuals who would like to work at the going market wage W_e who are unable to find work at that wage. There are some individuals who are "voluntarily unemployed" in the sense that they would be willing to work at a higher wage (e.g., at wage W_0), but at the wage W_e the value of their time spent in leisure or household production exceeds the value of their time spent in market work. (This aspect is evident from the rising labour supply curves in these two labour markets.)

The absence of involuntary unemployment implies that there are also no unexploited "gains from trade" that would mutually benefit employers and unemployed workers. For example, if there existed an unemployed worker who

FIGURE 13.1

(a) Temporary Wage Differences In Markets For Homogeneous Workers

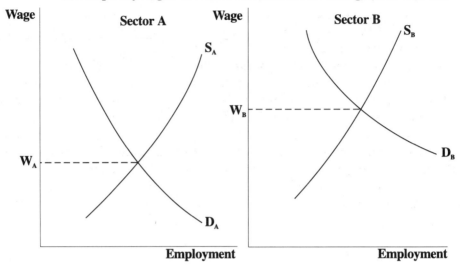

(b) Labour Market Equilibrium In Markets For Homogeneous Workers

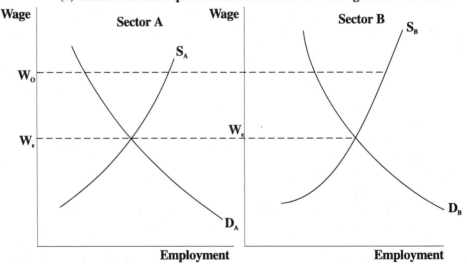

would be willing to work at a wage below the existing market wage, both the worker and the hiring firm would benefit from a job match.

A related implication of the basic competitive labour market equilibrium is that there will be no queues for or rationing of jobs. The predicted absence of queues follows from the assumed homogeneity of jobs (thus making all jobs equally satisfactory in terms of working conditions, job security, and so on) and the property of wages being equalized across jobs and sectors. As a consequence, there are no jobs worth waiting for or lining up for. Similarly, in the absence of queues, there is no need for firms to ration jobs.

Most of these predictions of the basic competitive model do not accord well with observed labour market behaviour. Wages do not appear to adjust quickly to "clear" the labour market (i.e., to equate labour demand and labour supply). Involuntary unemployment of workers appears to be a frequent, if not persistent, feature of many labour markets. Large and persistent differentials in the wages paid to apparently homogeneous workers appear to prevail. In addition, there is evidence that the jobs that pay higher wages are also often more desirable on other grounds, such as working conditions, job security, career prospects, and nonwage benefits. These observations have led some to hold the view that the economy is characterized by "dual labour markets" with good jobs and bad jobs rather than the equalization of wages across jobs predicted by the basic competitive model. In these circumstances, unemployed workers may queue for (or wait for) a good job rather than accept an available bad job, and employers offering good jobs may thus need to ration the limited supply of these jobs.

These apparent inconsistencies between the predictions of the basic competitive model (under the assumptions of full information, homogeneous workers, jobs, etc.) and observed behaviour has motivated considerable recent theoretical and empirical research. A central theme of this research is that much of the behaviour we observe in labour markets can be explained in terms of the responses of employers and employees to imperfect information, incentives, and risk and uncertainty. Although we do not provide a comprehensive treatment of these issues, this chapter does outline the main features of several of these recent developments. A common feature of these theories is that the wage performs functions in addition to "clearing" the labour market — the wage may also act as an incentive device, a risk-sharing mechanism, as part of a matching process, or as a signal of the productivity of the worker or the quality of the job. Because the wage rate serves more than one function, it need not — indeed generally will not — perform its market-clearing function fully. Like former U.S. President Gerald Ford who was reported to be unable to "walk and chew gum at the same time," the wage rate cannot continually equate labour demand and labour supply if it is also functioning as a risk-sharing, matching, or incentive device.

In this chapter we will (except when noted otherwise) deal with labour markets comprised of homogeneous workers and jobs. Heterogeneity on either the demand or supply side of the market introduces additional complexities to the analysis of market equilibrium, and these are examined in Part 5 of this book. There is considerable advantage in analyzing the role of imperfect information, incentives, and risk-sharing in the simplest possible labour market environment, which is that of homogeneous workers and jobs.

INFORMATION, SEARCH, AND MATCHING

Imperfect information is a common feature of many labour markets. Workers may lack information about the availability of jobs, the wages they pay, their location, the associated working conditions, and aspects such as employment security, risk of workplace injury, and career prospects. Employers may also be uncertain about potential job seekers, the wages at which they would accept employment, their productivity in the job, how they will interact with co-workers, and how long they will remain with the organization. If the economic environment were stable and unchanging, the amount of imperfect information would gradually decline over time. However, economic and labour market conditions are continually changing. Thus each period there is new information for employers, employees, and job seekers to acquire.

The labour market is called upon to perform the important task of matching workers with the available jobs and employers. How well this matching function is performed will determine the overall standard of living of the society, and the economic well-being of many of the individual participants. Imperfect information clearly makes this process of matching workers and jobs more difficult.

Imperfect information gives rise to a number of phenomena. *Search* refers to the activity of acquiring information on variables of interest prior to making a decision. Individuals seeking work may search to determine which jobs are available, their rates of pay, and the associated benefits prior to deciding which job to apply for or accept. Similarly, employers with unfilled jobs may search for potential employees prior to deciding which individual(s) should receive offers of employment.

Signaling occurs when the variable of interest is not observed until after the transaction takes place. Employers will often not know how productive an applicant or group of applicants will be until after — perhaps a considerable amount of time after — the hiring decision has been made. Similarly, workers may not know the nature of the working conditions — such aspects as job safety, relationships with management and co-workers, and employee morale — until after the decision to accept the job has been made. In these circumstances, both parties may look for variables they believe to be correlated with the variable of interest. For example, consumers uncertain about the quality of a product may use the price of the product or the warranty associated with the product as a "signal" of the product's quality. Similarly, employers may use the level of education of job applicants as a signal of their productivity, and workers may regard the level of remuneration offered by the employer as a signal of the quality of the working environment.

Signaling occurs when employers or employees are heterogeneous. Because our focus in this chapter is on labour market equilibrium in markets with homogeneous workers and jobs, the role of labour market signaling is discussed in Chapter 18 on Human Capital Theory, where the implications of education as a signal of worker productivity are examined. In this chapter we describe some of the implications of imperfect information and search for labour market equilibrium. Job search is also examined in Chapter 25 on Unemployment: Causes and Consequences.

As emphasized in an important early contribution by Stigler (1962), acquiring information is an economic activity with associated costs and benefits. For workers seeking employment, the benefits of additional search are associated with the possibility of obtaining a job with wages or working conditions preferred to those provided in a job known to be available. The costs of additional search include the opportunity cost of not working and the direct costs associated with sending applications, attending interviews, and so on. Rational individuals will continue searching as long as the incremental benefits from an additional period of job search exceed the additional costs; that is, until the marginal benefits of search equal the marginal costs. A similar rule applies to the search decision of firms.

Note that these decision rules imply that workers and firms will, in general, discontinue their search activities before they are fully informed. That is, optimal decision-making implies that it is typically worthwhile to acquire some information prior to making a decision (information for which the marginal benefits exceed marginal costs) but to not acquire all the available information. Because of diminishing returns to information acquisition, and possibly also rising marginal costs of information acquisition, workers and firms will discontinue their search activities prior to being fully informed.

Several implications follow from these optimal search decision rules. First, a labour market characterized by imperfect information will not "clear" instantaneously. While the process of *matching* workers and jobs proceeds, demand will not equal supply at each moment. Indeed, because search, information acquisition, and matching take time, unsatisfied demand (unfilled job vacancies) and unutilized supply (unemployed job seekers) will coexist at any point in time.

A related implication of imperfect information is that there generally will be a distribution of wage rates even in a labour market with homogeneous workers and jobs. This situation is illustrated in panel (a) of Figure 13.2, in contrast to the full information case shown in panel (b). Some employers will pay wages above the market average, both because they are not fully informed about wages offered by other firms and they may wish to expand their work force at a rapid rate, an aspect which is discussed further below. Similarly, other firms may offer wages below the market average. Because workers are not fully informed about the wages offered by all firms, some unemployed job seekers may accept employment at firms offering below average wages, an outcome that would not occur under full information.

A further, and closely related, implication is that under imperfect information employers possess some short-run monopsony power, even though the market is otherwise perfectly competitive, and employers are wage-takers in the long run.

FIGURE 13.2
WAGE DISTRIBUTIONS UNDER IMPERFECT
AND PERFECT INFORMATION

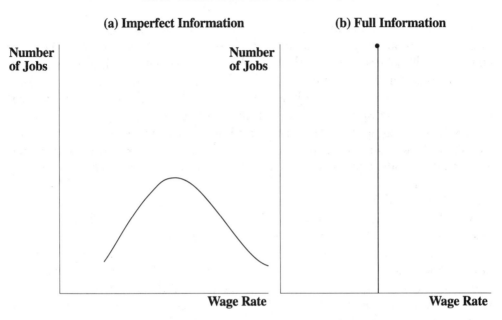

(a) Imperfect Information   **(b) Full Information**

This situation — referred to as *dynamic monopsony* — is most readily under-stood by assuming that workers have some information about the distribution of wage rates available in the market, i.e., the distribution illustrated in Figure 13.2(a), but do not, in the absence of search, know the wage rates offered by individual employers. In these circumstances, when an unemployed job seeker receives a job offer, she must decide whether to accept the offer or continue searching. She will be more likely to accept an offer that appears to be above average (based on her beliefs about the distribution of wage rates available in the market) than one that appears to be below average. Thus, from the perspective of the employer with unfilled job vacancies, paying a high wage (relative to what other firms are believed to be paying) increases the probability that an offer will be accepted and vice-versa for offering a low wage. In other words, the employer faces an upward sloping labour supply curve in that the higher the offered wage the greater is the acceptance rate of job offers, or hiring rate.

This discussion also indicates that in markets characterized by imperfect information the wage rate does not adjust instantaneously to equate labour demand and supply. Employers who find that their acceptance rate of job offers is low will, over time, adjust upward their offered wages, especially if it is very costly to leave jobs unfilled. Similarly, employers who discover that their offers are immediately accepted may lower their offered wages. These adjustments can be expected to occur gradually because it takes time for employers to become aware of how their offered wages compare to those available elsewhere, as well as to observe the rate at which their offers are accepted.

In summary, the response of workers and employers to imperfect information can account for several observed features of labour market behaviour:

- wages don't adjust instantaneously to clear the labour market;

- unsatisfied labour demand and unutilized labour supply coexist at any point in time;

- a distribution of wages, rather than a single wage, is observed, so that the wage rate is not necessarily equalized across homogeneous workers;

- some employers may pay above-average wages and others below-average wages for otherwise homogenous jobs;

- employers have some short-run discretion over the wage rate even though they may be wage-takers in the long run.

IMPLICIT CONTRACTS*

While search theory is concerned with the process of matching job vacancies and unemployed workers, implicit contract theory deals with issues that may arise when firms and workers are already engaged in a continuing employment relationship. In particular, implicit contract theory seeks to explain phenomena such as rigid wages and the use of quantity adjustments (layoffs and rehires) rather than wage adjustments to respond to variations in product demand.

Implicit contract theory is based on the view that wage and employment behaviour reflects risk-sharing between employers and employees. Risk-sharing arises because of differences in attitudes toward risk between workers and the owners of firms. Specifically, workers are believed to be more risk-averse than the shareholders of firms. These differences in attitudes toward risk create potential gains from trade; that is, both parties can benefit from a risk-sharing arrangement. Because workers dislike fluctuations in their incomes, they prefer an arrangement whereby they receive a somewhat lower average or expected income provided their income is sufficiently less variable (more certain). The owners of firms also prefer this arrangement because average or expected profits are higher (due to lower labour costs), albeit more variable because of the stabilization of workers' incomes. In effect, the employment relation involves two transactions: (1) provision of labour services by employees in exchange for payment by employers, and (2) provision of insurance services by employers in exchange for payment of an insurance premium (acceptance of a lower wage) by employees. For reasons discussed below, workers are generally unable to pur-

---

*more advanced material

chase insurance against the risk of income fluctuations in regular insurance markets such as those that exist for accident, property, and life insurance. However, the continuing nature of the employment relationship makes feasible the implicit purchase of income insurance from the employer.

Seminal contributions to implicit contract theory were made by Azariadis (1975), Baily (1974), and Gordon (1974). A large literature, much of it highly technical in nature, has subsequently developed. Our purpose in this section is to present the basic elements of implicit contract theory in as nontechnical a fashion as possible. Further details are available in surveys by Azariadis (1979), Azariadis and Stiglitz (1983), Hart (1983), and Rosen (1985).

Differences in attitudes toward risk between employers and employees provide the basis for both parties to benefit from a risk-sharing arrangement. Two reasons why workers may be more risk-averse than the owners of firms have been advanced. Perhaps the most significant factor is that for many workers their wealth consists largely of the value of their human capital, which cannot be diversified. In contrast, individuals whose wealth consists largely of financial capital can reduce the risk of a reduction in their wealth by holding a diversified portfolio; that is, by acquiring shares in (or income claims on) a variety of companies. It is not possible to diversify wealth holdings in the form of human capital because markets analogous to the stock market do not exist for buying and selling claims on the incomes of different individuals or groups of individuals. Such markets would constitute a form of slavery and would therefore be illegal, even if there were sufficient demand to make markets for trading in such claims viable. Workers, therefore, are in the awkward position of having most of their wealth in one risky asset — their human capital. As a consequence, they seek alternative ways of reducing the risk of fluctuations in the return on that asset, their employment income.

A second reason for differences in risk attitudes involves sorting according to innate risk preferences. Those who are venturesome — risk-neutral or perhaps even risk lovers — may be more likely to become entrepreneurs and thus the owners of firms. Cautious or risk-averse individuals may be more likely to become employees and wage earners.

If workers dislike the risk of fluctuations in their employment income, why don't they purchase income insurance from private insurance companies? Private markets for income insurance do not exist because of two phenomena — moral hazard and adverse selection — that may result in the selling of such insurance being an unprofitable activity, despite the demand that exists for the product. Moral hazard exists when individuals can influence the risk against which they are insured. For example, suppose workers are insured against reductions in their income associated with becoming unemployed. If the fact that they are insured affects their behaviour such that they are then more likely to be unemployed and collecting insurance — perhaps because those who become unemployed search longer for a better job when they are insured or because they become more willing to accept a job with a high risk of layoff than they would if they were not insured — then the profitability of selling such insurance is reduced, perhaps to the point at which selling such insurance is unprofitable.

Adverse selection occurs when the insurer cannot observe the risk that a particular insuree represents. The insurer thus charges each customer the same rate. However, the high-risk individuals are more likely and the low-risk individuals less likely to purchase insurance. Thus the average risk among those who purchase insurance will be higher than the risk for the population as a whole. The insurer will therefore earn less — perhaps incurring a loss — than would be expected on the basis of population risk statistics. Furthermore, raising its insurance rates may not increase profitability because fewer individuals will purchase insurance at the higher rates *and* those who decide not to purchase insurance because of the higher rates will be the customers facing the lowest risk. Thus, with each increase in its rates, the insurance company ends up selling to a smaller number of customers with a higher average risk. In these circumstances, there may be no price that would enable insurance to be sold at a profit.

Moral hazard and adverse selection may exist in any insurance market. In some cases — such as automobile insurance for most individuals and life insurance for individuals under 65 — the reduction in profitability due to moral hazard and adverse selection is small enough that private insurance markets continue to exist. In other cases — such as life insurance for individuals over 65 and income insurance in the labour market — the reduction in profitability is evidently large enough that insurance against these risks can't be purchased in the usual fashion. The central hypothesis of implicit contract theory is that employees purchase income insurance indirectly from the employer. The continuing nature of the employment relationship enables the employer to deal with the moral hazard and adverse selection problems. The firm provides insurance only to its own employees, thus avoiding adverse selection. In addition, the firm controls the probability of income loss due to layoff or wage and/or hours reduction, thus avoiding the moral hazard problem.

Implicit contract theory applies to situations in which there is a long-term attachment between the firm and its workers. Many economists have suggested that in labour markets with these characteristics wages do not adjust each period to equate demand and supply. In contrast, many product and asset markets behave like "Walrasian auction markets" in which the price adjusts each period to clear the market. According to implicit contract theory this difference in behaviour reflects risk-sharing in the labour market. The continuing nature of the employment relationship enables the firm to stabilize its employees' incomes over several periods by paying a wage above that which would exist with continuous market clearing when product demand conditions are weaker than normal and paying a wage below that which would exist with continuous market clearing when product demand conditions are stronger than normal.

The reasons for the importance of long-term attachments in the labour market are discussed subsequently. Specific human capital is perhaps the most fundamental factor. As discussed in Chapter 18 on Human Capital Theory, when employees acquire firm-specific human capital due to on-the-job training and experience, then both the employer and the employees subsequently earn rents. The worker receives a wage greater than the alternative wage and the firm earns a marginal revenue product greater than the worker's wage. The fact that both parties are earning rents provides the incentive for each to continue the employment relationship. A second reason for long-term attachments in the labour market is the cost of search and mobility discussed in the previous section. Because finding a new job is costly to employees and hiring new workers is costly to employers, both parties usually prefer to maintain the employment relationship.

Although employers and employees are assumed to be involved in a continuing employment relationship, the basic model of implicit contracts can be explained in a two-period setting. In the initial period, firms offer wage and employment contracts to workers and workers decide which firm's contract to accept. The wages and employment stipulated in these contracts may be contingent on the state of product demand realized in the second period. The contracts agreed to in the first period are then carried out in the second. Thus workers are mobile ex ante (in the initial period when they are choosing which firm's contract to accept) but immobile ex post (in the second period when the uncertainty about product demand conditions is resolved and the terms of the contract are carried out). The assumption of ex post immobility is intended to reflect the continuing nature of the employment relationship and the cost of severing that relationship.

These contractual arrangements are not formal written agreements — hence the term "implicit contract" — but rather represent understandings that govern the behaviour of firms and workers. As stated colourfully by Okun (1981, p. 89), "Employers . . . Rely heavily on the 'invisible handshake' as a substitute for the invisible hand that cannot operate effectively in the career labour market." The explicit contracts observed in the union sector may also reflect risk-sharing to some degree, but the purpose of the analysis is to explain behaviour in the unorganized sector.

To keep the analysis simple, we will assume that the workers are homogeneous, each with utility function $u(y)$ where $y = w \times h$ is income, w is the wage rate and h is hours worked. To focus on wages and employment, hours of work will be assumed to be constant at h. A worker is thus either employed, working h hours, or unemployed, working zero hours. With hours of work fixed, an employed worker's utility can be written in terms of the wage rate alone, $u(w)$. An unemployed worker receives utility $u(k)$, where k is the value of leisure time and any unemployment benefit received from the unemployment insurance program. (Workers on layoff receive no income from the employer.) Because workers are homogeneous, which workers are laid off (should layoffs be required) is randomly determined.

Let N_o be the number of workers attached to the firm; this is the number of workers who agreed to join the firm's "labour pool" given the contract offered in the first period. The firm's labour supply curve in the second period is thus shown by S in Figure 13.3. At wage rates equal to or greater than the reservation wage k, the firm can employ up to N_o workers. The firm cannot employ any workers at wage rates below k.

We will discuss the case in which perfect competition prevails in the output market so that the price of output is not affected by the output produced and hence labour utilized. The analysis for other product market structures is similar. The firm's labour demand schedule is the locus of points for which wages are just equal to the value of the marginal product of labour, which in turn equals the marginal product of labour times the price at which the extra output is sold. That is, $D = VMP_N = MPP_N P$ where MPP_N is the marginal product of labour, and P is the price of output.

There is uncertainty about the state of the product market. When product market conditions are strong, the product price and the demand for labour will be high. The opposite holds in weak product market conditions. For ease of exposition we will assume that there are only two possible states of the product market — the "good state" in which the product price is high at P_a and the "poor state" in which the product price is low at P_b — and that these are equally likely to occur. The assumption of two equi-probable outcomes does not affect the analysis in any way except to make the results easier to present. An important assumption, however, is that both the employer and employees observe the state which is realized. When this *symmetric information* assumption holds the two parties can make the implicit wage-employment contract contingent on the observed state. When *asymmetric information* exists — for example, if the firm has better information about the state of the product market than do the workers — this type of contingent contract may not be optimal because the firm will usually have an incentive to cheat. For example, if the wage rate depends on the state, the firm may want to claim that the poor state has been realized — whatever the actual state — in order to pay a lower wage. The nature of contracts under asymmetric information is discussed further below.

The behaviour of a continuously clearing labour market, as shown in Figure 13.3(a), provides a useful benchmark for examining the influence of risk-sharing on wages and employment. In the good state the wage rises to w_a and all the workers in the firm's labour pool are employed ($N = N_o$). In the poor state the wage falls to w_b (equal to the reservation wage k) and employment declines to N_b.

It is important to note that there is no involuntary unemployment with market clearing. In the poor state, the decline in demand produces excess labour supply and the wage rate declines. As the wage falls below k, some workers withdraw

FIGURE 13.3

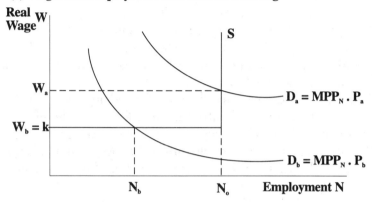

(a) Wages and Employment with Market Clearing

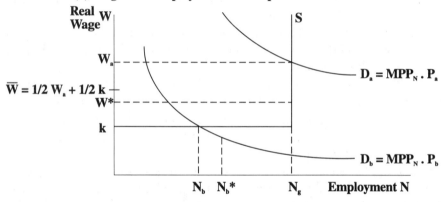

(b) Wages and Employment with Implicit Contracts

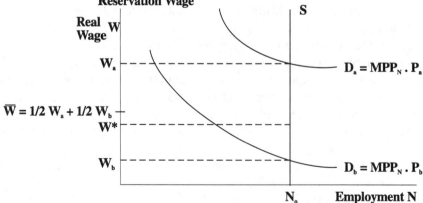

(c) Wages and Employment with Implicit Contracts and a Zero Reservation Wage

from employment. The result is an equilibrium at (w_b, N_b). Because the
ployed workers are paid the reservation wage k, unemployed workers do
envy the employed workers. Workers are indifferent between being employed
and unemployed, given the equilibrium wage rate and their value of leisure
time.

As is well known, in the absence of externalities and other sources of market
failure, the competitive market-clearing outcome produces an efficient allocation of resources. In the case analyzed here, the allocation of labour among firms
was determined ex ante. However, there remains the question of how to best
utilize each firm's labour pool. The allocation of workers shown in Figure 13.3(a)
— including the fact that N_bN_o workers are not employed in the poor state — is
efficient in the sense that it maximizes the value of output produced and income
generated in each state. However, with market clearing, workers face an uncertain income stream. Income may fluctuate because both the wage rate and
employment depend on the state which is realized. Because of this income
uncertainty, the market-clearing outcome is not, in general, a Pareto optimal
arrangement — i.e., it is possible through risk-sharing to make at least one of the
parties better off without making the other party worse off — even though it
does produce an efficient allocation of labour resources. An optimal arrangement takes into account both efficiency and risk-sharing considerations.

A Pareto optimal contract between a risk-neutral firm and risk-averse workers
has the following features: (1) the real wage rate is independent of the state
which is realized so that employed workers receive the same real wage whatever
the level of product demand, and (2) layoffs may occur in weak states of demand;
however, the number of workers laid off will be less than the number who would
be voluntarily unemployed under market clearing. Figure 13.3(b) shows an
example of an optimal wage-employment contract. Employed workers receive
the wage w^* in each state. Employment equals N_o in the good state and N_b^* in
the poor state.

Several features of the optimal contract should be noted. First, the contract
provides for a rigid real wage. Employed workers receive the same utility,
whatever the state of product demand and the overall price level. The contract
wage w^* is lower than the market-clearing wage in the good state and higher
than with market clearing in the poor state. Second, there are layoffs in the poor
state. However, the number of workers laid off $(N_b^*N_o)$ is less than the number
that would voluntarily withdraw from employment with market clearing (N_bN_o).
Third, there is involuntary unemployment in the poor state, which is why layoffs
are necessary. Employed workers receive utility $u(w^*)$ whereas unemployed
workers receive utility $u(k)$; laid-off workers would prefer to be employed given
that the wage does not decline in the poor state.

The way in which both parties may benefit from a risk-sharing arrangement
can be seen in Figure 13.3. Even though the firm is paying a higher wage and
employing more workers in the poor state than with market clearing, the savings

in labour costs in the good state are sufficiently large — given a contract wage w^* less than the expected wage w — that expected profits are higher under the implicit contract. The risk-neutral firm will therefore be better off with risk-sharing. Although their expected income is somewhat lower, risk-averse workers are also better off because their income is more certain. The reduction in the variability of workers' real income comes from two sources: real wage fluctuations are eliminated and more workers are employed in the poor state than with market clearing.

Workers' incomes are not fully stabilized, however. There remains some uncertainty due to the possibility of being laid off in weak demand conditions. In general, a contract that eliminates all income risk by stabilizing employment in addition to the wage is not optimal because in order for it to be in the firm's interest to employ all N_o workers in the poor state, the contract wage w^* would have to be lower than the reservation wage k. In these circumstances the workers would prefer not to be employed. Thus the optimal contract reduces, but doesn't necessarily eliminate, the income uncertainty that workers face.

The optimal contract represents a tradeoff between risk-sharing and production efficiency. On the basis of risk-sharing alone, the risk-neutral firm should absorb all the risk and the risk-averse workers should receive a fixed real income. For production efficiency alone, the real wage and employment should fluctuate as shown in Figure 13.3 (a). The optimal contract sacrifices some efficiency by employing more than N_b workers in the poor state. This inefficiency is reflected in the fact that the output produced by the additional $N_b N_b^*$ workers is less than the value of their leisure time. However, there is a benefit in terms of risk-sharing because employing these additional $N_b N_b^*$ workers in the poor state reduces the probability of a worker being laid off and not receiving the contract wage w^*. However, completely eliminating income uncertainty would be too costly in terms of production efficiency. The optimal contract strikes a balance between these two competing considerations.

Layoffs are a feature of the implicit contract because of the interaction between risk-sharing and the value placed by employees on time devoted to nonmarket activity. If the employees' reservation wage were zero — as would occur if there were no value placed on leisure time and no unemployment insurance benefits — the optimal contract would provide for a fixed real wage w^* and fixed employment of N_o, as illustrated in Figure 13.3 (c). In this case there is no tradeoff between production efficiency and risk-sharing, and employees' real income is completely stabilized. The higher the reservation wage k, the greater the tendency to employ layoffs in response to declines in demand, holding other influences constant, and the greater the tradeoff between production efficiency and risk-sharing.

Several conclusions emerge from this discussion of risk-sharing between the firm and its workers. Implicit contract theory can account for real wage rigidity, the use of layoffs to respond to reductions in demand, and the existence of involuntary unemployment. The optimal contract provides for a constant real wage and reduction in employment in the poor state. Because workers would prefer to be employed and earning the contract wage, the reduction in employment takes the form of layoffs. Those workers laid off are involuntarily unemployed in an ex post sense: given that the weak demand state occurs, that they are selected for layoff, and that the contract wage exceeds their reservation wage, they would prefer to be employed. However, the unemployment may be considered voluntary in an ex ante sense because the workers chose a wage-employment contract with some risk of layoff, and would make the same choice again in identical circumstances, given the Pareto optimal nature of the contractual arrangement. Thus the unemployment is involuntary in a rather limited sense.

Although implicit contract theory does provide a rigorous microeconomic explanation for wage rigidity, layoffs, and involuntary unemployment (in a restricted sense), the theory has been criticized for its inability to explain unemployment in excess of the amount that would be observed if wages adjusted each period to equate labour supply and demand (Akerlof and Miyazaki, 1980). Indeed, according to the basic implicit contract model discussed in this section, the number of workers involuntarily laid off in weak demand conditions is less than the number that would voluntarily withdraw from employment with market clearing. However, this implication of the model is closely related to the assumption that both parties observe the state of demand. When this symmetric information assumption is relaxed, optimal contracts may imply unemployment in excess of the amount that would occur with market clearing.

In many circumstances employers have more information than their employees on the true nature of the demand conditions they are facing. Thus it is not possible for the two parties to have wages and employment contingent on the state of demand, as was the case under symmetric information, because workers cannot verify the outcome. Contracts need to be contingent on some variable which both parties observe. In these circumstances, implicit contracts will generally involve layoffs rather than wage reductions in weak demand conditions. Contracts which provide for wage reductions in poor demand conditions will not work because the firm has an incentive to claim demand is weak whatever the true state. This is so because paying a lower wage is not costly to the firm. However, layoffs do impose costs on the employer because output is lower with fewer employees and the firm runs the risk of losing employees and has to incur rehiring costs. Rigid wage contracts that involve layoffs in poor states and rehires in good states discourage the firm from bluffing about the true nature of its product market, forcing the firm to reveal the true state of demand by its choice of employment. Thus under asymmetric information there is an additional rationale for wage rigidity and layoff.

EFFICIENCY WAGES

While implicit contract theory emphasized the role of wages and employment in risk sharing, efficiency wage theory focuses on the effect of wages on incentives and worker productivity. The central hypothesis is that firms may choose to pay wages above the market-clearing level in order to enhance worker productivity. Because higher wages also reduce labour demand, the consequence of this policy may be excess labour supply and unemployment.

There are several reasons why firms may benefit from paying a wage above the level necessary to attract labour. In less-developed countries, higher wages may result in better-fed and thus healthier and more productive workers. In developed economies wages may affect productivity in a variety of ways. Higher wages may improve worker morale, discourage shirking and absenteeism by raising the cost to workers of being fired, and reduce turnover. Firms may also prefer to pay high wages in order to reduce the threat of unionization or to obtain a larger and higher quality pool of job applicants.

The incentive to pay high wages will generally differ across firms and industries. Efficiency wages are most likely to be observed when other methods of enhancing productivity — such as supervision and monitoring of employees or the use of piece-rate compensation systems — are costly or ineffective. The impact of employee work effort on the quality and quantity of output is also an important factor. Shirking by employees can have disastrous effects in some jobs (such as the operator in a nuclear power station or the driver of a bus), while in others the consequences are much less severe. Similarly, in some production processes the work is highly interdependent, so that poor work effort by one employee affects the output of the entire group; whereas in other situations only that employee's output is affected. Other determinants of efficiency wages may also differ from one firm or industry to another. For example, turnover is more costly to some employers than others because of differences in hiring and training costs. As discussed in Chapter 14 on Union Growth and Incidence, some firms are more likely to become unionized than others because of factors such as size and capital intensity. Thus the incentive to pay high wages in order to discourage unionization will be stronger in some organizations than in others.

If each firm sets its wage rate above the market-clearing level, labour demand will be less than labour supply, resulting in involuntary unemployment. Figure 13.4(a) illustrates the equilibrium when each firm's optimal or efficiency wage is w^*. The unemployed wish to work at the prevailing wage; indeed, they are willing to work at wage rates below the efficiency wage w^*. However, firms will not lower the wage rate because doing so would reduce the productivity of their existing work force. The failure of wages to adjust to excess supply is a consequence of firm rather than worker behaviour.

Unemployment will result if wages are set above their market-clearing level in each sector. However, if there exist one or more sectors in which firms will pay the lowest wage necessary to attract labour, the consequence may be a dual labour market rather than unemployment. This point is illustrated in Figure 13.4(b) for the case in which there are two sectors in the economy, each employing the same type of homogeneous labour. Sector A consists of those firms which prefer to pay above-market wages in order to enhance productivity while sector B consists of employers who pay the market wage and who will lower the wage in the presence of excess supply. In the absence of efficiency wage effects, the equilibrium will be at (w_o, N_o) in each sector with the wage rate equalized across the two sectors $(w_0^A = w_0^B)$. However, because sector A firms set their wage at w_1^A,

FIGURE 13.4

(a) Labour Market Equilibrium with Efficiency Wages

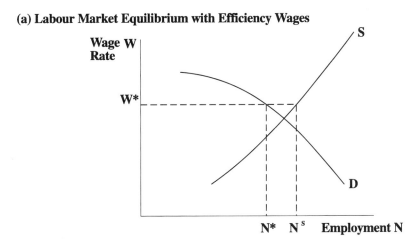

(b) Efficiency Wages and Dual Labour Markets

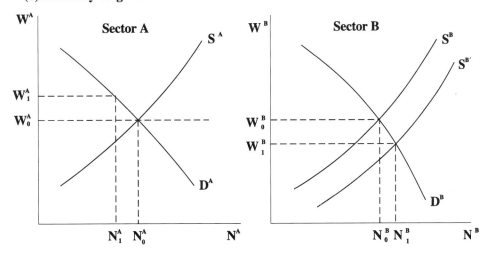

employment declines to N_1^A. The resulting increased labour supply to sector B lowers wages and increases employment in that sector. In these circumstances, efficiency wages result in a dual labour market, with high- and low-wage sectors for homogeneous workers. Wages differ not because of differences in the human capital characteristics of the workers in each sector but because of differences in the incentives facing firms.

Efficiency wages may cause unemployment even when there are firms — such as those in sector B in Figure 13.4(b) — which have no incentive to pay above-market wages. Some workers may search or wait for jobs in the high-wage sector rather than accept employment at low wages. This outcome is more likely the larger is the wage gap between the sectors, the greater is the turnover (and thus the probability of a job opening up) in the high-wage sector, and the more productive is unemployed search compared to employed search. Minimum wage laws may prevent the downward wage adjustment needed to absorb the additional labour supply created by efficiency wages in the high-wage sectors. In this case unemployment results from the interaction between efficiency wages and minimum-wage legislation.

The above analysis suggests that, in addition to being a source of wage rigidity and unemployment, efficiency wages may influence the occupational and industrial wage structure. Wages may differ not only because of differences in education, training, and other human capital characteristics of workers or because of job characteristics such as health and safety; but also because of such factors as the costs of monitoring and supervising employees, the importance of work effort to product quality and quantity, turnover costs, and the probability of union organization. That is, firms and industries may differ according to such factors as turnover costs, the ability and need for monitoring of work effort, and the importance of worker morale. Therefore we would expect to observe a distribution of wage rates across firms and industries. This prediction, and the empirical evidence relating to it, are examined in Chapter 20 on Wage Structures, where heterogeneity across firms and industries is taken into account.

Efficiency wage theory also implies that unemployment may act as a "worker discipline device" (Shapiro and Stiglitz, 1984). Employees are assumed to be motivated to work hard when they fear losing their job. Their work effort will thus be greatest when their wage is high relative to the next best alternative, as discussed above, and/or when unemployment is high. The greater the unemployment rate, the longer the expected search period facing a worker laid off and therefore the greater the cost of shirking on the present job.

In summary, a key assumption of the standard theory of labour demand is that the productivity of labour is technologically determined, and is therefore independent of the wage paid to workers. Efficiency wage theory relaxes this basic assumption, pointing to a variety of reasons why workers who are paid more will be more productive. This theory can account for a number of phenomena:

- some employers may rationally choose to pay homogeneous workers a wage above the market wage;

- the wage rate will not adjust to equate labour demand and supply, and this may result in involuntary unemployment;

- wages will not be equalized across homogeneous workers;

- there may be queues for the high-wage jobs, and these will be rationed by employers;

- there may be a dual labour market, with high-wage jobs in the primary sector (in which employers choose to pay above-market wages) and low wages in the secondary sector.

INSTITUTIONALISM, DUALISM, AND RADICALISM

Throughout this chapter we have emphasized the neoclassical approach to the analysis of labour market equilibrium. One of the central themes of recent research in the neoclassical tradition has been to demonstrate that much of observed labour market behaviour which appears initially to be inconsistent with the basic competitive model is in fact consistent with neoclassical theory, once account is taken of imperfect information, incentives, and uncertainty. This point has been illustrated in this chapter in our examinations of search, implicit contracts, and efficiency wages. There have always been schools of thought, however, that have de-emphasized, and at times attacked, the notion of market forces as being important factors in wage and employment determination. The severity of the criticism of the neoclassical approach has ranged from saying that economic factors play an important but minor role, to saying that they are distinctly subservient to other political-institutional factors, to the more radical critiques that argue that the economic framework simply masks the forces of power and class conflict that are the real determinants of wages and employment.

While it is always hazardous to try to categorize alternative perspectives and paradigms, they are labelled here is *institutionalism, dualism,* and *radicalism.* While there are some basic differences between these perspectives, they have as a common thread a critique of the neoclassical paradigm. In what follows an attempt will be made to briefly summarize each of these schools of thought with respect to their criticism of neoclassical theory, their own contribution, and their own strengths and shortcomings.

INSTITUTIONALISM

The institutionalism tradition in labour economics is one that plays down the economic forces and emphasizes the roles of individuals, institutions, custom and socio-political factors. It tends to emphasize descriptive realism as being more important than abstract, theoretical reasoning. The case-study and real-world observations are emphasized as being important in understanding the behaviour of the labour market.

The institutionalist tradition is evident in the work of Lester (1946) who attacked the assumptions of economics and its emphasis on marginal analysis as being unrealistic. He argued for empirical testing of the assumptions often employed in economics. For example, in responding to interviews and questionnaires, employers said that they did not use marginal analysis in their everyday business decisions and that they seldom responded to wage increases by reducing employment. Lester interpreted this as a rejection of conventional economic analysis.

Conventional economics was not without its defence. Machlup (1946) stressed the fact that it was the predictions and not the assumptions of theory that mattered, and that simplifying assumptions were necessary to deduce generalizations about labour market behaviour. For example, as discussed subsequently, the assumption of perfect mobility is a simplifying assumption that enables predictions about long-run wage differentials. It can be relaxed, and the implication for wage differentials examined. In addition, it may not be so unrealistic if one remembers that all that is required for the implications of economic analysis is that some workers at the margin of decision (not all workers) be mobile or, at times, have the threat of mobility.

The neoclassical paradigm can also be defended on the basis that it is operational and leads to empirically verifiable propositions. In defending the use of abstractions such as profit maximization on the part of firms and utility maximization on the part of individuals, Rottenberg (1956, p. 45) states: "complex motivation of real life does not destroy the truism of simple motivational behaviour in the abstract neoclassical model of the labour market. Other things equal, it can be a correct description of real life behaviour to say that workers make job choices with reference to relative prices."

The institutional tradition is also evident in the work of Arthur Ross (1956). He argued that wages are determined by "conscious human decision rather than by impersonal market forces" and that to understand union wage policy we must understand the internal policy aspect of the union — "what kind of organism it is, how it functions, and what is the role of leadership." Above all, Ross emphasized the interdependence of the wage structure and the importance of "equitable comparisons" as a criterion for wage determination.

The institutionalist tradition has continued in the writings of those who emphasized the importance of segmentation in labour markets. These writings in particular were the forefront of the more current work on dual labour markets. Kerr (1954), for example, coined the phrase "Balkanization of labour markets" to describe what he felt was a trend towards the increased segmentation of the labour market into a variety of noncompeting groups. In his view firms were becoming increasingly divorced or insulated from the competitive forces of the external labour market. They recruited from this labour market only at the lower "ports of entry." Most jobs were filled from internal promotion from the firm's "internal labour market." Consequently administered rules and internal company policies governing the internal labour market were more important than competitive economic forces that were important only for the seldom-used external labour market.

Dunlop (1957) also emphasized the importance of segmentation with the concepts of "job clusters" and "wage contours." Job clusters were occupational groupings with common wage-making characteristics so that wages within the clusters were highly interdependent. The clusters were linked together by technology, administrative rulings of the firm, or social custom. Wage contours referred to the stable wage relationships between certain firms that were linked together by similar product markets, labour markets, or custom. Thus the wage of an individual worker was dependent upon the job cluster of the occupation and on the wage contour of the firm. These clusters and contours defined the relevant comparisons for the actual wage determination process. Key wage settlements spilled over onto other jobs within the same cluster and onto other firms within the same contour. Thus, institutional forces which determined the nature of the job clusters, wage contours, key settlements, and spillover effects were seen as being of prime importance in the wage-determination process.

The institutionalist framework, with its impressive list of contributors, has given insight into the real-world behaviour of labour markets. It has been rich in classification and in describing the peculiarities of the labour market. However, it has not really provided an *analysis* of the basic forces that lie behind the everyday actions of labour and management. Consequently, it has seldom given rise to empirically testable implications that are capable of empirical confirmation or refutation. As such its contribution has not been cumulative. Rather, its insights have been incorporated into labour market analysis in a rather *ad hoc* fashion, often to explain particular anomalies.

In recent years, a literature has emerged applying Hirschman's (1970) concepts of exit and voice to the labour market. Exit refers to the traditional competitive response of leaving unsatisfactory market situations for more acceptable ones; voice refers to alternative means (e.g., complaint, protest, or suggestions) to improve unsatisfactory market situations. While the concepts of exit and voice can be subject to many of the criticisms of institutional economics, they do provide some interesting insights into labour market behaviour.

According to Hirschman, traditional economics regards exit, or the threat of exit, as the basic pressure that ensures that competition will provide goods and services efficiently: if they are not provided, then customers will exit and shop elsewhere. There are situations, however, where exit is not feasible, perhaps because it is a costly process or because of brand loyalty. In such circumstances, consumers may use voice ranging from constructive suggestion to severe protest. Most important, the possibility of exit reduces the necessity to use voice. From a policy perspective, Hirschman's analysis therefore implies that the usual economic solution of attaining efficiency through increased competition (exit) works against the possibility of attaining efficiency through mechanisms involving voice.

This has interesting implications for the analysis of labour markets since it suggests that increased competition and mobility (exit) — the panacea of economists — may reduce the use of voice as a way of handling labour market problems. As Freeman (1976) points out, collective bargaining can be viewed as the institutional embodiment of voice in the labour market. It therefore can serve as a viable alternative to the costly use of exit as a way of handling labour market problems.

Johnson's (1975) perspective on the impact of unions can also be viewed in the context of exit versus voice. Traditional economic analysis regards the union as a causal factor gaining higher wages for union members, largely through the threat of exit (withdrawal of services). Johnson raises the possibility, however, that causality works the other way; that is, for various reasons some firms and industries simply pay a greater-than-market wage, perhaps because it is important for them to reduce turnover and to have a queue of applicants from which to hire. Workers that are fortunate enough to be hired in this sector will not use exit to improve their situation because they cannot do better elsewhere. Because exit is not effective for them, they turn to voice in the form of collective bargaining and unionization to achieve other objectives (e.g., due process), knowing that they will be remaining in the firm for a long time. In essence, high wages reduce exit and encourage voice, raising the intriguing possibility that high wages lead to unionization rather than vice versa.

Clearly the exit, voice, and loyalty concepts raise some interesting issues, especially with respect to the advantages and disadvantages of various forms of exit as opposed to voice and loyalty as general ways of achieving efficiency and reconciling efficiency objectives with other objectives. In particular, they raise the possibility that many institutional arrangements can be regarded as efficient ways of utilizing voice in the labour market. These issues are discussed further in Part 4 on unions.

The efficiency wage literature is also relevant to the institutionalist tradition. One of the criticisms of neoclassical economic theory made by institutionalists was the prediction of a single equilibrium wage in a competitive labour market. Data collected by institutionalists and others indicated that a distribution of wages often existed in markets in which employers were sufficiently small relative to the market that they should not have any monopsony power (i.e., an essentially competitive labour market). Furthermore, this wage distribution was often found to be quite stable over time, suggesting that its existence could not be accounted for by factors such as imperfect information or shifts in demand and supply. As discussed previously, the efficiency wage literature can explain the existence of a distribution of wages in an otherwise competitive labour market if there are differences across firms in such factors as the costs of turnover, the ability to monitor work effort, and the ability of the firm to screen out the most productive workers. Because many of these factors are likely to remain stable over time, the associated wage distribution is also predicted to remain stable.

DUALISM

The dual or segmented labour market tradition characterizes the labour market as being segmented into two main parts, the primary or core, and the secondary or peripheral labour market. The characteristics of the primary and secondary labour markets are given in Doeringer and Piore (1971, p. 165):

> Jobs in the primary market possess several of the following characteristics: high wages, good working conditions, employment stability, chances of advancement, equity, and due process in the administration of work rules. Jobs in the secondary market, in contrast, tend to have low wages and fringe benefits, poor working conditions, high labour turnover, little chance of advancement, and often arbitrary and capricious supervision. There are distinctions between workers in the two sectors which parallel those between jobs: workers in the secondary sector, relative to those in the primary sector, exhibit greater turnover, higher rates of lateness and absenteeism, more insubordination, and engage more freely in petty theft and pilferage.

Of crucial importance in the analysis of segmented labour markets is the role of the internal labour market of individual firms in the primary sector. Within each internal labour market, well-developed hierarchies and stable employment relationships develop that are of mutual benefit to both management and workers. The job security and opportunities for career advancement that are so important to workers are also of value to management as ways of retaining a work force that has accumulated enterprise-specific skills and informal on-the-job training. In such circumstances, firms will try to reduce turnover cost by paying high wages, granting job security, and providing career advancement.

Within the internal labour market, the allocation of labour is determined by administrative rules and custom. Competitive forces, according to this perspective, have only a minor influence in determining broad limits within which administered and customary wage and employment policies are carried out.

Reasons for Segmentation

According to one view, segmentation may be historically determined and sustained by forces of custom and tradition. Well-established patterns emerge and these become accepted norms. As Doeringer and Piore (1971, p. 85) state with respect to the wage-determination process: "Any wage rate, set of wage relationships, or wage-setting procedure which prevails over a period of time tends to become customary; changes are then viewed as unjust or inequitable, and the work group will exert economic pressure in opposition to them." Thus segmentation becomes self-perpetuating through the forces of custom and tradition.

As discussed in the previous section, efficiency wage theory provides an explanation for dual labour markets which is consistent with rational behaviour on the part of profit-maximizing firms and homogeneous utility-maximizing workers. Dual labour markets may also arise from union wage bargaining, as discussed in Part 4 on unions. The higher wages may be negotiated via collective bargaining or be set by the employer to reduce turnover or encourage employee work effort. Because of the higher wage, employment is reduced below the level that would prevail in the absence of unionization or efficiency wage considerations. As a result, labour supply to other sectors is increased, reducing wages in these sectors. The consequence is a high-wage sector, in which workers are well paid because they are represented by a union or because paying a high wage is in the firm's interest, and a low-wage sector, in which union organization is difficult or in which firms have little incentive to pay above-market wages to reduce turnover, provide work incentives and so on.

Institutional and legal factors can also foster segmentation in labour markets. Unions, especially craft unions, can create barriers to entry into the primary (often unionized) labour market via such devices as the hiring hall, nepotism, and lengthy apprenticeship programs. Professional associations also create barriers to entry and prevent competition via occupational licensing and the control of paraprofessional substitutes. Discrimination can be an effective barrier preventing minority groups from moving from the secondary to the primary labour market.

Social legislation may foster segmentation, usually as an unintended side effect. For example, termination of employment legislation, experience rating under unemployment insurance, ceilings on the payroll tax used to finance social security — all of which result in quasi-fixed costs being associated with each employee — encourage firms to work their existing work force longer hours (to spread the fixed costs) rather than to hire additional employees which would result in additional fixed costs. The existing work force becomes protected from the competition of other workers because of these legislated fixed costs associated with hiring additional workers. These legislated quasi-fixed costs are compounded by the normal fixed costs associated with the recruitment, hiring, and training of new workers. To a certain extent employers may minimize these quasi-fixed costs by contracting out for work in the secondary labour market which does not have these high fixed costs, often because the legislation is inapplicable to small employers and to the self-employed. In essence, legislation that results in quasi-fixed costs of employment fosters segmentation by discouraging employers from hiring potential recruits from the secondary labour market, especially if their work habits are such that turnover may be high and hence the fixed costs are lost.

The impact of such legislation in fostering segmentation is illustrated most dramatically in the so-called "black markets" that are alleged to have developed as a way of avoiding the legislation. Especially in response to the difficulty of laying off workers, but also in response to minimum wage and other legislation, employers have at times responded by purchasing semifinished inputs from households or by contracting out to households to do specific tasks (e.g., sew on buttons, assemble a minor part). Veritable "cottage industries" develop which are not covered by legislation that creates fixed costs of employment, nor by safety laws, child labour laws, nor any other labour standards for that matter. In addition, the job of quality control, supervising, and monitoring the labour input is left up to the household, with firms simply buying or not buying the output. Markets develop and operate in unusual forms, and employers and employees adjust to legislation in subtle fashions!

Policy Implications of Segmented Labour Markets

Concerns for issues of public policy arise from the segmented labour market analysis for reasons of both distributive equity and allocative efficiency. Equity concerns arise because workers trapped in the secondary labour market tend to be disadvantaged workers, often caught in a vicious circle of poverty, and experiencing other problems associated with discrimination and poor working conditions. Efficiency concerns arise because of the costs associated with immobility of labour and barriers to entry.

As Smith (1976, p. 27) points out, as a general policy prescription the dual and segmented labour market analysis suggests a re-emphasis towards the structural aspects of labour demand and away from the structural aspects of labour supply and from macroeconomic full-employment policies. Macroeconomic full-employment policies to raise the level of aggregate demand are dismissed as helping mainly workers in the primary labour market, with at best a minor trickling-down effect on those in the secondary market, and a loosening of the barriers that prevent workers from moving from the secondary to the primary market. As Smith (1976, p. 28) concludes: "policies to maintain high, stable levels of aggregate demand have beneficial but limited effects in the dual theory."

Similarly, human capital policies that alter the structure of labour supply are not regarded by dual labour market analysts as viable ways of improving the earnings of workers in the secondary labour markets. Thus improved education, training, mobility, and labour market information — the traditional human capital factors — are dismissed largely on the grounds that they do little to improve earnings because they do not necessarily enable those in the secondary labour market to enter the primary labour market. In essence, improving the characteristics of workers via human resource development will do little if this is not accompanied by policies that alter the structure of labour demand so that these workers obtain jobs in the primary labour market.

After rejecting aggregate demand (full employment) and structural supply (human capital) policies, proponents of the dual labour market analysis suggest the structure of labour demand as being the most important determinant of the wages and working conditions of workers in the secondary labour market. The structure of labour demand is reflected in the characteristics of the industry, occupation, region, and firm in which workers find themselves. That is, the structure of labour demand in the secondary labour market reflects declining demand, low profitability, fierce competition in the product market, and competitive pressures from other sources of labour supply. Thus it is the characteristics of markets more than the characteristics of the workers themselves that are taken as being responsible for the low wages of those working in the secondary labour market.

Proponents of dual labour market analysis, therefore, recommend de-emphasizing human resource policies and recommend improvements in the characteristics of the markets in which low-wage workers find themselves. This could be done basically in two ways. One is to *extend* the protection, high wages and good working conditions of workers in the primary labour market to workers in the secondary labour market. This could be accomplished by wage fixing (e.g., minimum wages, equal pay laws), labour standards legislation, the encouragement of unionization and collective bargaining in the secondary labour market, and by policies such as unemployment insurance that would increase the bargaining power of workers in the peripheral sector. Wherever these policies have adverse side effects, perhaps on the employment opportunities of low-wage workers, they would have to be rectified by complementary programs, such as perhaps public employment.

A second device to improve the labour market situation in which low-wage workers find themselves is to break down the barriers that prevent movement from the secondary to the primary labour market. In particular, antidiscrimination laws may be effective, as would the removal of unnecessary educational or occupational licensing requirements. However, many of the protective features and job security of the primary labour market are regarded as desirable, and removing these features would be like "throwing out the baby with the bath water." Hence, the emphasis on extending the benefits of the primary labour market into the secondary labour market.

As discussed further in Chapter 20 on Wage Structures, the existence of dual labour markets due to efficiency wage considerations may provide a theoretical rationale for an "industrial strategy" designed to encourage and/or protect industries or firms which provide high-wage "good jobs." Such jobs are obviously desirable for workers — who receive a wage premium or economic rent relative to their best alternative wage — and are in the interest of firms — who choose to pay above-market wages — as a productivity-enhancing device. Thus society may be better off with more of such jobs, and fewer "bad jobs" in the secondary sector. The conditions under which this policy implication of efficiency wage and dual labour market theory is valid are currently the subject of considerable debate, and are discussed further in Chapter 20.

RADICALISM

An alternative perspective on labour market economics is found in the radical neo-Marxian writings as discussed, for example, in the works of Samuel Bowles (1972), Richard Edwards, Michael Reich and Thomas Weisskopf (1972), Herbert Gintis (1970, 1971), David Gordon (1972), Bennet Harrison (1972), and Howard Wachtel (1971, 1972). Probably the best summary of the radical perspective, giving an explicit comparison with the orthodox neoclassical and the dual labour market perspectives, is contained in Gordon (1972).

As Gordon (1972, p. 53) points out:

> Radical economic theory . . . has not yet been pulled together into a fully embellished theoretical system. Individual strands of radical economic analysis have begun to interweave, but many of the features of the eventual fabric have not become entirely clear. So far, much of the analysis has not even been precisely formulated, much less published; it continues to flow through conversations, letters and unpublished notes.

This makes a review of the radical perspective difficult; nevertheless, some common threads interweave through their analysis.

Following Marx, radicals emphasize that *economic classes* emerge as a result of the particular way in which productive activity is organized: under capitalism, the working class and capitalists emerge, with each group developing a strong subjective identification with its own class. Because of technological change and growth, an *economic surplus* develops over and above the subsistence level necessary to sustain and maintain the economic system. *Class conflict* arises over the division of this economic surplus. Under the current state of capitalism, this surplus tends to be appropriated by the capitalist owners of the means of production and they use the surplus to ensure their continued *power*. In particular, they use the instruments of the *state* (e.g., police, education, granting agencies, tax-transfer schemes) to ensure that effective power remains in their hands. These policies could include the granting of concessions to the working class, largely as a way to buy support, co-opt insurgence, and defuse protest.

Over time, however, capitalism develops *internal contradiction* — in essence, it contains the seeds of its self-destruction. In particular, competition, efficiency, and the specialization of labour are necessary for the survival of capitalism, but these very factors dehumanize and alienate the work force, strengthening class consciousness amongst exploited workers. Eventually, their plight will become intolerable and they will act in a concerted fashion to gain the effective power necessary to improve their position.

Radicals, therefore, emphasize the importance of class power in economic relationships. In particular, Gordon (1972, p. 64) argues that there are really two stages in the determination and distribution of income. First, the share of total income that goes to workers as opposed to capitalists depends on the relative power of the two groups. Second, once that share is determined by power relationships, then the amount that a given worker gets within the working class depends on the individual's productivity which in turn varies with the personal characteristics and capacities of the worker (as emphasized by human capital theory) and the characteristics of the job (as emphasized by dualism). Thus, to the conventional concepts of the quality of labour supply as emphasized by human capital theorists, and the structure of labour demand as emphasized by dual labour market analysts, radicals have added the concept of class power as an important factor in the determination and distribution of income. To cite Gordon (1972, p. 65): "The worker's final wage thus depends *both* on his individual productivity *and* on the relative power of the class to which he belongs. The radical theory thus combines the radical concept of class with orthodox notions of supply and demand."

As a general policy prescription, therefore, radicals tend to emphasize the importance of the development of class consciousness amongst workers and of fundamental changes in the power relationships between workers and capitalists. Specific policies to change these power relationships are not usually spelled out clearly, but what is clear is the radicals' belief that most of the policy recommendations that flow from traditional economic analysis, or from the dual labour market perspective, are ineffective without radical changes in power relationships.

AN OVERALL ASSESSMENT

Any overall assessment of the various competing paradigms — neoclassical, institutional, dual, and radical — is made difficult by the political overtones that surround any perspective. The matter is further complicated by the fact that the dual and new radical perspectives have not been subject to the test of time. And when they are judged, it is probably through the perspective of the reigning paradigm — neoclassical economics — with its emphasis on theory that yields implications capable of empirical acceptance or rejection. The extent to which this is a fair test is an open question, especially since the empirical testing itself is geared to the neoclassical paradigm with its emphasis on marginal analysis.

Each of the perspectives certainly does give insights into our understanding of labour market behaviour. Neoclassical economics, with its emphasis on income and relative prices as they affect various facets of labour supply and labour demand in alternative market structures, certainly has provided a systematic and consistent theoretical explanation for various aspects of observed labour market behaviour. This is especially the case when it has been modified to incorporate realistic factors such as uncertainty, transaction costs, incentive structures, and lack of information. Institutionalism, with its emphasis on the importance of individuals, institutions, custom, and socio-political factors, has reminded us of the importance of these factors, especially as they interact in that most peculiar of markets — the labour market. In addition, some perspectives of voice versus exit, for example, have suggested the possibility that certain institutional features of the labour market are rational efficient responses to peculiarities of the labour market. Dualism has continued the institutional tradition and has focused on the structure of labour demand that segments the labour market into essentially two noncompeting groups in the primary and secondary labour markets. However, much of the most recent literature on dual labour markets has been based on efficiency wage theory and is accordingly based on the neoclassical assumptions of profit-maximizing firms and utility-maximizing workers. Radicalism adds the importance of power relationships in a system of inherent class conflict.

While there certainly is no consensus as to which paradigm provides us with the greatest understanding of labour market behaviour, there is certainly more agreement that the insights from all of the paradigms have been useful. As a general proposition it is probably correct to say that the basic neoclassical economic paradigm has not been replaced — at least not yet — by any of the alternatives. Nevertheless, they have been useful in pointing out its possible weaknesses and in pressuring it to analyze institutional phenomena, labour market segmentation, and power relationships — all of which seem to be particularly important in labour markets. The extent to which the neoclassical paradigm is sufficiently adaptable remains an open — and interesting — question.

QUESTIONS

1. Discuss the implications of implicit contract theory for the following phenomena: wage rigidity, layoffs, and involuntary unemployment.

2. If workers are risk-averse, why might they enter into contractual arrangements that could involve the risk of layoffs as opposed to wage adjustments?

3. What is meant by efficiency wages and why do they arise? What are their implications for labour market equilibrium?

4. "By emphasizing descriptive realism, especially with respect to the assumptions of economics and the motivation of labour and management, institutional labour economics must be correct and therefore more useful in our understanding of labour market behaviour." Discuss.

5. "Because it assumes perfect information, perfect mobility, and zero transaction costs, conventional microeconomics has no applicability to the labour market." Discuss.

6. Discuss the meaning of the following terms as used in labour market analysis: internal versus external labour markets; job clusters; wage contours, key settlements, and spillover effects; and exit, voice, and loyalty.

7. Briefly outline the characteristics of the primary and secondary labour markets, and the characteristics of the workers in each. Is there a parallel in the literature on product markets as described in the field of industrial organization?

8. Discuss the main forces that give rise to the segmentation of the primary from the secondary labour market, according to the proponents of dual labour markets. Over time would we expect these forces to increase or decrease, and why?

9. Document a variety of internal administrative rules or company personnel policies that apply to the internal labour market. To what extent might these be an efficient response to the peculiarities of the labour market rather than simply inefficient institutional constraints?

10. Compare and contrast the *general* policy implications that flow from each of the following paradigms — neoclassical, institutional, dual, and radical. What would each say about the usefulness of the following specific policies:

    (a) minimum-wage legislation;

    (b) equal pay legislation;

    (c) unemployment insurance;

    (d) training programs;

    (e) job creation in the public sector.

11. "It is tempting to argue that each of the various labour market paradigms has a role and that different paradigms may be necessary to analyze different phenomena. However, it is intellectually unsatisfactory to utilize paradigms in such an *ad hoc* fashion so as to 'fit the facts' of each case. If a set of forces are important then they should apply as basic determinants of a variety of phenomena at a variety of levels." Discuss.

12. The use of voice has been suggested as an alternative to exit as a way of dealing with some labour market problems. Discuss various forms of exit and voice and discuss their costs and benefits.

13. "One of the problems of relying on voice is that the free market will provide an insufficient amount of voice because it has public good characteristics; that is, the benefits of voice are available to a wide group of persons (while the costs may be born by those who provide the voice) and the market does

not automatically extract payment from those who benefit, to reward those who bear the cost." Discuss the extent to which this is true. To the extent that it is true, are there institutional arrangements to make all those who benefit pay so as to reward those who bear the cost?

14. In his evaluation of the usefulness of the dual labour market approach in the Canadian context, David Smith (1976, p. 3) suggests that it highlights the following areas for further research: barriers to job choice arising from restrictions to entry into many types of jobs; the effect on job choice of behavioural traits induced by the work environment and peer groups; the extent to which on-the-job training is more of a socialization process than a costly investment, and the role of discrimination in this process; analysis of types of internal labour markets created by economic organizations and the integration of this analysis with economic theory developed primarily for external labour markets. Discuss briefly how our understanding of each of these areas could be furthered by the application of dual and/or conventional labour market analysis.

REFERENCES AND FURTHER READINGS

A. Information and Search

(See also references to Chapter 25 on Unemployment: Causes and Consequences)

Barron, J. and S. McCafferty. Job search, labour supply and the quit decision. *AER* 67 (September 1977) 683–691.

Burdett, K. A theory of employee job search and quit rates. *AER* 68 (March 1978) 212–220.

Gronau, R. Information and frictional unemployment. *AER* 61 (June 1971) 290–301.

Hasan, A. and S. Gera. *Job Search Behaviour, Unemployment and Wage Gains in Canadian Labour Markets.* Ottawa: Economic Council of Canada, 1982.

Johnson, W. A theory of job shopping. *QJE* 92 (May 1978) 261–278.

Lippman, S. and J. McCall. The economics of job search: a survey. *EI* 14 (June 1976) 155–189 and (September 1976) 347–368. Comment by G. Borjas and M. Goldberg 16 (January 1978) 119–125.

Lippman, S. and J. McCall (eds.). *Studies in the Economics of Search.* Amsterdam: North Holland, 1979.

Maki, D. *Search Behaviour in Canadian Job Markets.* Special Study No. 15. Ottawa: Economic Council of Canada, 1971.

McCall, J. Economics of information and job search. *QJE* (February 1970) 113–126. Comment by Peterson and reply (February 1972) 127–134.

Mortensen, D.T. Job search and labor market analysis. *Handbook of Labor Economics,* O. Ashenfelter and R. Layard (eds.). Amsterdam: North Holland, 1986.

Parsons, D. Quit rates over time: a search and information approach. *AER* 63 (June 1973) 390–401.

Phelps, E.S. et al. *Microeconomic Foundations of Employment and Inflation Theory.* New York: Norton, 1970.

Pissarides, C. Risk, job search and income distribution. *JPE* 82 (November/December 1974) 1255–1269.

Stephenson, S. The economics of youth job search behaviour. *R.E. Stats.* 58 (February 1976) 104–111.

Stigler, G.J. Information in the labour market. *JPE* 70 (October 1962) 94–105.

B. Implicit Contracts

Akerloff, G. and H. Miyazaki. The implicit contract theory of unemployment meets the wage bill argument. *R.E. Studies* 47 (January 1980) 321–338.

Arnott, R.J. and A.J. Hosios. Implicit contracts, labour mobility, and unemployment. *AER* 78 (December 1988) 1046–1066.

Azariadis, C. Implicit contracts and underemployment equilibria. *JPE* 83 (December 1975) 1183–1202.

Azariadis, C. On the incidence of unemployment. *R.E. Studies* 43 (February 1976) 115–126.

Azariadis, C. Implicit contracts and related topics: a survey. *the Economics of the Labour Market*, Z. Hornstein et al. (eds.). London: HMSO, 1979, 221–248.

Azariadis, C. and J.E. Stiglitz. Implicit contracts and fixed price equilibria. *QJE* 98 (Supplement 1983) 1–22.

Baily, M.N. Wages and employment under uncertain demand. *R.E. Studies* 41 (January 1974) 37–50.

Beaudry, P. and J. DiNardo. The effect of implicit contracts on the movement of wages over the business cycle: evidence from micro data. *JPE* 99 (August 1991) 665–688.

Bellante, D. and A.N. Link. Worker response to a menu of implicit contracts. *ILRR* 35 (July 1982) 590–599.

Bull, C. The existence of self-enforcing implicit contracts. *QJE* 102 (February 1987) 147–159.

Bull, C. Implicit contracts in the absence of enforcement and risk aversion. *AER* 73 (September 1983) 658–671.

Chary, V.V. Involuntary unemployment and implicit contracts. *QJE* 98 (1983) 107–122.

Cothren, R. Job search and implicit contracts. *JPE* 91 (June 1983) 494–504.

Farmer, R.E.A. Implicit contracts with asymmetric information and bankruptcy: the effect of interest rates on layoffs. *R.E. Studies* 52 (July 1985) 427–442.

Flanagan, R.J. Implicit contracts, explicit contracts, and wages. *AER* 74 (May 1984) 345–349.

Gamber, E.N. Long-term risk-sharing wage contracts in an economy subject to permanent and temporary shocks. *JOLE* 6 (January 1988) 73–99.

Gordon, D. A neoclassical theory of Keynesian unemployment. *EI* 12 (December 1974) 431–459.

Grossman, S. and O. Hart. Implicit contracts, moral hazard, and unemployment *AER* 71 (May 1981) 301–307.

Grossman, S.J. Implicit contracts under asymmetric information. *QJE* 98 (1983) 123–156.

Haley, J. Theoretical foundations for sticky wages. *Journal of Economic Surveys* 4 (1990) 115–155.

Hart, O.D. Optimal labour contracts under asymmetric information: an introduction. *R.E. Studies* 50 (January 1983) 3–35.

MacLeod, W.B. Implicit contracts, incentive compatibility, and involuntary unemployment. *Econometrica* 57 (March 1989) 447–480.

Mastuz, S.J. Implicit contracts, unemployment and international trade. *EJ* 96 (June 1986) 307–322.

Mastuz, S.J. The Heckscher-Ohlin-Samuelson model with implicit contracts. *QJE* 100 (November 1985) 1313–1329.

Mayers, D. and R. Thaler. Sticky wages and implicit contracts: a transactional approach. *EI* 17 (October 1979) 559–574.

Newbery, D.M. and J.E. Stiglitz. Implicit contracts, unemployment and economic efficiency. *EJ* 97 (June 1987) 416–430.

Okun, A.M. *Prices and quantities: a macroeconomic analysis.* Washington: The Brookings Institute, 1981.

Osano, H. and T. Inoue. Implicit contracts in the Japanese labour market. *Journal of the Japanese and International Economy* 2 (June 1988) 181–198.

Polemarchakis, H.M. Implicit contracts and employment theory *R.E. Studies* 46 (January 1979) 97–108.

Rosen, S. Implicit contracts: a survey. *JEL* 23 (September 1985) 1144–1175.

Samuelson, L. Implicit contracts with heterogeneous labor. *JOLE* 3 (January 1985) 70–90.

C. Efficiency Wages

Akerlof, G.A. and J.L. Yellen (eds.). *Efficiency Wage Models of the Labour Market.* Cambridge: Cambridge University Press, 1986.

Bar-Ilan, A. Monitoring workers as a screening device. *CJE* 24 (May 1991) 460–470.

Black, D.A. and J.E. Garen. Efficiency wages and equilibrium wages. *EI* 29 (July 1991) 525–540.

Carmichael, H.L. Efficiency wage models of unemployment — one view. *EI* 28 (April 1990) 269–295.

Cappelli, P. and K. Chauvin. An interplant test of the efficiency wage hypothesis. *QJE* 106 (August 1991) 769–787.

Copeland, B.R. Efficiency wages in a Ricardian model of international trade. *Journal of International Economics* 27 (November 1989) 221–244.

Hendricks, W.E. and L.M. Lawrence. Efficiency wages, monopoly unions and efficient bargaining. *EJ* 101 (September 1991) 1149–1162.

Kahn, C. and D. Mookherjee. A competitive efficiency wage model with Keynesian features. *QJE* 103 (November 1988) 609–645.

Kraft, K. The incentive effects of dismissals, efficiency wages, piece-rates and profit sharing. *R.E. Stats.* 73 (August 1991) 451–459.

Krueger, A.B. and L.H. Summers. Efficiency wages and the inter-industry wage structure. *Econometrica* 56 (March 1988) 259–293.

Lang, K. and S. Kahn. Efficiency wage models of unemployment: a second view. *EI* 28 (April 1990) 296–306.

Levine, D.K. Efficiency wages in Weitzman's share economy. *IR* 28 (Fall 1989) 321–334.

Lindbeck, A. and D.J. Snower. Efficiency wages versus insiders and outsiders. *European Economic Review* 31 (February–March 1987) 407–416.

Malcomson, J. Unemployment and efficiency wage hypothesis. *EJ* 91 (December 1981) 848–866.

Oi, W.Y. Employment relations in dual labour markets. *JOLE* 8 (January 1990) 124–149.

Ong, P.M. and D. Mar. Post-layoff earnings among semiconductor workers. *ILRR* 45 (January 1992) 366–379.

Raff, D.M.G. and L.H. Summers. Did Henry Ford pay efficiency wages? *JOLE* 5 (October 1987) 57–86.

Ramaswamy, R. and R.E. Rowthorn. Efficiency wages and wage dispersion. *Economica* 58 (November 1991) 501–514.

Shapiro, C. and J.E. Stiglitz. Equilibrium unemployment as a worker discipline device. *AER* 74 (June 1984) 433–444.

Strand, J. Unemployment and wages under worker moral hazard with firm-specific cycles. *IER* 32 (August 1991) 601–612.

Summers, L.W. Relative wages, efficiency wages, and Keynesian unemployment. *AER* 78 (May 1988) 383–388.

Weiss, A. *Efficiency wages: models of unemployment, layoffs, and wage dispersion.* Princeton: Princeton University Press, 1990.

Yellen, J. Efficiency wage models of unemployment. *AER* 74 (May 1984) 200–205.

D. Institutionalism, Dualism, and Radicalism

Albrecht, J. and S.B. Vroman. Dual labour markets, efficiency wages, and search. *JOLE* 10 (October 1992) 438–461.

Alexander, A. Income, experience and the structure of internal labour markets. *QJE* 88 (February 1974) 63–85.

Anderson, K.H., Butler, J.S. and F.L. Sloan. Labour market segmentation: a cluster analysis of job groupings and barriers to entry. *SEJ* 53 (January 1987) 571–590.

Barron, J. and M. Loewenstein. On employer-specific information and internal labor markets. *SEJ* 52 (October 1985) 431–445.

Bosanquet, N. and P. Doeringer. Is there a dual labour market in Great Britain? *EJ* 83 (June 1973) 421–435.

Boston, T.D. Segmented labour markets: new evidence from a study of four race-gender groups. *ILRR* 44 (October 1990) 99–115.

Bowles, S. Unequal education and the reproduction of the social division of labour. *The Capitalist System*, R.C. Edwards, M. Reich and T.E. Weisskopf, eds. Englewood Cliffs, N.J.: Prentice-Hall, 1972.

Bulow, J.I. and L.H. Summers. A theory of dual labour markets with application to industrial policy, discrimination, and Keynesian unemployment. *JOLE* 4 (July 1986) 376–414.

Cain, G. The challenge of dual and radical theories of the labour market to orthodox theory. *AER Proceedings* 65 (May 1975) 16–22.

Cain, G.G. The challenge off segmented labour market theories to orthodox theory: a survey. *Journal of Economic Literature* 14 (December 1976) 1215–1257.

Carnoy, M. and R. Rumberger. Segmentation in the U.S. labour market: its effects on the mobility and earnings of whites and blacks. *Cambridge Journal of Economics* 4 (June 1980) 117–132.

Dickens, W.T. and K. Lang. A test of dual labor market theory. *AER* 75 (December 1985) 792–805.

Dickens, W.T. and K. Lang. The reemergence of segmented labour market theory. *AER* 78 (May 1988) 129–134.

Doeringer, P. and M. Piore. *Internal Labour Markets and Manpower Analysis.* Lexington, Mass.: Health, 1971.

Doeringer, P. and M. Piore. Unemployment and the "dual" labour market. *Public Interest* 38 (Winter 1975) 67–79.

Donner, A. and F. Lazar. An econometric study of segmented labour markets and the structure of unemployment: the Canadian experience. *IER* 14 (1973) 312–237.

Dunlop, J. The task of contemporary wage theory. *The Theory of Wage Determination*, J. Dunlop, ed. New York: St. Martin's Press, 1957.

Edwards, R., M. Reich and T. Weisskopf. *The Capitalist System.* Englewood Cliffs, N.J.: Prentice-Hall, 1972.

Freeman, R. Individual mobility and union voice in the labour market. *AER Proceedings* 66 (May 1976) 361–368.

Gintis, H. Education, technology, and the characteristics of worker productivity. *AER Proceedings* (May 1971) 266–279.

Gordon, D. *Theories of Poverty and Underemployment.* Lexington, Mass.: Heath, 1972.

Gregory, R.G. and R.C. Duncan. Segmented labour market theory and the Australian experience of equal pay for women. *Journal of Post Keynesian Economics* 3 (Spring 1981) 403–428.

Harrison, B. *Education, Training and the Urban Ghetto.* Baltimore: Johns Hopkins, 1972.

Hirschman, A. *Exit, Voice and Loyalty.* Cambridge, Mass.: Harvard University Press, 1970.

Jones, S.R.G. Minimum wage legislation in a dual labour market. *European Economic Review* 31 (August 1987) 1229–1246.

Kennedy, B.R. Mobility and instability in Canadian earnings. *CJE* 22 (May 1989) 383–394.

Kerr, C. The balkanization of labour markets. *Labour Mobility and Economic Opportunity,* E. Wight Bakke et al., eds. Cambridge, Mass.: M.I.T. Press, 1954.

Lester, R. Shortcomings of marginal analysis for wage-employment problems. *AER* 36 (March 1946) 63–82.

Machlup, F. Marginal analysis and empirical research. *AER* 36 (September 1946) 519–554.

Magnac, T. Segmented or competitive labour markets. *Econometrica* 59 (January 1991) 176–177.

Malcolmson, J. Work incentives, hierarchy and internal labor markets. *JPE* 92 (June 1984) 486–507.

McDonald, I.M. and R.M. Solow. Wages and employment in a segmented labour market. *QJE* 100 (November 1985) 1115–1141.

McNabb, R. and G. Psacharopoulos. Further evidence on the relevance of the dual labor market hypothesis for the U.K. *JHR* 16 (Summer 1981) 442–448.

Meng, R. An empirical test for labour market segmentation of males in Canada. *IR* 24 (Spring 1985) 280–287.

Merrilees, W. Labour market segmentation in Canada: an econometric approach. *CJE* 15 (August 1982) 458–473.

Miyazaki, H. The rat race and internal labour markets. *Bell Journal of Economics* 8 (Autumn 1977) 394–418.

Neuman, S. and A. Ziderman. Testing the dual labour market hypothesis: evidence from the Israel labour mobility survey. *JHR* 21 (Spring 1986) 230–237.

Osberg, L., R. Muzany, R. Apostle and D. Clairmont. Job mobility, wage determination and market segmentation in the presence of sample selection bias. *CJE* 19 (May 1986) 319–346.

Osborne, M.J. Capitalist-worker conflict and involuntary unemployment. *R.E. Studies* 51 (January 1984) 111–127.

Osterman, P. An empirical study of labour market segmentation. *ILRR* 28 (July 1975) 508–523.

Piore, M.J. Fragments of a "sociological" theory of wages. *IRRA* (December 1972) 286–295.

Rebitzer, J.B. and L.T. Taylor. A model of dual labour markets when product demand is uncertain. *QJE* 106 (November 1991) 1373–1383.

Ross, A. *Trade Union Wage Policy.* Berkeley: University of California Press, 1956.

Rottenberg, S. On choice in labor markets. *ILRR* 9 (January 1956) 183–199. Comment by R. Lampman and reply, 9 (July 1956) 629–641.

Smith, David. *The Dual Labour Market Theory: A Canadian Perspective.* Kingston: Queen's University Industrial Relations Centre, 1976.

Taubman, P. and M. Wachter. Segmented labour markets. *Handbook of Labour Economics* Vol. 1. O. Ashenfelter and R. Layard (eds.). New York: Elsevier, 1986.

Thurow, L. *Poverty and Discrimination.* Washington: Brookings, 1969.

Van de Klundert, Th. On socioeconomic causes of "wait unemployment." *European Economic Review* 34 (July 1990) 1011–1022.

Wachtel, H. Capitalism and poverty in America: paradox or contradiction? *AER Proceedings* 62 (May 1972) 187–194.

Wachter, M. Primary and secondary labour markets: a critique of the dual approach. *BPEA* (No. 3, 1974) 637–694.

Wachter, M.L. and R.D. Wright. The economics of internal labour markets. *IR* 29 (Spring 1990) 240–262.

Williamson, O., M. Wachter and J. Harris. Understanding the employment relation: an analysis of idiosyncratic exchange. *Bell Journal of Economics* 16 (Spring 1975) 252–278.

PART 4

Unions

The three chapters in this part examine labour market equilibrium in situations where wages, employment, and working conditions are determined by collective bargaining between employers and unions. Previous chapters discussed the nature of labour market equilibrium when both firms and workers act on an *individual* basis. In these circumstances, the individual labour supply decisions of employees and the labour demand choices of employers interact to determine wages, employment, and other working conditions. In labour markets in which unions play a role, the distinguishing feature is that the conditions of employment are determined by *collective* rather than individual action; that is, the union negotiates with the employer on behalf of the workers as a group. As is the case in nonunion organizations, individual employees who are dissatisfied with wages and working conditions may quit or threaten to quit to seek employment elsewhere (the "quit threat"); however, in unionized establishments the main mechanism for bringing pressure to bear on the employer is the threat of the collective withdrawal of labour (the "strike threat").

Chapter 14 deals with the incidence of unionization — what determines which employers and employees are union members and which ones are nonunion. Differences in the extent of unionization across countries and changes in unionization over time — the growth and decline of collective bargaining — are also examined.

Chapter 15 deals with the theory of union behaviour, how unions and employers interact to determine wage and employment outcomes, and bargaining. This chapter provides the basic theory of behaviour in unionized labour markets, and how this behaviour differs from nonunion (competitive or monopsonistic) markets in which firms set wages unilaterally.

Chapter 16 examines the impacts of unions on a variety of labour market outcomes: the level and distribution of wages, employment, nonwage benefits, productivity, profitability of firms, and resource allocation. The empirical evidence discussed in this chapter indicates the numerous effects that unions have on the workers they represent, employees in the nonunion sector, employers and the shareholders of firms, and on society as a whole.

CHAPTER 14

Union Growth and Incidence

Unions are collective organizations whose primary objective is to improve the well-being of their members. In Canada, this objective is met primarily through collective bargaining with the employer. The outcome of this process is a collective agreement specifying wages, nonwage benefits such as those relating to pensions, vacation time, and health and medical expenses, and aspects of the employment relation such as procedures relating to hiring, promotion, dismissal, layoffs, overtime work, and the handling of grievances.

There are two basic types of unions. Craft unions represent workers in a particular trade or occupation; examples are found in the construction, printing, and longshoring trades. Industrial unions represent all the workers in a particular industry regardless of occupation or skill; examples are found in the automobile, steel, and forest industries. Some unions combine elements of both types.

In addition to their collective bargaining activities, unions play a role in social and political affairs. Close ties between unions and social democratic political parties are common, especially in Western Europe. In Canada the union movement provides financial and other support for the New Democratic Party. Unions also seek to influence the government in power, and accordingly have been involved in various forms of consultation and collaboration with governments and, in some cases, representatives of the business community. Although these and other political and social aspects of unions are important and deserve study, the major function of Canadian unions is to represent their members' interests in collective bargaining with employers, and it is this aspect of unions which is studied here.[1]

This chapter begins with a brief discussion of the nature and significance of unions and collective bargaining in Canada. The extent of unionization in the labour force, how this has changed over time, how it compares with other

---

1. For discussion of the social and political aspects of unions and collective bargaining see Banting (1986), Riddell (1986), and Lipset (1986).

countries, and the legal framework governing unionization and collective bargaining are described. We then examine the determinants of the extent and incidence of unionization in the economy: what factors result in some groups of workers being represented by a union while other groups of workers remain unorganized? The following chapters are devoted to understanding the economic consequences of unions and collective bargaining.

UNIONS AND COLLECTIVE BARGAINING IN CANADA

For a significant fraction of the Canadian labour force, wages and other conditions of employment are determined by collective bargaining. Table 14.1 provides several measures of the quantitative significance of unions and collective bargaining in the labour market. Over time, an upward trend is evident in all these measures. Union membership as a proportion of the civilian labour force increased from about 9 percent in 1920 to about 30 percent in 1990. Union membership as a proportion of (civilian) non-agricultural paid workers — a

Table 14.1
UNION MEMBERSHIP AND UNION DENSITY IN CANADA, 1920–1990

YEAR	UNION MEMBERSHIP (000S)	UNION MEMBERSHIP AS A PERCENTAGE OF CIVILIAN LABOUR FORCE	UNION MEMBERSHIP AS A PERCENTAGE OF NON-AGRICULTURAL PAID WORKERS
1920	374	9.4	16.0
1925	271	7.6	14.4
1930	322	7.9	13.9
1935	281	6.4	14.5
1940	362	7.9	16.3
1945	711	15.7	24.2
1951*	1029	19.7	28.4
1955	1268	23.6	33.7
1960	1459	23.5	32.3
1965	1589	23.2	29.7
1970	2173	27.2	33.6
1975	2884	29.9	35.6
1980	3397	30.2	37.0
1985	3666	29.8	38.1
1990	4031	29.9	36.2

*The survey was not conducted in 1950.

Sources: Labour Canada, *Directory of Labour Organizations in Canada*, 1990–91.
Reproduced with the permission of the Minister of Supply and Services Canada, 1993.

measure which excludes from consideration the self-employed and those employed in agriculture, who are generally not eligible for unionization — has increased from 16 to 36 percent over the same period. Both these measures tend to understate the extent of union organization in the economy because not all those covered by collective agreements are union members. (As is explained below, Canadian labour legislation provides that, once certified, the union is the exclusive bargaining representative for all employees in the bargaining unit, whether or not they are union members.) For example, data from Statistics Canada's 1990 Labour Market Activity Survey indicate that 37.6 percent of paid workers were covered by a collective agreement whereas only 33.1 percent were union members (see Table 14.3).

The influence of unions may extend further than is suggested by these measures of the extent of union organization or "union density." For example, the extent of unionization is much higher among non-office than office employees. The wages and benefits negotiated for production workers may set a standard for nonunion office employees in the same establishment. More generally, union agreements may influence the wages and working conditions of unorganized workers in the same industry, urban area, or region.

THE LEGAL FRAMEWORK

Union representation and collective bargaining in Canada are regulated by an elaborate legal framework. The evolution of this framework reflects the changing social attitudes toward the role of unions and collective bargaining in Canadian society, and has played a role in the increase in union organization evident in Table 14.1. In general terms, the law with respect to collective bargaining in Canada has passed through three main phases. (For a more detailed description of the evolution of Canadian labour legislation see Weiler, 1986.) In the first phase, the period mostly prior to Confederation, the law discouraged collective bargaining. Judges interpreted the common law to hold that collective action by employees constituted a criminal conspiracy. There were also other criminal and civil constraints on both individual and group action by workers. In the second phase, which began in the 1870s, the law was "neutral" with respect to collective bargaining. In particular, the Trade Unions Act of 1872, amendments to criminal law, and other legislative actions removed many of the restrictions on union formation and the collective withdrawal of labour. However, the law did not encourage or facilitate unionization. This neutral stance lasted in Canada until the enactment in 1944 of the National War Labour Order, Order-in-Council P.C. 1003, after which labour law facilitated union formation and, in turn, encouraged the spread of collective bargaining. P.C. 1003, which was partly modelled on the National Labor Relations Act (the Wagner Act) of 1935 in the United States, provided most private sector employees with the right to union representation and collective bargaining, established certification procedures, provided a code of unfair labour practices primarily intended to prevent employers from interfering with employees' right to union representation, and established a labour relations board to administer the law. Thus, in the post-World War II period, legislation encouraged collective bargaining.

These three phases in the history of collective bargaining applied primarily to the private sector. With the passage of the Public Service Staff Relations Act (PSSRA) in 1967 at the federal level and similar acts at provincial levels, governments encouraged collective bargaining and union formation in the public sector, which was also at that time an area of rapid growth in employment.

The Canadian labour relations policy which emerged in the 1940s had the following central features (Weiler, 1986):

1. workers who met the statutory definition of employee had the right to join and form unions;

2. collective bargaining rights were protected under unfair labour practices legislation, which prohibited acts by both employers and unions to discourage or interfere with the employees' prerogative to bargain collectively;

3. a system of defining appropriate bargaining units and certifying bargaining representatives was established;

4. once certified, the union became the exclusive bargaining representative of all employees in the bargaining unit;

5. unions and employers were required to bargain in good faith;

6. these rights and obligations were administered and enforced usually by a labour relations board, but in some cases in court.

This Wagner Act framework was combined with the traditional Canadian labour policy, expressed in the Industrial Disputes Investigation Act of 1907 and subsequent legislation, of regulating the use of work stoppages. These features included:

1. compulsory "cooling-off" periods/postpone of strikes and lockouts, coupled with compulsory mediation and/or conciliation procedures;

2. prohibition of strikes or lockouts during the term of the collective agreement coupled with a requirement that each collective agreement provide some alternative means for the resolution of grievances concerning the interpretation and application of the agreement;

3. while the content of the collective agreement was left largely to the parties, Ottawa and the provinces increasingly began to require certain items to be included in collective agreements, such as a recognition clause, a no-strike/no-lockout clause, a clause providing for a peaceful mechanism to resolve grievances arising during the term of the agreement, and provision for a date of termination of the agreement.

Another important aspect of the legal framework is the division of powers between the federal and provincial governments over labour legislation. The Constitution Act, 1867, has been interpreted by the courts as implying that jurisdiction over labour relations matters rests primarily with the provinces.

Table 14.2
UNION MEMBERSHIP OF NON-AGRICULTURAL WORKERS AS A PERCENTAGE OF NON-AGRICULTURAL WAGE AND SALARY EMPLOYEES: 1970–1986/87

	UNION DENSITY			CHANGE IN UNION DENSITY		
	1970	**1979**	**1986/87**	**1970–79**	**1979–86**	**1970–87**
COUNTRIES WITH SHARP RISES IN DENSITY						
Denmark	66	86	95	+20	+9	29
Finland	56	84	85	+28	+1	29
Sweden	79	89	96	+10	+7	17
Belgium	66	77	—	+11	—	—
COUNTRIES WITH 1970s RISES IN DENSITY; STABLE IN 1980s						
Germany	37	42	43	+5	+1	6
France	22	28	—	+6	—	—
Canada	32	36	36	+4	0	4
Australia	52	58	56	+6	−1	5
New Zealand	43	46	—	+3	—	3
Ireland	44	49	51	+5	+2	7
Switzerland	31	34	33	+3	−1	2
Norway	59	60	61	+1	+1	2
COUNTRIES WITH 1970s RISES IN DENSITY; DECLINE IN 1980s						
Italy	39	51	45	+12	−6	6
United Kingdom	51	58	50	+7	−8	−1
COUNTRIES WITH DECLINING DENSITY						
Austria	64	59	61	−5	+2	−3
Japan	35	32	28	−3	−4	−7
Netherlands	39	43	35	+4	−8	−4
United States	31	25	17	−6	−8	−14

Source: D.G. Blanchflower and R.B. Freeman. "Going different ways: unionism in the U.S. and other advanced OECD countries." *Industrial Relations*, Vol. 31(1) (Winter 1992) 56–79. Reprinted in *Labor Market Institutions and the Future Role of Unions*, edited by M. Bognanno and M. Kleiner, Basil Blackwell 1993. By permission.

Federal jurisdiction is limited to about 10 percent of the labour force — federal public servants, employees of federal Crown corporations and those employed in rail, air, shipping and truck transportation, banks, broadcasting, uranium mines, and grain elevators.

FACTORS INFLUENCING UNION GROWTH AND INCIDENCE

Inspection of Table 14.1 reveals that during the past six decades union growth in Canada has been substantial but erratic. In some periods union membership declined in absolute terms (1920–35), in others it increased absolutely but declined relative to the labour force (1955–65), while in others union membership grew substantially more rapidly than the labour force (1965–80; 1940–50). The causes of union growth — both the upward trend and the variations around the trend — have long been a subject of scholarly research and debate. There are also significant cross-sectional differences in the extent of unionization — across countries, industries, regions, and occupations — which call for an explanation.

Table 14.2 shows the extent of union organization in several countries. Union density in Canada is higher than in Japan and the U.S.A., but lower than in Australia and most European countries. These data also indicate that during the past two decades union density has been growing at a relatively rapid rate in some countries (especially the Scandinavian countries) yet declining in others (e.g., Japan and the United States). Canadian behaviour falls in a middle category: rising union density from 1970–79 followed by approximate stability during 1979–86/87. The divergent patterns displayed by Canada and the United States are particularly striking and are discussed in Exhibit 14.1.

The extent of union organization in Canada varies considerably by industry, occupation, region, and various individual employee or employer characteristics. Table 14.3 shows some of these variations. Unionization is highest in blue-collar occupations such as processing and machining, transportation and moving and materials handling and lower in many white-collar occupations. Industry differences are also substantial. Public administration (federal, provincial, and municipal public employees) and communications and public utilities are extensively unionized while wholesale and retail trade and finance, insurance and real estate have little unionization. Full-time workers are more likely to be unionized than part-time workers, and males are more unionized than females. (See also Exhibit 14.2 on the male–female unionization gap.) There are also important differences by age; the likelihood of being unionized increases sharply with age to 45–54 years of age, and then declines.

Explaining these differences in unionization — over time and across industries, regions, and occupations — is of considerable interest for its own sake. As discussed subsequently, this understanding is also important in interpreting research findings on the consequences of unions. Indeed, a central theme of recent research has been that union/nonunion status and the wage and nonwage

Table 14.3
UNION MEMBERSHIP AND COLLECTIVE AGREEMENT COVERAGE AS A PERCENT OF PAID WORKERS IN CANADA, 1990

	PERCENT OF PAID WORKERS WHO ARE:	
	UNION MEMBERS	**COVERED BY A COLLECTIVE AGREEMENT**
Both Sexes	33.1	37.6
16–24	15.4	19.7
25–34	32.4	36.8
35–44	39.8	44.4
45–54	44.0	48.3
55–64	40.7	45.4
Males	36.1	40.5
Females	29.7	34.3
Full-Time Employees	35.1	40.0
Part-Time Employees	23.1	27.6
By Occupation		
Managerial, Professional, Technical	36.7	42.7
Clerical	26.9	31.5
Sales	9.5	11.2
Service	28.3	32.3
Primary, except Mining	17.3	19.8
Processing, Machining	45.1	49.6
Transportation and Moving	39.8	42.4
Materials Handling	43.5	47.9
By Industry		
Agriculture	5.9	6.7
Mining	37.5	40.9
Construction	32.7	35.2
Manufacturing	39.3	43.5
Durable	40.4	44.4
Non-Durable	38.4	42.8
Transportation	47.9	51.5
Communications & Public Utilities	60.1	66.2
Wholesale Trade	12.4	14.2
Retail Trade	12.4	15.6
Finance, Insurance & Real Estate	9.0	13.1
Services	34.2	38.8
Public Administration	61.3	72.3

Source: Statistics Canada, unpublished data from the Labour Market Activity Survey 1990. Reproduced with the permission of the Minister of Industry, Science and Technology, 1993.

EXHIBIT 14.1 THE DIVERGENCE OF UNIONIZATION BETWEEN CANADA AND THE UNITED STATES

As illustrated in Figure 14.1, Canada and the United States displayed similar patterns of union growth from the early 1900s until the mid-1950s. Since that time, trends in unionization have diverged sharply, with union density declining steadily in the United States, but growing in Canada until the mid-1980s and subsequently declining modestly. As a consequence, a huge intercountry differential has emerged in the extent of union coverage — so that since the mid-1980s the fraction of the Canadian labour force represented by unions has been approximately double that of the United States.

A number of explanations have been advanced for the dramatic decline in union strength in the United States. Perhaps the most common explanation has to do with the changing structure of the economy and labour force. Specifically, most of the employment changes that have occurred in the past three decades — away from manufacturing and toward services; away from blue collar and toward white collar; away from male and full-time and toward female and part-time; away from large firms and towards small firms — represent declines in the relative importance of sectors which traditionally have been highly unionized and increases in the relative importance of sectors which traditionally had low union density. Thus if union density remained constant in each sector, or for each type of worker, the economy-wide extent of unionization would decline due to these structural shifts.

Another explanation involves changes in the U.S. legal regime (the laws, their interpretation, and their administration and enforcement) relating to unions and collective bargaining during the post-World War II period. A related view is that the decline in U.S. unionization can largely be attributed to the rise in management opposition, both legal and illegal, to unions. Increased management opposition to unions may be due to changes in the legal regime, a more competitive economic environment, and a substantial union-nonunion wage differential.

Another hypothesis is that there has been a reduction in the desire for collective representation because of the growth of substitute services. Governments have gradually provided more of the employment protection and nonwage benefits that were originally important factors underlying workers' desire for union coverage. In addition, employers have become increasingly sophisticated in their human resource practices and now provide services (e.g., grievance procedures) that workers previously only received in unionized firms.

A final hypothesis, which is perhaps the simplest and most profound, is that there has been in the United States a reduction in public sympathy toward unions and a reduction in workers' desire for collective representation.

Some evidence supporting each of these explanations has been put forward. However, beginning with Weiler (1983, 1984), the value of adopting a comparative Canada–U.S. perspective has increasingly been recognized. In particular, the many similarities between the economies of the two countries and their industrial relations systems results in a situation with elements of a controlled experiment, thus perhaps enabling some explanations of the decline in U.S. union strength to be rejected because they cannot account for the observed behaviour in Canada.

(continued)

Many of the structural changes described above (rise in service sector, in female and part-time, and in white-collar employment) also occurred in Canada. Thus it appears unlikely that the "structuralist" hypothesis can account for the behaviour of union density in both countries. Using comparable micro data on union incidence in the two countries, Riddell (1993) concludes that structural differences account for about 15 percent of the intercountry differential in union coverage; 85 percent is attributed to the fact that a Canadian worker with given characteristics is much more likely to be covered by a collective agreement than a U.S. worker with the same characteristics. The most important structural difference is the greater extent of public sector employment in Canada, which accounts for about seven percent of the unionization gap. (Public sector workers are much more likely to be unionized in both countries.)

Riddell (1993) also shows that underlying social attitudes towards unions, as measured by public opinion polls, are very similar in the two countries. Public attitudes toward unions have become less favourable during the last three decades on both sides of the 49th parallel, and to a similar degree. In addition, the impact of unions on wages in the private sector is very similar in the two countries. Thus, these hypotheses do not seem capable of explaining the observed behaviour in both societies. Equally unlikely is the "growth of union substitutes" view; Canadian governments have gone further than their U.S. counterparts in providing social and employment security.

These findings suggest that differences between Canada and the United States in the legal regime governing unions and collective bargaining may be able to explain the decline in the U.S. union movement and the divergent patterns of union growth in the two countries. Differences in such areas as certification and decertification procedures, bankruptcy and succession rights, first contract negotiation, and union security arrangements have been argued to be factors contributing to the intercountry differential in union coverage (Weiler, 1983, 1984). Institutional differences in the two countries' labour movements and industrial relations outcomes, detailed in Kumar (1992), also appear to be consistent with this perspective.

Sources: P. Weiler. Striking a new balance. *Harvard Law Review* 98 (December 1984) 351–420.

P. Weiler. Promises to keep. *Harvard Law Review* 96 (June 1983) 1769–1829.

W.C. Riddell. Unionization in Canada and the United States: a tale of two countries. *Small Differences That Matter: Labor Markets and Income Maintenance in Canada and the United States*, D. Card and R.B. Freeman (eds.). Chicago: University of Chicago Press, 1993.

effects of unions are jointly determined. That is, unionization may result in higher wages and improved working conditions, but also higher wages as well as other factors related to higher wages may make unionization more likely to occur.

Figure 14.1
UNION DENSITY IN CANADA AND THE UNITED STATES, 1920–90

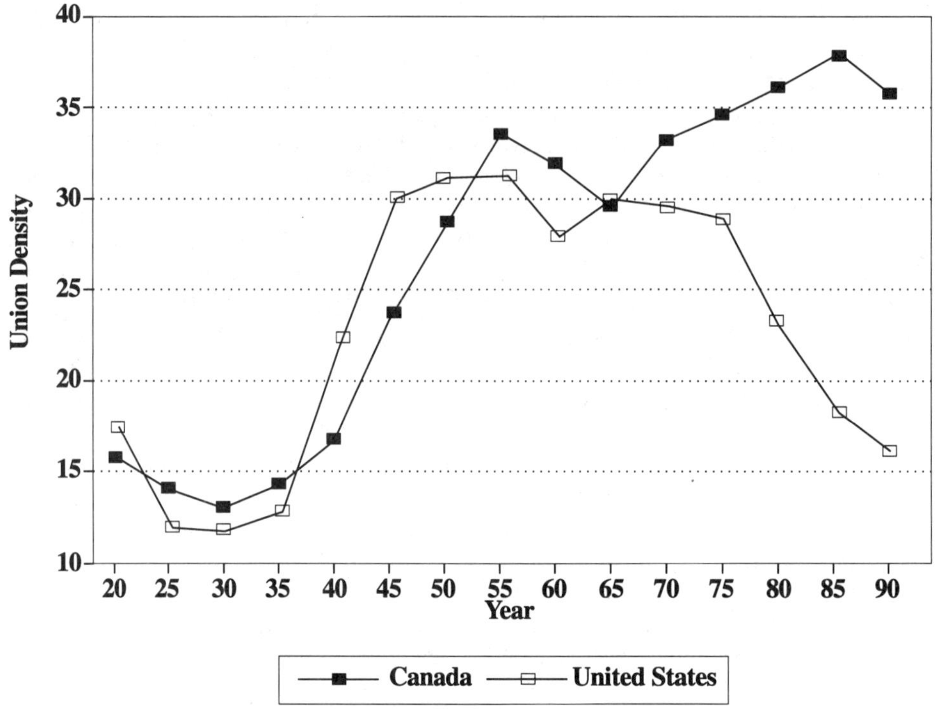

The growth and incidence of union organization can be analyzed using a demand and supply framework.[2] The demand for union representation emanates from employees and depends on the expected benefits and costs of union representation. The supply of union representation emanates from the organizing and contract administration activities of union leaders and their staff. These demand and supply forces are discussed in turn.

On the demand side the potential benefits of union representation may include higher wages and nonwage benefits, greater employment security (through protection from dismissal other than for cause) and protection from arbitrary treatment by management. Potential costs include direct costs such as union dues and time devoted to union activities, potential loss of income in the event of a work stoppage, and costs associated with the employer's actions, should the employer be willing and/or able to attempt to discourage unionization. Several factors may be net benefits or costs depending on the individual

---

2. This approach can be traced to Berkowitz (1954), Ashenfelter and Pencavel (1969) and Pencavel (1971). There is also a substantial institutional and historical literature on the causes of union growth.

employee's circumstances. Higher wages will benefit those who retain union jobs but may reduce employment opportunities of others. Unions typically alter procedures for determining promotions and layoffs — usually placing more emphasis on seniority — which will enhance some workers' opportunities and reduce those of others.

On the supply side, administering the contracts covering existing union members and organizing new (or previously nonunion) workplaces are costly activities. Unions need to allocate their scarce resources to the activities that are expected to yield the greatest return, measured in terms of the union's objectives. Success in organizing new workplaces and in maintaining membership and union representation in existing organizations will depend on a variety of factors including the resources devoted to contract administration and organizing drives, the activities of employers, and the economic and policy environment.

This demand and supply framework provides a useful conceptual model for analyzing the extent of union organization at a point in time and changes in unionization over time. The central concepts is that the extent of unionization depends on the choices made by employees, the organizational and contract administration activities of union leaders, and the actions of employers. Those enterprises in which a majority of the employees perceive substantial net benefits from being represented by a union and in which the actions of the employer will not significantly reduce these benefits will have a high demand for unionization, and vice versa. The supply of union organizational and contract administration effort will be high to those enterprises in which the per-worker cost of organization and representation is low. The interaction of these demand and supply forces determines an equilibrium extent of union organization at a point of time. Some enterprises will be organized while others — because of either low demand or high organizational costs — will remain unorganized. Changes over time in the factors influencing demand and supply will result in changes in union density.

EMPIRICAL EVIDENCE

This demand and supply framework was used as the basis for econometric analyses of union growth in the United States by Ashenfelter and Pencavel (1969) and the United Kingdom by Pencavel (1971). Subsequently, Bain and Elsheikh (1976a) carried out an extensive econometric study of union growth in Australia, Sweden, Great Britain, and the United States. Empirical studies by Swidinsky (1974), Bain and Elsheikh (1976b), Abbott (1982), and Kumar and Dow (1986) have applied and extended this approach in the Canadian setting. These studies employed aggregate data. More recently the demand and supply framework has been used to analyze micro data in which the union status of each individual is observed (Farber, 1983; Farber, 1990; Riddell, 1993). There have also been a number of U.S. and U.K. studies of the cross-sectional incidence of unionization according to such factors as industry, enterprise and personal characteristics, region, and legislative jurisdiction. Although this literature will not be reviewed in detail here, the main findings and their consistency with the demand and supply framework are briefly discussed.

Several factors suggested by the demand and supply framework have been investigated in empirical studies:

Social Attitudes Towards Unions and Collective Bargaining

Both the receptiveness of employees to union representation and the resistance of employers may be affected by the prevailing attitudes regarding the role of unions and collective bargaining in society, a factor which is, however, difficult to measure. In their U.S. study, Ashenfelter and Pencavel (1969) find that the percentage of Democrats in the House, a measure of pro-union sentiment, has a significant effect on U.S. union growth. Because society's attitudes toward unions tend to be reflected in labour legislation and its administration, measures of the legal regime (discussed below) capture social attitudes to some degree.

In the U.S., attitudes toward unions have become less and less favourable since the 1950s, leading some observers to explain the decline in unionization in that country in terms of less supportive social attitudes (e.g., Lipset, 1986). However, a very similar decline in favourable attitudes toward unions also occurred in Canada, thus casting doubt on this explanation (see Exhibit 14.1).

The Legislative Framework Governing Unionization and Collective Bargaining

Legislation governing the right of workers to act collectively and the machinery for administering and enforcing these rights are likely to exert an important influence on the demand for and supply of union representation. However, it must be recognized that labour relations legislation and its administration itself reflects certain unobservable factors such as society's attitudes toward unions; in the absence of good measures of social attitudes, it will be difficult to determine the independent impact of the legislation. Further, union growth and a legislative environment that is conducive to union growth may be joint manifestations of underlying societal attitudes. Without favourable attitudes, legislation may not have an independent influence.

The changes in the Canadian legislative framework described earlier have lowered the cost to employees of collective action and have increasingly restricted the employer's ability to attempt to discourage unionization. The former increases the demand for unions by raising the net benefits to workers of union representation; the latter increases the supply by lowering the costs of organizing and representing workers. The legislative changes are therefore predicted to have encouraged union growth. Although this trend in labour legislation and its administration has occurred gradually over time, particularly important changes took place in the 1870s (e.g., the Trade Unions Act of 1872), the early 1900s (the Industrial Disputes Investigation Act of 1907 contained an implicit form of union recognition by banning strikes or lockouts over recognition issues), the 1940s (the enactment of P.C. 1003 gave workers the right to form and join unions and made illegal a number of employer practices previously used to discourage collective bargaining; similar legislation was enacted in most provinces following the Second World War and the end of federal emergency powers), and in the 1960s (the extension to public sector employees of the right to collective bargaining).

Examination of Table 14.1 suggests that the changes in the legislative framework around 1944 and 1967 may have had a positive impact on union growth in Canada. Econometric studies by Abbott (1982) and Kumar and Dow (1986) support the view that labour legislation has had a significant effect on the union growth process. In addition, the divergent patterns of union growth in Canada and the United States since the 1960s has been argued to be associated to a considerable extent with differences between the two countries in labour relations legislation and its enforcement (Exhibit 14.1).

Other Economic and Social Legislation

The extent of union organization may also be affected by other economic and social legislation, although the direction of the effect is ambiguous. The setting (and raising) of employment standards such as those with respect to minimum wages, overtime premiums, statutory holidays, workplace health and safety, notice of layoff, and severance pay may narrow the gap in wages and working conditions between organized and unorganized workers, thus reducing the demand for unionization. On the other hand, such policies may have aided union growth by reducing competition from the lower labour costs that otherwise would have prevailed in the nonunion sector in the absence of extending such benefits to unorganized workers through employment standards. Social programs such as those respecting public pensions, medical and health care, and unemployment insurance may also have offsetting effects. Neumann and Rissman (1984) have suggested that the provision of certain welfare benefits by governments in the United States accounts for some of the decline in union strength there. In Canada, however, union density rose during the postwar period, an era of rapid expansion of the welfare state and significant increases in employment standards.

Aggregate Economic Conditions

Early institutional economists such as Commons (1918) emphasized the role of the business cycle in the growth of unions. Subsequent econometric studies have investigated the influence of such variables as the unemployment rate, the rate of growth of employment that is eligible for unionization, and the rate of price inflation. The general finding that the rate of union growth varies directly with the rate of growth of employment that is eligible for unionization is consistent with several hypotheses, including: (1) employer resistance to union formation is lowest when product demand is high and the labour market is tight, and (2) the ability of unions to secure wage and benefit increases (and thus the perceived net benefit of being represented by a union) is highest when there is excess labour demand. However, even in the absence of these cyclical effects, a positive relationship between union growth and growth in employment eligible for unionization is expected. Employment growth in existing organized enterprises will result in an increase in the unionized labour force unless the employees decide to decertify the union. In addition, some new enterprises and their employees will have characteristics (discussed below) which make them likely to become organized.

Although the above reasons suggest that union growth will be pro-cyclical in nature, the experience of the Great Depression — during and subsequent to which the union movement displayed strong growth, especially in the United States — led Ashenfelter and Pencavel (1969) to hypothesize that severe business contractions will raise worker discontent which in turn will be a spur to unionization, perhaps with a lag. Although U.S. studies generally find that union growth is positively related to the severity of past recessions, Canadian evidence on this issue is mixed (Abbott, 1982; Kumar and Dow, 1986).

Time series studies also generally find that union growth varies directly with the rate of price inflation. This finding is consistent with the hypothesis that unions are viewed by workers as an effective vehicle for maintaining real wages.

Industry and Enterprise Characteristics

As Table 14.3 shows, there are substantial differences across industries in the extent of union organization. Cross-sectional studies carried out in the United States and Great Britain have found that union density tends to be higher in industries with larger firms or establishments, in more concentrated industries, in industries with more capital-intensive production processes and in industries with more hazardous jobs (Ashenfelter and Johnson, 1972; Lee, 1978; Duncan and Stafford, 1980; Hirsch, 1982; Leigh, 1982; Hirsch and Berger, 1984).

These results can be interpreted using the demand-supply framework. The demand for union representation is likely to be higher in larger establishments because individual action becomes less effective the larger the group; and the need for formal work rules, communications, and grievance procedures is greater in larger organizations. In addition, the per-worker cost of union organizing is lower in larger establishments. Industries with hazardous jobs are also likely to be characterized by greater demand for a collective "voice" to represent workers' interests in the internal regulation of workplace health and safety.

Capital intensity may be associated with several effects. As with job hazards, capital-intensive production processes may entail greater need for a collective voice relating to the organization and flow of work — aspects such as the speed of assembly lines, scheduling of shifts, and overtime regulations. In addition, high capital intensity implies that labour costs are a small fraction of total costs. This tends to make labour demand more inelastic and thus increases the potential wage gains from unionization.

The finding that more concentrated industries tend to be more heavily unionized may reflect several factors. Concentrated industries typically have significant barriers to entry; there is thus less threat of nonunion competition in the form of new entrants. In addition, unions may enable workers to share in the excess profits or rents earned by the established firms. Organizing costs may also be lower in more concentrated industries.

Because of these differences across industries in the extent of union organization, changes over time in the economy's industrial structure will be accompanied by changes in union density, other things equal. In Canada and most other industrialized countries, significant changes in industrial structure have occurred in the postwar period: employment has grown rapidly in services-producing industries and the share of service sector employment increased substantially; goods-producing industries (manufacturing and construction) have experienced slower employment growth and a decline in the share of total employment; employment in agriculture both declined absolutely and relative to the share of total employment; and other primary industries (mining, forestry, fishing, and trapping) have experienced positive employment growth but declining shares of total employment.

These trends in the industrial composition of employment have had offsetting effects on union growth. Several factors tend to increase union density: the decline in employment in agriculture, an industry with little unionization; and the rapid growth in employment in public administration (employees of the federal, provincial, and municipal governments), a sector which has become extensively unionized since the mid-1960s. However, other changes in industrial structure have tended to reduce the extent of union organization in the economy: declining employment shares in manufacturing, mining, and forestry, sectors with above average unionization; and increases in the share of employment in trade, finance, insurance, real estate, and other private sector services, industries with low degrees of unionization.

Personal Characteristics

A number of personal characteristics affect the propensity to be a union member or to be represented by a union. Part-time workers and workers with intermittent labour force attachment are less likely to be organized. For these individuals, the net benefits of union representation — especially in the form of seniority provisions, pension benefits, and other forms of deferred compensation — are lower than is the case with full-time workers and employees with a permanent labour force attachment. The costs of organizing part-time workers and workers

EXHIBIT 14.2   GENDER DIFFERENCES IN UNIONIZATION

During the past two to three decades, significant changes have occurred in the extent of union organization of Canadian males and females. In the late 1960s, males were more than twice as likely to be unionized as females. For example, in 1966 union density of males was over 38 percent, whereas that of females was about 16 percent (Statistics Canada, 1992). However, the male–female unionization gap has subsequently narrowed considerably — reflecting the fact that female unionization has grown substantially whereas that of males has fallen slightly. The data shown below, based on a series of special surveys by Statistics Canada, show that the gender gap in unionization narrowed by almost half during the 1980s — from about 12 to 6 percentage points.

	EXTENT OF COLLECTIVE AGREEMENT COVERAGE		GENDER UNIONIZATION GAP
	MALE	FEMALE	
1981*	38.6	26.8	11.8
1984	46.0	36.6	9.4
1986	43.7	35.2	8.5
1988	37.6	28.7	8.9
1990	40.5	34.3	6.2

*union membership

   In an empirical study analyzing the 1981, 1984, and 1988 surveys, Doiron and Riddell (1992) find that the gender unionization gap is mainly due to differences between males and females in labour force characteristics; in particular, females possess fewer of each of the characteristics which make unionization likely. The occupation and industry distributions of females are the most important such characteristics, followed by job tenure and full-time work. Differences between males and females in the impacts of labour force characteristics on the probability of unionization contribute relatively little to the gap in union coverage.

   Over time, males and females have become more similar in their labour force characteristics (experience, education, part-time employment, etc.). Thus the gender gap in union coverage has narrowed.

   Because unionized workers typically earn more than their nonunion counterparts, this narrowing of the males–female unionization gap may have raised the earnings of women relative to those of men. Doiron and Riddell find that the male–female earnings differential would have widened by about 7 percentage points in the absence of the increase in female unionization relative to that of males.

(continued)

Gender differences in unionization have also declined substantially in the United States; however, in the U.S. this was brought about by a steep decline in male unionization while that of females remained approximately constant. Even and Macpherson (1992) find that the greater decline in male unionism is responsible for about one-fifth of the 13 percentage point decline in the gender wage gap which occurred between 1973 and 1987.

Sources: D. Doiron and W.C. Riddell. The impact of unionization on male–female earnings differences in Canada. UBC Discussion Paper 92-30, 1992.

W.E. Even and D.A. Macpherson. The decline in private sector unionism and the gender wage gap. *Journal of Human Resources*, forthcoming 1992.

Statistics Canada. *CALURA: Labour Unions 1989.* Ottawa: Statistics Canada, 1992.

with intermittent labour force attachments may also be high. Largely because they are more likely to work part-time and have tended to have less permanent attachment to the labour force, women are less likely to be represented by a union than men. (See Exhibit 14.2.)

As noted earlier, the extent of union organization is much higher among blue-collar than white-collar workers. Blue-collar workers are more likely to demand union representation because of less identification with management and possibly greater need for a collective voice in the determination of working conditions.

Age and experience have also been found to be related to the propensity to be organized in a number of studies (Antos, Chandler and Mallow, 1980; Farber and Saks, 1980; Duncan and Stafford, 1980; Hirsch and Berger, 1984). Although there are some conflicting results, this relationship appears to be concave; that is, the propensity to choose union representation at first increases, reaches a maximum, and subsequently decreases with age and/or experience. This result, which is also suggested by the tabulations in Table 14.3, is consistent with the view that, when members' preferences differ according to some characteristic like age, unions tend to represent most closely the preferences of members in the middle of the distribution of preferences (i.e., the median member).

An individual's position in the earnings distribution may also affect the net benefits of being represented by a union. As explained subsequently, unions tend to increase the wages of those at the lower end of the wage distribution by a relatively large amount and those in the upper tail of the wage distribution by a relatively small amount. Thus individuals whose earnings would be above average in the absence of a union are less likely to favour union representation.

Summary

In summary, the growth of unions over time and the extent of union organization across industries, regions, occupations, and employee characteristics are systematically related to a number of economic, social, and legal variables. These observed relationships are generally consistent with the demand-supply framework, which appears to provide a valuable conceptual device for understanding the growth and incidence of unions.

QUESTIONS

1. Account for the differential growth of unionization between Canada and the United States in recent years.

2. "Differential legislative structures may account for some of the differential union growth between Canada and the U.S.; however, that begs the question of why the legislative structures are different. That is, legislation should be regarded as an endogenous variable and its existence explained." Discuss.

3. Discuss differences in the concepts of union density based on union membership and the proportion of the work force that is covered by a collective agreement. What accounts for the differences in the magnitude in Canada? Which of the two best measures the significance of unions in the determination of wages and working conditions?

4. Outline the main legislative initiatives in Canada that have facilitated the development of unionization and collective bargaining.

5. "Any theory of union growth must be able to explain not only its long-run trend and short-run cyclical variation, but also why unionization varies across countries and across industries, regions, and occupations within a country." Discuss.

6. Given the determinants of unionization, discuss the future prospects for union growth in Canada.

7. The extent of unionization in Canada varies from one province to another. Discuss factors which could account for these differences.

8. Discuss the general factors that would explain the differences in union density across industries, occupations, and employee characteristics shown in Table 14.3.

9. Ashenfelter and Pencavel (1969) carry out an econometric analysis of the determinants of union growth in the U.S. between 1900 and 1960. Would their estimated equation over-predict, accurately predict, or under-predict the subsequent union growth in the U.S.? If you believe their equation would under- or over-predict the subsequent growth, what variable or variables may be omitted?

REFERENCES AND FURTHER READINGS

Abbott, M.G. An econometric model of trade union membership growth in Canada, 1925–1966. Princeton University, Industrial Relations Section, Working Paper 154, September 1982.

Abowd, J.M. and H.S. Farber. Job queues and the union status of workers. *ILRR* 35 (April 1982) 354–367.

Antos, J.R., M. Chandler and W. Mellow. Sex differences in union membership. *ILRR* 33 (January 1980) 162–169.

Ashenfelter, O. and G.E. Johnson. Unionism, relative wages and labor quality in U.S. manufacturing industries. *IER* 13 (October 1972) 488–508.

Ashenfelter, O. and J.H. Pencavel. American trade union growth: 1900–1960. *QJE* 83 (August 1969) 434–448.

Bain, G.S. and F. Elsheikh. *Union Growth and the Business Cycle: An Econometric Analysis.* Oxford: Blackwell, 1976a.

Bain, G.S. and F. Elsheikh. Trade union growth in Canada: a comment. *RI/IR* 31 (No. 3, 1976b) 482–490.

Banting, K.G. (ed.). *The State and Economic Interests.* Toronto: Royal Commission on the Economic Union and Development Prospects for Canada and University of Toronto Press, 1986.

Berkowitz, M. The economics of trade union organization and administration. *ILRR* 7 (July 1954) 575–592.

Bronars, S.G. and J.R. Lott. Why do workers join unions? The importance of rent seeking. *EI* 27 (April 1989) 305–324.

Carruth, A. and R. Disney. Where have two million trade union members gone? *Economica* 55 (February 1988) 1–20.

Chaison, G.N. and D.G. Dhavale. A note on the severity of the decline in union organizing activity. *ILRR* 43 (April 1990) 366–373.

Christie, V. and P. Miller. Attitudes towards unions and union membership. *Economic Letters* 30 (September 1989) 263–268.

Commons, J.E. et al. *History of Labor in the United States, I.* New York: Macmillan, 1918.

Disney, R. Explanations of the decline in trade union density in Britain: an appraisal. *British Journal of Industrial Relations* 28 (July 1990) 165–177.

Duncan, G. and D. Leigh. The endogeneity of union status: an empirical study. *JOLE* 3 (July 1985) 385–402.

Duncan, G. and F. Stafford. Do union members receive compensating wage differentials? *AER* 70 (June 1980) 335–371.

Farber, H.S. The determinants of the union status of workers. *Econometrica* 51 (September 1983) 1417–1437.

Farber, H.S. Right-to-work laws and the extent of unionization. *JOLE* 2 (July 1984) 319–352.

Farber, H.S. The decline of unionization in the United States: what can be learned from recent experience. *JOLE* 8 (January 1990) 75–105.

Farber, H. and D. Saks. Why workers want unions: the role of relative wages and job characteristics. *JPE* 88 (April 1980) 349–369.

Flanagan, R.J. Compliance and enforcement decisions under the National Labour Relations Act. *JOLE* 7 (July 1989) 257–280.

Freeman, R.B. and C. Ichniowski, (eds.). *When Public Sector Workers Unionize.* Chicago: University of Chicago Press, 1988.

Freeman, R.B. and J. Pelletier. The impact of industrial relations legislation on British union density. *British Journal of Industrial Relations* 28 (July 1990) 141–164.

Freeman, R.B. and M.M. Kleiner. Employer behavior in the face of union organizing drives. *ILRR* 43 (April 1990) 447–462.

Freeman, R.B. Unionism comes to the public sector. *JEL* 25 (March 1986) 41–86.

Freeman, R.B. Contraction and expansion: the divergence of private sector and public sector unionism in the United States. *JEP* 2 (Spring 1988) 63–88.

Freeman, R.B. and M.M. Kleiner. The impact of new unionization on wages and working conditions. *JOLE* 8 (January 1990) 8–25.

Goldfield, M. The causes of U.S. trade union decline and their future prospects. *Research in Political Economy* 7 (1984) 81–159.

Hirsch, B.T. The interindustry structure of unionism, earnings, and earnings dispersion. *ILRR* 36 (October 1982) 22–39.

Hirsch, B.T. and M.C. Berger. Union membership determination and industry characteristics. *SEJ* 50 (January 1984) 665–679.

Horn, H. and A. Wolinsky. Worker substitutability and patterns of unionization. *EJ* 98 (June 1988) 484–497.

Ichniowski, C. and J.S. Zax. Right-to-work laws, free riders and unionization in the local public sector. *JOLE* 9 (July 1991) 255–275.

Kokkelenberg, E.C. and D.R. Sockell. Union membership in the United States, 1973–1981. *ILRR* 38 (July 1985) 497–543.

Kumar, P. Union growth in Canada: retrospect and prospect. *Canadian Labour Relations*, W.C. Riddell (ed.). Toronto: Royal Commission on the Economic Union and Development Prospects for Canada and University of Toronto Press, 1986.

Kumar, P. and B. Dow. Econometric analysis of union membership growth in Canada, 1935–1981. *RI/IR* 41 (No. 2, 1986) 236–253.

Kumar, P. Industrial relations in Canada and the United States: from uniformity to divergence. Industrial Relations Centre, Queen's University, Working Paper 1991–1992.

Lee, Lung-Fei. Unionism and wage rates: a simultaneous equations model, with qualitative and limited dependent variables. *IER* 19 (June 1978) 415–434.

Leigh, J.P. Are unionized blue collar jobs more hazardous than nonunionized blue collar jobs? *JLR* 3 (Summer 1982) 349–357.

Leigh, D.E. The determinants off workers' union status: evidence from the National Longitudinal Surveys. *JHR* 20 (Fall 1985) 555–566.

Lipset, S.M. (ed.). *Unions in Transition: Entering the Second Century.* San Francisco: ICS Press, 1986.

Meltz, N. Labour movements in Canada and the United States: are they really that different? *Challenges and Choices Facing American Labor*, T. Kochan (ed.). Cambridge, Mass.: MIT Press, 1985.

Meltz, N.M. Interstate vs. Interprovincial differences in union density. *IR* 28 (Spring 1989) 142–158.

Moore, W.J., Newman, R.J. and L.C. Scott. Welfare expenditures and the decline of unions. *R.E. Stats.* 71 (August 1989) 538–542.

Neumann, G.R. and E.R. Rissman. Where have all the union members gone? *Journal of Labor Economics* 2 (April 1984) 175–192.

Pencavel, J.H. The demand for union services: an exercise. *ILRR* 24 (January 1971) 180–190.

Resnef, Y. Union decline: a view from Canada. *JLR* 11 (Winter 1990) 25–39.

Riddell, W.C. (ed.). *Labour-Management Cooperation in Canada.* Toronto: Royal Commission on the Economic Union and Development Prospects for Canada and University of Toronto Press, 1986.

Riddell, W.C. Unionization in Canada and the United States: a tale of two countries. *Small Differences that Matter: Labor Markets and Income Maintenance in Canada and the United States*, D. Card and R. Freeman (eds.). Chicago: University of Chicago Press, 1993.

Sheflin, N., L. Troy and C.T. Koeller. Structural stability in models of American trade union growth. *QJE* 96 (February 1981) 77–88.

Stepine, L.P. and J. Fiorito. Toward a comprehensive theory of union growth and decline. *IR* 25 (Fall 1986) 248–264.

Strauss, G., Gallagher, D.G. and J. Fiorito, (eds.). *The State of the Unions.* Madison: Industrial Relations Research Association, 1991.

Swidinsky, R. Trade union growth in Canada: 1911–1970. *RI/IR* 29 (No. 3, 1974) 435–450.

Voos, P.B. Union organization expenditures: determinants and their implications for union growth. *JLR* 8 (Winter 1987) 19–30.

Weiler, J. The role of law in labor relations. *Labour Law and Urban Law in Canada*, I. Bernier and A. Lajoie (eds.). Toronto: Royal Commission on the Economic Union and Development Prospects for Canada and University of Toronto Press, 1986.

Weiler, P. Striking a new balance: freedom of contract and the prospect for union representation. *Harvard Law Review* 98 (December 1984) 351–420.

Weiler, P. Promises to keep: securing workers' rights to self-organization under the NLRA. *Harvard Law Review* 96 (June 1983) 1769–1829.

Zax, J.S. and C. Ichniowski. Bargaining laws and unionization in the local public sector. *ILRR* 43 (April 1990) 447–462.

CHAPTER 15

Wage and Employment Determination Under Collective Bargaining

The previous chapter was concerned with explaining which organizations are unionized and which are nonunion. We now turn to the behavioural implications of unionization and collective bargaining. In order to understand and predict the consequences of collective bargaining we begin by examining union objectives and the constraints unions face in attempting to achieve those objectives. Then we examine the firm's objectives and constraints and how the two parties interact to determine collective bargaining outcomes. Much of this chapter focuses on the determination of wages and employment. Various nonwage aspects of employment — such as grievance procedures, seniority provisions, workplace health and safety, and fringe benefits — are discussed in the next chapter.

THEORY OF UNION BEHAVIOUR

The economic analysis of firms and households begins from the assumption that the decision-maker maximizes an objective function subject to the constraints imposed on the agent by the economic environment. Private sector firms are usually assumed to maximize profits subject to the production function (which summarizes the technical possibilities available to the firm), the demand conditions in the product markets, and the supply conditions in the markets for inputs. This is the theory that underlies the demand for labour discussed in Part 2 of this book. It is a theory rich in testable implications, and one which has generally been found to accord with the available evidence. Similarly, the household is assumed to maximize utility (which depends on the quantities consumed of various goods, including leisure) subject to the budget constraint (which depends on the prices of goods and the wage rate). This is the theory that underlies the demand for goods and services in product markets and the theory of labour supply discussed in Part 1 of this book. This theory is also rich in testable implications, and has generally been found to be consistent with the evidence on household behaviour.

Economists have generally agreed that an understanding of union behaviour and its consequences must likewise begin with a theory of union objectives. The union can then be assumed to maximize its objective or utility function subject to the constraints imposed on the union and its members by the economic environment. However, our ability to characterize union preferences has long been and remains controversial. The major aspects of this controversy were debated several decades ago by Dunlop (1944) and Ross (1948). Dunlop advocated an "economic" approach in which the union is modelled as attempting to maximize a well-defined objective function subject to labour market constraints. (The specific objective function suggested by Dunlop is discussed below). Ross criticized this approach and argued that union decision-making can only be understood by treating the union as a political institution. Although this debate has not been fully resolved, the modern approach to union behaviour recognizes some merit in both positions in that the union is modelled as attempting to maximize a well-defined objective function but attention is paid to the political nature of union decision-making.

UNION OBJECTIVES

Unions are collective organizations whose leaders represent members of the bargaining unit — the rank-and-file — in collective bargaining with the firm. Union objectives refer to the goals of the organization. These need not be identical to the objectives of the individual members or to those of the union leaders. Three factors influence the relationship among the preferences of the members, the union leaders and those of the union as a whole: (1) the information available to the rank-and-file about the available options, (2) the nature of the union's political decision-making process, and (3) the degree of homogeneity of the individual members' preferences. If the rank-and-file are well-informed and the union's decision-making processes are highly democratic, then the union leaders will make choices that have the support of a majority of the members. (Otherwise they will be quickly replaced). In these circumstances, the union's objectives are those which a majority of the rank-and-file would favour. However, asymmetric information (the members being less informed than the leaders) or imperfectly democratic decision-making processes may allow the union leaders to pursue their own objectives to some degree. In these circumstances the objectives of the members and those pursued by the union may differ.

The degree of homogeneity of individual members' preferences is another important factor. If the rank-and-file have very similar preferences it will be easy for the union leaders to determine the group's preferred choices. However, if preferences differ significantly across individuals or groups of individuals the task of the union leaders is more difficult and selecting an option that will be supported by a majority may require considerable skill and judgement.

Although various union objectives have been postulated, most involve two key aspects — the wages and employment of union members (abstracting from the various nonwage aspects of working conditions for the moment). This suggests a union objective or utility function whereby utility is a positive function of both the wage rate and the employment of union members (and which may also depend on other variables as discussed below). Combinations of the wage rate and employment which are equally satisfactory to the union — the union's indifference or iso-utility curves — must be downward sloping because a higher wage is needed to compensate for lower employment and vice versa. Furthermore, it is plausible to hypothesize that union preferences display a diminishing marginal rate of substitution between wages and employment; that is, holding utility constant, the union will be less willing to accept a wage reduction in return for a given increase in employment the higher the level of employment, and vice versa. This implies that the indifference curves have the convex shape displayed in Figure 15.1 (See the Appendix to Chapter 2 for a review of indifference curves.)

Union objectives are likely to depend on several additional variables. One is the overall price level, or the cost of living. Clearly what matters to the workers (and thus to the union leaders) is the real wage rate. A 10 percent increase (or decrease) in the nominal wage rate accompanied by a 10 percent increase (or decrease) in the price level leaves the workers' real income and therefore

Figure 15.1
UNION OBJECTIVES AND CONSTRAINTS

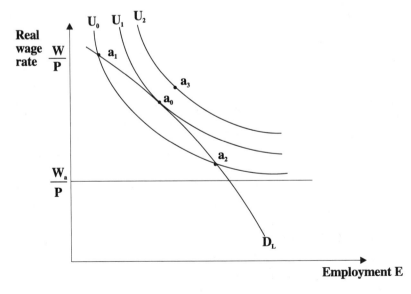

consumption and leisure possibilities unchanged. Union utility is also likely to depend on the alternative wage (i.e., the wage rate that union members could earn if employed elsewhere). Few workers will be willing to bear the costs of union representation (union dues and other costs discussed earlier) unless their compensation is equal to or greater than that available elsewhere for comparable work. For this reason the alternative (real) wage is shown as a lower bound to the union's indifference curves in Figure 15.1.

In using the term "employment" we have not distinguished between hours worked and number of employees. Implicitly hours worked are held constant, so that employment refers to the number of workers. This assumption is innocuous at this stage, but needs to be relaxed when examining issues such as worksharing versus layoffs in response to temporary reductions in demand. Also, we have not discussed how the size of the union membership is determined. Membership size will matter to both the union leadership and the existing union members — the former because a larger membership implies more dues revenue and possibly more power and prestige and the latter because a larger membership may imply reduced employment opportunities, depending on how scarce jobs are rationed when employment demand is insufficient.

SOME SPECIAL CASES

As noted earlier, various union objective functions have been suggested in the literature on union behaviour. Several of these are special cases of the utility function illustrated in Figure 15.1, and examining these may increase our understanding of union objectives.

Maximize the Wage Rate W

This would imply that the union places no weight on employment; i.e., that the indifference curves are horizontal straight lines as shown in Figure 15.2(a). Although this may seem implausible, the notion that, in negotiating wages, unions do not take into account the employment consequences of higher wages has a long tradition including the influential work of Ross (1948), and is often stated by union leaders today.

Maximize Employment E

This is the opposite extreme of the previous objective, and implies indifference curves which are vertical straight lines as shown in Figure 15.2(b).

Maximize the (Real) Wage Bill wE

This objective, proposed by Dunlop (1944), implies that the indifference curves are rectangular hyperbolae as shown in Figure 15.2(c). (wE = a constant is the equation of a rectangular hyperbola.) It has the advantage that the union

Figure 15.2
UNION OBJECTIVES: SOME SPECIAL CASES

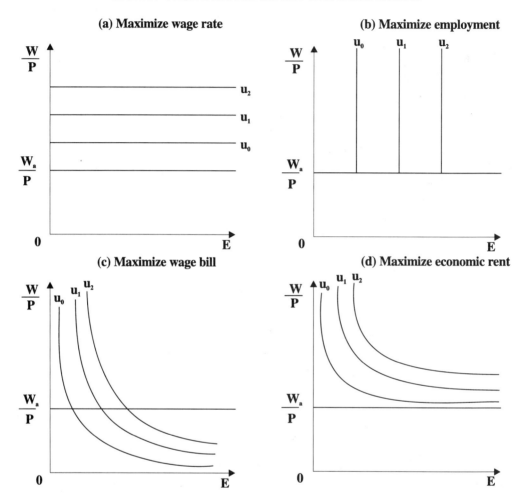

(a) Maximize wage rate

(b) Maximize employment

(c) Maximize wage bill

(d) Maximize economic rent

attaches weight to both wages and employment. The wage bill is, of course, total labour income. It may appear plausible that union members would attempt to maximize their total income, especially if there is some mechanism for sharing the income between employed and unemployed members. The main defect of this objective function is that is disregards the alternative wage; indeed it implies that the union may be willing to allow the wage to fall below W_a in exchange for higher employment.[1]

---

1. To avoid this difficulty, Dunlop (1944) suggested that the union maximize the wage bill subject to a "membership function" which shows the wage rate needed to attract a given number of union members. The membership function is analogous to a labour supply function.

Maximize (Real) Economic Rent ($w = w_a$)E

This objective is analogous to profit-maximization for a monopolist; the alternative wage is the opportunity cost to each union member so economic rent is analogous to total revenue (wE) minus total cost (w_aE). In this case the indifference curves are rectangular hyperbolae with respect to the new origin (w_a,O), as shown in Figure 15.2(d).

This objective is probably the most plausible of the group, in that it hypothesizes that the union members seek to maximize their "total return," given what they could earn elsewhere. Again, the objective is most plausible when there is some mechanism for sharing the income between employed and unemployed union members.

UNION PREFERENCES: SOME ADDITIONAL CONSIDERATIONS

The task of deriving union objectives from the preferences of the constituent members is simplest (though not necessarily simple) when: (1) members' preferences are homogeneous, (2) union leaders are constrained by democratic decision-making processes and the availability of information from pursuing their own objectives, and (3) union membership is exogenously determined. When these conditions do not hold, the economics of union behaviour is more complex.

Heterogeneous Preferences

The preferences of individual union members may differ for a variety of reasons. The union may represent several groups of employees with different occupations or skills. Industrial unions often contain highly skilled employees (usually a minority) in addition to semi-skilled and unskilled workers. Alternatively, workers performing similar jobs may differ according to a number of personal characteristics such as age, seniority, marital status, and alterative employment opportunities. Individual preferences may depend in a systematic way on these characteristics. For example, older workers may desire a greater emphasis on pensions and other forms of deferred compensation and less emphasis on current earnings in the total compensation package. Those with good alternative employment opportunities (high w_a's) may prefer a higher union wage than those with limited opportunities, for given levels of employment security. If layoffs take place in accordance with seniority, more senior workers may wish to place more emphasis on wages and less emphasis on employment in union objectives. Clearly these sources of heterogeneity may be important in particular settings and the economic analysis of union behaviour should be able to account for their implications.

The conditions under which heterogeneous individual preferences can be aggregated into a group or social objective function is a central issue in the theory of public choice. One well-known result is the median voter model which states that when choices over a single variable are made in a democratic fashion (i.e., by voting among alternatives) then under certain conditions the group's preferences will be identical to those of the median voter. The reason is that the outcome preferred by the median voter will defeat all others in a sequence of pairwise elections. In the union context this result implies that if, for example, the members' preferences with respect to the wage rate depend on the worker's seniority, the union's objective function will be that of the member with the median seniority (i.e., that individual for whom half the members are more senior and half the members are less senior).

The conditions under which the median voter model holds are fairly stringent. One key requirement is that the choice or voting set must be defined over a single variable (e.g., the wage in the union setting). Thus, with heterogeneous members, union preferences over multiple objectives (wages, employment, non-wage benefits, working conditions) cannot be derived from the assumption of pairwise voting among alternatives except under special conditions.

When the required conditions do not hold, a voting equilibrium will generally not exist; i.e., no single set of outcomes will defeat all others in a sequence of pairwise elections. In these circumstances, the organization's choice will depend on factors such as the order in which alternatives are presented for voting.

Clearly the assumptions of the median voter model do not apply precisely in the union setting. Unions typically do not choose among alternatives by conducting a sequence of pairwise votes. Most choices are made by union leaders who are elected periodically by the membership. In addition, the union leaders often face tradeoffs among several alternatives such as wages, nonwage benefits, and employment. Nonetheless when union members' preferences are heterogeneous the median voter model may predict union behaviour reasonably well. Union leaders wishing to remain in office will adopt positions that a majority of the membership support. When the members can be ordered by a preference-related factor such as age or seniority the choices of the median member, because of their central position in the distribution of preferences, may be more likely to receive majority support than any other set of choices.

Union Leaders vs. Union Members

Unions, like societies, generally do not make choices in a fully democratic fashion in the sense of conducting a referendum each time a decision must be made. Rather, union leaders are elected to implement policy and their position of authority may allow them to pursue their own interests to some extent. Furthermore, because they specialize in contract administration and collective bargaining with the employer, union leaders generally have greater access than members to information regarding the feasible options. In addition, the information that is received by the rank-and-file can be filtered to some degree by the leadership.

These observations suggest that union leaders may be able to pursue their own interests — such as maximizing dues revenue, expanding the union's membership and influence, and raising their own personal income. Nonetheless, the desire to remain in office (and perhaps other considerations) does constrain the leaders from deviating too far from the wishes of the membership. How binding these constraints are may well vary according to individual circumstances. A theory of union preferences based solely on the wishes of the members may not be able to account for all aspects of union behaviour in all settings. Nonetheless, it may provide a good approximation to observed behaviour.

Union Membership and Union Objectives

Changes in union membership may alter union preferences. For example if employment in the industry or firm is expanding rapidly, not only will membership size increase but the age and seniority of the median member will probably fall. The union's objectives, if the union represents members' preferences, will adjust to reflect the wishes of the younger, less senior worker. The opposite is likely to occur in declining sectors.

In formulating wage policy, the current union membership also influences the size of future membership and thus future union preferences. If the union pursues a high-wage policy, future membership will be reduced as employers substitute capital for labour and consumers substitute less expensive products. Thus union preferences may not be stable over time, but may depend on the wage policies adopted.

UNION CONSTRAINTS

The union cannot choose whatever wage and employment outcomes it desires, but most negotiate with the firm and must also take into account the consequences for its members of the outcomes of these negotiations. Initially we will assume that the two parties negotiate over the wage rate, leaving the firm free to choose the level of employment at whatever wage rate is agreed upon by collective bargaining. (The possibility of negotiating over both wages and employment is discussed subsequently.) In these circumstances, the firm will choose the level of employment that maximizes profits given the negotiated wage; i.e., will choose employment according to the labour demand curve.[2] It follows that when the

---

2. In the public and nonprofit sectors the labour demand curve is derived under different assumptions about the organization's objective; e.g., cost minimization. In these circumstances the demand curve shows the cost-minimizing level of employment for each wage rate.

firm can unilaterally determine employment the union is constrained by the firm's labour demand curve. That is, the firm's labour demand curve is analogous to a budget constraint for the utility-maximizing union.

In this situation, the union's preferred wage-employment outcome is the point a_o in previous Figure 15.1, where the iso-utility curve u_1 is tangent to the labour demand curve. Wage-employment combinations such as a_1 and a_2 are attainable but the union prefers a_o to a_1 and a_2. Outcomes such as a_3 are not attainable if the firm can choose employment.

The firm will prefer wage-employment outcomes at which profits (π) are higher to those at which profits are lower. The firm's iso-profit curves — combinations of wage rates and employment which yield equal profits — are shown in Figure 15.3. Note that lower curves imply higher profits; that is, $\pi_3 > \pi_2 > \pi_1 > \pi_o$. This is so, because for any given level of employment, say E^*, profits are higher when wages are lower, other things equal.

Each curve attains a maximum at the point at which the iso-profit curve and the demand for labour curve intersect. This is so because if we fix the wage at, say, w^*, then profits must be highest at c because D^L shows the profit-maximizing quantity of E at each wage. Thus, as the firm increases employment from e to f to c, π must increase (i.e., the firm must be moving to iso-profit curves with higher profit levels), reflecting the fact that each additional worker hired adds more to total revenue than to total cost (i.e., MRP > W where MRP is the "Marginal Revenue Product," the change in total revenue that results when one additional employee is hired). Similarly, as the firm increases employment from c to g to h, π must fall, reflecting the fact that for E > E^*, MRP > W (i.e., each additional employee adds less to total revenue than to total costs).

Figure 15.3
THE FIRM'S ISO-PROFIT CURVES

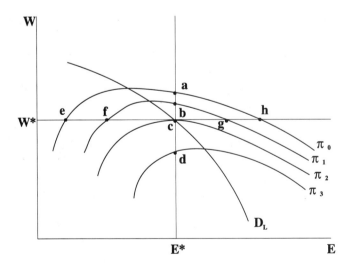

It is important to recognize that although the firm is maximizing profits at each point on the labour demand curve, the firm is not indifferent amongst these points because each corresponds to a different level of profits. The higher the wage rate, the lower the maximum attainable level of profits.

Because profits rise as the firm moves down and to the right along D_L, the firm will prefer lower points on D_L to higher points. However, the firm is constrained by the fact that it cannot pay wages below the alternative wage that employees could get elsewhere. Thus in figure 15.4 I_f (I for "ideal") is the firm's preferred (w,E) combination while I_u is the unions preferred (w,E) combination. The bargaining range is the interval $[w_a, w_u]$. The firm would not want the wage rate to fall below w_a, and the union would not want a wage higher than w_u because a higher wage would not be worth the reduction in employment. Within the bargaining range, however, higher wages make the union better off and the firm worse off (and vice versa).

The bargaining range may be further limited by the need for the firm to earn a normal rate of return (zero economic profits) in order to remain in business, a constraint analogous to the requirement that employees receive a wage at least equal to that available elsewhere. The iso-profit curve corresponding to zero economic profits is shown as π_o in Figure 15.4; thus if the wage were higher than w_o the firm would go out of business. The wage at which profits are zero (w_o) may be above or below the union's preferred wage, w_u. (In a competitive equilibrium w_o and w_a coincide, and there is no scope for raising wages without

Figure 15.4
THE FIRM'S AND UNION'S PREFERRED WAGE-EMPLOYMENT
OUTCOMES AND THE BARGAINING RANGE

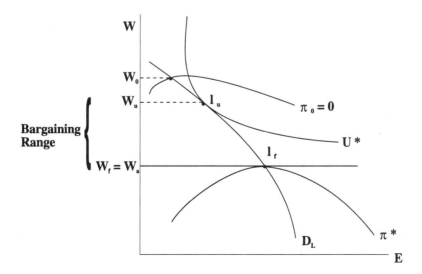

forcing some firms to shut down operations). When w_o is below w_u, the upper limit to the bargaining range is w_o.

The theory outlined thus far does not predict a unique wage and employment outcome, but rather predicts that the wage rate will lie somewhere in the bargaining range (w_a, w_u) with the corresponding level of employment being determined according to the labour demand function. To complete the theory we require an explanation of the determination of the bargaining outcome arrived at by the two parties. This requires a theory of bargaining, a subject taken up later. First we examine two closely related possibilities: (1) that the union may attempt to alter the constraint and (2) that the firm and union may bargain over both wages and employment, rather than leaving the latter to be unilaterally determined by the firm. In addition, the way unions can achieve their objectives by restricting supply is also discussed.

RELAXING THE DEMAND CONSTRAINT

Through collective bargaining with the employer, the union will attempt to reach a wage-employment outcome as close as possible to its preferred outcome I_u. However, unions can also enhance the set of available options by altering the constraint — either increasing labour demand or making the demand for labour more inelastic, so that increases in wages will have less severe consequences for employment. Various union practices attempt to achieve these results. With respect to the goal of making labour demand more inelastic, recall from Chapter 9 that the elasticity of labour demand depends on (1) the elasticity of product demand, (2) the elasticity of substitution between labour and other inputs, (3) the share of labour in total costs, and (4) the elasticity of labour supply of other inputs. Union activity is thus often directed at restricting substitution possibilities by consumers in product markets and by employers in labour markets, the latter including the substitution of nonunion for union labour in addition to the substitution of capital and other inputs for labour. Some restrictions come about through collective bargaining with the employer, while others are achieved by influencing public policy.

Because of the derived nature of labour demand, some attempts to alter the constraint focus on the product market. These include: support for quotas, tariffs, and other restrictions on foreign competition in order to reduce the number of substitutes for union-produced goods and services (observed, for example, in the Canadian auto, textile, and shoe industries); opposition to deregulation which generally enhances product market competition and may facilitate entry by nonunion competitors (observed, for example, in the communications and transportation industries); advertising of the product (e.g., union label on beer); and union attempts to organize all the firms in an industry in order to minimize substitution of nonunion-produced for union-produced goods and services.

Other union policies focus on the labour market. Unions are generally strong supporters of various forms of wage-fixing for nonunion labour such as minimum wage laws and "fair wage" provisions in government contracts. Raising the wage of nonunion workers reduces the potential for substitution of nonunion for union labour — both directly through arrangements such as contracting out of work by unionized employers and indirectly by the growth of nonunion firms at the expense of union firms. Advocating "fair wages" for unorganized employees and higher minimum wages for both organized and unorganized workers may also be viewed by the union movement as an effective method of raising the living standards of the labour force and increasing labour's share of national income. Union support for full employment policies may reflect similar desires; such support can also be viewed as an attempt to increase labour demand generally, including the demand for union labour, as well as reducing the size of the pool of unemployed workers competing for jobs and thus weakening the position of employed union members.

Attempts to alter the constraint can have offsetting effects; in these circumstances the union's optimal strategy will depend on the magnitude of the opposing effects. For example, raising the wages of unorganized workers limits the potential for substitution of nonunion for union labour, but may also reduce the incentive for these workers to organize. The growth paths adopted by early craft unions also illustrate opposing effects. Some craft unions remained narrowly focused, representing a small group of highly skilled workers. An advantage of this arrangement is that the earnings of these workers often account for a small fraction of total cost. Other craft unions expanded to cover semi-skilled and unskilled workers in the same enterprise and industry in order to minimize the substitution among different types of labour that would occur in response to an increase in the wage of the highly skilled group.

The potential for labour market substitution can also be reduced by restrictions on labour supply — such as limited immigration and control of entry through occupational licensing and apprenticeship arrangements. Restrictions on labour supply into the occupation or trade are most frequently employed by craft unions and professional associations such as those representing doctors and lawyers. Unions also attempt to limit the employer's substitution possibilities by negotiating contract provisions such as those relating to technological change, contracting out of work and work practices. This feature of collective bargaining involves the two parties in the negotiation of contracts relating to both wages and employment, a topic examined next.

EFFICIENT WAGE AND EMPLOYMENT CONTRACTS

Because unions generally care about the employment prospects of their members, we would expect the union to use its bargaining power to negotiate over employment as well as wages. Of course, the employer may resist this attempt because agreeing to certain levels of employment will have implications for the firm's profits. Nonetheless, if the union and its members care enough about employment, they may be willing to accept a lower wage in exchange for higher levels (or greater security of) employment. In these circumstances, the union and the firm may discover that an agreement with respect to both wages and employment is to their mutual advantage. Such an arrangement may be explicit (i.e., contained in the collective agreement) or implicit (i.e., an understanding that both parties respect).

In general, the firm and union can each benefit from negotiating a contract covering *both* wages and employment rather than having the firm choose employment subject to a fixed wage. That is, wage-employment outcomes on the labour demand curve imply unexploited "gains from trade" and thus are inefficient in the Pareto sense. (Recall that an outcome is said to be *Pareto-efficient* or *Pareto-optimal* if no individual or group of individuals can be made better off without making some individual or group of individuals worse off).

Figure 15.5
EFFICIENT AND INEFFICIENT WAGE-EMPLOYMENT CONTRACTS

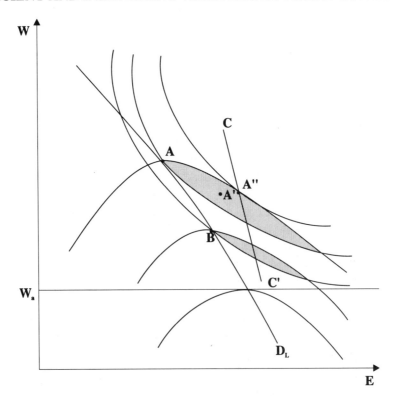

This is illustrated in Figure 15.5. Starting at points on the labour demand curve such as A and B, it is possible to make one or both parties better off by moving to outcomes in the respective hatched areas. At A′ both the union and firm are better off than at A, while at A″ the union is better off than, and the firm as well off as, at A. Note that at A″ it is not possible to move in any direction that does not make at least one party worse off. Thus A″ is a Pareto-efficient wage-employment outcome. It is characterized by the union's indifference curve being tangent to the firm's iso-profit curve. This is a necessary condition for an efficient wage-employment arrangement.

The reasoning underlying this condition is as follows. The slope of the indifference curve measures the union's willingness to trade off wages against employment, holding utility constant. That is, it shows the largest wage reduction the union would accept in exchange for an increase in employment of one unit. Similarly the slope of the iso-profit curve measures the firm's ability to substitute between wages and employment while holding profits constant. That is, it shows the minimum wage reduction the firm would require in exchange for an agreement to increase employment by one unit. When these slopes are not equal, unexploited gains from trade must exist. That is, by increasing employment by one unit and reducing the wage by more than the minimum the firm would require to maintain constant profits but less than the union would accept to maintain constant utility, both parties are made better off.

The locus of the Pareto-efficient wage-employment outcomes is called the "contract curve" and is shown as CC′ in Figure 15.5. When the union cares about wages and employment (i.e., if the indifference curves are downward sloping), the contract curve must lie to the right of the labour demand curve. This follows from the fact that the iso-profit curves are upward-sloping to the left of the labour demand curve, have a slope of zero where the two intersect, and are downward-sloping to the right of the labour demand curve. Thus the tangency conditions can only be met to the right of the labour demand function, where both the iso-profit and iso-utility curves slope downwards.

In summary, both the firm and the union can benefit from negotiating a wage-employment outcome on the contract curve. Of course, the two parties will have differing views about which *particular outcome* on the contract curve should be chosen. Moving up CC′, the union becomes better off and the firm worse off, and vice versa. As before, there is a bargaining range and where the parties end up within that range depends on their relative bargaining strengths. The lower limit of the bargaining range is determined as before; the firm would not want the wage rate to be lower than would be required to attract workers. The upper limit is determined by a zero economic profit condition; in general the union would not want wages and employment set at a level which caused the firm to cease operations.

Several implications follow from this analysis. Perhaps the most important implication is that the two parties have an incentive to negotiate wage-employment arrangements which may appear to be wasteful. Outcomes on the contract curve must involve the firm employing more labour than the firm would choose on its own, given the wage rate; i.e., hiring redundant personnel or spreading a given number of tasks among more workers. Such outcomes can be enforced by work rules of various kinds — arrangements which are often referred to as restrictive work practices or featherbedding. The above analysis shows why such practices may be to the mutual advantage of the employer and the union.

A second important implication is that, in the context of efficient contracts, the anticipated negative relationship (other influences being held constant) between wages and employment may not hold. The contract curve may be vertical or upward- or downward-sloping. Thus it need not be the case in this setting that higher wages imply reduced employment, and vice versa. A more powerful union will generally be able to achieve higher wages *and* employment than a less powerful union, other things equal. (The meaning of the term union power is discussed further later in this chapter.) Nonetheless, any particular union ultimately faces a tradeoff between wages and employment, even when the contract curve is upward-sloping, because the union is ultimately constrained by the need for the employer to earn a satisfactory level of profits in order to continue producing. Thus the union is constrained by the locus of wage-employment combinations that yield this satisfactory level of profits, illustrated by the iso-profit curve π_o in Figure 15.4, a locus which is downward-sloping to the right of the demand curve.

Obstacles to Reaching Efficient Contracts

While there is an incentive for firms and unions to bargain over both wages and employment and reach an outcome on the contract curve, there are also reasons for believing that outcomes on the labour demand curve may be more likely. One reason is that the information needed to recognize that there are unexploited gains from trade may not be available to both parties. For example, suppose the two parties are currently located at point A in Figure 15.5. As long as each has some information about the other side's preferences, then both will recognize that they are both better off at A′ than at A. What this requires is the firm knowing something about the union's willingness to trade off wages against employment, and similarly for the union. However, if neither party trusts the other then each may be unwilling to reveal information about their preferences to the other, perhaps because of a fear that the other will later use this information to their own advantage. Indeed, there may even be an incentive to misrepresent one's preferences in bargaining situations; i.e., not to "tell the truth, the whole truth, and nothing but the truth." In these circumstances it may be difficult for the two bargainers to recognize a Pareto-improving change.

Also, while an agreement over wages is easily enforced, an agreement about employment may be difficult (i.e., costly) to enforce. Some enforcement mechanism is needed to prevent the firm from "cheating" on the agreement; this follows from the fact that at each wage rate the firm's profits are highest at the level of employment given by the labour demand curve. Thus even though both sides are better off on the contract curve, given the negotiated wage rate the firm always has an incentive to cut employment.

If the demand for labour is expected to remain constant during the contract period (the period during which the collective bargaining agreement is in effect), then an agreement about employment should not be difficult to enforce. Presumably the union can keep track of its employed members and if the firm is caught cheating (e.g., not replacing a worker who quits, retires, or is fired) the union will act accordingly (e.g., a strike, slowdown, etc.). However many collective agreements cover a period of two or more years (almost all provide for a duration of at least one year) so that it is unlikely that labour demand will remain constant over the contract period. Thus an agreement concerning employment would have to specify the level of employment for each of the many sets of circumstances that may prevail in the future.[3] This is the case for such factors as the future state of the product market and the costs of other inputs for the firm. From the workers' point of view, important variables would be the cost of living and wages in other firms and industries. Clearly it may be very difficult and expensive to negotiate and enforce such a "contingent" agreement, depending on the different states of nature that may prevail.

The obstacles to negotiating efficient wage-employment contracts are often evident in concession bargaining. Unions have generally been reluctant to engage in concession bargaining unless job guarantees are part of the package, recognizing that without such guarantees the concessions may save few union jobs and may simply lead to higher firm profits. Firms are often reluctant to make explicit employment guarantees because of potential changes in product and labour market conditions.

Since an agreement covering all possible contingencies is unlikely to be workable, the two sides may try to approximate such an arrangement. This can be accomplished by tying employment to the output of the firm or to the use of other inputs. The most common such arrangement is to specify the minimum number of employees to be used for a given task; examples of such provisions are observed in railroads (size of freight train crews), airlines (number of pilots per aircraft), teaching (class size provisions), and musical performances (minimum size of orchestra). (Some of these provisions may also reflect other objectives such as safety.) Such rules may not be difficult to enforce; further, they have the advantage that they allow the firm to adjust employment in response to fluctuations in the demand for the firm's product, yet oblige the firm to hire more employees in each state of demand than the firm would choose on its own.

---

3. If D_L shifts, so does the set of Pareto-efficient (w, E) combinations. Even if an agreement for a specific level of employment could be enforced, there is no incentive for the two parties to reach such an agreement if the demand or supply of labour is expected to shift.

The practice of negotiating work rules that restrict the firm's ability to control employment can thus be interpreted as a mechanism for approximating an efficient contract. In effect the unconstrained demand curve D_L is replaced by a constrained demand curve D_L^C which shows the firm's profit-maximizing level of employment given the work rules implicitly or explicitly agreed to by the two parties (see Figure 15.6). Although this constrained demand curve (or "approximately efficient" contract curve) is unlikely to coincide exactly with the contract curve, for each outcome on D_L there is a Pareto-superior outcome on D_L^C.

Previously we noted that if the union is constrained by the labour demand curve, it is in the union's interest to shift the demand curve to the right or make it more inelastic. What wasn't clear was why the firm would accede to such requests. A primary insight of the analysis in this section is that featherbedding rules or restrictive work practices which force the firm to operate to the right of the unconstrained labour demand curve can be to the mutual advantage of both parties.

Efficient vs. Inefficient Contracts: Summary

Two alternative models of the determination of wages and employment in unionized settings have been outlined. The first — which can be termed the "labour demand curve model" — involves the two parties negotiating the wage rate and the firm unilaterally setting employment. The alternative model — the contract curve model — involves the two parties negotiating over both wages and employment. Which of these two models applies in any particular setting is

Figure 15.6
INEFFICIENT, APPROXIMATELY EFFICIENT,
AND EFFICIENT CONTRACTS

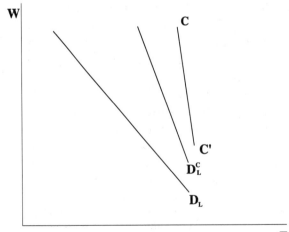

an empirical question. The observation that both parties can improve their welfare by negotiating both wages and employment suggests strong incentives to reach an outcome on the contract curve and provides an explanation of negotiated work rules as a rational response to these incentives. However, in many settings, employment is unilaterally determined by the firm and work rules appear to play a minimal role. This suggests that in many settings the costs of monitoring and enforcing an efficient contract are too high, given the variability in demand, to make this type of arrangement worthwhile.

MATHEMATICAL EXPOSITION*

Much of the above analysis can be stated succinctly using mathematics. Union preferences are given by $U(w, E; w_a)$ where $w = \frac{W}{P}$ is the real wage received by employed union members, w_a is the real alternative wage, and E is total employment of union members. Hours of work are assumed fixed for the purposes of simplicity. With the capital stock fixed in the short run, the firm's total revenue is a function of the variable factor and is written as $R(E)$. This notation is general enough to allow for various product market conditions. For example, if the firm is a perfect competitor in the product market then

$$R(E) = p \cdot Q = p \cdot f(E) \tag{15.1}$$

where p is the market price of the product and $Q = f(E)$ is the short-run production function. The case of a monopolist in the product market is

$$R(E) = p(Q) \cdot Q = p(f(E)) \cdot f(E) \tag{15.2}$$

where $p(Q)$ is the monopolist's product demand function.

Short-run or variable profits are given by

$$\Pi(E) = R(E) - W \cdot E \tag{15.3}$$

Since the aggregate price level P is exogenous to the individual firm, maximizing nominal and real profits are equivalent. Real profits are given by

$$\pi(E) = r(E) - w \cdot E \tag{15.4}$$

where $\pi = \dfrac{\Pi}{P}$ and $r(\cdot) = \dfrac{R(\cdot)}{P}$

Using Shepherd's lemma, the short-run labour demand function is obtained by differentiating the variable profit function with respect to employment:

$$r'(E) - w = 0 \tag{15.5}$$

where $r'(E) = \dfrac{d}{dE} r(E)$ is the marginal revenue product of labour.

---

*indicates more difficult material

When the firm chooses employment, the union's preferred wage-employment outcome is the solution to the following constrained maximization problem:

Max U(w, E; w_a)
{w,E}

subject to $r'(E) - w = 0$

Substituting in the constraint yields the unconstrained maximization problem:

Max U($r'(E)$, E; w_a)
{E}

The necessary condition for a maximum is:[4]

$$U_w r''(E) + U_E = 0 \tag{15.6}$$

where $U_w = \dfrac{\partial U}{\partial W}$ and $U_E = \dfrac{\partial U}{\partial E}$. This condition can be written as

$$-\frac{U_E}{U_W} = r''(E) \tag{15.7}$$

The left-hand side is the slope of a union iso-utility curve and the right-hand side is the slope of the labour demand function. Thus (15.7) states that the union's indifference curve must be tangent to the labour demand function, as shown in Figure 15.1.

The firm's iso-profit curves are defined by

$$r(E) - wE = \pi_0 \tag{15.8}$$

where π_0 is a constant. The slope of an iso-profit line is thus given by

$$\frac{dw}{dE} = \frac{(r'(E) - w)}{E} \tag{15.9}$$

Thus $\dfrac{dw}{dE} \gtreqless 0$ according to $r'(E) - w \gtreqless 0$ $\qquad$ (15.10)

That is, the iso-profit curves have a positive slope to the left of the labour demand curve, a slope of zero where they intersect the labour demand curve, and have a negative slope to the right of the labour demand curve. This is the shape shown in Figure 15.3.

The locus of Pareto-efficient contracts is obtained as the solution to the following constrained maximization problem:

Max (U(w, E; w_a)
{w,E}

---

4. It is straightforward to verify that the sufficient condition for a maximum requires that the union's indifference curves be more convex to the origin than the labour demand function.

subject to $r(E) - wE = \pi_0$

where π_0 is an arbitrary constant. The necessary conditions for a maximum simplify to

$$-\frac{U_E}{U_W} = \frac{(r'(E) - w)}{E} \tag{15.11}$$

The left-hand side is the slope of the union's iso-utility curves and the right-hand side is the slope of the firm's iso-profit curves. Thus the contract curve is the locus of points at which the iso-profit and iso-utility curves are tangent, as shown in Figure 15.5.

Because the iso-profit curves slope downward only to the right of the labour demand curve, the contract curve must lie to the right of the labour demand curve. This result can also be seen from rewriting (15.11).

$$r'(E) = w - E \cdot \frac{U_E}{U_W} < w \tag{15.12}$$

That is, efficient contracts are characterized by the marginal revenue product of labour being less than the wage rate.

The slope of the contract curve can be obtained by differentiating (15.11):

$$\frac{dw}{dE} = \frac{U_E + E\,U_{EE} + r''(E)\,U_w + (r'(E) - w)\,U_{wE}}{U_w - E\,U_{Ew} - U_{ww}\,(r'(E) - w)} \tag{15.13}$$

Because both the numerator and denominator of (15.13) may be either positive or negative, the contract curve may be upward- or downward-sloping or vertical.

EMPIRICAL APPLICATIONS

Compared to the large number of empirical studies of the product demand and labour supply behaviour of households and of the production and cost behaviour of firms, the number of studies of wage and employment determination in unionized labour markets is very limited. Nonetheless, following pioneering research by Farber (1978) and Dertouzos and Pencavel (1981), a number of studies have been carried out recently. The literature remains at an embryonic stage of development, and much remains to be discovered about union wage and employment policy.

Empirical applications of the theory of union wage and employment determination face several obstacles:

1. choosing an appropriate specification of union preferences,
2. the presence of two alternative models of wage and employment determination which are plausible on a priori grounds,

3. the need for a theory of firm-union bargaining to predict the specific wage-employment outcome chosen by the parties.

These are clearly major challenges. As noted previously, the issue of whether it is appropriate to characterize unions as attempting to maximize an objective function has long been a matter of scholarly debate. With respect to (2) and (3), the studies by Farber (1978), Dertouzos and Pencavel (1981), Pencavel (1984) and Carruth and Oswald (1985) assume that observed wage-employment outcomes lie on the labour demand curve and that the union is sufficiently powerful that it can achieve its preferred outcome, I_u in Figure 15.4.

In contrast, studies such as those by Abowd (1989) are based on the assumption of efficient contracts; i.e., that wage-employment outcomes are on the contract curve. Another group of studies — including Brown and Ashenfelter (1986), MaCurdy and Pencavel (1986), Eberts and Stone (1986), and Martinello (1989) — have addressed the question of whether observed wage and employment outcomes are more consistent with the labour demand or contract curve models. The most recent studies (Svenjar, 1986; Doiron, 1992) have begun the difficult task of incorporating bargaining into empirical models of wage and employment determination in unionized industries.

The Dertouzos and Pencavel study focuses on the International Typographical Union (ITU), a once-powerful closed shop union representing printers. The employers are mostly newspaper firms. Recently there have been important technological changes in the industry (computer typesetting, etc.) and employment has fallen drastically. The period studied, 1946–65, is before these drastic changes to the collective bargaining setting.

Dertouzos and Pencavel argue that the union has a dominant power position, and that this justifies the assumption that the union can determine the wage. The union's power derives mainly from the vulnerability of local newspaper firms to a prolonged period of nonproduction. This vulnerability results from several factors: (1) the product can't be stored, (2) both advertisers and subscribers have various alternative media they can turn to in the event of a strike, (3) it is often costly and difficult to regain former subscribers and advertisers after a strike.

For the union local examined in the most detail, the Cincinnati Post, Dertouzos and Pencavel's estimates imply that the union cares about both wages and employment, but that relatively more weight is put on employment than would be the case if the union's objectives were maximization of economic rents. Pencavel (1984) finds that larger union locals have preferences which are approximately those of rent maximization while smaller locals place more weight on employment.

Farber (1978) analyzes the wage and employment behaviour of the United Mine Workers in the U.S. bituminous coal industry over the period 1946–73. The union is a dominant force in the industry due to the fragmented market structure (large number of essentially competitive firms), and is thus assumed to be able to dictate the wage. However, the union is constrained by the position and shape of the labour demand curve which depend on substitution: (1) between coal and other fuels in the product market, (2) between labour and other inputs in coal production, and (3) between union-produced and nonunion-produced coal.

Farber's estimates imply that the union places a relatively high weight on employment relative to wages and benefits, higher than would be implied by wage bill or economic rent maximization. Other studies based on the labour demand model also find that unions place a relatively high weight on employment. Carruth and Oswald (1986) analyze the National Union of Mineworkers in the United Kingdom over the period 1950–80. Their estimates imply that the union places less weight on employment than its U.S. counterpart, the United Mine Workers, but nonetheless a high weight relative to that on wages. Martinello (1989) analyzes the behaviour of the International Woodworkers of America (IWA) in the British Columbia forest products industry. Although his primary objective is to compare the performance of the labour demand and contract curve models, his estimates for the labour demand model (assuming the union can set the wage) also imply that the union places a relatively high weight on employment.

On the basis of these studies based on the labour demand curve model, there appears to be agreement on several issues: (1) both employment and wages are important to unions, (2) union preferences display diminishing marginal rate of substitution between wages and employment, (3) union preferences are sensitive to the alternative wage, (4) unions generally place relatively more weight on employment than wages, (5) wage bill maximization is rejected as a union objective function, and (6) economic rent maximization is rejected in the majority of cases. In assessing these conclusions, it is important to keep in mind that only a small number of unions have been analyzed using modern econometric methods and that the conclusions are conditional on some key assumptions, in particular that the union can dictate the wage and the firm unilaterally sets employment.

The task of testing between the contract curve and labour demand models has been addressed in recent papers such as those by Brown and Ashenfelter (1986), MaCurdy and Pencavel (1986), Eberts and Stone (1986), Card (1986) and Martinello (1989). The first two studies are based on the ITU in the United States and the latter three studies are based respectively on New York public school teachers, U.S. airline mechanics, and forestry workers in the IWA in Canada. At this point there is not a consensus on the question of which model performs best. Most studies find that the contract curve model is more consistent with their data; however, the evidence is far from conclusive. In particular, whether the efficient bargaining model is better than alternative frameworks which allow the outcome to be to the right of the demand curve remains to be seen. The conditions under which outcomes are on the labour demand curve or the contract curve (or perhaps in between) is an important research topic.

Perhaps the most important aspect of union wage and employment determination that is not addressed by the above studies is the incorporation of firm-union bargaining in the empirical analysis. In general, the labour demand and contract curve models predict a range of possible wage-employment outcomes rather than a unique outcome. In order to determine the particular settlement the parties will reach we require a theory of bargaining, a subject to which we now turn.

THEORY OF BARGAINING*

Bargaining theory is concerned with predicting the outcome in any particular bargaining situation, and explaining what factors this outcome depends on. Because of the importance of bargaining in many aspects of life, bargaining theory has received considerable attention from social scientists. There is not yet, however, any single widely accepted theory of bargaining. In this section we examine some important contributions, with particular emphasis on their implications in the collective bargaining setting.

Although each bargaining situation may be unique, all share some common features: there is a set of possible outcomes, with minimum acceptable outcomes for each party; because agreement is voluntary, neither party will agree to an outcome worse than their minimum acceptable outcome. Whether the parties bargain over wages alone or over wages and employment, there is a set of feasible outcomes, with upper and lower limits representing the minimum outcomes acceptable to the union and the firm. This type of situation is illustrated in Figure 15.7(a). The point $d = (d_1, d_2)$ shows the utility of each party if no agreement is reached; this is referred to as the "disagreement" or "threat" point. Neither party will agree to an outcome worse than d and each can threaten to impose this outcome on the other by failing to agree. The bargaining set S consists of feasible outcomes in which one or both parties is better off than they would be without an agreement. This diagram illustrates the bargaining situation in terms of the parties' utilities because what ultimately matters to each party is the satisfaction obtained from an agreement. However, the parties do not bargain over utilities but rather over outcomes such as wages and employment and in some cases the analysis is best carried out in terms of these outcomes.

In bargaining there are incentives for both co-operation and conflict. The incentives for co-operation are evident from the fact that there are various feasible outcomes that make both parties better off than if no agreement is reached. As long as the employees' utility is at least as great as they can obtain elsewhere and the employer's profits at least as great as can be earned else-

---

*indicates more difficult material

where, both sides are better off continuing the employment relationship. Indeed, the two parties have an incentive to co-operate at any Pareto-inferior outcome, such as the point C in Figure 15.7(a), because both can gain from this process. The incentives for conflict arise because, once all Pareto-improvements have been realized, such as at the point A in Figure 15.7(a), actions which make one party better off result in the other becoming worse off. Further, the recognition by each party that the other benefits from reaching an agreement may create an incentive to threaten the other with non-agreement. For example, recognizing that the firm is better off with outcome f_1 than d_2 the union may threaten the firm with non-agreement unless the firm agrees to the outcome A. Alternatively the firm may threaten non-agreement unless the union agrees to B. Because of these incentives for both co-operation and conflict, most bargaining situations contain an uneasy mixture; on the one hand, recognition of the importance of reaching agreement and, on the other, attempts to mislead, out-manoeuvre or otherwise gain advantage over the other party.

SOLUTIONS TO THE BARGAINING PROBLEM*

A number of solutions to the bargaining problem have been proposed. These can be classified into two groups: theories of the bargaining process and theories of the bargaining outcome. The former not only predict the outcome but also model the process by which that outcome is reached. The latter model the bargaining outcome, by specifying a number of properties or axioms which the outcome should obey. Leading examples of each type are outlined below. The theories described here assume full information; each party knows the set of feasible payoffs or outcomes, the preferences of the other side, and therefore the set of feasible utility outcomes, S in Figure 15.7. However, this does not imply that there is no uncertainty; although each party is aware of the outcomes available to itself and the other party, it may well be uncertain about what the other will do. As is common in uncertain situations, the parties are assumed to maximize expected utility.

Nash's Bargaining Theory

Following the publication by Von Neuman and Morgenstern in 1944 of the *Theory of Games and Economic Behaviour*, the Princeton mathematician John Nash proposed a solution to the bargaining problem which is still widely used today. Nash (1950) specified four axioms that the solution should obey:

(A1) Pareto-efficiency: only Pareto-efficient bargains will be agreed to.

(A2) Symmetry: if the feasible bargaining set S is symmetric (i.e., if it doesn't matter which party's utility is measured on which axis), the solution will

---

*indicates more difficult material

Figure 15.7(a)
THE BARGAINING PROBLEM

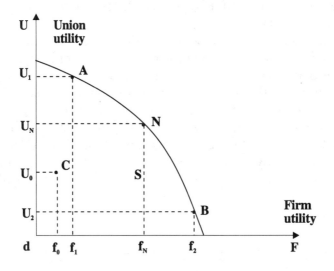

Figure 15.7(b)
THE NASH SOLUTION AND THE INDEPENDENCE
OF IRRELEVANT ALTERNATIVES AXIOM

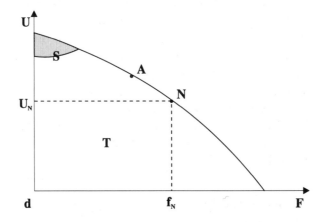

give equal utility increments (relative to the disagreement outcome) to each party.

(A3)   Transformation invariance: the solution is not altered by linear transformations of the utility function of either party.

(A4)   Independence of irrelevant alternatives: Suppose that the solution to the bargaining problem has been reached for a particular bargaining set S; for example, the point A in figure 15.7(b). Then the solution will not be altered if some outcomes other than A are unavailable (such as the shaded area in Figure 15.7(b)). That is, if A is the solution with the bargaining set S then A will also be the solution if the same two players face the bargaining set T (S less the shaded area).

The first axiom can be regarded as a natural consequence of the rationality of the two bargainers and the full information assumption. Why would they agree to an outcome such as C when both can be made better off at outcomes such as A and N? The second axiom can be viewed as stating that the bargaining power of each side is reflected in the nature of the set of feasible bargaining outcomes and the disagreement outcome. If the set of bargaining outcomes is symmetric, the two parties must be equally powerful because they both stand to gain or lose equally from agreement or non-agreement. Thus the solution should reflect this equality of bargaining power, and give each an equal gain from agreement.

The third axiom states that the units in which utility is measured should not matter. Utility functions simply represent an individual's preference ordering of outcomes, and any pair of functions that order outcomes in the same way are equally valid for this purpose. In expected utility theory, cardinal utility functions are needed; any linear transformation of a cardinal utility function will preserve the ordering and therefore be a valid utility function. The measurement of temperature is analogous; the fahrenheit and centigrade scales are equally valid measures of temperature, one being a linear transformation of the other. Because this axiom states that the actual numerical scale used to measure utility is arbitrary, its main consequence is to rule out interpersonal comparisons of utility.

The fourth axiom has probably been the most controversial. According to this axiom, if the parties chose the outcome A from all the alternatives in the set S, then the same outcome should be chosen from the set T because the only difference between the two situations is the absence of the outcomes in the shaded area, outcomes which have already been rejected in favour of A by the two bargainers. An analogy in the case of individual decision-making is the following. You go out to a restaurant for dinner, and are told by the waiter that there are three choices: chicken in mustard sauce, veal with a chanterelles sauce, and salmon with a cucumber sauce. You decide to order the veal. The waiter subsequently informs you that they are out of salmon. Would you want to change your order? Obviously not — you chose veal over salmon and chicken, so should choose veal over chicken alone. Whether this reasoning applies in a two-person bargaining situation is a more complex question — which is why the axiom has generated controversy.

Nash's (1950) remarkable result was that assumptions A1 to A4 imply a unique solution to any bargaining situation: the outcome that maximizes the product of the two parties' utility increments from the disagreement point; i.e., maximizes $(U - d_1)(F - d_2)$. The Nash solution is shown as the point N in Figure

15.7. Geometrically, the Nash outcome maximizes the area within the set S; i.e., the area $dU_N Nf_N$ in Figures 15.7(a) and (b) exceeds the area of any other rectangle contained in the bargaining set.

A number of other axiomatic theories of the bargaining outcome have been proposed; Roth (1979) provides a survey of these theories and Roth and Malouf (1979) and others have tested these competing theories in experimental settings. The Nash solution predicts the outcomes of some bargaining experiments well; however, in other situations the outcomes appear to contradict axiom A3, and a theory that assumes the parties make interpersonal utility comparisons predicts outcomes better than the Nash solution.

It can be debated whether the axiomatic approach constitutes a positive or normative theory of bargaining. The approach proceeds by stating properties that the solution ought to obey, and therefore may be more appropriate for the analysis of arbitration decisions (i.e., decisions about what the outcome ought to be) than collective bargaining outcomes. For the latter purpose, theories that explicitly model the bargaining process may be more appropriate. It should be noted, however, that although the axiomatic theories do not specify the process by which the outcome is reached, they may be consistent with one or several such processes. For example, in an early contribution to the theory of bargaining, Zeuthen (1930) proposed a model of the concession behaviour of each party. The assumed concession process results in the outcome that maximizes the product of the two utility increments, an outcome identical to the Nash solution. Recent contributions have provided more general theories of the bargaining process, and these are examined next.

Rubinstein's Bargaining Theory

In an important contribution, Rubinstein (1982) obtains a solution to the bargaining problem using some concepts of non-co-operative game theory. The type of bargaining situations analyzed by Rubinstein has the following characteristics. The two bargainers take turns making offers. If an offer is accepted, the process ends; if the offer is not accepted the other party makes a counter-offer in the next period. However, delay is costly to both sides. Each period the total "pie" to be shared by the two parties gets smaller. This situation is illustrated in Figure 15.8. As before, there is a bargaining set which shows the outcomes available to each party. In the first period this is the set bounded by $du_1 f_1$. However, if an agreement is not reached in the first period, the set of utility outcomes available in period two shrinks to that bounded by $du_2 f_2$. Similarly in period three the bargaining set becomes $du_3 f_3$, and in period four the only feasible outcome is the disagreement point $d = (u_4, f_4)$. Thus in this example the "pie" which the parties can share has shrunk to zero by period four. In general it may require delay of more than four periods in order to reach the situation where both parties gain nothing from an agreement (if, indeed, this situation is ever reached); the assumption that this outcome is reached quickly is made simply for ease of illustration.

As before, both parties are assumed to possess full information about the bargaining situation. That is, each knows the size of the "pie" available to be shared in the first period, the other party's preferences and the costs of delay (or disagreement) to both sides. Thus each can infer the nature of the bargaining set they will face in subsequent periods.

Each party is assumed to be concerned only with its own self-interest; i.e., it wants to obtain the highest possible utility from the bargain, and cares only about the outcome, not how it was reached. Also, each behaves rationally (and expects the other to do likewise) in the sense that neither will believe threats that would not be in the interest of the threatening party to carry out. For example, suppose it is the union's turn to make an offer in period 1. The union could offer $A = (u_A, f_A)$ and threaten to refuse to accept or make any future offers unless the outcome A is accepted. If the firm believed the threat, it would accept because f_A is better than f_4, which is what the firm would receive if the offer A is rejected and the union carries out the threat. However, the threat is not credible. Suppose the firm rejects the offer. In period 2 the bargaining set becomes du_2f_2 and if the firm offers the outcome B, the union will accept because u_B is better than u_3, the best outcome the union could obtain in the next period. For this reason, the firm knows that the union's threat in period 1 is not credible and it will therefore not believe it.

We can now turn to the solution. Suppose that in the initial period it is the union's turn to make an offer. Rubinstein's analysis predicts that the union will offer and the firm will accept the outcome $R = (u_R, f_R)$; that is, the Rubinstein

Figure 15.8(a)
RUBINSTEIN'S SOLUTION TO THE BARGAINING PROBLEM

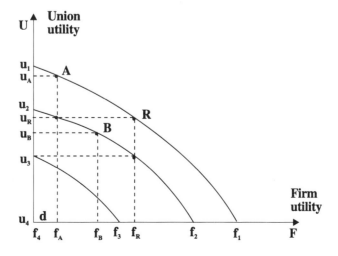

Figure 15.8(b)
THE EFFECT OF DELAY COSTS ON THE RUBINSTEIN OUTCOME

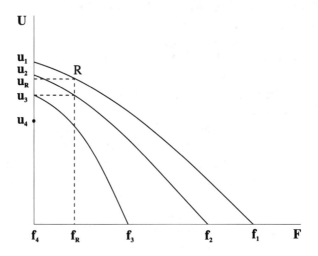

Figure 15.8(c)
THE EFFECT OF A PROPORTIONAL CHANGE IN DISAGREEMENT
COSTS ON THE RUBINSTEIN SOLUTION

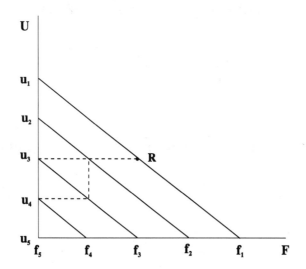

outcome is the point R in Figure 15.8. To see why this solution is predicted to occur, we proceed by backward induction. In the second period it will be the firm's turn to make an offer and in the third period the union's turn. In the fourth period there is nothing left to be shared. Now look at the situation facing the union in period 3. Both parties recognize that the best the firm can attain next period is the outcome f_4. Thus in period 3 the firm will accept any offer equal to or better than f_4. The union would therefore offer (u_3,f_4) and the firm would accept. Now go back to the situation facing the firm in period 2. The firm knows that the best the union can expect next period is u_3. Thus the firm won't offer more than u_3; i.e., will make an offer in the interval between (u_3,f_R) and (u_4,f_2). However, the firm also knows that next period the union will offer and the firm will accept (u_3,f_4). Thus in period 2 the firm will offer and the union will accept (u_3,f_R). Now go back to period 1 and the union's choice of offers. The same analysis leads to the conclusion that the union will offer and the firm will accept (u_R,f_R). Thus the point $R = (u_R,f_R)$ is the bargaining outcome; the costs of delay and the rational behaviour of the two parties result in an agreement that avoids incurring any disagreement costs.

Clearly delay costs play an important role in generating agreement in the Rubinstein model. The existence of these costs — and the fact that both parties are aware of them — gives each some power over the other. For this reason the relative magnitude of delay costs exerts an important influence on the negotiated outcome. This result is illustrated in Figure 15.8(b). The bargaining set du_1f_1 is identical to that in Figure 15.8(a), as are the delay costs to the firm (measured by the outcomes f_1, f_2, f_3, and f_4). However, the union's disagreement costs are less than in Figure 15.8(a), as indicated by the outcomes u_1, u_2, u_3, and u_4. As a consequence, the outcome predicted by Rubinstein's theory is more favourable to the union and less favourable to the firm.

It is important to recognize that *relative* disagreement costs determine the extent to which the negotiated outcome favours one party or the other. By holding constant the firm's disagreement costs and lowering the union's in the comparison between the situation in Figure 15.8(a) and (b), the union's relative delay costs declined. The consequence is a more favourable outcome for the union and a less favourable outcome for the firm. However, suppose the disagreement costs fall by the same proportion for both parties, leaving relative costs unchanged. This situation is shown in Figure 15.8(c). Two bargaining situations are depicted in the same diagram. In the first case the outer boundary of the bargaining set is u_1f_1 in the first period, u_2f_2 in the second period, and so on, reaching the disagreement point (u_5,f_5) in the fifth period. In the second bargaining situation, disagreement costs are proportionally higher for both parties; the boundary of the bargaining set is u_1f_1 in the first period, u_3f_3 in the second, and the disagreement point in the third period. In both cases the firm makes the first offer. The solution is unchanged by the proportionate increase in delay costs, illustrating the reason for emphasizing the role of relative disagreement costs in bargaining outcomes.

As a description of the collective bargaining process, the Rubinstein model is highly simplified. For example, in most bargaining situations either party can make an offer in any period. Nonetheless, this theory captures important aspects of collective bargaining, in particular the role played by the costs of delay or disagreement in bringing about a negotiated settlement. The negotiations process itself consumes scarce resources; thus the sooner an agreement is reached the lower the direct negotiations costs to each party. But the major disagreement costs are those associated with attempts of one side to bring pressure to bear on the other to make concessions. Although there are a variety of relatively inexpensive (and possibly also ineffective) methods for exerting pressure on the other party (e.g., work-to-rule actions, refusal to work overtime), the primary mechanism is the strike or lockout.

The strike involves the collective withdrawal of union labour in an attempt to halt or sharply curtail production. The lockout denies union members access to their normal job and source of income. Both types of work stoppage impose costs on the employer and the employees represented by the union (and also possibly on third parties). The firm loses profits and the workers lose income. In addition to the costs incurred during the strike or lockout, there may be costs of a more permanent nature. Clearly rational bargainers (and perhaps even irrational bargainers) will, in the negotiations process, take into account the magnitude of the costs that they and the other party will bear in the event of a strike or lockout. Rubinstein's theory confirms our intuition that the relative size of the costs of a work stoppage will be an important determinant of the negotiated settlement. Further, it is not necessary for a strike or lockout to occur in order for the costs associated with a work stoppage to influence the outcome.

UNION BARGAINING POWER

The term "bargaining power" arises frequently in discussions of firm-union bargaining. Like many terms that are widely used, the precise meaning is not always clear. In this section we contrast two possible meanings of union bargaining power. For simplicity we assume that the two parties negotiate over wages alone, leaving the firm to set employment. The extension of these ideas to the case of bargaining over wages and employment is straightforward.

One meaning of bargaining power is related to the elasticity of labour demand. A union facing an inelastic demand for labour can raise wages substantially with only minor adverse employment consequences. This notion of bargaining power is associated with the *willingness* to increase wages; other things equal, the more inelastic the demand for labour, the more willing is the union to raise the wage rate.

An alternative notion of bargaining power is associated with the *ability* to raise wages. According to this meaning of the term, a powerful union is one that can obtain a negotiated outcome close to its most preferred outcome. This aspect of union power depends not on the elasticity of labour demand but primarily on the relative costs of disagreement to each party.

Because these two aspects of bargaining power depend on different variables, it is possible for a union to be powerful according to one definition but weak according to another. This possibility is illustrated in Figure 15.9. Panel (a) shows a union that is willing to raise wages substantially but unable to do so; the negotiated wage w_c is considerably below the union's preferred wage w_u. Panel (b) shows a union that can obtain its preferred outcome but is not willing to raise wages substantially because of the employment consequences. Of course, in some circumstances a union may be powerful (or weak) according to both meanings. However, neither aspect of bargaining power alone ensures that the negotiated union wage will be substantially higher than the alternative wage.

UNION POWER AND LABOUR SUPPLY

Most unions attempt to improve the wages and working conditions of their members directly by bargaining with the employer, and the strike threat is their primary source of bargaining power. However, craft unions and professional associations can raise the wages and working conditions of their members indirectly by controlling labour supply. For these organizations, restrictions on the supply of labour can constitute an important source of union power.

Restricting Supply: Craft Unions

By restricting the supply of labour, unions can raise wages artificially relative to the situation where competition prevailed in the supply of labour. Craft unions generally have sought to raise the wage of their members by controlling entry into their craft through the apprenticeship system, discrimination, nepotism, high union dues, and such devices as the closed shop (the worker has to be a

Figure 15.9
ALTERNATIVE ASPECTS OF UNION POWER

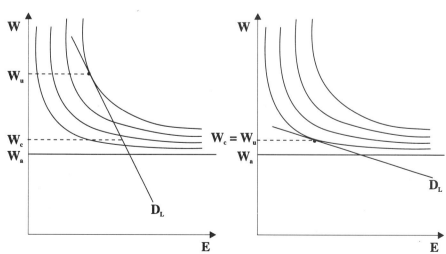

member of the union *before* being hired, i.e., the union controls the hiring hall). Because the supply of labour to the craft is reduced, wages will rise above the wage that would prevail in the absence of the craft restrictions.

Needless to say the craft union will try to control the whole trade; otherwise the benefits of the supply restriction would also go to nonunion craft workers. Hence the importance to the union of apprenticeship licensing, the closed shop, the union shop (the worker has to join the union after a probationary period of employment), or the agency shop (the worker must pay union dues but need not be a member of the union).

At the artificially high union wage rate, the quantity of labour supplied will exceed the quantity demanded. To the extent that the union controls entry, the scarce jobs can be rationed to the large number of workers who would like the jobs by such devices as discrimination, nepotism, or high union dues. To the extent that employers have a say in who is hired, they may ration jobs on the basis of discrimination or nepotism, but they may also ration on the basis of productivity-related factors such as education or training or experience. While these productivity-related factors obviously are useful requirements for employ-ers, they may be set *artificially* high because the employer has an excess supply of applicants, given the higher unionized wage.

Restricting Supply: Professional Associations

Professional associations can also behave much like craft unions in their control over labour supply, largely through the processes of occupational licensing and certification. This setting of standards is deemed necessary, ostensibly to protect the public interest in circumstances where it is important but difficult for the consuming public to judge the quality of the professional service. The job of governing the profession has usually gone to the profession itself — hence the term self-governing professions — because only members of the profession were deemed qualified to set standards. Traditionally, this has occurred in the stereotypic professions such as medicine and law, where an uninformed clien-tele purchases complex services from a self-employed professional. Increasingly, however, professionals are becoming employed on a salaried basis and the rationale for self-governing powers for salaried professionals can be questioned when the employer can provide an effective mediator between the consuming public and the salaried professional.

The techniques of occupational control utilized by professional associations usually involve occupational licensing or certification. Under certification only those with the professional certificate have the right to use the professional designation or title; that is, they have a *reserve-of-title certification*. However, other practitioners are allowed to practise in the profession, albeit they cannot use the certified title. Under licensing, only those with the professional licence can practise; that is, they have the *exclusive right-to-practise licence*. Clearly the exclusive right-to-practise licence involves more control over the occupation than the reserve-of-title certification; hence, it is sought after by most professional and quasi-professional groups.

Under licensing the supply of labour to the occupation is restricted only to those with the licence. Under certification the supply restrictions are less severe in the sense that others can practise in the profession, although competition is restricted because others cannot use the professionally designated certification. Whatever the impact on the quality of the service performed, the supply of labour to the occupation is restricted and wages rise accordingly. In this sense, professional associations like craft unions raise the wages of their members.

Although they receive higher wages, members of the profession may bear costs associated with the occupational licensing requirements. Such costs could include lengthy education periods as well as practical training periods, such as internship for doctors or articling for lawyers. There is a danger, however, in that the *incumbent* professionals may try to have the brunt of the licensing costs born by *new* entrants into the profession. This could be accomplished by having increasingly stringent education or training requirements as a qualification for entry into the profession, with so-called "grandfather" clauses exempting those who are already practising. In this fashion, the new entrants bear more of the licensing costs (they will do so as long as the wage premium outweighs these costs), the existing practitioners who control the profession benefit from the restricted competition, and the consuming public gets increasingly qualified practitioners, albeit at a higher cost. Restricting grandfather clauses, or having licensing requirements set by persons other than existing practitioners, could go a long way to correct this potential abuse of occupational licensing.

QUESTIONS

1. Discuss the reasons why the objectives of union leadership may differ from the objectives of the rank-and-file. Discuss the implications of such differences.

2. Draw the union indifference curves (iso-utility curves) for the following situations:

   (a) wages and employment are perfect substitutes.

   (b) there is a minimum wage below which union members will not reduce their wages in order to increase employment.

   (c) at the current level of employment within the union, unions attach more disutility to a given employment reduction than the utility they attach to a corresponding employment increase.

   (d) as their wealth increases, union members attach more weight to job security than to real income gains.

(e) the union objective is to maximize the wage bill.

(f) the union objective is to maximize total economic rent.

3. Discuss the effect of an increase in the alternative wage rate on the utility of the union.

4. Discuss why union members may differentiate in their preferences between hours of work and employment. What are the implications for union preferences?

5. Discuss the implications of unions being a democratic institution, reflecting the preferences of the median union voter. Why would the preferences of the median voter dominate in such a situation? What are the implications for minority rights within the union?

6. Discuss the error in the following statement: "Because the firm is maximizing profits at each point on its demand curve, it is indifferent amongst these points."

7. Discuss the constraints that limit the "bargaining range" between unions and employers.

8. "If unions engage in concession bargaining to maintain job security they should specifically bargain for a specific level of security to be associated with their wage concessions rather than simply agreeing to the concessions and allowing the firm to choose the level of employment." Discuss.

9. Discuss various ways in which unions may try to make the demand for labour more inelastic so as to minimize any adverse effect emanating from any wage increase.

10. Discuss why efficient wage and employment contracts are not on the firm's labour demand curve. That is, show how both labour and management potentially could be better off by agreeing to wage and employment combinations that are not on the firm's labour demand curve.

11. What determines where the parties will settle on the contract curve? Could the indeterminacy of where the parties will be on the contract curve ever prevent them from arriving at an efficient contract?

12. Discuss how certain featherbedding rules may be efficient and sustain themselves in a competitive market environment. Does this mean that firms will never try to "buy out" such seemingly inefficient practices?

13. Some trade unionists have argued that wage concessions have not preserved jobs. Use the material developed in this chapter to illustrate the circumstances under which this concern may be legitimate, and the conditions under which the concern is not correct.

14. Discuss the obstacles that unions and management may face in arriving at efficient wage-employment contracts.

15. Show that if the union's objective is to maximize economic rent and if the firm and the union negotiate an efficient contract, then the contract curve will be vertical and employment will equal the level that would occur if the employer could unilaterally set employment and were a wage taker facing the alternative wage.

16. "The main foe of any union is various forms of substitution, not the firm."

    (a) Explain the various forms of substitution that unions need to be aware of in formulating their bargaining objectives, and elaborate on the relationship between these forms of substitution and union bargaining power.

    (b) The United Mine Workers studied by Farber (1978) had two instruments for affecting the monetary returns to and employment of their members — the hourly wage rate and the output tax (tax per ton of coal produced), the proceeds from which were distributed to members. Under what circumstances would the use of these two instruments allow the union to achieve outcomes that could not be achieved by affecting the wage rate alone?

17. Indicate conditions under which one may expect unions to have a substantial impact on the wages of their members.

18. Craft unions and professional associations will affect the wages of their members in a different manner than will industrial unions. Discuss, and indicate why this may be the case.

19. If craft unions or professional associations are able to raise the wages in their trades above the competitive norm, excess supplies of applicants would result. Discuss how the scarce jobs may be rationed.

20. Self-government is an anachronism for salaried as opposed to self-employed professionals. Discuss.

21. Distinguish between occupational licensing and certification. Give examples of each.

22. If occupational licensing is necessary, the only group that could be entrusted not to abuse the powers of licensing would be the alumni of the profession, or at least members who were about to retire. Only they have knowledge about the profession, without having a self-interest in abusing the power of licensing. Existing practitioners, while knowledgeable about the profession, have a self-interest to restrict entry by putting unnecessarily costly entry requirements on new entrants into the profession. New entrants may not find it in their self-interest to put unnecessary restrictions on their entry into the profession, but they do not yet have sufficient knowledge about the profession and what is required for proper qualifications. The general public, while having an interest in ensuring quality performance

without unnecessary restrictions, may not be able to judge what requirements are necessary for professional competence. Only alumni have knowledge of the profession without having a self-interest in excessive quality
restrictions; consequently, they are the persons who should be entrusted
with the powers of occupational licensing, if it is necessary. Discuss.

23. Restrictions on the use of grandfather clauses would go a long way in
reducing the abuses of occupational self-licensing. Discuss.

24. Indicate why incumbent practitioners in a profession may want to put
excessive restrictions on entry into the profession.

REFERENCES AND FURTHER READINGS

Abowd, J.M. The effect of wage bargains on the stock market value of the firm. *AER*
79 (September 1989) 774–809.

Allen, S.G. Union work rules and efficiency in the building trades. *JOLE* 4 (April
1986) 212–242.

Atherton, W.N. *Theory of Union Bargaining Goals.* Princeton, NJ: Princeton University Press, 1973.

Blair, D.H. and D. Crawford. Labor union objectives and collective bargaining. *QJE*
99 (August 1984) 547–566.

Blanchflower, D.G., Millward, N. and A.J. Oswald. Unionism and employment behavior. *EJ* 101 (July 1991) 815–834.

Bronars, S.G. and D.R. Deere. The threat of unionization, the use of debt, and the
preservation of shareholder wealth. *QJE* 106 (February 1991) 231–254.

Brown, J. and O. Ashenfelter. Testing the efficiency off employment contracts. *JPE* 94
(July 1986) S40–S87.

Burda, M.C. Membership, seniority, and wage-setting in democratic labour unions.
Economica 57 (November 1990) 423–438.

Card, D. Efficient contracts with costly adjustment: short-run employment determination for airline mechanics. *AER* 76 (December 1986) 1045–1071.

Carruth, A.A. and A.J. Oswald. Miners' wages in post-war Britain: an application of a
model of trade union behavior. *EJ* 95 (December 1985) 1003–1020.

Carruth, A.A. and A.J. Oswald. On union preferences and labour market models: insiders and outsiders. *EJ* 97 (June 1987) 431–445.

Chapman, P.G. The Johnston wage bargaining model and trade union objectives. *Manchester School of Economics and Social Studies* 49 (December 1981) 310–318.

Clark, A. Efficient bargaining and the Macdonald-Solow conjecture. *JOLE* 8 (October
1990) 502–528.

Currie, J. Employment determination in a unionized public-sector labour market: the
case of Ontario's school teachers. *JOLE* 9 (November 1991) 45–66.

Dertouzos, J.N. and J.H. Pencavel. Wage and employment determination under
trade unionism: the international typographical union. *JPE* 89 (December 1981)
1162–1181.

Devereux, M.B. and B. Lockwood. Trade unions, non-binding wage agreements, and capital accumulation. *EER* 35 (October 1991) 1411–1426.

Doiron, D.J. Bargaining power and wage-employment contracts in a unionized industry. *IER* 33 (August 1992) 583–606.

Dowrick, S. Wage pressure, bargaining and price-cost margins in UK manufacturing. *Journal of Industrial Economics* 38 (March 1990) 239–267.

Dowrick, S. Union-oligopoly bargaining. *EJ* 99 (December 1989) 1123–1142.

Dunlop, J.T. *Wage Determination Under Trade Unions.* New York: Macmillan, 1944.

Eberts, R.W. and J.A. Stone. On the contract curve: a test of alternative models of collective bargaining. *JOLE* (January 1986) 66–81.

Farber, H.S. Individual preferences and union wage determination: the case of the united mine workers. *JPE* 86 (October 1978) 923–942.

Farber, H.S. The analysis of union behavior. *Handbook of Labor Economics*, O. Ashenfelter and R. Layard (eds.). Amsterdam: North Holland, 1986.

Grossman, G. Union wages, temporary layoffs, and seniority. *AER* 73 (June 1983) 277–290.

Grossman, G. International competition and the unionized sector. *CJE* 17 (August 1984) 541–556.

Gylfason, T. and A. Lindbeck. Union rivalry and wages: an oligopolistic approach. *Economica* 129–140.

Hall, R.E. and D.M. Lilien. Efficient wage bargains under uncertain supply and demand. *AER* 69 (December 1979) 868–879.

Hendricks, W.E. and L.M. Kahn. Efficiency wages, monopoly unions and efficient bargaining. *EJ* 101 (September 1991) 1149–1162.

Hicks, J.R. *The Theory of Wages.* 2nd ed. New York: Macmillan, 1963.

Hoel, M. Efficiency wages and local versus central wage bargaining. *Economic Letters* 30 (August 1989) 175–179.

Hoel, M. Union wage policy: the importance of labor mobility and the degree of centralization. *Economica* 58 (May 1991) 139–154.

Hollander, A. and R. Lacroix. Unionism, information disclosure and profit-sharing. *SEJ* 52 (January 1986) 706–717.

Johnson, G.E. Economic analysis of trade unionism. *AER* 65 (May 1975) 23–28.

Johnson, G.E. Work rules, featherbedding, and pareto-optimal union-management bargaining. *JOLE* 8 (January 1990) 237–259.

Jones, S.R.G. The role of negotiators in union-firm bargaining. *CJE* 22 (August 1989) 630–642.

Kaufman, B.E. and V.J. Martinez. Monopoly, efficient contract, and median voter models of union wage determination: a critical comparison. *JLR* 11 (Fall 1990) 401–423.

Kidd, D.P. and A.J. Oswald. A dynamic model of trade union behavior. *Economica* 54 (August 1987) 279–288.

Kotowitz, Y. and F. Mathewson. The economics of the union-controlled firm. *Economica* 49 (November 1982) 421–434.

Kuhn, P. Unions in a general equilibrium model of firm formation. *JOLE* 6 (January 1988) 62–82.

Layard, R. and S. Nickell. Is unemployment lower if unions bargain over employment? *QJE* 105 (August 1990) 773–788.

Lazear, E.P. A competitive theory of monopoly unionism. *AER* 83 (September 1983) 631–643.

Leigh, D.E. Union preferences, job satisfaction, and the union-voice hypothesis. *IR* 25 (Winter 1986) 65–71.

Lockwood, B. and A. Manning. Dynamic wage-employment bargaining with employment adjustment. *EJ* 99 (December 1989) 1143–1158.

MacDonald, G.M. and C. Robinson. Unionism in a competitive industry. *JOLE* 10 (January 1992) 33–54.

MacDonald, I.M. Customer markets, trade unions and stagflation. *Economica* 54 (May 1987) 139–154.

Malcomson, J.M. and N. Sartor. Tax push inflation in a unionized labour market. *European Economic Review* 31 (December 1987) 1581–1596.

Manning, A. The determinants of wage pressure: some implications of a dynamic model. *Economica* 58 (August 1991) 325–339.

Martin, D.L. *An Ownership Theory of the Trade Union.* Berkely, CA.: University of California Press, 1980.

Macurdy, T. and J. Pencavel. Testing between competing models of wage and employment determination in unionized markets. *JPE* 94 (July 1986) S3–S39.

Martinello, F. Wage and employment determination in a unionized industry: The IWA and the B.C. wood products industry. *JOLE* 7 (July 1989) 303–330.

McDonald, I. and R. Solow. Wage bargaining and employment. *AER* 71 (December 1981) 896–908. Also, comments and reply 74 (September 1984) 755–761.

Nash, J.F. The bargaining problem. *Econometrica* 18 (April 1950) 155–162.

Nickell, S. and S. Wadhwani. Employment determination in British industry: investigation using micro-data. *R.E. Studies* 58 (October 1991) 955–969.

Oswald, A. The microeconomic theory of the trade union. *EJ* 92 (September 1982) 576–595.

Oswald, A.J. The economic theory of trade unions: an introductory survey. *Scandinavian Journal of Economics* 87 (No. 2, 1985) 160–193.

Pemberton, J. A "managerial" model of the trade union. *EJ* 98 (September 1988) 755–771.

Pehkonen, J. Trade union objectives and the cyclical variability of wages and employment. *Scandinavian Journal of Economics* 92 (1990) 573–586.

Pencavel, J.H. The tradeoff between wages and employment in trade union objectives. *QJE* 99 (May 1984) 215–232.

Pencavel, J.H. Wages and employment under trade unionism: microeconomic models and macroeconomic applications. *Scandinavian Journal of Economics* 87 (No. 2, 1985) 197–225.

Rees, A. *The Economics of Trade Unions*, 2nd ed. Chicago, IL: University of Chicago Press, 1977.

Robinson, C. Union endogeneity and self-selection. *JOLE* 7 (January 1989) 106–112.

Ross, A.M. *Trade Union Wage Policy.* Berkeley: University of California Press, 1948.

Roth, A.E. *Axiomatic Models of Bargaining.* New York: Springer-Verlag, 1979.

Roth, A.E. and M. Malouf. Game theoretic models and the role of information in bargaining. *Psychological Review* 86(6) (November 1979) 574–594.

Rubinstein, A. Perfect equilibrium in a bargaining model. *Econometrica* 50(1) (January 1982) 97–110.

Sampson, A. Employment policy in a model with a rational trade union. *EJ* 93 (June 1983) 297–311.

Sampson, A. Efficient union bargains with a random arbitrator. *Economic Letters* 26 (1988) 99–102.

Scaramozzino, P. Bargaining with outside options: wages and employment in UK manufacturing. *EJ* 101 (March 1991) 331–342.

Spinnewyn, F. and J. Svenjar. Optimal membership, employment and income distribution in unionized and labor-managed firms. *JOLE* 8 (July 1990) 317–340.

Stewart, M.B. Union wage differentials, product market influences and the division of rents. *EJ* 100 (December 1990) 1122–1137.

Svenjar, J. Bargaining power, fear of disagreement and wage settlements: theory and empirical evidence from U.S. industry. *Econometrica* 54 (September 1986) 1055–1078.

Swidinsky, R. Bargaining power under compulsory unionism. *IR* 21 (Winter 1982) 62–72.

Ulph, A. The incentives to make commitments in wage bargains. *R.E. Studies* 56 (July 1989) 449–465.

Von Neumann, J. and O. Morgenstern. *Theory of Games and Economic Behaviour.* Princeton, N.J.: Princeton University Press, 1944.

Warren-Boulton, F.R. Vertical control by labour unions. *AER* 67 (June 1977) 309–322.

Zeuthen, F. *Problems of Monopoly and Economic Warfare.* London: G. Routledge and Sons, 1930.

CHAPTER 16

Union Impact on Wage
and Nonwage Outcomes

This chapter focuses on the impact of unions on various labour market outcomes. We begin by examining the impact of unions on the wages of their members and on the wages of other employees. The final part of the chapter discusses various nonwage consequences of unions — including their impact on fringe benefits, productivity, firm profitability, turnover (quits and layoffs), income distribution, and the allocation of resources.

UNION WAGE IMPACT

The impact of unions on wages has received more attention from economists than any other aspect of union behaviour. Most of the empirical research has been directed at measuring the union-nonunion wage differential, the (percentage) difference in wages between union and otherwise comparable nonunion workers. Our discussion of this research begins by outlining the relevant theoretical background and explaining some of the conceptual and measurement issues involved; we then proceed to a summary of the main empirical findings and a discussion of how these empirical results should best be interpreted.

Theoretical Background

In many respects, the basic theory of the union wage impact is similar to that underlying the impact of other wage-fixing arrangements such as minimum wages. However, there are some unique aspects of union wage determination and these will emerge in what follows. We first examine the union wage impact in the simplest setting, and subsequently introduce additional complications.

In the analysis of union wage and employment determination in the previous chapter, the union workers' alternative wage — which in many situations may be taken to be the nonunion wage — was taken as exogeneously given. At this stage, however, it is important to recognize the impact of unions not only on the wages of union members but also on the wages of others. This requires a general equilibrium model. Most of the basic principles can be illustrated with a two-sector model, as shown in Figure 16.1(a). The two sectors A and B can be thought of as two different industries or regions employing the same type of labour, which is assumed to be homogeneous. There is a sufficiently large number of firms and workers that, in the absence of unions, the labour market is competitive. The equilibrium wage in the absence of collective bargaining is therefore w_0. (If there are differences in the desirability of jobs in the two sectors, a compensating wage differential would exist in equilibrium, as discussed in Part 5 of this book; allowing for this possibility would not affect the analysis other than to make it more complex.) Now suppose a union organizes the workers in sector A and is able to raise the wage to w_u. Employment in sector A will therefore decline to E_1^A. (We are assuming here that the firm can unilaterally set employment; the possibility of wage-employment outcomes to the right of the labour demand curve is discussed below.) The workers who are unable to obtain (or retain) employment in sector A will search for jobs in sector B, increasing labour supply to that sector by $a = E_0^A - E_1^A$. The additional labour supply depresses the wage in sector B, resulting in a new equilibrium wage of w_n. Employment expands to E_1^B, an increase (the amount b in Figure 16.1(a)) which is smaller than the reduction in employment in sector A. The difference $(a - b)$ in Figure 16.1(a) is due to the upward-sloping supply of labour to the economy; as the market wage falls in sector B, some individuals drop out of the labour force, the market wage falling below their reservation wage.

In this economy the union-nonunion wage differential would be $d = w_u - w_N/w_N$. Note that this is greater than the difference between the union wage w_u and the wage union members would receive in the absence of unionization (w_0). The difference arises because the higher wages associated with unionization force more workers to seek employment in the unorganized sector, depressing wages there. In this respect, the impact of unions on wages in the two sectors is similar to that of firms in sector A paying efficiency wages, as discussed in Chapter 13.

According to this basic two-sector model, the magnitude of the union-nonunion differential depends on: (1) the elasticity of labour demand in each sector, (2) the ability of the union to raise wages in the organized sector, and (3) the elasticity of labour supply.

This basic model is consistent with the notion of a dual labour market. The primary sector (A) consists of firms or industries that have characteristics — such as establishment size, concentration in the product market, stability of employment, and the nature of production — that make unionization likely. The secondary sector (B) consists of firms or industries with the opposite characteristics. The reduction in employment in the primary sector (relative to what employment would be in the absence of wage increases associated with

Figure 16.1

(a) Union Wage Impact in a Two-Sector Model

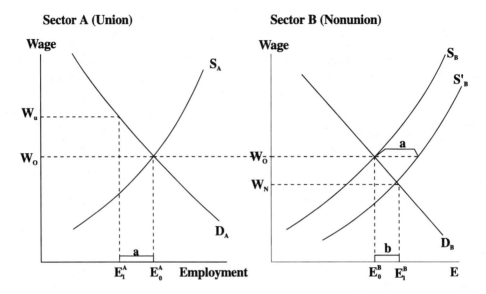

(b) Union Wage Impact with a Threat Effect

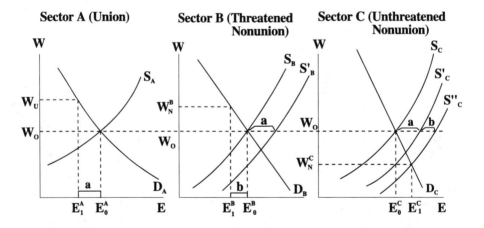

unionization) creates a widening of the wage gap between the two sectors as those who cannot obtain employment in the unionized sector crowd into the nonunion labour market, depressing wages even further in the secondary sector. These low wages may in turn be a source of many of the phenomena — such as high absenteeism and frequent turnover — that are characteristic of the secondary sector.

(c) Union Wage Impact with a Vertical Contract Curve

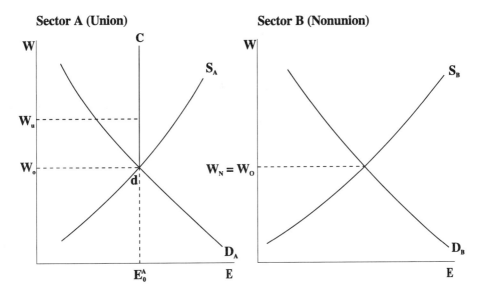

(d) Union Wage Impact with Wait Unemployment

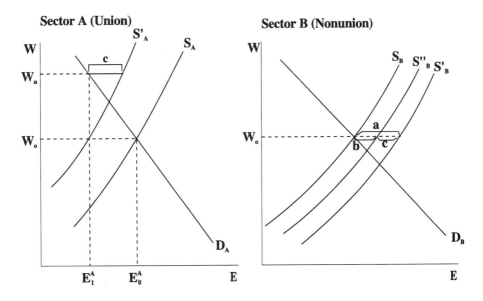

Contrary to the belief that union wage increases tend to depress the wage of unorganized workers is the perspective that nonunion firms may raise the wages of nonunion labour in order to compete with unionized firms for the work force. In addition, firms with both union and nonunion workers may raise the wages of nonunion labour in order to preserve traditional wage differentials. While this belief has intuitive appeal, especially because comparability appears to be so important in wage determination, it ignores the market factors — in particular the supply influx from workers who cannot get jobs in the unionized sector — that would enable employers to pay lower wages to nonunion labour. In essence, there may not be a need to compete with the unionized sector because there is a large pool of labour to hire that cannot get jobs in the high-wage unionized sector. It is true that the unionized sector may get the top applicants because of the high wages, and it is the case that the reserve of labour may dry up in periods of prosperity (hence compelling the nonunion sector to pay higher wages), but it is also true that those who cannot get jobs in the union sector are a supply influx into the nonunion sector, and this lowers wages in that sector.

Nonetheless it is the case that *some* nonunion employers may raise wages in order to reduce the threat of their employees choosing to become unionized. This threat effect will be greater if the union wage premium is high, the nonunion sector is easy to organize, the potential union is aggressive, or the employers have a strong aversion to unionization. In such circumstances, nonunion employees may be paid close to the union wage (perhaps even higher if the firm has a strong aversion to unionization) and provided with similar working conditions as union workers. Such employees could receive the benefits of unionization without bearing such costs as union dues or possible strike costs. However, such benefits may be precarious in that they hinge on the willingness of employers to provide them.

To illustrate the operation of the threat effect, we need to distinguish between those firms that choose to pay higher wages to avoid unionization — perhaps because of a strong aversion to unionization or because they are easy to organize — and those that will choose to pay the lowest wage needed to attract labour. This situation is shown in Figure 16.1(b). As before, the equilibrium in the absence of collective bargaining is the wage w_o in each sector. The establishment of a higher union wage w_u causes some employers (sector B) to raise wages to w_N^B, despite the supply of individuals willing to work at lower wages. (w_N^B may be above or below the union wage, depending on the employer's aversion to unionization.) Employment declines in sectors A and B; thus wages fall further in sector C than in the absence of the threat effect. In dual labour market analysis, sector C would be identified as the secondary labour market.

The impact of the threat effect on the union-nonunion wage differential is ambiguous. The nonunion wage is a weighted average of w_N^B and w_N^C with weights being the employment in each sector. This average nonunion wage may be higher or lower than the nonunion wage that would prevail in the absence of a threat effect, w_N in Figure 16.1(a).

Two additional theoretical points should be noted; these can also be explained using the basic two-sector model. One is the possibility that the firm and union will negotiate wage-employment outcomes to the right of the labour demand curve, for reasons discussed previously in Chapter 15. Figure 16.1(c) illustrates the case in which the parties negotiate outcomes on a vertical contract curve. (Recall from Chapter 15 that the contract curve may be upward sloping, downward sloping, or vertical.) In these circumstances, the higher union wage is not associated with reduced employment in the union sector and therefore with depressed wages in the nonunion sector. The union-nonunion differential in this case is identical to the difference between the union wage and the wage union members would earn in the absence of collective bargaining. If the contract curve is downward sloping, the higher union wage will result in some displacement of labour to the nonunion sector and therefore lower wages in that sector; however, these impacts will be smaller than will be the case if employment is on the demand curve.

The final complication to be discussed is that of "wait" or "queue" unemployment. The rationale behind this phenomenon can be understood by returning to the two-sector model illustrated in Figure 16.1(a). In that case it was assumed that all of the workers unable to obtain jobs in the union sector would seek employment in the nonunion sector. However, because union jobs pay relatively well, it may be rational to wait for a union job to turn up rather than seek employment elsewhere. This strategy is most likely to be sensible when there is rapid turnover in the union sector (so that union jobs become available fairly frequently) and when it is difficult to queue for a union job while employed. When these conditions apply, some individuals may choose to remain attached to the union sector, albeit unemployed. As a consequence, the increase in supply to the nonunion sector will be less than the reduction in employment in the union sector; compare Figures 16.1(a) and (d). The union-nonunion wage differential will be smaller than in the case of no queuing unemployment, though higher union wages nonetheless have some depressing effect on the wages of nonunion workers.

Higher union wages may also affect the demand for labour in the two sectors. This outcome is most likely to occur when the sectors constitute two groups of firms — possibly one group easy to organize and the other difficult to organize — in the same industry. Unless offset by higher labour productivity (a possibility discussed later) the union-nonunion wage differential may enable nonunion firms to expand their market share at the expense of union firms, thus shifting the demand for nonunion (union) labour to the right (left). The final equilibrium in this case will generally involve a smaller union-nonunion wage differential than that shown in Figure 16.1(a).

The above analysis of the union wage impact was based on several simplifying assumptions, including a homogeneous labour force and competitive labour markets (in the absence of collective bargaining). Several implications follow from relaxing these assumptions. With several different types of workers, changes in the wages of any one group will generally alter the wages of other groups. If union and nonunion workers are employed in the same firm or industry, an increase in union wages will increase (reduce) the demand for nonunion labour if the two groups are substitutes (complements) in production, and therefore increase (reduce) the wages of nonunion workers in that firm or industry. In addition, the wages of nonunion workers may be affected by various forms of wage-fixing such as minimum wage rates and fair wage provisions, many of which are supported by unions to reduce competition based on lower nonunion wages. In these circumstances the wage in the unorganized sector may not decline sufficiently to absorb the workers displaced by higher wages in the union sector. Jobs will be rationed in both sectors and there will be a number of individuals willing to work at existing wage rates but unable to find employment.

In summary, collective action (in particular the strike threat) will generally enable unions to raise wages relative to a hypothetical labour market equilibrium without unions, thus creating a wage differential between union and comparable nonunion workers. The magnitude of this differential depends not only on the ability and willingness of unions to raise wages but also on the impact on the wages of nonunion workers. Clearly the various factors discussed above can be at work simultaneously affecting the wages of nonunion workers, and since they do not all work in the same direction, economic theory does not provide unambiguous predictions of the effect of unions on the wages of nonunion workers. The wages of some nonunion workers may increase because of the increased demand for substitute nonunion labour, as a result of the threat effect, in response to legislative wage-fixing, or as part of the nonunion sector adjusts to restore old wage relativities. In contrast, some nonunion wages may decrease because of a decrease in the demand for complementary inputs, or because of the supply influx into the nonunion sector created by the reduced employment opportunities in the unionized sector. Ultimately, the impact of unions on the wages of nonunion workers is an empirical proposition.

Some Problems in Measuring the Union Wage Impact

Attempts to measure the union-nonunion wage differential face a number of interrelated problems, the most important being those of controlling for other wage-determining factors, accounting for the joint determination of union status and the union wage impact, and separating cause and effect.

In estimating a *pure* union-nonunion wage differential, it is important to control for other wage-determining factors so as to be able to attribute the differential purely to unionization. If, for example, unionized establishments tend to utilize more skilled labour than nonunionized establishments, it is crucial to control for these skill differences. Otherwise, the union-nonunion wage differential may simply reflect the skill differential, not a pure union impact on wages. Controlling for differences in labour quality between the union and nonunion sectors is especially important because, as indicated earlier, if unionized establishments pay a wage that is greater than the competitive wage, they may have a queue of applicants and may be able to hire the "cream of the crop" of applicants. Union firms faced with higher labour costs may also try to increase the job assignment, for example by increasing the pace of work; however, such adjustments are likely to be resisted by the union.

In these circumstances it is extremely important to control for differences in the characteristics of workers and job assignments. Otherwise these omitted variables are likely to impart an upward bias because the union-nonunion wage differential would reflect better quality workers and perhaps more onerous job assignments in the union sector in addition to the pure wage impact of unions. Empirical studies try to control for labour quality differences (for example, by including variables such as education, training, and experience in multiple regression analyses); however, some factors affecting labour quality (e.g., motivation, work effort, and reliability) may not be observed by researchers. Information relating to the pace of work and job assignments is also often unavailable, although some studies (e.g., Duncan and Stafford, 1980) have been able to control for these factors. Many studies are able to include only crude controls (such as occupation, industry, and region) for the characteristics of the job. Thus it is not clear that the available data enables researchers to *fully* control for differences in productivity-related factors, although as will be explained below the quality of the available data has steadily improved.

These issues illustrate the problem of selection bias in the context of the union wage impact. As discussed in Chapter 14, union status is the outcome of decisions made by individual workers, employers, and union organizers. The omission from the estimated earnings equations of unobserved variables which are related to wages (such as motivation, work ethic, reliability) will not bias the estimates of the union wage impact if these unobserved factors do not also influence the decisions relating to union status; that is, if the unobserved factors affecting wages do not also influence selection into the union sector. However, if factors unobserved by the researcher influence wages and selection decisions — for example, if ability is not observed by the researcher but is known by employers, perhaps because of screening procedures, and used to decide which applicants to hire — then the estimated union wage impact will be subject to "selection bias." Dealing with selection bias requires modeling the joint determination of union status and the union wage impact.

The issue of the selection process determining which workers enter the union sector illustrates the more general problem of sorting out cause and effect in studies of the union impact. Usually the argument is advanced that unions cause higher wages. However, there is the possibility that cause and effect also works in the opposite direction. That is, some firms may simply be high-wage firms, perhaps because of efficiency wage considerations relating to productivity and morale, perhaps because they are trying to reduce the turnover of workers with firm-specific human capital, or because they want to be known as a model employer, or because they want a queue of applicants from which to hire. Unionization may be more likely to emerge in high-wage sectors because they are easier to organize or because the workers may be more likely to demand union representation. Workers in these firms will be reluctant to leave because wages are so high. Knowing that they want to stay in the high-wage establishment, and knowing that they will probably stay there for a considerable period of time because of the high wages, they may turn to devices to try to improve the everyday work conditions of the job. One such device may be unionization with its emphasis on due process, regulating the work environment, and administered rulings that provide a degree of certainty and security for the unionized workers.

In Hirschman's (1970) terminology, the restriction of "exit" increases the use of "voice." In this case, voluntary quitting (exit) is reduced because of the high wages, and hence workers seek to have more of a say (voice) in their job by collective bargaining.

This possibility of cause and effect working in both directions suggests that some of the wage advantage of union establishments may not be due to unions, but may be attributed to other factors; in fact, unionism itself emerges because of the wage advantage. This suggests that econometric studies of the impact of unions should be based on simultaneous equation models, which allow wages to be a function of unionism as well as unionism to be a function of wages.

EMPIRICAL EVIDENCE ON UNION WAGE IMPACT

In varying degrees the numerous empirical studies of the wage impact of unions have attempted to deal with the measurement and modeling problems just discussed. Research has passed through several phases. Early studies based on aggregate data (e.g., industry or region) have given way to those using data based upon the individual worker as the unit of observation. Recent studies have attempted to account for the joint determination of union/nonunion status and the union wage impact and for unobserved quality differences between union and nonunion workers. Much of the research has been carried out in the United States and our summary reflects this fact. However, during the 1980s a number of data sets with information on the union status of individual employees became available in Canada. Accordingly, there is now a growing body of empirical literature on this important aspect of the Canadian labour market.

Early Studies

The classic work of H. Gregg Lewis (1963) both reviewed and reanalyzed the existing literature in the United States and provided new estimates up to 1958. While there was considerable variation in the estimated impact of the studies reviewed by Lewis, he attributed much of the variation to methodological differences. Lewis' estimate of the union-nonunion differential, based on his own work and a reanalysis of earlier studies, was approximately 10–15 percent for the U.S. economy as a whole, being larger in recessions and smaller in the boom phase of the business cycle. This estimate of a 10–15 percent differential was associated with an increase in the average union wage of 7 to 11 percent and a decrease in the average nonunion wage of approximately 3 or 4 percent.

Data availability imposed important limitations on these early studies. Because separate union and nonunion wage series were not available, the union-nonunion differential often had to be inferred from the average wage and the extent of unionization. This inference can be made only under certain restrictive assumptions; thus the estimates should be treated with caution.

Since Lewis' work there have been numerous additional studies, most utilizing new data sets where the unit of observation is the individual establishment or worker. Estimates of the union-nonunion wage differential based on individual cross-sectional data have tended to be higher than the 10–15 percent recorded by Lewis. These studies thus suggested that the union wage impact is higher than was previously believed. However, this conclusion is subject to the qualification that some of the difference between union and nonunion wages (after controlling for other wage-determining factors such as education, age, and experience) may be a reflection of reverse causality and/or unmeasured quality differences between union and nonunion workers. Recent research has focused on these issues.

Modeling Union Incidence and Impact

A number of recent studies have attempted to account for the joint determination of union status and the union wage impact. Ashenfelter and Johnson (1972), who were the first to formally deal with the simultaneity issue, used aggregate U.S. time series data. They found that higher wages make unionization more likely, and that the estimated union wage impact is considerably smaller when the two-way causality is taken into account. Most subsequent studies have applied simultaneous equation methods with individual data. Generally these studies model the wage determination process in the union and nonunion sectors and the selection process determining which workers are employed in each sector. Thus the researchers account for the possibility that the sample is not randomly generated with respect to union status; that is, some individuals have unobserved characteristics that make them more likely to be represented by a union than others.

Longitudinal Studies

Another recent development has been the availability of longitudinal or panel data which provide observations on the same individuals over time. With longitudinal data, person-specific characteristics which are conventionally unobservable to the researcher (e.g., reliability, work effort, and motivation) can be taken into account if these characteristics are constant for each individual over time. In these circumstances taking first differences (differences from one period to the next) in the data provides a natural way of removing the effects of such characteristics. This cannot be done with conventional cross-section estimates, which compare different individuals at the same point in time. If the unobserved variables are indeed constant over time, the use of longitudinal data allows the researcher to identify the union impact from the wage change for those individuals who *change* union status. Essentially the estimated differential is an average of the earnings gain of those who moved from nonunion to union jobs and the earnings loss of those who moved from union to nonunion jobs, after controlling for other observable factors.

Longitudinal analyses in the United States (e.g., Mellow, 1981; Mincer, 1983; Moore and Raisian, 1983; Jakubson, 1991) generally find that the estimated union-nonunion wage differential is about 10 percent, substantially smaller than suggested by cross-sectional estimates. Indeed, when cross-sectional analyses are carried out on the same data sets, the estimated differentials are approximately double those obtained using data on those who change union status. This finding could be due to unobserved (by the researcher) differences in worker quality between union and nonunion workers which are quantitatively significant; if true, this conclusion has important implications for the interpretation of not only observed earnings differences between organized and unorganized workers but also observed differences in labour productivity between the two sectors. However, the assumptions underlying the longitudinal estimates could also be invalid.

In addition, longitudinal data has some deficiencies. There is evidence of error in the measurement of union status which may bias downward the estimated union-nonunion wage impact (Freeman, 1984). Thus the estimated average wage impact of 10 percent may be too low. In addition, the number of individuals changing status is typically only 3–5 percent of the total sample, resulting in small and possibly unrepresentative samples. Furthermore, the facts that those individuals who change status may not be typical individuals and that changes in union status may themselves be associated with other unobservable factors suggest that it would be hazardous to rely exclusively on the estimates from longitudinal studies.

Summary of Average Impact in United States

Clearly important advances have taken place in the measurement of the union wage impact, a reflection of improved data sources and more sophisticated econometric methods. The accumulated U.S. evidence indicates that there is a statistically significant difference between the wages of union and otherwise comparable (to the researcher) nonunion workers — probably in the order of 10 to 20 percent for the economy as a whole after controlling for observed worker characteristics other than union status. In a major extension of his earlier work, Lewis (1985) reviews almost 200 empirical studies of the impact of unions on wages in the United States. He concludes that the average union-nonunion wage differential over the period 1967–79 was 15 percent.

However, there remains uncertainty about the magnitude of the union-nonunion wage differential when the joint determination of unionization and its impact is taken into account. Unfortunately the availability of better data and the application of more sophisticated techniques have not yet resulted in greater consensus about the magnitude of the average union wage impact. Indeed, the range of estimates produced by these studies is sometimes wide, a reflection both of differences in data sets and differences in model specification. Accordingly, recent research has attempted to determine which specification is most appropriate (Robinson, 1989; Jakubson, 1991; Green, 1991).

Canadian Evidence

Canadian studies of the average union wage impact are summarized in Table 16.1. Studies using aggregate industry data, plant level data, and individual data have been carried out. The estimated union impact generally differs substantially across individual workers; the estimates reported in Table 16.1 are for the average individual in the sample. The estimates are broadly similar to those obtained in comparable U.S. studies. Average differentials (for the samples used) generally lie in the 10–25 percent range. The highest estimates — those obtained by Robinson and Tomes (1984) and Robinson (1989) — appear to be due primarily to the nature of the sample (hourly paid workers). The estimated union wage impact is also sensitive to the econometric specification employed.

VARIATION IN THE UNION WAGE IMPACT

As would be expected on theoretical grounds, the estimated union-nonunion wage differential varies considerably across firms, industries, and workers. The average differential also varies over time. A number of generalizations about these variations emerge from the empirical literature, although for most generalizations there are exceptions in particular studies.

The union-nonunion wage differential tends to be larger when a high proportion of the relevant jurisdiction (industry, occupation, region) is organized. The relationship, however, is nonlinear; once a certain proportion of the jurisdiction is unionized, further increases have little additional effect on the differential. These findings can be interpreted in terms of two aspects of union power discussed previously: the elasticity of demand for union labour and the threat effect. As the fraction of the industry (or occupation or region) unionized

Table 16.1
ESTIMATES OF THE UNION/NONUNION WAGE DIFFERENTIAL IN CANADA

AUTHOR	TIME PERIOD	ESTIMATED DIFFERENTIAL (%)	CHARACTERISTICS OF SAMPLE; OTHER COMMENTS
1. Kumar (1972)	1966	17–23	Unskilled workers in manufacturing; aggregate data
2. Starr (1973)	1966	10–15	Unskilled male workers in Ontario manufacturing; disaggregate data (base wage rates at plant level)
3. Grant and Vanderkamp (1980)	1971	15	Individual data; annual earnings; "union member" includes membership in professional associations
4. MacDonald and Evans (1981)	1971–76	16	Aggregate industry data; 30 manufacturing industries
5. MacDonald (1983)	1971–79	20	Aggregate industry data; 30 manufacturing industries
6. Grant, Swidinsky and Vanderkamp (1987)	1969–71	12–14 (1969) 13–16 (1970)	Individual longitudinal data, annual earnings; "union membership" includes professional associations
7. Robinson and Tomes (1984) and Robinson (1989)	1979	24	Individual data on hourly paid workers
8. Simpson (1985)	1974	11	Microdata on wage rates for narrowly defined occupations
9. Kumar and Stengos (1985)	1978	10	Union and nonunion earnings data, by industry
10. Kumar and Stengos (1986)	1981	12	Individual data; hourly earnings, representative sample
11. Green (1991)	1986	15	Individual longitudinal data on males; hourly earnings

increases, the possibilities for the substitution of nonunion for union labour decline, reducing the elasticity of demand for union labour. However, as more of the industry is organized, the remaining nonunion firms may face a higher possibility that their work force will organize and thus will tend to raise their wages. The gap between union and comparable nonunion workers in the same jurisdiction will thus not continue to widen, and may even narrow.

Firm or establishment size has also been found to be positively related to both union and nonunion wages in the majority of studies. The relationship is considerably stronger for nonunion workers; thus the wage differential between union and comparable nonunion workers declines with firm or establishment size. This outcome may reflect several forces. Unions usually attempt to standardize wage rates across firms and occupations — to "take wages out of competition" — which implies raising wages more in smaller firms, given the positive relationship between wages and firm size in the absence of unionization. Further, the threat of union organization is generally higher (due to reduced organizational costs per prospective member) in larger firms. Thus larger nonunion firms are more likely to match union wage scales than their smaller counterparts.

Large differences in the estimated union-nonunion differential are usually found to exist across occupations and industries. The impact is higher for blue-collar workers than for white-collar workers; higher in non-manufacturing than manufacturing industries; and it can be exceptionally high in certain specific sectors such as construction. Skilled workers typically gain less from unionization than their semi-skilled and unskilled counterparts, although there are specific examples of skilled trades — usually organized in craft unions — that exhibit large earnings differentials. Simpson (1985) provides an analysis of skill differences in the union impact with Canadian data; some of his estimates are shown in Table 16.2. The estimated union-nonunion differentials decline almost monotonically with skill level; indeed, in the most skilled categories, nonunion wages exceed union wages. This outcome may largely reflect the fact that in many unions — especially industrial unions — skilled workers constitute a minority with little influence on the union's wage policy.

An alternative perspective on the relationship between union and nonunion wages and skill level is shown in Figure 16.2. (It should be understood that all other wage-determining factors are held constant.) Wages increase with skill level in both sectors; however, the increase is more gradual under collective bargaining, and at high skill levels the union may even fall below the nonunion wage. This type of relationship applies to a number of productivity-related characteristics (e.g., experience, education) in addition to skill. Simply stated, union wages are typically found to be less responsive to the personal characteristics of workers. The relationship between earnings and such factors as age, experience, education, and even marital status is weaker (i.e., age-earnings profile are flatter, albeit higher) in the organized sector. When fringe benefits —

Table 16.2
**ESTIMATES OF THE UNION-NONUNION PERCENTAGE WAGE
DIFFERENTIAL BY SKILL LEVEL, CANADA, 1974**

SKILL LEVEL	ALL INDUSTRIES	MANUFAC-TURING	NON-MANUFAC-TURING	PUBLIC SECTOR[a]	PRIVATE SECTOR
1 (Low)	33.2	26.5	33.7	12.0	32.9
2	28.9	17.1	29.6	2.9	28.7
3	18.1	11.0	22.6	2.3	19.1
4	17.9	3.3	23.4	−1.0	19.0
5	12.4	−1.9	14.9	4.8	10.2
6	13.1	14.1	9.5	−24.3	15.0
7 (High)	−10.7	−11.1	−14.2	−33.4	−9.4
All Levels	18.6	11.5	20.9	−10.6	18.9

Notes:
a. Includes workers in health and education as well as public administration.

Source: Wayne Simpson, The impact of unions on the structure of Canadian wages: an empirical analysis with microdata. *Canadian Journal of Economics* 18 (February 1985), 164–181.

which are typically of greater benefit to older, more experienced workers — are included together with wages this outcome is weaker, at least with respect to age and experience, but nonetheless remains.

U.S. studies generally find that union and nonunion wages are approximately equally responsive to gender. Lewis (1986) surveys 48 U.S. studies dealing with the impact of unions on male and female wages; the union-nonunion wage differential for females is generally estimated to be similar to that for males. However, in Canada there is evidence of a larger union-nonunion differential for females (Doiron and Riddell, 1992). Unionization thus has two offsetting effects on the average earnings of Canadian males and females. Females benefit less from unionization because they are less likely to be covered by a collective agreement (see Exhibit 14.2 in Chapter 14). However, unions raise female wages more than those of males. Doiron and Riddell (1992) find that these two effects tend to offset each other.

In the United States the union wage impact tends to be smaller in the public than in the private sector (Mitchell, 1983; Freeman, 1986). This finding may appear somewhat surprising, given that the elasticity of labour demand is generally estimated to be lower in the public sector. However, there are other important differences between the two sectors, and these may account for the

Figure 16.2
THE RELATIONSHIP BETWEEN UNION AND
NONUNION WAGES AND PERSONAL CHARACTERISTICS

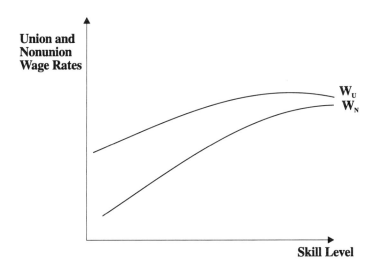

observed behaviour. Unlike their private sector counterparts, many U.S. public sector employees do not have the right to strike. Thus the *willingness* of public sector workers to raise wages may be higher than in the private sector, but their *ability* to do so may be lower. In addition, union-nonunion wage differentials for the private sector workers most alike the majority of those in the public sector — those in white-collar, service sector employment — are significantly lower than for employees in blue-collar, goods-producing jobs. Thus, once one controls for the nature of employment, the public-private sector differences in union impacts narrow considerably. Finally, for reasons discussed in Gunderson (1984), public sector unions may have a larger impact on fringe benefits and working conditions than their private sector counterparts.

Canadian evidence generally supports the conclusion that the union wage impact is smaller in the pubic sector. Robinson and Tomes (1984) find modestly lower differentials for hourly paid workers in the public sector (27 percent versus 34 percent in the private sector); however, only a minority of public sector workers are paid on that basis. As shown in Table 16.2, Simpson (1985) concludes that union-nonunion wage differentials are significantly lower in the public than private sector for each skill group. Indeed, the differentials for higher skill groups in the pubic sector are sufficiently negative that unionized public sector workers are estimated to earn *less* than comparable nonunion public sector workers, while unionized private sector employees earn substantially more than their nonunion counterparts. Riddell (1992) also finds that unions have a significantly larger wage impact in the private than in the public sector.

In his earlier work, Lewis (1963) concluded that unionism has probably lowered the wages of nonunion workers, this effect accounting for about 3 to 4 percentage points of the total estimated differential of 15 percent. Subsequent studies utilizing better data (Kahn, 1978, 1980) also reach this conclusion; however, the overall average effect masks considerable variability, with some groups such as nonunion white males receiving a large wage increase from unionism. This outcome suggests the operation of both the threat and labour supply effects of union wage increases, as analyzed previously.

Lewis also concluded that the union-nonunion wage differential varies cyclically, widening in recessions and narrowing in booms; that is, union wages are less responsive to variations in economic and labour market conditions than nonunion wages. This behaviour may partly result from the fact that collective agreements typically provide for fixed nominal wage rates — often for durations of two to three years — and these agreements overlap each other. At any point in time, only the fraction of union contracts being renegotiated are able to respond to changes in economic conditions. In the nonunion sector, agreements are implicit rather than explicit and wages can be adjusted in response to variations in economic conditions. However, even if union wage contracts were renegotiated frequently and did not provide for fixed durations (as is the case for example in Great Britain), union wages may well exhibit less cyclical variation than nonunion wages. For reasons discussed subsequently, unions are likely to prefer employment reductions to wage reductions as a method of responding to temporary reductions in labour demand.

Recent U.S. research has generally supported Lewis' conclusion that the union-nonunion wage differential varies counter-cyclically, although some conflicting results have been reported. Changes in the extent of cost-of-living-allowance clauses in collective agreements is one of several factors that may alter the cyclical behaviour of union wage rates. Because the price level varies pro-cyclically (in the absence of supply shocks), indexing union wages to the cost of living tends to increase their responsiveness to variations in aggregate economic conditions.

Canadian evidence on the time series behaviour of the union-nonunion wage differential is extremely limited. As can be seen from Table 16.1, nine years (1971–1979) is the longest period for which estimates are available. MacDonald (1983) finds the differential rose almost continuously during this period from 16.4 percent in 1971 to 22.8 percent in 1979. Because this was a period of generally rising unemployment and increasingly slack economic conditions, these results are broadly consistent with the hypothesis that the union wage impact varies counter-cyclically.

UNION WAGE IMPACT: CONCLUDING COMMENTS

Although exceptions can be found in particular studies, a number of generalizations emerge from the substantial body of empirical research on the wage impact of unions:

1. There is substantial evidence of a significant wage differential between union and nonunion workers who are otherwise comparable in terms of observable characteristics. Based on a relatively limited number of Canadian studies, the average differential appears to be approximately 10 to 25 percent, similar in magnitude to estimates based on U.S. data. The earnings gap between union and comparable nonunion workers varies across industries, firms, regions, occupations, and over time with aggregate economic conditions.

2. The overall impact of higher wages in the union sector is a somewhat lower average wage in the nonunion sector, consistent with additional labour supply to nonunion firms because of reduced employment opportunities in unionized enterprises. However, there is considerable variability in the effect on nonunion workers, with some gaining from higher union wages and others losing.

3. In unionized enterprises, wage differences among workers who differ according to various productivity-related personal characteristics — such as education, age, experience, skill, and marital status — are smaller than in nonunion enterprises, and evidently smaller than they would be in the absence of unionization. Thus the financial "return" to additional amounts of these attributes — e.g., years of education or experience — is generally lower with collective than individual bargaining.

4. The union-nonunion wage differential is generally higher in the private than the public sector, higher in small firms and enterprises, higher for blue-collar than white-collar workers, higher when a large proportion of the relevant jurisdiction is organized, and higher in recessions than booms.

Canadian empirical evidence is consistent with these conclusions; however, only a limited number of studies have been carried out with Canadian data and there is considerable scope for further research.

Clearly a great deal has been learned about the effects of unions on wages. Nonetheless, several unresolved issues remain:

1. On theoretical grounds there is reason to believe that unionized firms will respond to higher wages and the concomitant larger queue of applicants by hiring higher "quality" workers. The extent to which estimated union-nonunion differentials reflect this difference in labour quality rather than a pure differential remains largely unknown. U.S. evidence from longitudinal studies suggests that unobserved personal characteristics may account for a

substantial proportion — perhaps as much as one-half — of the estimated wage differential. However, the differences between longitudinal and cross-sectional analyses could be due to other factors.

2. Unionized firms may also respond to higher wages by attempting to increase the job assignment. Although empirical studies have generally been able to control for a variety of measured worker characteristics, few have been able to control for attributes of the job. Some studies have concluded that estimated union-nonunion differentials partly reflect compensating differentials for more hazardous jobs, faster work pace, more structured work settings, and less flexible working hours (Duncan and Stafford, 1980; Leigh, 1982).

3. Uncertainty exists regarding the extent to which earnings differences between union and comparable nonunion workers are caused by unions raising their members' wages as opposed to unions being more likely to exist in firms that would choose to pay relatively high wages even in the absence of unions.

UNIONS, WAGE DISPERSION, AND THE DISTRIBUTION OF INCOME

As documented by Freeman (1980, 1982) and Hirsch (1982) for the U.S. and Meng (1990) for Canada, there is significantly less wage dispersion among union workers than among comparable nonunion workers. This smaller variance is related to several aspects of union wage policy noted above. Unions tend to reduce wage differentials among workers that differ according to factors such as skill, age, experience, and seniority. Unions also attempt to standardize the wages of similar workers across establishments, especially those in the same industry or region. Further, unions have a larger impact on the wages of blue-collar than on white-collar workers. These policies imply that unions typically raise the wages of those at the lower end of the pay scale proportionally more than those at the upper end. As a result, the union wage distribution is both to the right and less dispersed than the nonunion distribution, as shown in Figure 16.3.

Although wage dispersion is substantially lower in the union than in the nonunion sector, it does not necessarily follow that unionism is associated with reduced wage dispersion in the economy as a whole. By creating a wage differential between union and comparable nonunion workers, unions also increase the variability of earnings relative to a hypothetical economy without collective bargaining. Because of these offsetting factors — lower wage dispersion within the union sector but greater inequality between union and nonunion workers — the direction of the overall impact of unionism on the wage distribution is indeterminate. Studies by Hyclak (1979, 1980), Freeman (1980), and Quan (1984) indicate that in the United States the net effect of unionism is to reduce wage dispersion. Unfortunately there is no Canadian evidence on this aspect of the impact of unions.

Figure 16.3
WAGE DISPERSION IN THE UNION AND NONUNION SECTORS

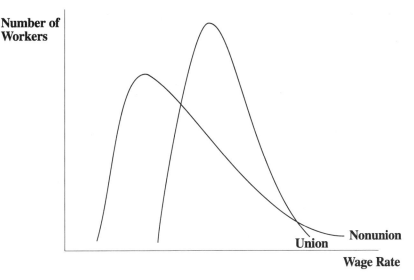

UNION IMPACT ON RESOURCE ALLOCATION AND ECONOMIC WELFARE

Because they alter wages and employment, unions affect the allocation of labour and other resources in the economy. The nature and magnitude of this effect depends on several factors discussed previously in the context of the union wage impact. As was done earlier in this chapter, the consequences of unions are studied by comparing the situation with unions to a hypothetical economy without unions.

Figure 16.4 shows the allocative consequences of the union wage impact in a simple two-sector general equilibrium model in which outcomes are on the labour demand curve and the supply of labour to the economy is completely inelastic. With homogeneous workers and competitive labour markets, the equilibrium in the absence of unions implies a common wage (w_0) in both sectors. Because of the higher union wage w_u, employment declines to E_1^u in the union sector and increases to E_1^N in the nonunion sector, reducing the wage there to w_N. With inelastic labour supply the increase in employment in the nonunion sector($E_0^N\, E_1^N$) equals the reduction in employment in the union sector ($E_1^u\, E_0^u$).

These changes in wages and employment result in an inefficient allocation of society's resources and a reduction in the total output and income generated by the economy. The magnitude of this allocative inefficiency can be illustrated using the fact that, because the labour demand curve is the marginal revenue product curve, the area under the labour demand curve equals the total value of

Figure 16.4
THE UNION WAGE IMPACT AND ALLOCATIVE INEFFICIENCY

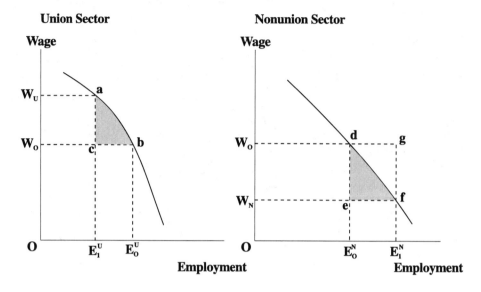

the output produced. Thus the value of the output produced by the union sector declines by E_1^u a b E_0^u as a result of the union wage impact, while the value of output produced by the nonunion sector increases by E_0^N d f E_1^N. Total income declines by E_1^u a b E_0^u minus E_0^N d f E_1^N, an amount equal to the sum of the two hatched areas cab and edf. (This follows from the fact that the area E_1^u c b E_0^u equals the area E_0^N d g E_1^N and the are def equals the area dgf, assuming a linear demand curve.)

An alternative way of viewing the losses is to distinguish between deadweight or real resource losses and transfer losses, the latter which represent a redistribution, not a loss, of output or income. For example, the loss of output E_1^u a b E_0^u in the union sector consists of two components: the loss of income E_1^u c b E_0^u to the displaced workers (which is a transfer loss since employers and ultimately customers are not paying that amount), and a real resource or deadweight loss abc which represents what consumers are willing to pay for the output over and above what the workers were willing to accept to produce the output. When the displaced workers move to the union sector ($E_1^u E_0^u = E_0^N E_1^N$), their income is reduced from E_1^u c b E_0^u in the union sector to E_0^N e f E_1^N in the nonunion sector. They therefore lose income of edgf. However, edf of that is a gain to the rest of society reflecting their value of the output E_0^N d f E_1^N over and above what they have to pay for the additional output E_0^N e f E_1^N. Therefore, the only real resource loss is dgf, which equals edf given a linear demand curve. Thus the total real output loss is abc plus def: the former represents what consumers are willing to pay for the loss of output in the unionized sector over and above what workers required to produce that output; the latter represents the loss associated with the fact that the output increase is valued less in the nonunion sector than the output loss in the union sector. The sum of these two components constitutes the welfare loss from union wage fixing in an otherwise competitive market.

This reduction in the value of total output is referred to as a "deadweight loss" because it is not offset by a benefit elsewhere. The misallocation of labour resources occurs because higher wages and reduced employment in the union sector push other workers into less-productive and more poorly paying jobs in the nonunion sector. As a result there are too few workers employed in the union sector and too many in the nonunion sector. The use of other resources is also affected. Relative to an efficient allocation of resources, firms will respond to the wage differential by utilizing excessively capital-intensive production techniques in the union sector and excessively labour-intensive techniques in the nonunion sector.

The existence and magnitude of a deadweight loss associated with the union wage impact depend on several factors discussed earlier: the elasticity of labour supply, the amount of queuing unemployment, and the extent to which firms and unions negotiate wage-employment outcomes to the right of the demand curve. Depending on these various factors, the magnitude of the reduction in total output may be lower or higher than that shown in Figure 16.4. For example, in the situation depicted in Figure 16.1(c) in which union wages and employment lie on a vertical contract curve, employment in the two sectors is identical to that which would occur in a competitive labour market equilibrium and no misallocation of labour resources results. In these circumstances, the higher union wage alters the distribution of income (increasing labour's share) but not total income. On the other hand, an additional output loss will occur if some individuals leave the labour force because the nonunion wage falls below their reservation wage and/or some individuals choose to remain unemployed hoping to receive a union job.

The potential allocative inefficiency associated with the union wage impact was illustrated using as a reference point an economy in a competitive equilibrium. From this starting point, any distortion such as monopoly in the product market or wage fixing in the labour market will result in a misallocation of resources and a reduction in total income. However, if the starting point is an economy with existing distortions, then it is not necessarily the case that adding another distortion (e.g., union wage impact) will cause a reduction in society's economic welfare. (This is the central proposition of the "Theory of Second Best" associated with Lipsey and Lancaster, 1956). The Canadian economy contains numerous distortions which would cause it to deviate from a first-best (Pareto-optimal) allocation of resources in the absence of unions; therefore it is an empirical question as to whether the union wage impact improves or worsens allocative efficiency.

Early estimates of the deadweight loss, such as that of Rees (1963), were based on fairly crude methods and did not take into account other distortions in the economy. These studies estimated the loss for the United States to be about 0.15 percent of GNP when the union-nonunion wage differential was 15 percent. More recently, De Fina (1983) reports estimates based on a computable general equilibrium model of the U.S. economy incorporating several existing distortions. De Fina estimates the deadweight loss to be about half that of earlier studies; when the union-nonunion wage differential is 15 percent, the loss is less than 0.1 percent of GNP, while a differential of 25 percent produces an estimated loss of 0.2 percent of GNP. The deadweight loss may be somewhat higher in Canada than in the United States owing to the higher union density. However, on the basis of the U.S. studies it appears unlikely to be substantially greater than 0.2 percent of GNP, or about 1300 million dollars per year at 1991 levels of national income, equivalent to about 50 dollars per person.

UNION IMPACT ON NONWAGE OUTCOMES

Unions bargain for a wide range of factors in addition to wages, and a more complete understanding of the economic and social consequences of unions will emerge from examining these other outcomes of collective bargaining. We begin by discussing the impact of unions on fringe benefits and proceed to working conditions, quit and layoff behaviour, and productivity and profitability.

Unions and Fringe Benefits

Nonwage or fringe benefits refer to employer payments such as those for pensions, medical and life insurance, vacations, holidays, and sick leave. Omitting these benefits understates both the cost of labour to the employer and the benefit received by the employee; thus total compensation is the relevant measure for both labour demand and supply decisions. (See Chapter 10 for a discussion of nonwage benefits and their effects on labour demand.) The omission of fringe benefits would not bias estimates of the impact of unions on the total compensation of union and nonunion workers if nonwage benefits were the same proportion of total compensation in both sectors. However, evidence indicates that nonwage benefits account for a greater proportion of total compensation among union than nonunion enterprises; i.e., unions tend to have a greater impact on fringe benefits than on wages. This implies that estimated union-nonunion differentials understate the total compensation gap between organized and unorganized workers.

Table 16.3 provides information on wage and nonwage compensation in Canada. Fringe benefits are divided into those which are legally required and those which are voluntary or negotiated. Legally required benefits account for similar proportions of total compensation in both sectors; however, voluntary/

Table 16.3
WAGE AND NONWAGE BENEFITS IN UNION AND
NONUNION FIRMS, CANADA, 1978[1]

COMPONENT OF COMPENSATION	UNION FIRMS		NONUNION FIRMS		UNION-NONUNION DIFFERENTIAL
	$/YR.	%	$/YR.	%	
Total Compensation	16,977	100.0	12,323	100.0	37.8
Pay for Time Worked[2]	13,192	77.7	10,159	82.4	29.9
Fringe Benefits (Total)	3,786	22.3	2,164	17.6	75.0
Legally Required[3]	1,043	6.1	651	5.3	60.2
Voluntary/Negotiated	2,743	16.2	1,513	12.3	81.3
Paid Absence[4]	1,634	9.6	1,085	8.8	50.6
Pension and Health[5]	699	4.1	248	2.0	181.9
Miscellaneous[6]	410	2.4	180	1.5	127.8

Notes:
1. Based on a survey of employers with 100 or more employees. Data are for wage earners only.
2. Incudes basic straight time pay, commissions, overtime, and shift premium pay.
3. Includes Canada and Quebec Pension Plans, workers' compensation, provincial medicare, and unemployment insurance.
4. Paid holidays, sick leave, and vacation pay.
5. Private pension and other benefit plans.
6. Includes COLA, bonuses, severance pay, and taxable benefits.

Source: Statistics Canada, *Employee Compensation in Canada*, 1980.

negotiated nonwage benefits account for a higher proportion of total compensation in unionized firms. In this sample, fringe benefits are 75 percent higher in unionized firms, while wage rates differ by about 30 percent between union and nonunion firms. The largest gap occurs with pension and health benefits. These data are not adjusted for other factors which may cause wages and nonwage benefits to differ between union and nonunion firms. However, empirical analyses with U.S. data (e.g., Freeman, 1981) indicate that significant differences between the two sectors remain after controlling for other factors.

There are numerous reasons why all parties — employees, employers, and governments — may prefer fringe benefits over wages as a form of compensation. (These are discussed in more detail in Chapter 10.) Workers may prefer fringe benefits because they are often not taxable or taxes are deferred; in addition there may be economies of scale and administrative simplicity associated with group purchases. Employers may benefit from the fringe benefits provided to employees to the extent that they facilitate the planning and operation of the production process, provide appropriate work incentives, and reduce turnover. In periods of wage control both parties may favour compensation via nonwage benefits. Governments may prefer fringe benefits (and hence grant favourable tax treatment) because private sector expenditures on items like pensions, workers' compensation, and unemployment insurance may reduce pressure for government expenditures in these areas.

While there are these reasons for the existence and growth of fringe benefits, the question remains as to why their magnitude is greater in the union than the nonunion sector. There are several theoretical reasons why this is so.

To the extent that union workers are made better off by unionization, they can afford to buy more of everything, including fringe benefits. Thus if the income elasticity of demand for nonwage benefits exceeds unity, union workers will devote a larger share of total income to fringe benefits than comparable nonunion workers. Compounding this effect is the fact that the effective price or cost of nontaxable fringe benefits falls as income (and therefore the marginal tax rate) rises.

The union's role in articulating its members' preferences regarding nonwage benefits may also affect the share of total compensation devoted to fringe benefits. As discussed previously, because of their political nature unions can be expected to represent the wishes of the median union voter. In contrast, nonunion employers will design their compensation packages to appeal to the marginal worker — the employee on the verge of joining or leaving the firm — who is likely to be younger and more mobile than the member of median age or seniority. Because the demand for such fringe benefits as pensions and life, accident and health insurance increases with age, seniority, and family responsibilities, compensation packages in unionized enterprises are thus expected to devote a greater share to these nonwage benefits.

Fringe benefits in the form of deferred compensation are also expected to be more prevalent in unionized establishments. As discussed in Chapter 23, employers may prefer deferred compensation because of favourable work incentive effects (employees will work diligently in order to retain their jobs and thus receive the deferred payment later in their careers) and because it reduces turnover (employees who quit lose some or all of their deferred compensation — for example, pensions and vacation rights). Deferred compensation will be attractive to employees if they are given a sufficiently high wage to compensate for some of it being deferred (and hence its receipt being uncertain), and/or if they are provided with sufficient guarantees that employers will ultimately pay the deferred wages. Such guarantees are more binding when they are provided in a collective agreement which, for example, protects against arbitrary dismissal. Thus deferred compensation arrangements are expected to be more acceptable to union than nonunion employees, other things equal.

Freeman's (1981) empirical analysis indicates that unionism increases fringe benefits both directly and indirectly (through higher levels of compensation for union workers). This suggests that the tendency of unions to represent the preferences of the median or inframarginal worker and the increased acceptability of deferred compensation to workers protected by a collective agreement help account for the increased significance of nonwage benefits in the union sector.

Union Impact on Working Conditions

In addition to their impact on wages and fringe benefits, unions affect a variety of working conditions. These changes may reflect both the collective bargaining objectives pursued by the union and the firm's response to the higher labour costs associated with unionism.

Union workplaces tend to be governed much more by various rules than comparable nonunion settings. Collective agreements typically stipulate such factors as the way grievances are to be handled, the role of company service, and the scheduling of worker hours. Nonunion enterprises, although not without their own bureaucratic procedures such as personnel policies, tend to be less rules-oriented, and to exhibit more worker and management flexibility.

In addition to a more structured work setting and less flexible hours, there is also evidence that unionized environments are characterized by a faster work pace. This may reflect the fact that unionism is more likely to occur in response to such working conditions or that employers are able to respond to the union wage effect by changing the conditions of work. Without being able to disentangle the true cause and effect, Duncan and Stafford (1980) estimate that about two-fifths of the union-nonunion wage differential reflects a compensating wage for these more onerous working conditions. This conclusion is relevant to interpreting empirical evidence not only on union-nonunion wage differences, but also on productivity differences between union and nonunion workers, discussed subsequently.

Union Impact on Turnover and Mobility

Unions also affect various aspects of labour market turnover and mobility — quits, layoffs, rehires, promotions, and terminations. Numerous U.S. studies (e.g., Freeman, 1980a, 1980b; Blau and Kahn, 1983) have found that the quit rate is significantly lower among union workers than comparable nonunion workers. In part this reflects the reluctance of workers to leave the higher-wage union environment. However, even after controlling for differences in wages, union workers quit less frequently than their nonunion counterparts.

There may be several explanations for this phenomenon. As discussed previously, deferred compensation is more prevalent in union enterprises; such benefits provide an extra inducement to remain with the employer. Causality may also operate in the other direction. Deferred compensation such as pensions and vacations tied to length of service is more attractive to individuals who expect to remain with the firm for a long time. Thus this type of compensation tends to act as a sorting or screening mechanism, attracting to union firms workers who are less likely to quit (and vice versa in nonunion firms).

The lower quit propensity among union workers can also be interpreted as reflecting their substitution of a "voice" mechanism for an "exit" mechanism (Freeman, 1980; Freeman and Medoff, 1984). Union employees dissatisfied with their working conditions can attempt to bring about improvements through collective bargaining, while the main mechanism available to dissatisfied workers in nonunion enterprises is the threat of seeking employment elsewhere.

Unions also affect the use of temporary layoffs, hours reductions, and wage reductions as mechanisms for responding to short-term fluctuations in demand. Collective agreements typically provide for a fixed nominal wage rate, possibly indexed to the cost of living but rarely indexed to other variables such as the state of the firm's product demand. With wages fixed, union firms rely more heavily on layoffs and recalls to respond to temporary fluctuations in demand than otherwise comparable nonunion firms, who make greater use of adjustments in wages and hours of work (Medoff, 1979; Blau and Kahn, 1983). In addition, seniority receives more weight in determining the order of layoffs and recalls in union firms. These differences in adjustment behaviour are consistent with the view that unions tend to represent the preferences of the median member. Having substantial seniority, such workers are unaffected by layoffs except in the case of extremely large reductions in the work force. In contrast, nonunion employers respond to the preferences of the marginal worker. Being younger and less senior, such workers would prefer that all workers take a reduction in wages and/or hours rather than some workers being laid off in reverse order of seniority.

Internal mobility also differs between union and nonunion employees and organizations. Promotion decisions depend much more on seniority in union than nonunion organizations, as do decisions about employee terminations (Medoff and Abraham, 1981; Blau and Kahn, 1983). As with layoffs, the greater weight given to seniority in these organizational decisions can be seen as reflecting the preferences of the median union member.

Union Impact on Productivity and Profitability

Although the impact of unions on various aspects of productivity has been debated for a long time, the subject received little quantitative analysis until recently. This recent literature has confirmed that unions can have both positive and negative effects on productivity, so that the overall impact depends on the magnitude of various offsetting factors.

In discussing the impact on productivity it is important to distinguish between the effects that occur directly as the outcome of collective bargaining and those that occur indirectly in the form of a behavioural response to higher wages. It is also important to recognize that unions may affect productivity not only in unionized firms but also elsewhere in the economy.

Economic theory predicts that firms will respond in several ways to the increase in labour costs associated with unionization. Relative to their nonunion counterparts, union firms will utilize more capital and other inputs and less labour per unit of output. Union employers will also attempt to hire more productive workers, and may try to increase job assignments and the pace of work.

Each of these responses to the union wage impact tends to raise labour productivity (output per worker) in the union sector and lower labour productivity in the nonunion sector. Thus a union-nonunion productivity comparison that does not control for differences in the amount of capital per worker, productivity-related worker characteristics, and the nature of the job assignment may well conclude that union workers are more productive than their nonunion counterparts. Although this conclusion would be correct, it would not necessarily follow that unions benefit society as a whole by raising worker productivity. For example, as discussed previously, the union wage impact and its consequences for the allocation of labour and other resources results in a reduction in society's total output and income, even though output per worker in the union sector rises. Similarly, society as a whole will, in general, not benefit from distributing the labour force such that unionized firms employ the more productive workers and nonunion firms the less productive. For these reasons, empirical analyses have attempted to determine whether unions and collective bargaining have direct effects on productivity in addition to those which arise as a consequence of the union wage impact. Such direct impacts, if positive, do imply a social benefit; if negative, they imply a social loss.

A common belief is that unions reduce productivity directly by inducing work stoppages and by negotiating restrictive work rules which compel the employer to use excessive amounts of union labour and/or prevent the employer from introducing technological innovations. However, as emphasized by Freeman and Medoff (1979, 1984), unions may also have positive effects on productivity by reducing turnover, improving morale and cooperation among workers, improving communications between labour and management, and by "shocking" management into more efficient practices.

Turnover is generally costly to firms because of expenses associated with hiring and training new workers to replace experienced personnel. While some turnover costs — for example, expenditures on job advertising — may not affect measured productivity, others — for example, existing personnel devoting less time to their own tasks and more to training new workers — do have this effect. In addition, lower turnover implies greater incentives for firms to engage in on-the-job training which enhances productivity. For these reasons, the lower quit rate in union enterprises is expected to result in higher productivity than in otherwise comparable nonunion enterprises. As discussed earlier, the differential in turnover can in turn be partly attributed to the union-nonunion wage differential and partly to the direct impact of collective bargaining.

Unions may affect employee morale, the amount of cooperation among workers, and communications between labour and management in several, possibly offsetting, ways. Employee morale may be enhanced through better wages, benefits and working conditions, by providing workers with a mechanism for expressing their collective preferences, and by negotiating grievance, dismissal, and other procedures which protect workers from arbitrary treatment by management. Better morale may in turn raise productivity, although it is also possible that grievance procedures can be abused by workers — resulting in management and employee time being devoted to trivial matters — and that making dismissal more difficult may reduce work effort and protect the incompetent. By making promotions much more dependent on seniority, unions may increase the amount of cooperation among workers because there will be less competition among employees for advancement. On the other hand, competition among workers for promotion may also enhance work effort and thus raise productivity. Unions also provide a formal mechanism for communication between workers and management, and these communications channels can be used to provide the firm with information about potential improvements in production techniques, product design, and the organization of the workplace. However, the union-management relationship can also become adversarial, with the consequence that there is less communication between the firm and its employees about potential improvements than in nonunion organizations.

Some analysts have also suggested that unionization — and the accompanying increase in labour costs — may shock the firm into adopting more efficient production and management techniques. This "shock effect" hypothesis assumes the existence of organizational slackness or inefficiency prior to unionization and is therefore consistent with satisficing behaviour on the firm's part, but not with optimizing behaviour (profit maximization and/or cost minimization). Although logically separate, the shock effect is difficult to distinguish empirically from the response of an optimizing firm to a change in relative factor prices — in this case, implementing a variety of changes in production techniques, hiring practices, and job assignments which result in the more efficient utilization of union labour.

In summary, economic theory suggests a variety of mechanisms through which unions may affect productivity. Because these mechanisms tend to both raise and lower productivity, the net impact is ambiguous. Empirical studies have found evidence of both positive and negative net impacts, suggesting that in some circumstances the productivity-enhancing mechanisms dominate while in others the opposite occurs. To the extent possible, the studies have attempted to control for such factors as differences in capital per worker and in productivity-related worker characteristics such as education and experience. The purpose, then, is to estimate the direct impact of unions on productivity, i.e., to exclude those effects which are a consequence of the union-nonunion wage differential.

Broadly based studies using aggregate data have generally produced inconclusive results. Using cross-sectional data for U.S. manufacturing industries, Brown and Medoff (1978) found a large positive union impact on productivity when certain strong assumptions were made regarding the technology of production in union and nonunion firms, but found no statistically significant impact under more general assumptions. With data for Canadian manufacturing industries, Maki (1983) obtains results which are even more sensitive to econometric specification, being positive when the Brown-Medoff assumptions are imposed but significantly negative under more general conditions. Adding to the conflicting results, Warren (1985) finds evidence of a negative union impact on productivity using aggregate U.S. time series data.

Empirical studies based on more disaggregated data generally appear to be much less sensitive to changes in specification. Using a large sample of individual U.S. firms, Clark (1984) estimates a small negative union impact on pro-ductivity. The growing number of econometric studies of individual industries provides evidence of both positive and negative impacts. In detailed studies of the U.S. cement industry, Clark (1980a, 1980b) finds that union labour was 6 to 8 percent more productive than nonunion labour, an amount which approximately offset the higher labour costs associated with the union wage impact. Large productivity effects have been estimated in residential and office construction (Mandelstamm, 1965; Allen, 1984), a sector in which the union-nonunion wage differential is also substantial. However, union and nonunion labour are estimated to be equally productive in the construction of schools (Allen, 1984). The difference may be due to the high degree of competition in residential and office construction relative to that for schools. Consistent with the hypothesis that product market competition plays an important role in the impact of unions on productivity is the general finding that there is no union-nonunion productivity differential in the provision of public sector services such as municipal libraries (Ehrenberg, Sherman and Schwarz, 1983) hospital services (Sloan and Adamache, 1984), and building department services (Noam, 1983).

Labour-management relations also appear to be an important determinant of labour productivity. The dramatic productivity decline in Canadian postal services was evidently caused by a sharp deterioration in labour-management relations (Read, 1982). U.S. studies of unionized automobile plants (Katz, Kochan and Gobeille, 1983) and paper mills (Ichniowski, 1984) find that productivity is negatively related to the number of grievances, a measure of labour-management relations.

Management response to unionization may also play a significant role. In his study of cement plants that became unionized, Clark (1980a) found that existing managers were replaced with new managers who were less paternalistic or authoritarian, and who placed more emphasis on controlling costs, setting production standards, and improving communication and monitoring. This adjustment can be interpreted either as a "shock effect" or as an optimizing firm's substitution of higher quality managerial input in response to an increase in the relative price of production labour.

In summary, the impact of unions on productivity, long a controversial issue, has received considerable attention recently. Evidence on the average economy-wide impact is inconclusive. However, in specific industries union labour has been found to be more productive, in some cases substantially so, than comparable nonunion labour. This productivity differential persists after controlling for differences in capital per worker and observed productivity-related worker characteristics, differences which can be attributed to the union-nonunion wage differential. However, the existence of a positive union impact on productivity is by no means assured. The direction and magnitude of the effect appears to depend on the management response to unionization, the quality of labour-management relations, and perhaps the degree of product market competition.

The conclusion that in some industries union labour is significantly more productive than comparable nonunion labour will no doubt surprise those who believe that, by restricting management flexibility and negotiating various work rules, unions reduce production efficiency. Sceptics are correct in claiming that several interpretations can be given to such evidence. Unions may organize the most productive firms in the industry, perhaps because these are the most profitable and have the greatest potential for wage gains. Alternatively, as discussed previously in the context of the union wage impact, there may be unobserved differences in worker characteristics and/or job assignments which account for observed productivity differences. However, it also may be the case that, for the various reasons discussed above, union representation makes otherwise comparable workers more productive.

One interesting consequence of a positive union productivity effect is that it may explain the coexistence of both union and nonunion firms in the same industry. In a perfectly competitive industry unions must either organize all the firms in the industry (and credibly threaten to organize potential entrants) or the cost-increasing consequences of the higher union wages must be offset by higher productivity. When barriers to entry or other factors enable firms to earn above-normal profits, it is not essential that union wage gains be offset by the higher productivity of union labour. In these circumstances unions may be able to capture some of the economic rents that would otherwise accrue to the firm's managers and/or owners. Empirical evidence indicates that this in fact occurs.

Despite differences in data sources, measurement of profitability and methodology, numerous U.S. studies are unanimous in concluding that union firms are less profitable than comparable nonunion firms. The available Canadian evidence, although somewhat inconclusive, is generally consistent with this finding (Maki and Meredith, 1986).

The conclusion that unions are associated with lower profits implies that owners and managers of firms are rational in generally opposing unionization, despite the possible existence of a positive union impact on labour productivity. In an interesting U.S. study, Ruback and Zimmerman (1984) find that the effect of unionization on firm profits is anticipated by stock market participants. Equity values fell significantly when an application for a union representation election was announced. The decline was substantially higher in those cases in which the union subsequently achieved certification, suggesting that investors were also able to anticipate the outcome of the representation vote.

QUESTIONS

1. Discuss the various wage structures and wage levels that unions can affect.

2. Discuss the various ways in which unions can affect the wages of nonunion labour.

3. Why may one expect the skill level of union workers to be different from the skill level of nonunion workers? What does this imply about union-nonunion wage differentials? Why may one expect high wages to cause unionism, as well as vice versa, and what does this imply about union-nonunion wage differentials?

4. Why may the union-nonunion wage differential not reflect the impact of unions on the wages of their members relative to what their wage would be in the absence of unionism?

5. If you expect the wage impact of unions to differ systematically with respect to certain conditions (e.g., extent of unionization in the industry or extent of concentration in the product market), how would you capture the interaction effect in a regression equation designed to measure the impact of unions?

6. The following union impact equation was estimated by Starr (1973, p. 56, Equation 11.3):

   $W = .361 \, CA + .00403 \, CA \cdot U + .00377 \, CA \cdot CR - .01228 \, CA \cdot RLC +$ other control variables where W is the male base wage rate in the establishment; CA represents the existence of a union in the establishment (coded one if the establishment has a collective agreement, zero otherwise); U represents the percentage of the industry that is unionized; CR represents industry concentration; and RLC is the ratio of wage labour cost to value added.

(a) Theoretically justify the use of each of these explanatory variables.

(b) What sign would you expect for each regression coefficient?

(c) Why are the explanatory variables entered in the equation in that particular fashion: that is, multiplied by CA?

(d) What is the effect on wages of being unionized as opposed to non-unionized?

(e) Evaluate this effect at the following mean values of the explanatory variables: $\overline{U} = 57.6$, $\overline{CR} = 7.8$, $\overline{RLC} = 36.1$.

(f) Compare this union impact in an establishment with the impact that would occur if the industry in which the firm operated was completely unionized, other things being equal.

(g) Compare the union impact of part (e) with the impact that would occur if the ratio of wage labour cost to value added were only 20 percent as opposed to the average of 36.1 percent.

7. Discuss why the union wage impact may vary according to such factors as industry, occupation, the stage of the business cycle, and the size of the firm.

8. Discuss the pros and cons of using longitudinal data to estimate the impact of unions.

9. Discuss why wage dispersion may differ between comparable union and nonunion workers.

10. Discuss the effect of unionization on the efficient allocation of resources. How would you measure these "welfare" effects?

11. Discuss the effect of unions on fringe benefits and working conditions.

12. Discuss the various ways in which unions may affect productivity.

13. Indicate how the average union nonunion wage differential, estimated from micro data based upon the individual as the unit of observation can be decomposed into two component parts: one part attributable to their different wage determining characteristics and the other due to different returns for the same characteristics.

14. In their article "The Two Faces of Unionism," R. Freeman and J. Medoff state:

"According to our recent research on the subject, treatment of unions as organizations without solidaristic or purposive values, whose sole function is to raise wages, is seriously misleading. While unions appear to raise wages above competitive levels, they also have significant nonwage effects which influence diverse aspects of modern industrial life. By providing workers with a voice both at the work place and in the political arena, unions can and do positively affect the functioning of the economic and social systems." Elaborate and assess this claim.

15. For each part, classify the statement as either true, partly true but incomplete, or false and explain your reasoning:

    (a) Unions tend to raise wages for unionized workers and lower wages for most nonunion workers. Thus unions tend to increase the amount of income inequality in the economy.

    (b) A powerful union is one for which the costs of a strike to the firm are very high, while a weak union is one for which costs of a strike are low.

    (c) Empirical research has found that unionized labour is more productive than comparable nonunion labour. This proves that unions increase the total value of output in the economy.

    (d) Unions can't raise wages in competitive industries because firms earn zero economic profits in equilibrium in these industries.

REFERENCES AND FURTHER READINGS

Abraham, K.G. and H.S. Farber. Returns to seniority in union and non-union jobs: a new look at the evidence. *ILRR* 42 (October 1988) 3–19.

Acs, Z.J. and D.B. Audretsch. Innovation in large and small firms: an empirical analysis. *AER* 78 (September 1988) 678–690.

Addison, J. and A. Barnett. The impact of unions on productivity. *BJIR* 20 (July 1982) 145–162.

Ahlseen, M.J. The impact of unionization on labor's share of income. *JLR* 11 (Summer 1990) 337–346.

Allen, S. Trade unions, absenteeism and exit voice. *ILRR* 37 (April 1984) 331–345.

Allen, S.G. Trade unionized construction workers are more productive. *QJE* 99 (May 1984) 251–274.

Allen, S.G. Unionization and productivity in office building and school construction. *ILRR* 39 (January 1986) 187–201.

Ashenfelter, O. and G.E. Johnson. Unionism, relative wages, and labour quality in U.S. manufacturing industries. *IER* 13 (October 1972) 488–508.

Barbezat, D.A. The effect of collective bargaining on salaries in higher education. *ILRR* 42 (April 1989) 443–455.

Becker, B.E. and C.A. Olson. Unionization and shareholder interests. *ILRR* 42 (January 1989) 246–262.

Belman, D. and J.S. Heywood. Union membership, union organization and the dispersion of wages. *R.E. Stats.* 72 (February 1990) 148–153.

Blackaby, D.H., Murphy, P.D. and P.J. Sloane. Union membership, collective bargaining coverage and the trade union mark-up for Britain. *Economic Letters* 36 (1991) 203–208.

Blakemore, A., J. Hunt and B. Kiker. Collective bargaining and union membership effects on the wages of male youths. *JOLE* 4 (April 1986) 193–211.

Blau, F.D. and L.M. Kahn. Unionism, seniority, and turnover. *IR* 22 (Fall 1983) 362–373.

Borjas, G. Job satisfaction, wages, and unions. *JHR* 14 (Winter 1979) 21–40.

Bronars, S.G. and D.R. Deere. The threat of unionization, the use of debt, and the preservation of shareholder wealth. *QJE* 106 (February 1991) 231–254.

Brown, C. and J. Medoff. Trade unions in the production process. *JPE* 86 (June 1978) 355–378.

Brown, C. and J. Medoff. The employer size-wage effect. *JPE* 97 (October 1989) 1027–1059.

Cable, J.R. and S.J. Machlin. The relationship between union wage and profitability effects. *Economic Letters* 27 (1991) 315–321.

Chowdhury, G. and S. Nickell. Hourly earnings in the United States: another look at unionization, schooling, sickness, and unemployment using PSID data. *JOLE* 3 (January 1985) 38–69.

Christensen, S. and D. Maki. The union wage effect in Canadian manufacturing industries. *JLR* 2 (Fall 1981) 355–368.

Clark, K. The impact of unionization on productivity: a case study. *ILRR* 33 (July 1980) 451–469.

Clark, K. Unionization and productivity. *QJE* 95 (December 1980) 613–640.

Clark, K.B. Unionization and firm performances: the impact on profits, growth, and productivity. *AER* 74 (December 1984) 893–919.

Connolly, R.A., Hirsch, B.T. and M. Hirschey. Union rent seeking, intangible capital, and market value of the firm. *R.E. Stats.* 68 (November 1986) 567–577.

De Fina, R.H. Unions, relative wages, and economic efficiency. *JOLE* 1 (October 1983) 408–429.

Doiron, D. and W.C. Riddell. The impact of unionization on male-female earnings differences in Canada. UBC Department of Economics Working Paper 92–30, 1992.

Duncan, G. and D. Leigh. Wage determination in the union and non-union sectors: a sample selectivity approach. *ILRR* 34 (October 1980) 24–34.

Duncan, G.M. and D.E. Leigh, The endogeneity of union status: an empirical test. *JOLE* 3 (July 1985) 385–402.

Duncan, G. and F. Stafford. Do union members receive compensating wage differentials? *AER* 70 (June 1980) 335–371.

Eberts, R. Union effects and teacher productivity. *ILRR* 37 (April 1984) 346–358.

Eberts, R.W. and J.A. Stone. Unionization and cost of production: compensation, productivity, and factor-use effects. *JOLE* 9 (April 1991) 171–185.

Ehrenberg, R.G., D.R. Sherman and J.L. Schwarz. Unions and productivity in the public sector: a study of municipal libraries. *ILRR* 36 (January 1983) 199–213.

Even, W.E. and D.A. MacPherson. The impact of unionism on fringe benefit coverage. *Economic Letters* 36 (May 1991) 87–91.

Fosu, A.K. Unions and fringe benefits: additional evidence. *JLR* 5 (Summer 1984) 247–254.

Freeman, R.B. Unionism and the dispersion of wages. *ILRR* 34 (October 1980) 3–23.

Freeman, R.B. The exit-voice trade-off in the labour market: unionism, job tenure, quits and separations. *QJE* 94 (June 1980) 643–673.

Freeman, R.B. The effect of unionism on worker attachment to firms. *JLR* 1 (Spring 1980) 29–62.

Freeman, R.B. The effect of unionism on fringe benefits. *ILRR* 34 (July 1981) 489–509.

Freeman, R.B. Union wage practices and wage dispersion within establishments. *ILRR* 36 (October 1982) 3–21.

Freeman, R.B. Longitudinal analyses of the effects of trade unions. *JOLE* 2 (January 1984) 1–26.

Freeman, R.B. Union density and economic performance: an analysis of U.S. states. *European Economic Review* 32 (March 1988) 707–716.

Freeman, R.B. Unionism comes to the public sector. *JEL* 24 (March 1986) 41–86.

Freeman, R.B. and M.M. Kleiner. The impact of new unionization on wages and working conditions. *JOLE* 8 (January 1990) 8–25.

Freeman, R. and J. Medoff. The two faces of unionism. *The Public Interest*. (Fall 1979) 69–93.

Freeman, R. and J. Medoff. The impact of the percentage organized on union and non-union wages. *R.E. Stats* 63 (November 1981) 561–572.

Freeman, R.B. and J.L. Medoff. The impact of collective bargaining: can the new facts be explained by monopoly unionism? *New Approaches to Labor Unions*, J.D. Reid, Jr. (ed.), Greenwich, CT: JAI Press, 1983.

Freeman, R.B. and J.L. Medoff. *What Do Unions Do?* New York, NY: Basic Books, 1984.

Graddy, D.B. and G. Hall. Unionization and productivity in commercial banking. *JLR* 6 (Summer 1985) 249–262.

Grant, E.K., R. Swidinsky and J. Vanderkamp. Canadian union-non-union wage differentials. *ILRR* 41 (October 1987) 93–107.

Grant, E.K. and J. Vanderkamp. The effects of migration on income: a micro study with Canadian data, 1965-71. *CJE* 13 (August 1980) 381–406.

Green, D. A comparison of estimation approaches for the union-nonunion wage differential. UBC Department of Economics Working Paper 91-13, May 1991.

Gunderson, M. The Public/Private Sector Controversy. *Conflict or Compromise: The Future of Public Sector Industrial Relations*, M. Thompson and G. Swimmer (eds.). Montreal: Institute for Research on Public Policy, 1984, 1–44.

Heywood, J.S. Who queues for a union job? *IR* 29 (Winter 1990) 119–127.

Hirsch, B.T. The interindustry structure of unions, earnings, and earnings dispersion. *ILLR* 36 (1982) 22–39.

Hirsch, B.T. Union coverage and profitability among U.S. firms. *R.E. Stats.* 73 (February 1991) 69–77.

Hirsch, B.T. Market structure, union rent seeking, and firm profitability. *Economic Letters* 32 (January 1990) 75–79.

Hirsch, B.T. and A.N. Link. Unions, productivity and productivity growth. *JLR* 5 (Winter 1984) 29–38.

Hirschman, A.O. *Exit, Voice and Loyalty*. Cambridge, MA: Harvard University Press, 1970.

Hyclak, T. The effect of unions on earnings inequality in local labour markets. *ILRR* 33 (October 1979) 77–84.

Hyclak, T. Unions and income inequality: some cross-state evidence. *IR* 19 (Spring 1980) 212–215.

Ichniowski, C. The effects of grievance activity on productivity. *ILRR* 40 (October 1986) 75–89.

Jakubson, G. Estimation and testing of the union wage effect using panel data. *R.E. Studies* 58 (October 1991) 971–991.

Johnson, G.E. The determination of wages in the union and non-union sectors. *BJIR* 15 (July 1977) 211–225.

Johnson, G.E. Changes over Time in Union-Non-union Wage Differential in the United States. *The Economics of Trade Unions: New Directions*, J. Rosa (ed.), Boston, MA: Kluwer-Nijhoff, 1984.

Jones, E.B. Union/non-union differentials: membership or coverage? *JHR* 17 (Spring 1982) 276–285.

Kahn, L.M. The effects of unions on the earnings of non-union workers. *ILRR* 31 (January 1978) 205–216.

Kahn, L.M. Union spillover effects on unorganized labour markets. *JHR* 15 (Winter 1980) 87–98.

Kahn, L.M. The effect of collective bargaining on production technique: a theoretical analysis. *JLR* 5 (Winter 1984) 1–12.

Kalachek, E. and F. Raines. Trade unions and hiring standards. *JLR* 1 (Spring 1980) 63–76.

Karier, T. Unions and monopoly profits. *R.E. Stats* 67 (February 1985) 34–42.

Katz, H.C., T.A. Kochan and K.R. Gobeille. Industrial relations performance, economic performance, and QWL programs: an interplant analysis. *ILRR* 37 (October 1983) 3–17.

Keefe, J.H. Do unions influence the diffusion of technology? *ILRR* 44 (January 1991) 261–274.

Killingsworth, M. Union-nonunion wage gaps and wage gains. *R.E. Stats* 64 (May 1983) 332–336.

Kuhn, P. Union productivity effects and economic efficiency. *JLR* 6 (Summer 1985) 229–248.

Kumar, P. Differentials in wage rates of unskilled labour in Canadian manufacturing industries. *ILRR* 26 (October 1972) 631–645.

Kumar, P. and T. Stengos. Measuring the union relative wage impact: a methodological note. *CJE* 18 (February 1985) 182–189.

Kumar, P. and T. Stengos. Interpreting the wage gap estimate from selectivity correction techniques using micro data. *Economic Letters* 20 (1986) 191–195.

Lee, Lung-Fei. Unionism and wage rates: a simultaneous equations model with qualitative and limited dependent variables. *IER* 19 (June 1978) 415–434.

Leigh, J.P. Are unionized blue collar jobs more hazardous than non-unionized blue collar jobs? *JLR* 3 (Summer 1982) 349–357.

Leigh, J.P. Unionization and absenteeism. *Applied Economics* 16 (February 1984) 147–157.

Lewis, H.G. *Unionism and Relative Wages in the United States: An Empirical Inquiry.* Chicago, IL: University of Chicago Press, 1963.

Lewis, H.G. Union relative wage effects: a survey of macro estimates. *JOLE* 1 (January 1983) 1–27.

Lewis, H.G. *Union Relative Wage Effects: A Survey.* Chicago: University of Chicago Press, 1985.

Linneman, P. and M.L. Wachter. Rising union premiums and the declining boundaries among noncompeting groups. *AER* 76 (May 1986) 103–108.

Lipsey, R.G. and J. Lancaster. The general theory of second best. *R.E. Studies* 24 (December 1956) 11–32.

Lovell, C.A.K., Sickles, R.C. and R.S. Warren. The effect of unionization on labor productivity: some additional evidence. *JLR* 9 (Winter 1988) 55–63.

Macdonald, G.M. The size and structure of union–non-union wage differentials in Canadian industry: corroboration, refinement and extension. *CJE* 16 (August 1983) 480–485.

Macdonald, G. and J. Evans. The size and structure of union–non-union wage differentials in Canadian industry. *CJE* 14 (May 1981) 216–231.

Machin, S.J. The productivity effects of unionization and firm size in British engineering firms. *Economica* 58 (November 1991) 479–490.

MacPherson, D.A. and J.B. Stewart. Unionism and the dispersion of wages among blue-collar women. *JLR* 8 (Fall 1987) 395–405.

Maki, D.R. The effects of unions and strikes on the rate of growth of total factor productivity in Canada. *Applied Economics* 15 (February 1983) 29–41.

Maki, D.R. Trade unions and productivity: conventional estimates. *RI/IR* 35 (No. 2, 1983) 211–225.

Maki, D. and S. Christensen. The union wage effect re-examined. *RI/IR* 35 (No. 2, 1980) 210–229.

Maki, D. and L. Meredith. The effect of unions on profitability: Canadian evidence. *RI/IR* 41 (No. 1, 1986) 54–68.

Mandelstamm, A.B. The effects of unions on efficiency in the residential construction industry: a case study. *ILLR* 18 (July 1965) 503–521.

Medoff, J.L. Layoffs and alternatives under trade unions in U.S. manufacturing. *AER* 69 (June 1979) 380–395.

Medoff, J.L. and K.G. Abraham. The role of seniority at U.S. workplaces: a report on some new evidence. Cambridge, Mass. *NBER* Working Paper No. 618, 1981.

Mefford, R.N. The effect of unions on productivity in a multinational manufacturing firm. *ILRR* 40 (October 1986) 105–114.

Mellow, W. Unionism and wages: a longitudinal analysis. *R.E. Stats* 63 (February 1981) 43–52.

Meng, R.A. Union effects on wage dispersion in Canadian industry. *Economic Letters* 32 (April 1990) 399–403.

Mincer, J. Union effects: wages, turnover, and job training. *New Approaches to Labor Unions*, J.D. Reid, Jr. (ed.), Greenwich, CT: *JAI Press*, 1983.

Mitchell, D. Union–non-union wage spillovers. *BJIR* 18 (November 1980) 372–376.

Mitchell, D.J.B. Unions and wages in the public sector: a review of recent evidence. *Journal of Collective Negotiations* 12 (4) (1983) 337–353.

Montgomery, M. Employment and unemployment effects of unions. *JOLE* 7 (April 1989) 170–190.

Montgomery, M. New evidence of unions and layoff rates. *ILRR* 44 (July 1991) 708–721.

Moore, W., R. Newman and J. Cunningham. The effect of the extent of unionism on union and non-union wages. *JLR* 6 (Winter 1985) 21–44.

Moore, W. and J. Raisian. Cyclical sensitivity of union/non-union relative wage effects. *JLR* 1 (Spring 1980) 115–132.

Moore, W.J. and J. Raisian. The level and growth of union/non-union relative wage effects, 1967–1977. *JLR* 4 (Winter 1983) 65–80.

Mulvey, C. and J. Abowd. Estimating the union/non-union wage differential: a statistical issue. *Economica* 47 (February 1980) 73–80.

Noam, E.M. The effect of unionization and civil service on the salaries and productivity of regulators. *New Approaches to Labor Unions*, J.D. Reid, Jr. (ed.), Greenwich, CT: *JAI Press,* 1983.

Parsley, C. Labor union effects on wage gains. *JEL* 18 (March 1980) 1–31.

Pearce, J.E. Tenure, unions, and the relationship between employer size and wages. *JOLE* 8 (April 1990) 251–269.

Pencavel, J.H. The distributional and efficiency effects of trade unions in Britain. *BJIR* 15 (July 1977) 137–156.

Pencavel, J. *Labor Markets Under Trade Unionism: Employment, Wages, and Hours.* Oxford and Cambridge, Mass.: Blackwell, 1991.

Pencavel, J.H. and C.E. Hartsog. A reconsideration of the effects of unionism on relative wages and employment in the United States, 1920–1980. *JOLE* 2 (April 1984) 193–232.

Pfeffer, J. and J. Ross. Unionization and income inequality. *IR* 20 (Fall 1981) 271–285.

Podgursky, M. Unions, establishment size and intra-industry threat effects. *ILRR* 39 (January 1986) 277–284.

Polachek, S.W. and E.P. McCutcheon. Union effects on employment stability: a comparison of panel versus cross-sectional data. *JLR* 4 (Summer 1983) 273–288.

Quan, N. Unionism and the size distribution of earnings. *IR* 23 (Spring 1984) 270–277.

Raisian, J. Union dues and wage premiums. *JLR* 4 (Winter 1983) 1–18.

Read, L.M. Canada Post, a case study in the correlation of collective will and productivity. *Research on Productivity of Relevance to Canada,* D.J. Daly (ed.), Ottawa: Social Science Federation of Canada (1982).

Rees, A. The effects of unions on resource allocation. *Journal of Law and Economics* 6 (October 1963) 69–78.

Riddell, W.C. Unionization in Canada and the United States: a tale of two countries. D. Card and R. Freeman, (eds.). *U.S. and Canadian Labor Markets.* Chicago: University of Chicago Press, forthcoming.

Robinson, C. The joint determination of union status and union wage effects: some tests of alternate models. *JPE* 97 (June 1989) 639–667.

Robinson, C. and N. Tomes. Union wage differentials in the public and private sectors: a simultaneous equations specification. *JOLE* 2 (January 1984) 106–127.

Rosen, S. Trade union power, threat effects and the extent of organization. *R.E. Studies* 36 (April 1969) 185–196.

Rosen, S. Unionism and the occupational wage structure in the United States. *IER* 11 (June 1970) 269–286.

Ruback, R. and M. Zimmerman. Unionization and profitability: evidence from the capital market. *JPE* 92 (December 1984) 1134–1157.

Schmidt, P. Estimation of a simultaneous equations model with jointly dependent continuous and qualitative variables: the union-earnings question revisited. *IER* 19 (June 1978) 453–465.

Schmidt, P. and R.P. Strauss. The effect of unions on earnings and earnings on unions: a mixed logit approach. *IER* 17 (February 1976) 204–212.

Shah, A. Job attributes and the size of the union/non-union wage differential. *Economica* 51 (November 1984) 437–446.

Simpson, W. The impact of unions on the structure of Canadian wages: an empirical analysis with micro data. *CJE* 18 (February 1985) 164–181.

Slichter, S.H., J.J. Healy and E.R. Livernash. *The Impact of Collective Bargaining on Management.* Washington, DC: Brookings Institution, 1960.

Sloan, F.A. and K.W. Adamache. The role of unions in hospital cost inflation. *ILRR* 37 (January 1984) 252–262.

Solnick, L. The effect of blue-collar unions on white-collar wages and fringe benefits. *ILRR* 38 (January 1985) 236–243.

Starr, G.E. *Union-Non-union Wage Differentials: A Cross-Sectional Analysis.* Toronto: Research Branch, Ontario Ministry of Labour, 1973.

Stewart, M. Relative earnings and individual union membership in the United Kingdom. *Economica* 50 (May 1983) 111–126.

Stewart, M.B. Union wage differentials, product market influences and the division of rents. *EJ* 100 (December 1990) 1122–1137.

Stewart, M.B. Union wage differentials in the face of changes in the economic and legal environment. *Economica* 58 (May 1991) 155–172.

Tauman, Y. and Y. Weiss. Labour unions and the adoption of new technology. *JOLE* 5 (October 1987) 477–501.

Viscusi, W.K. Strategic behavior and the impact of unions on wage incentive plans. *JLR* 3 (Winter 1982) 1–12.

Viscusi, W.K. Union, labour market structure, and the welfare implications of the quality of work. *JLR* 1 (Spring 1980) 175–192.

Voos, P. and L. Mishel. The union impact on profits: evidence from industry price-cost margin data. *JOLE* 4 (January 1986) 105–133.

Wadhwani, S. The effect of unions on productivity growth, investment, and employment: a report on some recent work. *British Journal of Industrial Relations* 28 (November 1990) 371–385.

Warren, R.S., Jr. The effect of unionization on labor productivity: some time series evidence. *JLR* 6 (Spring 1985) 199–207.

Wessels, W. The effects of unions on employment and productivity: an unresolved conflict. *JOLE* 3 (January 1985) 101–108.

Wunnava, P. and A.A. Okunade. Cross-sectional versus panel estimates of union wage effects. *Economic Letters* 35 (January 1991) 105–110.

PART 5

Wage Structures

Part 5 deals with the determinants of wages and various wage structures — by education, experience, occupation, region, industry, firm size, and between public and private sector workers and between males and females. The basic general determinants of wage structures are first analyzed and then used to explain why the different wage structures prevail and change over time.

The analysis in Part 3, and much of the analysis in Part 4, dealt with labour market equilibrium in a homogeneous labour market. This part of the book analyzes the determination of wages when workers and/or jobs are heterogeneous. The wage structure refers to the distribution of wages across workers according to characteristics such as gender, age, experience, education, industry, occupation, and region.

Chapter 17 develops the theory of compensating wage differentials. These refer to the wage differentials that exist *in equilibrium* to compensate workers for undesirable job characteristics such as unpleasant or unsafe working conditions, or desirable work place attributes such as valuable nonwage benefits or "perks." This theory is used extensively in the remaining chapters of Part 5; it provides the basic foundation for much of what follows. Chapter 17 also provides an application of the theory to the important area of work place safety.

Chapter 18 analyzes the acquisition of human capital in the form of education and training. There are large differences in earnings across workers with different amounts of education, experience, and training. The higher earnings of those with more education or training can be regarded as a compensating differential associated with the cost of acquiring the additional human capital. According to human capital theory, these individuals earn higher wages because they are more productive. This chapter also examines the possibility that education acts as a signal of ability, and is used by employers to sort out the more productive from the less productive, rather than directly enhancing the productivity of employees.

Chapter 19 examines wage structures by occupation and region. Reasons for interoccupational wage differences (e.g., between skilled and unskilled workers) are analyzed and then applied to explain why that wage structure has changed over time and over the business cycle. Geographic wage differences are analyzed, with particular attention to the extent to which they affect, and are affected by, geographic mobility — an important issue in a country like Canada. Issues of international migration are discussed, with particular attention to the problem of the "brain drain" and to the impact of immigration. Special attention is also given to the issue of the assimilation of immigrants into the Canadian labour market.

Chapter 20 examines wage structures by industry and firm size as well as across individual workers. Particular attention is paid to the issue of whether "efficiency wages" (i.e., high wages paid to induce positive labour market behaviour) can explain why some industries and firms consistently pay high wages. The implication of efficiency wages for industrial policy and trade policy to "protect good jobs" is also discussed, as is the wage impact of the trend towards deregulation. Special attention is given to an issue that has received considerable attention and policy concern — the growing wage polarization and wage inequality amongst individual workers. Reasons for the growing inequality are analyzed, with particular attention to the increasing returns to higher education and the role of free trade, global competition, and industrial restructuring.

Wage differences between workers in the public and private sector are analyzed in Chapter 21. Possible theoretical explanations for such wage differentials are first discussed, followed by an analysis of the empirical evidence to see if they prevail and how they have changed over time. The theory and evidence is then related back to the key policy issue of the pros and cons of curbing public sector wage settlements.

Special attention is also given to the issue of discrimination and male-female wage differentials (Chapter 22). The sources of discrimination are analyzed, as are alternative theories of discrimination, with particular attention to some of the feminist perspectives. Evidence on the male-female earnings gap is presented, emphasizing how it has changed over time and how it is affected by productivity related characteristics as well as discrimination. Alternative policy initiatives are discussed, including equal pay and pay equity (comparable worth) as well as equal employment opportunity legislation and employment equity (affirmative action). The emphasis is on their rationale as well as their expected and actual impact.

Chapter 23 deals with the design of optimal compensation systems for groups as diverse as auto workers, media superstars, and executives. Issues that are dealt with include: Do high wages ever pay for themselves by inducing positive work behaviour? Why do some salaries resemble tournament prizes? Why are some media superstars so highly paid while others are more like "starving artists"? Under what conditions might it be optimal to compress the salary structure within an organization to encourage co-operative behaviour and team work? Why are some executives so highly paid, and is this always tied to their impact on the performance of the firm? Why do deferred wages seem to prevail whereby workers are "underpaid" when younger in return for being "overpaid" when older? Why does mandatory retirement exist, and should it be banned? If it is banned, what impact will this have on various human resource practices such as training, promotion, and compensation?

In all of these areas, the emphasis — as it is throughout the book — is on using basic principles of economic theory to understand labour market behaviour. Empirical evidence on the determinants of that behaviour is presented. The theory and evidence is then tied back to important policy issues or practical issues pertaining to such factors as human resource practices. The ultimate intent is to show that labour market theory and evidence has relevance for our understanding of important practical issues and policy issues of the day.

CHAPTER 17

Compensating Wage Differentials

In previous chapters we discussed labour demand and various aspects of labour supply and how they interact to determine the wages and employment of a homogeneous group of workers. This chapter focuses on the wage *differentials* that arise when workers and jobs are not homogeneous, but differ because of positive characteristics like nonwage benefits and negative characteristics like risk.

PURPOSES OF WAGES AND WAGE STRUCTURES

The discussion of wage determination is complicated by the fact that wages and wage structures are called upon to serve a variety of purposes in our economy. Wage structures are the relative prices of labour that are utilized to allocate labour to its most productive and efficient use and to encourage human capital development — education, training, mobility, job search — into areas yielding the highest return. Wages are also the price that compensates workers for undesirable job characteristics and hence they ensure that the supply of and demand for such characteristics are in balance. In addition wages are an important component of family income and hence may have an important role to play in the achievement of an equitable distribution of income. Our macroeconomic objectives of full employment, price stability, and a viable balance of payments can also be affected by wage developments, just as wage determination is affected by the macroeconomic environment.

There is a complex nexus between wages and productivity. Usually we think of high wages being the result of high productivity that can come about, for example, through various forms of human capital formation, or through technological change. (With respect to the latter factor, wage increases are one way of distributing the gains of general productivity improvements; price reductions are another.) However, there is the possibility of cause and effect working the other way as well; that is, high wages may induce productivity improvements through improved morale, health and intensity of work effort, as well as by reductions in absenteeism and turnover. In addition wage increases may induce efficient changes elsewhere in the system — changes that should have occurred earlier but that would not occur without the shock induced by a wage change.

The matter is further complicated by the fact that wages and wage structures serve a social role, in that prestige is associated with wages; hence, the importance of *relative* positions in the wage hierarchy and the importance of key comparison groups. The social role of wages is further heightened by the fact that labour services are inseparable from the person who provides the services: hence, the importance of the human element in labour market analysis.

Clearly wages and wage structures are called upon to serve a variety of often conflicting roles in our economic system in particular, and in our social system in general. Perhaps this is why there is so much emotive conflict over various policies — such as minimum wages or wage controls or equal pay legislation — that affect wages.

THEORY OF COMPENSATING WAGES

Wages serve the purpose of compensating employees for undesirable working conditions or for negative or costly attributes associated with a particular job. Such conditions include an unsafe or unhealthy work environment, long commute time, undesirable working hours, lengthy training or other human capital requirements, and even the need for workers with discriminatory preferences to work in an integrated work environment. Wages also compensate — by being lower than they would otherwise be — for desirable working conditions such as flexible hours, stimulating tasks, and a pleasant working environment. As illustrated subsequently, many of these characteristics are associated with particular occupations, industries, regions, or firms and hence form part of the rationale for wage structures associated with these factors. In this chapter, the theory of compensating wages is illustrated with respect to compensating wages associated with the risk of injury or illness that can result from an unsafe or unhealthy work environment — an area where the theory has seen considerable development and empirical application.

Single Firm's Isoprofit Schedule

Figure 17.1a illustrates a single firm's isoprofit schedule, I, defined as the various combinations of wages and safety that the firm can provide and maintain the same level of profits. For the firm, both wages and a safe work environment are costly to provide; hence, the negative trade-off between the two; that is, the firm can provide more safety and maintain profits only if it can pay a lower wage. The isoprofit schedule exhibits a diminishing marginal rate of transformation

Figure 17.1
EMPLOYER'S ISOPROFIT SCHEDULES AND OFFER CURVE

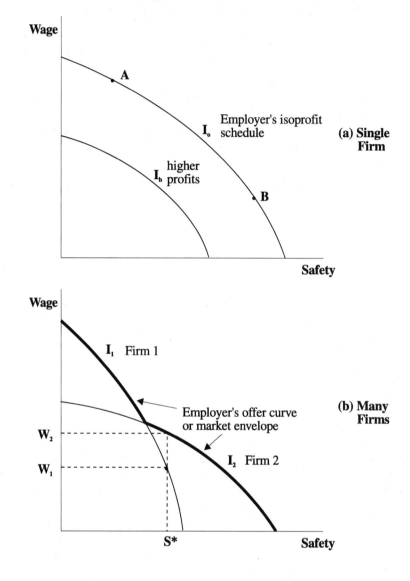

between wages and safety. That is, in the upper left portion, like at point A, where the firm is providing little safety, it can provide additional safety in a relatively inexpensive fashion with minor basic changes in its safety technology (e.g., better lighting, signs, guard rails). It is in a stage of increasing returns with respect to the provision of safety. In such circumstances, the firm can provide these inexpensive safety features without requiring much of a wage reduction in order to maintain a constant profit level; that is, the isoprofit schedule is relatively flat.

Conversely, in the bottom right segment as at point B, where the firm is providing considerable safety, it can provide additional safety only through the introduction of more sophisticated and costly safety procedures. It is in a stage of diminishing returns in the provision of safety, having already exhausted the cheapest forms of providing safety. That is, as the firm moves from left to right on the horizontal axis and provides additional safety, it will start with the cheapest forms of safety provision and then move to the more expensive. In such circumstances, it will require an even larger wage reduction to compensate for its additional safety costs, in order to maintain the same level of profits; that is, the isoprofit schedule will become steeper as more safety is provided and the firm moves from point A towards B.

Lower isoprofit schedules like I_h in Figure 17.1a imply higher levels of profits. That is, profits are the same at all points along I_h just as they are the same at all points along I_o. However, profits are higher along I_h than I_o. This is so because to the firm both wages and safety are costly to provide. Hence, providing both lower wages and lower safety means higher profits.

Different Firms with Different Safety Technologies

Different firms can have different abilities to provide safety at a given cost. Hence different firms may have differently shaped isoprofit schedules even for the same level of profit (e.g., even if they all operate at the competitive level of zero excess profits).

Figure 17.1b illustrates the isoprofit schedules for two firms (or industries, or occupations). Firm 1 exhibits rapidly diminishing returns to providing a safe work environment. Additional safety can only be provided if wages drop quite rapidly to compensate for the safety costs, if profits are to be maintained. This could be the case, for example, for firms in sectors that are inherently dangerous, such as mining or logging. In contrast, Firm 2 may be in an inherently safer industry and hence be able to provide high levels of safety without having to substantially lower wages to maintain profits. Competitive equilibrium requires that excess profits of Firms 1 and 2 are reduced to zero (i.e., $I_1 = I_2 = 0$).

The outer limits of the two isoprofit schedules (bold line), called the employers' *offer* curve or market *envelope* curve, show the maximum compensating wages that will be offered in the market for various levels of safety. Points within the envelope will not prevail in the market because they will always be dominated by points on the envelope. For example, for a given level of safety, S^*, Firm 2 is able to offer the wage W_2 and maintain its given level of profit, I_2. Firm 1, in contrast, can only offer the wage W_1 and continue to meet the zero profit condition. Hence, Firm 2's offer will dominate Firm 1's offer for levels of safety, like S*, to the right of the intersection of the isoprofit schedules; Firm 1's offer will dominate to the left of the intersection. In other words, workers would always go to employers that offer the highest compensating wage for every level of safety; hence, only points on the outer segments of the isoprofit schedules will prevail in a competitive market.

Single Individual's Preferences

As with any other items they value, individuals will have preferences for wages and safety. These preferences can be illustrated by a typical indifference or isoutility curve showing various combinations of wages and safety that yield the same level of utility, as depicted in the top panel, Figure 17.2a. The curvature of the isoutility curve illustrates a diminishing marginal rate of substitution between wages and safety. That is, at points in the upper left segment such as A where the individual does not have a very safe work environment, that individual would likely be willing to give up considerable wages to get a slightly safer work environment. Hence the indifference curve is steep. Conversely, at points such as B where the individual has a safer work environment, the individual may not be willing to give up much in the form of wages to obtain additional safety. Hence, the isoutility curve is relatively flat. Higher indifference curves, like U_h, indicate higher levels of utility since the individual has more of both wages and safety, both of which yield utility.

Different Individuals with Different Risk Preferences

Different individuals may have different risk preferences and hence may show a different willingness to give up safety in return for a compensating risk premium in the form of higher wages. Figure 17.2b illustrates a more risk-averse and a less risk-averse individual. As indicated at the point of intersection of the two schedules, the more risk-averse individual requires a larger compensating wage in return for giving up some safety and accepting a riskier work environment.

Equilibrium with Single Firm, Single Individual

Figure 17.3a illustrates the market equilibrium for a single firm and single individual as the point of tangency, E_c between the firm's isoprofit schedule, I_c and the individual's isoutility curve U_c. This outcome occurs because a perfectly competitive market yields the maximum worker utility subject to the firms' earning zero economic profits. The compensating wage, W_c, given the level of safety, S_c, is the highest the firm is able to offer for that level of safety, given the zero profit constraint. For movements to the left of S_c, the additional wage that the individual requires for accepting more risk (i.e., slope of the isoutility curve) is greater than what the firm is willing to give (i.e., slope of the isoprofit

Figure 17.2
WORKER ISOUTILITY CURVES

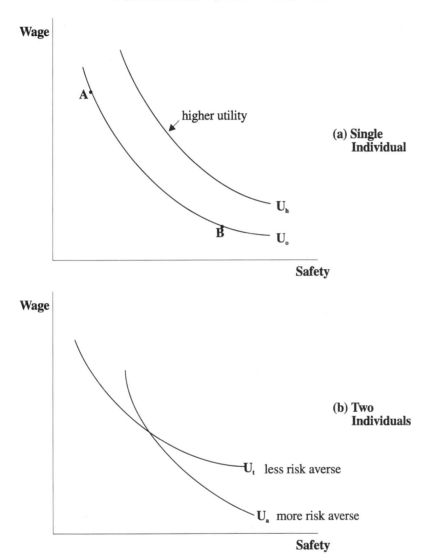

(a) **Single Individual**

(b) **Two Individuals**

schedule) in order to maintain the same competitive profit level. Conversely, for movements to the right, what the worker is willing to give up for additional safety is insufficient to compensate the employer so as to maintain the same competitive profit level. Thus, given the constraint I_c (due to the zero profit condition), worker utility is highest at E_c.

Higher isoutility curves are not attainable for the worker, because they would lie outside of the feasible choice set as dictated by the employer's isoprofit

Figure 17.3
MARKET EQUILIBRIUM

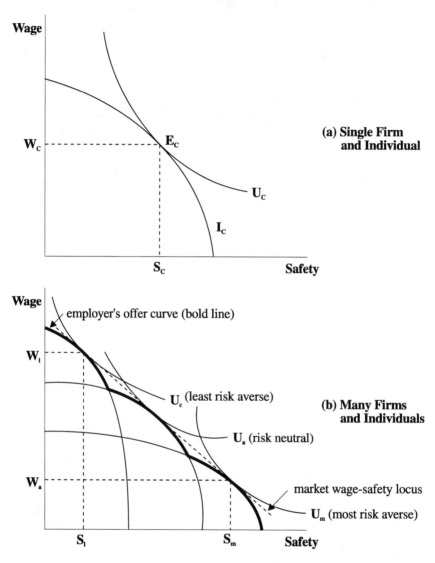

schedule, I_c, which gives the competitive level of profits. Higher isoprofit schedules would imply profits that are below the competitive level. Conversely, under
competitive conditions, individuals would not have to accept combinations of
lower wages and safety that would put them on isoutility curves below U_c
(and employers on isoprofit schedules that are closer to the origin, implying
profits above the competitive norm) because workers could move to firms that
would offer higher compensating wages and still maintain competitive profits.

Of course, if firms could offer lower combinations of wages and safety (e.g., if worker mobility were restricted or they did not know the risk to which they were exposed), then workers would be on lower isoutility curves and firms would have higher profits.

Equilibrium with Many Firms and Individuals

Figure 17.3b illustrates the market equilibrium that will prevail when there are many firms and individuals. Assuming perfect competition, individuals will sort themselves into firms (occupations, industries) of different risks (i.e., along the market envelope schedule) and receive differing compensating wages for the different levels of risk. The least risk-averse individual, for example, will enter a high-risk firm in order to get the higher wage, W_l, associated with the low level of safety, S_l. Conversely, the most risk-averse individual will enter the low-risk work environment and accept the lower wage, W_m, in order to have the safe work environment, S_m.

The set of tangencies between the various isoprofit and isoutility schedules gives the various equilibrium combinations of wages and safety that will prevail in the market. This is termed the wage-safety locus (dashed line in Figure 17.3b). The *slope* of that locus of wage-safety combinations gives the change in the wage premium that the market yields for *differences* in the risk of the job. The slope of that line can change for different levels of safety. It is determined by the interaction of workers' preferences and the firms' technology for safety, and these basic underlying determinants may change with the level of safety. The only restriction on the slope of the line is that it be negative, reflecting the fact that compensating wages are required for reductions in safety, given worker aversion to risk and the fact that safety is costly to firms.

The fact that the slope of the wage-safety locus can change magnitude but not direction means, for example, that the compensating wage premium required for *additional* risk may be very high in an already risky environment. Whether that is true, however, is an empirical proposition since it depends upon whether there are sufficient workers willing to take that risk in return for the higher wage, and whether firms could reduce the risk in ways that are less costly than paying a wage premium for workers to accept the risk.

In such circumstances of perfect information and competition (assumptions that will be relaxed later), the market will pay a compensating wage premium for undesirable working conditions, such as risk. This will induce workers to sort themselves into jobs depending upon their aversion to risk. It will also induce employers to adopt the most cost-effective safety standards (not necessarily the safest) since they can save on compensating wages by increasing their safety. In this fashion the need of employers to carry on production in a manner that may involve some risk is required to confront the equally compelling need of workers for desirable working conditions, and vice versa. The price that mediates these competing objectives is the compensating wage paid for the undesirable working conditions. Such prices are often termed "shadow" or "implicit" prices because they are embedded in the market wage rather than being attached explicitly to a job characteristic.

Obviously, the markets for each and every characteristic are likely to be "thin" (i.e., involve few buyers and sellers). For example, a worker who wants (and is willing to pay for) a job that is reasonably safe, close to home, and that allows flexible working hours, may not have a large set of jobs to choose from. Of course, if enough workers want those characteristics, then employers will find it is in their interest to provide them, because of the lower wages they can pay in return for providing the desired work characteristics. Nevertheless, it is easy to see that there may not be a large number of buyers and sellers for each and every characteristic, and hence the shadow price yielded by the market for each characteristic may not be the same as the competitive price that would prevail if there were many buyers and sellers for *each* characteristic.

ALTERNATIVE PORTRAYAL

The previous analysis was portrayed in terms of conventional indifference curve analysis with wages and safety — both of which are valued by employees — displayed on the axes. Following the insurance literature, the analysis is often portrayed in wage-*risk* space with risk portrayed on the horizontal axis. Since risk is negatively valued by employees this will obviously alter the portrayal; nevertheless, the same conclusions follow.

In Figure 17.4 the upward-sloping isoprofit schedules illustrate how firms can pay higher wages and maintain profits if risk can be increased because reductions in risk (increases in safety) are costly. The curvature of the isoprofit schedules exhibits diminishing returns; that is, as risk is reduced to very low levels in a firm it becomes extremely costly to reduce risk further and hence a larger wage reduction is necessary to maintain profits. The employers' offer schedule (bold line) indicates the maximum compensating wage that will be offered in the market for each level of risk. Points on employers' isoprofit schedules below that line will be dominated by points on the offer curve. Higher offer schedules yield lower profits since firms have to pay a higher compensating wage for every level of risk.

Worker indifference curves slope up because workers require a higher compensating wage to remain indifferent to the additional risk. Their curvature illustrates a diminishing marginal rate of substitution between wages and risk; that is, for a given individual to accept more risk and remain indifferent, an ever-increasing compensating wage is required to compensate for the increasing risk. Different individuals have different aversions to risk, as illustrated by the different indifference curves. As depicted at the intersection of the indifference curve for the most risk-averse (U_m) versus the least risk-averse (U_l) individual,

Figure 17.4
WAGE-RISK SPACE PORTRAYAL

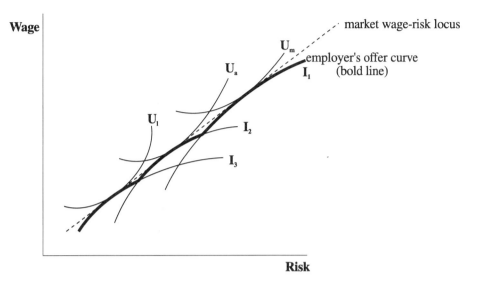

the risk-averse individual requires a higher compensating wage to accept additional risk.

Individuals will sort themselves into different firms (occupations, industries) on the basis of their willingness to accept risk in return for a compensating wage. The locus of tangencies between worker indifference curves and the employers' offer schedule will trace out the market wage-risk locus illustrating the varying wage premium the market will yield for different levels of risk. The market wage-risk locus can be of any configuration except that it slopes upwards, indicating the assumptions that workers require a compensating wage for additional risk, and reductions in risk are costly for employers.

EFFECT OF SAFETY REGULATION

Perfectly Competitive Markets

The effect of safety regulation depends critically upon whether markets are operating properly or not to compensate workers for occupational risk. In markets that are operating perfectly the effect of regulation may be perverse. This is illustrated in Figure 17.5a. For a single representative firm and individual, the competitive equilibrium is given at E_c, with the wage, W_c, being paid for the level of safety, S_c. In the absence of any external effects on third parties for which the market does not extract appropriate compensation, the level of safety, S_c, can be considered socially optimal in the sense that the price mechanism,

Figure 17.5
RESPONSES TO SAFETY STANDARD

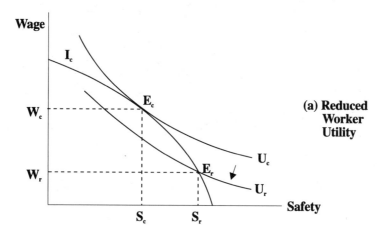

(a) Reduced
    Worker
    Utility

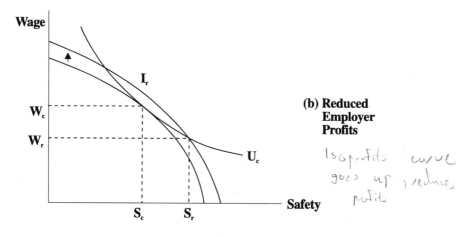

(b) Reduced
    Employer
    Profits

Isoprofits curve
goes up, reduces
profit

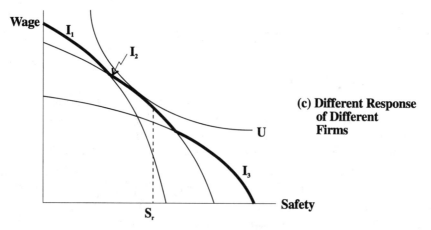

(c) Different Response
    of Different
    Firms

through compensating wages, ensures that neither employers nor employees want to move from the level of safety, S_c. By moving from S_c neither party can be made sufficiently better off to compensate the other party so that they are no worse off.

A regulatory agency that required the parties to increase the level of safety to S_r would actually make one or both parties worse off, in this particular situation of perfectly competitive markets. This is illustrated in Figure 17.5a, where the firm remains on its isoprofit schedule (it has to if this is the competitive level of zero excess profits) and the individual worker is on a lower level of utility U_r, with compensating wage W_r. The worker's level of utility under regulation is lower (i.e., $U_r < U_c$), because the worker is not voluntarily willing to accept the wage reduction from W_c to W_r in return for the safety increase from S_c to S_r; if given the choice the worker would return to E_c. In this particular example, the firm was no worse off after the regulation (i.e., on the same isoprofit schedule) but the worker was worse off (i.e., on a lower indifference curve). If the firm originally had excess profits it could absorb the cost increase associated with the safety regulation and not go out of business. In such circumstances the worker need not be worse off after the regulation. This is depicted in Figure 17.5b, where the firm moves to a higher isoprofit schedule, I_r, (with lower profits because it is experiencing both higher wage and safety costs) and the worker stays on the same isoutility curve. The worker still receives a reduction in the compensating wage from W_c to W_r, but this is exactly offset by the increase in safety, so that utility remains constant at U_c. In this case, the firm bears the cost of the safety regulation by taking lower profits. → *utility remains constant*

Realistically, the cost of the safety regulation will likely be borne by both workers in the form of compensating wage reductions and firms in the form of lower profits, their respective shares depending upon their relative bargaining powers.

Not all firms will be affected in the same fashion. For some, the safety regulation may be redundant since they are already meeting the standard. Others may be put out of business, unless their work force accepts wage concessions. This is illustrated in Figure 17.5c for three firms that are assumed to be operating under a competitive profit constraint so that excess profits are zero (i.e., $I_1 = I_2 = I_3 = 0$). For Firm 3, a uniform safety standard of S_r would be redundant since it is already providing a safer work environment than required by the standard; that is, the relevant portion of its isoprofit curve making up part of the market wage envelope always lies to the right of S_r. Firm 2 could meet the standard and stay in business only if its work force were willing to provide wage concessions (and hence receive a lower level of utility) for the firm to stay on its zero profit isoprofit schedule. (For simplicity only the isoutility curve for a representative worker of Firm 2 is shown.) Firm 1 would go out of business because its isoprofit schedule is below that of Firm 2 in the relevant region of the safety standard S_r; that is, for the minimal level of safety, S_r, Firm 2 could offer a higher compensating wage than Firm 1 and still stay in business. Workers in Firm 1 would go to Firm 2 if the standard were improved, even though they would prefer to be in Firm 1 with no standard. This competitive analysis, of course, assumes that Firm 2 can absorb the workers, with no adverse general equilibrium effects.

While this analysis suggests that the application of a uniform safety standard has the desirable effect of weeding out firms whose safety technology is such that they cannot meet the standard, it must be remembered that they were able to pay a sufficiently high wage to attract certain workers who were willing to accept the inherent risk. By imposing a uniform standard, the regulators are saying that firms that cannot meet the standard will not be able to operate no matter how much they are able to pay to have workers willingly accept the risk.

Imperfect Information

The previous analysis assumed perfect information about the level of safety involved. Such an assumption, however, may be unrealistic in situations where employers may have a vested interest in not disclosing information about health hazards, and where the latency period before an occupational disease shows up can be very long. In such circumstances, workers may think that they have a higher level of safety, and hence, utility, for the compensating wage they receive.

This situation is depicted in Figure 17.6. For a given compensating wage, W_a, associated with an actual level of safety, S_a, workers may perceive their level of safety to be greater at S_p. Their (mis)perceived level of utility would be U_p,

Figure 17.6
EFFECT OF IMPERFECT INFORMATION

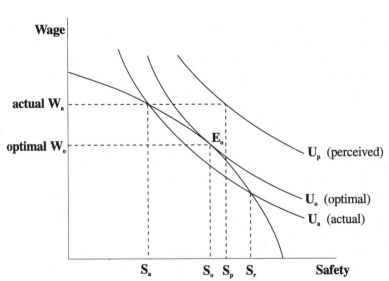

while their actual level would be U_a. In such circumstances, any imposed safety standard between S_a and S_r could improve workers' utility without making employers worse off since they would be on the same or a lower isoprofit schedule (lower schedules representing higher profits). The optimal standard would be at the point of tangency, E_o, between the employer's isoprofit schedule and the employee's highest attainable indifference curve.

Providing the parties the correct information would also lead to the optimal amount of safety because workers would increase their utility by moving to offers in the range of the offer curve between S_a and S_r. Ultimately, they would arrive at E_o, accepting the wage reduction from W_a to W_o in return for the increase in safety from S_a to S_o. With full information, market forces can also lead to the optimal amount of safety.

Rationale for Regulation

If a perfectly competitive market with proper information can lead to the socially optimal amount of safety at the workplace, why does regulation persist, especially since it runs the risk of making both employers and employees worse off? While no easy answer to this question exists, a number of plausible explanations can be given, in addition to the possibility that it may simply be a well-meaning mistake. First, information is not perfect and may be hidden by employers or improperly discounted by workers. Second, competition may not prevail for the buying and selling of each and every job characteristic, including safety in the work environment. Workers simply may not have much choice in the risk of their work. Third, workers who opt for risk in return for a high wage may not pay the full cost of their decisions to the extent that the state bears some of the medical or workers' compensation costs if the risk comes to fruition. Fourth, society may feel that workers ought not to have to sell their health and safety to make a living, even if they are willing to do so. If regulating safety means a lower compensating wage, then it may be best to try to take care of that problem through other income maintenance schemes. Fifth, the willingness of workers to accept some risk in return for a compensating wage premium may be dictated by the existing amount of risk to which they are currently exposed. If that risk is reduced by law for everyone, they may actually prefer the new level of risk even though earlier they were unwilling to accept the wage reduction to achieve the reduced risk. Although economists tend to regard preferences as given, preferences in fact may be influenced by our existing state. Lastly, there may be the feeling that the market simply may not yield the compensating wages necessary to ensure the socially optimal amount of safety.

EMPIRICAL EVIDENCE ON COMPENSATING WAGES

Obtaining empirical evidence on the existence of compensating wages for undesirable working conditions is difficult because of the problem of controlling for the myriad of positive and negative job characteristics to isolate the separate effect of one characteristic. This is compounded by the fact that situations of strong employee bargaining power often have high wages, good fringe benefits, and good working conditions. While there still could be a trade-off between wages and a better work environment, there is often insufficient variation in the data to identify the trade-off with any degree of precision.

In some cases the trade-off may be direct and explicit. For example, collective agreements or company pay policies may specify a premium for underground as opposed to above-ground mining, or for the night shift as opposed to the day shift. Premiums for an isolation allowance or even hazard pay may be explicitly mentioned with the explicit price stated.

In other circumstances, however, the compensating pay premium is not explicitly given but rather is embedded as part of an individual's wage rate and hence is an implicit or "shadow" price associated with the job characteristic. It can be estimated only by comparing wages in jobs that are similar except for the one characteristic. Given that jobs involve a multiplicity of characteristics this is not an easy task and usually requires fairly sophisticated statistical procedures.[1]

A number of such studies have estimated compensating wages for undesirable job characteristics such as the risk of being unemployed (Adams, 1985; Li, 1986), the uncertainty of pension receipts (Smith, 1981; Schiller and Weiss, 1980), the mandatory requirement to work overtime (Ehrenberg and Schumann, 1984), commuting time (Leigh, 1986), shift work (Kostiuk, 1990), retirement characteristics (Ehrenberg, 1980), and low pension benefits (Gunderson, Hyatt and Pesando, 1980). The most extensive literature in this area pertains to compensating wages for the risk of injury or death on the job. These studies have been plagued by a number of econometric problems that are typical of statistical work in economics.

For example, in addition to higher wages being paid to compensate for undesirable working conditions, cause and effect may work the other way as undesirable working conditions may result from lower wages — that is, low-wage individuals can't afford to buy a safe work environment and the forgone income from an accident is less to them. These income and substitution effects, respectively, cause them to buy less occupational health and safety. Failure to account for such reverse causality can lead to a simultaneous equation bias in the

---

1. Conventionally, this is done by multiple regression analysis which enables isolating the effect of one variable while controlling for the effect of the other variables included in the regression. In this case, wages are regressed on various wage-determining factors including job characteristics and human capital factors. The resulting regression coefficients for the job characteristics (e.g., risk) give the change in wages associated with a unit change in the job characteristic, holding constant the effect of all other variables included in the regression equation. The regression coefficients are therefore the shadow price (compensating wage premium) associated with changes in the job characteristics, such as risk. Such a wage equation is termed a hedonic wage equation.

estimates of compensating wages for work hazards. Simultaneous equation techniques to account for the two-way causality have been employed in only a few such studies (e.g., Chelius, 1979; McLean, Wendling and Neergaard, 1978).

Problems of "errors-in-variables" may also exist, especially in those studies that relate an *individual's* wage to an *aggregate* measure of risk such as the injury rate in the person's industry or occupation. Such aggregate measures of risk are likely to be subject to considerable random error as a proxy for the risk that an individual will face in that occupation. Econometrically, it can be shown that this will lead to an underestimation of the compensating wage paid for risk.

Omitted variable bias may also exist to the extent that crucial variables that affect wages and that are also correlated with risk are not controlled for in the measurement of the wage-risk trade-off. For example, bargaining power of workers (as represented by unionization) may lead to high wages and low risk. Failure to control for this factor may mean that there appears to be little or no wage premium paid for the risk. In reality the low wages in risky jobs may reflect the fact that workers with little bargaining power end up in jobs of low wages and high risk. The low wages in risky jobs may reflect the disproportionate absence of bargaining power of workers in those jobs, not the absence of a wage premium for risk. One method to control for such difficult-to-measure variables as preferences or an individual's bargaining power is to use longitudinal or panel data that follows the same individual over time so that such factors can be assumed to be relatively constant or fixed over time (e.g., Brown, 1980).

Sample selection bias problems may also prevail to the extent that wage-risk premiums are estimated on a subsample of risky jobs, perhaps because actuarial data is available only on risky occupations (e.g., Thaler and Rosen, 1975). In such circumstances, the risk premiums obtained from the subsample of risk-taking individuals may not give an accurate estimate of the compensating wage paid for a random worker.

More generally, account needs to be taken of the process via which individuals choose jobs, occupations, and careers. Specifically, those individuals who have the least aversion to an undesirable job characteristic (e.g., repetitive work) are, other things equal, the most likely to choose jobs with that characteristic. Thus the compensating wage needed to attract *additional* workers into an occupation may be higher than was needed to attract those already in the occupation. Recent research has tried to take into account this simultaneous determination of job risks and occupational choices (Garen, 1988; Godderis, 1988).

Such econometric problems are typical of the applied econometrics literature; hence, the econometric evidence on compensating wages for occupational risk is likely to be neither more nor less reliable than typical econometric estimates. Reviews of such evidence are contained in Smith (1979), Brown (1980), Gunderson and Swinton (1981), Digby and Riddell (1986), and Rosen (1986). The general consensus appears to be that compensating wage premiums are paid for work hazards and they increase with the seriousness of the risk; that is, they are larger for risks of death than for risks of injury and, for risks of injury, they are larger for permanent than for temporary injuries. In fact, for less serious injuries, compensating wage premiums are often found to not exist. A limited amount of evidence also suggests that the compensating wage premiums are reduced when the risk is partly covered by workers' compensation systems, and that compensating wage premiums are larger in union than in nonunion environments.

Canadian empirical evidence on compensating differentials for job related risks is extremely limited. (See Exhibit 17.1.) This is unfortunate given the importance of workplace health and safety and the many policy initiatives in this area (Digby and Riddell, 1986). The main obstacle facing researchers is lack of appropriate data. However, Meng (1989) utilizes a unique data set and finds evidence of significant compensating differentials for job risks in Canada.

POLICY IMPLICATIONS

A number of policy implications associated with compensating wage differences for undesirable job characteristics have already been discussed. In particular, competitive markets can yield the optimal amount of these characteristics, with compensating wages being the price that equilibrates markets. This ensures that the need of employers to carry on production in a manner that may involve undesirable working conditions is required to confront the need of workers for desirable conditions, and vice versa. This also implies, for example, that the optimal amount of safety is not zero; people are seldom willing to pay the price of attaining that otherwise desirable state.

In such competitive markets, regulations setting a uniform standard, such as a health and safety standard, run the risk of making the parties worse off, largely because compensating wages will adjust in a fashion that workers themselves would not have accepted for the improved working conditions. If there is imperfect information, or markets fail for other reasons, then regulation can make workers better off.

To the extent that wage premiums fully compensate workers for the expected risk of a job, then compensating them if the risk comes to fruition can involve double compensation. Of course, once this is anticipated by the parties, then the compensating wage premium itself will fall, and be paid only for the uncompensated risk. Thus, the compensating wage paid for the risk of injury or of being unemployed will be smaller, respectively, in situations of workers' compensation and unemployment insurance.

Empirical estimates of the compensating wage paid for the risk of death have been used to provide estimates of the "value of life." This phrase is somewhat of a misnomer since the estimates really reflect what people are willing to pay (as revealed through the market) to reduce the risk of death by a certain percentage.

EXHIBIT 17.1   THE VALUE OF SAFETY IN CANADIAN INDUSTRY

What is the value of an increase in safety in the workplace? Because compensating wage differentials associated with workplace risks are implicit prices associated with these negative job characteristics, this question can be addressed by empirical analysis which estimates these implicit prices. Meng (1989) utilizes a unique data source to obtain estimates of compensating wage differentials associated with several negative job characteristics in Canada. Information was available on the following job characteristics:

■ work under stressful conditions

■ work with machines

■ repetitive work (duties constantly repeated)

■ job requires physical exertion

These four job characteristics are expected to be positively related to hourly earnings (other things being equal). In addition, data on fatality rates (number of fatalities as a result of job-related accidents or disease) on detailed occupations were used to account for the risk associated with the job.

Meng finds that there is a significant and positive wage differential associated with the riskiness of the occupation, after controlling for other influences on earnings (such as education, labour market experience, union status, tenure on the current job, and other influences). This estimate is the implicit price associated with the riskiness of the occupation, or the slope of the market wage-risk locus (see Figure 17.4). The estimates indicate that, on average, a typical worker must be paid $2,712 (in 1981 dollars) more per year for an increase of 0.001 in the risk of a fatality on the job — that is, a firm with 1000 employees could save $2,712,000 in annual wage costs (in 1981 dollars) if it were able to reduce the risk of an occupation-related fatality by a probability of 1 in 1,000.

Meng also finds no evidence of compensating wage differentials for stressful working conditions, working with machines, repetitive work, and work requiring physical exertion. This result is similar to that found in several studies using U.S. and U.K. data — evidently, only the most adverse working conditions exert effects on earnings which are substantial enough to be measured in the (admittedly imperfect) data available to researchers.

Source: R. Meng. Compensating differences in the Canadian labour market. *CJE* 22 (May 1989) 413–24.

This is then extrapolated to arrive at the total price they would be willing to pay to reduce the risk of death to zero. These values are arrived at through impersonal transactions as people reveal how much in wages they are willing to give up to reduce their risk of death. These are not the values that people would attach to the *certainty* of saving a *specific* life, a value that would likely be incalculable.

For policy purposes such estimates have been used to compensate victims, or their survivors, for the loss of life or limb. They have also been used in decisions on how many resources to devote to life-saving activities or on the appropriate safety standards to set in industry. Rightly or wrongly, economics with its emphasis on trade-offs enters into many socially sensitive issues.

The notion that a job involves a set of positive and negative characteristics, each with its own implicit price, also has important implications for the analysis of wage differentials by such factors as occupation, industry, or region. As shown subsequently, the wage that is associated with each of these factors can have a component that reflects the compensating wage premiums paid for certain undesirable characteristics associated with each occupation, industry, or region. This will be important for understanding the existence and persistence of inter-occupation, interindustry, and interregional wage structures.

QUESTIONS

1. Suppose a simple economy with two kinds of workers — J (for "Jocks") workers who like physical labour and L (for "Lazy") workers who dislike physical labour. Further suppose that in this economy there are two kinds of jobs — O (for "Office") jobs which involve little or no physical labour and F (for "Forestry") jobs which involve substantial physical labour. The J workers will work for $6 per hour or more in the F jobs and $8 per hour or more in the 0 jobs, while the L workers will work for $6 per hour or more in the 0 jobs and $10 per hour or more in the F jobs. Neither group will work for less than $6 per hour. The L workers are indifferent between an F job at $10 per hour and 0 job at $6 per hour, while the J workers are indifferent between an F job at $6 per hour and an 0 job at $8 per hour.

   (a) Draw the supply of labour curve to the two sectors (O and F), assuming that there are 100 of each type of worker.

   (b) Suppose the demand for labour curves (which are derived from the demand for the products of the two sectors) intersect the labour supply curves at L = 150 and L = 50 in the F and O sectors respectively.

      (i) What is the equilibrium differential?

      (ii) Which workers are earning economic rent, and what is the amount of rent earned?

      (iii) Which workers are earning no economic rent?

2. A simple economy has two sectors, A and B, both of which use labour as an input in production. A jobs and B jobs are equally desirable from the point of view of workers. Also, all workers have the potential to do either job. However, a training period must precede employment in either job. For simplic-

ity, we will assume that the training is financed by the worker. Each period some workers retire in each sector, and new workers enter the labour force.

(a) If the training periods for the two jobs are equal in length and cost, what is the equilibrium differential in earnings between the two jobs?

(b) Assume the conditions given in (a), and assume further that the labour markets are initially in equilibrium. Then suppose there is an increase in demand for the products of sector A. What will happen in the two labour markets, both immediately and in the long run?

(c) Suppose job B requires a longer and more expensive training period than job A. Will there be an equilibrium earnings differential between the two jobs?

REFERENCES AND FURTHER READINGS

Adams, J. Permanent differences in unemployment and permanent wage differentials. *QJE* 100 (February 1985) 29–56.

Arnould, R.J. and L.M. Nichols. Wage-risk premiums and workers' compensation: a refinement of estimates of compensating wage differentials. *JPE* 91 (April 1983) 332–340.

Arthur, W. The economics of risks to life. *AER* 71 (March 1981) 54–64.

Atrostic, B.K. The demand for leisure and nonpecuniary job characteristics. *AER* 72 (June 1982) 428–440.

Brown, C. Equalizing differences in the labour market. *QJE* 94 (February 1980) 113–134.

Chelius, J. The control of industrial accidents. *Law and Contemporary Problems* 38 (Autumn 1979) 700–729.

Dardis, R. The value of a life: new evidence from the marketplace. *AER* 70 (December 1980) 1077–1082.

Diamond, C.A. and C.J. Simon. Industrial specialization and the returns to labor. *JOLE* 8 (April 1990) 175–201.

Digby, C. and W.C. Riddell. Occupational health and safety in Canada. *Canadian Labour Relations*, W.C. Riddell (ed.). Toronto: University of Toronto Press, 1986, pp. 285–320.

Dorsey, S. and N. Walzer. Workers' compensation, job hazards, and wages. *ILRR* 36 (July 1983) 642–654.

Duncan, G.J. and B. Holmlund. Was Adam Smith right after all? Another test of the theory of compensating wage differentials. *JOLE* 1 (October 1983) 366–379.

Dunn, L.F. Work disutility and compensating differentials: estimation of factors in the link between wages and firm size. *R.E. Stats.* 68 (February 1986) 67–73.

Eberts, R. and J. Stone. Wages, fringe benefits, and working conditions: an analysis of compensating differentials. *SEJ* 52 (July 1985) 274–280.

Ehrenberg, R. Retirement system characteristics and the compensating wage differentials in the public sector. *ILRR* 33 (July 1980) 470–484.

Ehrenberg, R. and P. Schumann. Compensating wage differentials for mandatory overtime. *EI* 22 (October 1984) 460–478.

Evans, R. and R. Weinstein. Ranking occupations as risky income prospects. *ILRR* 35 (January 1982) 252–259.

Fairris, D. Compensating wage differentials in the union and nonunion sectors. *IR* 28 (Fall 1989) 356–372.

Farber, S.C. and R.J. Newman. Regional wage differentials and the spatial convergence of worker characteristic prices. *R.E. Stats.* 71 (May 1989) 224–231.

Garen, J. Compensating wage differentials and the endogeneity of job riskiness. *R.E. Stats.* 70 (February 1988) 9–16.

Gerking, S. and W. Wierick. Compensating differences and interregional wage differentials. *R.E. Stats.* 65 (August 1983) 483–486.

Gunderson, M., D. Hyatt and J. Pesando. Wage-pension trade-offs in collective arguments. *ILRR* 46 (October 1992) 146–160.

Gunderson, M. and K. Swinton. *Collective Bargaining and Asbestos Dangers in the Workplace.* Ontario Royal Commission on Asbestos, 1981.

Goddeeris, J.H. Compensating differentials and self-selection: an application to lawyers. *JPE* 96 (April 1988) 411–428.

Gyourko, J. and J. Tracy. The importance of local fiscal conditions in analyzing local labor markets. *JPE* 97 (October 1989) 1208–1231.

Hamermesh, D.S. and J.R. Wolfe. Compensating wage differentials and the duration of wage loss. *JOLE* 8 (January 1990) 175–197.

Hersch, J. and W.K. Viscusi. Cigarette smoking, seatbelt use, and differences in wage-risk tradeoffs. *JHR* 25 (Spring 1990) 202–227.

Herzog, H.W. and A.M. Schjlottmann. Valuing risk in the workplace: market price, willingness to pay, and the optimal provision of safety. *R.E. Stats.* 72 (August 1990) 463–470.

Kennedy, B.R. Mobility and instability in Canadian earnings. *CJE* 22 (May 1989) 389–394.

Killingsworth, M.R. Heterogeneous preferences, compensating wage differentials, and comparable worth. *QJE* 102 (November 1987) 727–742.

Kostiuk, P.F. Compensating differentials for shift work. *JPE* 98 (October 1990) 1054–1075.

Leigh, J.P. Are compensating wages paid for time spent commuting? *AER* 18 (November 1986) 1203–1213.

Li, E.H. Compensating differentials for cyclical and noncyclical unemployment: the interaction between investors' and employees' risk aversion.

Low, S. and L. McPheters. Wage differentials and the risk of death: an empirical analysis. *EI* 21 (April 1983) 271–280.

Marin, A. and G. Psacharopoulos. The reward for risk in the labor market: evidence from the U.K. and a reconciliation with other studies. *JPE* 90 (August 1982) 827–853.

Marshall, J.M. Gambles and the shadow price of death. *AER* 74 (March 1984) 73–86.

McLean, R., W. Wendling and P. Neergaard. Compensating wage differentials for hazardous work: an empirical analysis. *Quarterly Review of Economics and Business* 18 (Autumn 1978) 97–107.

Meng, R. Compensating differentials in the Canadian labour market. *CJE* 22 (May 1989) 413–424.

Meng, R. Compensating wages for long-term job hazards in Canadian industry. *Economic Letters* 36 (July 1991) 331–336.

Olson, C. An analysis of wage differentials received by workers on dangerous jobs. *JHR* 16 (Spring 1981) 167–185.

Rea, S. Workmen's compensation and occupational safety under imperfect information. *AER* 71 (March 1981) 80–93.

Rosen, S. Hedonic prices and implicit markets: product differentiation in pure competition. *JPE* 82 (January 1974) 34–55.

Rosen, S. The theory of equalizing differences. *Handbook of Labour Economics*, Vol. 1. O. Ashenfelter and R. Layard (eds.). New York: Elsevier, 1986.

Schiller, B. and R. Weiss. Pensions and wages: a test for equalizing differences. *R.E. Stats.* 62 (November 1980) 529–538.

Smith, R. Compensating wage differentials and public policy: a review. *ILRR* 32 (April 1979) 339–352.

Smith, R.S. Compensating differentials for pensions and underfunding in the public sector. *R.E. Stats.* 63 (August 1981) 463–467.

Thaler, R. and S. Rosen. The value of saving a life: evidence from the labor market. *Household Production and Consumption*, N. Terleckyj (ed.). New York: National Bureau of Economic Research, 1975, pp. 265–302.

Viscusi, W. Kip. *Employment Hazards*. Cambridge, Massachusetts: Harvard University Press, 1979.

Viscusi, W. Imperfect job risk information and optimal workmen's compensation benefits. *J Pub Ec* 14 (December 1980) 319–338.

CHAPTER 18

Human Capital Theory: Applications to Education and Training

In Part 1 on labour supply we emphasized the *quantity* aspects of labour supply, ranging from family formation to labour force participation to hours of work. Labour supply also has a *quality* dimension encompassing human capital elements such as education, training, labour market information, mobility, and health. In addition, in the previous chapter on compensating wages, we indicated that compensating wages may have to be paid to compensate workers for the costly process of acquiring human capital like education or training. The hedonic wage approach developed in that chapter can be applied to costly attributes necessary to do a job, just as it can be applied to negative job attributes like risk.

While the economics of education and health economics are often the subject matter of separate courses and textbooks, they — along with training, job search, and mobility — have a common theoretical thread: that of human capital theory. This chapter presents the basic elements of human capital theory and applies it mainly to the areas of education and training. The acquisition of labour market information in connection with the job search process is discussed in Chapter 25 on unemployment. Mobility is discussed in Chapter 19 in the context of regional wage structures.

HUMAN CAPITAL THEORY

The essence of human capital theory is that investments are made in human resources so as to improve their productivity and therefore their earnings. Costs are incurred in the expectation of future benefits: hence, the term "investment in human resources." Like all investments, the key question becomes: is it economically worthwhile? The answer to this question depends on whether or not benefits exceed costs by a sufficient amount. Before dealing with the investment criteria whereby this is established, it is worthwhile to expand on the concepts of costs and benefits as utilized in human capital theory. In this chapter, only the basics are touched upon. A wealth of refinements and precise methodological techniques is contained in the extensive literature on human capital theory and its application.

In calculating the costs of human capital, it is important to recognize not only direct costs, such as books or tuition fees in acquiring university education, but also the opportunity cost or income forgone while people acquire the human capital. For students in university or workers in lengthy training programs, such costs can be the largest component of the total cost. The evaluation of these opportunity costs can prove difficult because it requires an estimation of what the people would have earned had they not engaged in human capital formation.

In addition, it is important to try to distinguish between the consumption and the investment components of human capital formation, since it is only the investment benefits and costs that are relevant for the investment decision. In reality this separation may be difficult or impossible — how does one separate the consumption from the investment benefits of acquiring a university degree? Nevertheless, the distinction must be made qualitatively, if not quantitatively, especially in comparing programs where the consumption and investment components may differ considerably.

A distinction must also be made between private and social costs and benefits. Private costs and benefits are those that accrue to the parties making the investment and as such will be considered in their own calculations. Social costs and benefits are all those that are accrued by society, including not only private costs and benefits but also any third-party effects or externalities that accrue to parties who are not directly involved in the investment decision. Training disadvantaged workers, for example, may yield an external benefit in the form of reduced crime, and this benefit should be considered by society at large even though it may not enter the calculations of individuals doing the investment.

A further distinction can be made between real costs and benefits as opposed to pecuniary or distributional or transfer costs and benefits. Real costs involve the use of real resources, and should be considered whether those resources have a monetary value or not. Pecuniary or transfer costs and benefits do not involve the use of real resources, but rather involve a transfer from one group to another: some gain while others lose. While it may be important to note the existence of such transfers for specific groups, it is inappropriate to include them in the calculation of social costs and benefits since, by definition, gains by one party involve losses by another. For example, the savings in unemployment insurance or social assistance payments that may result from a retraining program are worthy of noting, and for the unemployment insurance fund they may be a private saving, yet from the point of view of society they represent a reduction in a transfer payment, not a newly created real benefit.[1] Similarly, the installation of a retraining facility in a community may raise local prices for construction facilities, and this may be an additional cost for local residents; yet it is a pecuniary cost since it involves a transfer from local residents to those who raised the prices. While such a transfer may involve a loss to local residents, it is not a real resource cost to society as a whole, since it represents a gain for other parties.

From the point of view of the efficient allocation of resources, only real resource costs and benefits matter. Transfers represent offsetting gains and losses. However, from the point of view of distributive equity or fairness, society may choose to value those gains and losses differently. In addition, costs and benefits to different groups may be valued differently in the economic calculus.

Thus, in the calculation of the benefits from a training program, it is conceivable to weigh the benefits more for a poor disadvantaged worker than an advantaged worker. The appropriate weighting scheme obviously poses a problem, but it could be based on the implicit weights involved in other government programs or perhaps in the progressive income tax structure, or it could simply be based on explicit weights that reflect a pure value judgement.

Care must also be exercised in imputing a macroeconomic impact from investment programs. It is often tempting, for example, to multiply the benefits of some program to capture the multiplier effect as the investment sets up further rounds of spending throughout the economy. Or it is tempting to document the employment expansion that may accompany a particular investment program. The error in this reasoning occurs because it ignores the fact that the opportunities forgone, as resources were devoted to this particular investment rather than to some other one, also have a multiplier and employment creation effect. It is true that the multiplier effects may be different in magnitude (for example, if they involve different leakages into imports), and they may occur in different regions. While these factors may be worthy of note, they do not justify the imputation of a multiplier effect for all human capital investments.

EDUCATION

The main elements of human capital theory can be outlined by considering decisions relating to investment in education. This decision is illustrated in Figure 18.1 which shows alternative income streams associated with different levels of education: incomplete high school (10 years of education at age 16), high school completion (age 18), and university or college degree (age 22).

---

1. This discussion ignores the possible real resource costs involved in operating and financing unemployment insurance or social assistance programs. If savings in unemployment insurance or social assistance payments reduce the real resources required to operate and finance these programs, then a real externality is involved in addition to the transfer externality.

Figure 18.1
EDUCATION AND ALTERNATIVE INCOME STREAMS

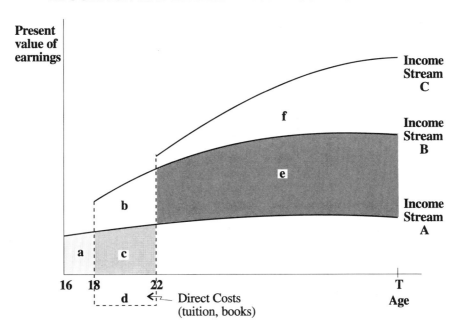

These three outcomes are used for illustration only; in general we may regard "years of education" as a continuous variable, each year being associated with a lifetime income stream. The earnings in each year are measured in present value terms to make them comparable across different time periods.

The shapes of the earnings streams (or "age-earnings profiles") reflect two key factors. First, for each profile, earnings increase with age but at a decreasing rate. This concave shape reflects the fact that individuals generally continue to make human capital investments in the form of on-the-job training and work experience once they have entered the labour force. This job experience adds more to their productivity and earnings early in their careers due to diminishing returns to experience. Second, the earnings of individuals with more years of education generally lie above those with fewer years of education. This feature is based on the assumption that education provides skills which increase the individual's productivity and thus earning power in the labour market. Because of the productivity-enhancing effect of work experience, individuals with more education may not begin at a salary higher than those in their age cohort with less education (and therefore more experience). Nonetheless, to the extent that education increases productivity, individuals with the same amount of work experience but more education will earn more, perhaps substantially more.

Which lifetime income steam should the individual choose? To address this question we will initially make several simplifying assumptions:

1. the individual doesn't receive any direct utility or disutility from the educational process;

2. hours of work (including work in acquiring education) are fixed;

3. the income streams associated with different amounts of education are known with certainty;

4. individuals can borrow and lend at the real interest rate r.

These assumptions are made to enable us to focus on the salient aspects of the human capital investment decision. The first assumption implies that we are examining education purely as an investment, not a consumption, decision. The second assumption implies that the quantity of leisure is the same for each income stream, so that they can be compared in terms of income alone. Assumption three allows us to ignore complications due to risk and uncertainty. The fourth assumption — often referred to as "perfect capital markets" — implies that the individual can base the investment decision on total lifetime income, without being concerned with the timing of income and expenditures. The consequences of relaxing these simplifying assumptions are discussed below.

In these circumstances, the individual will choose the quantity of education that maximizes the net present value of lifetime earnings. Once this choice is made, total net lifetime earnings (or human capital wealth) can be distributed across different periods as desired by borrowing and lending.

As illustrated in Figure 18.1, human capital investment involves both costs and benefits. The costs include both direct expenditures such as tuition and books and opportunity costs in the form of forgone earnings. For example, in completing high school the individual forgoes earnings equal to the shaded area 'a' associated with income stream A. The benefits of completing high school consists of the difference between earnings streams A and B, equal to the areas 'b + e' in Figure 18.1. For a high school graduate contemplating a university education, the additional costs include the direct costs (area 'd') and forgone earnings equal to area 'b + c', while additional benefits equal the earnings associated with income stream C rather than B (area 'f'). As Figure 18.1 is drawn, a university education yields the largest net present value of lifetime income. However, for another individual with different opportunities and abilities, and therefore different income streams, one of the other outcomes might be best.

The rule for optimal human capital investment can be expressed in a number of alternative, but equivalent, ways. Two common ways of stating this decision rule are illustrated in Figure 18.2. One is in terms of marginal rather than total costs and benefits: the individual should increase years of education until the present value of the benefits of an additional year of education equals the present value of the additional costs. In terms of Figure 18.1, for a high school graduate facing income stream B, the marginal benefits of a university education

Figure 18.2

(a) Marginal Benefits Equal Marginal Costs

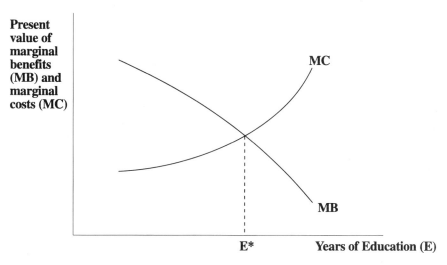

(b) Internal Rate of Return Equals Market Interest Rate

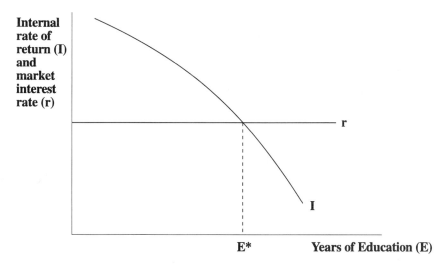

consist of the area 'f' while the marginal costs consist of the area 'b + c + d'. Marginal benefits generally decline with years of education due to diminishing returns to education and the shorter period over which higher income accrues. Marginal costs rise with years of education because forgone earnings increase with educational attainment. The point at which marginal benefits equal marginal costs yields the maximum net present value of lifetime earnings.

Human capital decisions, like those involving financial and physical capital, are also often expressed in terms of the rate of return on the investment. For any specific amount of education, the internal rate of return (i) can be defined as the implicit rate of return earned by an individual acquiring that amount of education. The optimal strategy is to continue investing as long as the internal rate of return exceeds the market rate of interest r, the opportunity cost of financing the investment. That is, if at a specific level of education i > r, the individual can then increase the net present value of lifetime earnings by acquiring more education, which may involve borrowing at the market interest rate r. Similarly, if i < r, the individual would increase lifetime earnings by acquiring less education. Because the present value of marginal benefits and marginal costs are generally declining and increasing functions, respectively, of years of education, the internal rate of return falls as educational attainment rises. The point at which i = r yields the optimal quantity of human capital.

Perhaps the most obvious implication of the theory is that human capital investments should be made early in one's lifetime. Educational investments made at later stages earn a lower financial return because forgone earnings increase with work experience and because of the shorter period over which higher income is earned. A related implication is that individuals who expect to be in and out of the labour force — perhaps in order to raise children — have less financial incentive to invest in education and will (other factors being equal) earn a lower return on any given amount of human capital investment.

This framework can be used not only to explain human capital investment decisions but also to predict the impact of changes in the economic and social environment and in public policy on levels of education. For example, changes in the degree of progressivity of the income tax system are predicted to alter levels of educational attainment. Because optimal human capital investment decisions are based on real after-tax income, an increase in the progressivity of the income tax system would shift down high income streams (such as C in Figure 18.1) relatively more than low income streams (such as A), thus reducing the demand for education. Similarly, policies such as student loans programs alter the total and marginal costs of education, and thus levels of educational attainment.

Not all individuals have or obtain sufficient information to make the detailed calculations needed to determine the optimal quantity of education. Nonetheless, people do take into account costs and benefits when making decisions, including those with respect to human capital investments. Consequently, as is frequently the case in economic analysis, models that assume rational decision-making may predict the behaviour of individuals quite well, especially the average behaviour of large groups of individuals. Optimization errors that result in a specific individual's choice of education deviating from the optimum level tend to offset each other and thus may have little effect on the average behaviour of large groups of individuals.

Decisions relating to investment in education are also complicated by the fact that the simplifying assumptions used in the above analysis may not hold in practice. The process of acquiring education may directly yield utility or disutility. The existence of this consumption component does not imply that the investment aspect is irrelevant; however, it does indicate that human capital decisions may not be based on investment criteria alone. Individuals who enjoy learning will acquire more education than would be predicted on the basis of financial costs and benefits, and vice versa for those who dislike the process of acquiring knowledge. The decision continues to be based on costs and benefits, but these concepts need to be broadened to include nonfinancial benefits and costs.

In addition to increasing one's future earnings, education may open up a more varied and interesting set of career opportunities, in which case job satisfaction would be higher among those with more education. The consequences may be even more profound; for example, the acquisition of knowledge may alter peoples' preferences and therefore future consumption patterns, possibly enhancing their enjoyment of life for a given level of income. In principle these aspects — to the extent that they exist — can be incorporated in the theory, but they clearly present challenges for measurement and empirical testing. Similarly, the returns to education are unlikely to be known with certainty so that investment decisions must be based on individuals' expectations about the future. Because some alternatives may be less certain than others, attitudes toward risk will also play a role. Risk-neutral individuals will choose the amount of education that maximizes the expected net present value of lifetime earnings, while risk-averse individuals will place more weight on expected benefits and costs that are certain than those that are uncertain.

Financing is generally an important aspect of any investment decision. In the case of human capital investments financing is particularly problematic because one cannot use the value of the human capital (i.e., the anticipated future earnings) as collateral for the loan. In contrast, machinery and equipment, land, and other physical assets can be pledged as loan collateral. There is therefore a fundamental difference between physical and human capital in terms of the degree to which "perfect capital markets" prevail. In the absence of subsidized tuition, student loan programs, and similar policies, the problems associated with financing human capital investments could prevent many individuals from choosing the amount of education that would maximize their net present value of lifetime earnings. Even in the presence of these policies, borrowing constraints may exert a significant influence on decisions regarding education.

This discussion of human capital theory has focused on the private costs and benefits of education because these are the relevant factors affecting choices made by individuals. However, the acquisition of knowledge may also affect third parties, in which case the social costs and benefits may differ from their private counterparts. These issues are discussed further below in the context of public policy towards education.

Education and Market Equilibrium

The income streams shown in Figure 18.1 represent hypothetical alternatives available to a given individual. In order to understand the relationship between income and education we need to study the interaction between individuals' preferences regarding different levels of education and employers' preferences with respect to workers with different amounts of human capital. This interaction determines market wage rates associated with different levels of education and employment of workers with various amounts of human capital. The nature of the market equilibrium is very similar to that analyzed in the previous chapter in terms of compensating wages for risk, the difference being that human capital is a positive attribute affecting the productivity of workers while risk is a negative characteristic associated with jobs. Because of the similarity, our treatment in this chapter will be brief.

Figure 18.3 shows indifference curves for two different types of workers (A and B workers) and isoprofit curves for two different types of firms (α and ß firms). Two types of each group are assumed purely for illustration. Employees differ in their preferences for education versus income because of such factors as differences in tastes for acquiring knowledge, learning ability, utility or disutility derived directly from the educational process, and need or ability to borrow to finance human capital investments. Because human capital is costly to acquire, both types of employee require a higher wage to induce them to obtain more education. Type A workers have a stronger preference for education, as reflected in the shape of their indifference curves U_A. Holding utility constant, a larger wage increase is required to induce B workers to undertake an additional year of education than is the case for A workers.

Firms' isoprofit curves will be upward-sloping, as depicted in Figure 18.3, if workers with more human capital are more productive. Holding profits constant, a higher wage can be paid to more educated employees due to their greater productivity. (Otherwise the isoprofit curves would be horizontal straight lines.) However, firms differ in the value of more educated workers to the production and marketing of their product. More educated workers are more valuable to the type α firms as reflected by the shape of their isoprofit curves I_α. Holding profits constant, α firms can pay a larger wage increase to workers with an additional year of education than is the case for ß firms.

In competitive equilibrium firms earn zero excess profits (normal rate of return). Thus if I_α and $I_ß$ are the isoprofit curves corresponding to zero profits, the outer limits of these two isoprofit curves (bold lines) constitute the employers' offer curve showing the maximum wage that can be paid for different levels of education consistent with competitive equilibrium. Market equilibrium

Figure 18.3
EDUCATION AND MARKET EQUILIBRIUM

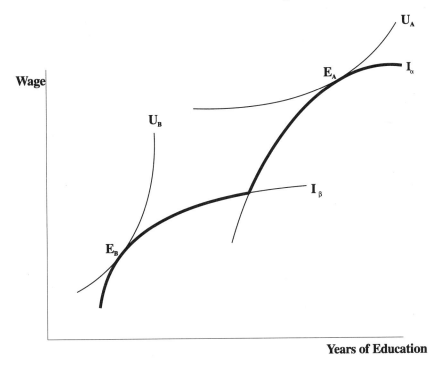

occurs at points E_A and E_B. This equilibrium has two important properties. First, workers who invest more in human capital receive a compensating differential for doing so. The magnitude of the compensating differential depends on both the preferences of different types of workers regarding income and education and the technology of production and market opportunities of firms. Second, employees with the strongest preferences for education are matched with employers for whom education is most valuable. This sorting equilibrium is Pareto-optimal. Both A and B types of workers are better off employed with the α and ß firms respectively than would be the case if they switched to the other firm type.

Empirical Evidence

Figure 18.4 shows age-earnings profiles for four educational categories of Canadian males: (a) eight years or less of elementary schooling (ELEM), (b) nine to thirteen years of elementary and secondary schooling (HS), (c) some post-secondary education but not a university degree (Some Post-HS), and (d) university degree (UNIV). Studies of the relationship between education and

Figure 18.4
EARNINGS BY AGE AND EDUCATION, CANADIAN MALES, 1979/81

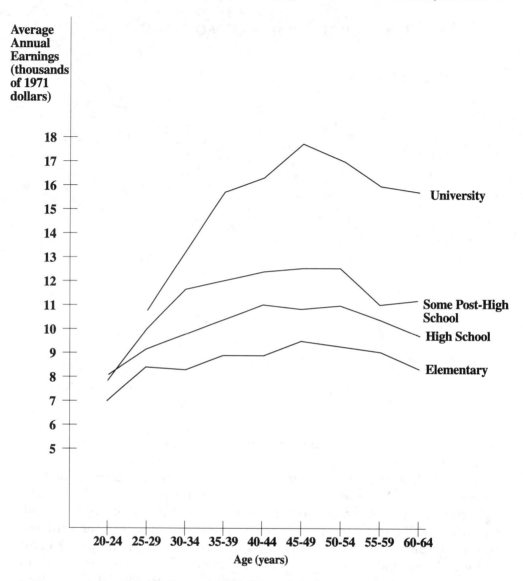

1. Earnings are average annual earnings of full-year, full-time males in 1971 dollars. Full-year, full-time workers are defined as those who worked 50 or more weeks and usually worked full weeks. Earnings are the average of 1979 and 1981 earnings.
2. Education categories are defined as follows:
   (a) ELEM: eight years or less of elementary schooling;
   (b) HS: nine to thirteen years of elementary and secondary schooling;
   (c) Some Post-HS: some post-secondary education, but not a university degree;
   (d) UNIV: university degree.

Source: M.D. Dooley, "The overeducated Canadian? Changes in the relationship among earnings, education, and age for Canadian men: 1971–81." *Canadian Journal of Economics* 19 (February 1986) 142–159.

earnings are often confined to males because of the empirical problems associated with the more intermittent labour force participation of females.

As these data indicate, there is a strong relationship between education and lifetime earnings. The income streams of those with more education generally lie above the streams of those with less education. Two additional patterns are evident. First, earnings increase with age and thus labour market experience until age 40 to 50 and then decline somewhat. As noted previously, this concave relationship between age and earnings is generally attributed to the accumulation of human capital in the form of on-the-job training and experience, a process which displays diminishing returns. Second, earnings increase most rapidly to age 40 or 50 for those individuals with the most education. Thus the salary differential between groups with different amounts of education is much wider at ages 40 to 50 than at ages 20 to 30.

Data on earnings by age and education can be used together with information on direct costs to calculate the rate of return on investments in education. Such calculations can be useful to individuals wishing to know, for example, whether a university education is a worthwhile investment. They can also be useful input into public policy decisions. In particular, efficient resource allocation requires that investments in physical and human capital be made in those areas with the greatest return.

Estimates of the rate of return to education are generally obtained by comparing different individuals at a point in time. After controlling for other observed factors that influence earnings, the differences in individuals' earnings are attributed to differences in educational attainment. However, there remains uncertainty regarding whether the differences in earnings may be due to differences in unobserved factors that influence earnings and which are related to the amount of education chosen. In particular, if those with greater earning ability — perhaps due to innate ability or to family background — choose higher levels of education, their earnings may have been greater even if they had not obtained more education. In these circumstances, some — perhaps all — of the measured return to education may be a return to unobserved ability. Later in this section we discuss empirical research which has addressed this difficult question of attempting to measure the pure impact of education on earnings.

Returns to Education During the 1960s and 1970s

Studies of the monetary returns to investments in education during the late 1960s and early 1970s generally concluded that rates of return were substantial, especially for an undergraduate university degree or less (Stager, 1972). Private after-tax returns were usually estimated to be in the 10–15 percent range — an excellent investment return, given that these are real (i.e., after taking inflation into account) after-tax returns. There is some evidence that rates of return increased during the 1960s. Mehmet's (1977) estimates indicate a significant increase in the rate of return to 22 percent in 1969 followed by a modest decline to 18 percent in 1971. During the 1970s the returns to education declined. Dooley (1986) examined changes in the relationship between earnings and education between the years 1971–73 and 1979–81 and found that the largest increases in earnings occurred for those groups with the least education. For those with some post-high school, earnings were approximately unchanged between 1971–73 and 1979–81, while for those with a university degree, earnings actually declined slightly. Using 1981 data, Vaillancourt and Henriques (1986) estimate private after-tax rates of return to university education in Canada ranging from 7 to 14 percent, depending on the region and length of degree program (three or four years). These estimates also indicate that monetary returns to education declined significantly during the 1970s.

In the U.S., the financial return to a university or college degree also declined during the 1970s. One explanation of this phenomenon was the substantial increase in the proportion of the population going to university, particularly the entry into the labour force of the "baby-boom" generation during the 1970s (Welch, 1979). Freeman (1976, 1980) argued that the demand for educated workers also declined, so that not all of the change in relative earnings could be attributed to temporary developments on the supply side.

Dooley (1986) examined these competing explanations in the Canadian setting using data for the period 1971–81. He concluded that the entry of the large baby-boom cohort during this period did lower earnings growth for this group, but that this demographic effect could not account for the observed narrowing of earnings differentials by level of education. Dooley's results thus suggest that demand-side forces may also have played a role.

Increasing Returns to Education in the 1980s

The decline in the monetary return to education during the 1970s — which gave rise to the phrase "overeducated American" — was reversed in the 1980s. This reversal was especially sharp in the United States, but also occurred — albeit to a lesser extent — in Canada (Freeman and Needels, 1991). The widening of earnings differences between the more educated and the less educated is part of a broader phenomenon of rising earnings inequality in general, a phenomenon which is discussed in more detail in Chapter 20. Empirical studies generally attribute the increased return to education to both demand and supply factors (Katz and Murphy, 1992; Murphy and Welch, 1992). On the demand side, the shift in employment out of heavy manufacturing industries and toward financial and business services has brought about a decline in demand for labour in semi-skilled and unskilled blue-collar jobs and an increase in demand for skilled, educated workers in the service sector. On the supply side, there were fewer educated baby-boomers graduating and entering the labour force during the 1980s.

In their comparative Canada–U.S. study, Freeman and Needels (1991) find that the returns to higher education did not increase as much in Canada as in the United States because of more rapid growth in the supply of university graduates in Canada compared to the U.S. In other words, the demand for more educated workers increased in both countries, but the greater supply response in Canada kept the relative earnings of the more educated from rising as quickly. In both countries industrial restructuring put downward pressure on the earnings of those with the least skill and education. Freeman and Needels find some evidence that less educated low-wage workers in the United States were more adversely affected by such forces as declining unionism and increased foreign competition than were their counterparts in Canada.

Table 18.1 shows estimates of the monetary return to education in Canada as of 1985 (based on data from the 1986 Census). Like most such estimates, these are obtained by comparing the earnings of individuals with different levels of education at a point in time, rather than following the same individuals over time. Other factors which might also account for earnings differences across individuals are taken into account using multi-variate regression analysis. The estimates shown are the private after-tax rates of return to the individual, taking into account such costs as tuition fees and forgone earnings.

Note that the rates of return generally decline with the amount of education already obtained, as would be expected from the basic economic theory of human capital (e.g., as shown in Figure 18.2). The highest returns come from completing high school, followed by incomplete high school (compared to only elementary education), and completion of a university degree (compared to completion of high school). As of the mid-1980s, females generally benefited more from additional education than males, a result that is consistent with the general finding that the gap between male and female earnings is largest at low levels of education and least at high levels of education.

Although rates of return to education are generally high, there are also large differences in these rates of return by the type of education obtained. The bottom part of Table 18.1 illustrates these differences for fields of study of Bachelor's degree holder. Those obtaining degrees in Medicine, Dentistry, and related fields earn the highest monetary return, while those graduating in Humanities and Fine Arts earn the lowest returns.

In concluding this discussion of the empirical evidence relating to education and earnings, several observations should be made. The returns to education discussed here are restricted to the private monetary benefits. Additional benefits to the individual — such as any enjoyment derived directly from acquiring knowledge, a more varied and interesting career, or even an enhanced ability to enjoy life — would increase the private returns to education. Further, the social returns to education may differ from the private returns for a variety of reasons. It is also important to note that estimates usually provide the *average* rate of return for all those making a particular educational investment. Policies such as

**Table 18.1
ESTIMATES OF THE PRIVATE RETURN TO
SCHOOLING IN CANADA, 1985**

LEVEL OF SCHOOLING	MALES	FEMALES
Secondary (incomplete)	20.7	18.6
Secondary (complete)	33.4	38.5
Community College or CEGEP	6.6	17.3
Bachelor's degree	8.3	18.8
Master's degree	6.5	0.1
Ph.D.	1.2	16.3

BACHELOR'S DEGREE BY FIELD OF STUDY	MALES	FEMALES
Education	9.8	16.3
Humanities and Fine Arts	0.7	5.5
Social Sciences	10.8	16.3
Commerce	19.6	23.9
Natural Sciences	10.6	16.3
Engineering and Applied Science	23.0	16.0
Health Sciences	9.2	26.6
Medicine, Dentistry, Optometry, Veterinary Medicine	30.8	28.8

Notes:
Rates of return by level of schooling are calculated relative to the next lowest level. For example, the return to Secondary (incomplete) is relative to Elementary, and the return to a Master's degree is relative to a Bachelor's degree.

Source: Vaillancourt, F. "Private and Public Monetary Returns to Schooling in Canada, 1985," Economic Council of Canada Working Paper No. 35, 1992. EC 26-1-35E. Reproduced with the permission of the Minister of Supply and Services Canada, 1993.

those relating to the allocation of resources should be based on *marginal*, not average, calculations; that is, for social efficiency funds should be allocated among various physical and human capital investments such that the social rate of return on the last dollar invested in each project is equal.

The calculated returns to education are based on cross-sectional data (different individuals at a point in time). This procedure will be accurate if the age-earnings profiles for each educational category are approximately constant over time, apart from overall earnings growth affecting all groups. To the extent that earnings differentials by education narrow or widen over time, the actual realized return to educational investments will be smaller or larger than that estimated on the basis of cross-sectional data.

Signaling, Screening, and Ability

The empirical analysis of the relationship between education and earnings is usually carried out by estimating an "earnings function" in which the dependent variable is (the logarithm of) earnings and the explanatory variables are various determinants of earnings such as age, experience, marital status, hours worked, and educational attainment (often measured by years of education). One potential determinant that is difficult to control for is ability, by which we mean ability in the workplace, not learning ability (though these may be correlated). If more able individuals are also more likely to invest in education, some of the estimated return to education may in fact be a return to innate ability. In other words, those who are more able would earn more even in the absence of education; we may be incorrectly attributing their higher earnings to education rather than to their innate ability.

A theoretical rationale for the potential importance of "omitted ability bias" is the hypothesis that higher education may act as a filter, screening out the more able workers rather than enhancing productivity directly. According to this "signalling/screening hypothesis," workers may use education to signal unobserved ability while firms use education to screen workers (Arrow, 1973; Spence, 1973, 1974). Such an outcome can arise when employers can't observe ability directly (perhaps even some time after hiring) and when the costs of acquiring education are lower for those workers with greater ability; otherwise, both able and less able workers will invest in education and employers' beliefs about the correlation between ability and education won't be confirmed. In these circumstances, education may yield a private return (more able individuals can increase their earnings by investing in education) but its social return consists only of its role in sorting the more from the less able rather than directly increasing productivity as in human capital theory.

The appendix to this chapter outlines a simple model in which education acts as a signalling or filtering device. In the equilibrium of this model, workers who obtain more education are more productive and receive higher earnings. Yet, by assumption, education does not affect worker productivity. Thus the signalling/screening hypothesis represents a potentially significant challenge to human capital theory, which attributes the higher earnings of the more educated to the productivity-enhancing effects of education.

Empirical tests of the signalling/screening hypothesis have not been conclusive (Riley, 1979; Lang and Kropp, 1986). Most would agree that the pure signalling model in which education has no impact on productivity does not appear capable of explaining observed behaviour (Rosen, 1977). Professional educational programs such as those in medicine, law, and engineering clearly are more than elaborate screening devices. However, educational programs may act as a filter to some degree. Furthermore, as discussed by Davies and MacDonald (1984), the informational role of education in terms of matching individuals' interests and abilities may be significant, albeit difficult to measure.

The best way to determine whether education does in fact enhance productivity and thus earnings would be to conduct an experiment. Different groups of individuals would be randomly assigned to different levels (and possibly types) of education, independent of their ability, family background, and other environmental factors. At a later date, the incomes of these groups would be compared. Because of the random assignment to groups, on average the only differences in earnings among the different groups would be due to the different levels and types of education. In the absence of such experimental evidence, economists seeking to obtain convincing evidence relating to the relationship between education and earnings have tried to find "natural experiments" which isolate the influence of education alone from the possible effects of unobserved ability factors.

One approach is to obtain data on identical twins who are genetically identical and have similar family backgrounds. Behrman, Hrubec, Taubman and Wales (1980) use a sample of male identical twins who were veterans of World War II. In their data, the simple relationship between education and income indicates that each additional year of schooling adds about 8 percent to annual earnings. However, when attention is focused on twins alone (i.e., the relationship between differences in education and differences in earnings for pairs of twins), the estimated return to an additional year of education falls to 2 to 2.5 percent. These estimates suggest that omitted ability may account for much of the estimated return to education. However, in this type of analysis measurement error in the amount of education obtained will result in downward bias in the returns to additional schooling. Thus considerable uncertainty about the true impact of education remained following the Behrman et. al. study.

More recently, Ashenfelter and Krueger (1992) obtained data on a sample of identical twins. At the same time, they obtained an independent source of information on education level in order to minimize the possible influence of measurement error. Their results show high returns to education: on average an additional year of education increases earnings by about 14 percent.

Another study uses the "natural experiment" associated with compulsory school attendance laws (Angrist and Krueger, 1991). Such laws generally require students to remain in school until their sixteenth or seventeenth birthday. However, because children born in different months start school at different ages, compulsory attendance laws imply that some children are required to remain in school longer than others. Of course, for those who remain in school longer than the minimum required period, such as those who obtain some post-secondary education, these laws will not influence the amount of education obtained.

However, for those who wish to leave school as soon as possible, compulsory attendance laws require students born in certain months to remain in school longer than those born in other months. Because month of birth is unlikely to be correlated with ability or family background, any variation in educational attainment associated with compulsory school attendance laws is likely to be randomly distributed in terms of ability and environmental background. Angrist and Krueger find that season of birth is indeed related to educational attainment in the U.S.; in particular, those born early in the year (and who therefore attain the legal drop-out age earlier in their education careers) have a slightly lower average level of education than those born later in the year. Furthermore, those who attend school longer because of compulsory schooling laws receive higher earnings. They estimate the impact of an additional year of school (due to compulsory attendance requirements) on the earnings of males to be an increase in earnings of 7.5 percent. Because ability is unlikely to be related to month of birth, this estimate should be free of any bias associated with unobserved ability. Of course, this study provides evidence regarding the relationship between schooling and earnings for levels of education around that of high school completion. Other natural experiments will be needed to obtain similar evidence for post-secondary education.

In summary, recent empirical studies utilize situations in which there are different levels of educational attainment for groups of individuals who should, on average, have the same level of unobserved ability. This evidence, although limited, provides convincing support for the basic assumption of human capital theory — that education enhances productivity and therefore earnings.

TRAINING

Like education, training is a form of investment in human capital in which costs are incurred in the present in the anticipation of benefits in the future. The benefits accrue because training imparts skills which raise the worker's productivity and thus value in the labour market.

In this section, we focus on some *economic* aspects of training rather than on an institutional description of training in Canada: the latter is discussed, for example, in Davies (1986) and Economic Council of Canada (1992). The main focus of our analysis is to shed light on the following questions: Who pays for training? Is a government subsidy warranted? How should training be evaluated?

Who Pays?

In his classic work on the subject, Becker (1964, pp. 11–28) distinguishes between *general* training and *specific* training. General training is training that can be used in various firms, not just in firms that provide the training. Consequently, in a competitive market, firms will bid for this training by offering a higher wage equal to the value of the training. Since competition ensures that the trainee reaps the benefits of general training in the form of higher earnings, then the trainee would be willing to bear the cost of training as long as benefits exceed costs. If a company were to bear the cost of such training they would still have to bid against other companies for the services of the trainee.

This argument is illustrated in Figure 18.5(a). In the absence of training, the individual can earn the alternative wage W_a equal to the value of marginal product without training (VMP_a). During the training period the value of the worker's output is VMP_t (which could be zero). After training the worker's value to *any* firm in this labour market rises to VMP^*. The costs and benefits are as shown. If the investment is worthwhile, the employee can finance the training and earn the benefits by being paid a wage equal to the VMP at each point in time; i.e., the worker receives $W_t = VMP_t$ during training and $W^* = VMP^*$ after the training period. The firm could incur the costs and hope to reap the benefits by paying the worker W_a before and after training. However, because the employee can earn W^* elsewhere, the firm's strategy won't work. Thus, in the absence of bonding arrangements (such as are used for limited periods in the armed forces for certain types of general training such as pilot training), general training will be financed by employees.

With specific training, however, the training is useful only in the company that provides the training. Consequently, other companies have no incentive to pay higher earnings for such training and the trainee would not bear the cost because of an inability to reap the benefits in the form of higher earnings. The sponsoring company, however, would bear the costs providing they exceed the benefits. In addition, the sponsoring company would not have to pay a higher wage for those persons with specific training since other firms are not competing for such trainees.

This case is also illustrated in Figure 18.5(a). The firm pays the alternative wage W_a during and after training, incurring costs of $W_a - VMP_t$ during the training period and receiving benefits of $VMP^* - W_a$ after the completion of training.

Specific human capital may also be a shared investment. This arrangement is particularly likely when there is some uncertainty about the continuation of the employment relation due to shifts in labour demand affecting the worker's VMP and shifts in supply affecting the alternative wage W_a. If the sponsoring company pays for the specific training, as shown in Figure 18.5(a), it faces the risk that the employee may quit at some point after the training period, thus reducing the employer's anticipated return on the investment. Because the worker is receiving no more than the alternative wage, the cost of quitting is low. Even a small increase in W_a could cause the worker to go elsewhere. However, the cost to the company is high because the trained employee is worth more to the firm than he or she is being paid. In these circumstances, the sponsoring company may pay trainees a wage premium to reduce their turnover and hence to increase the

Figure 18.5
COSTS, BENEFITS, AND FINANCING OF TRAINING

(a) Costs and Benefits of Training

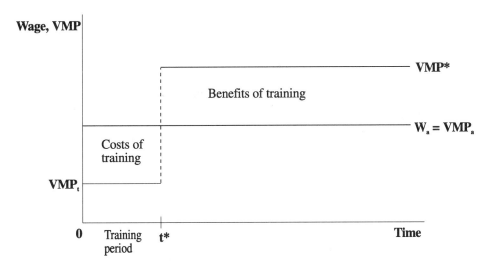

(b) Specific Training as a Shared Investment

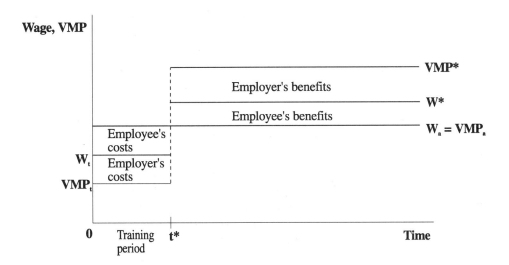

Figure 18.5 (continued)

(c) Earnings Growth with Gradual Training

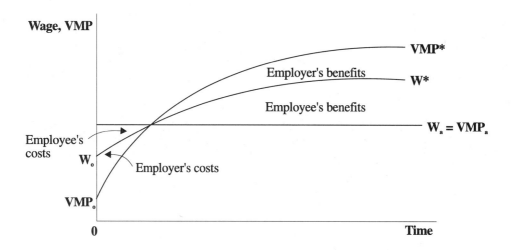

probability that the company will recoup its investment costs. To compensate for the wage premium, the firm may lower the wage paid during the training period, in which case the two parties share the costs and benefits of training.

This situation is shown in Figure 18.5(b). The firm pays $W_t < W_a$ during the training period and $W^* > W_a$ after training. Both parties incur costs and reap benefits as shown. The shared investment minimizes the risk that either party will wish to terminate the employment relation because both the employer and employee are earning rents after the completion of training. The employer's rents of $VMP^* - W^*$ each period reduce the risk that the employee will be laid off due to a decline in demand. Only when the worker's value to the firm falls below W^* will the firm consider layoffs. The employee's rents of $W^* - W_a$ each period reduce the risk that the employee will quit in response to an improvement in labour market opportunities. Only when the alternative wage rises above W^* will the employee consider quitting.

This analysis indicates that specific human capital investments provide an incentive for firms and workers to maintain their employment relationship in the face of external shocks to demand and supply. The return on the shared investment acts like "glue" keeping the two parties together. These incentives for long-term employment relationships and their consequences are discussed further in Chapter 13 in the context of implicit contracts, in Chapter 23 in the context of deferred compensation, and in Chapter 25 in the context of layoffs and unemployment.

In competitive markets, then, trainees will pay for general training whereas specific training investments may be paid by the sponsoring company or shared by the two parties. The form of payment may be subtle, as, for example, when trainees in an apprenticeship program forgo earnings by accepting a lower wage rate during the training period, or when companies forgo some output from workers when they provide them with on-the-job training.

The sharp distinction made in panels (a) and (b) of Figure 18.5 between the training and post-training periods may not hold in practice. Much on-the-job training is informal and takes place gradually as employees learn different facets of their job and its place in the overall organization. In these circumstances, earnings can be expected to rise gradually, as shown in Figure 18.5(c), rather than abruptly. The concave shape of the earnings profile reflects the assumption that there are diminishing returns to on-the-job training and work experience. As noted previously, this shape is typical of age-earnings profiles (see Figure 18.4).

In practice the distinction between general and specific training can be difficult to make. Training often contains elements of both. Even training that is geared to the specific production processes of a particular firm often contains elements that are transferable and the completion of such training can serve as a *signal* to firms that the trainee is capable of learning new skills even if the particular skills themselves are not transferable. On the other hand, general training that is provided in a particular firm may be somewhat more useful in the sponsoring company simply because the trainee is more familiar with that company. Nonetheless, the distinction between general and specific training is useful for conceptual purposes. Furthermore, although training usually contains a mixture of the two forms, occupations and industries will differ in the proportion of training that is general versus specific. Differences across occupations and industries in such factors as earnings growth and layoff and quit rates can therefore be explained in terms of differences in the amount of training and the proportions of general and specific training.

Appropriate Role of Government

If trainees will pay for and benefit from general training and sponsoring companies and employees will share the costs and benefits of specific training, why should governments be involved in the training process? In other words, are there situations when the private unregulated market does not provide a socially optimal amount of training?

This possibility may exist for trainees who cannot afford to purchase training (perhaps by accepting a lower wage during the training period) and who cannot borrow because of an inability to use their human capital (future earnings) as collateral for a loan. Subsidies to training the disadvantaged may be particularly appealing to taxpayers who prefer to support transfer programs that are associated with work activity. Concern about the working poor who work full-time but at a wage that is too low to yield a level of income above the poverty line also suggests the possibility of supporting training programs for disadvantaged workers.

The private market may also yield a less than socially optimal amount of training to the extent that training generates external (spillover, third-party) benefits that are not paid for in the market. In such circumstances firms and individuals would have no incentive to consider such benefits in their investment calculations and consequently may under-invest in training. Externalities may exist in the form of vacuum effects when trainees move up the occupation ladder and vacate a job that is filled by a member of the unemployed, or they may exist in the form of complementary multiplier effects as trainees reduce structural bottlenecks that previously resulted in the unemployment of related workers.

A sub-optimal amount of general training may also be provided by companies where on-the-job training is a natural by-product of their production process. Workers simply acquire training in their everyday work tasks. However, because it is difficult to know how much training they are acquiring and at what cost, they may be reluctant to pay for such training. In such circumstances, training may have public-good characteristics in that the training is available to all workers and yet it is difficult to exclude those who don't pay for the training. To be sure, only those who are willing to work for lower wages could be hired (and in this way non-payers are excluded). However, the indirect nature of the training makes it difficult for the purchasers to know how much training they are acquiring.

At the macroeconomic level, training may also yield benefits to the general public in the sense of helping the economy achieve such goals as growth, full employment, price stability, a viable balance of payments, and a more equitable distribution of income. As an anticyclical device to reduce the inflation-unemployment trade-off, training may be effective in absorbing some of the unemployed during a recession and in providing supplies of skilled workers when structural bottlenecks may otherwise lead to inflation.

APPENDIX: EDUCATION AS A FILTER

This Appendix outlines a simple theoretical model in which education acts as a signal or filtering device. The model is based on the seminal work of Spence (1974).

As discussed in Chapter 13, imperfect information is a common feature of many labour markets. Lack of information gives rise to at least two types of behaviour. One is search activity, whereby market participants expend time and effort to obtain information about variables of interest — such as the availability of jobs and the wages they offer — prior to making a decision. A second is signaling, which occurs when variables of interest are not observable (or are observable only at great cost) until after (perhaps a considerable amount of time after) a decision or transaction has taken place. In these circumstances, employers and employees may look for variables believed to be correlated with or related to the variables of interest. Such variables, which are observable prior to a decision or transaction being made, perform the role of being market signals. In the model described in this Appendix, education plays a role as a signal of the productivity of employees. This model is of importance in its own right because education may act, at least in part, as a signaling or sorting device and because it illustrates the more general phenomenon of signaling in labour and product markets.

In the model described here, education acts only as a signal; that is, we assume for simplicity and purposes of illustration that education has no effect on worker productivity. This assumption is made in part to keep the analysis as simple as possible, and in part to illustrate the proposition that job market signaling provides an alternative explanation of the positive correlation between education and earnings.

Employers in the model are assumed to not know the productivity of individual workers prior to hiring those workers. Even after hiring, employers may only be able to observe the productivity of groups of employees rather than that of each individual employee. However, employers do observe certain characteristics of prospective employees. In particular, they observe the amount of education obtained by the job applicant. Because employers are in the job market on a regular basis, they may form beliefs about the relationship between worker attributes such as amount of education and productivity. These beliefs may be based on the employer's past experience. In order for the employer's beliefs to persist, they must be fulfilled by actual subsequent experience. Thus, an important condition for market equilibrium is that employers' beliefs about the relationship between education and productivity are in fact realized.

If employers believe that more educated workers are more productive, they will (as long as these beliefs continue to be confirmed by actual experience) offer higher wages to workers with more education. Workers thus observe an offered wage schedule that depends on the amount of education obtained. In the model, we assume that workers choose the amount of education that provides the highest rate of return. Any consumption value of education is incorporated in the costs of acquiring education.

To keep the analysis as simple as possible, we assume that there are two types of workers in the economy. Low ability workers (Type L) have a marginal product of 1 (MP=1), and acquire y units of education at a cost of $y. High productivity workers (Type H) have a marginal product of 2, and acquire y units of education at a cost of $y/2. Note, as explained above, that the productivity or ability of workers is given and is independent of the amount of education obtained. Note also that the more able workers are assumed to be able to acquire education at a lower cost per unit of education obtained. This situation could arise because the more able workers acquire a specific amount of education more quickly, or because they place a higher consumption value on (or have a lower psychic dislike for) the educational process.

The assumption that more able workers have a lower cost of acquiring education is important. As will be seen, this is a necessary condition for education to act as an informative signal in the job market. If this condition does not hold, both low and high ability workers will acquire the same amount of education, and education will not be able to act as a signal of worker productivity.

To see what the market equilibrium might look like, suppose that employers' beliefs are as follows:

If $y < y^*$ then MP=1
If $y \geq y^*$ then MP=2

That is, there is some critical value of education (e.g., high school completion, university degree) and applicants with education less than this critical value are believed to be less productive, while applicants with education equal to or greater than this value are believed to be more productive.

In these circumstances, the offered wage schedule (assuming for the purposes of illustration that the labour market is competitive, so that firms will offer a wage equal to the expected marginal product) will be as shown in Figure 18.6. That is, applicants with education equal to or greater than y^* will be offered the wage w = \$2 and applicants with education less than y^* will be offered the wage w = \$1. In Figure 18.6 it is assumed that y^* lies between 1 and 2.

Also shown in Figure 18.6 are, for each type of worker, the cost functions C(y) associated with acquiring education. Note that low ability workers are better off by acquiring 0 units of education. This choice gives a net wage of \$1; because the cost of acquiring zero units of education is zero, the gross wage and net wage are equal in this case. In contrast, low ability workers would receive a net wage of w = \$2 $- y^* <$ \$1 if they were to acquire sufficient education to receive the higher wage offer given to those with education equal to or greater than y^*.

However, the high ability workers are better off by acquiring education level $y = y^*$. This choice yields a net wage of w = \$2 $- y^*/2 >$ \$1, whereas choosing $y = 0$ yields, for these individuals, a net wage of w = \$1.

Thus, given the offered wage schedule, if y^* lies in the range $1 < y^* < 2$, the low ability workers will choose $y = 0$ and the high ability workers will choose $y = y^*$. Thus employers' beliefs about the relationship between education and worker productivity will be confirmed. Those applicants with low education will in fact be the less productive, and those with higher education will be the more productive. Employers will therefore not have any reason to alter their beliefs, and therefore alter the offered wage schedule. Given the offered wage schedule,

Figure 18.6
OFFERED WAGE AND SIGNALING COST SCHEDULES

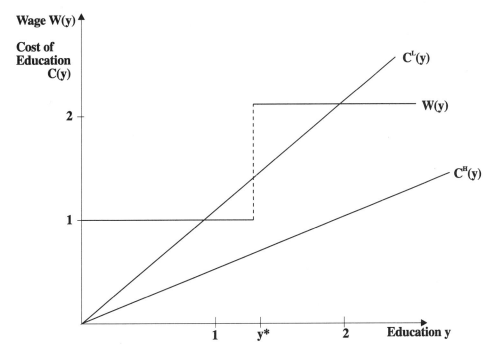

workers will continue to choose to acquire the educational "signal" such that the level of education is a good predictor (in this simple model it is a perfect predictor) of productivity. This outcome is a market equilibrium even though by assumption education does not increase the productivity of any individual worker; that is, education acts strictly as a signaling or sorting device in this case. Looked at from the outside, it might appear that education raises productivity because those with more education are more productive and receive higher earnings. However, this is not the case; education simply sorts the otherwise heterogeneous population into two distinct groups.

This simple model illustrates the central result of the theory of market signaling. This theory has been used to explain numerous other phenomena, such as the use of a high product price to signal the quality of the product, the use of product warranties to signal product quality, and the use by employers of an applicant's employment experience (e.g., number of jobs, amount of time spent unemployed) to signal worker quality.

Of course, we do not expect that education acts strictly as a filtering or signaling mechanism, as is the case in the simple model outlined in this Appendix. Most educational programs probably provide some skills and knowledge that raise the productivity of workers. However, it is possible that some forms of education or training act primarily as a signal, while other forms involve primarily human capital acquisition which raises productivity and earnings. The extent to which education serves as a signaling device versus a form of human capital acquisition is an interesting and important question.

QUESTIONS

1. Discuss the analogy between physical and human capital.

2. How would you evaluate the extent to which your acquiring a university education is a sound investment economically? Be precise in the information you would require and exactly what you would do with it.

3. You have been asked to evaluate an on-the-job training program in a particular company. Specify exactly what sort of information you require and what you would do with it.

4. The federal government supports a variety of human resource programs including education, training, mobility, labour market information, and health. Could you suggest any techniques that may be useful to suggest how resources should be allocated to the various functions?

5. Should governments subsidize human resource programs? If so, why? Be precise in your answer by indicating where, if anywhere, the private market may fail to yield a socially optimal amount of human resource development.

6. Assume that you are deciding whether or not to acquire a four-year university degree. Your only consideration at this moment is the degree as an investment for yourself. Costs per year are tuition fees $600, and books $100. The government also pays to the university an equivalent amount to your tuition fees to cover the real cost. If you didn't go to university, you would have earned $6,000 per year as an acrobat. With a university degree, however, you know that you can earn $10,000 per year as an acrobat. Because of the nature of your chosen occupation your time horizon for the investment decision is exactly 10 years after university; that is, if the investment is to be worthwhile, it must be so within a 10 year period after graduation. The market rate of interest is 5%. Would you make the investment in a degree?

7. Compare the virtues of on-the-job versus institutional training.

8. Give an example of each of the following as errors in a cost-benefit calculation: ignoring opportunity cost; failing to discount benefits; double counting; considering sunk cost with no alternative value; ignoring a real externality; considering a pecuniary externality; and ignoring consumption benefits.

REFERENCES AND FURTHER READINGS

A.   Human Capital Theory and Cost Benefit Evaluations

Additional readings in this area can be found in the bibliographies in the following entries.

Becker, G. *Human Capital.* New York: National Bureau of Economic Research, 1964.

Blaug, M. The empirical status of human capital theory. *JEL* 14 (September 1976) 827–856.

Bossiere, M., J. Knight and R. Sabot. Earnings, schooling, ability, and cognitive skills. *AER* 75 (December 1985) 1016–1030.

Bound, J., Z. Griliches and B. Hall. Wages, schooling and I.Q. of brothers and sisters: do the family factors differ? *IER* 27 (February 1986) 77–106.

Butler, R. The effect of education on wages — hedonic makes selectivity bias (sort of) simpler. *EI* 22 (January 1984) 109–120.

Coyte, P. Specific human capital and sorting mechanisms in labor markets. *SEJ* 51 (October 1984) 469–480.

Donaldson, D. and B. Eaton. Specific human capital and shared investments again. *CJE* 10 (August 1977) 474–475.

Dooley, M. The overeducated Canadian? Changes in the relationship among earnings, education, and age for Canadian men: 1971–1981. *CJE* 19 (February 1986) 142–159.

Eaton, J. and H. Rosen. Taxation, human capital, and uncertainty. *AER* 70 (September 1980) 705–715.

Garen, J. The returns to schooling: a selectivity bias approach with a continuous choice variable. *Econometrica* 52 (September 1984) 1199–1218.

Griliches, Z. Wages of very young men. *JPE* 84 (August 1976) 569–586.

Griliches, Z. Estimating the returns to schooling: some econometric problems. *Econometrica* 45 (January 1977) 1–22.

Heckman, J. and G. Sediacek. Heterogeneity, aggregation and market wage functions: an empirical model of self-selection in the labor market. *JPE* 93 (December 1985) 1077–1125.

Journal of Political Economy, Supplement, 70, Part 2 (October 1962).

Kenny, L. The accumulation of human capital during marriage by males. *EI* 21 (April 1983) 223–231.

Kiker, B. and R. Roberts. The durability of human capital: some new evidence. *EI* 22 (April 1984) 269–281.

Lee, L. Health and wages: a simultaneous equation model with multiple discrete indicators. *IER* 23 (February 1982) 199–222.

MacDonald, G. The impact of schooling on wages. *Econometrica* 49 (September 1981) 1349–1359.

MacDonald, G. A market equilibrium theory of job assignment and sequential accumulation of information. *AER* 72 (1982) 1038–1055.

McMahon, W. and A. Wagner. Expected returns to investment in higher education. *JHR* 16 (Spring 1981) 274–285.

Mincer, J. Investment in human capital and personal income distribution. *JPE* 66 (August 1958) 281–302.

Nitzan, S. and J. Paroush. Investment in human capital and social self protection under uncertainty. *IER* 21 (October 1980) 547–558.

Rosen, S. Human capital: a survey of empirical research, *Research in Labor Economics*, R. Ehrenberg (ed.). Greenwich, Conn.: JAI Press, 1977.

Rosenzweig, M., J. Morgan, A. Blinder. An exchange: On the appropriate specification of human capital models. *JHR* 11 (Winter 1976) 3–27.

Schultz, T. Investment in human capital. *AER* 51 (March 1961) 1–17.

Shaw, K. A formulation of the earnings function using the concept of occupational investment. *JHR* 19 (Summer 1984) 319–340.

Thurow, L. *Investment in Human Capital.* Belmont, California: Wadsworth, 1970.

Tomes, N. Religion and the rate of return on human capital: evidence from Canada. *CJE* 16 (February 1983) 122–138.

Tomes, N. The effects of religion and denomination on earnings and the returns to human capital. *JHR* 19 (Fall 1984) 472–488.

Vijverberg, W. Consistent estimates of the wage equation when individuals choose among income-earning activities. *SEJ* 52 (April 1986) 1028–1042.

Walsh, J. Capital concept applied to man. *QJE* (February 1935) 255–285.

Weisbrod, B. The valuation of human capital. *JPE* 69 (October 1961) 425–437.

Welland, J. Schooling and ability as earnings complements. *CJE* 13 (May 1980) 356–367.

Willis, R. Wage determinants: a survey and reinterpretation of human capital earnings functions. *Handbook of Labour Economics*, Vol. 1. O. Ashenfelter and R. Layard (eds.). New York: Elsevier, 1986.

B. Education

Albrecht, J. A procedure for testing the signaling hypothesis. *J Pub Ec* 15 (1989) 123–132.

Angrist, J.D. and A.B. Krueger. Does compulsory school attendance affect schooling and earnings? *QJE* 106 (November 1991) 979–1014.

Arrow, K.J. Higher education as a filter. *J Pub Ec* 2 (1973) 193–216.

Ashenfelter, O. and A. Krueger. Estimates of the economic return to schooling from a new sample of twins. Industrial Relations Section, Princeton University, Working Paper No. 304, July 1992.

Bartel, A. and F. Lichtenberg. The comparative advantage of educated workers in implementing new technology. *R.E. Stats.* (February 1987).

Behrman, J., Z. Hrubec, P. Taubman and T. Wales. *Socioeconomic Success: A Study of the Effects of Genetic Endowments, Family Environment, and Schooling.* Amsterdam: North Holland, 1980.

Bound, J., Z. Griliches and B.W. Hall. Wages, schooling and I.Q. Of brothers and sisters: do the family factors differ? *IER* 27 (February, 1986) 77–105.

Card, D. and A.B. Krueger. School quality and black-white relative earnings: a direct assessment. *QJE* 107 (February 1992) 151–200.

Chowdhury, G. and S. Nickell. Hourly earnings in the United States: another look at unionization, schooling, sickness, and unemployment. *JOLE* 3 (January 1985) 38–69.

Davies, J.B. Training and skill development. *Adapting to Change: Labour market Adjustment in Canada*, W.C. Riddell (ed.). Toronto: University of Toronto Press, 1986, 163–220.

Davies, J.B. and G.M.T. MacDonald. *Information in the Labour Market: Job-Worker Matching and Its Implications for Education in Ontario.* Toronto: Ontario Economic Council, 1984.

Dooley, M.D. The overeducated Canadian? Changes in the relationship among earnings, education, and age for Canadian men: 1971–1981. *CJE* 19 (February 1986) 142–159.

Economic Council of Canada. *A Lot to Learn: Education and Training in Canada.* Ottawa: Minister of Supply and Services Canada, 1992.

Ehrenberg, R.G. and D.R. Sherman. Employment while in college, academic achievement, and postcollege outcomes: a summary of results. *JHR* 22 (Winter 1987) 1–23.

Freeman, R.B. *The Overeducated American.* New York: Academic Press, 1976.

Freeman, R.B. The facts about the declining economic value of college. *JHR* 15 (1980) 124–142.

Freeman, R.B. and K. Needels. Skill differentials in Canada in an era of rising labor market inequality. National Bureau of Economic Research, Working Paper No. 3827, September 1991.

Globerman, S. Formal education and the adaptability of workers and managers to technological change. *Adapting to Change: Labour Market Adjustment in Canada*, W.C. Riddell (ed.). Toronto: University of Toronto Press, 1986, 41–70.

Griliches, Z. Estimating the returns to schooling: some econometric problems. *Econometrica* 45 (January 1977) 1–22.

Griliches, Z. and W.M. Mason. Education, income, and ability. *JPE* 80 (May/June, 1972).

Hause, J.C. Earnings profile: ability and schooling, *JPE* 80 (May/June 1972).

Harris, M. and B. Holmstrom. A theory of wage dynamics. *R.E. Studies* 44 (1982) 315–333.

Katz, L.F. and K.M. Murphy. Changes in relative wages, 1963–1987: supply and demand factors. *QJE* 107 (February 1992) 35–78.

Kodde, D.E. Uncertainty and the demand for education. *R.E. Stats.* 38 (August 1986) 460–467.

Krashinsky, M. The returns to university schooling in Canada: a comment. *CPP* 13 (June 1987) 218–221.

Lang, K. and D. Kropp. Human capital versus sorting: the effects of compulsory attendance laws. *QJE* 101 (August 1986) 609–624.

Mehmet, O. Economic returns on undergraduate fields of study in Canadian universities: 1961 to 1972. *RI/IR* 32 (1977) 321–339.

Murphy, K.M. and F. Welch. The structure of wages. *QJE* 107 (February 1992) 285–326.

Riley, J.G. Testing the educational screening hypothesis. *JPE* 87 (Supplement, 1979) S227–52.

Rumberger. R.W. The impact of surplus schooling on productivity and earnings. *JHR* 22 (Winter 1987) 24–50.

Spence, A.M. Job market signalling. *QJE* 87 (1973) 353–374.

Spence, A.M. *Market Signaling: Informational Transfer in Hiring and Related Screening Processes*. Cambridge, Mass.: Harvard university Press, 1974.

Stager, D. Allocation of resources in Canadian education. *Canadian Higher Education in the Seventies*, S. Ostry (ed.). Ottawa: Information Canada, 1972.

Stager, D. Economics of higher education: research publications in English in Canada, 1971 to 1981. *Canadian Journal of Higher Education* 12 (1982) 17–28.

Stapleton, D.C. and D.J. Young. Educational attainment and cohort size. *JOLE* 6 (1988) 330–361.

Taubman, P.J. and T.J. Wales. Higher education, mental ability and screening. *JPE* 81 (January 1973) 28–55.

Vaillancourt, F. Private and public monetary returns to schooling in Canada, 1985. Economic Council of Canada, Working Paper No. 35, 1992.

Vaillancourt, F. and I. Henriques. The returns to university schooling in Canada. *CPP* 12 (September 1986) 449–458.

Welch, F. Effects of cohort size on earnings: the baby boom babies' financial bust. *JPE* 87 (Supplement, 1979) 565–598.

Wilkinson, B.W. Elementary and secondary education policy in Canada: a survey. *CPP* 12 (December 1986) 535–572.

C. Training

Ashenfelter, O. Estimating the effect of training programs on earnings. *R.E. Stats.* 63 (February 1978) 47–57.

Barron, J.M., D.A. Black and M.A. Loewenstein. Employer size: the implications for search, training, capital investment, starting wages, and growth. *JOLE* 5 (January 1987) 76–89.

Bartel, A. and G. Borjas. Specific training and its effects on the human capital investment profile. *SEJ* 44 (October 1977) 333–341.

Block, F. (ed.). Evaluating manpower training programs. *Research in Labor Economics*, Supplement 1. Greenwich, Conn.: JAI Press, 1976.

Brown, J.N. Why do wages increase with tenure? On-the-job training and life cycle wage growth observed within firms. *AER* 79 (December 1989) 971–991.

Davies, J.B. Training and skill development. *Adapting to Change: Labour Market Adjustment in Canada*, W.C. Riddell (ed.). Toronto: University of Toronto Press, 1986, 163–219.

Donaldson, D. and B. Eaton. Firm-specific human capital: a shared investment or optimal entrapment? *CJE* 9 (August 1976) 462–472. Comment by S. Eastman and reply 10 (August 1977) 472–475.

Duncan, G. and S. Hoffman. On-the-job training and earnings by race and sex. *R.E. Stats.* (1979) 594–603.

Economic Council of Canada. *A Lot to Learn: Education and Training in Canada.* Ottawa: Minister of Supply and Services Canada, 1992.

Gunderson, M. Determinants of individual success in on-the-job training. *JHR* 8 (Fall 1973) 472–484.

Gunderson, M. Retention of trainees: a study with dichotomous dependent variables. *Journal of Econometrics* 2 (May 1974) 79–93.

Gunderson, M. Training subsidies and disadvantaged workers: regression with a limited dependent variable. *CJE* 7 (November 1974) 69–82.

Gunderson, M. The case for government supported training. *RI/IR* 29 (December 1974) 709–726.

Hashimoto, M. Firm specific human capital as a shared investment. *AER* 71 (June 1981) 475–482.

Hashimoto, M. Minimum wage effects on training on the job. *AER* 72 (December 1982) 1070–1087.

Hashimoto, M. and B. Yu. Specific capital, employment contracts and wage rigidity. *BJE* (Autumn 1980) 536–549.

Holtman, A. Joint products and on-the-job training. *JPE* 79 (July–August 1971) 929–931.

Maki, D. The direct effect of the occupational training of adults program on Canadian unemployment rates. *CJE* 5 (February 1972) 125–130.

Mincer, J. On-the-job training: costs, returns and some implications. *JPE* Supplement, 70, Part 2 (October 1962) 50–79.

Simpson, W. An econometric analysis of industrial training in Canada. *JHR* 19 (Fall 1984) 435–451.

Weiss, Y. The effect of labour unions on investment in training. *JPE* 93 (October 1985) 994–1007.

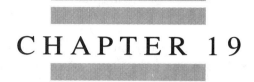

CHAPTER 19

Wage Structures:
By Occupation and Region

There are as many wage structures as there are ways of classifying workers. However, the most common structures are by occupation, region, industry, and firms, each of which also affect the wage structure across individual workers. Such wage structures are influenced by basic forces of supply and demand, or by institutional and noncompetitive factors that inhibit the forces of supply and demand. The basic determinants of each wage structure can be grouped under three categories: (1) compensating differences for nonpecuniary aspects, including possible human capital requirements; (2) short-run adjustments; and (3) noncompetitive factors.

OCCUPATIONAL WAGE STRUCTURES

As the *Canadian Classification and Dictionary of Occupations* — hereafter termed the CCDO — indicates, the term *occupation* denotes the "kind of work performed" (p.xiii); that is, "the term 'occupation' is used to refer to a number of jobs that have the same basic work content, even though they may be found in a number of different establishments or industries" (p.xv).

The occupational wage structure refers to the wage structure between various occupations or occupation groups. The occupation groups can be broadly defined, as for example, one of the 23 CCDO two-digit major groups such as managerial, clerical, sales, service, or processing. Or they can be one of the 496 four-digit codes for which Census data is often available, or a narrowly defined occupation, such as one of the 6700 seven-digit occupations, for example, typist (4113-126) or arc-welder (8335-138).

Figure 19.1 illustrates the occupational wage differential (skill differential) that exists between two hypothetical occupations, for example skilled welders and unskilled labourers. The demand schedules for each occupation are simply the aggregation of the demand schedules of the various firms that utilize each

EXHIBIT 19.1   ARE WORKERS IN THE SOVIET UNION PAID LIKE WESTERN WORKERS?

The Soviet wage system involved considerable centralized wage setting with state committees setting industry base wages. However, considerable variation is established through premiums for undesirable jobs and locations as well as for skill differentials amongst blue-collar employees and for job requirements for high-level positions. As well, local managers can respond to local labour market conditions by paying bonuses, premiums, piece rates, and reclassification payments. In essence, the centralized and decentralized mechanisms are designed to help meet planned staffing requirements and hence pay at least some attention to market focus.

However, the impact of market focus in wage determination can be blunted considerably by the fact that compensation can be based on political loyalty, with rewards, especially for higher level personnel, often in the form of privileges such as access to a car, housing space, health clinics, or special stores.

This combination of attention to market and nonmarket consideration suggests that Soviet workers may be paid somewhat like Western workers, but not completely so. This appears to be born out by data which indicate that education and experiences are recommended and that men earn more than women. However, returns to experience are lower and the occupational segregation of women is less prominent than in Western countries. As well, political loyalty is rewarded by higher compensation.

Adapted from: P. Gregory and J. Kohlhase. The earnings of Soviet workers: evidence from the Soviet interview project. *Review of Economics and Statistics* 70 (February 1988) 23–35.

type of labour. The schedules could be large or small, elastic or inelastic, depending on the circumstances.

The supply schedules for each of the occupations are upward-sloping, reflecting the fact that higher wages are required to attract additional workers into the occupations, away from other occupations, or from nonlabour market activities. The higher wages are often necessary to attract additional workers because of different worker preferences for the various occupations, or because of increasing costs associated with acquiring the skills necessary to do the work. Those who prefer the occupation or who have a natural talent for doing the work would enter at the lower wages; higher wages would be necessary to attract those who did not have a preference for the occupation or who could do the work only by a more costly acquisition of the skills.

The supply schedule, therefore, can be thought of as reflecting a ranking of actual and potential workers in the occupation, where those who most prefer the occupation or who have an innate talent for it are ranked at the beginning (and hence require a lower wage to enter), and those who least prefer the occupation or for whom the acquisition of the skills would be more costly are ranked at the

Figure 19.1
OCCUPATIONAL WAGE DIFFERENTIAL

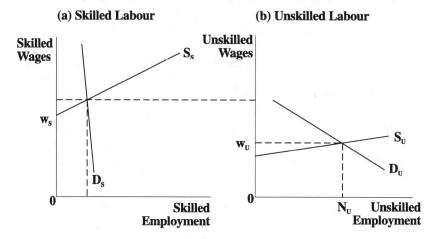

(a) Skilled Labour   **(b) Unskilled Labour**

end (and hence require the higher wage to enter). If all preferences were identical and there were no increasing costs associated with entering the occupation, then the supply schedule would be perfectly elastic at the going wage for the occupation.

The degree of elasticity of the supply schedule to each occupation also depends on the responsiveness of alternative sources of labour supply which, in turn, depends upon such factors as labour market information and mobility, training requirements, immigration, and the general state of the economy. For example, in the short run, the labour supply schedule to an occupation may be inelastic (and hence a demand increase would lead to more of a wage increase than an employment increase) for a variety of reasons: potential recruits to the occupation are not yet aware of the high wages; they have to be trained to enter the occupation; fulfilling some of the occupational demand through immigration takes time; and the labour market is presently tight, so there is no ready surplus of labour from which to draw.

In the long run, however, most of these factors would adjust to the new demands and the labour supply schedule to the occupation would be more elastic. It may not be perfectly elastic if the new sources of supply are more costly than the original sources. However, in the long run if all of the supply responses had sufficient time to adjust, and if the new sources of supply were not more costly than the original sources, then any inelasticity to the labour supply in an occupation would reflect the different preferences of workers, which would result in a need to raise wages to attract workers who do not have a strong preference for the occupation.

Reasons for Interoccupational Wage Differentials

Given the basic determinants of occupational supply and demand, economic theory predicts that the forces of competition will ensure an equal present value of net advantage at the margin in the long run across all occupations. If this does not prevail, then competition ensures that workers at the margin of decision will move from jobs of low net advantage to ones of high net advantage. This does allow for considerable variation in the occupational wage structure to reflect the main determinants of wage structures: the nonpecuniary aspects of the job (including human capital requirements); short-run adjustments; and noncompetitive forces.

For example, with respect to compensating differences for nonpecuniary aspects, occupations differ in their nonpecuniary characteristics, such as pleasantness, safety, responsibility, fringe benefits, seasonal or cyclical stability, and the certainty of their return (see Exhibit 19-2). As discussed in the previous chapter on compensating wage differentials occupational wage differentials may exist, therefore, to compensate for these nonpecuniary differences. As Adam Smith (1937, p. 100) observed over two hundred years ago:

> The five following are the principal circumstances which, so far as I have been able to observe, make up for a small pecuniary gain in some

---

EXHIBIT 19.2   STARVING ARTIST — MYTH OR REALITY?

Is there any truth to the stereotype of the starving artist — the person who gives up material gain in return for the creative rewards of an artist? In theory, this is a distinct possibility since low monetary rewards are consistent with compensating nonpecuniary aspects of a job such as fame and creative recognition. As well, artists may bequeath a legacy of art or fame to their heirs. Also, it simply may be an occupation that attracts risk-taking individuals who are willing to gamble for the "big break" that will lead to fame and fortune.

Contrary to the stereotypical view, empirical evidence indicates that artists tend to have similar earnings than otherwise comparable individuals in other occupations. However, there is more inequality within the artistic profession confirming the stereotype that some may have very low incomes while others have very high incomes. As well, it tends to be a "younger" profession, suggesting that those who don't make it often leave for other jobs.

The evidence therefore suggests that there is more myth than reality to the stereotype of the starving artist. It is true that there are a disproportionate number of artists at the low end of the earnings spectrum within the artistic profession. However, on average, artists earn about as much as otherwise comparable persons who chose more mundane jobs.

Source: R. Filer. The "starving artist" — myth or reality? Earnings of artists in the United States. *Journal of Political Economy* 94 (February 1986) 56–75.

employments, and counter-balance a great one in others: first, the agree-ableness or disagreeableness of the employments themselves; secondly, the easiness and cheapness, or the difficulty and expense of learning them; thirdly, the constancy or inconstancy of employment in them; fourthly, the small or great trust which must be reposed in those who exercise them; and fifthly, the probability or improbability of success in them.

Adam Smith's reference to the expense of learning how to do a job reminds us that part of the wage paid in a particular occupation may reflect a compensating wage premium to compensate workers for the costly acquisition of human capital necessary to do a particular job. A competitive wage premium would be one that just yields a competitive rate of return on the investment in human capital.

Individuals may also be *endowed* with skills that are valued in the market-place, in which case the wage premium that is paid to them reflects their natural embodiment of these skills rather than their costly acquisition of the skills. The individual may be endowed with a skill that is useful in a variety of occupations (e.g., strength, dexterity, or intelligence) or one that is unique to one or a few (e.g., ability to sing or put the puck in the net).

Short run adjustments factors may also affect the occupational wage struc-ture. For example, an increase in the demand for skilled workers, perhaps because of technological change, would lead to an outward shift in the demand schedule for skilled labour, as in Figure 19.1(a). In the short run, the supply schedule to the skilled occupation may be fairly inelastic because of a long training period required to do the skilled work. Thus wages would rise in this occupation and this would be a short-run signal for workers in other occupations to acquire the skills necessary to work in the skilled occupation.

In response to the short-run wage premium in the skilled occupation, new workers will enter this occupation and this increased supply will reduce the wage premium until it is just large enough to restore an equality of net advan-tage at the margin; that is, until no further gains in net advantage are possible. The new equality of net advantage may or may not be associated with the old occupational wage differential. This depends on the preferences of workers in the occupations (i.e., on the slopes of the supply schedules to each occupation). If the skilled occupation had an upward-sloping supply schedule, and therefore higher wages were necessary to attract the additional workers, the new skill differential may be greater than before. The skilled workers who were already in the occupation (intramarginal workers) would receive an additional rent because they were willing to work in the occupation at the old wage but now receive the new higher wage that was necessary to expand the number of workers in the occupation.

Noncompetitive factors can also affect the occupational wage structure. For example, occupational licensing on the part of professional associations, by regulating entry into the profession, can effectively lead to a reduction in the supply of labour to the occupation and hence artificially high salaries. Unions, especially craft unions through the hiring hall and apprenticeship requirements, can also exclude people from some occupations and crowd them into others, raising wages in the former and depressing them in the latter. Other policies may work directly on wages rather than indirectly through the supply of labour. Minimum wage legislation would affect low-wage occupations and equal pay for equal work laws would affect wages in female-dominated occupations.

Evidence on Changing Occupational Wage Structure

The occupational wage structure or skill differential in Canada narrowed between 1931 and 1951 and remained roughly constant or even widened slightly between 1951 and 1971 (Meltz and Stager, 1979). This occurred in response to changes in the basic forces of supply and demand that determine the occupational wage structure. In other words, the narrowing of the occupational wage structure until 1971 occurred because there was an increase in the demand for unskilled workers relative to skilled workers, and an increase in the supply of skilled workers relative to unskilled workers. The increase in the demand for unskilled relative to skilled workers reflects the shift from skilled craft production to less skilled work under automation and mechanization at that time. The increase in the supply of skilled workers reflects such factors as increased education and skilled restrictions on immigration. At the same time there was a "drying up" of pools of unskilled workers that had earlier been readily available from various sources: unskilled immigration, unskilled farm labour, and unskilled youths.

Since the 1970s, many of these forces have reversed, leading to a widening of the occupational wage structure or skill differential. There has been an increase in the demand for skilled workers reflecting the fact that recent technological changes are putting a premium on skills and flexibility and on the managerial, technical, professional, and administrative jobs at the top of the occupational hierarchy. Conversely, the industrial restructuring has been away from the blue-collar jobs in manufacturing, especially the heavy-manufacturing "smoke-stack" industries. As these workers have been displaced down the occupational ladder, they have put pressure on the lower-wage jobs. This pressure has been compounded by the fact that these lower-wage jobs are also subject to foreign competition from low-wage countries. As well, the supply of skilled workers has been limited somewhat by restrictions on immigration as well as a slow growth in enrolments in higher education.

These various pressures from the demand and supply side of the labour market have given rise to a widening of the occupational wage structure or skill differential since the 1970s. Evidence on this is discussed more extensively in

Chapter 20, in the last section on wage polarization and growing wage inequality across individual workers. Suffice it to say at this stage that the occupational wage structure has contributed to that growing wage inequality across individuals both because the occupational wage structure itself has widened, and because more people are in high-wage and low-wage jobs rather than the middle-wage jobs.

REGIONAL WAGE STRUCTURES AND GEOGRAPHIC MOBILITY

The existence of pronounced regional wage differentials is a well-documented fact in Canada as in most economies. As with most wage structures, however, what is less well known are the reasons for the differentials and the extent to which a pure geographic wage differential would persist in the very long run. The problem is compounded by the fact that regional wage structures tend to reflect a variety of other wage structures, notably those by industry, occupation, and personal characteristics of the workers; hence, it is difficult to isolate what could be considered a pure interregional wage differential, independent of other wage differentials.

Reasons for Regional Wage Differentials

Economic theory predicts that the forces of competition would ensure an equality of net advantage at the margin associated with identical jobs in each region. That is, if the nonwage aspects of employment (including the probability of obtaining employment) were the same, competition would ensure that wages for the same job would be the same across all regions. Competitive forces could operate in the form of workers moving from the low-wage regions to the high-wage regions, or in the form of firms moving from high-wage regions to low-wage regions to minimize labour cost. As workers leave the low-wage region, the reduced supply would raise wages, and as firms enter the low-wage region, the increased demand would also increase wages. The converse would happen in the high-wage region and the process of the mobility of labour and/or capital would continue until an equality of net advantage were restored in the long run.

To the extent that different people have different geographic preferences, the supply of labour for a given occupation in a particular region may not be perfectly elastic. That is, for a region to expand its workforce, higher wages may have to be paid to attract workers from other regions. This could obviously be the case in the short run, but it may also be the case in the long run to the extent that geographic preferences exist. In such circumstances, some workers may be receiving location rents in the sense that they would prefer to remain in a particular region even if wages were lower.

The equality of net advantage at the margin predicted by economic theory does allow for considerable variation in the interregional wage structure; however, as with interoccupational wage differentials that variation would have to reflect compensating differences, short-run adjustments, or noncompetitive factors.

The compensating differences are most obvious and they include wage compensation for such factors as cost of living, remoteness, and climate. They may even include compensation for nonpriced externalities such as pollution and congestion. Short-run interregional wage differentials could also be present and, in fact, serve as the signals that are necessary to induce the mobility that will lead to a long-run equilibrium.

Noncompetitive factors may also affect the regional wage structure. Not only is geographic mobility hindered by the barriers that arise because of the high direct and psychic costs of moving, but also it can be hindered by artificial barriers and by public policies that have an indirect, and perhaps unintentional, impact on mobility.

Occupational licensing, for example, can hinder geographic mobility. Some provinces may not recognize training done in other jurisdictions, others may have residency requirements, and others may even require citizenship. In addition, trade unions that control hiring (i.e., have the "hiring hall") can require that people who have worked in the zone be hired first, hence discouraging others from entering the zone. While these policies do achieve other objectives, they also reduce the effectiveness of interregional mobility as a force to reduce interregional wage differentials.

Social transfer programs that try to reduce income disparities will obviously have different regional impacts and this could reduce regional mobility. In such circumstances governments are faced with a basic dilemma: as they use transfer programs to reduce income inequality, they reduce part of the incentive for people to engage in regional mobility, unless the transfer programs are conditional upon geographic mobility. Regional expansion policies can encourage capital to move into low-income areas, but this also reduces the mobility of labour from that region and hence reduces some of the market forces that would reduce inequalities. Obviously these policies can achieve other important objectives — noticeably, perhaps, a more equitable distribution of regional income — but they do have the effect of reducing the effectiveness of geographic mobility as a force to reduce geographic wage differences.

Migration Decision

As the previous section indicates, economic theory predicts that the forces of competition would serve to reduce pure regional wage differentials so that they reflect compensating differences, short-run adjustments, or noncompetitive factors. Those forces of competition were the movement of capital from high- to low-wage areas, and the movement of labour from low- to high-wage areas.

This latter movement — geographic mobility or migration — can be treated as a human capital decision, the determinants of which are discussed in Chapter 18. That is, geographic mobility will occur as long as the marginal benefits exceed the marginal costs to the individuals making the decision. Benefits can include such factors as expected income (which in turn can be a function of earnings and job opportunities as well as transfer payments), climate, and the availability of social services. Costs can include the usual costs of information-seeking and moving, as well as the psychic costs associated with geographic moves. In addition the migration decision may depend on one's ability to finance the move — a problem that is particularly acute in human capital theory because one cannot legally use human capital (e.g., the larger expected earnings) as collateral to finance any necessary loan.

The human capital framework provides a variety of empirically testable propositions concerning the mobility decision. Other things being equal, relative to older workers, younger workers would tend to engage in more job mobility because they have a longer benefit period from which to recoup the costs, their opportunity cost (forgone income) of moving is lower as are their direct costs if they do not have a home to sell or family to move, and their psychic costs are probably lower because they have not established long ties in their community. In addition, they may not be locked in by seniority provisions or pensions that are not portable. Mobility should be from areas of high unemployment towards areas of low unemployment. It should also increase at the peak of a business cycle when job opportunities are more abundant, and decrease in a recession when jobs are scarce. Mobility in and out of Quebec is likely to be lower than in other regions of Canada because of language and cultural differences. In addition distance is an inhibiting factor because it increases moving costs and uncertainty as well as the psychic costs associated with being uprooted from one's familiar environment.

Empirical evidence[1] tends to verify the implications of migration as a human capital decision. As well, the process of migration tends to reduce the regional disparities that induce the migration decision. For example, empirical evidence form the U.S. indicates that the wage differential between the North and the South had disappeared by the 1970s, after controlling for the differences in cost of living and the human capital endowments of the work force (Bellante 1979; Bishop, Formby, and Thistle 1982; Loelho and Ghali 1971, 1973; Farber and Newmen 1987). This wage convergence occurred as firms relocated to the low-wage South (thereby reducing the demand for labour in the North and

---

1. In the Canadian context, this is evident in the work of Courchene (1970), 1974), Grant and Vanderkamp (1976), Laber and Chase (1971), Mansell and Copithorne (1986), and in the extensive work of John Vanderkamp (1968, 1971, 1972, 1976, and 1986).

increasing it in the South) and as workers left the South for jobs in the high-wage North (thereby increasing the supply of labour in the North and reducing it in the South).

International Migration, Brain Drain, and Impact of Immigration

The decision to migrate internationally can be viewed as a human capital decision that depends upon benefits and costs. In this decision, the noneconomic constraints are even more important, however, because immigration is strictly regulated and hence subject to political decisions of the host economy. These political decisions, however, often are a function of the economic environment.

In Canada, for example, immigration currently is regulated through a point system whereby a large percentage of the points are given for such factors as education, training, previously arranged employment, and possession of skills that are in high demand and short supply. Clearly the control of immigration is highly job-oriented and it can be turned on and off depending upon economic conditions.

Immigration into Canada also tends to be pro-cyclical. That is, in prosperous times immigration tends to be large, and in periods of recession and high unemployment it is small. This is attributed to the fact that in periods of high unemployment, immigration officials may tighten up on the influx, and new immigrants may be less prone to enter. The importance of the economic environment is further illustrated by the fact that new immigrants have chosen to settle in the more prosperous regions and urban centres of Canada.

Form a policy perspective, one of the most interesting and controversial aspects of international migration is the issue of the so-called *brain drain*. The problem arises because countries, especially less-developed ones, may lose their most-skilled labour to the more-developed countries. It is the skilled workers that tend to leave because they can afford to do so, they have the knowledge of foreign opportunities (perhaps acquired while studying abroad), and they can usually amass sufficient points to enter the host country.

The problem is especially acute with countries that heavily subsidize the education and training of their workers. In such circumstances, the home countries bear the cost of the education, and the skilled emigrant reaps all of the benefits in the form of higher earnings in the host country. In many circumstances these skilled workers are the very persons that the developing economy can least afford to lose.

The brain drain often occurs when students from less-developed economies go to more-developed economies for advanced education and training. The psychic costs associated with the cultural and environmental change have already been incurred, and the direct cost of education or training is often borne by the home country. By definition, the income opportunities are greater in the more-developed host country and hence the temptation to stay.

For countries that are net losers of highly skilled labour, there are some possible remedies, but they all involve other problems. Having the individuals themselves pay for their own education and training would at least ensure that they were paying the costs for the benefits they could receive if they emigrate. However, especially in developing economies, this may hinder large numbers of otherwise poor people from acquiring the education or training. Placing an "exit tax" equal to the amount of the human capital cost borne by the state is theoretically possible, but practically difficult to enforce, and it may be politically unpopular as it becomes compounded with issues of human rights. Recruiting your own students abroad and encouraging their return is a policy that is utilized, but it does have costs and it may be vacuous if viable job opportunities cannot be provided. Providing job opportunities at high wages, of course, is easier said than done, especially in countries that have followed a conscious policy of minimizing wage differentials, possibly for equity reasons. Clearly, as in so many elements of public policy a variety of delicate trade-offs are involved.

By augmenting the supply of labour, immigration may reduce the wages of domestic labour relative to what it would have been in the absence of immigration. This may have a differential impact on different regions and different types of labour, depending upon the location decisions of immigrants and their skill distribution. For example, if immigrants tend to be less skilled, then they would constrain wages at the lower end of the wage spectrum. If they are more skilled, then they would constrain wages more at the higher end.

The impact of immigration on the wages and employment of the domestic work force is complicated by the fact that immigrants may also be a source of demand, thereby increasing domestic wages and employment opportunities. As well, to the extent that immigration may be used to fill labour demand bottlenecks that are not filled by the domestic work force, then this may have a complementary effect on the demand for domestic labour, thereby increasing their wage and employment opportunities. To a certain degree, immigration may also be an alternative to trade in goods; that is, if free trade facilitates the movement of goods, then there may be less incentive for labour itself to migrate. For example, Canadians can import goods produced by labour-intensive countries that have low wages, or we can "import" the labour itself through immigration. In essence, imported goods embody the otherwise "imported" labour and hence serves as an alternative to immigration as a source of low-wage labour. The impact of immigration on domestic wages and employment, therefore, must be considered as relative to alternatives such as the importation of goods that may also affect domestic wages and employment. For these various reasons, the impact of immigration on the wages and employment of domestic-born workers is more complicated than the simple supply side effects would imply. Obviously, it is necessary to appeal to the empirical evidence to ascertain that impact.

The empirical evidence tends to be somewhat mixed, albeit most studies find that immigration does not have a substantial net impact on the wages and employment of the domestic-born work force. This is so in studies that compare domestic wages and employment in cities with large influxes of immigrants compared to cities with small influxes.[2] Borjas, Freeman and Katz (1992) use a different methodology that involves simulating the impact of immigration on labour supply, and then the effect of this increase of labour supply on wages and wage structures. For the U.S. as a whole, they found that immigration did depress domestic wages somewhat, especially at the lower wage level, since this is the level at which much of the immigrant labour competed. They suggest that their economy-wide results may differ from those based on intercity comparisons because the potential wage-depressing effect in specific cities may be offset by changes in the migration behaviour of domestic labour and by the demand increases associated with the immigration.

Assimilation of Immigrants

One of the more interesting areas of recent research in labour economics involves the assimilation of immigrants into the labour market. The issue is whether immigrants fully assimilate in the sense that their earnings eventually catch up to those of comparable domestic-born workers. A related issue is whether assimilation is occurring faster or slower for recent immigrants. The "vertical mosaic" aspect of Canadian immigration requires that immigrants would have a reasonable chance of being assimilated throughout the occupational ladder and not be segregated into jobs at the lower rungs.

Three different but interrelated effects have been identified in the literature. The *entry effect* is the wage difference between immigrants and domestic-born workers who are otherwise comparable in terms of observable wage-determining characteristics such as age, education, training, experience, and gender. It is expected to be negative for immigrants reflecting the loss of country-specific human capital and job networks, as well as differences in language and culture. It may also reflect discrimination against new immigrants, especially if they are visible minorities.

The *assimilation effect* is the growth in earnings of immigrants associated purely with their being in the new host country. It is over and above any growth in earnings associated with their aging or acquisition of skills like education or training — skills that also affect the earnings of domestic-born workers. The assimilation effect is expected to be positive as immigrants acquire the (conventionally unobserved) characteristics of job networks, language skills, and cultural aspects that may affect their earnings. It may also reflect a dissipation of

---

2. Recent studies include Altonji and Card (1991), Butcher and Card (1991), Card (1990), and Lalonde and Topel (1991). These studies also refer to earlier studies on this issue.

discrimination, at least to the extent that immigrants who have been in the host country for a longer period are likely to be less discriminated against than recent arrivals.

The implied time-to-equality occurs when the positive assimilation effect eventually overcomes the negative entry effect, so that the earnings of immigrants catch up to the earnings of otherwise comparable domestic-born workers. After that point, the earnings of immigrants may even exceed the earnings of otherwise comparable domestic-born workers to the extent that immigrants are a select group in terms of conventionally unobserved characteristics such as motivation and initiative — characteristics that may have influenced them to make the immigration decision in spite of the hardships.

Cohort effects may also exist, reflecting differences in the earnings capacity of different cohorts of immigrants who arrived in different periods (e.g., immigrants who arrived in 1980–1985 as opposed to 1945–1955). The pure period-of-immigration cohort effects are over and above the effect of differences in the education or training of the groups, the later differences being controlled for in the statistical analysis. The cohort effects often capture differences in the conventionally unobservable factors that can vary across different immigrant cohorts, depending upon the time period when they entered the country. These are often labelled as "quality" differences in the literature, although it should be kept in mind that these are over and above the effect of conventionally observed wage-determining factors of age, education, training, or gender. The cohort effects may reflect changes in immigration policy, or perhaps differences in the country-of-origin mix or the legacy of the state of the labour market at the time of immigration, depending on whether the later factors are controlled for in the statistical analysis.

Figure 19.2 illustrates these different effects. The domestic-born earnings profile is drawn flat, since it is independent of the years-since-immigration of the immigrants as given on the horizontal axis. The immigrants and domestic-born workers are assumed to be comparable in terms of conventionally observable wage-determining characteristics like age, education, training, and gender.

The negative entry effect is exhibited by the fact that the immigrant earnings profile *starts* below that of comparable domestic-born workers when immigrants first enter (i.e., when years-since-migration is zero). The positive pure assimilation effect is exhibited by the positive wage growth for immigrants associated with each year that they are in the host country (i.e., for each year since immigration). The catch-up or time-to-equality is given by T, when the profiles cross.

Empirical evidence for Canada and the United States tends to confirm the existence of negative entry effects and positive assimilation effects. For example, in Bloom and Gunderson (1991), immigrants to Canada tended to earn about 17 percent less than comparable domestic-born workers when they first entered Canada, but their earnings grew three-quarters of one percent (i.e., 0.77 per-

Figure 19.2
EARNINGS PROFILES FOR IMMIGRANTS AND DOMESTIC BORN

(a) All Immigrants

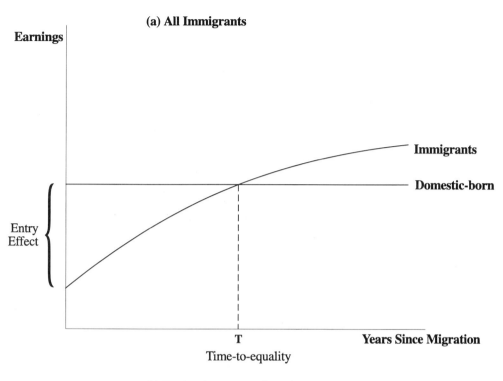

(b) Early and Recent Cohorts

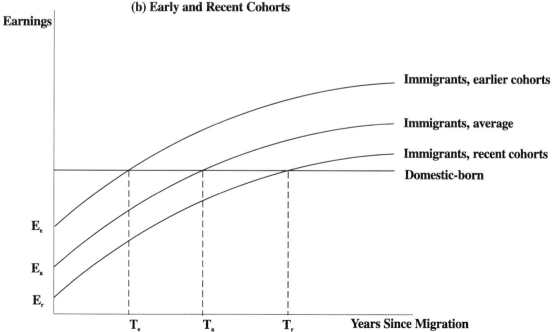

cent) per year faster, so that by 21 years they had caught up to the earnings of comparable Canadian-born workers.

However, the pattern is different for different cohorts of immigrants, depending upon the period in which they immigrated. Specifically, more recent cohorts of immigrants appear to be taking longer to assimilate. This is indicated in the bottom panel, Figure 19.2(b). To a certain extent this may reflect changes in immigration policy away from a labour market and skills orientation (of the 1950s and 1960s) and towards an emphasis on family reunification, human rights, and refugee status (in the 1970s and 1980s). It may also reflect differences in the composition or country of origin of immigration away from European immigrants (whose culture, labour market skills, and language may have facilitated immigration) and towards immigrants from Asia, Africa, and Latin America, where these differences may have been greater and where discrimination may be more prominent. Assimilation may also have been more difficult in recent periods, reflecting the continuous upward drift in unemployment.

QUESTIONS

1. Discuss the interrelationship between occupational wage structures and other wage structures, particularly wage differentials by industry, region, and personal characteristics. Why does the interrelatedness make it difficult to talk about such a thing as a pure occupational wage differential? Illustrate by an example.

2. Discuss the various factors that determine the shape of the labour supply schedule for an occupation. Give an example of an occupation that may have an elastic supply schedule and one that may have an inelastic one, and indicate why this is so. If both occupations received an equal increase in the demand for their labour, what would happen to the skill differential between the two occupations?

3. Discuss the expected impact on the occupational wage structure of each of the following policies:

   (a) an increase in unionization;

   (b) an increase in public subsidies to education, training, job search, and mobility;

   (c) wage-price controls;

   (d) child labour laws;

   (e) laws governing the minimum age one can leave school;

   (f) a reduction in rural-urban migration;

   (g) a reduction in immigration;

(h) an exogenous influx into the labour force of younger workers and married women;

(i) more strict control of entry into the medical profession.

4. Just as in computing the aggregate price index it is important to net out the effect of quality change over time, so in computing the aggregate wage index we should net out changes in the quality of labour and in the characteristics of jobs. Otherwise we won't know how much, if any, an increase in the aggregate wage level is attributable to quality change or to changes in the work environment as opposed to a pure wage increase. How do you think these factors — labour quality and job environment — have changed over time, and how would this affect our aggregate wage level?

5. Discuss the competitive forces that should work to reduce interregional wage differentials.

6. The economic model of migration has been criticized for predicting that the forces of competition would reduce regional disparities. Some have argued that disparities would increase as the best people leave for the better job alternatives, and as capital flows to the already expanding sector. Discuss.

7. It could be argued that market forces will not prevent cities from growing beyond a socially optimal size because pollution or congestion externalities associated with city growth are not accounted for through the market mechanism. Discuss the extent to which wage adjustments may reflect these externalities. Will the wage adjustments ensure a socially optimal city size?

8. Discuss the variety of artificial barriers that can impede geographic mobility in a country like Canada. Should these barriers be removed?

9. Discuss the ways in which government social transfer programs may decrease or increase geographic mobility. To the extent that they decrease mobility, should they be removed? Are there ways of mitigating their adverse effects on geographic mobility?

10. Discuss the extent to which the migration decision is a human capital decision. To the extent that it is a human capital decision, what does this tell us about the determinants of migration? Derive some empirically testable hypotheses from the human capital model of migration.

11. Discuss what is meant by the brain drain, why it arises, and possible solutions to the problem.

REFERENCES AND FURTHER READINGS

A. Occupational Wage Structures

Bernhardt, D. and D. Backus. Borrowing constraints, occupational choice, and labor supply. *JOLE* 8 (January 1990) 145–173.

De Meza, D. Wage uncertainty, expected utility and occupational choice. *IER* 25 (October 1984) 757–762.

Filer, R. The "starving artist" — myth or reality? *JPE* 94 (February 1986) 56–75.

Gleicher, D. and L. Stevens. *A Classical Approach to Occupational Wages.* London: Praeger, 1991.

Headen, A. Wage, return to ownership, and fee responses to physician supply. *R.E. Stats.* 72 (February 1990) 30–37.

Jones, J. and W. Walsh. Salary determination in the national hockey league. *ILRR* 41 (July 1988) 592–604.

Kumar, P. Differentials in wage rates of unskilled labor in Canadian manufacturing industries. *ILRR* 26 (October 1972) 631–645.

Kumar, P. and M. Coates. Occupational earnings, compensating differentials and human capital: an empirical study. *CJE* 15 (August 1982) 442–457.

McCall, B. Occupational matching: a test of sorts. *JPE* 98 (February 1990) 45–69.

Meltz, N. and D. Stager. Trends in the occupational structure of earnings in Canada 1931–1971. *CJE* 12 (May 1979) 312–315.

Orazem, P. and J. Matilla. Occupational entry and uncertainty. *ILRR* 68 (May 1986) 265–223.

Reder, M. The theory of occupational wage differentials. *AER* 45 (December 1955) 833–852.

B. Regional Wage Structures and Geographic Mobility

Barro, R. and X. Sala-l-Martin. Convergence across States and regions. *BPEA* 1 (1991) 107–158.

Beesm, P. and R. Eberts. Identifying productivity and amenity effects in interurban wage differentials. *R.E. Stats.* 71 (August 1985) 443–452.

Bellante, D. The North–South differential and the migration of heterogeneous labor. *AER* 69 (March 1979) 166–175.

Bishop, J., J. Furmby and P. Thistle. Convergence of the South and Non-South income distributions. 1969–1979. *AER* 82 (March 1992) 262–272.

Drewes, T. Regional wage spillover in Canada. *R.E. Stats.* 59 (May 1987) 215–223.

Courchene, T. Interprovincial migration and economic adjustment. *CJE* 3 (November 1970) 550–576.

Courchene, T. *Migration, Income and Employment: Canada 1965–1968.* Montreal: Howe Research Institute, 1974.

Farber, S. and R. Newman. Accounting for South/Non-South real wage differentials and for changes in those differentials over time. *R.E. Stats.* 59 (May 1987) 215–223.

Farber, S. and R. Newman. Regional wage differentials with spatial convergence of worker characteristic prices. *R.E. Stats.* 71 (May 1989) 224–231.

Gerking, S. and S. Wierick. Compensating differences and interregional wage differentials. *R.E. Stats.* 65 (August 1983) 483–486.

Grant, E.K. and J. Venderkamp. *The Economic Causes and Effects of Migration: Canada 1965–1971.* Ottawa: Economic Council of Canada, 1976.

Laber, G. and R. Chase. Interprovincial migration in Canada as a human capital decision. *JPE* 79 (July/August 1971) 795–804.

Leolho, P. and M. Ghali. The end of the North-South wage differential. *AER* 61 (Dec. 1971) 932–937, and comment and reply 63 (September 1973) 757–762.

Mansell, R.L. and L. Copithorne. Canadian regional economic disparities: a survey. *Disparities and Interregional Adjustment.* Toronto: University of Toronto Press, 1986, pp. 1–51.

Newman, R. Dynamic patterns in regional wage differentials. *SEJ* 49 (July 1982) 246–254.

Roback J. Wages, rents and amenities. *EI* 26 (January 1988) 23–46.

Sahling, H. and S. Smith. Regional wage differentials: has the South risen again? *R.E. Stats.* 65 (February 1983) 131–134.

Tremblay, C. Regional wage differentials: has the South risen again? *R.E. Stats.* 68 (February 1986) 175–178.

Vanderkamp, J. Interregional mobility in Canada: a study of the time pattern of migration. *CJE* 1 (August 1968) 595–608.

Vanderkamp, J. Migration flows, their determinants and the effects of return migration. *JPE* 79 (September/October 1971) 1012–1031.

Vanderkamp, J. Return migration: its significance and behaviour. *Western Economic Journal* 10 (December 1972) 460–466.

Vanderkamp, J. The role of population size in migration studies. *CJE* 9 (August 1976) 508–516.

Vanderkamp, J. The efficiency of the interregional adjustment process. *Disparities and Interregional Adjustment*, K. Norrie (ed.). Toronto: University of Toronto Press, 1986, pp. 53–108.

C. International Migration, Brain Drain, and the Impact of Immigration

Ault, D., G. Rautman and T. Stevenson. Some factors affecting mobility in the labor force for academic economists. *EI* 20 (January 1982) 104–132.

Altonji, J., and D. Card. The effects of immigration on the labor market outcomes of less-skilled natives. In J. Abowd and R. Freeman (eds.), *Immigration, Trade, and the Labor Market.* Chicago: University of Chicago Press, 1991, pp. 201–234.

Berry, R.A. and R. Soligo. Some welfare aspects of international migration. *JPE* 77 (September/October 1969) 778–794.

Blomqvist, A. International migration of educated manpower and social rates of return to education in LDCs. *IER* 27 (February 1986) 165–174.

Boehm, T., H. Herzog and A. Schlottmann. Intraurban mobility, migration and tenure choice. *R.E. Stats.* 73 (February 1991) 59–58.

Borjas, G., R. Freeman and L. Katz. On the labour market effects of immigration and trade. *The Economic Effects of Immigration in Source and Receiving Countries*, G. Borjas and R. Freeman (eds.). Chicago: University of Chicago Press, 1992.

Butcher, K.F., and D. Card. Immigration and wages: Evidence from the 1980s. *AER* 81 (May 1991) 292–296.

Canto, V. and F. Udwadia. The effect of immigration quotas on the average quality of migrating labor and income distribution. *SEJ* 52 (January 1986) 785–793.

Card, D. The impact of the Mariel boatlift on the Miami labor market. *ILRR* 43 (January 1990) 245–257.

Da Vanzo, J. Repeat migration in the United States: who moves back and who moves on? *R.E. Stats.* 65 (November 1983) 552–559.

Dunlevy, J. and D. Bellante. Net migration, endogenous incomes and the speed of adjustment to the North-South differential. *R.E. Stats* 65 (February 1983) 66–75.

Ethier, W. International trade and labor migration. *AER* 75 (September 1985) 691–707.

Ethier, W. Illegal immigration: the host country problem. *AER* 76 (March 1986) 56–71.

Falaris, E. Migration and regional wages. *SEJ* 48 (January 1982) 670–686.

Falaris, E. A nested logit migration model with selectivity. *IER* 28 (June 1987) 429–443.

Falaris, E. Migration and the wages of young men. *JHR* 23 (Fall 1988) 514–534.

Farber, S. Post-migration earnings profiles: an application of human capital and job search models. *SEJ* 49 (January 1983) 693–705.

Grant, E.K. and J. Vanderkamp. The effects of migration on income: a micro study with Canadian data 1965–1971. *CJE* 13 (August 1980) 381–406.

Green, A. *Immigration and the Postwar Canadian Economy*. Toronto: Macmillan, 1976.

Greenwood, M. Research on internal migration in the United States. *JEL* 13 (June 1975) 397–433.

Greenwood, M. and G. Hunt. Migration and interregional employment redistribution in the United States. *AER* 74 (December 1984) 957–969.

Greenwood, M. and J. McDowell. Differential economic opportunity, transferability of skills, and immigration to the United States and Canada. *R.E. Stats.* 73 (November 1991) 612–623.

Grossman, J. Illegal immigrants and domestic employment. *ILRR* 37 (January 1984) 240–251.

Herzog, H., Jr., and A. Schlottmann. Labor force migration and allocative efficiency in the United States: the roles of information and psychic costs. *EI* 19 (July 1981) 459–475.

Herzog, H., Jr., and A. Schlottmann. Migrant information, job search and the remigration decision. *SEJ* 50 (July 1983) 43–56.

Hunt, J. and J. Kau. Migration and wage growth: a human capital approach. *SEJ* 51 (January 1985) 697–710.

Johnson, G. The labour market effects of immigration. *ILRR* 33 (April 1980) 331–341.

Kleiner, M.R. Gay and K. Greene. Barriers to labor migration: the case of occupational licencing. *IR* 21 (Fall 1982) 383–391.

Kwok, V. and H. Leland. An economic model of the brain drain. *AER* 72 (March 1982) 91–100.

LaLonde, R., and R. Topel. Labor market adjustments to increased immigration. In J. Abowd and R. Freeman (eds.), *Immigration, Trade, and the Labor Market.* Chicago: University of Chicago Press, 1991, pp. 235–259.

Mincer, J. Family migration decisions. *JPE* 86 (October 1978) 749–776.

Mueser, P. A note on simultaneous equations models of migration and employment growth. *SEJ* 52 (October 1985) 516–522.

Nakosteen, R. and M. Zimmer. Migration and income: the question of self-selection. *SEJ* 46 (January 1980) 840–851.

Rauch, J. Reconciling the pattern of trade with the pattern of migration. *AER* 81 (September 1991) 775–796.

Robinson, C. and N. Tomes. Self-selection and interprovincial migration in Canada. *CJE* 15 (August 1982) 474–502.

Schlottmann, A. and H. Herzog. Home economic conditions and the decision to migrate: new evidence for the U.S. labor force. *R.E. Studies* 48 (April 1982) 950–961.

Schwartz, A. Migration, age and education. *JPE* 84 (August 1976) 701–720.

Scott, A. and H. Grubel. The international flow of human capital. *AER* 56 (May 1966) 268–274. Comment by N. Aitken and reply 58 (June 1968) 539–547.

Sjaastad, L. The costs and returns of human migration. *JPE* 70 (October 1962) S80–S93.

Todaro, M. A model of labour migration and urban unemployment in less developed countries. *AER* 59 (March 1969) 138–140.

Topel, R. Local labor markets. *JPE* 94, Part 2 (June 1986) S111–S143.

Webb, M. Migration and education subsidies by governments: a game-theoretic analysis. *J Pub Ec* 26 (March 1985) 249–262.

Wong, K. The economic analysis of internal migration: generalization. *CJE* 19 (May 1986) 357–362.

D. Assimilation of Immigrants

Bloom, D. and M. Gunderson. An analyses of the earnings of Canadian immigrants. In *Immigration, Trade and the Labor Market*, J. Abowd and R. Freeman (eds.). Chicago: University of Chicago Press, 1991, pp. 321–342.

Borjas, G. The earnings of male Hispanic immigrants in the United States. *ILRR* (April 1982).

Borjas, G. Assimilation, changes in cohort quality, and the earnings of immigrants. *JOLE* 3 (October 1985) 463–489.

Borjas, G. Self-selection and the earnings of immigrants. *AER* 77 (September 1987) 531–553.

Borgas, G. Immigrant and emigrant earnings. *EI* 27 (January 1989) 21–37.

Borjas, G. *International Differences in the Labor Market Performance of Immigrants.* Kalamazoo, Michigan: W.E. Upjohn Institute for Employment Research, 1988.

Borjas, G. Immigration and self-selection. In *Immigration, Trade and the Labor Market.* Edited by J. Abowd and R. Freeman. Chicago: University of Chicago Press, 1991, pp. 29–76.

Chiswick, B. The effect of Americanization on the earnings of foreign-born men. *JPE* 86 (October 1978) 897–921.

Kossondji, S. Immigrant worker assimilation. *JHR* 24 (Summer 1989) 494–527.

Chiswick, B. Is the new immigration less skilled than the old? *JOLE* 4 (April 1986) 168–192.

Chiswick, B. Immigration policy, source countries, and immigrant skills: Australia, Canada, and the United States. In *The Economics of Immigration*, L. Boher and P. Miller (eds.). Canberra: Australian Government Printing Service, 1988, pp. 163–206.

Chiswick, B. and P. Miller. Earnings in Canada: the roles of immigrant generations, French ethnicity and language. *Research in Population Economics*, Edited by T. Paul Schultz, 6 (1988) 183–228.

Kossondji, S. Immigrant worker assimilation. *JHR* (Summer 1989) 494–527.

Meng, R. The earnings of Canadian immigrant and native-born males. *Applied Economics* 19 (1987) 1107–1119.

Nakamura, A. and M. Nakamura. Effects of labour-market-entry unemployment rates on the earnings of U.S. and Canadian immigrant and native workers. In *Immigration, Language and Ethnicity: Canada and the United States*, B. Chiswick (ed.). Washington: American Enterprise Institute, 1992.

CHAPTER 20

Wage Structures: By Industry, Firm Size, and Individuals

While wage structures by occupation and region tend to receive the most attention, wage structures also exist by industry, firm size, and across different individuals — the latter being associated with the issue of wage polarization or growing wage inequality. To a large extent, changes in these wage structures also reflect changes in the basic underlying determinants of all wage structures — compensating differences for nonpecuniary aspects, short-run demand changes, and noncompetitive factors.

INTERINDUSTRY WAGE DIFFERENTIALS

As the Canadian *Standard Industrial Classification* manual indicates, industrial designations refer to the principal kind or branch of economic activity of establishments in which individuals work. The industries may be broadly defined sectors, as, for example, agriculture, mining, manufacturing, construction, transportation, trade, and public administration. Or they may be more narrowly defined. For example, within the nondurable manufacturing sector there are food and beverage, tobacco, rubber, textile, and paper industries. The interindustry wage structure indicates the wage differential between industries, however broadly or narrowly defined.

Theoretical Determinants of Interindustry Wage Differentials

The average wage in an industry will reflect a variety of factors, including the occupational composition and personal characteristics of the work force, and the regional domination of some industries. Consequently, pure interindustry wage differentials are difficult to calculate because they reflect other wage structures, notably those by occupation, region, and the personal characteristics of the work force.

If we could net out the effect of these other wage structures, however, so that we were comparing wages across industries for the same occupation, the same type of worker, and the same region, then the pure interindustry wage differential would reflect only the different characteristics of the industry. Many of these different wage-determining characteristics are similar to the ones analyzed as determinants of the interoccupational wage structure — nonpecuniary aspects of the jobs in each industry, short-run adjustments, and noncompetitive forces. If there are no differences in any of these aspects, competitive forces should ensure that all industries pay the same wage for the same type of labour. The basic determinants of interindustry wage differentials, therefore, can be categorized according to the basic determinants of any wage structure: nonpecuniary characteristics of the industry, short-run demand changes, and noncompetitive factors.

Interindustry wage differences, for example, may exist to compensate for differences in the nonpecuniary aspects of the work in different industries. Such factors could include unpleasant or unsafe work conditions, or seasonal or cyclical employment. Workers in the construction industry, for example, may receive a wage premium in part to compensate for risk as well as the seasonal and cyclical nature of much of the work.

Interindustry wage differences may also reflect short-run demand changes. The wage differences are the price signal necessary to encourage reallocation from declining sectors to expanding ones. Different industries are affected in different fashions by various demand changes emanating from such factors as technological change or free trade and global competition. For example, freer trade will lead to an expansion of export-oriented industries and to a contraction of industries most affected by imports. This should lead to a wage premium in the expanding export sector, which in turn is the price signal to encourage the reallocation of labour from the declining import sector to the expanding export sector.

There may also be a variety of noncompetitive factors that could affect the interindustry wage structure. Monopoly rents may differ by industry and these may be captured, in part, by workers, especially if they are unionized. Even in declining industries, or industries subject to severe import competition, the rents going to workers could increase if they were able to appropriate a larger *share* of the declining rents. This could occur, for example, if the declining nature of the industry meant that there was no threat of entry, and the firm had no alternative use for its plant and equipment, and therefore no viable threat to relocate if labour appropriate a larger share of the rents. In essence, labour is still able to "milk a dead cow," at least for a time.

Other noncompetitive factors can also affect the interindustry wage differential. Minimum wage laws obviously affect low-wage industries such as personal services (laundries, hotels, restaurants, and taverns), retail trade, and parts of nondurable manufacturing, especially clothing. Equal pay laws will affect industries with a high proportion of females (and which also often are low-wage industries), for example nondurable manufacturing retail trade, services and finance, insurance and real estate). Fair wage laws and wage extension laws tend to be most important in the construction sector. Occupational licensing and unionization tend to be most prominent in high-wage industries — in part, of course, because they may have raised wages in those industries. In summary, basic economic theory suggests that, in the long run, competitive forces should eliminate pure interindustry wage differences. To the extent that they exist, interindustry wage differences should reflect only such factors as compensating wages for the nonpecuniary characteristics of the industry, short-run demand changes, or noncompetitive factors.

Efficiency Wages and Interindustry Wage Differences

As discussed previously in Chapter 11 (Wages and Employment in Competitive Labour Markets) firms may rationally pay wages above the market clearing level so as to improve morale, reduce turnover and absenteeism, elicit effort, reduce shirking, reduce the threat of unionization, and to have a queue of applicants so as to reduce recruiting and hiring costs. It may be efficient for the firm to pay wages above the market clearing wage because of its positive effects on labour productivity or on reducing costs in these other areas. Such higher wages are rents to the workers who receive them in the sense of being payments in excess of the market clearing wage and in excess of their next-best-alternative job, assuming it is at the market clearing wage. In fact it is the receipt of this rent or excessive wage payment that elicits the desirable behaviour on the part of workers. Although they are payments above the market clearing wage, these are competitive wage differentials in the sense that they are consistent with the profit maximizing behaviour of firms and will not be eliminated by the forces of competition.

The desirability of paying efficiency wages may well differ by industry thereby giving rise to persistent interindustry wage differentials.[1] For example, industries that utilize expensive capital equipment or for which the training costs make turnover costly, may find it profitable to pay such efficiency wages to have a trustworthy work force and low turnover. Such interindustry wage differentials may also persist over time if the efficiency rationales for paying a pure wage premium also persist over time.

---

1. In theory, efficiency wages could vary by any of the factors that give rise to wage structures, such as occupation or region as well as industry. In practice, however, the theoretical rationales for efficiency wages suggest that they will differ by industry and by firm size and not so much by occupation and region. Hence, the efficiency wage literature tends to be associated with the literature on wage differences by industry and firm size, and not by the other factors that give rise to wage structures.

Such high-wage jobs also correspond to the notion of "good jobs" as opposed to "bad jobs." The good jobs are those in industries that pay wages in excess of the market clearing wage. Workers in those jobs are fortunate in the sense that the efficiency wage premium is a pure rent, not a compensating wage premium paid for undesirable working conditions or costly human capital. It is a rent or wage premium designed explicitly to make it a good job so that workers will not want to lose the job. As such, they are likely to exhibit good work characteristics with respect to such factors as absenteeism, turnover, and honesty. Obviously, to the extent that there is disutility associated with having to deliver such good work characteristics, then the higher wage would simply be a compensating wage premium associated with such characteristics. The essence of efficiency wages, however, is that they are rents or pure wage premiums and not compensating wages.

Efficiency wage premiums are different from conventional rents in that they are not extracted from the firm through, for example, the monopoly power of a union. Rather, they are voluntarily paid by the firm to serve a positive human resource management function such as eliciting loyalty or honesty or reducing turnover. Workers are paid the marginal product of their labour; that marginal product is simply enhanced by the productivity inducing aspects of the higher wage. For workers, however, it is like a conventional rent in that the wage premium is greater than any disutility they experience from their actions (increased loyalty and honesty, reduced turnover, absenteeism or shirking) that they undertake as a result of the efficiency wage premium.

Workers who are displaced from those "good" jobs (perhaps because of industrial restructuring or import competition) would loose those rents if they are displaced to jobs that may simply pay the market clearing wage. The loss of such efficiency wages may be one of the reasons for the large wage losses that are often experienced by workers who are displaced because of a plant closing or permanent job loss.

The existence of efficiency wages may also provide a theoretical rationale for industrial policies designed to protect the so-called "good wage" jobs (Dickins and Lang, 1988; Katz and Summers, 1989). Such policies could include subsidies to "high-tech" or other industries where such jobs may prevail, or even minimum wage laws that reduce employment in low-wage jobs that are unlikely to embody any efficiency wage. Tariffs could also protect good jobs in industries that pay efficiency wages; however, in developed countries, high-wage industries tend to be export-oriented ones and low-wage industries tend to be subject to import competition. Therefore, freer trade to enhance both exports and imports should increase the number of "good jobs" related to low-wage jobs; that is, free trade and not protectionism would be the best policy to enhance the number of jobs that pay efficiency wage premiums, especially if workers can be reallocated from the lower-wage import-competing jobs to the higher-wage export-oriented jobs.

In summary, countries may want to protect or encourage jobs that pay efficiency wages because they are good jobs in the sense that firms want to pay the high wages (i.e., it is in their profit maximizing interest to do so) and workers obviously want to receive such high wages because they are rents in excess of their next-best-alternative wage. Countries that happen to have such jobs are fortunate, just like countries that happen to have endowments of natural resources.

While efficiency wages can provide a *theoretical* rationale for industrial policy to protect certain good jobs, this should not be taken as a justification for all such policies. Most are likely instituted to protect monopoly rents or to provide subsidies or protection that is politically motivated and likely comes at the expense of consumers having to pay higher product prices. At a practical level, it may be extremely difficult to distinguish efficiency wages from pure rents so as to target the protection and encouragement only of the jobs that pay efficiency wages. Issues of equity and fairness may also be involved since the jobs that pay efficiency wages are likely to be high-wage jobs for reasons other than efficiency wages.

Furthermore, there is legitimate debate over the existence and magnitude of pure efficiency wages. It is difficult to statistically control for all of the other factors that influence wages, and yet this is crucial in this area since one would expect firms to be able to hire the best workers in terms of both observable characteristics (the influence of which can be controlled for through statistical techniques) and unobservable factors (the influence of which is difficult to control for). As such, the efficiency wages may simply reflect a return to an unmeasured characteristic. As well, competition on the part of workers for such good jobs should put a cap on the magnitude of the efficiency wage that is necessary to ensure that these are regarded as good jobs. There are also likely to be other compensation mechanisms besides paying efficiency wages to elicit the desired behaviour on the part of workers. One such mechanism, discussed subsequently in the chapter on optimal compensation systems, is deferred compensation whereby workers are "underpaid" when young and "overpaid" when older as a way to encourage them to want to stay with the firm.

In spite of these theoretical and empirical problems in using industrial policies to encourage the "good" jobs that pay efficiency wages, the potential policy importance of the area merits that more consideration and research be given to the topic. This is especially the case because the encouragement and protection of efficiency wages is subject to a degree of policy control. Sensible policy application, however, requires answers to a number of questions: Can efficiency wage premiums be precisely identified so as to target policies to encourage such jobs? What will be the ultimate outcome if all jurisdictions (countries, provinces, states, regions, etc.) compete for such jobs? Are there ways of dealing with the distributional consequences if the already high-wage jobs effectively become protected or subsidized? If industrial policies are used to protect or encourage efficiency wages, what will happen when employers and employees try to have other jobs protected and encouraged in the same fashion? Answers to these and other questions would be useful to shed light on this important policy issue.

Empirical Evidence on Interindustry Wage Differences, Efficiency Wages, and the Impact of Deregulation

Systematic evidence on the existence of pure interindustry wage differences is difficult to compile because of the problem of controlling for other factors that affect the wage difference across industries. Different industries, for example, utilize different skill mixes, and they are concentrated in different regions and are composed of firms of different sizes. Hence, interindustry wage differences often confound wage differences by other factors such as occupation, region, or plant size.

With the advent of microeconomic data sets using the individual worker as the unit of observation, however, it has been possible to use econometric techniques to control for the influences of these different factors, and to thereby isolate a pure interindustry wage differential. The econometric results indicate that interindustry wage differences do reflect the conventional determinants of wage structures: nonpecuniary differences in the jobs, short-run demand changes, and noncompetitive factors. However, the studies also indicate that even after controlling for these factors a pure interindustry wage differential or efficiency wage seems to prevail[2]. Some industries simply pay higher wages than appear necessary to compensate for the nonpecuniary aspects of the job, or to meet short-run demand changes, or because of noncompetitive factors. As well, this interindustry wage pattern appears quite stable over time; high-wage industries tend to remain high-wage industries.

That these are rents or payments above the market clearing wage is supported by other indirect evidence. Workers who change industries are likely to gain or loose the full amount of the rent premium associated with that industry, suggesting it is the industry and not their own individual characteristics that gave rise to the wage premium. As well, the wage premium goes to workers throughout the occupational spectrum in an industry (e.g., secretaries, blue-collar workers, and managers) suggesting that it is something in the industry itself that generates the premium. The higher wages also tend to be associated with lower quit rates and large queues of applicants, and this would not prevail if the premium were simply a compensating premium for skills or nonpecuniary aspects of the job. While these observations suggest that the wage premiums are rents, it is more

---

2. Dickens and Katz (1987), Katz and Summers (1989), Krueger and Summers (1987, 1988) and Murphy and Topel (1987) for U.S. data, and Gera and Grenier (1991) for Canadian data. These studies also review the earlier studies that documented a substantial interindustry wage differential based on aggregate industry data.

difficult to determine whether they reflect efficiency wages or the results of noncompetitive factors associated with each industry.

The interindustry wage structure tends to be quite stable over time and across different countries (Katz and Summers, 1989; Krueger and Summers, 1988). That is, high-wage industries tend to be high-wage industries for considerable periods of time, and they tend to be high-wage industries in countries with very different labour market structures and institutional features. Industries that pay efficiency wages tend to be capital intensive, and typically are composed of large firms with considerable market power and ability to pay. These industries are also usually highly unionized and have a work force of above average education and with low quit rates.

Based on Canadian data, Table 20.1 illustrates the magnitude of the pure interindustry wage differential (i.e., efficiency wage) that prevails after controlling for the effect of a wide range of other factors that can influence wages across industries. Clearly the differences are substantial. High-wage industries like tobacco products and mineral fuels respectively pay wage premiums of 33.4 percent and 2.5 percent above the competitive norm (i.e., above the average wage paid to workers of comparable wage-determining characteristics in other industries). Conversely, the service industries tend to pay wages considerably below the competitive norm — 31.8 percent below in religious organizations, and 20.3 percent below in accommodation and food services.

These extreme examples of the largest differences above and below the competitive norm also highlight the potential difficulty of fully controlling for the other factors that could affect interindustry wage differences. For example, some of the wage premium in tobacco products could reflect a compensating premium for the uncertainties associated with the restructuring of that industry. Some of the wage disadvantage in accommodation and food services may reflect the fact that some of the compensation may occur in the form of "tips." Some of the wage disadvantage in religious organizations may reflect perceived compensation in nonmonetary forms.

Whether these interindustry wage premiums reflect noncompetitive factors or efficiency wages, there is empirical evidence to suggest that industry wage premiums will be dissipated by international competition. This has been documented in studies that tend to find a negative correlation between average industry wage levels and the extent of import penetration into the industry.[3] It has also been documented in a number of recent studies that relate individual wages to measures of import penetration into the industry in which the individuals work.[4]

---

3. See, for example, Grossman (1986, 1987), Lawrence and Lawrence (1985) and Ravenga (1992).
4. See, for example, Dickens and Lang (1988), Freeman and Katz (1991), and Katz and Summers (1989).

Table 20.1
PURE INTERINDUSTRY WAGE DIFFERENTIALS, CANADA 1986[a]

INDUSTRY	WAGE PREMIUM	INDUSTRY	WAGE PREMIUM
Primary Industries		**Construction**	
Mineral fuels	25.5	Special trade	17.0
Mining services	22.7	General contractors	11.4
Forestry	18.9	Related services	−2.4
Metal mines	18.9		
Nonmetal mines	13.7	**Transportation, Communication, Utilities**	
Quarries and pits	0.1		
Fishing and trapping	−9.5	Utilities	14.4
		Communications	10.5
Manufacturing		Transportation	8.7
		Storage	6.3
Tobacco products	33.4		
Petroleum and coal	20.8	**Trade**	
Chemicals	14.4		
Paper	12.0	Wholesale trade	3.8
Primary metals	11.5	Retail trade	−11.1
Wood products	8.7		
Transport equipment	7.2	**Finance, Insurance and Real Estate**	
Rubber and plastics	7.1		
Printing and publishing	6.4	Insurance carriers	13.7
Nonmetallic minerals	4.8	Finance	9.3
Metal fabricating	3.7	Insurance and real estate agencies	−4.1
Electrical products	2.6		
Machinery	0.8	**Other Services**	
Miscellaneous manufacturing	−3.4		
Food and beverages	−3.5	Business management	4.8
Clothing	−8.1	Education	−1.0
Leather	−10.0	Miscellaneous	−2.8
Furniture and fixtures	−14.4	Health and welfare	−3.1
Textiles	−19.0	Amusement/recreation	−11.4
		Personal services	−16.7
		Accommodation/food	−20.3
		Religious organizations	−31.8

a. Percent wage differential between the wage in each particular industry and the average two-digit industry wage, after controlling for the effect of other wage-determining factors. Data is from Statistics Canada's 1986 Labour Market Activity Survey. The industries are listed in descending order of the magnitude of the industry wage premium within each major industry group.

Source: S. Gera and G. Grenier, Interindustry wage differentials and efficiency wages: some Canadian evidence. In *Canadian Unemployment*, S. Gera (ed.). Ottawa: Economic Council of Canada, 1991, pp. 6–8.

There is also empirical evidence indicating that industries that are concentrated in the sense that they are dominated by a small number of firms in the product market tend to pay higher wages, presumably because of their monopoly profits. As well, industries that are regulated pay higher wages, presumably because the higher wage costs can be passed to consumers in the form of rate increases without there being much of a reduction in the demand for the regulated service and hence in the derived demand for labour. Ehrenberg (1979), for example, found that workers in the regulated telephone industry in New York were paid about 10 percent more than comparable workers in other industries. Shackett and Trapani (1987) found regulatory wage premiums of around 10 percent for males and 15 to 20 percent for females in regulated industries like trucking, airlines, banks, utilities, and insurance in the mid-1970s in the U.S.

Conversely, the deregulation that has occurred in recent years has led to substantial reductions in wages in those industries. Rose (1987), for example, found that the union wage premium was cut almost in half after deregulation of the trucking industry in the United States. The wages of unionized truck drivers fell by almost $4,000 or 14 percent as a result of deregulation. Prior to deregulation, she indicates that unions had captured two-thirds of the regulatory rents or excess profits generated by the regulatory protection. In such circumstances, it is not surprising that unions tend to strongly oppose deregulation. This is especially the case because the new firms that enter the industry after deregulation tend to be nonunion, and the ones that fail after deregulation tend to be union operations.

In the case of the airline industry, Card (1986) also found that wage reductions occurred after deregulation, albeit the magnitudes were smaller than in trucking. This was attributed in part to the fact that the large airline carriers were still able to maintain a monopoly position because of their control over key airline terminals as a result of the "hub-and-spoke" system.

In summary, the empirical evidence strongly suggests that industries that are subject to greater competitive pressures on the product market (e.g., import competition, deregulation, more competing firms) tend to pay lower wages. The evidence also suggests the existence and persistence of pure interindustry wage differentials or rents that are consistent with the payment of efficiency wages. More empirical work is necessary, however, to document the precise magnitude of these wage differences and their basic determinants.

INTERFIRM WAGE DIFFERENCES AND FIRM SIZE

Wage differences may also exist across firms within the same industry and region and for the same occupational group. As with other wage structures, economic theory suggests that such wage differences should reflect the basic determinants of wage structures: nonpecuniary differences in the nature of the jobs, short-run demand changes, and noncompetitive factors. Firms that pay high wages, for example, may have poor working conditions, or short-run demand increases, or noncompetitive conditions such as a monopoly position in the product market or a union in the labour market, or they may find it in their interest to pay efficiency wages above the competitive norm.

One off the puzzling empirical regularities that seems to prevail is the positive relationship between wages and firm size, with larger firms tending to pay higher wages (see the references to this section). This relationship has been somewhat of a puzzle to economists because there is little theoretical reason to expect such a relationship once one has controlled for the effect of other factors that influence wages.

Certainly other factors that influence wages may be correlated with firm size, giving the appearance that it is the size of the firm that affects wages. For example, large firms tend to be unionized and hence may pay a union wage premium. Even if they are not unionized, they may pay that premium to minimize the threat of being organized. Large firms may be more likely to have monopoly profits that may be shared with labour. They may also be more likely to be capital intensive and hence employ more skilled workers, thereby confounding the occupational wage differential with a firm size wage differential. Some of the wage advantage in large firms may be a compensating wage premium for their more structured and formalized work requirements.

In such circumstances, the wage premium in large firms arises because large firms are more likely to have these wage-enhancing characteristics: unionization, skilled workers, rents, and undesirable working conditions. It is not firm size *per se* that matters, but rather the correlation of firm size with these other wage-determining factors.

These characteristics may be conventionally observable factors, as were the previously mentioned ones, or they could be conventionally unobservable factors that are difficult to control for in statistical procedures that isolate the pure effect of firm size on wages. Such unobservable factors could include motivation or skills that are not measured by years of education or training. Large firms may attach a premium to these characteristics because of their importance in the capital intensity and interrelated production of such firms. It may also be harder for large firms to get rid of their "mistakes" given their (usually) more formal personnel systems and discipline procedures. Having the best workers (even if more expensive) may also be more important to large firms since monitoring and supervision may be more difficult. Having the best workers may be an important way of conserving on scarce managerial time, and this may be especially important for large firms if they employ the most talented managers so as to amortize that talent over the large firm size. To the extent that large firms tend to employ workers with conventionally unobserved characteristics that attract a wage premium, then some of the firm size-wage effect may simply be the wage premium for these conventionally unobserved factors.

Large firms may also be more likely to pay efficiency wages. As discussed previously, these are wage premiums above the market clearing wage. They are voluntarily paid by firms in order to encourage loyalty and reduce costs associated with turnover, shirking, monitoring, and supervision. These may be more important for large firms, given their capital intensity and their possibly higher costs of monitoring and supervision. In essence, a high-wage policy may pay more for large firms than smaller firms. In such circumstances, the wage premium need not be a premium for workers who have better observable or even unobservable characteristics. It can be a pure rent or "prize" that elicits the positive behaviour on the part of workers so as not to risk loosing the prize. Large firms may find it more cost effective to pay such prizes rather than devote resources to continuous monitoring and supervision. This is analogous to the cost effectiveness of setting a large penalty (i.e., the loss off the efficiency wage) when it is costly to detect or prosecute malfeasant behaviour.

The empirical evidence (cited in the references) has not disentangled the relative importance of each of these factors contributing to the positive relationship between firm size and wages. Nevertheless, the evidence suggests that most of these factors contribute in some way. That is, large firms pay higher wages because of differences in observable and conventionally unobservable characteristics of their work force. Even when these factors are accounted for, large firms still seem to pay a wage premium, at least some of which may be an efficiency wage.

INDIVIDUAL WAGE DIFFERENCES AND GROWING WAGE INEQUALITY

The wages of individual workers reflect the previously discussed wage structures since the wages of individuals reflect their occupation, region, industry, and the firm in which they work. Individual wages also reflect such factors as human capital, discrimination, and union status — attributes that are discussed in other chapters.

A number of recent studies have documented a growing wage inequality amongst individual workers. This inequality has been manifest in a variety of forms: wage polarization, increasing returns to education, and greater wage inequality in general.

Wage Polarization

Wage polarization[5] has been associated with the decline in unionized, blue-collar, reasonably well-paid jobs in heavy manufacturing "smokestack" industries

---

5. Canadian evidence on wage polarization is provided in Economic Council of Canada (1990, 1991) and Picot, Miles and Wannell (1990). In the U.S., the wage polarization phenomenon is emphasized in Bluestone and Harrison (1988).

like iron and steel. This "declining middle" in the occupational and industrial wage distribution has been displaced by jobs at the more extreme ends of the pay scale. These include the higher paid jobs in managerial, professional, technical, and administrative positions in industries like financial and business services — the new "information-oriented" economy. The middle-wage jobs have also been displaced by jobs at the low end of the pay scale, especially in personal services and retail trade. This is part of the deindustrialization or industrial restructuring that has been occurring away from middle-wage manufacturing and towards the service economy, with its more polarized wage distribution.

Increasing Returns to Education

A related development involves increasing returns to higher education, reflecting the premium that these new jobs place on higher education. The empirical evidence indicates that the returns to education in the United States generally increased slightly in the 1960s, decreased in the 1970s, and then increased dramatically in the 1980s.[6] This reflects a combination of demand factors (e.g., the demand for skilled, educated workers rising in the 1980s) and supply factors (e.g., the supply of educated baby-boomers flooding the labour market in the 1970s and fewer graduates entering in the 1980s). This flooding of the labour market with college graduates in the 1970s, with the resulting depressing effect on the returns to education, gave rise to the phrase, "the overeducated American," a phenomenon that reversed in the 1980s.

Freeman and Needels (1991) indicate that the returns to higher education also increased in Canada in the 1980s; however, the increase was not as large as in the United States. Their evidence indicates that the returns to higher education in Canada did not increase as fast as in the United States because of a faster growth of the supply of university graduates in Canada relative to the United States. In other words, the demand for more highly educated persons increased, but the supply response kept the economic returns to higher education in check more in Canada than in the United States.

Freeman and Needels (1991) also indicate that other factors contributed to the slower growth of economic returns to education in Canada, albeit these effects are not as important as the supply response, and they are more difficult to measure. These factors include the greater strength of unions, the faster economic growth, and the stronger trade balance in Canada. In other words, in the United States less educated low-wage workers were more adversely affected by the decline of unionism, the slower economic growth, and the effect of import competition.

---

6. Freeman and Needels (1991), Katz and Murphy (1992), and Murphy and Welch (1992).

Growing Wage Inequality

Other studies have documented the growing wage inequality by comparing wage differences across individual workers.[7] A common measure of wage inequality, for example, is the "90–10 differential," which involves a comparison of the average wage of high-wage workers (those in the 90th decile such that 90 percent of the work force has lower wages) and in the 10th decile (where only 10 percent have lower wages). This simple summary figure apparently gives similar results as do more complicated measures of wage dispersion that reflect inequality throughout the wage distribution (Katz and Murphy, 1992, p. 45).

The empirical evidence indicates that wage inequality has increased throughout the 1970s and especially the 1980s. While there is not a consensus on the relative contribution of the different supply and demand side factors that have contributed to the growing inequality, there is reasonable agreement that demand side factors were most important, especially in the 1980s. That is, the technological change and industrial restructuring that occurred was biased in favour of more skilled and educated workers. As well, the increased international competition and imports tended to adversely affect low-wage workers, while increased exports positively affect higher-wage workers.

While these demand side factors were the most important determinants of the growing wage inequality, they were augmented by other interrelated institutional and supply side factors, many of which also contributed to the previously discussed increasing returns to education that occurred over the 1980s. Specifically, the declining unionization that occurred in the United States contributed to wage inequality since unions tend to reduce wage disparity. Increases in the legislated minimum wage did not keep up with inflation or average wage growth so that wages at the low end would not increase for that reason. On the supply side, there was a slower growth in the influx of educated college grads so that higher wages were not constrained by that factor in the 1980s. In contrast, in the 1960s and 1970s, supply side factors notably the influx of the large cohort of baby-boomers, did serve to constrain the wages of persons in those large cohorts.[8]

Supply and demand changes, as well as legal and institutional factors, also seem to explain the tendency towards greater wage equality that occurred in the earlier period of the 1940s and 1950s in the United States. Golden (1992) indicates that the compression of the individual wage structure at that time occurred because of the confluence of a variety of factors: increases in the demand for unskilled workers to work on automated production processes;

---

7. Blackburn, Bloom and Freeman (1990), Borjas, Freeman and Katz (1992), Bound and Johnson (1992), Katz and Murphy (1992), and Murphy and Welch (1992).
8. Murphy and Welch (1992) and Welch (1979).

increases in the supply of more educated skilled workers; and increases in the minimum wage and union power.

International Trade, Wages, and Wage Inequality

Given the increased importance of international competition and free trade, the impact of trade on wages merits additional comment. Since much of our international competition comes from low-wage countries who specialize in more labour-intensive production, then import competition could be expected to lower wages and to increase wage inequality. The latter would occur if low-wage labour was disproportionately displaced by import competition, and high-wage labour was positively affected by their being disproportionately involved in the production of our more capital-intensive exports.

As discussed previously in the section on interindustry wage differences, industries subject to greater import competition (as measured, for example, by the import penetration or share of imports in the industry output) tend to pay lower wages. Those studies also tended to find that export-oriented industries tend to pay higher wages. Since the import competition is greatest in the lower-wage industries, given the comparative advantage of low-wage, labour-intensive countries, then this should also serve to increase interindustry wage differences and wage inequality.

This impact of trade on wage inequality also tends to be found in the studies on individual wage inequality, when they relate individual wages to measures of trade.[9] The methodology for doing so typically involves estimating the domestic labour supply equivalent (at different skill levels) of the import penetration, and then simulating the impact of this labour supply increase on the wage structure and hence individual earnings. Since the import competition tends to adversely affect low-wage workers, and the export expansion tends to increase the wages of skilled high-wage workers, the increased trade tends to contribute to the growing wage inequality that is occurring.

This is not meant to imply that free trade is undesirable even for labour. The benefits of free trade are well known, and many of these benefits will accrue to labour in the form of lower product prices and hence increased *real* wages. There will be winners and losers, even if the net gains are positive. Unfortunately, workers who already are disadvantaged in the sense of having low wages are likely to be losers, at least from trade with low-wage countries. This highlights the importance of labour adjustment policies to facilitate the reallocation of labour from the declining low-wage import competing sectors to the expanding higher-wage export sectors. It also highlights the fact that when we are competing with countries whose wages are less than one-fifth of those in Canada, we

---

9. Borjas, Freeman and Katz (1991), Katz and Murphy (1992), and Murphy and Welch (1991, 1992).

cannot compete on the basis of low wages, but rather have to compete on the basis of a high value-added strategy.

QUESTIONS

1. "In a competitive economy, there can be no such thing as a pure interindustry wage differential in the long run." Discuss.

2. If efficiency wage structures prevailed in an industry, would you expect self-employed individuals to also receive an efficiency wage premium? If not, could you use this theoretical expectation to test for whether pure interindustry wage premiums reflect efficiency wages?

3. Assume that efficiency wage premiums do prevail and that the "good jobs" that pay the efficiency wages premiums can be identified. Does this justify a government industrial strategy to protect and perhaps even subsidize the existence of such jobs? What if all countries follow such a strategy and compete for these good jobs?

4. If it is cost effective for certain industries to pay efficiency wages, would we expect such wage premiums to prevail for all occupations in those industries? What would one expect if the monitoring costs were higher amongst the less skilled occupations and greater amongst the more skilled occupations?

5. "Efficiency wages cannot prevail in the long run because there would always be a queue of qualified applicants for the jobs that paid the efficiency wage premiums. In such circumstances, the employer could increase the hiring standards or the job requirements and the wage premium would therefore reflect a compensating wage for the greater skill requirements or the more difficult requirements of the job." Discuss.

6. "We should be willing to engage in free trade with low-wage countries like Mexico, but not with high-wage countries like the United States. This is so because the import competition from low-wage countries is likely to displace the domestic 'bad jobs' that do not pay efficiency wages, but the import competition from high-wage countries is likely to displace the 'good jobs' that pay efficiency wages." Discuss.

7. Is there any theoretical reason to believe that the efficiency wage premium itself should be related to the average wage level of the industry, net of the efficiency wage component? In other words, it is theoretically possible that low-wage jobs may also pay an efficiency wage premium?

8. Why might efficiency wage premiums be more prevalent in large firms as opposed to small firms?

9. Why might large firms pay higher wages than small firms? Why might this relationship prevail even after controlling for the effect of other wage-determining factors such as differences in the skill distribution and in working conditions?

10. Why has wage inequality amongst individual workers increased throughout the 1970s and especially the 1980s? Why has this pattern been less prevalent in Canada as opposed to the United States?

11. What impact should free trade have on wage inequality in Canada? Does this differ depending upon whether we trade with the United States or Mexico?

12. "The dual labour market literature has amply documented that industries with poor working conditions are also low-wage industries. Therefore, compensating premiums do not exist and traditional labour market analysis must be replaced by an analysis grounded in the realities of the system." Discuss.

13. "Even if laws or unions can fix wages in an industry above the competitive rate, the resulting queue of applicants will enable employers to hire better quality labour or change the nonpecuniary aspects of the job so as to restore an equality of net advantage." Discuss.

14. Why might we expect, roughly speaking, that the supply of labour to an occupation is more inelastic than the supply of labour to an industry?

15. Consider an economy of only two industries, one high-wage and the other low-wage. Each employs the same number of workers, equally divided between two groups, skilled and unskilled. Skilled workers are identical with each other in the two industries and so are unskilled workers. The low-wage industry pays its skilled labour $10.00 per hour, and its unskilled labour $5.00 during both a recession and an expansion; that is, it has a rigid wage structure. The high-wage industry pays its skilled labour $20.00 per hour in both a recession and an expansion, but it pays unskilled labour $5.00 during recession and $15.00 during expansion; that is, only its skilled wages are rigid over the business cycle.

    For the whole economy, calculate the interoccupational wage differential (ratio of skilled to unskilled wages) in both the recession and the expansion. Calculate the interindustry wage differential (ratio of high to low industry) for both skilled and unskilled workers in both the recession and the expansion. Why does the interoccupational wage differential contract in the expansion and the interindustry differential widen?

16. "Even if we knew the determinants of the overall interindustry wage structure, and why it changed, this would be of little policy value since there is not much we could or should do to change that structure. It is the result of the interplay of a variety of market forces and what we want to do is to make

sure that the market forces are legitimate (e.g., nondiscriminatory, not artificial). Once that is done, then the interindustrial wage structure is what it is, and nothing more should be done." Discuss.

REFERENCES AND FURTHER READINGS

A. Interindustry Wage Structures

Bell, L. and R. Freeman. The causes of increasing interindustry wage dispersion in the United States. *ILRR* 44 (January 1991) 275–287.

Bell, L. and R. Freeman. The facts about rising industrial wage dispersion in the U.S. *IRRA* 39 (December 1986) 331–337.

Black, D. and J. Garan. Efficiency wages and equilibrium wages. *EI* 29 (July 1991) 525–540.

Christofides, L., R. Swidinsky and D. Wilton. A micro-economic analysis of spillovers within the Canadian wage determination process. *R.E. Stats.* 62 (May 1980) 213–221.

Davidson, C. and M. Reich. Income inequality: an inter-industry analysis. *IR* 27 (Fall 1988) 263–286.

Dickens, W. and L. Katz. Industry wage differences and industry characteristics. *Unemployment and Structure of Labor Markets*, K. Lang and J. Leonard (eds.) Oxford: Basil Blackwell, 1987.

Dickens, W. and K. Lang. Why it matters what we trade: a case for active policy. *The Dynamics of Trade and Employment,* L. Tyson, W. Dickens, and J. Zysman (eds.). Cambridge: Ballinger, 1988, 87–112.

Drago, R. Efficiency wages: an experimental analysis. *Australian Economic Papers* 29 (June 1991) 68–84.

Ehrenberg, R. *The Regulatory Process and Labor Earnings.* New York: Academic Press, 1979.

Freeman, R. B. and L. F. Katz. Industrial wage and employment determination in an open economy. *Immigration, Trade, and the Labor Market*, J. Abowd and R. Freeman (eds.). Chicago: University of Chicago Press, 1991, 235–259.

Gera, S. and G. Grenier. Interindustry wage differentials and efficiency wages: some Canadian evidence. In *Canadian Unemployment*, S. Gera (ed.) Ottawa: Economic Council of Canada, 1991.

Groshen, E. Sources of intra-industry wage dispersion. *QJE* 106 (August 1991) 869–884.

Grossman, G. The employment and wage effects of import competition. *Journal of International Economic Integration* 2 (Spring 1987) 1–23.

Grossman, G. Imports as a cause of injury: The case of the U.S. steel industry. *Journal of International Economics* 20 (May 1986) 201–223.

Helwege, J. Sectoral shifts and interindustry wage differentials. *JOLE* 10 (January 1992) 55–84.

Heywood, J. Labor quality and the concentration — earnings hypothesis. *R.E. Stats.* 68 (May 1986) 342–346.

Holzer, H., L. Katz and A. Kreuger. Job queues and wages. *QJE* 106 (August 1991) 739–768.

Howell, D. Production technology and the interindustry wage structure. *IR* 28 (Winter 1989) 32–50.

Johnson, N., R. Sambharya and P. Bobko. Deregulation, business strategy, and wages in the airline industry. *IR* 28 (Fall 1989) 419–430.

Katz, L. F. and L. H. Summers. Can inter-industry wage differentials justify strategic trade policy? *Trade Policies for International Competitiveness*, R.C. Feenstra (ed.). Chicago: University of Chicago Press, 1989.

Katz, L. and L. Summers. Industry rents: evidence and implications. *Brookings Papers: Microeconomics* (1989) 209–290.

Kaufman, B. and P. Stephen. Determinants of interindustry wage growth in the seventies. *IR* 26 (Spring 1987) 186–194.

Kleitzer, L. Industry wage differentials and wait unemployment. *IR* 31 (Spring 1992) 250–269.

Kreuger, A. and L. Summers. Reflections on the inter-industry wage structure. *Unemployment and the Structure of Labor Markets*, K. Lang and J. Leonard (eds.). Oxford: Basil Blackwell, 1987.

Krueger, A. and L. Summers. Efficiency wages and the interindustry wage structure. *Econometrica* 56 (March 1988) 259–293.

Kumar, P. Differentials in wage rates of unskilled labour in the Canadian manufacturing industries. *ILRR* 26 (October 1972) 631–645.

Lawrence, C. and R. Lawrence. Manufacturing, wage dispersion: an end game interpretation. *BPEA* 1 (1985) 47–106.

Murphy, K. and R. Topel. Unemployment, risk and earnings. In *Unemployment and its Structure of Labour Markets*, Edited by K. Lang and J. Leonard. Oxford: Basil Blackwell, 1987.

Ramaswamy, R. and R. Rowthorn. Efficiency wages and wage dispersion. *Economica* 58 (November 1991) 501–514.

Revenga, A. Exporting jobs: The impact of import competition on employment and wages in U.S. manufacturing. *QJE* 107 (February 1992) 255–284.

Rose, N. Labor rent sharing and regulation: evidence from the trucking industry. *JPE* 6 (December 1987) 1146–1178.

Shackett, J. and J. Trapani. Earnings differentials and market structure. *JHR* (Fall 1987) 518–531.

Thaler, R. Interindustry wage differentials. *JEP* 3 (Spring 1989) 181–193.

Thornicroft, K. Airline deregulation and the airline labour market. *JLR* 10 (Spring 1989) 163–181.

Vanderkamp, J. Industrial mobility: some further results. *CJE* 10 (August 1977) 462–472.

B. Interfirm Wage Differences and Firm Size

Baron, J., D. Black and M. Loewenstein. Employer size: the implications for search, training, capital investments, starting wages and wage growth. *JOLE* 5 (January 1987) 76–89.

Brown, C., J. Hamilton and J. Medoff. *Employers Large and Small.* Cambridge: Harvard University Press, 1990.

Brown, C. and J. Medoff. The employer size–wage effect. *JPE* 97 (October 1989) 1027–1059.

Davis, S. and J. Haltiwanger. Wage dispersion between and within U.S. manufacturing plants, 1963–86. *BPEA* (1991) 115–180.

Dunn, L. Work disutility and compensating differentials: estimation of factors in the link between wages and firm size. *R. E. Stats.* 68 (1986) 67–83.

Evans, D. and L. Leighton. Why do smaller firms pay less? *JHR* 24 (Spring 1989) 299–318.

Garen, J. Worker heterogeneity, job screening, and firm size. *JPE* 93 (August 1985) 715–739.

Groshen, E. Five reasons why wages vary among employers. *IR* 30 (Fall 1991) 350–381.

Groshen, E. Sources of intra-industry wage dispersion: how much do employers matter? *QJE* 106 (August 1991) 869–885.

Idson, T. and D. Feaster. A selectivity model of employer-size wage differentials. *JOLE* 8 (January 1990) 99–122.

Kappelli, P. and K. Chauvin. An interplant test of the efficiency wage hypothesis. *QJE* 106 (August 1991) 769–787.

Kruse, D. Supervision, working conditions, and the employer size–wage effect. *IR* 31 (Spring 1992) 229–249.

Mellow, W. Employer size, unionism and wages. *Research in Labor Economics*, R. Ehrenberg (ed.), Greenwich, Conn.: JAI Press, 1983. pp. 253–282.

Oi, W. Heterogeneous firms and the organization of production. *EI* 21 (April, 1983) 147–171.

Pearce, J. Tenure, unions, and the relationship between employer size and wages. *JOLE* 8 (April 1990) 251–269.

Rebitzer, J. and M. Robinson. Employer size and dual labor markets. *R.E. Stats.* 73 (November 1991) 710–715.

Strand, J. The relationship between wages and firm size. *IER* 28 (February 1987) 51–68.

Schmidt, C. and K. Zimmermann. Work characteristics, firm size and wages. *R.E. Stats.* 73 (November 1991) 705–710.

C. Individual Wage Differences and Growing Wage Inequality

Beach, C. The "vanishing" middle class?: evidence and explanations. Kingston: Queens Papers in Industrial Relations, 1988.

Blackburn, M. What can explain the increase in earnings inequality amongst males? *IR* 29 (Fall 1990) 441–456.

Blackburn, M. D. Bloom and R. Freeman. The declining position of less skilled American males. *A Future of Lousy Jobs? The Changing Structure of U.S. Wages*, G. Burtless (ed.). Washington, D.C.: Brookings Institute, 1990.

Bluestone, B. and B. Harrison. *The Great U-Turn: Corporate Restructuring and the Polarization of America.* New York: Basic Books, 1988.

Borjas, G., R. Freeman and L. Katz. On the labor market effects of immigration and trade. *The Economic Effects of Immigration in Source and Receiving Countries*, G. Borjas and R. Freeman (eds.). Chicago: University of Chicago Press, 1992.

Bound, J. and G. Johnson. Changes in the structure of wages in the 1980s: an evaluation of alternative explanations. *AER* 82 (June 1992) 371–392.

Burtless, G. (ed.) *A Future of Lousy Jobs.* Washington, D.C., Brookings Institute, 1990.

Connelly, R. A framework for analysing the impact of cohort size on education and labor earnings. *JHR* 21 (Fall 1986) 543–562.

Dooley, M. and P. Gottschalk. Earnings inequality amongst males in the United States: trends and the effect of labor force growth. *JPE* 92 (February 1984) 59–89.

Economic Council of Canada. *Good Jobs, Bad Jobs.* Ottawa: Supply and Services, 1990.

Economic Council of Canada. *Employment in the Service Economy.* Ottawa: Supply and Services, 1991.

Freeman, R. The effect of demographic factors on the age-earnings profile in the United States. *JHR* 14 (Summer 1979) 289–318.

Freeman, R. and K. Needels. Skill differentials in Canada in an era of rising labor market inequality. National Bureau of Economic Research working paper 3827, 1991.

Golden, C. and R. Margo. The great compression: the wage structure in the United States at mid-century. *QJE* 107 (February 1992) 1–34.

Grubb, W. and R. Wilson. Sources of increasing inequality in wages and salaries, 1960–80. *MLR* 112 (April 1989) 3–13.

Horrigan, M. and S. Horrigan. The declining middle class thesis. *MLR* 111 (May 1988) 3–13.

Katz, L. and K. Murphy. Changes in relative wages, 1963–1987: supply and demand factors. *QJE* 107 (February 1992) 35–78.

Katz, L. and A. Ravenga. Changes in the structure of wages: the United States versus Japan. *Journal of Japanese and International Economies.* 3 (December 1989) 522–553.

Kennedy, B. Mobility and instability in Canadian earnings. *CJE* 22 (May 1989) 283–294.

Leonard, J. and L. Jacobson. Earnings inequality and job turnover. *AER* 80 (May 1990), 298–302.

Murphy, K. and F. Welch. The role of international trade in trade differentials. *Workers and Their Wages*, M. Kosters (ed.). Washington, D.C.: American Institute Press, 1991, pp. 39–69.

Murphy, K. and F. Welch. The structure of wages. *QJE* 107 (February 1992) 285–326.

Picot, G., J. Males and T. Wannell. Good jobs/bad jobs and the declining middle, 1967–1986. Ottawa: Statistics Canada Analytical Studies Branch, 1990.

Rosenthal, N. The shrinking middle class: myth or reality? *MLR* 8 (March 1985) 3–10.

Welch, F. Effects of cohort size on earnings: the baby boom babies financial bust. *JPE* 87 (October 1979) S65–S98.

CHAPTER 21

Public–Private Sector Wage Differentials

The issue of public–private sector wage differentials is a topic of interindustry wage determination since the public sector is simply one of many industries. However, the public sector is singled out for special attention for various reasons: it is a large sector; it is the subject of policy concern mainly because of strikes and wage settlements; its impact may spill over into the private sector; and it has peculiarities that make wage determination somewhat unique.

As our earlier discussion indicated, the theoretical determinants of interindustry wage differentials include compensating adjustments for nonpecuniary differences, short-run adjustments, and noncompetitive factors. Just as these broad categories provided a convenient framework for analyzing the determinants of interindustry wage differentials, they also are convenient for categorizing the theoretical determinants of the particular interindustry wage differential examined in this chapter — the public–private sector wage differential.

The public sector can be broadly defined to include education, health, and government enterprises, as well as the more narrowly defined government employment at the federal, provincial, and municipal levels. Thompson and Ponak (1992) indicate that the broadly defined public sector in Canada in the mid-1980s employed almost 3 million workers, or nearly one-third of the work force. Of that 3 million approximately 31 percent were in the health sector, 26 percent in education, 17 percent in government enterprises, and 26 percent in the federal, provincial, and municipal governments. Clearly, the public sector is a large and important employer.

THEORETICAL DETERMINANTS OF PUBLIC–PRIVATE WAGE DIFFERENTIALS

Compensating Adjustments for Nonpecuniary Differences

Interindustry wage differentials may reflect compensating adjustments for differences in the nonpecuniary aspects of employment across industries. With respect to the private and public sectors, nonpecuniary advantages may exist with respect to such factors as job security, fringe benefits, and political visibility. To the extent that these advantages prevail in the public sector, we would expect a correspondingly lower wage to compensate for the advantages. However, to the extent that these advantages are dissipating over time, we may also expect public sector wages to rise relative to those in the private sector to compensate for the loss of these nonpecuniary advantages.

Job security often is discussed as being more prevalent in the public than in the private sector. Theoretically this may be the case because job security could be necessary to prevent the abuses of political patronage. That is, without a modicum of job security, civil servants could be replaced each time a new political party came into power, or whenever a politician wanted to gain favour by granting patronage in the form of civil service jobs. To avoid this potential abuse, and to ensure a degree of continuity in the public sector work force, a degree of job security may be granted. Larger elements of the public sector also may be able to easily provide such job security because their size gives them a portfolio of jobs within which to reallocate their work force.

As with job security, there are theoretical reasons to suggest that the public sector may provide more liberal fringe benefits. This occurs because in the public sector there are not adequate checks to prevent employers from saving on *current* wage costs by granting liberal retirement benefits and pension schemes, the cost of which may be borne by *future* generations of taxpayers. Fringe benefits that are payable in the future, such as early retirement plans and employer-sponsored pensions, can be viewed as deferred wages in the sense that workers would willingly give up some current wages for these future benefits. Such deferred compensation systems may be a way for the public sector to shift costs to future generations of taxpayers; however (as discussed in the later chapter on optimal compensation systems) deferred compensation also may exist as part of an optimal compensation system to ensure honesty and work effort. Such a system may be especially important in the public sector, to the extent that other mechanisms, such as the threat of dismissal or the monitoring of output, are not available.

In the private sector there is a built-in check to ensure that employers are constrained in their granting of such deferred benefits: eventually they have to meet the obligation of paying for them. However, in the public sector, unless such benefits are fully funded, their costs will be borne by future taxpayers. To the extent that they have little or no say in the current political process, there is no automatic mechanism to prevent public sector employers from saving on current wage costs by granting liberal deferred cost items, such as early retirement or substantial pensions or pensions indexed for future inflation. The only check is the possibility that future generations may not honour such commitments made by their predecessors, or the possibility that local property values may fall to the extent that they reflect the future tax obligations associated with such future cost commitments.

Additional nonpecuniary advantages offered by the public sector could include political visibility, access to control over political rules, and the opportunity to provide public service. For some, these factors may be valued for their own end; for others they may be valued as a means to other objectives. For example, some workers may regard a period of public sector employment as a low-wage apprenticeship period that provides them with inside knowledge, access to power sources, and contacts in the political arena. These factors may be of immense value (and hence lead to higher remuneration) in future private sector jobs as consultants, lobbyists, or simply partners in firms that do business with the government or that would benefit from inside political information.

For others, public sector employment may provide "the nonpecuniary satisfaction of doing good" — to use a phrase utilized by Reder (in Hamermesh 1975, p. 28). This could be a reason, for example, for people to do volunteer work for churches or charities, or for some to accept lower salaries in such nonprofit institutions. It may be a more prevalent phenomenon throughout the public sector, especially in teaching or health care. The term "public service" means just that to many persons.

The nonpecuniary factors that we have discussed — job security, fringe benefits, and political visibility — generally are ones for which there are theoretical reasons to believe that they would be greater in the public than private sector. Certainly conventional wisdom and casual empiricism seems to suggest this to be the case. To the extent that it is true, the public sector would be expected to have lower wages to compensate for these factors.

While these nonpecuniary factors may be greater in the public than private sector, there are reasons to believe that the gap may be dissipating over time. Certainly this seems to be the case with job security and it may be the case with fringe benefits to the extent that the public sector may have reached diminishing returns with respect to fringe benefits it can provide, and the private sector is now catching up. Even with respect to the opportunity to do public service, attitudes seem to be changing so that public sector jobs are done in return for pay, like most other jobs. To the extent that these nonpecuniary factors have diminished in the public sector, one would expect the wages of public sector workers to rise relative to those of their private sector counterparts.

Short-Run Adjustments

Interindustry wage differentials may reflect a short-run disequilibrium situation. In fact the disequilibrium wage serves as the signal for new entrants, and this supply response should restore the interindustry wage structure to its long-run equilibrium level.

This scenario could be relevant to the public sector. In essence, the expansion of the public sector that occurred, especially in the 1950s and 1960s, would lead to increases in the demand for public sector labour and, hence, to increases in their wages. The higher wages would enable public employers to recruit the necessary work force associated with the expansion of the public sector.

In recent years, however, there are forces at work to suggest that the rapid expansion of the public sector may have ended. Certainly political pressures are mounting as taxpayers show opposition to an ever-expanding role of the public sector; the pressure now is for public sector retrenchments. In addition, basic demographic changes may be reducing the demand for public services, especially in education as the baby boom has left the education sector, and in health as family size reduces. Again, this may change in the future as health care for the aged becomes an issue.

Empirical evidence for Canada,[1] in fact, indicates that the public sector grew substantially during the 1950s and 1960s, but levelled off after 1970. The fastest growth in recent years has been in provincial and local administration and in health and welfare.

Noncompetitive Factors

Interindustry wage differentials may also reflect noncompetitive factors and it is in this area that the differences between public and private sector labour markets become most pronounced. Specific features that have important implications for public–private wage differentials and their changes over time revolve around the fact that the public sector can be characterized as subject to various peculiarities including a political constraint rather than a profit constraint, possible monopsony, an inelastic demand for labour, and a high degree of new unionization. Each of these factors will be discussed in turn, with an emphasis on their implications for public–private wage differentials and their changes over time.

Political Rather than Profit Constraint

The public sector usually is not subject to a profit constraint as traditionally exists in the private sector. Rather, the profit constraint is replaced by an ultimate political constraint and there is the belief that the political constraint is less binding. This occurs because taxpayers are diffuse, often ill-informed, and

---

1. See, Bird (1978), Gunderson and Riddell (1991) and Picot (1986).

can exert pressure only infrequently and with respect to a package of issues. Public sector managers also are diffuse in the sense that lines of responsibility are not always well defined and "buck-passing" can prevail. Workers in the public sector, on the other hand, are often portrayed as a unified, distinct interest group providing direct pressure for wage increases. A more serious imperfection of the political process arises from the fact, discussed earlier, that *future* taxpayers have little or no representation in *today's* political process.

While these features of the political constraint suggest the possibility that it is less binding than the profit constraint, there still are constraining influences on wage settlements in the public sector. Taxpayer scrutiny is now extremely strong and in fact politicians may gain by appearing to be cost-conscious guardians of the public purse by reducing wage costs. Public sector workers may also be called upon to set an example of moderate wages to curb inflation as was evident in recent wage controls applied to the public sector. Even the alleged diffuse nature of management in the public sector may work against employees; it can be difficult to win wage gains if one doesn't know with whom one is bargaining.

Perhaps the most important feature of the political constraint that works against public sector workers is that, during a strike, tax revenues keep coming in to the public sector even though wage expenditures are reduced and public services are not provided. This is in contrast to the private sector where employers are under considerable pressure to settle because they are losing customers and sales revenues during a strike. To be sure, there can be pressure from taxpayers to have the wage savings go into general tax revenues or into tax reductions because services are not provided. Nevertheless, the fact remains that this political pressure is less stringent than the profit constraint when firms lose customers and revenues during a strike.

Monopsony

To the extent that the government sector is often the dominant employer in particular labour markets, governments may utilize their monopsony power to pay lower wage rates than if they behaved competitively. Political forces may pressure them to act as model employers and not utilize their monopsony power; nevertheless, there is also pressure to be cost conscious and this may lead them to exercise their monopsony power.

In fact, the empirical evidence cited earlier in the chapter on monopsony suggested that for at least two elements of the public sector — teaching and nursing — there was some evidence of monopsony. The extent to which these results can be generalized for other elements of the public sector, or for the teaching and nursing professions as a whole, remains an unanswered empirical question.

To the extent that monopsony power exists and is exercised in the public sector, public sector wages would be lower than they would be in the absence of monopsony. However, the pressure of monopsony also means that, for a range of wage increases, unions would not have to be concerned about employment reductions. In fact, as illustrated in the chapter on monopsony, wage increases may actually lead to employment increases, suggesting the possibility of a substantial union wage impact in monopsonistic labour markets.

Inelastic Demand for Public Sector Labour

The possibility of a substantial union impact on wages in the public sector is furthered by the possibility that the demand for labour in the public sector may be wage inelastic. In such circumstances unions could bargain for substantial wage increases without worrying about large reductions in the employment of their members. The inelasticity of the demand for labour may occur because noncompetitive forces restrict the utilization of substitute inputs as well as substitute services.

Many of the services produced in the public sector are so-called essential services that are not provided in the private sector, and when they are so provided, they are often under governmental regulation. Since consumers (taxpayers) are unable to substitute other services for those provided by the public sector, their demand for these services may be relatively price (tax) inelastic. In essence, wage increases can be passed on to taxpayers in the form of tax increases for the essential services without taxpayers reducing their demand for these services and, hence, the derived demand for labour.

This is strengthened by the fact that there are often few good inputs to substitute for public sector labour as it becomes expensive. This may reflect the nature of the public sector production function, but it may also reflect the fact that the public sector is heavily professionalized, and professional labour tends to have a degree of control over the utilization of other inputs, including nonprofessional labour.

While these factors suggest that the demand for public sector labour would be wage inelastic, other factors are at work in the opposite direction. Specifically, the high ratio of labour cost to total cost in many public services suggests that the demand would be elastic. In addition, substitute inputs and services certainly could be utilized, especially in the long run. Even the essentiality of the service may work against public sector employees by restricting their right to strike and reducing public sympathy during a strike. In addition, the employers may stall in the bargaining process, knowing that the essential elements of the service will be maintained and, as indicated earlier, tax revenues are still forthcoming.

Since theory does not indicate unambiguously whether the demand for public sector labour would be elastic or inelastic, we must appeal to the empirical evidence. The evidence from various U.S., studies indicates that the demand for labour in the public sector is inelastic, and that the more essential the service, the greater the inelasticity (Gunderson and Riddell, 1991 p. 167).

Unionization

A final noncompetitive factor — unionization — may also affect the wages of public sector workers relative to their private sector counterparts. This is especially the case when the unionization is coupled with the other factors discussed in this chapter — the inelastic demand for public sector labour, the possible monopsony in some public sector markets, and the absence of a profit constraint.

In Canada, this recent growth and high degree of unionization in the public sector is documented, for example, in Ponak and Thompson (1992). Some sectors, like the federal civil service, are organized almost to their full potential. Others, like teaching and health, are now predominantly unionized.

Competitive Floor But Not Ceiling

The previously discussed factors — absence of a profit constraint, monopsony, inelastic demand, and the high degree of recent unionization — all suggested that wage determination may be different between the public and private sectors, and that noncompetitive factors may give rise to an interindustry wage differential between the public and private sectors. Most of the factors, with the exception of monopsony, would appear to favour a noncompetitive wage advantage in the public sector, although a variety of subtle constraining influences are present.

However, there is a stronger reason to expect the wage advantage to be in favour of public sector workers. This is so because the forces of competition would ensure that wages in the public sector would not fall much below wages in the private sector for comparable workers. If they did, the public sector would not be in a competitive position to recruit labour, and would experience problems of recruitment, turnover and morale. That is, competition would ensure a competitive *floor* on wages in the public sector.

Theses same competitive forces, however, need not provide an effective *ceiling* on wages in the public sector. To be sure, if public sector wages exceeded wages of comparable workers in the private sector, there would be queues of applicants and excessive competition for public sector jobs. Nevertheless, employers in the public sector are not under a profit constraint to respond to these disequilibrium signals; they need not lower their relative wage. They *may* not pay an excessive wage if the political constraint is a binding cost constraint; nevertheless, there is no guarantee that this will be the case. In essence, the forces of competition ensure a floor, but they need not provide an effective ceiling on wages in the public sector. Hence the potential bias, in theory, for wages in the public sector to be in excess of wages in the private sector for comparable workers.

EMPIRICAL EVIDENCE

Empirical studies of wage differentials between the public and private sector have tried to control for other legitimate wage-determining factors by multiple regression analysis or by choosing groups that are similar except that one group is in the public sector and the other is in the private sector. A decomposition procedure based on multiple regression analysis is first discussed; it may be skipped by the reader not familiar with regression analysis, albeit the empirical evidence is self-explanatory.

Decomposition of Public–Private Wage Differentials: Mathematical Exposition

The overall average earnings differential between public and private sector workers can be decomposed into two component parts: a portion due to differences in the endowments of wage-determining factors between the two sectors, and a portion attributable to a pure interindustry wage differential. The latter can be termed an economic rent or surplus payment because it represents payment for being in a particular industry, after controlling for the effect of the usual wage-determining factors.

This decomposition technique — which will also be utilized in the subsequent chapter on male–female wage differentials — can be illustrated formally. Let y denote the dependent variable (earnings), X the set of usual wage-determining explanatory variables (human capital factors such as education and training, and control variables for such factors as occupation and region), and b the set of regression coefficients estimated from an earnings equation based on individuals as units of observation. In such an earnings equation, the X's can be regarded as the endowments of wage-determining factors and the b's as the pay structure associated with differences in these endowments. The regression coefficient for the education variable, for example, indicates the change in earnings that results from an additional year of education — it is the monetary reward or pay structure associated with the acquisition of additional education.

Separate earnings equations can be estimated for each of the public and private sectors, respectively denoted by the subscripts g (government or public) and c (private or competitive). That is, for the public sector

$$y_g = \Sigma b_g X_g \tag{21.1}$$

and for the private sector

$$y_c = \Sigma b_c X_c \tag{21.2}$$

In regression analysis, the mean of the dependent variable is equal to the regression coefficients times the mean values of the explanatory variables. Therefore,

$$\overline{y}_g = \Sigma b_g \overline{X}_g \text{ and} \tag{21.3}$$

$$\overline{y}_c = \Sigma b_c \overline{X}_c \tag{21.4}$$

Subtracting Equation 21.4 from Equation 21.3 gives the average earnings differential between public and private sector workers: that is,

$$\bar{y}_g - \bar{y}_c = \Sigma b_g \bar{X}_g - \Sigma b_c \bar{X}_c \qquad (21.5)$$

Subtracting and adding $\Sigma b_c \bar{X}_g$ to the right-hand side (and therefore not changing the equality) yields

$$\bar{y}_g - \bar{y}_c = \Sigma b_g \bar{X}_g - \Sigma b_c \bar{X}_g + \Sigma b_c \bar{X}_g - \Sigma b_c \bar{X}_c \qquad (21.6)$$

which, after collecting terms, yields

$$\bar{y}_g - \bar{y}_c = \Sigma(b_g - b_c)\bar{X}_g + \Sigma b_c(\bar{X}_g - \bar{X}_c) \qquad (21.7)$$

The first term on the right-hand side is the differences in pay structure (b's) between public and private sector workers, evaluated for the same wage-determining endowments ($\bar{X}_g$'s). This represents the pure interindustry wage differential because it arises solely from differences in the way public and private sector workers are paid for the same wage-determining characteristics. The second term on the right-hand side of Equation 21.2 is the difference in the endowments of wage-determining characteristics (X's) between the public and private sector workers, evaluated according to the private sector pay structure (b_c's). Since the private sector pay structure is usually regarded as the competitive norm, it makes sense to evaluate differences in the wage-determining endowments according to that norm.

Obviously it is the first term on the right-hand side of Equation 21.7 that is important for policy purposes since it reflects the economic rent or surplus payment associated with being in one sector as opposed to the other sector. If the public sector pays its workers more than the private sector (that is, the b_g's are greater than the b_c's) for similar levels of education, training, and other wage-determining factors (that is for the $\bar{X}_g$'s), then this will result in a positive surplus payment, $\Sigma(b_g - b_c)\bar{X}_g$. The second expression, on the other hand, is not of policy concern since it represents a differential payment for different endowments of legitimate wage-determining factors, such as human capital. These two components — a surplus payment, and a payment for differences in productivity-related endowments — together make up the overall average earnings differential, $\bar{y}_g - \bar{y}_c$, between public and private sector workers.

Estimates of this average earnings differential and its two component parts for full-year, full-time workers are provided in Table 21.1, based on 1970 and 1980 Canadian census data. The public sector is the industry designation of public administration and the private sector is the manufacturing industry.

Clearly workers in the public sector on average have higher earnings than workers in the private sector (column 1). However, much of the differential comes about because workers in the public sector have greater endowments of wage-determining factors such as education (column 2). The pure surplus payment or economic rent associated with being employed in the public as opposed to the private sector (column 3) is 6.2 percent for males and 8.6 percent for females in 1970, falling slightly to 4.2 for males and rising slightly to 12.2 percent

Table 21.1
PUBLIC SECTOR EARNINGS ADVANTAGE AND ITS DECOMPOSITION
(% over private sector earnings)

YEAR AND GENDER	OVERALL DIFFERENTIAL	AMOUNT ATTRIBUTED TO	
		ENDOWMENTS	RENT
1970			
Males	9.3	3.1	6.2
Females	22.3	13.7	8.6
1980			
Males	19.1	14.8	4.2
Females	27.2	15.0	12.2

Source: 1970 figures are from Gunderson (1979b) and 1980 figures are from Shapiro and Stelcner (1989).

for females in 1980. Since there are over twice as many males as females in the public administration category, this would yield a pure public sector earnings advantage of approximately 7 percent overall in both 1970 and 1980. These results generally are in line with those reported in Smith's (1977b) comprehensive analysis, which summarizes and updates her application of this decomposition technique in a variety of U.S. studies.

Other Empirical Evidence

Gunderson and Riddell (1991) summarize other empirical evidence on public–private compensation differences. The Canadian evidence[2] suggests that there is a pure public sector wage advantage, likely in the neighbourhood of 5 to 10 percent. It is larger for females than males and for low-wage workers as opposed to high-wage workers; in fact, it is likely to be negative for high-wage workers. This highlights a policy dilemma in that reducing the premium would disproportionately affect workers who are already disadvantaged in the labour market — women and low-wage workers.

The public sector wage premium appears to be dissipating somewhat over time, especially for males, although it appears to be increasing somewhat for

2. Canadian studies based in microdata sets with the individual worker as the unit of observation include Daniel and Robinson (1980), Gunderson (1978, 1979, 1984) and Shapiro and Stelcner (1989).

EXHIBIT 21.1   ARE SEVERE CHECKS NECESSARY TO CURB PUBLIC SECTOR WAGES?

The empirical evidence discussed in this section is a useful input into the important policy question of whether severe checks are necessary to curb public sector wages, given that a pure rent or wage premium over and above that paid to comparable private sector workers seems to prevail. The evidence suggests, however, that strong public policy intervention is not necessary, at least at this stage. This is so for a variety of reasons.

The public sector wage advantage is not large, probably in the neighbourhood of 5 to 10 percent, and this tends to be based on data prior to more recent policies of retrenchment and restraint in the public sector. Since the public sector premiums tend to be largest for low-wage workers and women, then any policy to curb these premiums would disproportionately fall on such workers who are already relatively disadvantaged in the labour market.

The wage advantage also tends to be dissipating over time. Usually high settlements that get reported in the press and media often reflect "catch-ups" and tend to be followed by low settlements. Overall, aggregate public sector settlements tend to be similar to those in the private sector, and they respond similarly to changes in economic conditions. As well, public sector settlements tend not to have any significant spillover effects to settlements in the private sector.

females. Public sector premiums also tend to be quite volatile, with unusually high settlements often reflecting a "catch-up," and dissipating fairly quickly over time as they are followed by smaller settlements.

The public sector premium in fringe benefits is likely to be greater than the wage premium, although data is limited on this point. The wage premium tends to be largest at the provincial and municipal level and smallest at the federal level. Given the high degree of unionization in the public sector, much of the wage premium likely reflects the premium that typically goes to unions.

Aggregate wage settlements[3] in the public and private sectors were fairly similar over the 1970s and 1980s. Wage settlements in both sectors tended to be quite responsive to market forces and to changes in aggregate demand. Furthermore, public sector settlements tended not to have spillover effects to wage settlements in the private sector.

These empirical generations suggest that severe policies are not necessary, at least at this stage, to restrain public sector wage settlements (Exhibit 21.1).

3. Canadian studies based on aggregate wage equations (augmented Phillips curves) include Auld, Christophides, Swidinsky and Wilton (1980), Auld and Wilton in Conklin, Courchene and Jones (1985), Cousineau and Lacroix (1977), Riddell and Smith (1982), and Wilton (1986).

EXHIBIT 21.2   WHAT DO PUBLIC SECTOR WAGES AND ENVIRONMENTAL ISSUES HAVE IN COMMON?

The empirical evidence discussed in this section and in Exhibit 21.1 suggests that severe policies to curb public sector settlements are not necessary, at least at this stage. There is one area, however, where constant vigilance is required.

Neither market nor political forces are likely to provide a sufficient check to ensure that current taxpayers do not shift some of the costs of public sector settlements to future generations of taxpayers through deferred compensation arrangements. Such deferred compensation could include pensions, job security, and seniority-based wage increases, all of which tend to be more prominent in the public sector as opposed to the private sector.

While checks on "shifting the bill" to future generations of taxpayers are necessary, the problem is that the current taxpayers have little incentive to put such checks in place. Reducing the deferred compensation package would mean that they would likely have to incur higher current costs, which falls on their tax bill. In contrast, deferred compensation falls on the tax bill of future generations of taxpayers.

It is in that sense that public sector compensation and environmental issues have something in common. Current generations may have insufficient economic incentives to preserve the environment for future generations, since preserving the environment can be costly for the existing populace. What is needed are mechanisms whereby the current generations are required to pay for the full costs of their decisions. This applies to public sector compensation decisions as well as to environmental decisions. The irony is that the current population has little incentive to put such mechanisms in place.

EXHIBIT 21.3   ARE THERE QUEUES FOR GOVERNMENT JOBS?

One of the best measures of whether a job is "overpaid" and whether incumbent workers therefore receive economic rents, is whether there are queues of qualified applicants for such jobs. This is so because queues provide a good "bottom-line" measure of the total compensation of the job (wages and fringe benefits) relative to the requirements and working conditions of the job. That is, as an alternative to trying to estimate, for example, pure public-private wage differentials after controlling for the effect of other legitimate wage-determining characteristics, it may be desirable to estimate queues of qualified applicants.

Krueger (1988) does so based on a number of different U.S. data sets. He finds that there are substantial queues for federal government jobs. As well, as the ratio of federal earnings to private sector earnings increases, the application rate for federal government jobs increases, as does the quality of such applicants.

Source: A. Krueger. The determinants of queues for federal government jobs. *Industrial and Labor Relations Review* 41 (July 1988) 567–581.

Nevertheless, constant vigilance is required especially to prevent current taxpayers from passing public sector costs to future taxpayers in the form of deferred compensation (Exhibit 21.2).

While these generalizations are based on Canadian studies, they also tend to apply to the United States[4], with the possible exception that public sector wage premiums tend to be largest at the federal level in that country. As well, there is U.S. evidence indicating that there are substantial queues of qualified applicants for public sector jobs, and the greater the wage premium, the larger the queue (Exhibit 21.3).

QUESTIONS

1. Since the public sector is simply one of many industries, public–private sector wage differentials are really a special topic of interindustry wage differentials. Discuss.

2. Are there any *theoretical* reasons to believe that job security and fringe benefits may be different in the public and the private sector?

3. Indicate how the various determinants of the public–private sector wage differential may have *changed* during the 1950s and 1960s in Canada and how this may have affected the wage differential over that period.

4. Cousineau and Lacroix (1977, p. 113) indicate: "because of its particular characteristics as an employer, the public sector is particularly 'vulnerable' to the unionization of its employees and to union pressure." Discuss.

5. The fact that wage determination in the public sector ultimately depends on political factors may well work against workers in the public sector, even though we traditionally think that this puts them in a better position than when wages are under the ultimate profit constraint in the private sector. Discuss.

6. Criticisms of public sector wages as being too high often implicitly assume that private sector wages are the correct norm. What is so virtuous about private sector wages given that they can reflect market imperfections, unequal bargaining power, and a variety of noneconomic constraints? Discuss.

7. The potential upward bias for wages in the public sector relative to the private sector occurs because the market provides a reasonably effective floor on wages in the public sector but it does not necessarily provide an effective ceiling. Discuss.

---

4. Reviews of U.S. studies include Ehrenberg and Schwarz (1986), Gunderson (1980), and Wise (1987). In a more recent U.S. study, Moulton (1990) indicates that the pure wage advantage of federal government workers is only 3 percent after controlling for detailed occupations and location cost of living differences.

8. Based on your knowledge of the determinants of the elasticity of demand for labour, would you expect the demand for public sector labour to be inelastic or elastic and why? Why might one expect the elasticity to differ across different elements of the public sector?

9. Differences in the elasticity of demand for labour between the public and the private sector, by themselves, are not sufficient conditions for a wage differential between the two sectors. Discuss.

10. Ehrenberg (1973) provides empirical estimates of the wage elasticity of demand for public sector labour in the U.S., ranging from –1.00 in public welfare to –.28 in health, fire, and protection services. What does this imply about the expected employment impact of a wage change in each of these sectors?

11. What are the pros and cons of using queues as a measure of rents or excess payments paid to workers in the public sector?

12. (a) What variables would you utilize as determinants of wages in a micro wage equation (individual workers is the unit of observation) to be estimated separately for public and private sector workers?

    (b) Why would you want to estimate *separate* wage equations for public and private sector workers rather than a single wage equation for both, with a dummy variable for public as opposed to private sector workers?

    (c) How would you test to see if the underlying wage determination process is different in the two sectors?

    (d) Based on this regression analysis, illustrate formally how one could decompose the overall average earnings differential between the public and the private sector into two components: a portion due to an economic rent or pure surplus payment, and a portion attributable to different endowments of productivity-related factors.

    (e) Which is the portion that public policy is concerned with and why?

    (f) Discuss some possible pitfalls in employing this methodology to estimate a pure public–private sector wage differential.

13. Decompose the overall average public–private sector wage differential by adding and subtracting $\Sigma b_g \bar{X}_c$ to the right-hand side of Equation 21.5. Is this also a reasonable decomposition technique?

14. Just as with measuring a pure impact of unions on relative wages, it is difficult to measure a pure public–private sector wage differential. It is extremely important, but difficult, to control for differences in the quality of labour, and to consider wage spillover effects between the two sectors. Discuss.

15. Estimating a pure public–private sector wage differential is difficult because the choice of sector itself is endogenously determined in that individuals may sort themselves into sectors on the bases of unobservable characteristics that influence wages. In such circumstances, the conventional regression procedure for decomposing the public–private sector wage differential into rents and differences due to observable differences in endowments of wage-determining characteristics does not control for differences in unobservable characteristics. Illustrate the appropriate econometric procedures for correcting for the biases that may result from the differences in unobservable characteristics between workers in the public and private sectors.

16. In Canada in the year 1975, the rate of increase in base wages in collective agreements was 20.0 percent in the provincial public sector and 21.3 percent in the health-education-welfare sector, compared to 14.4 percent in the private sector. These large increases in particular elements of the public sector apparently triggered public reaction. Whey might it be a bit hazardous to react to specific public sector increases in a particular year; that is, why might one expect public sector wage settlements to be more volatile than private sector wages?

REFERENCES AND FURTHER READINGS

Abbott, M. and T. Stengos. Alternative estimates of union–nonunion and public–private wage differentials in Ontario, 1981. Kingston: Queen's University Industrial Relations Centre, 1987.

Auld, D., L. Christofides, R. Swidinsky and D. Wilton. A microeconomic analysis of wage determination in the Canadian public sector. *JPubEc* 13 (June 1980) 369–388.

Asher, M. and J. Popkin. The effect of gender and race differentials on public–private wage comparisons: a study of postal workers. *ILRR* 38 (October 1984) 16–25.

Bartel, A. and D. Lewin. Wages and unionism in the public sector: the case of police. *R.E. Stats.* 63 (February 1981) 53–59.

Belman, D. and J. Heywood. Public wage differentials and the public administration industry. *IR* 27 (Fall 1988) 385–393.

Belman, D. and J. Heywood. Government wage differentials: a sample selection approach. *Applied Economics* 21 (April 1989) 427–438.

Belman, D. and J. Heywood. Direct and indirect effects of unionization and government employment on fringe benefit provisions. *JLR* 12 (Spring 1991) 111–122.

Belman, D. and J. Heywood. The effect of establishment and firm size on public wage differentials. *Public Finance Quarterly* 18 (April 1990) 221–235.

Bird, R. The growth of the public service in Canada. *Public Employment and Compensation in Canada.* D. Foot (ed.) Toronto: Butterworths, 1978.

Borjas, G. Labor turnover in the U.S. federal bureaucracy. *JPubEc* 19 (November 1982) 187–202.

Borjas, G. Wage determination in the federal government. *JPE* 88 (December 1980) 1110–1147.

Christensen, S. Collective bargaining in provincial public administration. *RI/IR* 36 (No. 3, 1981) 616–628.

Conklin, D., T. Courchene and W. Jones (eds.) *Public Sector Compensation.* Toronto: Ontario Economic Council, 1985.

Cousineau, J.-M. and R. Lacroix. *Wage Determination in Major Collective Agreements in the Private and Public Sectors.* Ottawa: Economic Council of Canada, 1977.

Cromwell, B. Dedicated taxes and rent capture by public employees. *IRRA* 42 (December 1989) 169–176.

Currie, J. Employment determination in a unionized public sector labor market: the case of Ontario's school teachers. *JOLE* 9 (January 1991) 45–66.

Currie, J. and S. McConnell. Collective bargaining in the public sector: the effect of legal structure on dispute costs and wages. *AER* 81 (September 1991) 693–718.

Daniel, M. and W. Robinson. *Compensation in Canada: A Study of the Public and Private Sectors.* Ottawa, Conference Board, 1980.

Delaney, J., P. Fenstle and W. Hendricks. Police salaries, interest arbitration and leveling effect. *IR* 23 (Fall 1984) 417–423.

Dunson, B. Pay experience and productivity: the government-sector case. *JHR* 20 (Winter 1985) 153–160.

Edwards, L. and F. Edwards. Public unions, local government structure and the compensation of municipal sanitation workers. *EI* 20 (July 1982) 405–425.

Ehrenberg, R. The demand for state and local government employees. *AER* 63 (June 1973a) 366–379.

Ehrenberg, R. Municipal government structure, unionization and wages of fire fighters. *ILRR* 27 (October 1973b) 36–48.

Ehrenberg, R. Heterogeneous labor, minimum hiring standards, and job vacancies in public employment. *JPE* 81 (November/December 1973a) 1442–1459.

Ehrenberg, R. and G. Goldstein. A model of public sector wage determination. *Journal of Urban Economics* 2 (1975) 223–245.

Ehrenberg, R. and J. Schwarz. Public sector labour markets. *Handbook of Labour Economics*, Vol. 1, O. Ashenfelter and R. Layard (eds.). New York: Elsevier, 1986.

Elliott, R. and P. Murphy. The relative pay of public and private sector employees, 1970–1984. *Cambridge Journal of Economics* 11 (June 1987) 107–132.

Ferland, G. La politique de remunération dans les secteurs publics et parapublics au Québec. Salary policy in Québec's public and parapublic sectors. *RI/IR* 36 (No. 3, 1981) 475–498.

Feville, P., J. Delaney and W. Hendricks. Police bargaining, arbitration and fringe benefits. *JLR* 6 (Winter 1985) 1–20.

Freeman, R. Unionism comes to the public sector. *JEL* 24 (March 1986) 41–86.

Freeman, R. and C. Ichniowski. *When Public Sector Workers Unionize*. Chicago: University of Chicago Press, 1988.

Gelb, A., J. Knight and R. Sabot. Public sector employment, rent seeking and growth. *EJ* 101 (September 1991) 1186–1199.

Gonzalez, R., S. Mehay and D. Deno. Municipal residency laws: effects on police employment, compensation and productivity. *JLR* 12 (Fall 1991) 439–452.

Grosskopf, S., K. Hayes and T. Kennedy. Supply and demand effects of underfunding pensions on public employee wages. *SEJ* 51 (January 1985) 745–753.

Gunderson, M. Data on public sector wages in Canada. *Public Employment and Compensation in Canada*, D. Foot (ed.). Toronto: Butterworths, 1978a, 107–126.

Gunderson, M. Public–private wage and nonwage differentials in Canada: some calculations from published tabulations. *Public Employment and Compensation in Canada*, D. Foot (ed.). Toronto: Butterworths, 1978b, 167–188.

Gunderson, M. Wage determination in the public sector: Canada and the U.S. *Labour and Society* 4 (January 1979a) 49–70.

Gunderson, M. Earnings differentials between the public and private sectors. *CJE* 12 (May 1979b) 228–242.

Gunderson, M. Public sector compensation in Canada and the U.S. *IR* 19 (Fall 1980) 257–271.

Gunderson, M. The public/private sector compensation controversy. In *Conflict or Compromise*, M. Thompson and G. Swimmer (eds.). Ottawa: Institute of Research on Public Policy, 1984.

Gunderson, M. and C. Riddell. "Provincial Public Sector Payrolls." In *Provincial Public Finances*, M. McMillan (ed.). Toronto: Canadian Tax Foundation, 1991.

Gyourko, J. and J. Tracy. An analysis of public- and private-sector wages allowing for endogenous choices of both government and union status. *JOLE* 6 (April 1988) 229–253.

Hamermesh, D. (ed.). *Labor in the Public and Non-Profit Sectors*. Princeton, N.J.: Princeton University Press, 1975.

Hartman, R. *Pay and Pensions for Federal Workers*. Washington, D.C.: Brookings Institute, 1983.

Hundly, G. Public- and private-sector occupational pay structures. *IR* 30 (Fall 1991) 417–434.

Hunter, W. and C. Rankin. The composition of public sector compensation. *JLR* 9 (Winter 1988) 29–42.

Ichniowski, C., R. Freeman and H. Lauer. Collective bargaining laws, threat effects and the determinants of police compensation. *JOLE* 7 (April 1989) 191–209.

Ippolito, R. Why federal workers don't quit. *JHR* 22 (Spring 1987) 281–299.

Katz, H. Municipal pay determination: the case of San Francisco. *IR* 18 (Winter 1979) 44–58.

Krueger, A. Are public sector workers paid more than their alternative wage? Evidence from longitudinal data and job queues? *When Public Sector Workers*

Unionize, R. Freeman and C. Ichniowski (eds.). Chicago: University of Chicago Press, 1988.

Krueger, A. The determinants of queues for federal jobs. *ILRR* 41 (July 1988) 567–581.

Lewis, H.G. Union/nonunion wage gaps in the public sector. *JOLE* 8 (January 1990) S260–S328.

Linneman, P. and M. Wachter. The economics of federal compensation. *IR* 29 (Winter 1990) 58–76.

Lovejoy, L. The comparative value of pensions in the public and private sectors. *MLR* 111 (December 1988) 18–26.

Mehay, S. and R. Gonzalez. The relative effect of unionization and interjurisdictional competition on municipal wages. *JLR* 7 (Winter 1986) 79–93.

Moore, W. and R. Newman. Government wage differentials in a municipal labor market. *ILRR* 45 (October 1991) 145–153.

Moore, W. and J. Raisian. Government wage differentials revised. *JLR* 12 (Winter 1991) 13–33.

Moulton, B. A re-examination of the federal private wage differential in the United States. *JOLE* 8 (April 1990) 270–293.

Olson, C. The impact of arbitration on the wages of firefighters. *IR* 19 (Fall 1980) 325–339.

Pedersen, P. Wage differentials between the public and private sectors. *JPubEc* 41 (February 1990) 125–145.

Perloff, J. and M. Wachter. Wage comparability in the U.S. postal service. *ILRR* 38 (October 1984) 26–35.

Picot, G. *Canada's Industries: Growth and Jobs Over Three Decades*. Ottawa: Statistics Canada, 1986.

Porter, F. and R. Keller. Public and private pay levels in large labor markets. *MLR* 104 (July 1981) 22–26.

Preston, A. The nonprofit worker in a for-profit world. *JOLE* 7 (October 1989) 438–463.

Proulx, P. Comment établir des comparaisons de la rémunération entre les secteurs publics et privés? *RI/IR* 35 (No. 2, 1980) 202–208.

Quinn, J. Postal sector wages. *IR* 18 (Winter 1979) 92–96.

Quinn, J. Wage differentials among older workers in the public and private sectors. *JHR* 14 (Winter 1979) 41–62.

Riddell, W.C. and P. Smith. Expected inflation and wage changes in Canada. *CJE* 15 (August 1982) 377–394.

Robinson, C. and N. Tomes. Union wage differentials in the public and private sectors. *JOLE* 2 (January 1984) 106–127.

Saunders, G. The impact of interest arbitration on Canadian federal employees' wages. *IR* 25 (Fall 1986) 320–327.

Shapiro, D. and M. Stelcner. Canadian public-private sector earnings differentials, 1970–1980. *IR* 28 (Winter 1989) 72–81.

Smith, S. Pay differentials between government and private sector workers. *ILRR* 29 (January 1976) 179–197. Also comment by W. Bailey and reply 31 (October 1977) 78–87.

Smith, S. Government wage differentials by sex. *JHR* 11 (Spring 1976) 185–199.

Smith, S. Are postal workers over- or under-paid? *IR* 15 (May 1976) 168–176.

Smith, S. *Equal Pay in the Public Sector: Fact or Fantasy.* Princeton, N.J.: Industrial Relations Section, 1977a.

Smith, S. Government wage differentials. *Journal of Urban Economics* 4 (July 1977b) 248–277.

Smith, S. Public/private wage differentials in metropolitan areas. *Public Sector Labor Markets*, P. Mieszkowski and G. Peterson (eds.). Washington, D.C.: Urban Institute, 1981.

Smith, S. Prospects for reforming federal pay. *AER* 72 (May 1982) 273–277.

Spizman, L. Public employee unions: a study in the economics of power. *JLR* 1 (Fall 1980) 265–273.

Stager, D. Lawyers' earnings in the Canadian private and public sectors. *RI/IR* 43 (No. 3, 1988) 571–589.

Stelcner, M., J. van der Gagg and W. Vijverberg. A switching regression model of public–private sector wage differentials in Peru. *JHR* 24 (Summer 1989) 545–559.

Subbaro, A. Impasse choice and wages in the Canadian federal service. *IR* 18 (Spring 1979) 233–236.

Thompson, M. and J. Cairnie. Compulsory arbitration: the case of British Columbia teachers. *ILRR* 29 (October 1973) 3–17.

Thompson, M. and A. Ponak. Restraint, privatization and industrial relations in the 1980s. In *Industrial Relations in Canadian Industry*, R. Chaykowski and A. Verma (eds.). Toronto: Dryden Press, 1992.

Utgoff, K.C. Compensation levels and quit rates in the public sector. *JHR* 18 (Summer 1983) 394–406.

van der Gagg, J. and W. Vijverherg. A switching regression model for wage determination in the public and private sectors of a developing country. *R.E. Stats.* 70 (May 1988) 244–252.

Weisbrod, B.A. Non-profit and proprietary sector behavior: wage differentials among lawyers. *JOLE* 1 (July 1983) 246–263.

Wiatrowski, W. Comparing employee benefits in the public and private sectors. *MLR* 111 (December 1988) 3–8.

Williams, C. and G. Swimmer. The relationship between public and private sector wages in Alberta. *RI/IR* 30 (No. 2, 1975) 217–226.

Wilton, D. Public sector wage compensation. In *Canadian Labour Relations*, W.C. Riddell (ed.). Toronto: University of Toronto Press, 1986.

Wise, D. (ed.) *Public Sector Payrolls.* Chicago: University of Chicago Press, 1987.

CHAPTER 22

Discrimination and Male–Female Earnings Differentials

The economic analysis of discrimination provides a good application for many of the basic principles of labour market economics. In addition, it indicates the limitations of some of these tools in an area where noneconomic factors may play a crucial role. In fact, many would argue that discrimination is not really an economic phenomenon, but rather is sociological or psychological in nature. While recognizing the importance of these factors in any analysis of discrimination, the basic position taken here is that economics does have a great deal to say about discrimination. More specifically, it can indicate the labour market impact of the sociological or psychological constraints and preferences. More important, economics may shed light on the expected impact of alternative policies designed to combat discrimination in the labour market.

Although the focus of this analysis is on gender discrimination[1] in the labour market, it is important to realize that discrimination can occur against various groups and in different markets. Discrimination can occur in the housing market, product market, capital market, and human capital market (e.g., education and training), as well as in the labour market. It can be based on a variety of factors including race, age, language, national origin, sexual preference, or political affiliation, as well as on sex.

In order to better understand the economics of gender discrimination, the reasons, sources, and forms of sex discrimination are first discussed. Various theories of discrimination are then presented and empirical evidence is given to document the existence of sex discrimination in the labour market. The chapter concludes with a discussion of alternative policies to combat discrimination in the labour market.

---

1. The terms sex discrimination and gender discrimination are used interchangeably in this chapter, although the former is often used to denote biological differences and the latter to denote cultural and socially determined differences.

DISCRIMINATION: REASONS AND SOURCES

Reasons for Discrimination

Labour market discrimination against females may result because males have a preference for working with or buying from fellow males. This prejudice would be especially strong against females in supervisory positions or in jobs of responsibility.

Discrimination may also occur because of erroneous information on the labour market worth of females. Such erroneous information could come from females who consistently underestimate their own capacity in the labour market or it could come from employers who consistently underestimate the productivity of females. Erroneous information on the part of employers may come from the subjective reports of co-workers or supervisors, or it may be based on sex-biased test scores.

Because information on individual workers is extremely costly to acquire, employers may judge individual females on the basis of the average performance of all females. Statistical and signalling theories of discrimination are discussed in Aigner and Cain (1972), Lundberg and Startz (1983), and Phelps (1972). If efficiency wage premiums are paid to reduce costly turnover, individual females who expect to stay in the labour market may find it difficult to credibly signal their intent. In such circumstances, employers may judge them as having the average turnover of all women (which employers perceive to be higher than that of males) and hence may not place them in the jobs that pay the efficiency wage premiums. Although, from the employers' viewpoint, such statistical judgment may be efficient, it could be inequitable for many individual females.

Males may also discriminate for reasons of job security. To protect their high-wage jobs from low-wage female competition, males would use the forces of governments, unions, and business cartels to ensure that power remains in their hands and is used to further their own ends.

Obviously, discrimination can occur for any, or all, of the previously mentioned reasons of preference, erroneous information, statistical judgement, or job security. As we shall see later, the effectiveness of policies designed to combat discrimination often depends on the reasons for the discrimination.

Sources of Discrimination

Labour market discrimination can come from a variety of sources. *Employers* may discriminate in their hiring and promotion policies as well as their wage policies. To a certain extent the forces of competition would deter employers from discriminating since they would forgo profits by not hiring and promoting females who are as productive as their male counterparts. However, profit-maximizing firms do have to respond to pressure from their customers and male employees. For this reason they may be reluctant to hire and promote females, or they may pay females a lower wage than equally productive males. In addition, firms in the large not-for-profit sector (e.g., government, education, hospitals) may be able to discriminate without having to worry about losing profits by not hiring and promoting females.

In addition to discrimination on the part of employers, *male co-workers* may also discriminate for reasons of prejudice, misinformation or job security. Representing the wishes of a male majority, craft unions may discriminate through the hiring hall or apprenticeship system, and industrial unions may discriminate by bargaining for male wages that exceed female wages for the same work. Another potential source of discrimination is *customers* who may be reluctant to purchase the services of females or who may not patronize establishments that employ females, especially in positions of responsibility.

THEORIES OF LABOUR MARKET DISCRIMINATION

Alternative theories of sex discrimination in the labour market can be classified according to whether they focus on the demand or supply side of the labour market, or on noncompetitive aspects of labour markets.

Demand Theories of Discrimination

Demand theories of discrimination have the common result that the demand for female labour is reduced relative to the demand for equally productive male

---

EXHIBIT 22.1   DISCRIMINATION OR PRODUCTIVITY DIFFERENCES?: EVIDENCE FROM BASEBALL CARDS

Conventionally, the extent of discrimination is measured by estimating the earnings differential between, for example, males and females or blacks and whites, after controlling for productivity differentials that could be regarded as legitimate determinants of wages. It is extremely difficult, however, to control for productivity differentials. Even if the extent of discrimination can be estimated, it is difficult to determine the extent to which it emanates from customers, co-workers, or employers.

The market for baseball cards gets around many of these problems. It is a highly competitive market with competitive prices for the cards and it is possible to link the price to objective measures of productivity (i.e., performance of the players). The evidence, indicates that the price of cards is 10 to 13 percent lower for nonwhite players than for white players of the same productivity (i.e., performance). This can be taken as evidence of pure *customer* discrimination since neither employers nor co-workers systematically influence the price.

Source: C. Nardinelli and C. Simon. Customer racial discrimination in the market for memorabilia: the case of baseball. *QJE* 105 (August 1990) 575–595.

labour. The decreased demand for female labour would reduce the employment of females and, unless the supply of female labour is perfectly elastic, the decreased demand would reduce the wages of females relative to the wages of equally productive males. How does discrimination lead to a reduction in the demand for female labour?

According to Becker (1971, p. 14), employers act as if $W_f(1 + d_f)$ were the net wage paid to females, male co-workers act as if $W_m(1 = d_m)$ were the net wage they receive when working with females, and consumers act as if $P_c(1 + d_c)$ were the price paid for a product sold by a female. In all cases, the discrimination coefficient, "d," represents the cost, in percent, associated with hiring, working with, or buying from females. For example, a firm that has a discrimination coefficient of .10 and that can hire all the female labour it wants at $6.00 per hour, would act as if it paid $6.00 (1 + .10) = $6.60 each time it hired a female. Clearly such discrimination reduces the demand for females relative to equally productive males.

Arrow (1973) gives a neoclassical theory of discrimination based on the firm's desired demand for labour when the firm maximizes utility (rather than simply profits) and the firm's utility is increased by employing fewer workers from minority groups. Again, this results in a reduced demand for female labour and lower female wages.[2]

The demand for female labour relative to male labour also depends on information concerning their relative productivity. To the extent that employers consistently underestimate the productivity of females, they would correspondingly reduce their demand for female labour. This misinformation on the part of employers may be due to their own ingrained prejudices as well as to erroneous information fostered by male customers and co-workers.

Supply Theories of Discrimination

Supply theories of discrimination have the common result that the supply of female labour is increased by discrimination or, conversely, that the female asking wage is reduced by discrimination.

---

2. Arrow's (1973) neoclassical theory of discrimination can be presented formally as follows. Assume that the firm maximizes utility, U, which is a positive function of its profits, Π, and a negative function of the number of females, F, it employs (discrimination). That is $U = u(\Pi, F)$ where the partial derivatives are denoted by $U_\pi > 0$ and $U_F < 0$ and were $\Pi = P \cdot Q(M + F) - W_m M - W_F F$. For simplicity we have assumed that output, Q, is produced by perfectly substitutable male and female labour of equal productivity so that the output is denoted by the production function $Q(M + F)$. Utility maximizing firms will hire females until $U_F = 0$ which, from the utility function, implies $U_\pi \Pi_F + U_F = 0$. Since $\Pi_F = PQ_F - W_F$, utility maximization implies $U_\pi(PQ_F - W_F) + U_F = 0$, which upon rearrangement yields $W_F = PQ_F + U_F/U_\pi$. Thus female wages will be less than their marginal revenue product, PQ_F, since $U_F < 0$ and $U_\pi > 0$. In addition, female wages will be less than male wages if males are paid a wage equal to their marginal revenue product.

The crowding hypothesis, as formalized by Bergmann (1971), implies that females tend to be segregated into "female-type" jobs. The resulting abundance of supply lowers their marginal productivity and hence their wage. Thus, even if females are paid a wage equal to their marginal productivity, their wage will be less than male wages that are not depressed by an excess supply.

In a similar vein, dual labour market theory posits two separate and distinct labour markets. The primary or core labour market (unionized, monopolistic, expanding) provides secure employment at high wages. The secondary or peripheral labour market (nonunionized, highly competitive, declining) is characterized by unstable employment at low wages. Men tend to be employed in the primary labour market, women in the secondary labour market. Prejudice on the part of the dominant group and its desire to exclude female competition will prevent the entry of women into the primary labour market. Unions, occupational licensing, discriminatory employment tests, and barriers to education and training can all work against the entry of females into the core labour market. Female immobility may also result because of women's stronger ties to the household and their tendency to move to the places of their husbands' employment. For these various reasons, females tend to be crowded into jobs in the secondary labour market with its concomitant unstable employment and low wages. Their low wages and undesirable working conditions in turn create high absenteeism and turnover which further depress wages in the secondary labour market.

Females' attitudes of their own labour market worth also may induce them to lower their asking wage when seeking employment. Because of conditioning in a male-dominated labour market, females may erroneously underestimate their own labour market worth. This would tend to lower female asking (reservation) wages; that is, the female labour supply schedule would be shifted vertically downwards. Employers would naturally foster these attitudes because they lead to lower labour costs. Supply theories of discrimination also emphasize the importance of preferences in determining various decision with respect to education, training, hours of work, working conditions, and occupational choice — all of which can influence the job that women take and the pay they receive. The importance of these preferences are emphasized, for example, in Butler (1982) Daymont and Andrisani (1984) and Filer (1983, 1985, 1986). Of course, the extent to which these "choices" reflect preferences or discriminatory constraints, perhaps arising from outside of the labour market, is an interesting and important question.

Noncompetitive Theories of Discrimination

In theory, male–female wage differentials for equally productive workers are inconsistent with competitive equilibrium. As long as females could be paid a wage lower than that of equally productive males, firms that do not have an aversion to hiring females would increase their profits by hiring females. The resultant increased demand for females would bid up their wages, and the process would continue until the male–female wage differential is eliminated. Firms that do not have an aversion to hiring females would be maximizing profits by employing large numbers of females; firms that have an aversion to hiring females would be forgoing profits by employing only males. According to competitive theory, discrimination leads to a segregation of males and females, not wage differentials. It also implies that discrimination should be reduced over time since firms that discriminate will go out of business as they forgo profits to discriminate.

Some would argue that these predictions of competitive economic theory are at variance with the facts. Male–female wage differentials seem to persist, and discrimination does not appear to be declining over time. What then are some of the factors that may explain this persistence in the face of competitive forces?

Arrow (1973) attributes the persistence of male–female wage differentials to costs of adjustment and imperfect information. Even if they do acquire the information that profits can be increased by hiring low wage females to do certain jobs performed by higher wage males, firms cannot immediately replace their male workers by an all-female work force. There are fixed costs associated with recruiting and hiring; consequently, firms want to spread these fixed costs as much as possible by retaining their existing work force. Firms may replace male turnover by new female recruits, but they would be reluctant to immediately replace their male work force. In essence, Arrow implies that the long run may be a very long time.

Queuing theories based on efficiency wages may also explain the persistence of male–female wage differentials for equally productive workers. He argues that some firms pay efficiency wages that are wages greater than competitive market wages in order to reduce turnover, improve morale, or secure the advantages of always having a queue form which to hire workers. Other firms pay greater than competitive wages because of union or minimum wage pressure, or because of a necessity to share monopoly or oligopoly profits with their workers. The resultant higher-than-competitive wage enables these high-wage firms to hire from a queue of available workers. Rationing of the scarce jobs may be carried out on the basis of nepotism or discrimination. Thus, discriminatory wage differentials may exist in a long-run competitive equilibrium when profit-maximizing firms minimize labour costs by paying high wages in order to reduce turnover or improve morale, or because they face institutional constraints such as unions or minimum wage legislation.

As an alternative explanation for the persistence of discrimination, it can be argued that discrimination comes mainly from noncompetitive sectors, mainly the government and trade unions. Reflecting majority wishes, governments discriminate both in their own hiring practices and in the provision of education and training. In the case of sex discrimination, this would reduce the skill endowment of minorities and reduce the number of professional-managerial females who would hire fellow females. Also reflecting the wishes of a male majority, unions could discriminate against females, especially in apprenticeship and training programs. Social pressure can foster discrimination, especially when profits can be made through discrimination. Majority groups may use the power inherent in governments, unions, and business monopolies to further their own ends through discrimination and segregation.

A more radical perspective regards discrimination as deliberately fostered by employers as part of a conscious policy of divide-and-conquer (Reich 1978; Roemer 1979). Employers will consciously pit workers against each other and utilize a reserve of unemployed workers as ways of disciplining the work force and reducing working class solidarity and power. In this perspective, group power is more important than the economic forces of supply and demand in determining wages.

Monopsony is another noncompetitive factor that can affect female wages relative to male wages, mainly in two ways. First, if females tend to be employed by monopsonists, their wages will be reduced correspondingly. Because of their immobility (tied to household and to husbands' places of employment), females may not have the effective threat of mobility necessary to receive a competitive wage. Second, monopsonists may try to differentiate their work force so as to pay the higher wage rate only to some employees. One obvious way of differentiating workers is by sex. Thus, the company may pay higher wages to male employees and yet not have to pay these higher wages to its female employees who do the same work, simply because the female employees do not consider themselves direct competitors with male employees doing the same job. Employers find it in their interest to foster this attitude, since it enables them to maintain the wage differentials based on sex. The formal monopsony model of sex discrimination with separate male–female labour supply schedules was discussed earlier in Chapter 12.

The concept of systemic (not to be confused with systematic) discrimination has been advanced to explain discrimination that may be the unintended byproduct of historically determined practices. It may result, for example, from informal word-of-mouth recruiting networks (the "old-boy system") that perpetrate the existing sex composition of the work force. Or, it may result from certain job requirements pertaining to height or strength that may no longer serve a legitimate purpose. In situations where wages are institutionally set above the competitive norm, such job requirements may simply serve to ration the scarce jobs and serve little or no function related to productivity.

Productivity Differences: Choice or Discrimination?

Female wages may differ from male wages because of productivity differences that arise from differences in the human capital endowments and differences in the absenteeism and turnover of males and females. The human capital endowments could include *acquired* attributes such as education, training, labour market information, mobility and labour market experience, as well as more *innate* characteristics such as intelligence, strength, perseverance, or dexterity. In general there is little reason to believe the endowment of these *innate* characteristics to differ in an important manner between the sexes, and in an increasingly mechanized society the innate characteristic for which there may be the greatest difference — physical strength — takes on reduced importance. It is in the area of acquired human capital endowments that productivity differences may arise.

Human capital formation will occur until the present value of the marginal benefits (usually in the form of increased earnings) equals the present value of the marginal cost (both direct cost and the opportunity cost of the earnings forgone while acquiring human capital). Because of their dual role in the household and in the labour market, women traditionally have a shorter expected length of stay in the labour market. Consequently, they have a reduced benefit period from which to recoup the costs of human capital formation. In addition, their time in the labour market tends to be intermittent and subject to a considerable degree of uncertainty, thus creating rapid depreciation of their human capital and preventing them from acquiring continuous labour market experience. For this reason it may be economically rational for females (or firms) to be reluctant to invest in female human capital formation that is labour market oriented.

The human capital decision therefore may reflect rational choice, but it may also reflect discrimination as well as rational choice subject to discriminatory constraints. Females may be discriminated against in the returns they receive for acquiring human capital as well as in borrowing to finance the cost of human capital formation. In addition, family and peer group pressures may close off certain avenues of human capital formation. Young girls, for example, may be conditioned to become nurses and not doctors, or secretaries and not lawyers.

Most important, female responsibility for household tasks (see Exhibit 22.2) may reflect discrimination more than choice. Whatever the reason, females tend to acquire less labour-market-oriented human capital than males, and consequently their wages and employment opportunities are reduced in the labour market. This occurs even if in the labour market there is no discrimination on the part of firms, co-workers, or customers. Differences in wage and employment opportunities may reflect productivity differences, which in turn may come about because of rational economic choice as well as discrimination prior to entering the labour market.

Productivity and hence wage differences may also occur because of differences in the absenteeism and turnover of females. This is especially the case if female turnover occurs primarily as a result of leaving the labour force (which would reduce earnings as explained earlier), and male turn-over occurs primarily

---

EXHIBIT 22.2   DO DIFFERENCES IN FAMILY COMMITMENTS AFFECT THE PAY GAP BETWEEN MALE AND FEMALE MANAGERS?

Based on survey data of middle-managers in a large Canadian organization, Canning (1991) determined that female managers spend about four times as much on household work than do male managers. Only 15 percent of the females, but 69 percent of the males, thought that their own careers were more important to their families than were the careers of their spouses. Ten percent of the females had professional or managerial husbands who would be willing to move for the sake of the career of their wife; in contrast, 28 percent of the males had professional or managerial wives who would be willing to move for the sake of the career of their husband.

Clearly, family commitments and constraints are more important to female managers than they are to male managers. Based on statistical analysis of the determinants of the earnings of male and female managers, these differences in family commitment accounted for about 20 percent ($900) of the earnings gap that prevailed between otherwise similar male and female managers.

Source: K. Cannings. Family commitments and career success: earnings of male and female managers. *Relations Industrielles/Industrial Relations* 46 (No. 1, 1991) 141–153.

---

to move to a higher paying job. The empirical evidence on differences in quit rates and turnover rates by sex tends to be mixed. Some studies, especially those based on earlier data, tend to find quit and turnover rates to be higher for women than men so that women accumulate less tenure or seniority at a particular job (Donohue 1988; Hall 1982; Ureta 1992). However, other studies find that after controlling for the effect of other variables that affect turnover (e.g., wages, education, occupation, age) there is no significant difference in the turnover rates of men and women (Blau and Kahn 1981; Viscusi 1980; Meitzen 1986). In other words, to the extent that women have higher turnover rates then men, it tends to be because women are employed in occupations and low-wage jobs that have high turnover. As well, most studies find that any differences in turnover rates tend to be declining over time, and that the wage gap is smaller when comparisons are made between males and females with the same continuity in the labour forces (Light and Ureta 1990, 1992).

The absenteeism of women tends to be much higher than that of men, but this difference also becomes negligible after controlling for the effect of other determinants of absenteeism, notably the presence of small children in the household. Boulet and Lavallée (1984 p. 67), for example, cite Canadian evidence indicating that while total absenteeism was only slightly higher for females than males without children or with school-age children only, it was about seven times higher for females than males with pre-school age children. In addition to causing lower wages, higher absenteeism and turnover may be the result of low wages: cause and effect work in both directions to reinforce each other.

Differences in household responsibilities and in the division of labour within the household can be quite pronounced. Gunderson and Muszynski (1990, p. 26) for example, cite Canadian evidence indicating that when both the husband and wife work full-time in the labour market, the wife tends to work an average of 16 hours per week on household tasks, while the husband averages about seven hours.

Differences in household responsibilities can lead to a situation where women develop a comparative advantage in household tasks and men develop a comparative advantage in labour market tasks, leading to each specializing in their area of comparative advantage. This is reinforced by discriminatory behaviour against women in labour market tasks. The specialization, in turn, leads to cumulative effects whereby women become less prone to accumulate the continuous labour market experience that leads to higher earnings. This leads them to be especially vulnerable in case of divorce or if they return to the labour market after a prolonged absence due to childraising.

The fact that the household responsibility of females is so important in these aspects of productivity makes it more understandable that there will be increased pressure for a more equitable division of labour in the household as well as more daycare facilities for children. In addition, in empirical work that attempts to control for productivity differences between males and females, we should be careful in interpreting the productivity adjusted wage and employment patterns as ones that are free of discrimination. The discrimination may simply be occurring outside of the labour market.

Feminist Perspectives

Feminist perspectives on economics in general may also shed light on alternative theories of discrimination. Examples include Bergman (1986, 1989), Cohen (1985), Ferber and Nelson (forthcoming), and Waring (1988), as well as references cited in those studies. While such perspectives are continuously being articulated, they tend to have a number of common elements. Many of these are similar to the segmented labour market and radical perspective discussed earlier.

Exhibit 22.3 contrasts certain elements of differences between conventional economics and feminist perspectives on economics. Certainly not all feminists would share the feminist perspective, just as not all economists would share this stylized view of traditional economics. As well, many economists would argue that the conventional tools of economics can be applied to most of the approaches advocated by the feminist perspective.

Nevertheless, the feminist perspective highlights a set of critical issues that are important for understanding discrimination. The issues are sufficiently important and complex that our understanding of them could be enhanced by

---

EXHIBIT 22.3 FEMINIST VERSUS TRADITIONAL ECONOMIC PERSPECTIVES

Traditional Economics	Feminist
■ Rational choice and optimization	■ Power, tradition, dependence, coercion
■ Market exchange	■ Human interactions and processes
■ Self-interest, individualism	■ Human relations
■ Budget constraints and endowments	■ Constraints from broader society and family
■ Often emphasis on material needs and objectives	■ Broader social and nonmaterial needs and goals
■ Objectivity, reasoning, and analytical inquiry	■ Additional emphasis on subjectivity and holistic approach
■ Mathematical statistical approach	■ Alternative methodologies
■ Dominated by males	■ Male dominance in field of economics determines the issues and methodologies

---

a variety of alternative approaches — feminist perspectives, conventional economics suitably modified, as well as other approaches.

EVIDENCE ON MALE–FEMALE EARNINGS DIFFERENTIALS

Empirical studies have employed a variety of techniques to estimate the male–female earnings differential and to see how much of the differential reflects discrimination. The results are quite varied, in part because of different data sources and methodology, but also in part because of an emphasis on different aspects of discrimination. Some studies focus on wage discrimination within the same establishment and occupation; others involve measures of sex discrimination that also reflect the crowding of females into low-wage establishments, occupations, and industries; and other studies utilize measures of earnings differentials that could reflect discrimination outside of the labour market, perhaps in educational institutions or in households.

Figure 22.1 indicates the ratio of the average earnings of females relative to males in Canada, *before* controlling for the effect of other factors also believed to influence earnings. On average, females tend to earn about 55 to 60 percent of what males earn, with the ratio being more like 65 percent or slightly higher

Figure 22.1
FEMALE–MALE EARNINGS RATIO
(Canada 1967–89)

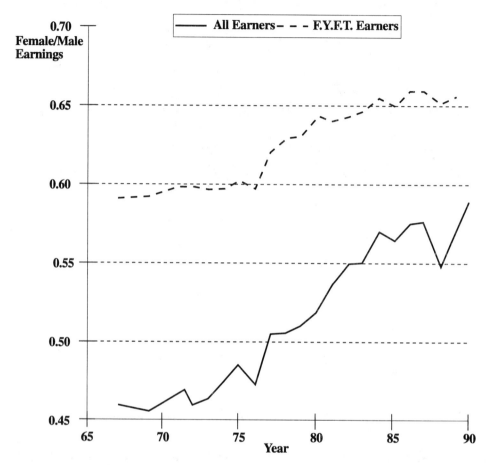

Source: Statistics Canada, *Earnings of Men and Women, 1988*, No. 74–401, and subsequent
updates. Ottawa: Supply and Services, 1989.

for full-year, full-time earners. The ratio has been increasing steadily but very
slowly since the 1960s.

Gunderson and Riddell (1991) discuss how the ratio of female to male earnings in Canada varies by other factors. The ratio is much larger (i.e., the gap is
much smaller) after controlling fully for differences in hours worked. Specifically, it is more in the neighbourhood of 75 to 80 percent when comparisons are
made between the hourly wages of men and women. The ratio is also much
higher for younger workers under the age of 24 (above 80 percent), for single
persons (almost 90 percent), and for university graduates (above 70 percent).

These raw or unadjusted earnings ratios do not simultaneously control for the variety of other factors besides gender that can influence wages. Such factors include age, education, race, training, labour market experience, seniority, marital status, health, hours of work, city size, region, quality of schooling, absenteeism, and number of children. The individual's industry and occupation can also affect earnings, but these are more likely to be channels through which discrimination can occur, rather than exogenous wage determining variables.

Econometric studies, however, have estimated the separate impact of many of these other determinants of wages, with the intent of isolating a pure male–female wage differential that remains after controlling for the effect of productivity-related factors that are regarded as "legitimate" or nondiscriminatory determinants of the wage gap. The net or adjusted wage gap that remains is then taken as a measure of discrimination. The most common procedure is to estimate equations of the determinants of earnings, separately for males and females. The overall average earnings gap is then decomposed into a component attributable to differences in the endowments of wage-determining characteristics and another attributable to discrimination, defined as differences in the economic returns to the same wage-determining characteristics.[3]

A number of studies have reviewed the evidence based on the econometric studies of gender discrimination.[4] While there are exceptions to almost every statement, a number of generalizations do emerge from the evidence on gender differences in earnings.

Even after controlling for a wide-range of wage-determining variables, a pure wage gap appears to remain that is attributable to discrimination. The gap is narrowed considerably after controlling for the effect of these other determinants

---

3. The methodology — first utilized with U.S. data by Oaxaca (1973) — is identical to that which was formally presented in the previous chapter to decompose the public–private sector wage differential. Separate earnings equations are estimated for males and females. Subtracting the female from the male earnings equations, evaluated at the mean of the explanatory variables, yields
$$\bar{y}_m - \bar{y}_f = \Sigma b_m \bar{X}_m - \Sigma b_f \bar{X}_f$$
Subtracting and adding $\Sigma b_f \bar{X}_m$ from the right-hand side, and collecting terms, yields
$$\bar{y}_m - \bar{y}_f = \Sigma b_m (\bar{X}_m - \bar{X}_f) + \Sigma (b_m - b_f) \bar{X}_m$$
That is, the average male–female earnings differential can be decomposed into two parts: a portion attributable to differences in endowments of wage-determining characteristics (i.e., differences in the explanatory variables evaluated according to the male pay structure), and a portion attributable to different wage payments for the same characteristics (i.e., differences in the regression coefficients or returns to the same characteristics). The latter term is taken as a measure of wage discrimination or unequal pay for the same characteristics. Based on Canadian data, such equations have been estimated in Robb (1978), Gunderson (1979), Shipiro and Stelcner (1981, 1987), Miller (1987), and Foot and Stager (1989).
4. Reviews of Canadian and U.S. studies are contained in Gunderson (1985a, 1989) and of U.S. studies in Treiman and Hartmann (1981), Willborn (1986), Cain (1986), and Blau and Ferber (1987a).

EXHIBIT 22.4   ARE WOMEN WHO ARE DISCRIMINATED AGAINST MOST LIKELY TO REPORT DISCRIMINATION?

Conventionally, one would expect persons who are most subject to discrimination to also be most likely to report themselves as being subjected to discrimination. However, based on both Canadian and U.S. evidence, Peter Kuhn (1987) finds that women who are most subject to objective measures of wage discrimination are the least likely to subjectively self-report that they are subject to discrimination. The objective measure of discrimination is the statistical evidence of discrimination based on the decomposition analysis from male and female earning equations as discussed in the text. The subjective measure of perceived discrimination is whether the woman self-reports on a confidential survey that she is subject to discrimination. Kuhn's analysis indicates that younger, well-educated women, for example, are more likely to report discrimination, even though they appear to be less subject to wage discrimination.

Kuhn attributes this difference in reporting behaviour to possible differences in nonstatistical aspects of discrimination. For example, younger, more educated women may work in mixed occupations where male comparator jobs are more readily observed, or they may be more perceptive of the subtle forms of discrimination, or they may have not yet sorted themselves into jobs where they perceive less discrimination, or they have greater expectations of nondiscriminatory norms.

Barbezat and Hughes (1990) also indicate that employees may be less likely to engage in discrimination in ways that can be measured and used against them, if they feel that such information is more likely to be reported by particular groups such as younger and more educated women. Consequently, the degree of statistically measured discrimination is likely to be smallest against such groups precisely because they are more likely to report discrimination.

Sources: P. Kuhn. Sex Discrimination in labour markets; the role of statistical evidence. *American Economic Review* 77 (Sept 1987) 567–583.
D. Barbezat and J. Hughes. Sex Discrimination in Labour Markets; the role of statistical evidence: comments. *American Economic Review* 80 (March 1990) 277–286.

of wages; nevertheless, a residual wage gap appears to remain. As well, it is not clear that it is appropriate to control for the effect of many of these other factors, since differences in these factors themselves may reflect discrimination and, in fact, may be a channel through which discrimination operates. This is the case, for example, with respect to such factors as hours worked, training, labour market experience, seniority, and type of education.

Discrimination and inequality of responsibilities and opportunities originating from outside of the labour market are likely to be a more important source of male–female wage inequality than is discrimination from within the labour market itself. This is especially the case with respect to the unequal division of labour with respect to household responsibilities (especially childraising) which can manifest itself in different labour market behaviour pertaining to such factors as hours worked, absenteeism, turnover, mobility, career interruption, training, and occupational choice. This highlights the limited scope of policies that focus only on labour market discrimination.

Wage differences arising from differences in the occupational distribution between males and females is likely to be a more important contributor to the overall wage gap, than are wage differences between men and women within the same occupation, especially within the same establishment. This highlights the importance of policies to reduce occupational segregation and to encourage the occupational advance of women. Such initiatives include equal employment opportunity policies, as well as the provision of education and training that is labour market oriented. It also highlights the limitations of conventional equal pay legislation involving comparisons between men and women within the same job in the same establishment, since pay differences are likely to be small in those circumstances.

In addition to being disproportionately employed in low-wage occupations, women tend to be employed in low-wage establishments and industries. This highlights the further limitations of equal pay policies, since comparisons are allowed only within the same establishment and therefore also the same industry.

The productivity-adjusted wage gap tends to be smaller in the union sector as opposed to the nonunion sector, and in the public sector as opposed to the private sector. Within the private sector, it tends to be smallest when product markets are competitive highlighting that competitive market forces can dissipate discrimination.

The gap seems to be declining slightly over time with some of the decline coming from improvements in the productivity-related characteristics of women, and some coming from a reduction in discriminatory pay differences for the same characteristics.

The Ontario *Green Paper on Pay Equity* (1985, p. 10) provides what can be considered a representative summary picture of the magnitude of the male–female earnings gap and the factors that contribute to that gap. They indicate that in the early 1980s, the ratio of female to male average annual full-time earnings was 0.62, implying an unadjusted gap of 0.38. That gap was attributed to the following factors: 0.16 to differences in hours worked; 0.05 to 0.10 to differences in such factors as experience, education, and unionization; 0.10 to 0.15 to occupational segregation; and 0.05 to narrowly defined wage discrimination.

POLICIES TO COMBAT SEX DISCRIMINATION

Because of the variety of sources and forms of discrimination, policies to combat discrimination have also tended to take on a variety of forms. The main thrusts of public policy have been in three areas: equal pay legislation (including pay equity or equal pay for work of equal value); equal employment opportunity legislation (including employment equity or affirmative action); and policies

EXHIBIT 22.5   ARE FEMALE MANAGERS AS LIKELY TO BE PROMOTED AS ARE MALE MANAGERS?

Discrimination can occur in the form of differential promotion opportunities as well as unequal wages for the same work. Based on a survey of large Canadian corporations, Kathy Cannings found that female managers earned on average 87 percent of the pay of male managers. However, they were only 80 percent as likely to be promoted in any given year of their career with the organization.

Some of this difference in promotion opportunities can be explained by differences in such factors as formal education and productivity-related characteristics. However, even after controlling for differences in these factors, female managers were significantly less likely to be promoted than were male managers.

Source: K. Cannings. Managerial promotion: the effects of socialization, specialization, and gender. *Industrial and Labor Relations Review* 42 (October 1988) 77–88.

designed to facilitate female employment and to alter attitudes and preferences. Each of these will be discussed in turn.

Conventional Equal Pay Legislation

All Canadian jurisdictions have conventional equal pay legislation which requires equal pay for equal work within the same job and within the same establishment. The courts and enforcement agencies have generally interpreted equal work as work that is *substantially similar*, with minor differences being allowed, especially if offset by differences in other aspects of the work performed. For example, the work could be considered the same even if males do occasional heavy lifting, especially if females do other occasional tasks not done by males.

The scope of conventional equal pay legislation is limited by the fact that it deals with only one aspect of discrimination — wage discrimination within the same job within an establishment — and as our earlier discussion of the empirical evidence indicated, this is probably a quantitatively small aspect of discrimination. In addition, the enforcement of the law can be difficult because it usually relies upon individuals to complain and because there are problems of interpreting what is meant by equal work, although the courts have interpreted this rather broadly.

Equal Value or Comparable Worth Legislation

In part because of the limited potential for conventional equal pay policies to narrow the overall male–female earnings gap, further legislative initiatives have been advocated on the pay side in the form of equal pay for work of equal value. The phrases equal value, pay equity, and comparable worth are usually used interchangeably, although equal value is often used internationally, pay equity is usually used in Canada, and comparable worth is used most often in the United States.

Existence of Equal Value Legislation

In Canada, equal value or pay equity legislation or regulation exists in all jurisdictions except for Alberta and Saskatchewan (Table 21.1). It has been in place in Quebec since 1976, although most cases in that province have tended to be conventional equal work cases with the comparisons being made between males and females in the same job. In most Canadian jurisdictions, pay equity legislation is restricted to the public sector either by law or by practical application. In some jurisdictions a complaint is required before the process will be instigated. In other jurisdictions, a so-called proactive system is in place whereby employers are required to initiate gender-neutral job evaluation plans and to make the appropriate wage adjustments whether or not a complaint has been or there is *prima facie* evidence of discrimination.

Only Ontario, however, has both a proactive system and a requirement to be applied also to the private sector. In fact, these two characteristics make Ontario's equal value legislation the most extensive in the world.

In the United States, by the mid-1960s over a dozen states had passed comparable worth legislation covering state employees; however, these laws were rarely enforced (Ehrenberg and Smith 1987b). Only a few states such as Washington State and Minnesota had active comparable worth policies. At the local level a larger number of cities, counties, or school districts (mainly in California, Washington, and Minnesota) were experimenting with or studying the concept; however, it has yet to be adopted on a widescale basis in the United States. Equal value initiatives have not been prominent in Europe in spite of the fact that a number of European countries have made a token adoption of such policies. In Australia, a form of comparable worth was adopted in 1972 when their national wage awards made by their wage tribunals (which set wages for the majority of workers) eliminated wage differentials across occupations that were designated as predominantly male and predominantly female. However, this decision, which substantially raised the wages of females, was based on legislative fiat not on the results of job evaluation procedures, and hence is not a comparable worth procedure in the conventional sense.

Equal Value Procedures

Equal value policies generally require an equality of pay between jobs of equal value where value is determined by a job evaluation scheme that is free of gender bias. Such schemes generally involve comparisons between jobs that are designated as predominantly male and predominantly female, where gender

Table 22.1
EXISTENCE OF COMPARABLE WORTH IN VARIOUS CANADIAN JURISDICTIONS, 1989

JURISDICTION	YEAR	PRIVATE SECTOR	ENFORCEMENT
Comparable Worth			
Quebec	1976	Yes[a]	Complaints based
Federal	1977	Yes[a]	Complaints based
Manitoba	1985	No	Proactive[b], plans
Yukon	1986	No	Complaints based
Ontario	1987[c]	Yes	Proactive[b] and complaints
Newfoundland	1988	No	Collective bargaining[d]
Nova Scotia	1988	No	Proactive[b]
Prince Edward Is.	1988	No	Proactive and complaints
New Brunswick	1989	No	Proactive[b]
N.W. Territories	1990	No	Complaints based
British Columbia	1990	No	Collective bargaining[d]
Conventional Equal Pay[e]			
Alberta	1957	Yes	Complaints based
Saskatchewan	1952	Yes	Complaints based

Notes:
a. Almost all cases have been in the public sector.
b. Employers are required to initiate gender-neutral job evaluation plans and to adjust wages in female-dominated jobs to ensure equal pay for work of equal value, whether or not a complaint has been made or there is *prima facie* evidence of discrimination.
c. The legislation was passed in 1987 to commence on January 1, 1988. Wage adjustments are to commence no later than January 1, 1990 in the public sector and no later than January 1, 1991 in the private sector for larger employers. Smaller private sector employers must begin adjustments in subsequent years.
d. Legislation has not been passed, but the government has committed itself to pay equity for its civil service, through the collective bargaining process.
e. Promised for the private sector in 1991 or 1992.

Source: Updated from Morley Gunderson and Roberta Robb (1991a, 1991b). More detailed analysis is given in Nan Weiner and Morley Gunderson (1990).

predominance is designated, for example, as involving seventy percent or more of either sex. The next step involves the determination of the *factors* such as skill, effort, responsibility, and working conditions that are believed to be the important determinants of the value or worth of a job. For the various jobs, job evaluators then usually assign point scores for each of these factors. The points are then summed to get a total point score for each job (which implies equal weights for a point score for each of the factors) or different weights or ranges of scores could be assigned to the scores of each factor on an *a priori* basis. Predominantly female jobs are then compared to predominantly male jobs of the same total point score, and the wages adjusted to those of the predominantly male jobs. In situations where this equal value approach has been applied, wages in female-dominated jobs typically have been only 80 to 90 percent of wages in male-dominated jobs of the same job evaluation point scores. The jobs can be in different occupations as long as they have the same job evaluation point scores. One case in the federal jurisdiction, for example, involved a comparison of predominantly female librarians with predominantly male historical researchers.

Economic versus Administrative Concepts of Value

This procedure highlights the fact that equal pay for work of equal value involves an *administrative* concept of value, where value is determined by the *average* value of the *inputs*, such as skill, effort, responsibility, and working conditions, that are involved in a job. This is in contrast to the *economic* concept of the value of the marginal product of labour whereby value is determined by the value of the *output* produced by an *additional* unit of labour. The administrative concept of value is akin to the notion of value-in-use (i.e., the average value of inputs) while the economic concept involves value-in-exchange (i.e., marginal contribution to the value of output). According to the economic concept, inputs that are in abundance in supply may have little value-in-exchange (and hence command a low market wage) even though they may have a high value-in-use because they involve substantial average inputs of skill, effort, responsibility, and working conditions. This is analogous to the diamond-water paradox whereby diamonds have a high value-in-exchange (because of their scarcity) but water has a low value-in-exchange (because of its abundance) in spite of its high value-in-use.

The equal value concept in fact explicitly rejects the notion that market forces should be the prime determinant of the value and hence pay for jobs. This rejection is based on the belief that market forces reflect discrimination. Even if market forces enable employers to hire workers in predominantly female jobs at rates of pay that are lower than those paid to workers in predominantly male jobs, proponents of comparable worth would argue that such an outcome is socially unacceptable because it reflects discriminatory segregation and the systematic undervaluation of female-dominated jobs. The belief is that females should not have to leave female-dominated jobs to get the same pay as in male-dominated jobs that require the same inputs of skill, effort, responsibility, and working conditions.

In essence, the primacy of market forces of supply and demand is rejected in that the value and hence remuneration of a job is not deemed to be low simply because there is an abundance of labour willing to do the work, or because there is little demand for that type of labour. This is in contrast to the economic emphasis on the market forces of supply and demand to determine the value and hence remuneration of a job.

While there is a sharp contrast between the economic and comparable worth concepts of value, the two concepts do not have to be diametrically opposed to each other. Skill, effort, and responsibility are scarce resources and hence will receive a market premium, just as they are assigned point scores by job evaluators. Similarly, compensating wage premiums are paid in the market for undesirable working conditions, just as they can be assigned point scores by job evaluators. In fact, it is possible to use the premiums that the market yields for the various factors of skill, effort, responsibility, and working conditions, based only on the male-dominated jobs, and then to apply those premiums to weigh the scores of the female-dominated jobs to arrive at the value of those jobs had they been paid the same market premium as the male-dominated jobs. This procedure is termed a "policy capturing approach" in that it is a policy that simply captures the pricing mechanism that is used for the male-dominated jobs and extends it to the female-dominated jobs. Comparable worth approaches can also pay attention to market forces by allowing exceptions for occupations that are in scarce supply.

Clearly market forces do not have to be ignored by the principle of comparable worth, but it is equally clear that the comparable worth concept of value is different from the economic concept of value. The former is an administrative concept of value based upon job evaluation procedures; the latter is based upon market forces of supply and demand.

Rationale for Equal Value Initiatives

Equal value initiatives have been rationalized on the grounds of being able to deal with *both* wage discrimination and occupational segregation — the latter on the grounds that comparisons can be made across different occupations as long as they are of the same value as determined by job evaluation procedures. This is important because occupational segregation generally is believed to be a more important contributor to the overall earnings gap than is wage discrimination, and conventional equal pay can only deal with the latter. Equal value policies have also been rationalized on the grounds of securing redress for those women who do not want to leave predominantly female jobs to get the same pay as predominantly male jobs of the same value as determined by job evaluation procedures.

Scope of Equal Value Initiatives

The scope of equal value policies is potentially large because it enables comparisons across occupations, unlike conventional equal pay policies which enable comparisons only within the same occupations. However, the scope of equal value policies will be limited in a complaints-based system by the fact that it is difficult for individuals or even groups to lodge a complaint under such a complicated procedure. For individuals, the fear of reprisals may also deter complaints. For this reason, supporters of comparable worth have advocated a "proactive system-wide" procedure whereby employers would be required to have a bona fide job evaluation system in place to help achieve pay equity. Even if such a system were in place, the scope of comparable worth would be limited by the fact that comparisons can be made only within the same establishment. This means that comparable worth could not reduce that portion of the overall earnings gap that reflects the segregation of females into low-wage establishments or industries. Based upon U.S. data, for example, Johnson and Solon (1986) indicate that these restrictions mean that even if comparable worth completely eliminated the relationship between wages and the sex composition of an occupation, it would reduce the average male–female earnings gap by only a small amount.

Design Features of Comparable Worth

If comparable worth becomes more prominent as a policy initiative, there are a large number of program design features that will have to be worked out to facilitate the practical implementation of the policy.[5] Such design features include: the definition of gender predominance for comparing male-dominated with female-dominated jobs; the job evaluation procedure for establishing the value of a job; the procedure for relating the job evaluation point scores to the pay of jobs; the procedure for adjusting pay in undervalued jobs and in overvalued jobs; the definition of establishment and of pay itself; the appropriate exemptions if any; and the optimal enforcement procedure including whether a complaints-based or more proactive, system-wide procedure should be followed. These and other design features can have a substantial impact on the scope of equal value initiatives and hence on their positive and negative consequences.

Pay Equity Example

A hypothetical example can illustrate many of the issues associated with the application of pay equity. In Figure 22.2 the job evaluation points for each job are plotted on the horizontal axis, and the pay for each job is plotted on the vertical axis. The upper line is the pay line for the male-dominated jobs (e.g., 70 percent or more males) and the bottom line is the pay line for the female-dominated jobs (e.g., 70 percent or more females). The points around each line are the particular male or female jobs in the establishment. The points illustrate the combination of pay and points associated with each job. The pay lines could have been simply drawn or "eyeballed" to fit the scatter of points, or they could be estimated by statistical techniques such as regression analysis. They could be

---

5. These design and implementation features are discussed in more detail in Gunderson (1989) and Gunderson and Robb (1991a, 1991b).

Figure 22.2
PAY EQUITY CASE

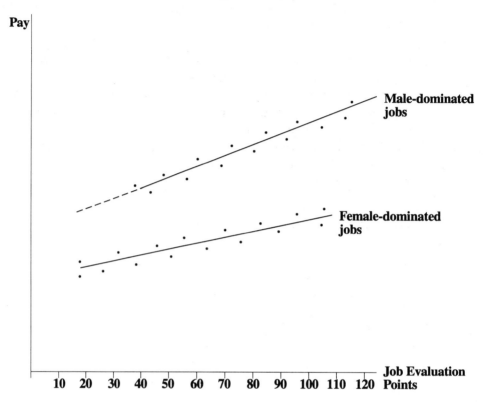

straight lines, as in this example, or nonlinear if the relationship between pay and points is nonlinear. The slopes could be the same, in which case the same absolute pay difference would be associated with different job values, and therefore the percentage difference would be smaller in the jobs of high pay and points. Or the slopes could be different, as in this example where the higher paying jobs are associated with higher absolute wage differences, albeit a possible constant percentage differential.

A number of observations are worth noting. The male pay line is above the female pay line indicating that male-dominated jobs tend to receive higher pay than female-dominated jobs of the same value, where value is measured by job evaluation points. Typically, the female pay line is 80 to 90 percent of the height of the male pay line, indicating that female-dominated jobs tend to be paid 80 to 90 percent of the pay in male-dominated jobs of the same job evaluation score. (They are drawn further apart here to simply avoid clutter in the diagram.)

Some of the female-dominated jobs at the lower end of the job evaluation scores do not have direct male comparator jobs of the same value. This raises the issue of whether projection of the male pay line would be allowed (the dashed line segment) to provide hypothetical male comparators (sometimes terms "phantom jobs") that indicate what such male jobs would have been paid, based on the relationship between pay and points in the other male jobs. In fact, there may even be an issue of whether extrapolation of the male pay line is allowed within the sample points of the male-dominated jobs, or whether a male comparator job of comparable job evaluation points is required before a comparison can be made.

Although not shown in this particular diagram, it is possible that some female "dots" (pay for points) could lie above the male pay line, or above a male "dot" of the same value. This would simply indicate that some female jobs could pay more than male jobs (hypothetical or actual) of the same job evaluation score. Similarly, a male "dot" could lie below the female line. Such outlying dots are not likely to occur, but they are possible. It is also possible to have multiple male or female "dots" at the same point score. This would simply indicate that unequal pay for work of the same value can exist within female-dominated jobs or within male-dominated jobs.

The diagram also illustrates alternative wage adjustment procedures that are possible. The "line-to-line" procedure would raise the female pay line to the male pay line, thereby eliminating the systematic differences in the pay between male-dominated and female-dominated jobs. The random deviations in pay that previously prevailed around the female pay line would remain, just as they remain for male-dominated jobs around the male pay line. Some female jobs would be "overvalued" (i.e., above the male pay line), and others would be "undervalued" (i.e., below the male pay line) just as they were overvalued and undervalued relative to the previous female pay line, and just as some male-dominated jobs are overvalued and others undervalued relative to the male pay line.

The "point-to-line" procedure, in contrast, would raise all the points around the female pay line to the male pay line. This would remove both the systematic differences as evidenced by differences in the heights of the pay lines, as well as the random deviations that previously prevailed around the female pay line. Random deviations for male jobs around the male pay line would still prevail, unless it became a company policy to try to remove such random deviations. This would be hampered, however, by the fact that pay equity legislation invariable prohibits the wages in any job to be reduced as a result of pay equity, in which case male points above the male pay line could not be brought down to the male pay line.

Other adjustment procedures are also possible. A "point-to-point" procedure would raise the female point to the closest male comparator point (as is specified in the Ontario legislation). This would not require the estimation of pay lines, but it would require decision rules for the choice of particular male comparators jobs. An average pay line could be used that is an amalgam of the male and female lines, or a corridor approach could be used whereby the female pay line is raised only to a lower boundary of the male pay line. A single average adjustment could be calculated, with the same amount then allocated to each female-dominated job.

These are meant to be simply illustrative of the technical and administrative issues that can be associated with the implementation and design of pay equity. They can have a substantial impact on the magnitude of the awards that are made, and hence are likely to be subject to considerable debate if the policy becomes more extensively applied.

Equal Employment Opportunity Legislation

In contrast to equal pay and equal value policies which deal with pay discrimination, equal employment opportunity legislation is designed to prevent discrimination in recruiting, hiring, promotion, and dismissals. In general, provincial equal employment opportunity legislation is part of the Human Rights Code of each province. Usually, complaints concerning discrimination in employment are made by the individual party to the Human Rights Commission. An officer of the Commission tries to reach a settlement by conciliation. If this is unsuccessful, a board of inquiry investigates and makes a final decision. Most provinces have appeal procedures to regular courts. Although the Commissions generally act only on complaints, on a more informal basis they can and do persuade employers to increase their quotas of female employees.

In spite of the fact that this procedure is time-consuming and cumbersome, equal employment opportunity legislation does have the virtue of increasing the demand for female labour at the recruiting, hiring, and promotion stages. This, in turn, should increase female wages *and* employment. Unlike equal pay laws which increase female wages at the expense of their employment, equal employment opportunity legislation would work through the market to increase both the wages and employment of females. However, equal employment opportunity legislation is likely to be more beneficial to new recruits or women seeking to change their jobs; it may do little to help incumbent females in their existing jobs.

Some have argued that equal pay and equal employment legislation are complementary, in that one is useless without the other. The argument is that without equal employment opportunity legislation, equal pay would result in employers refusing to recruit, hire, and promote females. Similarly, without equal pay, equal employment opportunity legislation would result in employers hiring females but paying them lower wages than equally productive males. However, this ignores the economic argument that the increased demand for females resulting from equal employment opportunity legislation would work through the market to increase female wages. Equal pay may be a natural byproduct of equal employment opportunities.

Affirmative Action or Employment Equity

While equal employment opportunity legislation is designed to remove discriminatory barriers so as to ensure an equality of *opportunity*, affirmative action is a more intrusive form of intervention, focusing on *results* and not just opportunities. The rationale is that an equality of opportunity is insufficient to compensate for the legacy of a cumulative history of discrimination as well as systemic discrimination that may be the unintended byproduct of certain personnel policies. True equality may require some preferential treatment for women, at least on a temporary basis, to ensure an equality of starting points. The hope is that temporary affirmative action programs may break the circle of entrapment whereby women are relegated to low-wage dead-end jobs, which in turn foster labour market behaviour that keeps them in such jobs.

Affirmative action legislation tends to be called employment equity legislation in Canada — a phrase coined by the 1984 Abella Commission — in part to differentiate it from the earlier U.S. initiatives which tended to be criticized on the grounds of requiring rigid quantitative quotas for the hiring of the target groups. In Canada, legislated employment equity currently exists only in the federal jurisdiction, which covers about 10 percent of the Canadian work force. It is embodied in the Employment Equity Act of 1986. A separate employment equity program was also adopted at that time for federal contractors as part of a federal contract compliance program. Employment equity programs have also been adopted in particular cities or municipalities and can be utilized on a voluntary basis without being considered a form of reverse discrimination by the courts.

The federal programs were to apply to four designated groups which were identified by the Abella Commission as particularly susceptible to systemic discrimination. The four designated groups are: women, visible minorities, disabled persons, and aboriginal people.

Employment equity essentially involves four steps. First, an internal audit is conducted within the firm, usually based on a survey, to determine the internal representation of the designated groups and their position within the firm. Second, this internal representation is compared to the external availability of these groups in the relevant external labour market, as given for example by census data. Third, targets or goals are established to achieve an internal representation of the designated groups that is similar to their availability in the relevant external labour market. Fourth, a plan and timetable for achieving these targets is established.

In the United States, affirmative action has a longer history. It is part of the federal contract compliance program under Executive Order 11246 of 1965, and subsequent orders. It can also be imposed by the courts as a form of redress under the general antidiscrimination legislation: Title VII, the Equal Employment Opportunity Provision, of the Civil Rights Act of 1964. When first instituted, affirmative action was targeted towards blacks, but in the 1970s it was also directed towards women.

Policies to Facilitate Female Employment

In addition to the various direct policies such as equal pay and equal employment opportunities legislation, and affirmative action, there are a variety of policies that can indirectly affect the employment opportunities and wages of women. These could be labelled *facilitating* policies, in that they are generally designed to expand the range of choices open to women and hence to facilitate their participation in labour market activities in a nondiscriminatory fashion.

Currently, women tend to have the primary responsibility for household tasks, especially the raising of children. This may reflect historical tradition, comparative advantage (which itself may be historically determined), or discrimination within the household. Whatever the origin, this responsibility can create tensions over the equitable division of labour within the household when both parties engage in labour market work.

To minimize some of this tension, and to expand the employment opportunities available to women, facilitating policies have been suggested. Improved availability of daycare, flexible working hours, part-time jobs, and childbirth leaves are examples of such facilitating policies. In the interests of equality they would apply to males as well as females, so as to maximize the opportunities of households to allocate their labour amongst various market and nonmarket alternatives.

Certainly such policies are not without their costs, which could be substantial in some areas such as subsidized daycare. Consequently there is legitimate debate over the efficiency of such policies. What is less contentious, from the perspective of economics at least, is that at a minimum such facilitating policies be allowed to emerge in response to market demands. This may require, for example, that daycare expenses be tax deductible or that quality regulations do not restrict excessively the availability of daycare. Or it may require that government policies do not discourage flexible working arrangements by increasing quasi-fixed costs of employment which may encourage employers to utilize only a core of male workers.

To a large extent these institutional features — flexible hours, part-time jobs, childcare leaves and daycare arrangements — are emerging as endogenous responses to the increased labour market role of women. Their emergence, however, may be subject to impediments reflecting discrimination, government policies designed to achieve other objectives, or simply the slow operation of market forces subject to inertia and fixed costs.

Considerable debate emerges over the appropriate role of public policy in this area. Some would argue that such facilitating policies should be discouraged in order to preserve the traditional sex division of labour. Others argue that public policies should be neutral by simply removing impediments to the emergence of these facilitating policies. Others argue for a more active role — subsidized daycare, extended childcare leave — to compensate for discrimination elsewhere in the system, including past discrimination.

The role of protectionist labour standards policies takes on an interesting light in the context of policies designed to facilitate the labour market work of women. Such protectionist policies could include special provisions requiring employers to provide free transportation for women on night shifts, or prohibiting the employment of women in some specific occupations, or restrictions on the hours of work for females. Some may argue that such protective devices would enable women to take on some jobs they would otherwise be reluctant to do. Most would argue that such protective devices simply protect male jobs by reducing the employment opportunities of females and by perpetuating the stereotype of women as in need of outside protection. If such protection is desirable, it should be provided to all workers, male and female, through labour standards laws or collective agreements.

This suggests another institutional arrangement for facilitating the labour market work of females — unionization. The rationale for increasing the role of women in unions is really two-fold. The first is to have women share in the wage and job security gains that unions obtain for their members. In order to obtain these gains, it is not only necessary for women to unionize; they must aspire to powerful positions within the union or at least ensure that their minority rights are guaranteed. The second rationale for increasing the role of women in unions is to ensure effective monitoring of equal pay, equal employment opportunities, and affirmative action.

Females may be more willing and able to press for equal pay and employment opportunities through the union both because the apparatus is available (shop steward, grievance procedure) and because reprisal by the company is less likely when the worker is protected by a union.

Alter Preferences and Attitudes

Economics traditionally regards tastes and preferences as exogenously given, and then inquires into what happens to the demand for something when such factors as relative prices and income change. In the area of sex discrimination, this is a legitimate inquiry. Equal pay and equal employment opportunity legislation can be viewed as policies designed to raise the "price" of discrimination to those who discriminate. The price in this case is the legal costs which include court costs and the expected fine. The contention is that raising the price of discrimination will reduce the quantity demanded.

In the area of discrimination, however, some have suggested the utilization of public policies designed to alter the basic tastes and preferences that give rise to discrimination, since tastes and preferences of employers, co-workers, and customers are an important source of discrimination. The dominant group, for example, may find it important to foster uniform preferences for discrimination and to punish nondiscriminators. Preferences are also shaped by the media, the education system, and the socio-economic system in general. Preferences are not immutable over time, nor are they likely to be the same for all firms, employees, or customers. For this reason, many antidiscrimination policies are designed to alter basic tastes, preferences, and attitudes.

In practice, of course changing basic tastes and attitudes is a difficult task. Specific policies have been instituted and include: removal of sex stereotyping in schoolbooks, guidance programs, and television; provision of information on the labour market performance and attachment of females; and politicization of women to raise their own group awareness. The process will be slow, but many argue that the results at least will be long lasting.

IMPACT OF POLICY INITIATIVES[6]

While there is considerable empirical evidence on the magnitude of the male–female wage differential and its determinants, there is much less evidence on the impact of various policy initiatives.

Conventional Equal Pay Policies

The Canadian evidence on the impact of equal pay policies is so far restricted to conventional equal pay policies which require equal pay for *equal work*, rather than the broader concept of work of equal value. Econometric studies in the area of equal pay for equal work have found those policies not to have had any impact on closing the male–female earnings gap (Gunderson 1975, 1985b). This lack of impact likely reflected the limited scope for such policies, since they could only deal with male–female wage differences within the same occupation and establishment, and such differences are unlikely to be very large compared to differences across occupations and establishments. As well, reliance on a complaints procedure may deter enforcement since individuals may be reluctant to complain.

In the United States, conventional equal pay policies are combined with equal employment opportunity initiatives under Title VII of the Civil Rights Act; therefore, the separate impact of equal pay is difficult to disentangle, especially because both should serve to raise the wages of women. The econometric studies of that legislation (reviewed in Gunderson 1989a) tends to find somewhat inconclusive results. Some studies find positive effects on the earnings or women, some find no effect, and when the effects are positive they tend to be small.

---

6. Much of the evidence from this section is summarized in more detail in Gunderson (1989a).

In contrast, more positive effects of equal pay policies are found in Britain (Zablaza and Tzannatos, 1985a, 1985b) where it was incorporated into the process of centralized collective bargaining that occurs in that country. This suggests that conventional equal pay policies may be more effective in reducing the earnings gap when it is incorporated by the private parties into more formal systems of centralized collective bargaining, rather than legislative systems relying on complaints in the more market oriented decentralized systems of wage determination and collective bargaining that prevails in North America.

Affirmative Action

A number of econometric studies (summarized in Gunderson 1989a) have evaluated the impact of affirmative action under the federal contract compliance of Executive Order 11246 in the United States. That legislation did benefit the minority groups to which it was targeted — mainly black males in the earlier years of the program. Much of that benefit,however, came at the expense of losses to other minority groups — mainly white females — to which the program was not targeted in the earlier years. However, when the affirmative action programs began to be more targeted towards females, they too benefited from them. The legislative initiatives also tended to be more effective in an expanding growing economy than in a recession (Beller 1982) and when it was aggressively enforced (Beller 1982; Leonard 1984b, 1985; Osterman 1982).

Comparable Worth

The impact of comparable worth (pay equity) programs has also been analyzed in a number of studies, usually based on the application of the policy in a number of public sector jurisdictions in Canada and the United States[7]. Average wage adjustments of $3,000 to $4,000 per recipient are common, although there is considerable variability in the adjustments. However, such adjustments only go to the portion of the work force that is in female-dominated jobs that are undervalued. When the costs are amortized over the whole work force of the particular public sector, they are in the neighbourhood of 4 to 8 percent of payroll costs. Typically, in those public sector work forces, the ratio of female earnings to male earnings (in all public sector jobs, not just those that receive the adjustment) is around 0.78 before the comparable worth adjustment and 0.84 after the adjustment. This increase of 0.06 in the earnings ratio implies that comparable worth closes about one-third (i.e., 0.06/0.22) of the earnings gap of 0.22 (i.e. 1.00–0.78).

---

7. The studies are reviewed in Gunderson and Riddell (1992a, 1992b). The jurisdictions where the policy has been applied in the public sector include Manitoba, Ontario, Minnesota, Iowa, Michigan, Minnesota, Washington State, and the City of San Jose in California.

These estimates of the impact of comparable worth are based on particular public sector cases where the policy has been applied. They are also based on estimates of the gap within a particular public sector employer, and this gap of around 0.22 is only a little over half of the earnings gap that prevails in the economy as a whole. Simulations of the potential economy-wide impact of comparable worth in the United States have estimated that the policy would close at most 8 to 20 percent of the overall gap of 0.41 (Johnson and Solon 1986) or 15 to 20 percent of the gap (Aldrich and Buchelle 1986). The policy can only close a portion of the overall gap because much of the gap reflects the segregation of females into low-wage firms and industries, and comparable worth does not involve comparisons across different firms or industries.

In summary, comparable worth would likely close about one-third of the pay gap within particular elements of the public sector. For the economy as a whole, it would likely close a smaller portion, perhaps 8 to 20 percent of the gap. This highlights that the policy can close a substantial portion of the earnings gap, but most would still remain even after a comprehensive application of the program.

Employment Effects of Comparable Worth

Ehrenberg and Smith (1987) simulate the potential adverse employment that would result from the widespread application of comparable worth to the public sector in the United States. Based upon the magnitude of the wage adjustments that occurred in a number of comparable worth cases in the United States, they assume that comparable worth would lead to a 20 percent increase in the wages of females in the state and local public sector. Based upon existing estimates of the elasticity of demand for public sector labour, this in turn would yield an employment reduction of only 2 to 3 percent. These adverse employment effects are small because of the existing lack of substitutability between males and females in the public sector. Whether this would be true in the long run when there is sufficient time to substitute away from female-intensive activities remains an unanswered question. Aldrich and Buchelle (1986) also estimate that comparable worth wage increases of 10 to 15 percent in the *private* sector would lead to employment reductions of about 3 percent. Kahn (1992) finds no adverse employment effects of comparable worth in the city of San Jose, California.

Gregory and Duncan (1981) also estimate the employment effect that resulted from the Australian pay awards emanating from their wage tribunals which set wages for the majority of the work force. Between 1972 and 1978 those awards led to an increase in the ratio of female to male wages from 0.774 to 0.933. This in turn led to a statistically significant but quantitatively small reduction in the growth of female employment relative to male employment. That is, female employment growth averaged 3.0 percent per year as opposed to the 4.5 percent that would have occurred without the substantial wage increase. Most of the reduced growth occurred in the manufacturing and service sector, not in the public sector where employment growth was relatively insensitive to the wage increases. While these various studies suggest either a small or nonexistent adverse employment effect from fairly substantial comparable worth wage increases, Killingsworth's (1990) results suggest otherwise, based on data from Minnesota, California, and Australia. He tends to find that the long-run adverse employment effects are small (certainly relative to the normal growth of female employment), but this occurred because the long-run wage increases from comparable worth were small. If they had been larger, then the adverse employment effect would have been substantial because he estimates a very elastic demand for female labour. In essence, Killingsworth tends to find a small adverse employment effect because of a small wage effect combined with a large elasticity of demand for female labour. The other studies tended to find a small adverse employment effect because of a substantial wage effect but a small elasticity of demand for female labour. Clearly, more work is necessary to resolve the important differences in this area.

SUMMARY OBSERVATIONS

Although it is hazardous to draw a conclusion about the impact of the various policy initiatives from such a limited number of studies, some tentative generalizations can be made. The effect of equal pay and equal employment opportunity policies tends to be inconclusive, although there is some evidence that affirmative action policies in the United States have been effective for the target groups to which they have been applied. Comparable worth policies have reduced the earnings gap and resulted in considerable awards in the few situations where they have been applied, albeit on a comprehensive basis their scope will be limited by an inability to deal with the considerable portion of the wage gap that reflects the segregation of females into low-wage establishments and industries. The evidence on the adverse employment effects from comparable worth adjustments tends to be inconclusive.

It should be kept in mind, however, that estimating the impact of such policy initiatives is extremely difficult, especially given the other dramatic changes that are occurring simultaneously in the female labour market. This is especially the case since the policy initiatives themselves may be a function of the policy problem they are designed to eliminate, and the initiatives may be related to other unobservable factors that are the true cause of the changes in labour market behaviour. Such problems are not unique to these policy initiatives; they apply to estimating the impact of almost any social program.

QUESTIONS

1. "If females are paid a wage equal to their marginal productivity, then sex discrimination in the labour market does not exist." Discuss.

2. "Male–female wage differentials for equally productive workers will not persist in the long run because the forces of competition would remove these differentials." Discuss

3. If we adjusted the gross male–female wage differential for all of the productivity-related factors that influence wages (e.g., education, experience, absenteeism, turnover, etc.) and found that the wage gap would be zero if males and females had the same productivity-related characteristics, would this indicate the absence of gender discrimination?

4. What impact would equal pay laws have on the wages and employment of female workers? What impact would equal employment opportunity laws (fair employment laws) have on the wages and employment of female workers?

5. For policy purposes does it matter if the reason for discrimination is prejudice, erroneous information, or job security? Does it matter if the main source of discrimination is employers, co-workers, or customers?

6. Based on the utility maximization theory given in footnote 2, show that male wages will exceed their marginal revenue product when nepotism exists whereby, other things being equal, the firm's utility is increased when it employs more males.

7. Based on the decomposition technique discussed in footnote 3, indicate how an alternative decomposition could be utilized by subtracting $\Sigma b_m \overline{X}_f$ rather than $\Sigma b_f \overline{X}_m$ from the right-hand side. Which decomposition is correct?

8. Would you expect the overall, unadjusted male–female pay gap to be increasing or decreasing over time, and why?

9. "Comparable worth" is akin to the elusive notion of a "just price," where the price is determined independent of the basic forces of demand (reflecting what people are will to pay for a service) and supply (reflecting what people are willing to accept to provide the service). Discuss.

10. Discuss the pros and cons of the following design or implementation features of comparable worth: (a) defining gender dominance as 60 percent as opposed to 70 percent of either sex, (b) constraining the male and female pay lines to have the same slope as opposed to different slopes, (c) allowing projection of the male payline outside of the sample range of the data in cases where the female-dominated job otherwise have no male comparator groups, (d) following a point-to-line, versus a line-to-line versus a point-to-point wage adjustment procedure.

11. Would you expect discrimination to be greater in the public sector or in the private sector and why? Would you expect it to be greater in union or nonunion sectors and why?

12. Based upon your knowledge of the various components of the overall male–female wage, discuss the potential scope of the various possible policy initiatives.

13. In designing a viable equal value policy, what steps might you take to minimize any inherent conflict between administrative concepts of value (as implied by job evaluation procedures) and economic concepts of value (as implied by market-determined wages)?

14. Assume that you have used a job evaluation procedure to establish the point scores for a number of male-dominated jobs and a number of female-dominated jobs in your organization. Indicate how you would estimate a pay line between point scores and pay for each of the predominantly male and predominantly female jobs. What is the appropriate functional form for such pay lines? Should they each have an intercept? If the pay in the undervalued female-dominated jobs is to be adjusted should the female pay line be raised to the male pay line (line-to-line adjustment) or should the pay in each female job be raised to the male pay line (point-to-line adjustment) or to some other value such as an average line or the pay of the closest male job? Should interpolation of the pay lines be allowed, within the sample range of the data? Should extrapolation of the pay lines be allowed outside of the sample range of the data?

REFERENCES AND FURTHER READINGS

(References here refer mainly to gender discrimination and theoretical issues pertaining to discrimination in general.)

Abella, R., (Commissioner). *Report of the Commission on Equality in Employment.* Ottawa: Supply and Services, 1984.

Abowd, J. And M. Killingsworth. Sex discrimination, atrophy, and the male-female wage differential. *IR* 22 (Fall 1983) 387–402.

Agarwal, N. and Harish Jain. Pay discrimination against women in Canada. *ILR* 117 (March–April 1978) 169–178.

Aigner, D. and G. Cain. Statistical theories of discrimination in labour markets. *ILRR* 30 (January 1972) 175–189.

Akerlof, G. Discriminatory, status-based wages among tradition-oriented stochastically trading coconut producers. *JPE* 93 (April 1985) 265–276.

Albelda, R. Occupation segregation by race and gender, 1958–1981. *ILRR* 39 (April 1986) 404–411.

Aldrich, M. The gender gap in earnings during World War II. *ILRR* 42 (April 1989) 415–429.

Aldrich, M. and R. Buchele. *the Economics of Comparable Worth.* Cambridge: Ballinger Publishing Company, 1986.

Arrow, K. The theory of discrimination. *Discrimination in the Labour Market*, O. Ashenfelter and A. Rees (eds.). Princeton, N.J.: Princeton University Press, 1973.

Ashenfelter, O. and T. Hannan. Sex discrimination and product market competition: the case of the banking industry. *QJE* 101 (February 1986) 149–174.

Becker, G. *The Economics of Discrimination*. Chicago: University of Chicago Press, 1971.

Beller, A. Occupational segregation by sex: determinants and changes. *JHR* 17 (Spring 1982) 371–392. Also, comments and reply 20 (Summer 1985) 437–444.

Beller, A. Changes in the sex composition of U.S. Occupations, 1960–1981. *JHR* 20 (Spring 1985)235–250.

Bergmann, B. The effect on white incomes of discrimination in employment. *JPE* 79 (March/April 1971) 294–313.

Bergmann, B. *the Economic Emergence of Women*. New York: Basic Books, 1986.

Bergmann, B. Does the market for women's labour need fixing? *JEP* 3 (Winter 1989) 43–60.

Blau, F. and A. Beller. Trends in earnings differentials by gender, 1971–1981. *ILRR* 41 (July 1988) 513–529.

Blau F. and L. Kahn. Race and sex differences in quits by young workers. *ILRR* 34 (July 1981) 563–577.

Blau, F. and L. Kahn. The gender earnings gap: learning from international comparisons. *AER* 82 (May 1992) 533–538.

Blau, F. and M. Ferber. Discrimination: empirical evidence from the United States. *AER* 77 (May 1987) 316–320.

Blau, F. and M. Ferber. Career plans and expectations of young women and men: the earnings gap and labour force participation. *JHR* 26 (Fall 1991) 581–607.

Block, W. and M. Walker (eds.). *Discrimination, Affirmative Action and Equal Opportunity*. Vancouver: Fraser Institute, 1982.

Bloom, D. and M. Killingsworth. Pay discrimination research and litigation: the use of regression. *IR* 21 (Fall 1982) 318–339.

Borjas, G. The measurement of race and gender wage differentials: evidence from the federal sector. *ILRR* 37 (October 1983) 70–91.

Borjas, G. and S. Bronors. Consumer discrimination and self employment. *JPE* 97 (June 1989) 581–605.

Borooah, V. and K. Lee. The effect of changes in Britain's industrial structure on female relative pay and employment. *EJ* 98 (September 1988) 818–832.

Brown, C. and J. Peckman (eds.). *Gender in the Workplace*. Washington, D.C.: Brookings Institute, 1987.

Buchele, R. and M. Aldrich. How much difference would comparable worth make? *IR* 24 (Spring 1985) 222–233.

Burt, S. Voluntary affirmative action: does it work? *RI/IR* 41 (No. 3, 1986) 541–551.

Butler, R. Estimating wage discrimination in the labour market. *JHR* 17 (Fall 1982) 606–621.

Butler, R. Direct estimates of the demand for race and sex discrimination. *SEJ* 49 (April 1983) 975–990.

Cain, G. The economic analysis of labour market discrimination: a survey. *Handbook of Labour Economics*, Vol. 1. O. Ashenfelter and R. Layard (eds.). New York: Elsevier, 1986.

Cannings, K. Family commitments and career success: earnings of male and female managers. *RI/IR* 46 (No. 1, 1991) 141–158.

Cannings, K. The earnings of female and male middle managers: a Canadian case study. *JHR* 23 (Winter 1988) 34–56.

Cannings, K. Managerial promotion: the effects of socialization, specialization and gender. *ILRR* 42 (October 1988) 77–88.

Cannings, K. and C. Montmarquette. The attitude of subordinates to the gender of superiors in a managerial hierarchy. *Journal of Economic Psychology* 12 (December 1991) 707–724.

Carlson, L. and C. Swartz. The earnings of women and ethnic minorities, 1959–1979. *ILRR* 41 (July 1988) 530–546.

Chaykowski, R. (ed.). *Pay Equity Legislation: Linking Economic Issues and Policy Concerns.* Kingston: Queen's University Industrial Relations Centre, 1989.

Chiplin, B. An alternative approach to the measurement of sex discrimination: an illustration from university entrance. *EJ* 91 (December 1981) 988–997.

Chiplin, B. and P. Sloan. Personal characteristics and sex differentials in professional employment. *EJ* 86 (December 1976) 729–745.

Cohen, M. The razor's edge invisible: feminism's effects on economics. *International Journal of Women's Studies.* 8 (May 1985) 286–296.

Corcoran, M. and G. Duncan. Work history, labour force attachment, and earnings differences between races and sexes. *JHR* 14 (Winter 1979) 3–20.

Cotton, J. On the decomposition of wage differentials. *R.E. Stats.* 70 (May 1988) 236–243.

Cox, D. Panel estimates of the effects of career interruptions on the earnings of women. *EI* 22 (July 1984) 386–403.

Cymrot, D. Does competition lessen discrimination? Some evidence. *JHR* (Fall 1985) 605–611.

Daymont, T. and P. Andrisani. Job preferences, college major and the gender gap in earnings. *JHR* 19 (Summer 1984) 408–428.

Dolton, P. and G. Makepeace. Sample selection and male-female earnings differentials in the graduate labour market. *Oxford Economic Papers* 38 (July 1986) 317–341.

Dolton, P. and G. Makepeace. Marital status, childrearing and earnings differentials in the graduate labour market. *EJ* 97 (December 1987) 897–992.

Donohue, J. Determinants of job turnover of young men and women in the United States: A hazard rate analysis. *Research in Population Economics* 6 (1988) 257–301.

Ehrenberg, R. and R. Smith. Comparable worth wage adjustments and female employ-ment in the state and local sector. *JOLE* 5 (January 1987) 43–62.

Eichengreen, B. Experience and the male-female earnings gap in the 1890s. *Journal of Economic History* 44 (September 1984) 822–834.

Elberts, R. and J. Stone. Male-female differences in promotions: EEO in public educa-tion. *JHR* 20 (Fall 1985) 504–521.

England, P. The failure of human capital theory to explain occupational sex segrega-tion. *JHR* 17 (Summer 1982) 358–370. Also, comments and reply 20 (Summer 1985) 437–444.

Evans, S. and B. Nelson. *Wage Justice: Comparable Worth and the Paradox of Techno-cratic Reform.* Chicago: University of Chicago Press, 1989.

Evan, W. Career interruptions following childbirth. *JOLE* 5 (April 1987) 255–277.

Evan, W. and D. Macpherson. The gender gap in pensions and wages. *R.E. Stats.* 72 (May 1990) 259–265.

Ferber, M. and J. Nelson (eds.). *Beyond Economic Man: Feminist Theory and Economics.* Chicago: University of Chicago Press, 1993.

Ferber, M. and J. Spaeth. Work characteristics and the male-female earnings gap. *AER* 74 (May 1984) 260–264.

Fields, J. and E. Wolf. The decline of sex segregation and the wage gap, 1970–80. *JHR* 26 (Fall 1991) 608–622.

Filer, R.K. Sexual differences in earnings: the role of individual personalities and tastes. *JHR* 18 (Winter 1983) 82–99.

Filer, R. The role of personality and tastes, in determining occupational structure. *ILRR* 39 (April 1986) 412–424.

Fishback, P. and J. Terza. Are estimates of sex discrimination robust? The use of never-marrieds. *EI* 27 (April 1989) 271–285.

Foot, D. and D. Stager. Intertemporal market effects on gender earnings differentials: lawyers in Canada, 1970–80. *Applied Economics* 21 (August 1989) 1011–1028.

Fuchs, V. *Women's Quest for Economic Equality.* Cambridge, Ma.: Harvard Univer-sity Press, 1988.

Fuchs, V. His and hers: gender differences in work and incomes, 1959–1979. *JOLE* 4 (July 1986) S245–S272.

Gerhart, B. Gender differences in current and starting salaries. *ILRR* 43 (April 1990) 418–433.

Gerhart, B. and N. Cheikh. Earnings and percentage female: a longitudinal analysis. *IR* 30 (Winter 1990) 10–20.

Gerlack, K. A note on male-female wage differences in West Germany. *JHR* 22 (Fall 1987) 584–592.

Gill, A. The role of discrimination in determining occupation structure. *ILRR* 42 (July 1989) 610–623.

Goldberg, M. Discrimination, nepotism, and long-run wage differentials. *QJE* 79 (May 1982) 307–320.

Goldberger, A. reverse regression and salary discrimination. *JHR* 19 (Summer 1984) 293–318.

Goldin, C. Monitoring costs and occupational segregation by sex: a historical analysis. *JOLE* 4 (January 1986) 1–27.

Goldin, C. *Understanding the Gender Gap: An Economic History of American Women.* Oxford: Oxford University Press, 1990.

Goldin, C. and S. Polachek. Residual differences by sex: perspectives on the gender gap in earnings. *AER* 77 (May 1987) 143–151.

Green, C. and M. Ferber. Employment discrimination: an empirical test of forward versus reverse regression. *JHR* 19 (Fall 1984) 557–569.

Greenhalgh, C. Male-female wage differentials in Great Britain: is marriage an equal opportunity? *EI* 90 (December 1980) 751–755.

Gregory, R. Jobs and gender: a legal approach to the Australian labour market. *Economic Record* (1991) 20–40.

Gregory, R. and R. Duncan. Segmented labour market theories and the Australian experience of equal pay for women. *Journal of Post Keynesian Economics* 2 (Spring 1981) 403–428.

Grenier, G. The effects of language characteristics on the wages of Hispanic-American males. *JHR* 19 (Winter 1984) 35–52.

Griffin, P. The effect of affirmative action on labor demand. *R.E. Stats.* 74 (May 1992) 251–260.

Gronan, R. Sex-related wage differentials and women's interrupted labor careers — the chicken or the egg. *JOLE* 6 (July 1988) 277–301.

Groshen, E. The structure of the female/male differentials. *JHR* 26 (Summer 1991) 457–472.

Gunderson, M. Male-female wage differentials and the impact of equal pay legislation. *R.E. Stats.* 57 (November 1975) 426–470.

Gunderson, M. Decomposition of male-female earnings differentials: Canada 1970. *CJE* 12 (August 1979) 479–484.

Gunderson, M. Discrimination, equal pay and equal opportunities in the labour market. In *Work and Pay: The Canadian Labour Market.* Edited by C. Riddell. Toronto: University of Toronto Press for the Mcdonald Commission, 1985a.

Gunderson, M. Spline function estimates of the impact of equal pay, legislation: the Ontario experience. *RI/IR* 40 (No. 4, 1985b) 775–791.

Gunderson, M. Male-female wage differentials and policy responses. *JEL* 27 (March 1989a) 46–117.

Gunderson, M. Implementation of comparable worth in Canada. *Journal of Social Issues* 45 (No. 4, 1989b) 209–222.

Gunderson, M. and L. Muszynski. *Women and Labour Market Poverty.* Ottawa: Canada Advisory Council on the Status of Women, 1990.

Gunderson, M. and W.C. Riddell. Economics of women's wages in Canada. *International Review of Comparative Public Policy* 3 (1991) 151–176.

Gunderson, M. and W.C. Riddell. Comparable worth: Canada's experience. *Contemporary Policy Issues* 10 (July 1992) 85–94.

Gunderson, M. and R. Robb. Equal pay for work of equal value: Canada's experience. *Advances in Industrial and Labour Relations* 5 (1991a) 151–168.

Gunderson, M. and R. Robb. Legal and institutional issues pertaining to women's wages in Canada. *International Review of Comparative Public Policy* 3 (1991) 129–150.

Gyimah-Brempong, K., R. Fichtenbaum and G. Willis. The effect of college education on the male-female wage differential. *SEJ* 58 (January 1992) 790–804.

Haberfeld, Y. and Y. Shenhav. Are women and blacks closing the gap? *ILRR* 44 (October 1990) 68–82.

Hall, R. The importance of lifetime jobs in the U.S. economy. *AER* 72 (September 1982) 716–724.

Hartmann, H.I. (ed.). *Comparable Worth: New Directions for Research*. Washington: National Academy Press, 1985.

Hartman, M. and O'Farrell (eds.) *Pay Equity: Empirical Inquiries*. Washington: National Academy Press, 1989.

Heckman, J. and B. Payner. Determining the impact of federal anti-discrimination policy on the economic status of blacks. *AER* 79 (March 1989) 138–177.

Hersch, J. Male-female differences in hourly wages: the role of human capital, working conditions and housework. *ILRR* 44 (July 1991) 746–759.

Hill, A. and M. Killingsworth (eds). *Comparable Worth: Analyses and Evidence*. Ithaca: ILR Press, 1989.

Hirsch, J. and J. Stone. "New and improved" estimates of qualification discrimination. *SEJ* 52 (October 1985) 484–491.

Hodson, R. and P. England. Industrial structure and sex differences in earnings. *IR* 25 (Winter 1986) 16–32.

Holmes, R. Male-female earnings differentials in Canada. *JHR* 11 (Winter 1976) 109–112.

Horrigan, M. and J. Markey. Recent gains in women's earnings: better pay or longer hours? *MLR* 113 (July 1990) 11–17.

Hunt, A. (ed.). *Women and Paid Work*. New York: St. Martin's Press, 1988.

Johnson, G. and G. Solon. Estimates of the direct effects of comparable worth policy. *AER* 76 (December 1986) 1117–1125.

Jones, F. On decomposing the wage gap: a critical comment on Blinder's method. *JHR* 18 (Winter 1983) 126–130.

Jones, J. and W. Walsh. Product market imperfections, job content differences and gender employment discrimination at the managerial level: some evidence from the Canadian manufacturing sector in 1971 and 1981. *CJE* 24 (November 1991) 844–858.

Kahn, L. Customer discrimination and affirmative action. *EI* 29 (July 1991) 555–571.

Kahn, S. Economic implications of public sector comparable worth: the case of San Jose California. *IR* 31 (Spring 1992) 270–291.

Killingsworth, M. Comparable worth in the job market: estimating its effects. *MLR* 108 (July 1985) 39–41.

Killingsworth, M. Heterogeneous preferences, compensating wage differentials and comparable worth. *QJE* 102 (November 1987) 727–742.

Killingsworth, M. *The Economies of Comparable Worth.* Kalamazoo, Mich.: W.E. Upjohn Institute, 1990.

Korenman, S. and D. Neumark. Marriage, motherhood and wages. *JHR* 27 (Spring 1992) 233–255.

Kuhn, P. Sex discrimination in labor markets: the role of statistical evidence. *AER* 77 (September 1987) 567–583. Also, comments by E. Barbezet and J. Hushes and by W. Evans, and reply 80 (March 1990) 277–297.

Lang, K. A language theory of discrimination. *QJE* 101 (May 1986) 363–382.

Lazear, E. and S. Rosen. Male-female wage differentials in job ladders. *JOLE* 8 (January 1990) S106–S123.

Leonard, J. Antidiscrimination or reverse discrimination: the impact of changing demographics, Title VII and affirmative action on productivity. *JHR* 19 (Spring 1984a) 145–174.

Leonard, J. Employment and occupational advance under affirmative action. *R.E. Stats.* 66 (August 1984b) 377–385.

Leonard, J. The impact of affirmative action on employment. *JOLE* 2 (October 1984c) 439–464.

Leonard, J. Affirmative action as earnings redistribution: the targeting of compliance reviews. *JOLE* 3 (July 1985) 363–384.

Leonard, J. What promises are worth: the impact of affirmative action goals. *JHR* 20 (Winter 1985) 3–20.

Leonard, J. Women and affirmative action. *JEP* 3 (Winter 1989) 61–75.

Lewis, G. Gender and promotions. *JHR* 21 (Summer 1986) 406–419.

Light, A. and M. Ureta. Gender differences in wages and job turnover among continuously employed workers. *AER* 80 (May 1990) 293–397.

Light, A. and M. Ureta. Panel estimates of male and female job turnover behaviour. *JOLE* 10 (April 1992) 156–181.

Lindsay, C. and M. Maloney. A model and some evidence concerning the influence of discrimination on wages. *EI* 26 (October 1988) 645–680.

Lundberg, S. and R. Startz. Private discrimination and social intervention in competitive labor markets. *AER* 73 (June 1983) 340–347.

Madden, J. Gender differences in costs of displacement. *AER* (May 1987) 246–251.

Manchester, J. and D. Stapleton. On measuring the progress of women's quest for economic equality. *JHR* 26 (Summer 1991) 562–580.

Maki, D. and I. Ng. Effects of trade unions on the earnings differential between males and females. *CJE* 23 (May 1990) 305–311.

McCrate, E. Gender differences: the role of endogenous preferences and collective action. *AER* 78 (May 1988) 235–239.

McCarthy, D. and J. Turner. Sex discrimination in pension compensation. *IRRA*. 42 (December 1985) 129–138.

Meitzen, M. Differences in male and female job quitting behavior. *JOLE* 4 (April 1986) 151–167.

Milgrom, P. and S. Oster. Job discrimination, market forces and the invisibility hypothesis. *QJE* 102 (August 1987) 453–476.

Miller, P. The wage effect of occupational segregation in Britain. *EJ* 97 (December 1987) 885–896.

Miller, P. Gender differences in observed and offered wages in Canada 1980. *CJE* 20 (May 1982) 225–244.

Miller, P. and S. Rummery. Male-female wage differentials in Australia: a reassessment. *Australian Economic Papers* 30 (June 1991) 50–69.

Mincer, J. Intercountry comparisons of labor force trends and of related developments. *JOLE* 3 (January 1985) S1–S32.

Mincer, J. and Ofek. Interrupted work careers: depreciation and restoration of human capital. *JHR* 17 (Winter 1982) 3–24.

Montgomery, E. and W. Wascher. Race and gender wage inequality in services and manufacturing. *IR* 26 (Fall 1987) 284–290.

Moore, R. Are male-female earnings differentials related to life-expectancy-based pension cost differences? *EI* 25 (July 1987) 389–401.

Nakamura, A. and M. Nakamura. *The Second Paycheck*. Toronto: Academic Press, 1985.

Nardinelli, C. and C. Simon. Customer racial discrimination in the market for memorabilia: the case of baseball. *QJE* 105 (August 1990) 575–595.

Neumark, D. Employers' discriminatory behaviour and the estimation of wage discrimination. *JHR* 23 (Summer 1988) 279–295.

Oaxaca, R. Male-female wage differentials in urban labour markets. *IER* 14 (October 1973) 693–709.

Olson, C. and B. Becker. Sex discrimination in the promotion process. *ILRR* 36 (July 1983) 624–641.

O'Neill, J. The trend in the male-female wage gap in the United States. *JOLE* 3 (January 1985) S91–S116.

O'Neil, J., M. Brien and J. Cunningham. Effects of comparable worth policy: evidence from Washington State. *AER* 79 (May 1989) 305–309.

Ontario, *Green Paper on Pay Equity*. Toronto: Attorney General's Office, 1985.

Orazem, P. and J.P. Mattila. The implementation process of comparable worth: winners and losers. *JPE* 98 (February 1990) 134–152.

Osterman, P. Affirmative action and opportunity: a study of female quit rates. *R.E. Stats.* 64 (November 1982) 604–613.

Paglin, M. and A. Rufolo. Heterogeneous human capital, occupational choice, and male-female earnings differentials. *JOLE* (January 1990) 123–144.

Parsons, D. Specific human capital: an application to quit rates and layoff rates. *JPE* 80 (November/December 1972) 1120–1143.

Phelps, E. The statistical theory of racism and sexism. *AER* 62 (September 1972) 659–661.

Peterson, J. The challenge of comparable worth: an institutionalist view. *Journal of Economic Issues* 24 (June 1990) 605–612.

Phipps, S. Gender wage differences in Australia, Sweden and the United States. *Review of Income and Wealth* 36 (December 1990) 365–379.

Polachek, S. Occupational self selection in human capital approach to sex differences in occupational structure. *R.E. Stats.* 63 (February 1981) 60–69.

Polachek, S. Occupational segregation: a defence of human capital predictions. *JHR* 20 (Summer 1985) 437–440.

Preston, A. Women in the white-collar nonprofit sector. *R.E. Stats.* 72 (November 1990) 560–568.

Ragan, J. and S. Smith. The impact of differences in turnover rates on male-female pay differentials. *JHR* 16 (Summer 1981) 343–365.

Raymond, R., M. Sesnowitz and D. Williams. Comparable worth and factor point analysis in state government. *IR* 31 (Winter 1992) 195–215.

Rea, S. Jr. The market response to the elimination of sex-based annuities. *SEJ* 54 (July 1987) 55–63.

Reich, M. Who benefits from racism? *JHR* 13 (Fall 1978) 524–544.

Remick, H. (ed.). *Comparable Worth and Wage Discrimination*. Philadelphia: Temple University Press, 1984.

Robb, R.E. Earnings differentials between males and females in Ontario. *CJE* 11 (May 1978) 350–359.

Robb, R.E. Equal pay policy. In *Towards Equity*. Ottawa: Supply and Services, 1985.

Robb, R. Equal pay for work of equal value. *CPP* 13 (December 1987) 445–461.

Roemer, J. Divide and conquer: micro-foundations of a Marxian theory of wage discrimination. *Bell Journal of Economics* 10 (Autumn 1979) 695–706.

Riach, P. and J. Rich. Measuring discrimination by direct experimental methods. *Journal of Post Keynesian Economics* 14 (Winter 1991–92) 143–150.

Sandell, S. and D. Shapiro. Work expectations, human capital accumulation, and the wages of young women. *JHR* 15 (Summer 1980) 335–353.

Schotter, A. and K. Weigelt. Asymmetric tournaments, equal opportunity laws, and affirmative action, *QJE* 107 (May 1992) 511–539.

Schrank, W. Sex discrimination in faculty salaries. *CJE* 10 (August 1977) 411–433.

Schwab, S. Is statistical discrimination efficient? *AER* 76 (March 1986) 228–234.

Shapiro, D. and L. Shaw. Growth in the labor force attachment of married women: accounting for changes in the 1970s. *SEJ* (October 1983) 461–473.

Shapiro D. and M. Stelcner. Male-female earnings differentials and the role of language in Canada, Ontario and Quebec, 1970. *CJE* 14 (May 1981) 341–348.

Shapiro, D. and M. Stelcner. Earnings differentials among linguistic groups in Quebec. *IR* 21 (Fall 1982) 365–375.

Shapiro D. and M. Stelcner. Language legislation and male-female earnings differentials in Quebec. CPP 8 (Winter 1982) 106–133.

Shapiro, P. and M. Stelcner. The persistence of the male-female earnings gap in Canada, 1970–1980: the impact of equal pay laws and language policy. *CPP* 13 (December 1987) 462–476.

Simpson, W. Starting even? Job mobility and the wage gap between young single males and females. *Applied Economics* 22 (June 1990) 723–737.

Smith, J. and M. Word. Women in the labour market and in the family. *JEP* 3 (Winter 1989) 9–23.

Sorensen, E. Measuring the pay disparity between typically female occupations and other jobs. *ILRR* 42 (July 1989) 624–639.

Sorensen, E. The crowding hypothesis and comparable worth. *JHR* 25 (Winter 1990) 55–89.

Sorensen, E. Implementing comparable worth: a survey of recent job evaluation studies. *AER* Proceedings 76 (May 1986) 364–367.

Sorensen, E. Effect of comparable worth policies on earnings. *IR* 26 (Fall 1987) 227–239.

Spurr, S. Sex discrimination in the legal profession. *ILRR* 43 (April 1990) 406–417.

Stewart, M. and C. Greenhalgh. Work history patterns and the occupational attainment of women. *EJ* 94 (September 1984) 493–519.

Swimmer, G. Gender based differences in promotions of clerical workers. *RI/IR* 45 (No. 2, 1990) 300–310.

Treiman, D. and H. Hartman (eds.). *Women, Work and Wages: Equal Pay for Jobs of Equal Value.* Washington: National Academy of Sciences, National Research Council, 1981.

Tzannatos, E. The long-run effects of the sex integration of the British labour market. *R.E. Studies* 15 (1988) 5–18.

Tzannatos, P. and A. Zabalza. the anatomy of the rise of British female relative wages in the 1970s. *BJR* 22 (July 1984) 177–194.

Ureta, M. The importance of lifetime jobs in the U.S. economy, revisited. *AER* 82 (March 1992) 322–335.

Uri, N. and J. Mixon. Effects of U.S. affirmative action programs on women's employment. *Journal of Policy Modeling* 13 (Fall 1991) 367–382.

Viscusi, K. Sex differences in worker quitting. *R.E. Stats.* 62 (August 1980) 388–398.

Weiner, N. and M. Gunderson. *Pay Equity: Issues, Options and Experiences.* Toronto: Butterworths, 1990.

Weiss, Y. and R. Gronau. Expected interruption in labour force participation and sex related differences in earnings growth. *R.E. Stats.* 48 (October 1981) 607–620.

Welch, F. Affirmative action and its enforcement. *AER* 71 (May 1981) 127–133.

Willborn, Steven. *A Comparable Worth Primer.* Lexington: D.C. Heath, 1986.

Woring, M. *If Women Counted: A New Feminist Economics.* San Francisco: Hooper and Row, 1988.

Wright, R. and J. Ermisch. Gender discrimination in the British labour market. *EJ* 101 (May 1991) 508–522.

Zabalza, A. and Z. Tzannatos. The effect of Britain's anti-discrimination legislation on relative pay and employment. *EJ* 95 (September 1985a) 79–99; and comments and reply *EJ* 98 (September 1988) 812–843.

Zabalza, A. and Z. Tzannatos. *The Effects of Legislation in Female Employment and Wages.* Cambridge: Cambridge University Press, 1985b.

Zolokar, N. Male-female differences in occupational choice and the demand for general and occupation-specific human capital. *EI* 26 (January 1988) 59–74.

CHAPTER 23

Optimal Compensation Systems, Deferred Compensation, and Mandatory Retirement

Many features of our compensation system in particular (and personnel policies in general) seem peculiar and inefficient. For example, promotion and pay are often based on seniority and not necessarily merit. Workers are often required to retire from specific jobs even though they could carry on many of the necessary functions. Certain individuals receive astronomical pay, certainly relative to what they would receive in their next best alternative activity, even though they appear to be only slightly better performers than others. In many circumstances, pay seems to resemble a prize rather than payment equal to the marginal revenue product of the employee to the firm. Retroactive cost of living adjustments are often made to the pensions of retired workers, even though neither the firm nor the incumbent employees have a formal obligation to the retired workers. In many circumstances wages appear to be in excess of the amount necessary to recruit and retain a viable work force to the firm.

To some, these examples are simply taken as illustrations of the fact that labour markets do not operate in the textbook fashion predicted by conventional economics. They are regarded as inefficient pay practices or the results of constraints imposed by unions. In recent years, however, basic economic principles have been applied to help explain, in part at least, the existence of various institutional features of labour markets, including compensation systems. An efficiency rationale was sought to explain their survival value. This is part of the trend away from simply regarding these institutional features as exogenously given and examining their impact. It is part of the trend to try to explain these institutional features (i.e., to regard them as endogenous) using basic principles of economics.

AGENCY THEORY AND EFFICIENCY WAGE THEORY

In many instances the theoretical framework that has been applied is part of what is termed principal-agent theory. That perspective deals with the problem of designing an efficient contract between the principal (worker) and the agent (employer) when there are incentives to cheat. The problems are particularly acute when monitoring costs are an issue or there is asymmetric information (i.e., one party has better information than the other and there is no incentive to reveal that information). In such circumstances the parties may try to design contracts that elicit "truth-telling" and that ensure incentive compatibility so that no party has an incentive to cheat on the contract.

In such circumstances the pay of workers need not equal their marginal revenue product to the firm at each and every point in time (a condition economists label the "spot" or "auction market"). Rather their *expected* pay would equal their expected productivity over their expected lifetime with the firm. Such circumstances are often characterized by explicit contracts like collective agreements that provide a degree of guarantee of the receipt of payment. Or they are often characterized by implicit contracts that rely largely upon reputation. Where reputation is weak, a compensating wage may be paid for the risk that the contract may be broken. Economists term such markets "contract" markets to highlight the long-term contract nature of the arrangements.

The growing literature on efficient compensation systems is also related to the literature on "efficiency wages" (discussed previously) which argues that wages may affect productivity as well as wages being paid according to the value of the marginal product of labour in the firm; that is, causality can run both ways. This efficiency wage literature has had a number of applications. The economic development literature has often emphasized the subsistence wage necessary for basic levels of nutrition. The Marxist literature has emphasized the subsistence wage necessary for the reproduction of the labour force. As discussed in the chapter on the impact of unions, much of the new literature on the impact of unions has emphasized that unions may arise in response to high wages (in addition to the conventional view that unions *cause* higher wages) and that such unions can have a positive influence on productivity. As discussed previously, one possible response to minimum wage legislation (or any other form of wage fixing) is for employers to be "shocked" into more efficient practices; again higher wages induce higher productivity. High pay also enables employers to hire from a queue of applicants and this may serve as a worker discipline device, making the cost of layoff high to the worker, especially if there is a pool of unemployed. The personnel literature, and many private personnel practices, has long recognized the potential efficiency gains of paying high wages (see Exhibit 23.1). Pay (and especially perceptions about the fairness of pay) has long been recognized in the personnel literature as potentially being able to affect

---

EXHIBIT 23.1   DO HIGH WAGES PAY FOR THEMSELVES?

On January 14, 1914, Henry Ford overnight doubled the going wage rate to $5.00 per day. As a result there were hugh queues of applicants, so much so that rioting developed and the lines of job applicants had to be controlled with fire hoses.

Worker performance and productivity improved dramatically. Within a year, turn-over fell to one-eighth, absenteeism to one-quarter, and dismissals to one-tenth of their previous level.

Henry Ford touted the high wage policy — it was termed profit sharing — as good business, rather than altruism. It certainly did have the effect of improving work effort and productivity, and it reduced the need for costly supervision. This raises the possibility that high wages may "pay for themselves," at least to a degree.

It is interesting to note that women and "girls" were excluded from eligibility for the $5.00 day, largely on the ground that they were not supporting a family. As well, a Sociological Department with 150 inspectors was established, with the power to exclude workers from eligibility if they engaged in excessive drinking, gambling, untidiness, and consumption of unwholesome food!

Source: D. Raff and L. Summers. Did Henry Ford pay efficiency wages? *Journal of Labor Economics* 5 (October 1987) S57–S86.

---

productivity. In all of these examples the causality runs in the direction of pay affecting productivity. This supplements the conventional economic emphasis that pay is given in return for the value of the marginal product of labour to the firm. This two-way causality must be recognized in the design of any optimal compensation system.

ECONOMICS OF SUPERSTARS

Optimal compensation systems must also consider the fact that the value of the marginal product of labour may reflect the size of the market as well as the contribution of an employee to the productivity of others. As emphasized by Rosen (1981) small differences in the skill input of certain individuals may get magnified incredibly in the value of the marginal product of the service consumed by the public or co-workers in certain circumstances. This could be the case, for example, in a collectively consumed consumption good like a concert or T.V. performance where the size of the audience does not usually detract from the ability of others to watch. In such circumstances of a common demand, people are willing to pay to hear the best person because the additional price they have to pay to hear the best is shared amongst a large group of collective recipients. The additional *per person* cost of hearing the best person is small because it is shared amongst the large audience. Hence, "the best" person may command a superstar salary that is astronomical relative to the next-best person even though the superstar's ability or skill may only be marginally better than that of the next-best person. Small differences in skill get magnified into large differences in the value of the marginal product of the service when that service can be consumed by a large audience that can share the cost.

Similarly, if an individual can have a positive effect on the productivity of other workers in the job hierarchy then the value of the marginal product of that person's labour to the organization may be high, even if that positive effect is small for each person. A small positive effect on each person can cumulate to a large total effect when the number of affected persons in the organization is large. Hence, we may expect certain executives or talented individuals to receive a much larger salary than the next most talented individual, even though there may not be much difference in the skills of the two individuals. Small differences in skill get magnified into large differences in the value of an individual to an organization when those skills can improve the productivity of a large number of persons. We would also expect such talented individuals to be drawn to large organizations where the benefits of their talent could be spread over larger numbers.

SALARIES AS TOURNAMENT PRIZES

As pointed out in Lazear and Rosen (1981), executive salaries often seem to resemble prizes to the winners of contests rather than compensation in return for the value of the marginal product of one's services. This is starkly illustrated in the situation of a company that has a number of vice-presidents, all of whom are of roughly comparable ability. However, the one who gets promoted, presumably on the basis of being the best, gets a huge increase in salary. The salary difference between the president and the vice-president seems to reflect a prize for winning a contest more than differences in ability or the value of the person to the organization. That is, the president is probably slightly more capable than the vice-president and yet the salary difference seems much more than the ability difference.

Such a compensation system may be efficient, however, if the organization is only able to *rank* its executives according to the relative value of their contribution to the organization. That is, the organization may be able to say that individual A is better than B who is better than C, and so forth. However, the organization is not able to assess the precise contribution of each individual or by how much the contribution of one individual exceeds that of another. If the organization is able to assess the productivity of the group, then it could pay individual wages equal to the *average* productivity of each person. However, this compensation system may not create much incentive to perform since the individual's pay would be based on the productivity of the group, not the individual. In fact, it may create perverse incentives for individuals, who themselves know that they have below average productivity, to enter the group and get the average pay. This adverse selection procedure (a common problem in markets with asymmetric information) could lead to a situation where the bad workers drive out the good workers.

In order to create performance incentives in such markets, organizations may pay a top salary that resembles a prize to the winner in the group. Even if most are paid a wage equal to the average productivity of the group, there will be an incentive to perform in order to be promoted to win the prize. The organization can judge the winner in order to award the prize because, by assumption, the organization can rank individuals on the basis of their performance, even though it cannot gauge their individual productivity.

Such compensation in the form of prizes may be particularly important for senior persons who otherwise have few promotional opportunities left in their career. There may still be an incentive to perform if one of those promotions means the prize of a large salary increase.

Such compensation systems in the form of prizes can also create perverse incentive structures. For example, they may discourage co-operative behaviour since that could lead to one's competitor receiving the prize, unless of course the organization is able to identify and reward such "team-players." Such compensation systems may even encourage individuals to disrupt the performance of their competitors for the prize so as to enhance their own probability of winning the prize. The firm is also limited in the extent to which it can use prizes since risk-adverse individuals would not enter a contest for which the runner-ups (e.g., vice-presidents) would receive nothing. Hence, it must balance the needs to provide an incentive to win a large prize against the needs to guarantee some payment to get people to enter the contest.

EFFICIENT PAY EQUALITY AND EMPLOYEE COOPERATION

In establishing the pay structure within an organization, trade-offs are involved in determining the optimal degree of inequality of pay, as well as competitive and cooperative behaviour amongst employers.[1] Some dispersion of inequality of pay is necessary to provide an incentive to perform well in order to advance up the hierarchy and perhaps ultimately reach the top. However, too much inequality or dispersion may be inefficient in that it can discourage cooperative behaviour and teamwork if only a few can rise to the top. It may even encourage sabotage to prevent your fellow employees, if they are competitors, from being promoted if that reduces your own chance of promotion. As well, a large degree of salary inequity may not be attractive to persons with a degree of risk-aversion, or who are motivated by concepts of fairness.

---

1. Much of the material in this section is based on Lazear (1979).

The challenge for compensation specialists and human resource managers is to establish the optimal degree of pay equality or salary compression to provide performance incentives and yet to encourage teamwork and cooperative behaviour if that is important. The optimal degree of pay equality likely differs across different situations. A greater degree of pay equality or salary compression will be efficient if workers affect each other's output. In such circumstances, cooperative behaviour and teamwork have a greater potential to enhance group productivity, while sabotage has a greater potential to do the opposite.

Different pay incentive schemes also may be appropriate for different levels within an organization, as well as for different organizational structures. For example, at higher levels within an organization the potential for cooperative or sabotaging behaviour may be greater than at lower levels. As well, the types of persons who achieve those higher levels may be disproportionately "hawkish" and willing to engage in aggressive competitive behaviour. In such circumstances, it may be sensible to pay higher-level managers bonuses on the basis of individual or group output, rather than paying them on the basis of their relative performance. The latter could encourage noncooperative behaviour and even sabotage, on their own part as well as on the part of their peer with whom they compete.

Firms may also try to discourage noncooperative behaviour or sabotage by keeping such opposing contestants at arm's length, for example, by having them each head relatively autonomous separate business units. Selecting a president from "within" a company also may be more feasible if that person is chosen from a group of vice-presidents who head autonomous units whose relative performance can be evaluated easily, and who cannot affect the relative performance of the other competing vice-presidents through noncooperative behaviour or sabotage. If such opportunistic behaviour is possible, then it may be necessary to choose a president from the "outside," albeit this may dilute the incentive of inside vice-presidents because of the reduced probability for internal promotion.

The optimal degree of competition and cooperation amongst employees is also likely to differ across firms and individuals. Some firms may get higher output, more innovation, and better quality by fostering competition; others may do better by fostering cooperation amongst employees. Some workers may thrive on competition; others may be devoured by it and do their best work in a cooperative environment. In such circumstances it is incumbent upon human resource managers to appropriately "mix and match" such workers. Workers themselves cannot always be relied upon to sort themselves into their appropriate environment, since there may be an incentive for the more "hawkish" competitive employees to disguise that trait, enter the more "dovish" cooperative work environment, and then dominate that environment by their noncooperative behaviour. In such circumstances, it may be rational for human resource managers to use personality traits as a hiring criterion so as to appropriately mix and match employees on the basis of the importance of competitive versus cooperative behaviour.

These examples highlight that human resource policies which at first glance may appear to be inefficient (e.g., egalitarian wage policy, paying attention to personality in the hiring decision) may be efficient practices to encourage cooperative behaviour amongst employees (see also Exhibit 23.2). In essence, basic economic principles may be able to explain the existence of a variety of human resource practices and compensation policies, including egalitarian wage structures within a firm, personality as a criterion in hiring, the establishment of semiautonomous business units, and promotion from within the firm as opposed to from the outside. Explaining the variation in the use of such practices across time and across different work environments is an important new area of research in labour economics. It also highlights the interrelationship of labour economics with practical issues of human resource management and compensation practices.

EXECUTIVE COMPENSATION

The issue of executive compensation has received increased attention and public scrutiny in recent years. Executive compensation has increased relative to the

---

EXHIBIT 23.2   WHY IS THERE ACADEMIC TENURE?

Historically, academic tenure was perceived to be necessary to ensure freedom of expression in the university environment. Other rationales have been offered, including the possibility that it encourages researchers to undertake potentially risky but innovative research projects, since unsuccessful projects will not lead to job loss.

Carmichael (1988) offers an alternative explanation in that such job security is necessary to provide an incentive for more senior professors to hire junior professors who are better than themselves. Without such job security, professors would be reluctant to hire their potential replacement.

Tenure is prominent in the academic field because the recruiting method is unusual in that the search and hiring decisions are done usually by the employees themselves (i.e., by professors). In a sense, the university can be thought of as a workers' cooperative, run by the workers themselves! In order to provide an incentive to hire the best young people, academics are granted the job security of tenure after meeting certain competency requirements, usually after about six years as an assistant professor.

Again, what appears to be an inefficient work rule may well be an efficient practice, given the policy of recruiting and hiring being done by professors themselves. On the other hand, it may also be a convenient rationale for those of us with tenure!

Source: L. Carmichael. Incentives in academics: why is there academic tenure? *Journal of Political Economy* 96 (June 1988) 453–472.

average pay of workers, and it is much higher in the United States (where the attention has focused) in relation to other countries like Japan, whose economic performance has exceeded that of the United States. Stories abound in the popular press of skyrocketing executive pay that has occurred in spite of the poor performance of their companies, suggesting that executive pay is not tied to performance.

The previous discussion of tournaments and superstars suggests that high executive salaries may be motivated in part by the need to pay "prizes" to the top persons, as a way to induce incentives amongst other executives so as to win the "tournament prize." Making sure that the top person gets the job is especially important for a top executive who is responsible for an extremely large organization. Correct decisions in these circumstances, for example, can have hugh ramifications for the performance of large numbers of workers. In that vein, the incentive that is important is not the incentive for the executive to work hard or to acquire managerial skills, but rather it is to make sure that the executive with the right talent gets matched with the organization that needs and values that talent the most. Paying a person a million dollars as opposed to half a million dollars a year is not likely to make her work any harder or be more productive. It may not even be necessary to have such a large "prize" to improve the incentives of potential candidates to be promoted and "win" the prize. However, it may be necessary to ensure that the appropriate person goes to the organization where her talent is valued most. Such organizations are likely to bid for that talent if they can identify it as being of particular use to their organization. This need to pay high salaries as prizes may be tempered by the negative ramifications of excessive inequality if, for example, it induces noncooperative behaviour or feelings of resentment if such high salaries violate norms of fairness.

Popular impressions to the contrary, the empirical evidence tends to suggest that the compensation of top executives *is* positively related to the performance of their firm (Antle and Smith 1986; Murphy 1985; Jensen and Murphy 1990, and references in those studies). Whether this link is sufficiently strong to induce appropriate incentives, or is in fact unnecessarily and excessively strong, is subject to considerable controversy. It is also not clear that the underlying determinants of executive compensation have *changed* sufficiently in recent years to explain the rapid *increase* in their relative compensation that appears to have occurred.

The link between executive compensation and firm performance is complicated by the institutional arrangements whereby executive compensation is set. Executive compensation usually consists of a base salary as well as stock options — the latter of which provide an incentive for the CEO to improve the stock market value of their firm. Typically, the compensation process involves the Board of Directors approving the executive pay recommendation made by a compensation committee. The Boards of Directors may have an incentive to award high executive salaries because they themselves are usually executives, and they are often appointed by the CEO whose salary they are determining. Shareholders may not be able to provide an effective check because of the complexities of evaluating the magnitude of the compensation package. As a result of these institutional arrangements for setting executive pay, some reforms have focused on making the process more open to the scrutiny of shareholders to serve as an effective check.

DEFERRED WAGES

Evidence

Compensation often appears to be deferred in the sense that wages are above an individual's productivity for the more senior employees (Figure 23.1(a)). Alternatively stated, wages rise with seniority or experience, independent of the individual's productivity. Abraham and Medoff (1982), for example, review a large number of empirical studies and find that such deferred compensation systems are prominent. Topel (1991) also provides econometric evidence indicating a substantial wage effect of seniority, albeit other econometric studies find it to be small (Abraham and Farber 1987, Altonji and Shakotko 1987, Ruhm 1990).

The deferred wage aspect would be even larger when one considers the fact that pension accruals become important in the pre-retirement years. That is, for every additional year an individual works, that individual experiences not only a wage increase but also pension increments that are a cost to the employer but a compensation benefit to employees. Pension benefits themselves also can be considered part of the deferred compensation package.

Associated Characteristics

Obviously such deferred compensation systems will exist only in situations where there is some long-term commitment on the part of the firm to continue the contracted arrangements. Otherwise the firm has an incentive to dismiss the employee at the time when the workers' pay begins to exceed their productivity (at the point of intersection, S_B of the pay and productivity schedules of Figure 23.1(a)). Firms that did so, however, would develop a bad reputation and have to pay a higher regular compensating wage in return for the risk that employees may not receive their deferred wage.

Of course, firms that find themselves with an aging work force with accumulated seniority and few younger workers (perhaps because of a decline in their hiring due to a decline in their product demand), may find it "profitable" to declare bankruptcy and hence sever their deferred wage obligations, especially if they can relocate elsewhere and hire younger workers at a wage less than their productivity. Obviously, if this were known by the new recruits, then a

Figure 23.1
WAGE PRODUCTIVITY PROFILES

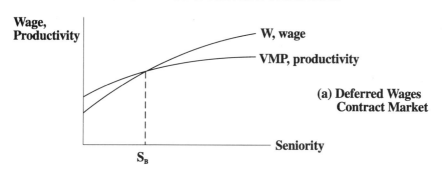

(a) Deferred Wages
Contract Market

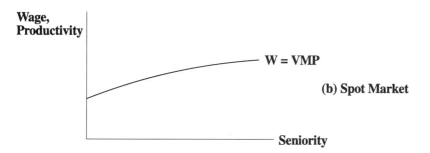

(b) Spot Market

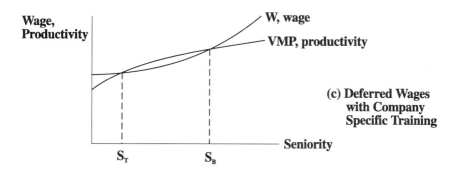

(c) Deferred Wages
with Company
Specific Training

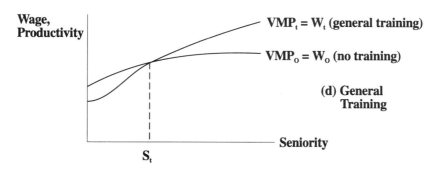

(d) General
Training

compensating wage would be required for accepting employment in a firm that deliberately restructured to avoid its deferred compensation obligations. However, the true rationale for such bankruptcy can be difficult to detect, especially in situations of asymmetric information where the firm has more information than employees on the true profitability of the firm.

Given the fact that firms have a short-run incentive to renege on the deferred wage obligation when that obligation becomes due, such a compensation system will prevail only when there are certain checks and balances to ensure that the deferred compensation is paid. Such checks and balances include the reputation of the firm, the payment of wages related to seniority, and the fact that unions can provide a degree of "due process" to ensure that their members are not arbitrarily dismissed. These checks and balances are characteristic of the longer term "contract market," where a longer term employment relationship is expected, rather than the "spot" or "auction" market, where wages would equal the workers' productivity at each and every point in time (Figure 23.1(b)).

Rationales for Deferred Compensation

Such deferred compensation systems may serve a number of important functions and hence have an efficiency rationale. As emphasized by Lazear (1979), they may ensure honesty and work effort because the worker wants to stay with the company to get the deferred wage. It is like the worker posting a bond (equal to the excess of the worker's productivity over wages when young) that will be returned (in the form of wages in excess of productivity) when the worker has accumulated seniority with the firm. The returning of the bond (deferred compensation) will be done conditional upon the worker having performed satisfactorily after a period over which past performance can be observed. Workers whose performance has not been satisfactory can be dismissed, subject to the checks and balances discussed previously. The greater honesty and work effort that results from such a system enhances productivity and hence provides the means for such workers to receive greater compensation over their expected work life.

Such a compensation system also reduces the need for constant everyday monitoring of the productivity of workers. With deferred wages monitoring can be made on a periodic, retrospective basis (i.e., examining past performance) with the worker being promoted and hence receiving the deferred wage conditional upon satisfactory performance (again subject to the previously discussed checks and balances). In contrast, under the spot market system (Figure 23.1(b)), where wages are equal to productivity at each and every point in time, productivity and performance must be assessed or known on a regular basis in order to compensate accordingly. The paradigm of the spot market is the piece-rate system, where payment is made directly according to output — a system that can exist only where output is easily measured (e.g., sales, logs cut, buttons sewn).

Deferred wages may also reduce unwanted turnover and hence enable firms to recoup their quasi-fixed hiring and training costs that are usually incurred early in the employee's career with the company. Such costs to the firm are illustrated in Figure 23.1(c), by the area above the value-of-marginal product, VMP, line and below the wage line. In order to recoup these costs and amortize them over a longer work life of their employees, firms may want to defer compensation in order to provide employees with an incentive to stay with the firm.

The training here is *company-specific* in the sense that it is usable mainly in the company providing the training. As discussed in the previous chapter on human capital theory, if the training were *generally usable* in other firms then the employee would have the incentive to pay for the training, often in the form of accepting a lower wage during the training program. In this case the trainee's productivity and wage during training ($VMP_t = W_t$ in Figure 23.1(d)) would fall below what their productivity and wage would have been in the absence of training ($VMP_o = W_o$). After the training period, the employee's productivity would increase and, since general training is usable anywhere, so would the trainee's wage (i.e., $VMP_t = W_t$ and both lie above $VMP_o = W_o$ in the post-training period). In the case of general training, where the employee bore the cost of accepting a lower wage during the training period, the company would not have incurred any quasi-fixed costs, and hence would have no incentive to pay a deferred wage, at least for that reason. In such circumstances the wage profile would equal the productivity profile (i.e., $W_t = VMP_t$) as in a spot market. If deferred wages were desirable for other reasons, then they could be superimposed upon the general-training wage profile.

Deferred wages may also exist so as to provide workers with a financial interest in the solvency of the firm. Under the pure spot market of Figure 23.1(b), except for the transactions costs of finding a new job, the employee would be indifferent as to whether the firm went bankrupt since employees would get a wage equal to their productivity elsewhere. With deferred wages, as in Figure 23.1(a), however, the employee stands to lose the deferred wage portion if the firm goes bankrupt. Thus deferred wages, including pension obligations, provide employees with an interest in the financial solvency of the firm. The loss of deferred wages, for example, may be one reason for the larger income losses that are incurred by older workers when they lose their jobs and are displaced to their next-best-alternative employment.

Public sector employers may prefer deferred wages because this arrangement passes costs to future generations of taxpayers, the only check being the willingness of the future generations to honour these commitments. Such deferred compensation may come in the form of more liberal pension benefits or greater job security so that one's *expected* wage (i.e., wage times the probability of receiving the wage) is higher. Deferred wages may also be legitimately needed more in the public than private sector to the extent that honesty is more important in the public sector and to the extent that monitoring and the measurement of current output is more difficult.

Deferred wages may also facilitate the synchronization of income and expenditures over the life cycle. That is, the shortfall of wages below productivity for junior people can be regarded as a form of forced savings, paid out later in the form of wages in excess of productivity in later years. Whether this corresponds to their years of greatest expenditure needs and whether saving in the form of deferred wages is better than private saving through capital markets is an open question. Deferred wages may also correspond to a sense of equity and fairness to employees so that they receive wage increases commensurate with seniority, even if their productivity does not increase or even declines.

Clearly there are a variety of possible rationales for deferred wages that can confer mutual benefits to both employers and employees. Even when the initial benefits would go mainly to employers (e.g., increased work incentives, reduced monitoring costs, reduced unwanted turnover) such benefits would provide the means for them to compensate employees for any adverse effects of deferred wages. Employees could be compensated in the form of a higher lifetime wage profile and more job security. As mentioned previously, employees may also benefit from the lifetime employment aspects of the contract market, from the

---

EXHIBIT 23.3 DO WORKERS ALSO PREFER DEFERRED WAGES?

While there are good reasons why *employers* prefer deferred wages (i.e., rising wage profiles), it is less obvious why *workers* should also prefer them. This is especially the case since having a positive discount rate, given the uncertainties of the future, implies that people should discount the future and hence not prefer to give up current wages in return for future wages.

Loewenstein and Sicherman (1991) surveyed individuals and found that they actually *preferred* the rising wage profiles. They maintained this preference even when they were told that they could have higher lifetime income by having the money sooner and investing it.

This "irrational" preference may occur as people regard deferred wages as a form of forced saving or because they get satisfaction from anticipating future consumption. Whatever the reason, individuals seem to prefer to given up current wages in return for future wages. For that reason they would willingly accept deferred wage profiles, which are also preferred by employers.

Source: G. Loewenstein and N. Sicherman. Do workers prefer increasing wage profiles? *Journal of Labor Economics* 9 (January 1991) 67–84.

periodic rather than everyday monitoring, from the due process and seniority rules that provide security to ensure that deferred wages are paid, and from the possible synchronizing of income and expenditures and the sense of equity and fairness that may be associated with wages based on seniority.

RATIONALE FOR MANDATORY RETIREMENT

To Enable Deferred Wages

While there clearly can be efficiency and equity reasons for deferred wages to exist, such a contractual arrangement requires a termination date for it to exist. Otherwise, employers run the risk of paying wages in excess of productivity for an indefinite period (as illustrated in Figure 23.1(a)). In such circumstances a contractual arrangement involving deferred wages could not persist.

Lazear (1979) argues that mandatory retirement provides the termination date for such contractual arrangements with deferred wages, and he provides empirical evidence indicating that mandatory retirement provisions are more prevalent in situations of deferred wages. The mandatory retirement date, MR in Figure 23.2, provides an equilibrium condition, enabling the expected present value of the wage stream to be equal to the expected present value of the productivity stream to the firm. Without that termination date to the existing contractual arrangement it would be difficult or impossible for the deferred wage contract to exist since the firm would be paying a wage in excess of productivity for an indefinite period. Mandatory retirement is the institutional rule that gives finality to the existing contractual arrangement. Viewed in this light, it is an endogenous institutional feature of labour markets that arises for

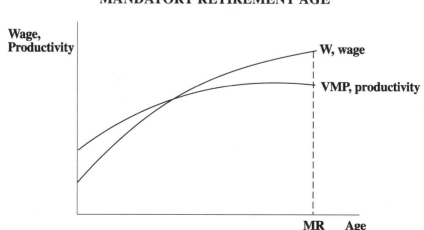

Figure 23.2
MANDATORY RETIREMENT AGE

efficiency reasons to allow deferred compensation systems to exist; it is not an exogenously imposed constraint as is often perceived.

As discussed in the earlier chapter on retirement, the phrase "mandatory retirement" is somewhat of a misnomer since it does not mean that the person is required to retire from the *labour force*. It means that the existing contractual arrangement is now terminated, and the employee must renegotiate an alternative arrangement with another employer or with the same employer (the latter being possible under compulsory but not automatic retirement). Presumably any new arrangement would be for a wage commensurate with the worker's productivity and this may involve a substantial wage drop since deferred wages no longer would be paid. The wage drop need not reflect a decline of productivity with age; rather it occurs because deferred wages are in excess of productivity for senior employees (in return for wages being below productivity in their early years with the company). In fact, this emphasizes that mandatory retirement need not exist because productivity (correctly or incorrectly) is perceived to decline with age. The appropriate response to such a phenomenon, if it existed, would be wage reductions, rather than requiring that the person must leave the job. Mandatory retirement could be an efficient compensation rule even if productivity continues to increase throughout a worker's working life (as it in fact does in the diagram).

Other Possible Rationales for Mandatory Retirement

In addition to enabling a deferred wage compensation system, other possible rationales may exist for mandatory retirement. It may facilitate worksharing by opening up promotion and employment opportunities for younger workers. While this may be true in particular firms and industries, as indicated in our previous discussion of worksharing, caution should be used in applying this argument to the economy as a whole so as to avoid the "lump-of-labour fallacy." That is, there need not be a fixed number of jobs in the economy so that the job held by an older worker would mean that a younger worker would not have a job. The younger worker may not have that *particular job*, but by working longer the older worker may be generating other jobs through increased consumption expenditures and through affecting wage levels.

Mandatory retirement may also create a greater degree of certainty about when an employee will retire from a particular job. For employers, this may facilitate planning for new staffing requirements, pension obligations and medical and health expenditures, the latter of which can be quite high for older workers. For workers, the advent of mandatory retirement may encourage pre-retirement planning and preparation of eventual retirement, preparation they may otherwise postpone indefinitely if the retirement date were not fixed. Retirement may also be facilitated by the fact that pensions, as a form of deferred compensation, are prominent in situations of mandatory retirement.

Mandatory retirement also reduces the need for the monitoring and evaluation of older workers and it enables them to "retire with dignity" since they are retiring because of a common retirement rule, not because they are singled out for incompetence. Less productive workers can be "carried" to the retirement age by employers or co-workers because that may be a short period and it is known with certainty. In contrast, without mandatory retirement, the performance of older workers will have to be monitored and evaluated, and dismissals may occur more frequently.

Arguments Against Mandatory Retirement

The arguments against mandatory retirement tend to rest on human rights issues, with mandatory retirement being regarded as a form of age discrimination. The appearance of age discrimination is further enhanced by the fact that employment standards or human rights legislation often has an age limit, not covering workers who are 65 years of age and older. Workers beyond that age have no formal resource to the law to prevent age discrimination. This apparent sanctioning of mandatory retirement provisions certainly makes the provisions appear as a form of age discrimination. Stereotypes that give rise to discrimination are likely to be as prominent against older workers as against women, racial groups, or any other visible minorities.

Even if certain rules may be efficient, they may not be socially justifiable if they discriminate against individuals. Thus even if mandatory retirement creates certainty for employers and employees, and it facilitates retirement with dignity, if this occurs at the expense of some who want to continue working, and are capable of working, it still may be regarded as discriminatory against those individuals.

The abolition of mandatory retirement has also been supported on the grounds of improving the viability of public and private pensions, a growing concern as the aging population puts increasing pressure on pension obligations at the same time as a declining work force provides fewer persons paying into pension funds. If workers continue to work beyond what was the former retirement age they will be less of a drain on pensions. Of course, this argument assumes that those who continue to work will not draw full pension income, an assumption that is seldom explicitly stated. In fact, many unions tend to be concerned that banning mandatory retirement is simply a veiled excuse to reduce private and public pension obligations.

A Middle Ground

Given the pros and cons of mandatory retirement it is not surprising that the debate over banning mandatory retirement is conducted at an emotive level. The real question, however, is: to what extent should legislation ban parties from entering into contractual arrangements (like mandatory retirement) that inhibit their freedom at some time in the future, presumably in return for other benefits such as job and promotion opportunities when they are younger, pensions when they retire, and the benefits of a deferred compensation scheme? Clearly legislation sanctions some contracts like marriage contracts and loan contracts that can inhibit freedom at certain points in time, presumably because of the other benefits they bring. Yet the state does forbid other contracts like indentured service or separate pay or promotion lines for females or racial groups, even if the individuals themselves are willing to enter into such contractual arrangements. Not all contractual arrangements are sanctioned by the state, especially if people can be misinformed or if they can be exploited because of a weak bargaining position.

One possible solution to prevent the worst abuses of mandatory retirement and yet to preserve its benefits would be to remove the age ceiling on antidiscrimination legislation but to exempt bona fide collective agreements or employees covered by a bona fide pension plan. This would enable people to enter into arrangements involving mandatory retirement if they had the protection of a union. Presumably the union would be informed of the trade-offs involved and it would represent the interests of its older workers, because all of its members could expect to become older and subject to any constraints like mandatory retirement. Exempting employees covered by a bona fide pension plan would ensure that when workers become subject to the mandatory retirement constraint, they would retire with an adequate pension.

Such exemptions, however, would allow mandatory retirement to continue much in its present form since most workers who are currently subject to mandatory retirement are also protected by a collective agreement and/or a pension plan. However, it would prevent the potential abuse of the employer who engages in age discrimination or who requires mandatory retirement but does not provide a pension.

Impact of Banning Mandatory Retirement

As indicated in Gunderson (1983) whatever the pros and cons of mandatory retirement as a work rule, banning it can be expected to have a variety of implications on various aspects of the employment relationship. To the extent that deferred compensation is no longer as feasible, then the wages of younger workers will rise and the wages of older workers with more seniority will fall, as wage profiles will now have to more closely approximate productivity profiles (i.e., the profiles will be more like Figure 23.1(b) then 23.1(a)). The wages of older workers may drop quite substantially to compensate for any increase in age-related fringe benefit costs like medical and disability insurance. Those adjustments, in turn, may bring charges of age discrimination in pay practices. Until such wage adjustments occur, however, older workers may receive substantial "windfall gains" from the abolition of mandatory retirement, as their wages could now exceed their productivity for an indefinite period of time (i.e., beyond MR in Figure 23.2).

Not only will the wage profile "tilt" downwards towards the productivity profile but the whole productivity profile and wage profile will drop to the extent that productivity was enhanced by the deferred wage system, for example, because of enhanced work incentives and reduced monitoring. New efforts will also be made to restore the deferred wage system in a fashion that does not rely upon mandatory retirement for its existence. For example, periodic bonuses may be more prominent. If deferred wages are continued, employee voluntary buy outs or "golden handshakes" may become more prominent to encourage voluntary retirement. Actuarial reductions in pension benefits may also be used to encourage people to voluntarily retire. Pensions themselves may become less prominent to the extent that they were a part of the deferred compensation scheme.

Monitoring and evaluation of workers will likely increase to the extent that it is less feasible to use periodic evaluations based upon past performance. In addition, they will be more necessary to detect early signals of performance in younger workers and to provide documentation in the case of dismissals or wage reduction of older workers so as to avoid age discrimination charges.

The employment and promotion opportunities of younger workers may be reduced in particular jobs when older workers continue past the usual retirement age. The extent to which other job opportunities may open up because of the spending activities of those who remain in the work force, or because of their effect on wage levels, is an empirical unknown. To the extent that deferred wages are not as prominent, promotion will take on a different meaning, since it will not mean promotion to receive the deferred wage. If wages are now equal to the worker's value of marginal product at each and every stage of the life cycle, then promotion by itself will be less important.

Training will decrease to the extent that deferred wages can no longer exist to discourage the unwanted turnover of those who receive company-specific training. However, training may increase to the extent that the expected benefit period, without mandatory retirement, is longer. Forecasting and planning will be more difficult given the greater uncertainty about retirement ages.

Jobs may be redesigned to accommodate the older workers who stay, but they may also be designed to encourage them to leave. Some dismissals of older workers who otherwise may have been "carried" to the mandatory retirement age will be inevitable, and this will likely increase the amount of litigation over unjust dismissal cases. Layoffs of older workers may also be more prominent to the extent that seniority declines as a work rule that was coexistent with deferred compensation.

These are meant to be simply illustrative of the sorts of adjustments that may occur if one aspect of the intricate employment relationship is changed, in this case if mandatory retirement is abolished. The adjustments may be minor if they are going on already in anticipation of mandatory retirement being banned, or if alternative procedures can arise to enable deferred wages, or if workers continue their trend towards earlier retirement. Nevertheless, we do not know if that trend will continue especially as the proportion of the white-collar work force grows, as retirement plans have more lead-time to change, as larger numbers of co-workers postpone retirement, and if pensions dissipate. In addition, as discussed in the chapter on retirement, Canadian public pension plans do not have a tax-back feature as exists in the United States, so that Canadian workers may be more prone to postpone retirement than their American counterparts.

Clearly, the current debate over the pros and cons of banning mandatory retirement is not one that is likely to be resolved easily, and if it is resolved in favour of banning mandatory retirement, a number of repercussions are likely to follow. Hopefully, the application of some basic principles of economics can narrow the focus of the debate, though it may also make the issues more complicated than they appear at first glance. At the very least, these principles should provide us with information on the repercussions of a policy change and the alternative and complementary policies that may be desirable.

QUESTIONS

1. Discuss the pros and cons of efficiency wages as opposed to deferred compensation as compensation systems to deter shirking and encourage work effort.

2. If Henry Ford decided to pay efficiency wages almost "overnight," what discrete change in the work environment all of a sudden made it efficient to pay such wage premiums? In other words, if they truly were efficiency wages, would not one expect an abrupt change in the work environment that would have prompted such a high-wage strategy? If it was an efficient strategy after January 14, 1914, why was it not an efficient strategy prior to that time? Or is this simply an example of an efficient innovation being developed?

3. What should happen to the pay of superstars if there is technological change that enables the very best to reach a wider audience?

4. If deferred compensation prevails and firms face an exogenous, unanticipated permanent reduction in their demand conditions, then they may have a strong incentive to engage in permanent layoffs or even bankruptcy. Discuss.

5. While there may be advantages to deferred compensation, there are other ways besides mandatory retirement to enable deferred compensation. Discuss.

6. If mandatory retirement exists in part to enable deferred compensation, discuss the implication of banning mandatory retirement for different elements of the personnel function, including compensation policies.

7. "The relevant question is not: are you for or against mandatory retirement? Rather, it is: under what conditions should governments prohibit private parties from entering into arrangements like mandatory retirement?" Discuss.

8. "Banning mandatory retirement is not likely to be important to enable low-wage workers to continue working and avoid poverty in retirement, because such workers are unlikely to be subject to mandatory retirement." Discuss.

9. "Any theory of executive compensation must be able to consistently explain two stylized facts — the increase in executive compensation that has occurred relative to average pay, and the higher executive compensation in the U.S. relative to other countries." Discuss.

10. "Some have argued that stock options are an ineffective component of executive compensation since there is no negative penalty if the value of the stock declines. This ignores the fact that the decline imposes an opportunity cost or forgone income associated with the fact that the stock did not rise or even stay the same." Discuss.

REFERENCES AND FURTHER READINGS

Abraham, K. and H. Farber. Job duration, seniority and earnings. *AER* 77 (June 1987) 279–297.

Abraham, K. and J. Medoff. Length of service and the operation of internal labour markets. *IRRA* (December 1982) 308–318.

Acharya, S. Maximizing market value of a firm to choose dynamic policies for managerial hiring, compensation, firing and tenuring. *IER* 33 (May 1992) 373–398.

Akerlof, G. Labor contracts as partial gift exchange. *QJE* (November 1982) 543–570.

Akerlof, G. and J. Yellen. The fair wage-effort hypothesis and unemployment. *QJE* 105 (May 1990) 255–284.

Akerlof, G. and L. Katz. Workers' trust funds and the logic of wage profiles. *QJE* 104 (August 1990) 525–536.

Allen, S. and R. Clark. Unions, pension wealth and age-compensation profiles. *ILRR* 39 (July 1986) 502–518.

Altonji, J. and R. Shakotko. Do wages rise with job seniority? *R.E. Studies* 54 (July 1987) 437–460.

Antle, R. and A. Smith. An empirical investigation into the relative performance of corporate executives. *Journal of Accounting Research* 24 (Spring 1986).

Arnott, R. and J. Stiglitz. Labor turnover, wage structures and moral hazard: the inefficiency of competitive markets. *JOLE* 3 (October 1985) 434–462.

Bar-Ilan, A. Monitoring workers as a screening device. *CJE* 24 (May 1991) 460–470.

Barro, J. and R. Barro. Pay, performance, and turnover of bank CEOs. *JOLE* 8 (October 1990) 448–481.

Baumol, W. Entrepreneurship: productive, unproductive and destructive. *JPE* 98 (October 1990) 893–921.

Bishop, J. The recognition and reward of employee performance. *JOLE* 5 (October 1987) S36–S56.

Blakemore, A., S. Low and M. Ormiston. Employment bonuses and labor turnover. *JOLE* 4 (October 1987) S124–S135.

Blanchflower, D. and A. Oswald. Profit related pay. *EJ* 98 (September 1988) 720–730.

Blinder, A. (ed.). *Paying for Productivity.* Washington: Brookings Institute, 1990.

Blinder, A. and D. Choi. A shred of evidence on theories of wage stickiness. *QJE* 105 (November 1990) 1003–1015.

Boher, G. Incentive contracts and performance measurement. *JPE* 100 (June 1992) 598–614.

Brown, J. Why do wages increase with tenure? *AER* 79 (December 1989) 971–991.

Bull, C., A. Schotter and K. Weigelt. Tournaments and piece rates: an experimental study. *JPE* 95 (February 1987) 1–33.

Bulow, J. and L. Summers. A theory of dual labor markets with application to industrial policy, discrimination and Keynesian unemployment. *JOLE* 4 (July 1986) 376–414.

Burkhauser, R. and J. Quinn. Is mandatory retirement overrated? *JHR* 18 (Summer 1983) 337–358.

Carmichael, L. Does rising productivity explain seniority rules for layoffs? *AER* 73 (December 1983) 1127–1131.

Carmichael, L. Reputations in the labor market. *AER* 74 (September 1984) 713–725.

Carmichael, L. Incentives in academics: Why is there tenure? *JPE* 96 (June 1988) 453–472.

Carmichael, L. Self-enforcing contracts, shirking, and life-cycle incentives. *JEP* 3 (Fall 1989) 65–83.

Carruth, A. and A. Oswald. *Pay Determination and Industrial Prosperity.* Oxford: Oxford University Press, 1989.

Chaykwoski, R. and C. Beach. Prizes in an industrial union environment. *JLR* 6 (Spring 1985) 181–198.

Clark, R. and A. McDermed. Earnings and pension compensation. *QJE* 101 (May 1986) 341–362.

Clark, R. and N. Ogawa. Effect of mandatory retirement on earnings profiles in Japan. *ILRR* 45 (January 1992) 258–266.

Curme, M. and L. Kahn. The impact of the threat of bankruptcy on the structure of compensation. *JOLE* 8 (October 1990) 419–447.

Donaldson, D. and B.C. Eaton. Person-specific costs of production: hours of work, rates of pay, labour contracts. *CJE* 17 (August 1984) 441–449.

Drago, R. and G. Turnbull. The incentive effects of tournaments with positive externalities amongst workers. *SEJ* 55 (July 1988) 100–106.

Dye, R. Optimal length of labor contract. *IER* 26 (February 1985) 251–270.

Eaton, C. and W. White. The economy of high wages. *Economica* 50 (May 1983) 175–182.

Ehrenberg, R. and M. Bognanno. Do tournaments have incentive effects? *JPE* 98 (December 1990) 1307–1324.

Ehrenberg, R., R. Chaykowski and R. Ehrenberg. Determinants of the compensation and mobility of school superintendents. *ILRR* 41 (April 1988) 386–401.

Fama, E. Time, salary, and incentive payoffs in labor contracts. *JOLE* 9 (January 1991) 25–44.

Frank, R. Are workers paid their marginal products? *AER* 74 (September 1984) 549–571.

Gaynor, M. and M. Pauly. Compensation and productive efficiency in partnerships. *JPE* 98 (June 1990) 544–573.

Gibbons, R. and K. Murphy. Optimal incentive contracts in the presence of career concerns. *JPE* 100 (June 1992) 468–505.

Gibbons, R. and K. Murphy. Relative performance evaluation of chief executive officers. *ILRR* 43 (February 1990) S30–S51.

Gibbons, R. Piece-rate incentive schemes. *JOLE* 5 (October 1987) 413–429.

Glammerino, R. and E. Nosel. Wage smoothing as a signal of quality. *CJE* 23 (February 1990) 159–174.

Gomez-Mejla, L. and D. Balkin. Effectiveness of individual and aggregate compensation strategies. *IR* 28 (Fall 1989) 431–445.

Gunderson, M. and J. Pesando. Eliminating mandatory retirement: economics and human rights. *CPP* 6 (Spring 1980) 352–360.

Gunderson, M. Mandatory retirement and personnel policies. *Columbia Journal of World Business* (Summer 1983) 8–15.

Hamermesh, D. The costs of worker displacement. *QJE* 102 (February 1987) 51–76.

Hamlen, W. Superstardom in popular music. *R.E. Stats.* 73 (November 1991) 729–732.

Hashimoto, M. Bonus payments, on-the-job training, and lifetime employment in Japan. *JPE* 87 (October 1979) 1086–1104.

Hashimoto, M. and J. Raisian. Employment tenure and earnings profiles in Japan and the United States. *AER* 75 (September 1985) 721–735. Comment by R.C. Clark and N. Ogawa and reply, 82 (March 1992) 336–354.

Hutchens, R. Seniority, wages and productivity. *JEP* 3 (Fall 1989) 49–64.

Hutchens, R. Delayed payment contracts and a firm's propensity to hire older workers. *JOLE* 4 (October 1986) 439–457.

Hutchens, R. A test of Lazear's theory of delayed payment contracts. *JOLE* 5, Part 2 (October 1987) S153–S170.

Industrial and Labor Relations Review, Special Issue on Compensation Policies, 43 (February 1990).

Ippolito, R. Encouraging long-term tenure: wage tilt or pensions. *ILRR* 44 (April 1991) 520–535.

Jensen, M. and K. Murphy. Performance pay and top management incentives. *JPE* 98 (April 1990) 225–264.

Johnson, W. The social efficiency of fixed wages. *QJE* 100 (February 1985) 101–118.

Kanemoto, Y. and W.B. MacLeod. The ratchet effect and the market for secondhand workers. *JOLE* 10 (January 1992) 85–98.

Kaufman, R. The effects of Improshare on productivity. *ILRR* 45 (January 1992) 311–322.

Kleiner, M. and M. Bouillon. Providing business information to production workers. *ILRR* 41 (July 1988) 605–617.

Kraft, K. The incentive effect of dismissals, efficiency wages, piece rates and profit sharing. *R.E. Stats.* 73 (August 1991) 451–459.

Krueger, A. Ownership, agency and wages: an examination of franchising in the fast food industry. *QJE* 106 (February 1991) 75–101.

Kuhn, P. Wages, effort and incentive compatibility in life-cycle employment contracts. *JOLE* 4 (January 1986) 28–49.

Kuhn, P. and J. Roberts. Seniority and distribution in a two-worker trade union. *QJE* 104 (August 1989) 485–506.

Lakhani, H. The effect of pay and retention bonuses on quit rates in the U.S. army. *ILRR* 41 (April 1988) 430–438.

Lang, K. Why was there mandatory retirement? *JPubEc* 39 (June 1989) 127–136.

Lapp, J. Mandatory retirement as a clause in an employment insurance contract. *EI* 23 (April 1985) 69–92.

Lazear, E. Why is there mandatory retirement? *JPE* 87 (December 1979) 1261–1284.

Lazear, E. and S. Rosen. Rank-order tournaments as optimum labor contracts. *JPE* 89 (October 1981) 841–864.

Leigh, D. Why is there mandatory retirement? An empirical re-examination. *JHR* 19 (Fall 1984) 512–531.

Leonard, J. Carrots and sticks: pay, supervision and turnover. *JOLE* 5 (October 1987) S136–S152.

Levine, D. Just-cause employment policies when unemployment is a discipline device. *AER* 79 (September 1989) 902–905.

Light, A. and M. Ureta. Gender differences in wages and job turnover among continuously employed workers. *AER* 80 (May 1990) 293–297.

Loewenstein, G. and N. Sicherman. Do workers prefer increasing wage profiles? *JOLE* 9 (January 1991) 67–84.

MacDonald, G. The economics of rising stars. *AER* 78 (March 1988) 155–166.

MacLeod, W.B. and J. Malcomson. Reputation and hierarchy in dynamic models of employment. *JPE* (August 1988) 832–854.

Medoff, J. and K. Abraham. Experience, performance and earnings. *QJE* 95 (December 1980) 703–736.

Montgomery, J. Equilibrium wage dispersion and interindustry wage differentials. *QJE* 106 (February 1991) 163–179.

Murphy, K.J. Incentives, learning and compensation. *Rand Journal of Economics* 17 (Spring 1986) 59–76.

Murphy, K. J. Corporate performance and managerial remuneration. *Journal of Accounting and Economics* 7 (1985) 11–42.

Nakamura, M. and A. Nakamura. Risk behavior and determinants of bonus versus regular pay in Japan. *Journal of Japanese and International Economics* 5 (June 1991) 140–159.

Nalbantian, H. (ed.). *Incentives, Co-operation and Risk Sharing.* Totowa, N.J.: Littlefield, Adams, 1987.

Parsons, D. The employment relationship: job attachment, work effort, and the nature of contracts. *Handbook of Labour Economics* Vol. 1. O. Ashenfelter and R. Layard (eds.). New York: Elsevier, 1986.

Putterman, L. and C. Skillmore, Jr. The incentive effects of monitoring under alternative compensation schemes. *International Journal of Industrial Organization* 6 (March 1988) 109–119.

Rosen, S. The economics of superstars. *AER* 71 (December 1981) 845–858.

Ruhm, L. Do earnings increase with job seniority? *R.E. Stats.* 72 (February 1990) 143–147.

Rupert, P. Bonus payments and pay flexibility in Japan. *International Economic Journal* 5 (Autumn 1991) 31–38.

Schuster, M. and C. Miller. An empirical assessment of the age discrimination in employment act. *ILRR* 38 (October 1984) 64–74.

Shapiro, D. and S. Sandell. Age discrimination in wages and displaced older men. *SEJ* 52 (July 1985) 90–102.

Siow, A. Are first impressions important in academic? *JHR* 26 (Spring 1991) 236–255.

Siow, A. and B. O'Flaherty. On the job screening, up or out rates, and firm growth. *CJE* 25 (May 1992) 346–368.

Stern, S. Promotion and optimal retirement. *JOLE* 5 (October 1987) S107–S123.

Tachibanki, T. Further results on Japanese wage differentials: nenko wages, hierarchical position, bonuses, and working hours. *IER* 23 (June 1982) 447–462.

Topel, R. Specific capital, mobility and wages: wages rise with job seniority. *JPE* 99 (February 1991) 145–176.

Viscusi, W. Self-selection, learning-induced quits, and the optimal wage structure. *IER* 21 (October 1980) 529–546.

Viscusi, W. Moral hazard and merit rating over time: an analysis of optimal intertemporal wage structures. *SEJ* 52 (April 1986) 1068–1079.

Weitzman, M. Some macroeconomic implications of alternative compensation systems. *EJ* 93 (December 1983) 763–783.

Williams, N. Re-examining the wage, tenure and experience relationship. *R.E. Stats.* 73 (August 1991) 512–516.

Yellen, J. Efficiency wage models of unemployment. *AER* 74 (May 1984) 200–205.

PART 6

Unemployment and Inflation

Part 6 deals with unemployment and inflation. These are subjects that are frequently topics of discussion, and are the source of much public debate. As stated by Nobel Laureate James Tobin in 1972: "Unemployment and inflation still preoccupy and perplex economists, statesmen, journalists, housewives and everybody else. The connection between them is the principal domestic economic burden of presidents and prime ministers, and the major area of controversy and ignorance in macroeconomics."

Although the study of inflation and unemployment is largely the domain of courses in macroeconomics, there are also important labour market aspects to these phenomena. This part of the book focuses on these labour market aspects, leaving to courses in macroeconomics a more thorough treatment of these subjects.

Chapter 24 examines the meaning of unemployment, and describes how unemployment is measured in Canada and elsewhere. The salient features of Canada's experience with unemployment and inflation are also described in this chapter.

Chapter 25 examines the causes and consequences of unemployment. Understanding the causes of unemployment is important to determining what policies are most likely to be effective in reducing underutilized labour resources.

Chapter 25 draws heavily on several topics covered previously in Chapter 13, such as efficiency wages and implicit contracts. Some readers will wish to review that material at this time.

Chapter 26 deals with the relationship among wage inflation, price inflation, and unemployment. This chapter also examines a variety of policies — such as wage and price controls and changes to wage setting arrangements — that have been proposed and/or employed an attempt to reduce inflation without exacerbating unemployment.

CHAPTER 24

Unemployment: Meaning and Measurement

With the possible exception of the Consumer Price Index, no single aggregate statistic receives as much attention as the unemployment rate. To a considerable extent this attention reflects concern about the hardship which unemployment may impose on affected individuals and their families and the waste of human resources that may be associated with unemployment. In addition, the unemployment rate is used as a measure of the overall state of the economy and of the degree of tightness (excess demand) or slack (excess supply) in the labour market. However, in spite of this evident interest, there is often confusion about what is meant by our unemployment statistics. This chapter examines the meaning and measurement of unemployment. The causes and consequences of unemployment and the role of public policy in this area are discussed in the following chapter.

MEASURING UNEMPLOYMENT

The unemployed are generally defined as those who are not currently employed and who indicate by their behaviour that they want to work at prevailing wages and working conditions. In Canada, measurement of the unemployed can be obtained from the Labour Force Survey, the Census, or from the number of claimants of the Unemployment Insurance program. While there are these alternative measures, and different ones can be useful for different purposes, the unemployment measure according to the Labour Force Survey is the one that receives the most attention: it is the one that is so often quoted by the press and utilized for general policy purposes.

Labour Force Survey

The Labour Force Survey is conducted monthly by Statistics Canada. According to the Labour Force Survey (LFS), people are categorized as unemployed if they did not have work in the reference period but they were available for and actively seeking work. There are two exceptions to this general principle. The unemployed also includes persons who were available for work but who were not seeking work because they were on temporary lay-off from a job to which they expect to be recalled, or they had a new job to start within four weeks.

Apart from these exceptions, the general principle is that to be counted as unemployed, one has to be available for and actively seeking work. That is, it is necessary to engage in active job search to be classified as unemployed. Active job search is defined in the LFS as having looked for work sometime during the previous four weeks. Individuals who are not employed and who are not actively seeking work (such as students, persons engaged in household work, those permanently unable to work, and retirees) are classified as being out of the labour force.

Persons are categorized as employed if they did any work for pay or profit, including unpaid work on a family farm or business, or if they normally had a job but were not at work because of such factors as bad weather, illness, industrial dispute, or vacation. The employed plus the unemployed make up the labour force, and the unemployment rate is defined as the number of unemployed divided by the labour force.

Census

The Canadian Census, conducted every ten years since 1871, also provides information on the unemployed, at least since 1921. (A smaller census containing more limited information is also conducted in between the ten-year census.) Rather than being based on a sample of the population, the Census is based on the whole relevant population, with detailed information coming from approximately one-third of this population. Consequently, Census data is more comprehensive and provides detailed information on such factors as unemployment by industry and occupation, as well as a wealth of demographic and economic information on the respondents. For this reason the Census data has proven useful in detailed statistical and econometric analysis of labour market behaviour. However, the usefulness of the Census data is limited by the fact that it is conducted infrequently and its reliability on labour force issues may be questioned because of the fact that it covers so many issues in addition to labour force activity. In addition, there is a time lag of about two years before the Census data are available, whereas the Labour Force Survey results are available within a few weeks of the survey.

Unemployment Insurance Claimants

A third data source — unemployment insurance claimants — can be used to obtain estimates of the unemployed. Comparisons of the number of unemployed according to unemployment insurance figures versus Labour Force Survey figures are given in Levesque (1989).

While the number of unemployed according to the Labour Force Survey may be similar to the number claiming unemployment insurance or registered in an unemployment insurance office, the two series generally differ. The number of unemployment insurance claimants may exceed the Labour Force Survey number because some people may be collecting unemployment insurance but are not actively seeking work. These people could range from persons on maternity or sick leave who are legally entitled to collect unemployment insurance, to persons who traditionally do seasonal work and collect unemployment insurance for part of the year, to outright cheaters who simply don't bother looking or who may even be working illegally. Such persons may be collecting unemployment insurance, but they need not indicate, in the confidential Labour Force Survey questionnaire, that they are actively seeking work.

On the other hand, the number of unemployed according to the Labour Force Survey could exceed the number of unemployment insurance claimants. New job seekers, for example, may not be eligible for UI, or the long-term unemployed may have exhausted their benefit period. Some, such as the self-employed, may not be covered by unemployment insurance; others may not register because then they would have to accept a job. They (or their parents who may be the respondents) may still indicate on a Labour Force Survey questionnaire that they are actively seeking work.

Clearly the two series need not be identical since they indicate somewhat different things. Because, from a policy perspective, what is usually wanted is an indicator of the number of persons who are available for work and actively seeking work, then the Labour Force Survey figures are the ones that are commonly used.

Canadian Experience

As Figure 24.1 illustrates, Canada's unemployment rate fluctuated widely during the 1921–91 period, to a considerable extent because of cyclical fluctuations in business activity. With the onset of the Great Depression in 1929, the unemployment rate soared from about 3 percent to almost 20 percent. A lengthy period of declining unemployment followed, and during World War II unemployment fell to very low levels. Following the war, unemployment remained low until the recession of 1957–58. As the economy recovered from the 1957–58 recession, the unemployment rate gradually declined, reaching 3.4 percent in 1966. Since that time the overall trend in unemployment has been upward, although

Figure 24.1
UNEMPLOYMENT IN CANADA, 1921–91

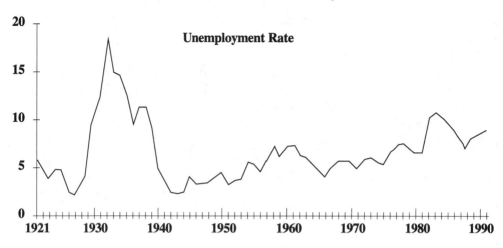

fluctuations around the trend have continued to occur. The most significant increases occurred in the recession of 1974–75, in the severe recession of 1981–82 which resulted in the unemployment rate reaching 11.8 percent, the highest level since the Great Depression, and in the recession of 1990–92. During the economic expansion which followed the 1981–82 recession, the unemployment rate gradually declined to a low of 7.5 percent in 1989. However, by 1992 the unemployment rate was back above 11 percent.

The unemployment rate is the most commonly used measure of aggregate labour market activity and the degree of labour force utilization. Two alternative measures are the labour force participation rate (the ratio of the labour force to the population of working age, or source population) and the employment rate (the ratio of employment to the source population). These statistics are shown for selected years in Table 24.1. (The years 1956, 1966, 1973, 1981, and 1989 correspond to the peak years of the business cycle in the postwar period.) Because they focus on somewhat different aspects of labour market activity, these three measures need not always move together. For example, since 1966 the unemployment rate has risen sharply, yet at the same time both the fraction of the source population employed and the fraction in the labour force have risen rather than fallen. Additional perspective on these divergent patterns is provided by the fourth column of Table 24.1; the rate of growth of employment between 1966 and 1981 was the most rapid of the postwar period. The combination of rapid employment growth and rising unemployment rates reflected several phenomena, including the substantial number of youths and women who entered the labour force, particularly after the mid-1960s. Although

Table 24.1
LABOUR FORCE PARTICIPATION, EMPLOYMENT AND
UNEMPLOYMENT, CANADA, 1946–89[1]

YEAR	LABOUR FORCE PARTICIPATION RATE	EMPLOYMENT RATE	UNEMPLOYMENT RATE	RATE OF GROWTH OF EMPLOYMENT[2]
1946	55.0	53.1	3.4	
				1.8
1956	53.5	51.7	3.4	
				2.5
1966	57.3	55.4	3.4	
				3.0
1973	59.7	56.4	5.5	
				2.9
1981	64.8	59.9	7.5	
				2.0
1989	67.0	62.0	7.5	

Notes:
1. All statistics are based on the former Labour Force Survey prior to 1966 and the Revised Labour Force Survey from 1966 on.
2. Civilian employment; growth rates are averages of compound annual increases from the level in the year in the row one line above to the level in the year one line below.

Source: Statistics Canada, *The Labour Force*, various issues, and calculations by the authors.

employment grew rapidly between 1966 and 1981, the growth in labour supply was even more substantial; thus the unemployment rate rose.

Which of these three measures is used for analyzing aggregate labour market activity will depend on the purpose of the analysis. The employment and labour force participation rates focus on the fraction of the source population which is employed and in the labour force respectively, while the unemployment rate measures the fraction of the labour force which is out of work and searching for work. Because this chapter deals with unemployment, the main aggregate statistic referred to is the unemployment rate. However, in many circumstances the employment and labour force participation rates may provide useful additional information on labour market developments.

Hidden Unemployment/Marginal Labour Force Attachment

The measurement of unemployment raises some difficult and controversial issues. Hidden unemployment or marginal labour force attachment refers to

EXHIBIT 24.1 INTERNATIONAL DIFFERENCES IN UNEMPLOYMENT

Canada and the United States measure employment, unemployment, and labour force participation in almost identical ways, using very similar monthly surveys (the Labour Force Survey and Current Population Survey respectively). Measures used in other countries differ in a variety of ways; however, the unemployment rate statistics shown in Table 24.1a have been adjusted so that they are comparable to the U.S. and Canadian concepts.

As these data indicate, the unemployment rate has varied substantially over time within most countries, and also varies considerably across countries. During the 1950s and 1960s, unemployment rates in North America were substantially higher than those in Australia, Japan, and most European countries. However, after the first OPEC oil price shock in 1973, unemployment rose sharply in several European countries (e.g., France, Germany, Netherlands, and the U.K.) and then escalated further during the 1980s. Australia followed a similar pattern. Japan, as well as a few European countries (e.g., Sweden), did not experience this dramatic rise in unemployment during the 1970s and 1980s.

Thus Canadian and American unemployment rates, which were unusually high by international standards during the first two decades of the postwar period, were no longer so during the 1980s. Indeed, unemployment rates in North America declined substantially during the long period of economic expansion following the 1982–83 recession, whereas those in many European countries remained stubbornly high. Japan and Sweden stand out as countries that were somehow able to adjust to the economic shocks of the 1970s and 1980s without experiencing dramatic increases in unemployment. Explaining these large differences in international experience is an important challenge for social scientists.

situations in which individuals may be without work, yet they desire work but are not classified as unemployed according to the official statistics. Such individuals are attached to the labour force to some degree, but are not sufficiently strongly attached that they are actively seeking work. One important example — especially during recessions and in regions where few jobs are available — is the "discouraged worker" phenomenon. This refers to individuals who are not employed, who may wish to work at prevailing wages, but who are not actively seeking work because they believe that no work is available. Other examples include: individuals still awaiting recall after more than six months on layoff; forms of underemployment, such as individuals working fewer hours than they desire to work or normally work; and those temporarily employed in jobs that do not utilize their skills or training. Each of these examples illustrates the difficulties involved in making a satisfactory distinction between "employment," "unemployment," and "out of the labour force."

Table 24.1a
LABOUR FORCE, EMPLOYMENT, AND UNEMPLOYMENT, 1959–90

YEAR	UNITED STATES	CANADA	AUSTRALIA	JAPAN	FRANCE	GERMANY	ITALY	NETHER-LANDS	SWEDEN	UNITED KINGDOM
				Unemployment Rate (percent) Approximating U.S. and Canadian Concepts						
1959	5.3	5.5	NA	2.3	1.5	2.0	4.7	NA	1.6	2.7
1960	5.4	6.4	NA	1.7	1.4	1.1	3.6	NA	1.7	2.2
1961	6.5	6.6	NA	1.5	1.2	.6	3.2	NA	1.5	1.9
1962	5.4	5.4	NA	1.3	1.4	.6	2.8	NA	1.5	2.7
1963	5.5	5.1	NA	1.3	1.5	.4	2.3	NA	1.6	3.3
1964	5.0	4.3	1.4	1.2	1.2	.4	2.6	NA	1.5	2.4
1965	4.4	3.6	1.3	1.2	1.5	.3	3.4	NA	1.2	2.1
1966	3.7	3.3	1.6	1.4	1.6	.3	3.6	NA	1.5	2.2
1967	3.7	3.8	1.8	1.3	2.0	1.3	3.3	NA	2.1	3.2
1968	3.5	4.4	1.8	1.2	2.6	1.1	3.5	NA	2.2	3.2
1969	3.4	4.4	1.8	1.1	2.3	.6	3.4	NA	1.9	3.0
1970	4.8	5.6	1.6	1.2	2.5	.5	3.2	NA	1.5	3.0
1971	5.8	6.1	1.9	1.2	2.7	.6	3.2	NA	2.5	3.9
1972	5.5	6.2	2.6	1.4	2.8	.7	3.7	NA	2.7	4.2
1973	4.8	5.5	2.3	1.3	2.7	.7	3.6	3.0	2.4	3.2
1974	5.5	5.3	2.6	1.4	2.8	1.5	3.1	3.5	2.0	3.1
1975	8.3	6.9	4.8	1.9	4.0	3.3	3.4	5.0	1.6	4.5
1976	7.6	7.1	4.7	2.0	4.4	3.3	3.8	5.3	1.6	5.9

(continued)

Table 24.1a (continued)
LABOUR FORCE, EMPLOYMENT, AND UNEMPLOYMENT, 1959–90

YEAR	UNITED STATES	CANADA	AUSTRALIA	JAPAN	FRANCE	GERMANY	ITALY	NETHER-LANDS	SWEDEN	UNITED KINGDOM
				Unemployment Rate (percent) Approximating U.S. and Canadian Concepts						
1977	6.9	8.0	5.6	2.0	4.9	3.4	4.0	4.8	1.8	6.3
1978	6.0	8.3	6.2	2.3	5.2	3.2	4.1	5.0	2.2	6.2
1979	5.8	7.4	6.2	2.1	5.9	2.8	4.3	5.0	2.0	5.3
1980	7.0	7.4	6.0	2.0	6.3	2.8	4.3	5.9	2.0	6.9
1981	7.5	7.5	5.7	2.2	7.4	3.9	4.8	8.7	2.5	10.4
1982	9.5	10.9	7.1	2.4	8.1	5.5	5.3	10.1	3.1	11.2
1983	9.5	11.8	9.9	2.7	8.3	6.7	5.8	11.3	3.4	11.7
1984	7.4	11.2	8.9	2.7	9.7	7.0	5.8	11.3	3.1	11.6
1985	7.1	10.4	8.2	2.6	10.2	7.0	5.9	9.4	2.8	11.1
1986	6.9	9.5	8.0	2.8	10.4	6.5	7.4	9.6	2.6	11.1
1987	6.1	8.8	8.0	2.9	10.5	6.2	7.7	9.9	1.9	10.2
1988	5.4	7.7	7.2	2.5	10.0	6.2P	7.8	9.2	1.6	8.5
1989	5.2	7.5	6.1	2.3	9.4	5.6P	7.7	8.5P	1.3	7.0P
1990	5.4	8.1	6.9	2.1	9.0	5.1P	6.9P	7.9P	1.5	6.8P

Notes:
1. P stands for preliminary estimate
2. NA means "not available"

Source: U.S. Department of Labor, Bureau of Labor Statistics.

As discussed by Kaliski (1984) and Akyeampong (1987), some evidence on the quantitative significance of hidden unemployment in Canada is available. Since 1979 an annual supplement to the Labour Force Survey has identified those who want work, are available for work, but are not seeking work for "personal" or "economic" reasons. These "persons on the margin of the labour force" typically number between one-quarter and one-third of the officially unemployed. That is, if these individuals were classified as unemployed rather than out of the labour force, the unemployment rate would be more than 20 percent higher.

Whether individuals who desire work but are not seeking work should be classified as unemployed is a controversial issue. The observation that they are without work and want work (assuming the response to the survey is accurate) may argue for including them among the unemployed. According to this view, the fact that they are not actively searching for work may simply be a rational way to spend their time, for example if they believe that no jobs are available. Similarly, waiting more than 6 months for recall may be sensible behaviour rather than wishful thinking, especially if the chances of recall are good and the job to which the individual expects to return is attractive relative to potential alternatives. In contrast, the absence of active job search may indicate a low degree of labour force attachment, justifying the classification "out of the labour force."

These phenomena respond to changes in aggregate economic conditions. The number of discouraged workers increased substantially during the 1981–82 recession and subsequently declined as the economy gradually recovered. An increase, albeit smaller was also observed during the 1990–92 recession (Akyeampong, 1992). The number of individuals awaiting recall after more than six months displayed a similar pattern. As a consequence, the increase in the unemployment rate in recessions is smaller than would be the case if those desiring but not seeking work were classified as unemployed. Similarly, the decline in the unemployment rate during the recovery is smaller than would be the case if these individuals were included in the unemployed.

Some evidence on the significance of underemployment is also available from the Labour Force Survey. For example, in 1990–91 about 25 percent of those working part-time would have preferred full-time work. Once again, the measured unemployment rate may understate the extent of unutilized labour services, especially in recessionary periods.

The major lesson to be learned from this discussion is *not* that the official unemployment rate is a poor measure of unemployment, but rather that the concept of unemployment is not sufficiently well defined that any single measure will suit all purposes. For many purposes the official unemployment measure will be appropriate; however, in other situations this should be supplemented by measures of hidden unemployment and underemployment.

;OUR FORCE DYNAMICS

Since the early 1970s, research by economists has emphasized the dynamic nature of the labour force. The Labour Force Survey provides a snapshot at a point in time, an estimate of the stock of persons in each labour force state. However, even if the magnitudes of these stocks remain approximately constant from one period to another, it would be a mistake to conclude that little change had taken place in the labour force. In fact, as Figure 24.2 illustrates, the flows between the three labour force states (employment, unemployment, out of the labour force) are large in comparison to the stocks and the gross flows are huge in comparison to net flows. For example, on average over the 1976–91 period between any pair of months the number of unemployed declined only slightly

Figure 24.2
LABOUR MARKET STOCKS AND FLOWS, 1976 TO 1991

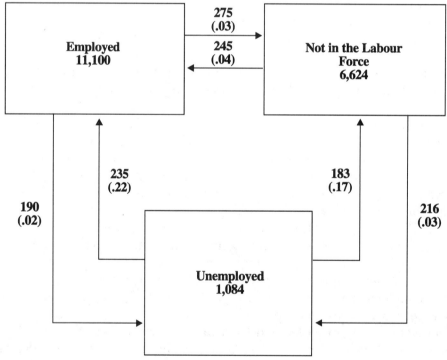

Notes:
1. All numbers are in thousands.
2. All flows and stocks are averages of monthly values from 1976 to 1991.

Source: Stephen R.C. Jones. "Cyclical and Seasonal Properties of Canadian Gross Flows of Labour." This article first appeared in Vol. XIX:I March 1993 of *Canadian Public Policy—Analyse de Politiques*. By permission.

(by 12,000). However, 235,000 individuals, 22 percent of those unemployed in an average month, became employed by the following month. This large gross flow from unemployment to employment was offset by a movement of 190,000 individuals in the opposite direction; the average net flow from unemployment to employment was thus 45,000 (235,000 minus 190,000). Similarly, 183,000 individuals, 17 percent of those unemployed, left the labour force between the average pair of months. However, this change was more than offset by the 216,000 individuals who entered the labour force and actively sought work. The net addition to unemployment of 33,000 due to the movements between unemployment and not in the labour force thus partially offset the net decline of 45,000 associated with the flows between unemployment and employment.

Below each flow (in parentheses) is the average probability of moving from the origin state to the destination state. For example, on average over the 1976–91 period, the probability of an unemployed individual being employed in the following month was 0.22; the probability of moving from employment to unemployment was 0.02.

In summary, even though the number of unemployed changes only marginally from one month to the next, many of the individuals unemployed in one month are no longer unemployed in the next month, with about 22 percent having obtained work and about 17 percent no longer seeking work. The picture that emerges from these data is that of a highly dynamic labour force in which there is a great deal of movement among employment, unemployment, and outside of the labour force each month.[1]

---

1. Gross flows data are subject to some limitations; see Jones (1993).

Table 24.2
DECOMPOSITION OF UNEMPLOYMENT BY REASON, 1975–1989

	NATIONAL UNEMPLOYMENT RATE	REASON FOR SEPARATION			
		JOB LOSERS	JOB LEAVERS	RE-ENTRANTS	NEW ENTRANTS
		(Percentage points)			
1975–79	7.6	3.6	1.7	1.9	0.5
1980–81	7.5	3.8	1.5	1.9	0.4
1982–83	11.4	6.8	1.6	2.5	0.5
1984–87	10.0	5.6	1.7	2.3	0.5
1988–89	7.6	4.0	1.6	1.9	0.3

Source: Surendra Gera (ed.). *Canadian Unemployment*. Ottawa: Economic Council of Canada, 1991.

Table 24.2 provides data on the significance of the various flows into unemployment. The picture of unemployment consisting almost entirely of individuals who lost their job (either temporarily or permanently) is clearly misleading; such job losers account for 40 to 60 percent of all unemployment. In recessions (such as during 1981–82) the absolute and relative importance of job losers increases, and the relative importance of job leavers declines. New entrants typically account for only 4 to 6 percent of unemployment; however, re-entrants (those moving from out of the labour force to unemployed *and* who previously were in the labour force) constitute a substantial proportion of the unemployed.

Incidence and Duration of Unemployment

Because of the dynamic nature of labour force movements, the understanding of unemployment requires information on the flows between various labour force states in addition to the stocks in each state at any point in time. Useful measures for this purpose are the incidence of unemployment — the proportion of individuals who become unemployed in any period — and the duration of unemployment — the length of time spent in the unemployed state before obtaining employment or leaving the labour force.[2] For any particular group of workers, incidence measures the probability of a member of the group becoming unemployed, while duration measures the length of time the individual can be expected to remain unemployed. Using the data in Figure 24.2, the incidence of unemployment for those employed in the average month was 1.7 percent (190,000 out of 11,100,000) while the incidence of unemployment among those out of the labour force in the average month was 3.7 percent (245,000 out of 6,624,000). The overall incidence among those either employed or not in the labour force was 2.5 percent.

The amount of unemployment at any point in time is affected by both the incidence and duration. Indeed, the unemployment rate (UR) can be expressed as the product of the incidence (I) and duration (D).[3]

$$UR = I \cdot D \tag{24.1}$$

Table 24.3 shows the unemployment rate and the incidence and duration of unemployment for various age-sex groups in Canada in 1980, chosen because this was an average year in terms of unemployment. The most striking feature of

---

2. This is sometimes referred to as the duration of a "completed spell of unemployment" to distinguish it from the duration of an "interrupted spell" (i.e., a spell in progress). For further discussion of these duration concepts see Beach and Kaliski (1983).

3. When the labour force is in a steady state equilibrium (i.e., the fraction of the labour force in each state and the proportions flowing between states are constant) this relationship is exact. Otherwise the relationship is only approximately correct. Equation (24.1) is an example of the steady-state identity that the stock at any point in time equals the flow times average duration.

Table 24.3
INCIDENCE AND DURATION OF UNEMPLOYMENT BY
AGE AND SEX, CANADA, 1980

GENDER OR AGE GROUP		UNEMPLOYMENT RATE		SPELLS OF UNEMPLOYMENT	
		ACTUAL	CALCULATED[1]	INCIDENCE	DURATION (MOS.)
Men	15–19	17.1	16.2	8.0	2.0
	20–24	11.5	11.0	4.2	2.6
	25–44	5.2	5.2	1.9	2.8
	45–64	4.2	4.4	1.7	2.6
	Average	6.9	6.8	2.7	2.5
Women	15–19	15.3	13.6	7.1	1.9
	20–24	10.7	9.6	4.2	2.3
	25–44	7.0	6.4	3.0	2.2
	45–64	5.8	5.8	2.5	2.3
	Average	8.4	7.7	3.6	2.2
Overall Average		7.5	7.1	3.1	2.3

Notes:
1. Calculated unemployment rate is the product of incidence and duration. Rounding may
   result in small differences.

Source: Economic Council of Canada, *In Short Supply: Jobs and Skills in the 1980's*
Ottawa: Supply and Services, Canada, 1982, Table 6.2.

the data in Table 24.3 is that the groups with the highest unemployment rates
(males and females 15–19 years of age) have the lowest duration (2.0 and 1.9
months respectively). The very high youth unemployment rate is associated with
a high probability of becoming unemployed rather than an unusually long time
being required to find a job or to exit from the labour force. The data for
individuals 20–24 years of age are less striking but tell a similar story. For these
individuals, duration is about average but incidence is significantly above average, resulting in an unemployment rate well above that of most other groups.
Older workers, on the other hand, have the lowest unemployment rates within
each sex group. This reflects the fact that older workers are less likely to become
unemployed; those that do become unemployed require a slightly longer than
average period to find employment or leave the labour force. Females have a

higher unemployment rate than males but suffer shorter spells of unemployment; that is, the higher unemployment rate of females is associated with a higher probability of becoming unemployed rather than with longer spells of unemployment.

Changing Perspectives on Unemployment

The data summarized in Figure 24.2 and in Tables 24.2 and 24.3 suggest several conclusions:

1. the labour force is highly dynamic, with large flows into and out of unemployment each period;

2. about half the flow into unemployment in an average year is due to individuals losing their job; the remainder is associated with job leavers, new entrants, and re-entrants;

3. the average duration of unemployment in an average year is approximately 2 to 2 1/2 months;

4. the age-sex groups with the highest unemployment rates have the shortest average unemployment durations, but the highest incidence of unemployment.

These conclusions are not peculiar to Canada, but have been reported in empirical studies carried out in various countries during the 1970s and early 1980s. (For Canada see Economic Council of Canada, 1976, 1982; Hasan and de Broucker, 1982; Beach and Kaliski, 1983. For the U.S. see Hall, 1970, 1972; Feldstein, 1973; Marston, 1976; and Clark and Summers, 1979.)

These general findings have important implications for the role of policies to deal with unemployment. The large amount of turnover in the labour force, the fact that job losers account for only about half the flow into unemployment, and the relatively short average duration of unemployment (especially among the groups with the highest unemployment rates) suggest that much unemployment may be caused not by a general shortage of jobs but by employment instability — brief spells of employment followed by periods of job search and/or exit from the labour force. In particular, the short observed duration of unemployment was regarded as evidence that most people could find an acceptable job fairly quickly. Feldstein (1973) used the term "the new unemployment" to describe this view of unemployment being primarily associated with rapid turnover and employment instability; this contrasted with the "old view" in which the unemployed were regarded as a stock of individuals without work for a lengthy period while waiting for a business upturn. The new view suggested that policies aimed at reducing turnover and employment instability may be more successful in achieving lower levels of unemployment than policies aimed at increasing the number of jobs.

Subsequent research, such as that of Clark and Summers (1979) and Akerlof and Main (1980) for the U.S., and Hasan and de Broucker (1982), and Beach and Kaliski (1983) for Canada, has indicated that the "new view" of unemployment contains important elements of truth but is an overly simplistic picture of unemployment. Three conclusions have emerged from this recent research; taking these into account results in a "modified new view." First, although the average duration of completed spells of unemployment is fairly short, much unemployment is accounted for by those suffering lengthy spells of unemployment. An example may help to illustrate this point. Suppose that five individuals become unemployed, four for spells of one month each and the fifth for eight months. The average duration of unemployment for this group is 2.4 months, but two-thirds of all the unemployment (8 out of 12 months) is associated with the one long spell.

A second important consideration is that not all periods of unemployment end in employment. As Figure 24.2 suggests, a significant number of individuals end their search for work by withdrawing from the labour force. Some may leave the labour force to return to school, raise children, or work in the home; however, as discussed previously, some may want paid work but have stopped searching for work because they believe no work is available or for other "economic reasons." By not classifying as "unemployment" the period during which these individuals wanted work but were not searching for work, the duration of unemployment may be considered to be understated. For example, consider an individual who becomes unemployed, searches for work for two months, gives up and stops searching for three months, and then resumes the job search, finding work after a further two months. According to the official statistics, this person would have experienced two spells of unemployment, each lasting a brief two months. However, it could be argued that the individual experienced a single bout of unemployment lasting seven months. Even if this argument is rejected (i.e., the active job search requirement for being considered unemployed is retained), it is clear that, when many spells of unemployment end in withdrawal from the labour force, the fact that the average duration of unemployment is fairly short does not imply that most of the unemployed can find an acceptable job fairly quickly. In Canada in 1980, 44 percent of all bouts of unemployment ended in withdrawal from the labour force and 56 percent ended in employment (Economic Council of Canada, 1982, Table 6-9). A related observation is that spells of unemployment which end in employment are, on average, longer than those ending in withdrawal from the labour force. Thus the average length of time required to successfully find a job is understated by the duration statistics such as those reported in Table 24.3.

The third modification is related to the distribution of unemployment across the population. The implications of a specific unemployment rate differ according to how widely the unemployment is distributed. For example, an unemployment rate of eight percent could mean that eight percent of the labour force is unemployed all year (i.e., the distribution of unemployment is highly concentrated) or that all of the labour force is unemployed for eight percent of the year (i.e., the distribution of unemployment is widely dispersed). The high rate of labour force turnover and the short average duration of unemployment led some analysts to believe that unemployment must be widely distributed among the labour force. In fact, this conclusion turns out not to be correct: unemployment is highly concentrated among a minority of the labour force. The reason for these apparently contradictory findings is that even though the average duration of unemployment is short, a minority of individuals suffer repeated spells of unemployment, sometimes interrupted by periods out of the labour force. For these individuals — the "chronically unemployed" — obtaining employment, especially durable employment, appears to be a difficult task.

Developments During the 1980s and 1990s

Recent behaviour of unemployment has been dominated by cyclical changes: the recession of 1981–82, recovery and expansion from 1983–89, and the recession of 1990–92. During 1983 the unemployment rate reached 11.8 percent, its highest level of the postwar period (Figure 24.1). Only by 1989, following a long period of economic growth, had the unemployment rate returned to its pre-recession level. However, during the recession of 1990–92, unemployment again climbed above 11 percent. The 1980s and early 1990s were thus characterized by higher unemployment rates at each point in the business cycle than was experienced in the 1960s and 1970s.

Some changes in the nature of unemployment were also evident. These are illustrated in Table 24.4. Both the incidence and duration of unemployment increased during the 1981–82 downturn; as is common in recessions, incidence rose more quickly and dramatically than duration. However, during the subsequent recovery the incidence of unemployment not only returned quickly to its pre-recession level, but also fell well below that level. On the other hand, the duration of unemployment remained stubbornly high as the expansion proceeded, and even in the peak years of the business cycle, 1988–89, was still significantly above the pre-recession level. Thus, it appears that unemployment during the latter part of the 1980s and early 1990s was more concentrated on a smaller subset of the population than was the case prior to the 1981–82 recession. At the same time, those who became unemployed took longer, on average, to find employment (or, in some cases, leave the labour force).

Table 24.4 also shows that the proportion of unemployed who are long-term unemployed (where long term is defined as six months or longer) was also significantly higher in the latter part of the 1980s and early 1990s than in previous periods. This finding also holds when other definitions of long-term unemployment (e.g., unemployment lasting longer than one year) are used (Gera, 1991).

Table 24.4
INCIDENCE AND DURATION OF UNEMPLOYMENT AND LONG-TERM UNEMPLOYMENT, SELECTED PERIODS, 1990–91

YEAR	UNEMPLOY-MENT RATE (PERCENT)	INCIDENCE OF UNEMPLOY-MENT (PERCENT)	AVERAGE DURATION OF UNEMPLOY-MENT (WEEKS)	PROPORTION UNEMPLOYED SIX MONTHS OR LONGER (PERCENT)
1976–79	7.6	27.6	14.7	14.9
1980–81	7.5	26.1	14.9	15.3
1982–83	11.4	30.8	19.6	23.9
1984–87	10.0	24.9	21.0	24.7
1988–89	7.6	22.0	18.1	20.2
1990–91	9.2	24.0	18.2	20.9

Notes:
Based on data from Statistics Canada, *The Labour Force*, versus issues. Incidence calculated using the approximation Unem Rate = Incidence ∘ Duration (in months).

For these reasons, the late 1980s and early 1990s saw a shift in emphasis away from concern about employment instability and high turnover in the labour force toward unemployment that is longer term in nature. This shift took place in Canada but was especially evident in Europe, where the problem of long-term unemployment was particularly severe.

UNEMPLOYMENT RATE AS A SUMMARY STATISTIC

Much of the previous discussion suggests that the unemployment rate may not be telling all that we think it tells us, or perhaps used to tell us, about such things as the aggregate state of the economy, the extent to which the labour market is in equilibrium, the tightness or looseness of the labour market, or the extent of hardship in the population. It is not that the unemployment rate is wrong; rather it may be asking too much of a single measure to indicate all these things, especially when dramatic demographic changes have occurred in the labour force, and institutional changes have occurred in such things as unemployment insurance, pensions, and social assistance.

There are reasons to suggest that the unemployment rate may understate the degree of hardship and loss in the economy, and there are reasons to suggest that it may overstate the hardship and loss. For these reasons, complementary indicators have been suggested to supplement the aggregate unemployment rate.

EXHIBIT 24.2 THE MEASUREMENT OF UNEMPLOYMENT AND THE CANADA–UNITED STATES UNEMPLOYMENT DIFFERENTIAL

Throughout most of the postwar period, unemployment rates in Canada and the United States followed very similar trends, as is illustrated in Figure 24.3. Average unemployment rates were nearly equal in the two countries during the 1950s and 1960s, and only slightly higher in Canada during the 1970s. However, since 1982 the Canadian unemployment rate has been two to three percentage points higher than in the United States.

The emergence of a large Canada — U.S. unemployment gap has sparked much speculation about and research into its causes. Initially many observers argued that the gap reflected the more severe economic downturn in Canada in 1981–82. However, the persistence of the differential throughout the subsequent expansion and into the 1990–92 recession suggests the emergence of a permanent or structural difference between the two countries rather than a short-run cyclical phenomenon.

In their analysis of the Canada—U.S. unemployment differential, Card and Riddell (1993) point out that intercountry differences in other measures of aggregate labour force activity actually converged during the 1970s and 1980s, in contrast to the divergence observed in the unemployment rate. This development is illustrated in Figure 24.4, which shows Canada–U.S. differences in the nonemployment rate (one minus the employment rate), the unemployment rate, and the labour force participation rate. Traditionally, labour force participation in Canada was lower than that in the U.S.; however, by the late-1970s Canada's participation rate had caught up to the U.S. rate, and during the 1980s consistently exceeded the U.S. rate. Similarly, although intercountry differences in the unemployment and nonemployment rates followed the same cyclical pattern, the gap in nonemployment actually closed during the 1980s, at the same time the unemployment gap widened. Thus a comparison of relative employment rates and participation rates paints a much brighter picture of the Canadian labour market during the 1980s than a comparison of unemployment.

These aggregate data suggest that the Canadian and U.S. labour forces became more alike during the 1980s in terms of the amount of time spent in the labour force and the amount of time spent employed. Yet they became less alike in terms of the amount of time spent unemployed. How can these apparently contradictory trends be explained? The answer lies in the way nonworking time is spent. In their analysis of Canadian and U.S. micro data, Card and Riddell find that much of the intercountry unemployment differential can be attributed to a rise in the fraction of nonemployment time spent unemployed (i.e., searching for work) rather than out of the labour force. This finding illustrates the fact that alternative measures of labour force activity can sometimes appear to indicate contradictory trends. It also suggests that an understanding of the emergence of a substantial Canada–U.S. unemployment gap must account for the rise in the likelihood of a nonemployed Canadian being engaged in active job search (rather than being out of the labour force) compared to a nonemployed American.

Source: David Card and W. Craig Riddell. "A Comparative Analysis of Unemployment in Canada and the United States." *Small Differences That Matter: Labor Markets and Income Maintenance in Canada and the United States*, D. Card and R.B. Freeman, (eds.). Chicago: University of Chicago Press, 1993.

Figure 24.3
UNEMPLOYMENT RATES IN U.S. AND CANADA, 1953–1990

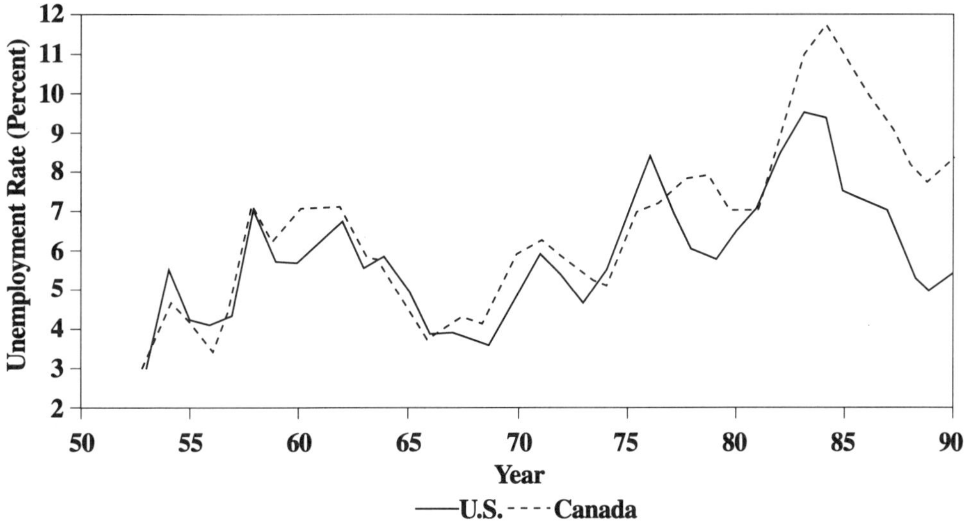

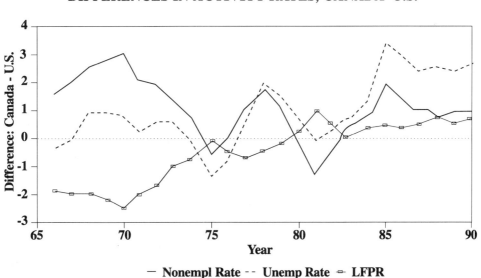

Elements Understating Hardship and Loss

The reported unemployment rate may understate what could be labelled a true unemployment rate because it does not indicate the hidden unemployed. As previously discussed, the hidden unemployed include discouraged workers who

Figure 24.4
DIFFERENCES IN ACTIVITY RATES, CANADA–U.S.

gave up looking for work and who would therefore be counted as not in the labour force, as they engaged in such activities as household work, early retirement, or even school. The number of discouraged workers increases in periods or places of high unemployment. In addition, the hidden unemployed may even include people who drop out of the population that is being surveyed. For example, to the extent that crime increases during periods of high unemployment, these people may be engaging in (unreported) criminal activity, or if apprehended they may be in criminal institutions. In neither case would they be part of the labour force. Or, to the extent that vagrancy increases in times of high unemployment, these people usually would not be considered part of the population that is surveyed. Our unemployment figures do not reflect the loss or hardship of these hidden or disguised unemployed.

In addition, our aggregate unemployment figures by themselves do not indicate the extent to which some workers have added themselves to the labour market to maintain a declining family income associated with a high unemployment rate. In other words, they do not indicate the loss of household production or the loss of education on the part of these added workers who may be compelled by the high unemployment rate to leave the household or educational institutions or retirement.

Nor does the aggregate unemployment rate capture the fact that many employed people may be involuntarily employed only part-time. People are recorded as employed if they did any work during the survey reference period. The fact that some may want to do more work is not reflected. Underemployment can also take the form of individuals working temporarily in jobs that do not utilize their skills and abilities — such as the Ph.D. chemist driving a taxi until a job as a chemist opens us.

In addition, a high unemployment rate by itself does not indicate the degree of hardship associated with the fact that most measures of income inequality are greater during periods of high unemployment. Hollister and Palmer (1972) and Maslove and Rowley (1975), for example, indicate that the poor, more than the non-poor, are adversely affected by high unemployment even if this means reduced inflation. On the labour market side this can occur for a variety of reasons: the poor are most likely to be unemployed; secondary workers from poor families are more likely to drop out of the labour force; the wages of the poor fall the most or rise the slowest; and their hours of work are reduced during periods of high unemployment. The unemployment rate by itself does not reflect the fact that poverty is exacerbated in periods of high unemployment.

Research in Canada, the U.K., and the U.S. has established a positive correlation between unemployment and a variety of social ills including mental and physical illness, suicide, mortality, crime, drug abuse, child abuse, and social unrest. (See, for example, Brenner, 1973; Brenner and Mooney, 1983; Kelvin and Jarrett, 1984; Platt, 1984; Stern, 1983.) Although the causal mechanisms remain a matter of debate (in particular, some of the causation may run from these social ills to unemployment in addition to causation in the opposite direction), these findings indicate that there are important social costs associated with unemployment in addition to the economic costs such as lost output and income. These social costs have significant implications for both the unemployed and for society as a whole through publicly funded institutions such as hospitals, police, and social services.

These various factors suggest that the unemployment rate may understate the true loss and hardship in the economy. In addition, to the extent that these factors worsen in periods of high unemployment — hidden unemployment, the number of added workers, involuntary part-time employment, poverty and various social ills such as mental and physical illness, crime and suicide all increase — then this underestimate of the loss is most severe in periods of high unemployment, when conditions already are severe.

Elements Overstating Hardship and Loss

There are also reasons to believe, however, that the unemployment rate may be overstating the degree of hardship and loss in our economy, especially relative to the situation in the 1950s and early 1960s. Many of these reasons hinge on the fact that our unemployment increasingly has fallen on younger workers and married women, and for a variety of reasons there is a belief — held by some, but not all — that the hardship and loss associated with such unemployment is not as severe as that which is associated with the unemployment of males in families where only the male otherwise works.

Increasingly our unemployed consist of females and younger workers, as they have expanded their participation in labour force activities and increasingly are subject to a high incidence or likelihood of being unemployed. (See Chapter 5 for details of the changing patterns of labour force participation and Table 24.3 in this chapter for differences in the incidence of unemployment across age-sex groups.) To the extent that younger workers and women have alternative sources of income to rely upon, and to the extent that the economic well-being of the family is more dependent upon the employment of males, then there is the possibility that the economic hardship associated with a given unemployment rate is not as great now as when males dominated the labour market. This perspective, of course, can be challenged — and is being challenged increasingly — as the participation of women in labour market activities and their contribution to family income is becoming regarded as permanent rather than secondary. As a more general proposition, with the growing proportion of families where both the husband and wife work, the unemployment of either may result in less hardship than when males tended to dominate the labour market and became unemployed. For example, since 1975, a period in which the aggregate unemployment rate generally exceeded seven percent, seventy to seventy-five percent of families with one family member unemployed had at least one family member employed. Even in 1983, when the unemployment rate reached 11.8 percent, 70 percent of families with one member unemployed had at least one member employed and 23 percent of families with one member unemployed had two or more members employed (Kaliski, 1985, Table 2-8).

This possibility of reduced hardship is buttressed by the fact that many of the unemployed may be looking for part-time jobs. This could be the case especially for married women who also maintain prime responsibility for household tasks. It may also be the case for students who go to school part-time. The unemployment rate counts all searchers equally, whether they are seeking part-time or full-time work.

The hardship associated with being unemployed may also be reduced, or at least cushioned, by the availability of unemployment insurance benefits. Although many changes have been made to Canada's unemployment insurance system during the past several decades, the most substantial changes were those made in 1971–72. These significantly increased the coverage of the UI program, reduced the number of weeks of previous work experience required to qualify for UI benefits, increased the number of weeks for which benefits could be received, and raised the benefit rate. As a result of these and other changes to unemployment insurance and other social programs, Canadians are more protected from the loss of income due to unemployment than was the case in earlier periods.

These possibilities are not meant to play down the hardship associated with being unemployed. It is a serious loss to the economy and more importantly to the individuals and families involved. Nevertheless it is important to be aware that a six percent unemployment rate in the mid 1970s probably would not inflict the hardship that an identical rate would have inflicted on families in the 1950s. Whether an average unemployment rate of over six percent in the 1970s inflicts more hardship than an average rate of slightly over five percent in the 1960s, or four percent in the 1950s, remains an open question.

Complementary Measures

Because of these problems associated with a single aggregate unemployment estimate, some have suggested using complementary measures. In general, these measures are suggested to supplement rather than replace the information contained in the aggregate unemployment rate estimate. For example, Statistics Canada now reports eight supplementary measures of unemployment in addition to the official unemployment rate (Devereaux, 1992). These are:

- R1 — Persons unemployed 14 or more weeks as a percentage of the labour force.

- R2 — Unemployment rate of persons heading families with a child or children under age 16.

- R3 — Unemployment rate excluding full-time students.

- R4 — Unemployment rate including full-time members of the Canadian Armed Forces.

- R5 — Official unemployment rate.

- R6 — Unemployment rate of the full-time labour force.

- R7 — Unemployment rate including discouraged workers and other persons "on the margins" of the labour force.

- R8 — Underutilization rate based on hours lost through unemployment and underemployment.

- R9 — Unemployment rate of the part-time labour force.

These various alternative measures of unemployment are listed in order of their rank (from lowest to highest) in 1977, the year when data were available for all calculations. Measures R1 and R2 focus on groups for whom unemployment may present particular economic hardship, while R3 and R4 reflect groups with varying degrees of labour force attachment. R6 and R7 incorporate aspects of hidden unemployment or the under-utilization of labour. R8 is a measure based on hours, whereas the official unemployment rate is based on numbers of individuals. R9 is the unemployment rate for a particular group, part-time workers.

Labour Market Disequilibrium

The unemployment rate is also often used as a measure of the overall state of the economy and of the degree of disequilibrium in the labour market. A low unemployment rate is associated with a booming economy and with excess demand in the labour market while a high unemployment rate is associated with a recession and with excess supply in the labour market. However, this begs the question of what is meant by a "low" or "high" unemployment rate. Clearly some reference point is needed. However, as our discussion of the Canadian experience indicated, the level of unemployment that might be considered normal has changed substantially over time.

A useful reference point is the unemployment rate at which the economy-wide labour market is in equilibrium; that is, at which aggregate demand for labour equals aggregate supply of labour. This concept is often referred to as the "natural" or equilibrium unemployment rate and is discussed in more detail in Chapter 26. There is aggregate excess demand in the labour market if the unemployment rate is below the natural rate, and excess supply if the unemployment rate exceeds the natural rate; indeed, that is what is meant by an equilibrium unemployment rate. However, because the natural rate is not directly observed, some care is required in interpreting increases or decreases in the unemployment rate as indicating increased slack or tightness in the labour market. A decrease in the unemployment rate could indicate either a tighter labour market or a decline in the natural rate; the former is associated with excess demand in the labour market while the latter is not. Relying solely on the unemployment rate as an indicator of the state of the economy could result in macroeconomic policy mistakes.

Perhaps the best example of a Canadian macroeconomic policy error caused in part by excessive reliance on the unemployment rate as a measure of the overall state of the economy occurred in the early 1970s. During the recession of 1970–71, unemployment climbed to over 6 percent, a level which was high by the experience of the 1950s and 1960s and which was considered indicative of substantial slack in the economy. When inflation fell somewhat, the authorities set out on an expansionary path. Although employment and output grew rapidly, the unemployment rate remained stubbornly above 6 percent during 1972 and 1973. The "high" level of unemployment was interpreted as indicating the need for further stimulus. The result was an overly expansionary policy during this period which, combined with the 1973 OPEC oil price shock, led to a substantial increase in inflation and ultimately to wage and price controls in 1975–78. In retrospect, the overly expansionary policy can be attributed in part to undue reliance on the unemployment rate as a measure of excess demand or supply and a concomitant failure to recognize that the natural unemployment rate had risen sharply during the early 1970s due mainly to the changing demographic structure of the labour force and the 1971–72 revisions to the unemployment insurance system.

Again, the lesson of this discussion is not that the unemployment rate is unsuitable as a measure of the state of the labour market, but rather that it should be used in conjunction with other indicators. In this case, although the unemployment rate remained approximately constant (at about 6.3 percent) between 1971 and 1973, employment grew very rapidly and the job vacancy rate more than doubled during this period, indicating an increasingly tight labour market.

QUESTIONS

1. (a) Discuss the various ways in which the number of unemployed typically are measured in Canada.

   (b) Which is the measure that receives the most publicity?

   (c) Discuss the strengths and weaknesses of the various measures.

2. Discuss the extent to which our aggregate unemployment rate measure may understate and overstate the degree of hardship associated with a given level of unemployment.

3. (a) Explain the meaning of the following measures of labour force utilization: (i) labour force participation rate, (ii) employment rate, and (iii) unemployment rate.

   (b) For each of these measures, describe circumstances under which that measure would be preferred to the other two.

   (c) Under what conditions will these three measures all move in the same direction? Under what conditions would they move in opposite directions?

4. Explain the concept of hidden unemployment. Discuss the pros and cons of including the hidden unemployed in our unemployment statistics.

5. Using the March issues of *The Labour Force* since 1979, construct an annual time series of the "discouraged worker" phenomenon for 1979 to the present. During which years were there relatively more discouraged workers? Why?

6. Explain the distinction between gross flows and net flows among various labour force states.

7. Discuss the differences between static and dynamic measures of unemployment.

8. In discussing the nature of unemployment, some analysts stress a shortage of jobs while others emphasize labour force turnover and employment instability.

   (a) Elaborate these competing perspectives on unemployment.

   (b) Describe research findings which attempt to distinguish between these competing perspectives. Which view is correct?

9. Why might we expect younger workers to have higher unemployment rates than older workers, and what does this imply about the changes in our aggregate unemployment rate over time?

10. Historically in Canada, female unemployment rates have been considerably lower than male unemployment rates; however, by the mid-1970s they began to exceed male rates.

    (a) Why might this be the case?

    (b) What does this imply about the time pattern of our aggregate unemployment rate?

11. Why is it that the groups with the highest unemployment rates also have the shortest average duration of unemployment?

12. (a) Indicate some of the uses for which our aggregate unemployment rates are often used.

(b) Why might this single measure increasingly be inadequate for these purposes?

(c) What are some of the alternatives and what are their strengths and weaknesses?

13. What factors might account for the rise in duration of unemployment during the latter part of the 1980s and the associated decline in incidence?

REFERENCES AND FURTHER READINGS

Akerlof, G. and B. Main. Unemployment spells and unemployment experience. *AER* 70 (December 1980) 885–893.

Akerlof, G. and B. Main. An experience-weighted measure of employment and unemployment durations. *AER* 71 (December 1981) 1003–1011.

Akyeampong, E.G. Persons on the margins of the labour force. *The Labour Force* (April 1987) 85–142.

Akyeampong, E.B. Discouraged workers. *Perspectives on Labour and Income* (Autumn 1989) 64–69.

Akyeampong, E.B. Discouraged workers — where have they gone? *Perspectives on Labour and Income* (Autumn 1992) 38–44.

Beach, C. and S. Kaliski. Measuring the duration of unemployment from gross flow data. *CJE* 16 (May 1983) 258–263.

Beach, C. and S. Kaliski. The impact of recession on the distribution of annual unemployment. *RI/IR* 41 (No. 2, 1986) 317–328.

Blanchard, O. and P. Diamond. The cyclical behavior of the gross flow of U.S. workers. *BPEA* (1990) 85–155.

Brenner, H. *Mental Illness and the Economy.* Cambridge: Harvard University Press, 1973.

Brenner, H. and A. Mooney. Unemployment and health in the context of economic change. *Social Science and Medicine* 17 (No. 16, 1983) 1125–1138.

Card, D. and W.C. Riddell. A comparative analysis of unemployment in Canada and the United States. *Small Differences That Matter: Labor Markets and Income Maintenance in Canada and the United States*, D. Card and R.B. Freeman (eds.). Chicago: University of Chicago Press, 1993.

Clark, K. and L. Summers. Labor market dynamics and unemployment: a reconsideration. *BPEA* (No. 1, 1979) 13–60.

Economic Council of Canada. *People and Jobs: A Study of the Canadian Labour Market.* Ottawa: Information Canada, 1976.

Economic Council of Canada. *In Short Supply: Jobs and Skills in the 1980s.* Ottawa: Supply and Services Canada, 1982.

Feldstein, M. The economics of the new unemployment. *Public Interest* 33 (Fall 1973) 3–42.

Gonul, F. New evidence on whether unemployment and out of the labor force are distinct states. *JHR* 27 (Spring 1992) 329–361.

Gower, D. Time lost: an alternative view of unemployment. *Perspectives on Labour and Income* (Spring 1990) 73–77.

Gramlich, E. The distributional effects of higher unemployment. *BPEA* (No. 2, 1974) 293–342.

Green, C. Labour market performance from an employment perspective. *CPP* 3 (September 1977) 315–323. Comment by S, Kaliski 3 (Autumn 1977) 515–517.

Hall, R. Why is the unemployment rate so high at full employment? *BPEA* (No. 3, 1970) 369–402.

Hall, R. Turnover in the labour force. *BPEA* 3 (1972) 709–756.

Hasan, A. and P. De Broucker. Duration and concentration of unemployment. *CJE* 15 (November 1982) 706–734.

Hollister, R. and J. Palmer. The implicit tax of inflation and unemployment. *Redistribution to the Rich and Poor*, K. Boulding and M. Pfaff (eds.). Belmont, Calif.: Woodworth, 1972.

Hurd, M. A compensation measure of the cost of unemployment to the unemployed. *QJE* 97 (September 1980) 225–244.

Jackson, G. Alternative concepts and measures of unemployment. *The Labour Force* (February 1987) 85–120.

Jones, S.R.G. The cyclical and seasonal behavior of Canadian gross flows of labour. *Canadian Public Policy* 19 (March 1993).

Jones, S.R.G. and W.C. Riddell. The measurement of labour force dynamics with the labour market activity survey: the LMAS filter. UBC Department of Economics Discussion Paper No. 91-17, 1991.

Kaitz, H. Analysing the lengths of spells of unemployment. *MLR* 93 (November 1970) 11–20.

Kaliski, S.F. Why must unemployment remain so high? *CPP* 10 (June 1984) 127–141.

Kaliski, S.F. Trends, changes and imbalances: a survey of the Canadian labour market. *Work and Pay: The Canadian Labour Market*. W.C. Riddell (ed.). Toronto: University of Toronto Press, 1985, 77–140.

Kelvin, P. and J. Jarrett. *Unemployment: Its Social Psychological Effects*. Cambridge: Cambridge University Press, 1984.

Levesque, J.M. Unemployment and unemployment insurance: a tale of two sources. *Perspectives on Employment and Income* 1 (Winter 1989) 49–53.

Marston, S. Employment instability and high unemployment rates. *BPEA* (No. 1, 1976) 169–210.

Maslove, A. and J. Rowley. Inflation and redistribution. *CJE* 8 (August 1975) 399–409.

Monthly Labor Review. Measuring Unemployment. 102 (March 1979) 13–52.

Norwood, J.L. The measurement of unemployment. *AER* 78 (May 1988) 284–288.

Platt, S. Unemployment and suicidal behaviour: a review of the literature. *Social Science and Medicine* 19 (No. 2, 1984) 93–115.

Romer, C. Spurious volatility in historical unemployment data. *JPE* 94 (February 1986) 1–37.

Rosen, C. Bibliography: hidden unemployment and related issues. *MLR* 96 (March 1973) 31–37.

Stern, J. The relationship between unemployment, morbidity and mortality in Britain. *Population Studies* 37 (1983) 61–74.

Summers, L.H. *Understanding Unemployment,* Cambridge, MA: MIT Press, 1990.

CHAPTER 25

Unemployment: Causes and Consequences

The previous chapter discussed the meaning and measurement of unemployment and described the salient aspects of the Canadian experience. In this chapter we examine the causes and consequences of unemployment and the role of public policy in this area.

TYPES OF UNEMPLOYMENT

Economists usually distinguish among several types of unemployment. These differ according to their causes, their consequences, and the policies that are likely to be effective in dealing with them. The most common categories are frictional, structural, deficient demand, and seasonal unemployment. Each of these will be discussed in turn. As will become evident in the ensuing discussion, the distinctions are often not clear-cut, either conceptually or practically.

Frictional Unemployment

Frictional unemployment is associated with normal turnover in the labour force; it can thus be thought of as unemployment that would prevail even in a well-functioning labour market. Change is a pervasive feature of modern economies. As discussed previously, in each period some individuals enter the labour force to search for work while others leave to return to school, retire, or work in the home. Similarly, new jobs open up in some firms and disappear in others. As a consequence, unemployed workers and unfilled job vacancies will coexist at any point in time. This unemployment exists even if jobs and workers are potentially matched in the sense that the unemployed workers are qualified for the available jobs and are willing to fill these jobs; they simply have to be brought together and this process takes time. For this reason, frictional unemployment is often associated with job search activity within a given labour market. Because suitable unfilled vacancies exist for the frictionally unemployed, this type of unemployment will typically be of short duration; however, the optimal duration will depend on the benefits and costs of continued search, as discussed below.

Unfilled job vacancies and unemployed job seekers coexist because of imperfect information. Time and money are required for the unemployed to discover the available jobs, their rates of pay, and their working conditions. Such is also the case for employers to identify applicants and determine their suitability. The process of matching job seekers with job vacancies yields benefits — both to the individual employers and employees involved and to society — but it also involves costs. By improving the flow of information, it may be possible to reduce the amount of frictional unemployment. However, it is unlikely that such unemployment could be eliminated, nor is it necessarily the case that reducing frictional unemployment would be desirable. In many circumstances the benefits of search and the acquisition of information by employers and job seekers will exceed the costs. In such cases it is both privately and socially optimal for the parties to engage in such activity. Furthermore, the additional benefits of reducing frictional unemployment — perhaps by reducing the time needed to locate job vacancies — may not exceed the additional costs associated with bringing about this change. In these circumstances it would not be desirable to implement policies intended to reduce frictional unemployment.

Frictional unemployment may also take the form of temporary layoffs. Even in the absence of fluctuations in aggregate economic activity, individual firms and industries are affected by changes in product demand due to changes in international economic conditions, consumer demand, weather, work stoppages, and so on. Employers may respond to reductions in demand in a variety of ways: inventory accumulation, wage and/or hours reduction affecting most or all of the work force, or temporary layoffs affecting a portion of the work force. The combination chosen will depend on a variety of factors discussed below. For some firms and industries temporary layoffs are a common form of adjustment to changing circumstances. As explained previously, individuals who have been laid off and are expecting to be recalled to their former employer are classified as unemployed even though they may not be searching for work. Temporary layoffs typically account for 7 to 12 percent of unemployment in Canada (Kaliski, 1985). About 20 percent of workers on temporary layoff search for work, perhaps because they are uncertain about being recalled or perhaps because they are using the opportunity to try to find a better job. Workers on temporary layoff are considered to be frictionally unemployed in the sense that the fluctuations in individual markets which give rise to layoffs are part of the normal "frictions" of a decentralized market economy. However, temporary layoffs are conceptually different from other forms of frictional unemployment in that job vacancies for these workers do not necessarily exist at the time of the survey; rather, the unemployed are waiting for their previous jobs to reappear.

Temporary layoffs in response to fluctuations in demand can be an efficient arrangement from the perspective of both employers and employees if other forms of adjustment (inventory accumulation, wage and/or hours reductions) are costly, and if the costs to workers of brief periods of temporary unemployment (and reduced income) are not too high.

Structural Unemployment

Structural unemployment results when the skills or location of the unemployed are not matched with the characteristics of the job vacancies. Unemployed workers and job vacancies are considered to be in different labour markets, either by virtue of geography or because they do not coincide in terms of qualifications and characteristics. The analogy that is often used is that of matching square pegs to round holes. As is the case with frictional unemployment, structural unemployment is characterized by the coexistence of unemployed workers and job vacancies. However, in this case, successful matching of workers and jobs requires more than acquisition of information. In particular, employees or employers will either have to relocate — when there are job vacancies in one region and unemployed workers in another — or alter their characteristics or requirements (in the case of mismatching along occupational and skill lines). Proposed solutions to structural unemployment usually involve improving the human capital characteristics of the workers by education or training programs or encouraging labour mobility and job search in other regions. They could also include adapting the characteristics of the jobs themselves by altering entrance requirements, rearranging the basic job components to adapt to available skills, and even job enrichment and job enlargement. Regional development policies which attempt to expand employment opportunities in areas with high levels of structural unemployment have also been used in Canada.

One of the most difficult issues relating to structural unemployment involves determining whether a particular situation is temporary or permanent. For example, suppose that there is an increase in demand in one industry, occupation or region, giving rise to job vacancies in that sector and a decrease in demand in another industry, occupation or region, giving rise to unemployed workers in that sector. If these changes in demand are permanent in nature, the resulting unemployment is clearly structural and the best course of action will usually be to adjust to the structural change, perhaps via retraining or relocation. However, if the changes in demand are temporary in nature, the resulting unemployment is more frictional than structural and the best course of action may well be to simply wait for demand to return to normal levels in each sector. In these circumstances, costly activities such as relocation or retraining are inappropriate because they yield negligible (perhaps zero) benefits. Unfortunately it is not always evident in advance which situations are temporary and which are permanent. From the perspective of individuals adversely affected by economic change, this uncertainty implies that it may not be clear whether investments such as retraining or relocation are worthwhile. In these circumstances, some individuals may wait for the uncertainty to be resolved before taking action. Providing it is based on a careful assessment of the likelihood of demand returning to previous levels, and not on "wishful thinking," waiting for the resolution of uncertainty can be viewed as a productive activity analogous to the acquisition of information associated with job search. Waiting can result in the acquisition of information about future developments while job search can result in acquiring information about current job opportunities.

When the changes giving rise to structural unemployment are clearly permanent in nature, the best course of action — both from the perspective of the individuals involved and from that of society — usually will be to adjust to the altered circumstances. An exception occurs for people near the end of their working lives. For these individuals, the benefits of retraining or relocation may not justify the costs, and alternatives such as early retirement may be best, given the available options.

The distinction between frictional and structural unemployment can be blurred conceptually, and in practice it can be difficult to clearly delineate the two. Nevertheless, there are differences in degree if not in kind. Frictional unemployment is a result of the matching process; structural unemployment occurs when there is mismatching. Frictional unemployment is associated with job search in an individual labour market; structural unemployment involves more costly solutions, ranging from retraining for a job within the individual labour market to job search and relocation in other labour markets. Frictional unemployment is associated with a productive activity — the acquisition of information regarding job opportunities and applicants — and thus policies designed to reduce frictional unemployment are not necessarily desirable. Structural unemployment that is clearly permanent in nature is not associated with a productive activity and actions to reduce this type of unemployment should be taken, providing the benefits from doing so exceed the costs. What the two have in common is that they are related to the characteristics of the workers and the job, not to the general state of aggregate demand in the economy. Even here, though, the distinction between demand-deficient unemployment and nondemand-deficient unemployment (frictional, structural) may become blurred as the characteristics of the workforce and of jobs themselves may be related to the aggregate state of the economy.

Demand-Deficient Unemployment

Demand-deficient unemployment exists when there is insufficient aggregate demand in the economy to provide jobs. It is not a matter of workers engaging in normal job search or lacking the correct skills or being in the wrong labour market; rather it is a matter of insufficient aggregate demand to generate sufficient job vacancies. Therefore, job vacancies would fall short of the number of unemployed job seekers. That is, defining labour supply as employed plus unemployed, and labour demand as employed plus vacancies, then insufficient demand, defined as being less than supply, would imply vacancies being less than the number unemployed.

Demand-deficient unemployment is usually associated with adverse business cycle conditions; hence, the term cyclical unemployment is often used. However, it may also be associated with a chronic (as opposed to short-term cyclical) insufficiency of aggregate demand as occurred, for example, in the Great Depression of the 1930s. Since the cause of such unemployment is a deficiency of aggregate demand, its cures usually involve macroeconomic policies to increase consumption, investment, exports, or government spending, or to decrease imports and taxes. Monetary, fiscal, and exchange rate policies are the traditional macroeconomic instruments.

Seasonal Unemployment

Seasonal unemployment is often associated with insufficient demand in a particular season. In this sense it can be considered as demand-deficient unemployment; nevertheless, it is different in the sense that it is not a shortage of aggregate demand for the economy as a whole, but rather a shortage of demand in a particular season. The patterns are usually predictable over the year and specific to particular industries. For example, seasonal unemployment is usually prevalent in the winter months in construction, agriculture, and the tourist trade. Seasonal fluctuations in labour supply may also occur, the most significant example being the large number of college and university students who enter the labour force during the summer.

Seasonal unemployment is analogous in several respects to temporary layoffs, and is therefore often included under the rubric of frictional unemployment. Both seasonal unemployment and temporary layoffs involve the workers returning to their previous job or employer. To the extent that layoffs or seasonal unemployment are anticipated, employees generally will receive some compensation ex ante in the form of a wage premium to compensate them for the probability of being unemployed.

To a certain extent seasonal jobs could be adapted to employ more people over the slow season. Some construction jobs could be modified so as to be more continuous throughout the year. Tourist facilities can likewise adapt, for example, as winter ski areas become summer schools in other sports or crafts. In some industries like agriculture, migrant labour often moves to other labour markets, following the seasonal patterns of demand.

Clearly the fact that such adaption of jobs and workers can and does go on illustrates that seasonal unemployment is not immutable. It is related to the costs and benefits of reducing such unemployment, and the private market will respond to changes in these costs and benefits. For example, to the extent that the cost to employers and employees of seasonal unemployment is reduced by the availability of unemployment insurance benefits, then one may expect more seasonal unemployment.

Aggregate unemployment rates are often referred to as seasonally adjusted or unadjusted. The seasonally unadjusted figures simply show the unemployment rate as it is in a particular month, unadjusted for seasonal fluctuations. The seasonally adjusted figures show the unemployment rate that would have prevailed had the particular month not been associated with unusually high or low seasonal demand conditions. Those who are seasonally unemployed are not removed from the figures; they are simply averaged in over the year.

Involuntary Unemployment and Wage Rigidity

Another distinction often made is that between voluntary and involuntary unemployment. According to the most common definition of these terms, individuals are involuntarily unemployed if they are willing to work at the going wage rate for their skills or occupation but they are unable to find a job. Voluntary unemployment exists when a job is available but the worker is not willing to accept it at the existing wage rate. These definitions are based on the willingness to work at the prevailing wage, rather than on the reason for becoming unemployed, such as job loser or job leaver. Nonetheless, many job losers may be involuntarily unemployed and many job leavers voluntarily unemployed.

Cyclical unemployment is often viewed as being largely involuntary in nature because evidence suggests that many of those unemployed in economic downturns would accept work, even at wages below the prevailing wage. This phenomenon could be associated with a failure of the wage to adjust to a decline in aggregate demand, creating excess supply of labour at the existing wage rate. For this reason, involuntary unemployment and wage rigidity are often regarded as being closely linked. Explaining why wages may not adjust to "clear" the labour market is a puzzle that has challenged economists for many years. Leading responses to that challenge are examined in the next section.

EXPLANATIONS OF UNEMPLOYMENT

The economic theory of frictional, structural, and demand-deficient unemployment has received considerable attention, especially in the past three decades. To a considerable extent this interest has been motivated by the increase in unemployment that occurred in many countries during the 1970s and 1980s. In addition, many economists believed that the theory of unemployment suffered from a lack of clear microeconomic foundations similar to that which provides the basis for the theory of household or firm behaviour. Significant recent developments include search theory, implicit contracts, efficiency wages, and insider-outsider theories. The relationship between unemployment insurance and unemployment has also been extensively investigated. This section discusses these and related developments and their contribution to understanding unemployment. Parts of this section build on the discussion of search and matching, implicit contracts and efficiency wages contained in Chapter 13, and some readers may wish to review that material at this time.

Search Unemployment

The job search and matching process is associated with imperfect information on both sides of the labour market. Unemployed workers are not aware of all available jobs, their rates of pay, location, and working conditions. Employers with job vacancies are not aware of all individual workers and their characteristics. If both sides were fully informed, the process of matching workers and jobs could take place in a few days, if not hours. However, because the acquisition of information about job opportunities and job applicants takes time, unemployment and unfilled vacancies coexist.

Following the important early contributions of Stigler (1962) and Phelps et al. (1970), the economics of job search has received considerable attention in the past two decades. To illustrate the main ideas we focus primarily on the job search process of employees. Employer search is discussed in Lippman and McCall (1976) and in the Canadian context in Maki (1971, Chapter 3).

Job search is an economic decision in that it involves both costs and benefits. There are two main aspects to the decision: (1) determining whether it is worthwhile *initiating* the job search process, and (2) once begun, determining when to *discontinue* the process. For employees, initiating a job search or continuing to search rather than accepting the first job offer that comes along may yield benefits in the form of a superior job offer involving higher wages or better working conditions. Similarly, by continuing to search rather than filling the job with the first "warm body" that becomes available, the firm may obtain benefits in the form of more suitable or more qualified applicants. In both cases the magnitude of these benefits is uncertain; their size depends on the employee's expectations regarding the probability of receiving a superior job offer or on the firm's expectations regarding the quality of future applicants. Those engaged in search must make their decisions based on these expectations: i.e., on the information available ex ante. The realized or ex post benefits may turn out to be larger or smaller than the expected benefits.

The expected benefits have to be weighed against the costs of search. These include both the direct costs — such as the firm's costs of advertising positions and interviewing applicants and the employee's costs of sending applications and travelling to interviews — and the indirect or opportunity costs. For employees the opportunity costs are measured by the best alternative use of their time devoted to job search. For those who quit their previous job to search for a better one their opportunity cost would be their previous wage. For others the opportunity cost would be measured by the best job offer received so far or the wage that could be earned in some job that is known to be available. It could also be an individual's implicit home wage or the value of their time doing household work. For employers the opportunity cost of continuing to search is the difference between the value of the output that would be produced by the best known applicant and the wages that would be paid to that applicant.

Not all job seekers are unemployed; some individuals will search for a better job while employed. Employed search has the advantage of being less costly. However, it may also be less effective because of the difficulties associated with contacting potential employers and following up promising opportunities while working, especially if employed on a full-time basis. In some circumstances individuals may choose to quit their current job in order to search for a better one, while in other circumstances employed search will be the preferred method.

The central assumption in the economic analysis of job search is that individuals will search in an optimal fashion: that is, will choose their search activities in order to maximize their expected utility. Note, however, that the assumption of optimal search behaviour does not imply that all unemployed job seekers have *chosen* to be unemployed or that all search unemployment is voluntary in nature. Individuals who quit their previous job in order to search for a new job — the "job leavers" in Table 24.2 — may have chosen unemployment. Rational employees would not do so unless the expected utility of searching for a new job exceeds the expected utility of remaining in the existing job. However, as Table 24.2 indicates, about half the unemployed are "job losers"; many of these individuals may have preferred to remain in their previous job. Nonetheless, given that they have lost their previous job, rational individuals will carry out their job search in an optimal fashion. In the case of a person who is risk-neutral, this assumption implies that the individual will choose the amount of search activity that maximizes the net expected benefit (expected benefits minus expected costs). Risk-averse employees will also take into account the costs and benefits of search, but will attach a greater weight to benefits and costs which are certain compared to those which are uncertain. The basic principles of optimal job search are most easily explained in the context of risk-neutral searchers. However, very similar principles and conclusions follow when individuals are risk-averse.

In order to maximize the net expected benefits of job search, employees should continue searching until the marginal expected benefit of search equals the marginal expected cost. This condition is simply another example of the rule that the net benefit of any activity is maximized by expanding the activity to the

Figure 25.1
OPTIMAL JOB SEARCH

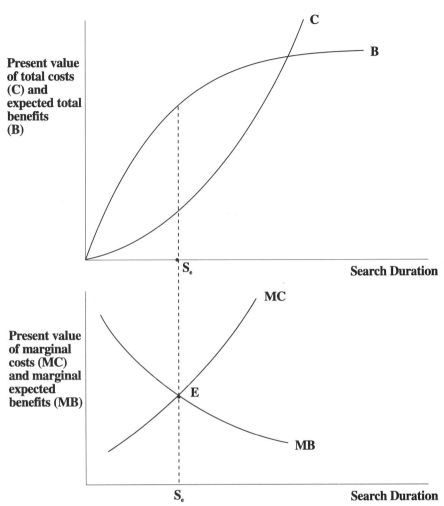

point at which its marginal benefit equals its marginal cost. Figure 25.1 shows the way in which the benefits and costs of search are likely to be related to the amount of time devoted to job search. The case in which search is worthwhile is illustrated (i.e., total expected benefits exceed total costs). For low levels of search, the marginal costs of search are fairly low because low-cost, usually informal, search processes can be used. For example, friends and relatives can be contacted, want ads examined, and perhaps a few telephone calls made. As the search continues, however, more costly processes are often necessary to acquire additional labour market information. For example, it may be necessary to apply directly to a company or to sign up with an employment service. In some cases it may even be necessary to change locations or to quit working if one already has a job. For these reasons the marginal cost of search probably rises with the amount of job search undertaken, as depicted in Figure 25.1.

The marginal benefits of search, on the other hand, probably are a declining function of the amount of search undertaken. One starts out the search process with an examination of the most promising alternatives and then continues examining further activities in the hope of finding an even better one. Obviously, a better alternative may occur; however, one may encounter diminishing returns with respect to the additional information. As search continues it becomes less likely — but still possible — that a better offer will be received simply because there are fewer options left to examine.

Given the costs and benefits of additional search, the rational individual will acquire labour market information and engage in search until the point E, where the marginal expected benefits equal the marginal cost. To the left of E, the benefits of additional search exceed the cost and hence additional search is worthwhile; to the right of E, the benefits of additional search are not worth the costs. In this sense E will be an equilibrium with S_e being the optimal amount of expected search activity. Note, however, that S_e is the *expected* amount of search required to maximize net benefits. The actual amount of search undertaken in any particular situation may turn out to differ from S_e. For example, a lucky individual who receives an extremely good job offer during the first week will experience an actual search duration less than S_e because the best course of action will be to accept the job and discontinue the search. Similarly, an unlucky individual may have to search longer than anticipated, or accept a lower paying job than expected.

The conditions for optimal search can alternatively be stated in terms of a "stopping rule": the individual should choose a minimum acceptable wage and search until a job paying this wage or better is found. The minimum acceptable wage is often referred to as a "reservation wage," although this concept is not identical to the reservation wage discussed in Chapter 5 on labour force participation. Choosing a minimum acceptable wage is equivalent to choosing an expected search duration. On average, given the distribution of wage offers in the market and the rate at which firms can be contacted, the individual will require the expected search duration to find an acceptable job. The minimum acceptable wage is therefore chosen to equate the marginal benefits and marginal costs of search.

Factors Determining Optimal Search

This framework is useful to illustrate how the optimal amount of search depends on the various factors underlying the cost and benefit schedules and how search activity will respond to shifts in these schedules. For example, factors affecting the expected benefits of job search are the dispersion of wage offers (for otherwise comparable jobs) and the expected duration of the job. If there is little variation in the attractiveness of different jobs in terms of wages and working conditions, there is little point in continuing to search once an offer has been received. In contrast, if some jobs are much more desirable than others, the optimal strategy may be to continue searching even after receiving an average job offer. Similarly, if some potential jobs are more desirable than others it will be rational to devote more time to search the longer is the anticipated duration of employment because of the longer period over which the benefits of a superior job can be enjoyed.

These two factors may help explain the observed differences in the average duration of unemployment across age and sex groups, as summarized in Table 24.3. Because of their limited work experience, youths typically face a less dispersed distribution of potential wages than older workers. In addition, many youths try out a variety of jobs before settling on a career; thus, on average, their expected duration of employment is much lower than is the case for adult workers. For both these reasons, there is less incentive for youths than for adult workers to continue searching in the expectation of obtaining a better job. These considerations are consistent with the observation that the average duration of unemployment of youths is significantly lower than that of adults. Similarly, because adult women tend to exit from and re-enter the labour force more frequently than adult men, the expected duration of employment for women is lower than that for men, at least on average. This may help account for the fact that the average duration of unemployment for adult women is lower than that for adult men, as shown in Table 24.3.

Shifts in the schedules may affect the number of searchers as well as the expected duration of search. For example, an increase in expected total benefits or a decrease in total costs will imply a larger number of people for whom search is worthwhile. As a consequence the following may be observed: more entry and re-entry into the labour force, an increase in the number of quits as more employed workers seek a new job, and more search by the employed.

Other factors affecting the benefits and costs of employee search are the institutional mechanisms for disseminating labour market information, the number of employers with job vacancies, the rate at which employers make job offers, the value of "leisure" time (time spent not searching or working), the number of other searchers competing for the available jobs, and the occupational and regional segmentation of labour markets. Social and labour market policies can also alter the costs and benefits of search and hence its amount and duration. Unemployment insurance and portable pensions would reduce the total and marginal costs of search and hence increase the number of searchers and the optimal duration of search. Improving the arrangements for disseminating labour market information would increase the total benefits and, in most cases, the marginal benefits as well. Thus the number of job seekers would increase but the expected duration would fall. The total amount of search unemployment may either increase or decrease, depending on which effect is larger.

Aggregate economic conditions can also affect the schedules and hence the amount and duration of job search. In recessions the expected benefits of search decline because there are fewer unfilled vacancies and more job losers competing for the available jobs. The downward shift in the benefit schedule can be

---

EXHIBIT 25.1   MANDATORY NOTICE AND UNEMPLOYMENT

Employment protection laws take a variety of forms, including severance or redundancy pay, mandatory notice, and requirements to establish adjustment programs to assist displaced workers. In Canada, mandatory notice is the primary legal obligation on employers who wish to reduce their work force for "economic" reasons. All jurisdictions (federal, provincial, and the territories) have some form of advance notice requirement for individual layoffs or dismissals. Several provinces have additional requirements for mass layoffs. In addition, under Canadian common law, "reasonable" notice must be provided to employees dismissed for economic reasons.

Does mandatory notice reduce unemployment? By giving employees time to search prior to the termination of their existing job, some may be able to move from one job to another, without an intervening period of nonemployment, thus reducing the incidence of unemployment. Others may be able to begin the search process during the notice period, and to thereby reduce the duration of unemployment.

The impact of mandatory notice on unemployment has been examined by Jones and Kuhn (1992) using a sample of workers laid off due to plant closures in Ontario. They find that even small amounts of notice are quite helpful in reducing the number of workers who experience some unemployment following displacement (the incidence of unemployment). For example, giving notice of less than one month reduces the proportion of workers in a shutdown who experience unemployment from 92 percent to 76 percent. However, Jones and Kuhn also find that there are few, if any, additional gains from providing notice of more than one month, and they even find some evidence that notice of more than six months can be harmful to workers.

Unfortunately, there appears to be little scope for using advance notice to reduce the long-term unemployment that results from many mass layoffs. In particular, Jones and Kuhn find that, no matter how much notice is given, about 30 percent of workers remain unemployed one year after shutdown. The reason appears to be that, except in small amounts, pre-displacement search is significantly less effective than post-displacement search in obtaining reemployment, a conclusion also reached in several U.S. studies of displaced workers (e.g., Swaim and Podgursky, 1990; Ruhm, 1991).

Source: Stephen R.G. Jones and Peter Kuhn. "Mandatory Notice and Unemployment." McMaster University, Department of Economics, Working Paper No. 92-15, July 1992.

expected to result in fewer employed workers quitting their jobs to search for a new job, which explains why the quit rate varies pro-cyclically. The tendency for quits to decline in recessions is evident from Table 24.2 which shows that the proportion of job leavers fell sharply during the 1981–82 recession and subsequently gradually increased as the economy recovered. The decline in the benefits of search will also result in less entry into and more withdrawal from the labour force, as fewer of the unemployed find search worthwhile. This "discouraged worker" phenomenon was discussed previously in the context of hidden unemployment. Because of this phenomenon the labour force participation rate also varies pro-cyclically. As a further consequence, in a recession the rise in unemployment will be less than the decline in employment; the opposite occurs as the economy recovers from the recession.

The job search paradigm may also help explain why wages adjust slowly to excess supply or demand in the labour market. As noted above, the optimal search strategy involves choosing an acceptance wage and searching until a job offering this wage or better is found. Thus wages are not adjusted downward even if an acceptable job is not found quickly. Over time, unemployed searchers will revise downward their acceptance wages if they discover that their initial beliefs regarding the distribution of wage offers and the rate at which job offers are made were too optimistic, as would be the case if there were more labour supply or less demand than originally anticipated. However, because this learning process takes time, the adjustment of wages to excess demand or supply will occur less quickly than in the presence of complete information.

Empirical Evidence

A significant amount of job search takes place. Table 25.1 provides information on job search activity during 1991. In most years about 10 to 15 percent of the labour force is engaged in job search at any point in time. About two-thirds of searchers are unemployed and one-third employed. On average, three to five percent of employed workers are searching for a new job. As indicated in Table 25.1, a much larger proportion of those employed part-time are engaged in job search than is the case for those employed full-time. Groups that are more likely to engage in employed search are men, part-time workers, the highly educated and those in managerial, professional, and service occupations (Hasan and Gera, 1982).

Workers can engage in a variety of search activities ranging from informal discussions with friends and relatives to more formal activities such as contacting private or public employment agencies. In addition, varying degrees of success can be associated with each of these activities. Table 25.1 indicates the most common search activities used in Canada according to a recent Labour Force Survey. The most popular measure was to contact employers directly, followed by looking at ads and using a public employment agency. Most searchers used more than one method.

Table 25.1
SEARCH ACTIVITY OF THE EMPLOYED AND UNEMPLOYED, 1991

GROUP AND SEARCH ACTIVITY		NUMBER USING ACTIVITY (THOUSANDS)	PERCENT USING ACTIVITY[1]
Employed:	Full-time	389	3.8
	Part-time	217	10.7
Unemployed:	Did not search	90	6.4
	Searched for full-time work	1144	80.7
	Searched for part-time work	183	12.9
Search activity:	Contacted employers directly	923	65.1
	Used public employment agency	534	37.7
	Looked at ads	769	54.3
	Used other methods	487	34.4

Note:
1. For search activities, the sum of the methods does not equal the total because many individuals use more than one method.

Source: Statistics Canada, *Labour Force Annual Averages*. Ottawa: Statistics Canada, 1992. Reproduced with the permission of the Minister of Industry, Science and Technology, 1993.

Using an activity, however, does not mean that it will be successful. In his study of the search activity of the unemployed in Canada, Maki (1971) found that informal search activities, such as contacting friends and relatives and local employers, were more successful than formal activities such as using a private or public employment agency.

Hasan and Gera (1982) examine the search behaviour of both the employed and the unemployed in Canada. They conclude that, on average, employees who quit to search for a new job experience a wage increase while workers who are laid off experience a wage decrease. This finding is consistent with the theory of optimal search discussed above: job leavers would not quit to search unless they expected to be able to find a better job, whereas job losers need not anticipate being able to improve on their previous job even if they pursue an optimal search strategy.

In summary, labour market search is an important activity among the unemployed, some of the employed, and many employers. Most fundamentally, the economic analysis of job search helps explain the coexistence of unemployed workers and unfilled vacancies and the process through which these are matched. More specifically, search theory offers insights into several phenomena, including the duration of unemployment, the cyclical behaviour of quits and labour force participation, the sluggish adjustment of wages to changes in economic conditions, and the consequences of public policies such as unemployment insurance and the provision of labour market information.

Implicit Contracts

While search theory is concerned with unemployment that is associated with the process of matching job vacancies and unemployed workers, implicit contract theory deals with unemployment that may arise when firms and workers are already engaged in a continuing employment relationship. In particular, implicit contract theory seeks to explain phenomena such as rigid wages and the use of quantity adjustments (layoffs and rehires) rather than wage adjustments to respond to variations in product demand.

The basic elements of implicit contract theory are discussed in detail in Chapter 13. Here we summarize the key features of the theory and describe its implications for unemployment.

Implicit contract theory is based on the view that wage and employment behaviour reflects, in addition to payment for labour services provided, risk-sharing between employers and employees. The need and opportunity for risk-sharing arise because firms are believed to be less risk-averse than employees, and employees are unable to purchase insurance against human capital risks in private insurance markets. In addition, unlike those whose wealth consists mainly of financial capital, workers aren't able to hold a diversified portfolio of human capital investments (i.e., acquire a variety of skills or insure against the vagaries of the business cycle by acquiring skills that will be in demand whatever the state of the economy). Thus, employers may provide a natural source of risk-sharing for employment-related income risks. As a consequence, both parties can gain from an arrangement whereby workers receive a lower average wage than would occur without risk-sharing and firms smooth out to some extent the fluctuations that would otherwise occur in worker incomes.

Implicit contract theory deals with an environment in which firms face fluctuations in product demand. These variations in product demand lead to variations in the demand for labour. With market clearing, wages and employment would be low in poor states of product demand and high in good states of demand. These variations in wages and employment under market clearing imply that workers face the risk of fluctuations in their incomes. A central assumption of implicit contract theory is that prior to the state of demand being realized, firms and their employees enter into implicit arrangements regarding the future behaviour of wages and employment. As discussed in more detail in Chapter 13, these implicit arrangements reflect both production efficiency and risk-sharing.

The basic model of implicit contract theory assumes that workers are risk-averse, firms are risk-neutral, and both parties are fully informed about the fluctuations in product demand that may occur in the future. This basic model predicts that the firm will maintain a rigid real wage in the face of fluctuations in product demand, in contrast to the market clearing outcome in which wages would fall in poor states of demand and rise in good states. Fluctuations in employment, when they occur, take the form of layoffs in poor states of demand. Workers' incomes are less uncertain under these implicit contracts than under market clearing. Because the contract wage is less than the average wage that would occur with market clearing, firms' expected profits are higher than in the absence of implicit contacts. Thus both parties gain from these risk-sharing arrangements.

Note, however, that it will generally not be in the mutual interest of the two parties to fully stabilize workers' incomes, i.e., to have both fixed real wages and fixed employment. This implication arises because the optimal contract trades off production efficiency and risk-sharing considerations. On the basis of productive efficiency alone, workers should be laid off when the value of their marginal product is less than the wage. On the basis of risk-sharing alone, both real wages and employment should be stabilized. The optimal contract provides the best (Pareto-optimal) combination of these two competing considerations. Real wages are stabilized (i.e., the risk of wage fluctuations is eliminated) and firms employ more workers in poor states of demand than would be the case under market clearing (i.e., the risk of employment fluctuations is reduced, but not eliminated). Both these features of the optimal contract reduce production efficiency and thus the total amount of income to be shared between the employer and its employees. However, production efficiency is not ignored, which is why some workers are laid off in poor states of demand.

Several implications follow from this form of risk-sharing between the firm and its employees. In particular, implicit contract theory can account for real wage rigidity, the use of layoffs to respond to reductions in demand, and the existence of involuntary unemployment. The optimal contract provides for a constant real wage and reduction in employment in the poor state. Because workers would prefer to be employed and earning the contract wage, the reduction in employment takes the form of layoffs. Those workers laid off are involuntarily unemployed in an ex post sense: given that the weak demand state occurs, that they are selected for layoff, and that the contract wage exceeds their reservation wage, they would prefer to be employed. However, the unemployment may be considered voluntary in an ex ante sense because the workers chose a wage-employment contract with some risk of layoff, and would make the same choice again in identical circumstances, given the Pareto-optimal nature of the contractual arrangement. Thus the unemployment is involuntary in a rather limited sense.

Although implicit contract theory does provide a rigorous microeconomic explanation for wage rigidity, layoffs, and involuntary unemployment (in a restricted sense), the theory has been criticized for its inability to explain unemployment in excess of the amount that would be observed if wages adjusted each period to equate labour supply and demand (Akerlof and Miyazaki, 1980). Indeed, according to the basic implicit contract model discussed in this section, the number of workers involuntarily laid off in weak demand conditions is less than the number that would voluntarily withdraw from employment with market clearing. However, this implication of the model is closely related to the assumption that both parties observe the state of demand. When this symmetric information assumption is relaxed, optimal contracts may imply unemployment in excess of the amount that would occur with market clearing.

In many circumstances employers have more information than their employees on the true nature of the demand conditions they are facing. Thus it is not possible for the two parties to have wages and employment contingent on the state of demand, as was the case under full information, because workers cannot verify the outcome. Contracts need to be contingent on some variable which both parties observe. In these circumstances, implicit contracts will generally involve layoffs rather than wage reductions in weak demand conditions. Contracts which provide for wage reductions in poor demand conditions will not work because the firm has an incentive to claim demand is weak whatever the true state. That is so because paying a lower wage is not costly to the firm. However, layoffs do impose costs on the employer because output is lower with fewer employees and the firm runs the risk of losing employees and has to incur rehiring costs. Rigid wage contracts that involve layoffs in poor states and rehires in good states discourage the firm from bluffing about the true nature of its product market, forcing the firm to reveal the true state of demand by its choice of employment. Thus under asymmetric information there is an additional rationale for wage rigidity and layoff.

Efficiency Wages

Another explanation of wage rigidity and unemployment which has received considerable attention in recent years is the notion of "efficiency wages." While implicit contract theory emphasized the role of wages and employment in risk-sharing, efficiency wage theory focuses on the effect of wages on incentives and worker productivity. The central hypothesis is that firms may choose to pay wages above the market-clearing level in order to enhance worker productivity. Because higher wages also reduce labour demand, the consequence of this policy may be excess labour supply and unemployment.

The basic features of efficiency wage theory are described in Chapter 13, where its implications for labour market equilibrium and wage differentials are discussed. Here we focus on the theory's implications for unemployment.

The central assumption of efficiency wage theory is that firms may prefer to pay "above market" wages because doing so enhances worker productivity. This productivity-enhancing effect may arise because of improved morale, reduced turnover, selection of the most productive employees from applicant pools, or because being paid more than their alternative wage encourages work effort and reduces absenteeism and shirking.

For these reasons, the wage rate paid by the firm affects profits in two ways. Higher wages enhance productivity and thus output and total revenue. However, higher wages also raise labour costs and thus total costs. The employer will set the wage rate to maximize total profits. This optimal (from the perspective of the firm) or "efficiency" wage has several important properties which are illustrated in Figure 25.2. First, the efficiency wage (W^* in panel a of Figure 25.2) exceeds the competitive market wage W_0. Because of the productivity-enhancing effect of higher wages, the firm finds it profitable to pay an above market wage. In addition, the firm will not wish to lower the wage below W^* even in the presence of excess supply because doing so will reduce total revenue (due to lower worker productivity) more than it reduces total cost. Another property of the efficiency wage W^*, at least in the basic efficiency wage model, is that it is independent of the state of demand, so that fluctuations in demand (such as from D^* to D' in Figure 25.2(a)) lead to fluctuations in employment at a constant real wage.

If each firm sets its wage rate above the market-clearing level, labour demand will be less than labour supply, resulting in involuntary unemployment. Figure 25.2(a) illustrates the equilibrium when each firm's optimal or efficiency wage is W^*. The unemployed are willing to work at wage rates below the efficiency wage W^*. However, firms will not lower the wage rate because doing so would reduce the productivity of their existing work force. The failure of wages to adjust to excess supply is a consequence of firm rather than worker behaviour.

Unemployment will result if wages are set above their market-clearing level in each sector. However, as discussed in Chapter 13, if there exist one or more sectors in which firms will pay the lowest wage necessary to attract labour, the consequence may be a dual labour market rather than unemployment. This point is illustrated in Figure 25.2(b) for the case in which there are two sectors in the economy, each employing the same type of homogeneous labour. Sector A consists of those firms which prefer to pay above-market wages in order to enhance productivity while sector B consists of employers who pay the market wage and who will lower the wage in the presence of excess supply. In the absence of efficiency wage effects, the equilibrium will be at (W_0, N_0) in each sector with the wage rate equalized across the two sectors ($W_0^A = W_0^B$). However, because sector A firms set their wage at W_1^{A*}, employment declines to N_1^A. The resulting increased labour supply in sector B lowers wages and increases employment in that sector. In these circumstances, efficiency wages result in a dual labour market, with high- and low-wage sectors for homogeneous workers. Wages differ not because of differences in the human capital characteristics of the workers in each sector but because of differences in the incentives facing firms.

Figure 25.2

(a) Labour Market Equilibrium with Efficiency Wages

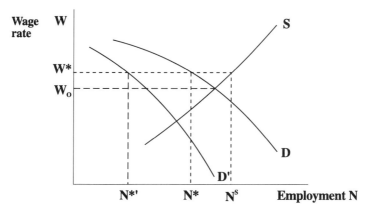

(b) Efficiency Wages and Dual Labour Markets

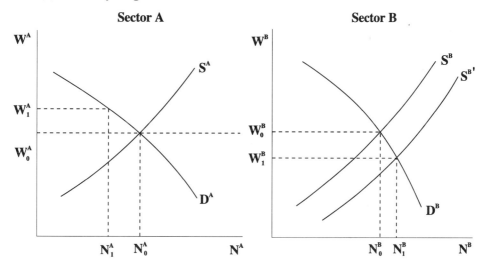

Efficiency wages may cause unemployment even when there are firms — such as those in sector B in Figure 25.2(b) — which have no incentive to pay above-market wages. Some workers may search or wait for jobs in the high-wage sector rather than accept employment at low wages. This outcome is more likely the larger is the wage gap between the sectors, the greater is the turnover (and thus the probability of a job opening up) in the high-wage sector, and the more productive is unemployed search compared to employed search. Minimum wage laws may prevent the downward wage adjustment needed to absorb the additional labour supply created by efficiency wages in the high-wage sectors. In this case unemployment results from the interaction between efficiency wages and minimum-wage legislation.

Efficiency wage theory also implies that unemployment may act as a "worker discipline device" (Shapiro and Stiglitz, 1984). Employees are assumed to be motivated to work hard when they fear losing their job. Their work effort will thus be greatest when their wage is high relative to the next best alternative, and/or when unemployment is high. The greater the unemployment rate, the longer the expected search period facing a worker laid off and therefore the greater the cost of shirking on the present job.

Insider-Outsider Theory

In the presence of excess supply, labour market equilibrium is restored by a decline in the real wage, and vice versa for excess labour demand. Persistent unemployment suggests that this natural equilibrating mechanism is not operating, or not operating sufficiently quickly. For this reason, explanations of persistent unemployment typically focus on the wage determination process, and in particular on reasons why wages may not adjust to eliminate excess supply or demand. Both the implicit contract and efficiency wage theories provided explanations for real-wage rigidity and unemployment that are consistent with rational behaviour on the part of employers and employees. Insider-outsider theory is another explanation of wage inflexibility and unemployment, one that has received considerable attention, especially in the context of the persistence of high unemployment in several European countries. (References include Lindbeck and Snower, 1986, 1987, 1988; and Solow, 1985.)

The central theme of insider-outsider theory is that wage setting is determined by bargaining between the employer and its existing work force (the "insiders") with unemployed workers (the "outsiders") exerting little, if any, influence on the outcome. This view is based on the proposition that it is costly for the firm to replace all or some of its existing work force with new workers recruited from among the ranks of the unemployed. These costs give incumbent workers bargaining power which they can use to raise their wages, even in the presence of excess labour supply in the form of unemployed workers willing to work at lower wages.

Turnover costs are one potentially important source of insider power. As discussed in Chapter 10, there are often costs associated with recruiting and hiring new workers. There may also be significant costs associated with dismissal of existing employees (severance pay, requirements for advance notice). Firm-specific training and work experience is another factor giving insiders power over outsiders. As discussed in Chapter 18, in the presence of firm-specific human capital, the employer is not indifferent between an experienced incumbent and an otherwise identical individual without the specific training.

In these circumstances, the fact that there exist unemployed job seekers willing to work at lower wages may not be enough of an incentive for employers to attempt to replace some or all of the existing work force with new employees.

Another potential source of insider power stressed by Lindbeck and Snower (1988) is the ability of incumbent employees to not cooperate with or even to harass new hires, especially if new hires are replacing otherwise incumbent workers.

Analytically, the insider-outsider theory is similar to wage determination under collective bargaining, as discussed in Chapter 15. Through bargaining between the employer and incumbent "insiders," the wage will be set above the market-clearing level, thus generating less employment than would occur in the absence of insider power. This reduced employment in sectors in which workers possess insider power will generate unemployment in the labour market as a whole unless wages fall enough in other sectors (sectors in which workers do not have insider power, perhaps because turnover costs and firm-specific training are unimportant) to absorb the excess supply.

Insider-outsider theory has been used to explain the persistence of high unemployment, combined with substantial real-wage growth among employed workers, observed in many European countries since the 1970s. The persistence of unemployment, including the phenomenon of "hysteresis" — according to which the level of unemployment may drift upward or downward rather than tending to return to its natural or equilibrium level — is discussed further in Chapter 26.

Sectoral Shifts and Unemployment

The rapidity of structural adjustment (due to technological and economic change) in the economy may also affect the aggregate level of unemployment. Macroeconomists typically view aggregate fluctuations as arising from shocks to aggregate demand or supply. However, as emphasized by Lilien (1982), shifts in the sectoral composition of demand (by region, industry, or occupation) can raise the equilibrium level of unemployment, as time is required for labour and other resources to be reallocated to other sectors.

This sectoral shift hypothesis can be illustrated as follows. Consider two economies in which labour demand and labour supply are growing at equal average rates. If labour demand is growing at the same rate in each industry (or region or occupation) in economy A, but growing at above-average rates in some industries and below-average rates in other industries in economy B, the latter economy would have a higher natural unemployment rate due to the necessity for labour to be re-allocated from those industries which are growing slowly (perhaps even declining) to those industries which are growing rapidly. For this reason, Lilien (1982) includes the variance of employment growth across U.S. industries as an additional variable explaining movements in the equilibrium level of unemployment in that country. Lilien's model was estimated for Canada by Samson (1985) and Charette and Kaufman (1987). Both studies find a relationship between the variance of employment growth across industries and the unemployment rate.

Although periods of high dispersion of employment growth across industries are associated with periods of high unemployment, there is considerable disagreement as to whether this relationship confirms the structural shifts hypothesis. The problem arises because a standard macroeconomic model of the business cycle based on aggregate demand shocks is also likely to generate a positive relationship between the variance of employment growth across industries and unemployment (Abraham and Katz, 1986). In particular, some industries are more cyclically sensitive than others, so that shocks to aggregate demand will have a differential effect across industries, producing a rise in the dispersion of employment growth across industries at the same time as unemployment increases. Thus, it is difficult to separate cause and effect; in particular, it is difficult to determine whether sectoral shifts exert an independent influence on unemployment, or whether aggregate shocks cause both a rise in unemployment and a cyclically induced increase in the dispersion of employment growth across industries. In a Canadian study which attempts to separate out the cyclical from noncyclical components of changes in sectoral composition of demand, Neelin (1987) finds that noncyclical shifts in the variance in employment growth across industries do not have an independent effect on unemployment. Rather, the causation runs in the reverse direction; economy-wide shocks which influence the aggregate unemployment rate also cause shifts in the industrial composition of employment. However, Neelin does find evidence supporting the sectoral shifts hypothesis for changes in the regional composition of labour demand. Evidence reported in Gera (1991, chapter 4) also suggests that increasing regional imbalances played an important role in the rise in unemployment in Canada during the 1970s and 1980s.

Osberg (1991) analyzes micro data on the interindustry mobility of Canadian workers during the 1980–86 period. He finds that interindustry mobility falls sharply during economic downturns, in contrast to the implications of the sectoral shifts hypothesis. Murphy and Topel (1987) report similar findings for the U.S. The tendency for recessions to "chill" the process of intersectoral labour mobility may contribute to the slow recovery from economic downturns.

Structural unemployment will increase if there is a mismatch between industries and regions experiencing growth in labour demand and those experiencing growth in labour supply. However, there are incentives for employers to adjust their hiring to the available labour supply, as well as for workers to acquire characteristics in demand by employers. Beach and Kaliski (1986) examine changes in the demographic and industry characteristics of employment in detailed Canadian industries during the 1966–83 period, when most of the growth in labour supply involved entry by youths and women. They find that changes in industry employment structure generally accommodated the changes on the supply side; most industries increased their proportions of women and young workers employed, and industries which employed women intensively grew more rapidly than average. Thus the interaction between changes in labour demand and those in the composition of labour supply facilitated employment growth and prevented the increase in structural unemployment that would occur in the absence of adjustments on both the demand and supply sides of the labour market.

Although the evidence supporting the sectoral shifts hypothesis is limited, the message that some economic shocks may have more severe consequences for unemployment than others (holding constant the overall magnitude of the shock) is an important one. The 1981–82 recession had a much larger impact in the Western provinces (especially Alberta and British Columbia) than in central Canada, whereas the 1990–92 downturn adversely affected Ontario more than it did the Western provinces. Because of the additional need for interregional re-allocation of labour resources, adjustment to these shocks is more difficult than adjustment to shocks that have an even impact across regions.

Other features of economic shocks may also contribute to the ease or difficulty of adjustment. During the 1980s and 1990s, a considerable amount of economic dislocation involved losses in high-wage, manufacturing jobs and growth in lower wage service sector employment. In addition, there was substantial displacement among older workers with substantial job tenure. These features of displacement appear to contribute to a slow process of adjustment and high levels of unemployment. Osberg (1991) finds a very strong relationship between job tenure and interindustry labour mobility, consistent with the hypothesis that workers accumulate specific human capital on the job as discussed in Chapter 18. Workers displaced from high-wage jobs are also generally found to have longer periods of nonemployment than comparable workers displaced from low-wage jobs (e.g., Kruse, 1987).

The extent to which unemployment is associated with sectoral shifts and the process of labour re-allocation is important for policy purposes. Policies which facilitate adjustment to change (training, mobility assistance) are more likely to be useful in periods characterized by substantial sectoral adjustment, whereas macroeconomic stabilization policies are more likely to be useful in response to aggregate shocks.

UNEMPLOYMENT INSURANCE AND UNEMPLOYMENT

The relationship between unemployment insurance and unemployment has been the subject of considerable theoretical and empirical research. Unemployment insurance is intended to provide workers with protection against the risk of income loss due to unemployment. However, the unemployment insurance program may also affect the incidence and duration of unemployment by altering the incentives facing workers and firms. In this section we examine the influence of unemployment insurance on labour supply, job search, layoffs, seasonal unemployment, and inter-regional mobility. A brief description of the main features of Canada's unemployment insurance system is also provided.

Because economic circumstances continually change, risk and uncertainty are often present. As discussed previously in the context of implicit contract theory, labour market risks are especially significant because workers are generally unable to diversify their human capital wealth. In addition, comprehensive private insurance markets which would enable most workers to purchase insurance against the risk of unemployment (and possibly other sources of loss of labour market income) have generally failed to emerge despite the demand that evidently exists for such insurance. According to the economic theory of risk and insurance, this absence of private insurance markets is probably due to moral hazard and adverse selection. In the context of unemployment insurance, moral hazard implies that individuals with insurance are more likely to become or remain unemployed. Adverse selecton in this context implies that the purchasers of insurance will be those who are most likely to become unemployed. Relative to a situation without moral hazard and adverse selection, these two effects reduce the profitability of selling insurance, and may result in a situation in which the sale of insurance is not profitable at any price.

The experience with Career Guard, an insurance plan to protect executives who might be fired, illustrates these effects. (See "Insurance Against Being Fired: A Plan that Just Didn't Fly" *Financial Post*, May 29, 1983, p. 1). Although this insurance policy didn't cover executives fired within six months of purchasing insurance, the entrepreneurs who started Career Guard nonetheless discovered that a very high proportion of those who purchased insurance were dismissed by their employers subsequent to the six-month qualifying period. It appears that Career Guard failed primarily because of adverse selection — those executives who knew they were likely to be dismissed were the main purchasers of insurance, and the insurer couldn't distinguish high-risk from low-risk customers. Moral hazard may also have played a role.

Moral hazard and adverse selection are generally present in any insurance situation. Whether they prevent the emergence of private insurance markets depends on their magnitudes. For example, in life insurance, moral hazard is not a serious problem because few individuals will take their own lives in order to collect insurance. Insurance companies often reduce the already small amount of moral hazard by not paying insurance in the event of suicides or murder by a beneficiary. Adverse selection is minimized by such means as not selling insurance to some high-risk groups (e.g., sky divers), charging differential fees to others (e.g., smokers), and requiring medical examinations for certain individuals seeking insurance.

In the absence of private sector provision, governments in many countries have introduced unemployment insurance (UI) as part of social policy. Canada's unemployment insurance system was established in 1940, following the recommendations of the Royal Commission on Dominion-Provincial Relations and the resolution, through a constitutional amendment, of the difficulties which had resulted in an earlier attempt to introduce UI being declared outside the legislative authority of the federal government. The UI fund is financed by premiums collected from employers and employees. Coverage is compulsory for those groups that come under the Act, a feature which significantly reduces the effects of adverse selection. However, as the empirical evidence reviewed below indicates, moral hazard effects do occur.

Canada's unemployment insurance system has evolved significantly over time. Between 1940 and 1971 there were gradual and modest changes in the eligibility, coverage, and benefit and financing provisions of the UI Act. Dramatic changes to these key features of the act were made in 1971/72, including a substantial expansion in coverage, an increase in the benefit rate (benefits as a proportion of previous earnings), a reduction in the minimum period of employment required to qualify for benefits, an increase in the maximum benefit period, and the introduction of extended benefits in regions with high unemployment. With these changes, UI covered a much larger proportion of the labour force (about 90 percent) and became considerably more "generous." Making UI benefits taxable offset this increased generosity to some degree. The 1971/72 changes were also partially reversed by revisions made in 1978/79. Nonetheless, the UI system of today contrasts sharply with that of pre-1971 in terms of coverage and the generosity of its provisions. These dramatic changes have provided a rather unique "social experiment" for studying the impact of unemployment insurance on labour market behaviour. Accordingly, many of the Canadian empirical studies contrast pre- and post-1971/72 behaviour or pre- and post-1978/79 behaviour. Of course, the 1971/72 and 1978/79 changes did not constitute a controlled experiment. As is so frequently the case in empirical research, controlling for other factors which may have affected behaviour represents a major challenge.

On the basis of economic theory, changes in the provisions of the unemployment insurance system can be expected to affect labour force behaviour in several ways. Indeed, examining the impact of such changes provides a useful test of the theories of unemployment discussed previously in this chapter and of the analysis of the incentive effects of alternative income maintenance schemes discussed in Chapter 4. The following effects are discussed in turn: incidence and duration of search unemployment, temporary layoffs, employment instability and seasonal unemployment, labour force participation and labour supply, and interregional labour market mobility.

Analysis of the relationship between unemployment and unemployment insurance is complicated by the fact that the UI program has several key parameters, including the benefit rate, the minimum employment period to qualify for benefits, the maximum duration of benefits, the relation between weeks of previous employment and weeks of potential benefits, and the coverage of the labour force. Much of the empirical literature focuses on the consequences of changes in the benefit rate, and our discussion reflects this emphasis. However, other "program parameters" are also important, and we will discuss evidence regarding their impacts when this evidence is available.

Unemployment insurance can affect the incidence and duration of search unemployment by altering the costs and benefits of job search. Several cases need to be considered, depending on whether the individual is (1) employed, (2) unemployed and eligible for UI benefits, and (3) not eligible for benefits. For the employed, an increase in the benefit rate makes unemployed search more attractive relative to employed search; as a consequence, the incidence of unemployment is predicted to rise. Evidence appears to support this prediction. Kaliski (1985) notes that the ratio of employed to unemployed search rose sharply following the introduction in 1978/79 of a reduced benefit rate and tighter qualification requirements.

For the unemployed who are eligible for benefits (because of a previous spell of employment), an increase in the benefit rate lowers the marginal cost of search. According to the theory of optimal search, the expected or average duration of job search will increase. This prediction has been extensively tested in the U.S., the U.K., and to a lesser extent in Canada. Cousineau (1985) summarizes various Canadian and U.S. studies and Atkinson et al. (1984) and Atkinson and Micklewright (1991) critically evaluate U.S. and U.K evidence. Early Canadian studies using aggregate data concluded that the 1971/72 changes to the UI Act — which increased the benefit rate from 43 to 67 percent, among other changes — raised the average duration of unemployment by 1-1/2 to 2 weeks (Green and Cousineau, 1976; Rea, 1977; Maki, 1977; Lazar, 1978). This increase of approximately 20 percent in duration corresponds to an increase in the unemployment rate of one to one and one-half percentage points. More recent studies in Canada and the U.S. employ micro data which allow for more careful testing of UI effects. Using gross flow data, Beach and Kaliski (1983) find that the 1978/79 revisions — which lowered the benefit rate from 67 to 60 percent and tightened qualification requirements — resulted in an increase in the flow from unemployment to employment and a reduced duration of unemployment for all age-sex groups. Studies by Glenday and Alam (1982) and Ham and Rea (1987), using micro data on the employment and unemployment experience of individuals, find significant effects of UI entitlements on the duration of job search.

Changes in the benefit rate may also affect the search behaviour of those who are not eligible for UI. Such individuals may be entering the labour force for the first time, they may be re-entering after an extended absence from the labour force, or they may have exhausted their benefits. In these circumstances it may be rational to accept employment quickly, even temporary work, in order to subsequently qualify for UI benefits. In effect, a job has two components: the income received directly and a "ticket" entitling the worker to benefits in the event of unemployment. Consequently, for individuals ineligible for benefits, an increase in the benefit rate does not affect the costs of search but lowers the marginal benefit of search, thus reducing optimal search duration.

There is no direct evidence on this effect in Canada, but it is likely to be small. Consider, for example, a 10 percent increase in the benefit rate. For individuals who qualify for UI benefits, this change has a significant effect on the costs of job search. However, for those not eligible for benefits, the *additional* incentive to accept a job quickly is not large. Furthermore, this response, to the extent that it occurs, is confined to new entrants, some re-entrants, and UI exhaustees, a small fraction of the unemployed.

In summary, unemployment insurance affects job search behaviour in several ways. For workers who qualify for UI, a more generous benefit structure lowers the cost of job search, thus raising average search duration, and makes unemployed search more attractive relative to employed search, thus increasing the incidence of unemployment. Empirical evidence generally supports these predictions of the theory of optimal search behaviour. The impact on search duration has been most extensively investigated. Although most studies find that UI benefits do affect the duration of job search, the studies are not always robust to changes in specification (Atkinson et al. 1984; Atkinson and Micklewright, 1991). Possibly offsetting these effects to some extent, a more generous benefit structure may reduce search duration for those currently ineligible for UI. This latter effect is likely to be small, in which case the overall impact of a more generous benefit structure will be to increase search duration. Empirical studies of the 1971/72 UI changes conclude that the overall duration of unemployment did increase significantly.

The impact of UI on layoffs can be analyzed using implicit contract theory. As discussed previously in Chapter 13 and in this chapter, optimal risk-sharing arrangements may provide for layoffs in adverse product market conditions when the value of the marginal product of labour falls below the workers' reservation wage. An increase in the benefit rate raises the reservation wage, thus increasing the number of states in which layoffs will occur and increasing the number of workers laid off in each adverse state. These predictions can be illustrated by shifting up the labour supply curve S in Figure 13.2(b).

The theory of compensating wage differentials discussed in Chapter 17 is also relevant to the analysis of layoffs. Although all firms and industries experience fluctuations in demand due to cyclical, seasonal, and other factors, these variations are much larger for some employers than others. Firms can adjust to these disturbances in various ways: through inventories, by altering the time lag between orders and shipments, by diversifying product lines, by saving work for slack periods, by adjusting wages and/or hours of work, and by layoffs and rehires. Firms will choose the least-cost method of adjustment. Unemployment insurance lowers the cost of adjusting through layoffs relative to other methods because, in the absence of UI, employers who rely extensively on layoffs will have to pay a compensating wage to attract employees. In this way a higher benefit rate may contribute to larger variations in employment in response to cyclical and seasonal fluctuations in demand.

The financing of the UI program is also relevant. Under "experience rating," employers' UI premiums are related to the amount of benefit payments to their employees. Firms which rely heavily on layoffs will pay a penalty in the form of higher UI premiums, thus providing an incentive to utilize other forms of adjustment. Canada's unemployment insurance system is not experience-rated, a feature which has been the subject of considerable policy discussion (Kesselman, 1983; Riddell, 1985). As a consequence, industries with stable employment patterns cross-subsidize those with unstable employment. In the United States, various states have experience rating in their unemployment insurance programs.

The relationship between UI and employment instability has been investigated in a number of empirical studies. Much of this research has been carried out in the United States, utilizing differences that exist across states in the degree to which UI is experience-rated. Studies by Feldstein (1978), Topel (1983), and Saffer (1982, 1983) indicate that incomplete experience rating has a significant impact on the incidence of layoffs; i.e., the lower the degree of experience rating the greater the use of layoffs, ceteris paribus. In Canada, Kaliski (1976) found that for the majority of industries and provinces, the trend toward reduced seasonality in employment moderated, or even reversed, after 1971. The increase in seasonal fluctuations in employment was particularly large in construction, one of the industries most heavily subsidized by UI. Glenday and Alam (1982) conclude that the regional extended benefits (longer entitlement periods in regions with high unemployment) contribute to the amount of seasonal unemployment and short-term employment instability in high unemployment regions.

The choice between hours reductions and layoffs has received special policy attention. As discussed in Reid (1985) and Riddell (1985), UI may bias employers and employees toward the use of layoffs as opposed to reductions in hours of work. This is so because workers on layoff qualify for UI benefits while those whose unemployment takes the form of reduced hours do not. This bias can be offset in one or two ways: (1) have UI premiums experience-rated so that firms pay a penalty for adjustments in the form of layoffs or (2) introduce a UI-assisted worksharing program. Canada has followed the latter route. As described in Chapter 4 and Reid (1985) the Canadian program basically provides UI benefits of two-thirds of each employee's lost wages for voluntarily giving up an average of one day of work per week so that other workers in the firm would not have to be laid off.

Unemployment insurance also affects employment and unemployment through its impact on labour force participation. This aspect was analyzed in

Chapter 4 using the income-leisure choice model, and the reader may wish to review that discussion at this time. The analysis applies to situations in which employees can adjust their weeks of employment and nonemployment in response to the incentives inherent in the UI system. The many individuals for whom a job entails being employed throughout the year do not fit in this category. The relevant group consists of those who work less than a full year either because the worker quits or is laid off after a certain period of employment or because the job itself is short-term in nature, as is the case in much seasonal work.

Our earlier analysis showed that more generous UI will decrease weeks worked by those who, prior to the change, worked more than the minimum number of weeks required to qualify for benefits, whereas more generous UI will increase weeks of employment for those who previously did not qualify. Most of the latter group would have been out of the labour force prior to the change; for these individuals, higher UI benefits make labour force participation sufficiently attractive to obtain at least enough employment to qualify for benefits. (In Chapter 4 we compared equilibria with and without unemployment insurance. The analysis of equilibria with different UI benefit levels is very similar.)

These two responses have offsetting effects on employment, reducing weeks worked by those with relatively strong labour force attachment and increasing weeks worked by those with little or no labour force attachment. Total employment may therefore increase or decrease. However, the impact on labour force participation and unemployment is unambiguous. Labour force participation is predicted to rise because of the entry by those who now wish to obtain enough work to qualify for UI. Those who previously qualified for benefits do not exit from the labour force, although they do work fewer weeks. Measured unemployment increases for both groups: those who previously qualified for UI because they now spend less time employed and more time unemployed, while those previously not eligible spend more time employed and unemployed and less "not in the labour force."

These predictions of the income-leisure choice model refer to desired combinations of employment and nonemployment given the individual's preferences and constraints. Of course, not all workers can achieve their desired allocation of time to work and nonmarket activity, especially in the short run. However, as discussed in Chapter 17, firms have an incentive to offer jobs with more desirable and fewer undesirable characteristics because doing so reduces the compensating wage that must be paid in equilibrium. Thus if short-term employment is desirable for some workers, jobs with these features can be expected to emerge. In this way, changes in the UI system may affect not only employee behaviour but also the structure of labour demand.

Glenday and Alam (1982) conclude that UI benefits contribute to the amount of employment instability; in particular, the regional differentiation of benefits reinforces the short-term and seasonal nature of much of the employment in high unemployment regions.

A variety of Canadian empirical evidence is available on these labour supply effects. Early studies by Green and Cousineau (1976), Rea (1977), and Sharir and Kuch (1978) found that the 1971–72 changes to the UI program increased aggregate labour force participation, especially among groups with lower average rates of labour force attachment. More recently, Phipps (1990, 1991a, 1991b) used the income-leisure choice framework discussed in Part 1 of this book (in particular, in Chapter 4) to analyze the impacts of UI parameters on labour force participation and employment. An important implication of Phipps' research is that changes in the UI program designed to increase work incentives (i.e., to increase labour supply) may have little effect if there are constraints on the demand side of the labour market (e.g., insufficient jobs in depressed regions or in economic downturns).

Green and Riddell (1993) examine the labour supply effects of UI coverage of older workers, a group generally considered to have a low degree of labour force attachment. In particular, they study the impact of a 1976 change in UI regulations which disenfranchised workers between 65 and 70 years of age, who were covered by the UI program prior to this change. They find that a large proportion of these individuals withdrew from the labour force on the removal of UI eligibility.

Corak (1992) uses UI administrative data covering the period 1971–90 to examine patterns of employment and nonemployment among participants in the UI program. He finds a high degree of repeat use of UI. For example, during the latter part of the 1980s, about 80 percent of the UI claimants in any year had previously received UI, with 40 to 50 percent having experienced five or more previous claims. The extent to which this degree of repeat use is due to the structure of labour demand (for example, the importance of seasonal work) or due to individuals adjusting their labour supply to the parameters of the UI program is an important question.

The regional extended benefit structure brought about by the changes made to the UI program in the late 1970s provides a relatively strong incentive for individuals in high unemployment regions to work at least 10 weeks, and thereby qualify for up to 42 weeks of benefits. In their study of Canada–U.S. unemployment (see Exhibit 24.1), Card and Riddell (1993) find evidence that during the 1980s Canadians increasingly adjusted their labour supply to the parameters of the UI program. In particular, for both males and females the distributions of annual weeks of employment show "spikes" at 10 to 14 weeks (the minimum weeks required to qualify for UI), and the magnitudes of these spikes increased during the 1980s. As noted above, such behaviour could arise both because some individuals who would otherwise (i.e., in the absence of UI eligibility) not participate in the labour force work enough weeks to qualify for UI and because some individuals who would otherwise work more weeks reduce their labour supply.

Unemployment insurance may also affect the inter-regional mobility of the labour force. The regional extended benefits, in particular, appear to retard interregional mobility by providing benefits for longer periods in regions with high unemployment (Winer and Gauthier, 1982; Vanderkamp, 1986).

In summary, the unemployment insurance system has numerous effects on labour force behaviour, including the incidence and duration of search unemployment, layoffs, annual patterns of employment and unemployment, labour force participation, and interregional mobility. A substantial amount of empirical research has been devoted to estimating the size of these effects. Although there are some offsetting influences, the overall impact of a more generous UI benefit structure is to increase unemployment. This does not imply that the 1971–72 changes were undesirable or that subsequent changes which to some extent "tightened up" the UI program were desirable; it simply points out that the changes affected our aggregate unemployment rate and that this should be considered in interpreting this statistic. More generally, the tradeoffs inherent in the unemployment insurance system need to be recognized in the design of UI financing and benefits. The more generous the benefit structure, the greater the insurance value of the program but also the larger the adverse incentive effects and the amount of induced unemployment. Optimal UI design must strike a balance between these social costs and benefits.

QUESTIONS

1. Indicate the main types of unemployment, their causes, and the usual policies discussed to curb each type of unemployment.

2. Frictional unemployment is optimal. Discuss.

3. In the late 1950s and early 1960s there was considerable debate over the relative importance of structural versus demand deficient unemployment. Much of the debate waned during the relatively full employment period of the mid- and late-1960s. What does this seem to imply about the existence of structural unemployment?

4. There is some renewed interest in the issue of structural unemployment, associated with the demographic changes in the labour force, and with the high unemployment that has occurred even during periods of inflation. Discuss why this may be the case.

5. Indicate some basic differences between frictional and structural unemployment. If possible, provide examples of when it may be unclear as to whether to categorize a worker as frictionally or structurally unemployed.

6. All unemployment can be categorized as frictional, structural, or demand-deficient. Discuss.

7. Discuss the various components of the cost and benefits of job search for employees. How might the availability of unemployment insurance alter those components? Discuss other factors that may have had important effects on the costs and benefits, and hence duration, of job search.

8. Discuss the costs and benefits of job search for younger workers and females. Relate these factors to their unemployment duration.

9. In recent years, private employment agencies have taken over much of the placement function of the public employment agency. Discuss the pros and cons of this phenomenon.

10. Discuss the implications of job search theory for the following phenomena: unemployment duration, the cyclical behaviour of quits, and the phenomenon of wage rigidity over the business cycle

11. Discuss the implications of implicit contract theory for the following phenomena: wage rigidity over the business cycle, labour hoarding, and layoffs as opposed to wage reductions.

12. If workers are risk-averse, why might they enter into contractual arrangements that could involve the risk of layoffs as opposed to minor wage adjustments?

13. Why don't we observe private insurance companies selling unemployment insurance? How could employers provide such insurance?

14. Discuss the distinction between voluntary and involuntary unemployment.

15. Discuss the importance of asymmetric information in implicit contract theory.

16. Compare and contrast the implications of efficiency wages and insider-outsider models for unemployment.

17. Discuss the various design features (policy parameters) of unemployment insurance that can affect the benefits from such insurance and hence the behaviour of recipients.

18. Discuss the various elements of labour market behaviour that can be affected by unemployment insurance.

19. Discuss the relationship between compensating wages for the risk of unemployment and unemployment insurance.

20. Discuss the pros and cons of experience rating in unemployment insurance.

REFERENCES AND FURTHER READINGS

Abraham, K.G. Structural/frictional vs. deficient demand unemployment: some new evidence. *AER* 73 (September 1983) 708–724.

Abraham, KG. and L.F. Katz. Cyclical unemployment: sectoral shifts or aggregate disturbances? *JPE* 94 (1986) 507–522.

Addison, J.T. and P. Portugal. Job displacement, relative wage changes, and duration of unemployment. *JOLE* (July 1989) 281–302.

Akerlof, G. and H. Miyazaki. The implicit contract theory of unemployment meets the wage bill argument. *R.E. Studies* 47 (January 1980) 321–338.

Akerlof, G.A. and J.L. Yellen (eds.). *Efficiency Wage Models of the Labour Market.* Cambridge: Cambridge University Press, 1986.

Albrecht, J.W. and B. Axell. An equilibrium model of search unemployment. *JPE* 92 (1984) 824–840.

Ashenfelter, O. and J. Ham. Education, unemployment, and earnings. *JPE* 87 (October 1979) S99–S116.

Ashenfelter, O. and D. Card. Unemployment in Canada and the U.S. *Economica* 53 (1986) 171–196.

Atkinson, A., J. Gomulka, J. Micklewright and N. Rau. Unemployment benefit, duration and incentives in Britain: how robust is the evidence? *JPubEc* 23 (February/ March 1984) 3–26.

Azariadis, C. Implicit contracts and underemployment equilibria. *JPE* 83 (December 1975) 1183–1202.

Azariadis, C. On the incidence of unemployment. *R.E. Studies* 43 (February 1976) 115–126.

Azariadis, C. Implicit contracts and related topics: a survey. *The Economics of the Labour Market*, Z. Hornstein et al. (eds.). London: HMSO, 1979, 221–248.

Azariadis, C. and J.E. Stiglitz. Implicit contracts and fixed price equilibria. *QJE* 98 (Supplement 1983) 1–22.

Baker, G.M. and P.K. Trivedi. Estimation of unemployment duration from grouped data: a comparative study. *JOLE* 3 (April 1985) 153–174.

Baily, M.N. Wages and employment under uncertain demand. *R.E. Studies* 41 (January 1974) 37–50.

Ball, L. Insiders and outsiders: a review essay. *Journal of Monetary Economics* 26 (1990) 459–469.

Barnes, W. Job search models, the duration of unemployment and the asking wage: some empirical evidence. *JHR* 10 (Spring 1975) 230–240.

Barron, J., R. McAfee and P. Speaker. Unemployment insurance and the entitlement effect: a tax incidence approach. *IER* 27 (February 1986) 175–186.

Barron, J. and S. McCafferty. Job search, labour supply and the quit decision. *AER* 67 (September 1977) 683–691.

Barron, J. and W. Mellow. Unemployment insurance: the recipients and its impact. *SEJ* 47 (January 1981) 606–616.

Beach, C.M. and S.F. Kaliski. On the design of unemployment insurance: the impact of the 1979 amendments. *CPP* 9 (June 1983) 164–173.

Beach, C.M. and S.F. Kaliski. Structural unemployment: demographic change or industrial structure? *CPP* 12 (June 1986) 356–367.

Bean, C.R., et al. The rise in unemployment: a multi-country study. *Economica* 53 (1986) 1–22.

Beaudry, P. and J. DiNardo. The effect of implicit contracts on the movement of wages over the business cycle: evidence from micro data. *JPE* 99 (August 1991) 665–688.

Benjamin, D. and L. Kochin. Searching for an explanation of unemployment in inter-war Britain. *JPE* 87 (June 1979) 441–478. Also comments 90 (April 1982) 369–436.

Bester, H. Incentive-compatible long-term contracts and job rationing. *JOLE* 7 (April 1989) 238–255.

Betcherman, G. Labour market imbalances in Canada: 1966–1983. *RI/IR* 41 (1986) 802–815.

Blanchard, O. and P. Diamond. The Beveridge curve. *BPEA* (1989) 1–60.

Blank, R. and D. Card. Recent trends in insured and uninsured unemployment: is there an explanation? *QJE* 107 (1991) 1157–1189.

Blau, D, and P. Robins. Job search, wage offers, and unemployment insurance. *JPubEc* 29 (March 1986) 173–198.

Blau, D. and P. Robins. Job search outcomes for the employed and unemployed. *JPE* 98 (June 1990) 637–655.

Broadway, R. and A. Oswald. Unemployment insurance and redistributive taxation. *JPubEc* 20 (March 1983) 193–210.

Brechling, F. The incentive effects of the U.S. unemployment insurance tax. *Research in Labor Economics* R. Ehrenberg (ed.) Greenwich, Conn.: JAI Press, 1977.

Brown, E. and H. Kaufold. Human capital accumulation and the optimal level of unemployment insurance provision. *JOLE* 6 (October 1988) 493–514.

Burdett, K. and B. Hool. Layoffs, wages and unemployment insurance. *JPubEc* 21 (August 1983) 325–358.

Burdett, K. and J.I. Ondrich. How changes in labour demand affect unemployed workers. *JOLE* 3 (January 1985) 1–10.

Burgess, P.L. and S.A. Low. Pre-unemployment job search and advance job loss notice. *JOLE* 10 (July 1992) 258–287.

Carmichael, H.L. Efficiency wage models of unemployment: one view. *EI* 28 (1990) 269–295.

Charette, M.F. and B. Kaufman. Short-run variation in the natural rate of unemployment. *Journal of Macroeconomics* 9 (1987) 417–427.

Card, D. and W.C. Riddell. A comparative analysis of unemployment in Canada and the United States. *Small Differences That Matter: Labor Markets and Income Maintenance in Canada and the United States*, D. Card and R.B. Freeman (eds.). Chicago: University of Chicago Press, 1993.

Chesher, A. and T. Lancaster. The estimation of models of labour market behaviour. *R.E. Studies* 50 (October 1983) 609–624.

Corak, M. Repeat users of the unemployment insurance program. *Canadian Economic Observer* (January 1992) 3.1–3.25.

Cousineau, J.M. Unemployment insurance and labour market adjustments. *Income Distribution and Economic Security in Canada*, F. Vaillancourt (ed.). Toronto: University of Toronto Press, 1985, 187–214.

Cousineau, J. and C. Green. Structural unemployment in Canada: 1971–1974. Did it worsen? *RI/IR* 33 (No. 2, 1978) 175–192.

Deere, D.R. Unemployment insurance and employment. *JOLE* 9 (October 1991) 307–324.

Diamond, P. Mobility costs, frictional unemployment and efficiency. *JPE* 89 (August 1981) 798–812.

Donner, A. and F. Lazar. An econometric study of segmented labour markets and the structure of unemployment: the Canadian experience. *IER* 14 (June 1973) 312–327.

Eaton, C. and P. Neher. Unemployment, underemployment and optimal job search. *JPE* 83 (April 1975) 355–376.

Eaton, C. and W. White. The economy of high wages. *Economica* 50 (May 1983) 175–182.

Ehrenberg, R. and R. Oaxaca. Unemployment insurance, duration of unemployment and subsequent wage gain. *AER* 66 (December 1976) 754–766.

Feldstein, M. The economics of the new unemployment. *Public Interest* 33 (Fall 1973) 3–42.

Feldstein, M. Unemployment compensation: adverse incentives and distributional anomalies. *National Tax Journal* 37 (June 1974) 231–244.

Feldstein, M. The importance of temporary layoffs. *BPEA* (No. 3, 1975) 725–745.

Feldstein, M. Temporary layoffs in the theory of unemployment. *JPE* 84 (October 1976) 937–958.

Feldstein, M. The effect of unemployment insurance on temporary layoff unemployment. *AER* 68 (December 1978) 834–846.

Feldstein, M. and J. Poterba. Unemployment insurance and reservation wages. *JPubEc* 23 (February/March 1984) 141–168.

Follman, D.A., M.S. Goldberg and L. May. Personal characteristics, unemployment insurance, and the duration of unemployment. *Journal of Econometrics* 45 (1990) 351–366.

Gamber, E.N. Long-term risk-sharing wage contracts in an economy subject to permanent and temporary shocks. *JOLE* 6 (January 1988) 83–99.

Gibbons, R. and L.F. Katz. Layoffs and lemons. *JOLE* (October 1991) 351–380.

Glenday, G. and J. Alam. The labour market experience of individuals: unemployment insurance and regional effects. Mimeo. Toronto: York University, 1982.

Gordon, D. A neoclassical theory of Keynesian unemployment. *EI* 12 (December 1974) 431–459.

Gottschalk, P. and T. Maloney. Involuntary terminations, unemployment, and job matching: a test of job search theory. *JOLE* 3 (April 1985) 109–123.

Gramlich, E. The distributional effects of higher unemployment. *BPEA* (No. 2, 1974) 293–342.

Green, C. Labour market performance from an employment perspective. *CPP* 3 (September 1977) 315–323. Comment by S. Kaliski 3 (Autumn 1977) 515–517.

Green, C. and J.-M. Cousineau. *Unemployment in Canada: The Impact of Unemployment Insurance.* Ottawa: Economic Council of Canada, 1976.

Green, D.A. and W.C. Riddell. The economic effects of unemployment insurance in Canada: an empirical analysis of UI disentitlement. *JOLE* 11 (January 1993).

Gronau, R. Information and frictional unemployment. *AER* 61 (June 1971) 290–301.

Grubel, H., D. Maki and S. Sax. Real and insurance induced unemployment in Canada. *CJE* 8 (May 1975) 174–191. Comments by S. Kaliski and reply 8 (November 1975) 600–605.

Hall, R. The rigidity of wages and the persistence of unemployment. *BPEA* (No. 2, 1975) 301–350.

Ham, J.C. and S.A. Rea, Jr. Unemployment insurance and male unemployment duration in Canada. *JOLE* 5 (July 1987) 325–353.

Hamermesh, D. *Jobless Pay and the Economy.* Baltimore: Johns Hopkins University Press, 1977.

Hamermesh, D. Unemployment insurance and labor supply. *IER* 21 (October 1980) 517–546.

Hansen, B. Excess demand, unemployment, vacancies and wages. *QJE* 84 (February 1970) 1–23.

Hart, O.D. Optimal labour contracts under asymmetric information: an introduction. *R.E. Studies* 50 (January 1983) 3–35.

Hasan, A. and S. Gera. *Job Search Behaviour, Unemployment and Wage Gains in Canadian Labour Markets.* Ottawa: Economic Council of Canada, 1982.

Heckman, J. and G. Borjas. Does unemployment cause future unemployment? *Economica* 47 (August 1980) 247–284.

Helwege, J. Sectoral shifts and interindustry wage differentials. *JOLE* 10 (January 1992) 55–84.

Hey, J. and K. Mavronmaras. The effect of unemployment insurance on the riskiness of occupational choice. *JPubEc* 16 (December 1981) 317–342.

Holzer, H.J. Search method used by unemployed youth. *JOLE* 6 (January 1988) 1–20.

Hosios, A.J. Unemployment and recruitment with heterogeneous labor. *JOLE* 3 (April 1985) 175–187.

Howitt, P. and R.P. McAfee. Costly search and recruiting. *IER* 28 (1987) 89–107.

Johnson, W. A theory of job shopping. *QJE* 92 (May 1978) 261–278.

Jones, S.R.G. Reservation wages and the cost of unemployment. *Economica* 56 (May 1989) 225–246.

Jones, S.R.G. The relationship between unemployment spells and reservation wages as a test of search theory. *QJE* 103 (1988) 741–765.

Jovanovic, B. Work, rest, and search: unemployment, turnover, and the cycle. *JOLE* 5 (April 1987) 131–148.

Kaitz, H. Analysing the lengths of spells of unemployment. *MLR* 93 (November 1970) 11–20.

Kaliski, S. Unemployment and unemployment insurance: testing some corollaries. *CJE* 9 (November 1976) 705–712.

Kaliski, S. Trends, changes and imbalances: a survey of the Canadian labour market. *Work and Pay: The Canadian Labour Market*, W.C. Riddell (ed.). Toronto: University of Toronto Press, 1985, 77–140.

Katz, A. (ed.). The economics of unemployment insurance: a symposium. *ILRR* 30 (July 1977) 431–526.

Katz, L.F. Efficiency wage theories: a partial evaluation. *NBER Macroeconomics Annual*. Cambridge, MA: MIT Press, 1986.

Katz, L.F. Some recent developments in labour economics and their implications for macroeconomics. *Journal of Money, Credit, and Banking* 20 (August 1988) 507–530.

Keeley, M.C. and P.K. Robins. Government programs, job search requirements, and the duration of unemployment. *JOLE* 3 (July 1985) 337–362.

Kennan, J. Equilibrium interpretations of employment and real wage fluctuations. *NBER Macroeconomics Annual*. Cambridge, MA: MIT Press, 1988.

Kesselman, J.R. *Financing Canadian Unemployment Insurance*. Toronto: Canadian Tax Foundation, 1983.

King, I.P. A natural rate model of frictional and long-term unemployment. *CJE* 23 (August 1990) 523–545.

Kuhn, P. Mandatory notice. *JOLE* (April 1992) 117–137.

Lancaster, T. Econometric methods for the duration of unemployment. *Econometrica* 47 (July 1979) 939–956.

Lazar, F. The impact of the 1971 unemployment insurance revisions on unemployment rates: another look. *CJE* 11 (August 1978) 559–569.

Li, E.H. Compensating differentials for cyclical and noncyclical unemployment: the interaction between investors' and employees' risk aversion. *JOLE* 4 (April 1986) 277–300.

Lilien, D. The cyclical pattern of temporary layoffs in United States manufacturing. *R.E. Stats* 62 (February 1980) 24–31.

Lilien, D. Sectoral shifts and cyclical unemployment. *JPE* 90 (August 1982) 777–793.

Lindbeck, A. and D.J. Snower. *The Insider-Outsider Theory of Employment and Unemployment*. The MIT Press: Cambridge, 1988.

Lindbeck, A. and D.J. Snower. Cooperation, harassment, and involuntary unemployment. *AER* 78 (1986) 167–188.

Lindbeck, A. Efficiency wages versus insiders and outsiders. *EER* 31 (1987) 407–416.

Lippman, S. and J. McCall. The economics of job search: a survey. *EI* 14 (June 1976) 155–189 and (September 1976) 347–368. Comment by G. Borjas and M. Goldberg 16 (January 1978) 119–125.

Lippman, S. and J. McCall (eds.). *Studies in the Economics of Search*. Amsterdam: North Holland, 1979.

Lipsey, R. Structural and deficient demand unemployment reconsidered. *Employment Policy and the Labour Market*, A. Ross (ed.). Berkeley: University of California Press, 1965, 210–255.

MacCallum, J. Unemployment in Canada and the United States. *CJE* 20 802–822.

MacLeod, W.B. and J.M. Malcomson. Implicit contracts, incentive compatibility, and involuntary unemployment. *Econometrica* 57 (March 1989) 447–480.

Maki, D. *Search Behaviour in Canadian Job Markets*. Special Study No. 15. Ottawa: Economic Council of Canada, 1971.

Maki, D. Regional differences in insurance-induced unemployment in Canada. *EI* 13 (September 1975) 389–400.

Maki, D. Unemployment benefits and the duration of claims in Canada. *Applied Economics* 9 (1977) 227–236.

Malcomson, J. Unemployment and efficiency wage hypothesis. *EJ* 91 (December 1981) 848–866.

Marshall, R.C. and G.A. Zarkin. The effect of job tenure on wage offers. *JOLE* 5 (July 1987) 301–324.

Marston, S. The impact of unemployment insurance on job search. *BPEA* (No. 1, 1975) 13–60.

Marston, S. Employment instability and high unemployment rates. *BPEA* (No. 1, 1976) 169–210.

Marston, S. Two views of the geographic distribution of unemployment. *QJE* 100 (February 1985) 57–79.

McCall, J. Economics of information and job search. *QJE* (February 1970) 113–126. Comment by Peterson and reply (February 1972) 127–134.

Medoff, J.L. Layoffs and alternatives under trade unions in U.S. manufacturing. *AER* 69 (June 1979) 380–395.

Milbourne, R.D., D.D. Purvis and W.D. Scoones. Unemployment insurance and unemployment dynamics. *CJE* 24 (November 1991) 804–826.

Moorthy, V. Unemployment in Canada and the United States: the role of unemployment insurance benefits. *Federal Reserve Bank of New York Quarterly Review* (Winter 1989) 48–61.

Mortensen, D.T. Job search and labor market analysis. *Handbook of Labor Economics*, O. Ashenfelter and R. Layard (eds.). Amsterdam: North Holland, 1986.

Neelin, J. Sectoral shifts and Canadian unemployment. *R.E. Stats.* 69 (1987) 718–732.

Nickell, S. The effect of unemployment and related benefits on the duration of unemployment. *EJ* 89 (March 1979) 34–49.

Nickell, S. Estimating the probability of leaving unemployment. *Econometrica* 47 (September 1979) 1249–1266.

Nickell, S. Education and lifetime patterns of unemployment. *JPE* 87 (October 1979) S117–S132.

Nickell, S. A picture of male unemployment in Britain. *EJ* 90 (December 1980) 776–794.

Nickell, S. The determinants of equilibrium unemployment in Britain. *EJ* 92 (September 1982) 555–575.

Ohashi, I. Cyclical variations in wage differentials and unemployment. *JOLE* 5 (April 1987) 278–300.

Okun, A.M. *Prices and Quantities*. Washington: The Brookings Institution, 1981.

Osberg, L. Unemployment insurance in Canada: a review of the recent amendments. *CPP* 5 (Spring 1979) 223–235.

Osberg, L. Unemployment and inter-industry labour mobility in Canada in the 1980s. *Applied Economics* 23 (1991) 1707–1717.

Parsons, D. Unemployment, the allocation of labour, and optimal government intervention. *AER* 70 (September 1980) 626–635.

Phelps, E.S. et al. *Microeconomic Foundations of Employment and Inflation Theory*. New York: Norton, 1970.

Phipps, S. Quantity constrained responses to UI reform. *EJ* (1990) 124–40.

Phipps, S. Equity and efficiency responses of unemployment insurance reform in Canada: the importance of sensitivity analyses. *Economica* 58 (May 1991a) 199–214.

Phipps, S. Behavioral response to UI reform in constrained and unconstrained models of labour supply. *CJE* 24 (February 1991b) 34–54.

Pissarides, C.A. Search intensity, job advertising, and efficiency. *JOLE* 2 (January 1984) 128–143.

Rea, S. Unemployment insurance and labour supply: a simulation of the 1971 Unemployment Insurance Act. *CJE* 10 (May 1977) 263–278.

Reder, M. The theory of frictional unemployment. *Economica* 36 (February 1969) 1–28.

Rees, A. An essay on youth joblessness. *JEL* 24 (June 1986) 613–628.

Reid, F. Reductions in work time: an assessment of employment sharing to reduce unemployment. *Work and Pay: The Canadian Labour market*, W.C. Riddell (ed.). Toronto: University of Toronto Press, 1985, 141–170.

Reid, F. Combatting unemployment through work time reductions. *CPP* 12 (June 1986) 275–285.

Riddell, W.C. (ed.). *Work and Pay: the Canadian Labour Market*. Toronto: University of Toronto Press, 1985.

Rosen, S. Implicit contracts: a survey. *JEL* 23 (September 1985) 1144–1175.

Ruhm, C.J. Advance notice and postdisplacement joblessness. *JOLE* 10 (January 1992) 1–32.

Saffer, H. Layoffs and unemployment insurance. *JPubEc* 19 (October 1982) 121–130.

Saffer, H. The effects of unemployment insurance on permanent and temporary layoffs. *R.E. Stats.* (1983) 647–652.

Sampson, L. A study of the impact of sectoral shifts on aggregate unemployment in Canada. *CJE* 18 (August 1985) 518–530.

Shapiro, C. and J.E. Stiglitz. Equilibrium unemployment as a worker discipline device. *AER* 74 (June 1984) 433–444.

Sharir, S. and P. Kuch. Contribution to unemployment of insurance-induced labour force participation: Canada 1972. *Economics Letters* 1 (1978) 271–4.

Solow, R. On theories of unemployment. *AER* 70 (March 1980) 1–11.

Solow, R.M. Insiders and outsiders in wage determination. *Scandinavian Journal of Economics* 87 (1985) 411–428.

Solow, R.M. *The Labour Market as a Social Institution*. Cambridge, MA: Basil Blackwell, 1990.

Stigler, G.J. Information in the labour market. *JPE* 70 (October 1962) 94–105.

Summers, L.H. *Understanding Unemployment*. Cambridge, MA: The MIT Press, 1990.

Swaim, P.L. and M.J. Podgursky. Advance notice and job search: the value of an early start. *JHR* 25 (Spring 1990) 147–178.

Topel, R.H. On layoffs and unemployment insurance. *AER* 73 (September 1983) 541–559.

Topel, R.H. Equilibrium earnings, turnover, and unemployment: new evidence. *JOLE* 2 (October 1984) 500–522.

Van de Berg, G.J. Search behavior, transitions to nonparticipation and the duration of unemployment. *EJ* 100 (1990) 842–865.

Vanderkamp, J. The effect of out-migration on regional employment. *CJE* 3 (November 1970) 541–549.

Vanderkamp, J. The efficiency of the interregional adjustment process. *Disparities and Interregional Adjustment*. K. Norrie (ed.). Toronto: University of Toronto Press, 1986, 53–108.

Wadsworth, J. Unemployment benefits and search effort in the UK labour market. *Economica* 58 (February 1991) 17–34.

Warner, J., C. Poindexter and R. Fearn. Employer-employee interaction and the duration of unemployment. *QJE* 94 (March 1980) 211–234.

Winer, S.L. and D. Gauthier. *Internal Migration and Fiscal Structure*. Ottawa: Economic Council of Canada, 1982.

Wolcowitz, J. Dynamic effects of the unemployment insurance tax on temporary layoffs. *JPubEc* 25 (November 1984) 35–52.

Yaniv, G. Unemployment insurance benefits and the supply of labor of an employed worker. *JPubEc* 17 (February 1982) 71–88.

Yellen, J. Efficiency wage models of unemployment. *AER* 74 (May 1984) 200–205.

CHAPTER 26

Wage Changes, Price Inflation, and Unemployment

Previous chapters dealt with a variety of wage *structures*, such as the occupational, industrial, and regional wage structures. In this chapter, we deal with the aggregate wage *level* and its determinants, in particular how it is affected by other aggregate variables such as the price level and unemployment. Often this topic is analyzed at the macroeconomic level of aggregation of the economy as a whole; however, since the early 1970s, increased attention has been paid to its microeconomic foundations — that is, how aggregate wage changes and movements in aggregate employment and unemployment are the results of decisions made by many individual workers, unions, and firms. In addition, empirical analysis of the determinants of wage changes is often made at the more disaggregate level of the industry, or for individual wage contracts, as well as the economy as a whole.

The focus of our analysis is on the labour market dimensions of the issue. Since this is essentially a topic of macroeconomics, a more complete treatment is left to texts and courses in macroeconomics.

CANADIAN EXPERIENCE

Table 26.1 summarizes Canada's long-term experience with respect to inflation, unemployment, and related economic aggregates. (See also Figure 24.1). In order to abstract from the effects of business cycles, the table shows average performance between the postwar cyclical peak years of 1956, 1966, 1973, 1981, and 1989. Average data for the period prior to 1947 are also included for historical perspective. However, because this period includes both the Great Depression and World War II, the averages mask considerable variability.

Overall economic performance was excellent during the first two decades of the postwar period. Two bursts of inflation occurred during this period, one immediately after the war and the second during the Korean War (1950–53).

Table 26.1
AGGREGATE ECONOMIC TRENDS, CANADA, 1927–89

| | GROWTH RATE[1] OF | | | ANNUAL AVERAGE | |
PERIOD	REAL INCOME PER CAPITA[2]	PRODUC- TIVITY[3]	EMPLOY- MENT[4]	INFLATION RATE[5]	UNEM- PLOYMENT RATE[6]
1927–46	2.2	2.1	1.4	0.1	8.1
1947–56	2.6	3.5	1.8	4.3	3.2
1957–66	2.4	2.1	2.5	2.0	5.5
1967–73	3.9	2.5	3.0	4.4	5.2
1974–81	1.7	0.1	2.9	9.7	7.3
1982–89	2.4	1.6	1.7	5.3	9.8

Notes:
1. Growth rates are averages of compound increases from the level in the year before the period specified to the level in the final year of the period specified.
2. Real GNE divided by population.
3. Real GNE per person employed.
4. Civilian employment; minor non-comparabilities in series occur in 1946 and 1966.
5. As measured by the Consumer Price Index.
6. Minor non-comparabilities occur in 1946 and 1966.

Sources: M.C. Urquhart (ed.) *Historical Statistics of Canada*, Toronto: Macmillan, 1965; Bank of Canada, *Bank of Canada Review*, various issues; Statistics Canada, *National Income and Expenditure Accounts*, various issues.

Otherwise, inflation rates were low by the standards of both previous and subsequent experience and displayed no obvious trend until they began to creep upward in the mid-1960s. Unemployment rates were also low by the standards of other periods, although they did rise significantly during the 1957–58 recession and remained above normal levels until the early 1960s. Productivity and real income per capita grew at rapid rates from 1947 to 1973 compared to rates attained earlier and subsequently. Employment also grew rapidly.

A marked deterioration in economic performance is evident after 1973. The coexistence of high inflation and high unemployment — a phenomenon often referred to as "stagflation" — during the 1974–81 period was the most dramatic departure from the past. Prior to the 1970s, increases in unemployment were generally accompanied by decreases in the rate of inflation and vice versa. A sharp decline in productivity growth also occurred and real income growth slowed considerably. Only employment growth remained healthy.

The final period is dominated by the onset of and recovery from the severe recession of 1981–82. Unemployment rose dramatically during the recession and the rate of inflation fell substantially. Productivity growth recovered somewhat and real income growth increased. Employment growth remained healthy by historical standards, but lower than the very rapid pace set in the 1960s and 1970s. Unemployment declined gradually during the long period of economic growth from 1983 to 1989, reaching its pre-recession level of 7.5 percent at the end of this period.

Inflation and unemployment vary with the state of the business cycle, unemployment rising in recessions and declining in booms and inflation generally falling in recessions and rising in booms. However, as the data in Table 26.1 indicate, there are also large variations in the average levels of inflation and unemployment from one business cycle to another. The rate of inflation averaged only two percent during the decade 1957–66 but almost 10 percent during 1974–81. Similarly, the unemployment rate averaged 3.2 percent during the first decade of the postwar period (1947–56) but almost 10 percent during 1982–89. Explaining these variations has been the subject of a great deal of economic research.

A more detailed summary of recent experience is provided in Table 26.2. Two measures of wage inflation, both based on large unionized employers, and one measure of price inflation are shown. All major collective agreements in force includes deferred increases and cost-of-living-allowance (COLA) increases negotiated in previous periods but occurring in the year noted; new non-COLA settlements includes only settlements reached during the period noted (some of which may take effect in subsequent periods). Other measures of wage and price change tell a very similar story of events during these two decades.

Several aspects of the 1967–90 experience should be noted. Inflation crept upward during the late 1960s, with the rate of price increase falling below four percent only during the brief 1970–71 recession (brought on by the restrictive monetary policy introduced in 1970 — see the drop in money supply growth in column one of Table 26.2). Real wage growth (the rate of wage change less the rate of change of prices) was very high during this period. Despite rapid employment growth (except during the 1970–71 recession), the unemployment rate also moved upward and by 1971–72 exceeded six percent, a high level by the standards of the previous postwar period. After 1970–71 Canada, together with most other Western economies, embarked on an expansionary path, as is evident from the high rates of money supply growth during 1971–73. Employment grew extremely rapidly during 1972–74 but only a modest decline in unemployment took place. Wage and price inflation climbed sharply; by 1975 wage increases exceeded 19 percent and price inflation had reached almost 11 percent. The 1973 OPEC oil price increase contributed to the higher rates of inflation while at the same time causing a recession in many Western countries in 1974–75. In part because Canada is a producer as well as consumer of oil, this recession was milder here than in many countries; nonetheless, employment growth declined

Table 26.2
INFLATION, UNEMPLOYMENT, AND RELATED ECONOMIC AGGREGATES, CANADA, 1967–90

| | RATE OF CHANGE[1] OF | | | | | | |
YEAR	MONEY SUPPLY (M1) (1)	CON-SUMER PRICE INDEX (2)	WAGE RATES[2] (3)	EM-PLOY-MENT (4)	NEW NON-COLA SETTLE-MENTS[3] (5)	REAL WAGE SETTLE-MENTS[4] (6)	UNEM-PLOY-MENT RATE (7)
1967	9.8	3.6	—	2.9	8.4	4.8	3.8
1968	4.4	4.0	8.0	1.9	7.5	3.5	4.5
1969	7.1	4.6	9.1	3.1	7.6	3.0	4.4
1970	2.4	3.3	8.5	1.1	8.7	5.4	5.7
1971	13.0	2.9	7.8	2.3	7.9	5.0	6.2
1972	14.4	4.8	7.4	3.0	8.8	4.0	6.2
1973	14.5	7.5	8.3	5.0	11.0	3.5	5.5
1974	9.3	10.9	12.3	4.2	14.7	3.8	5.3
1975	14.0	10.8	14.2	1.7	19.2	8.4	6.9
1976	8.0	7.5	14.3	2.1	10.9	3.4	7.1
1977	8.5	8.0	11.2	1.8	7.9	–0.1	8.1
1978	10.1	9.0	6.8	3.5	7.1	–1.9	8.3
1979	6.9	9.1	7.8	4.1	8.8	–0.3	7.4
1980	6.2	10.1	9.7	3.0	11.1	1.0	7.5
1981	3.3	12.4	11.5	2.7	13.4	1.0	7.5
1982	–0.1	10.9	10.3	–3.5	10.1	–0.8	11.0
1983	10.1	5.7	4.7	0.5	5.6	–0.1	11.8
1984	3.4	4.4	3.6	2.4	3.5	–0.9	11.2
1985	4.3	3.9	3.5	2.6	3.7	–0.2	10.5
1986	5.0	4.2	3.4	2.8	3.5	–0.7	9.5
1987	12.9	4.4	3.9	2.9	3.9	–0.5	8.8
1988	6.1	4.0	4.3	3.2	4.4	0.4	7.8
1989	4.6	5.0	5.3	2.0	5.3	0.3	7.5
1990	–2.1	4.8	5.8	0.7	6.1	1.3	8.1

1. Year-to-year percentage change.
2. All major collective agreements in force.
3. Compound annual average percentage increase in base wage rates in major collective agreements over the duration of the contract, new non-COLA settlements only. Contracts containing COLA clauses are excluded because the value of these increases is not known at the time the contract is signed.
4. Column (5) minus column (2).

Sources: Bank of Canada, *Bank of Canada Review*, various issues; Labour Canada, *Major Wage Settlements*, various issues; Statistics Canada, *The Labour Force*, various issues.

noticeably and unemployment rose. The 1974–75 experience provided a clear example of stagflation — increased inflation and unemployment.

Faced with the continuing escalation in wage and price increases and the deteriorating demand conditions associated with the world recession, the federal government introduced the Anti-Inflation Program (AIP) in October 1975. The key ingredients in the program were mandatory wage and price (or profit) controls and gradual reduction in the growth in aggregate demand through fiscal and monetary restraint.

During the 1975–78 AIP, wage settlements declined substantially; price increases also fell, although by a smaller amount. As discussed later in this chapter, most empirical studies attribute some of this decline in inflation to the controls program. The reduction in demand due to fiscal and monetary restraint also played a role. Because of the slack economic conditions, unemployment rose to over 7 percent in 1976 and over 8 percent in 1977 and 1978.

The combination of the 1974–75 world recession and the 1975–78 AIP resulted in some moderation of inflationary pressures. Nonetheless, in the late 1970s inflation began to move upward again in Canada and elsewhere. In 1979 the second major OPEC oil price increase occurred. By the end of the decade rates of price inflation exceeded 10 percent in both the United States and Canada. A variety of aspects of economic behaviour suggested that expectations of continuing inflation were deeply entrenched.

The consequence of these developments was the adoption of a much more determined anti-inflationary stance by the monetary authorities in the United States and subsequently in Canada, as is indicated by the sharp drop in money supply growth in 1981 and 1982. The general results of this policy change are evident from Table 26.2. Initially, the impact of the reduced growth in aggregate demand fell primarily on output and employment: employment declined by over 3 percent in 1982 and unemployment soared to 11 percent. Wage and price inflation moderated only slightly during the first year of the recession. However, as the impact of the severe recession became more widely felt, inflation began to decline dramatically. The adoption in 1982 of the federal "6 and 5" wage restraint program and subsequent related provincial programs which limited the wage increases of public sector employees may also have played a role in the decline in wage increases. By 1984 the rates of wage and price inflation were the lowest since the 1960s. The restrictive monetary policy, possibly with some additional contribution from the public sector wage control programs, had clearly achieved its objective of ending a deeply entrenched inflation. Equally clear was the enormous cost of such a policy — high unemployment, business failures, mortgage foreclosures, and reduced output.

During the long period of economic growth from 1983–89, the Bank of Canada continued to follow a policy of reducing inflation. Indeed, during this period the Bank of Canada increasingly adopted the view that its primary objective should be that of price stability; i.e., zero inflation. Pursuing this policy generally meant a less expansionary monetary stance, and thus higher interest rates and lower employment growth. The purpose of the policy was to prevent a return to the high inflation rate era of the 1970s and early 1980s. Unemployment fell gradually during this period of economic expansion, the unemployment rate reaching by 1989 its pre-recession level of 7.5 percent. The policy of inflation reduction met with considerable success. In the peak year of the business cycle (1989), the rate of inflation was 5.0%, a modest level by the standards of the previous two decades.

The 1990–92 recession resulted in further downward pressure on prices, so that by 1992 the annual rate of inflation had fallen to the one to two percent range. However, employment growth slowed sharply and the unemployment rate rose to over 11 percent. Once again the high cost of achieving price stability following a period of deeply entrenched inflation was apparent to many Canadians.

As this brief review of the Canadian experience indicates, inflation and unemployment have been dominant policy concerns for the past three decades. They remain major challenges facing policymakers today. We begin our discussion of the labour market dimensions of inflation and unemployment with an examination of the determinants of wage changes.

DETERMINANTS OF WAGE CHANGES

Wage changes play a significant role in the inflationary process. Labour costs are an important component of total cost and reductions in the rate of price increase are unlikely to occur, other than temporarily, without accompanying reductions in the rate of wage increase. For this reason, policies to control inflation are typically directed to achieving reductions in wage as well as price increases. The first step in the design of such policies is to understand the factors that influence wage determination.

Unemployment Rate

In his classic article based on data from the United Kingdom for the period 1861–1957, Phillips (1958) estimated a negative relationship between aggregate money wage changes and unemployment in the economy as a whole — a relationship now generally referred to as the Phillips curve, as depicted in Figure 26.1(e). Much subsequent research has been devoted to the theoretical underpinnings of this relationship as well as to empirical investigations of other determinants of aggregate wage changes.

Lipsey (1960) explained the negative relationship on the basis that (1) the unemployment rate is a measure of the overall excess demand or supply in the aggregate labour market, and (2) the rate of wage change is a function of the amount of excess labour demand or supply. Aggregation over individual labour markets, each with possibly different amounts of excess demand or supply, is a key aspect of this explanation.

Figure 26.1 illustrates the main elements of Lipsey's theory. At any point in time some labour markets are characterized by excess demand for labour and others by excess supply, as shown in panel (a) of Figure 26.1. Basic economic theory predicts that wages will be rising in those markets characterized by positive excess demand and falling in markets with negative excess demand (excess supply). The rate at which wages adjust is assumed to be related to the amount of excess demand. This relationship may take various forms; for purpose of illustration it is shown in Figure 26.1(b) as being linear throughout (heavy line). Alternatively (shown by the dashed line), it could be linear with a kink at the origin, if wages rise more quickly in response to positive excess demand than they fall in response to negative excess demand, or even nonlinear (not shown). Note that the amount of excess demand in market i, D_i-S_i is expressed relative to the size of the market, $D_i - S_i/S_i$, because excess demand of 25 workers has a different impact in a market with 50 workers than in a market with 500 workers.

The aggregate rate of wage change $\dot{W}$ is simply the weighted sum of the wage changes in each individual market, the weights α_i being employment in market i:

$$\dot{W} = \sum_i \alpha_i \dot{W}_i \tag{26.1}$$

Because the rate of wage change in market i is related to the amount of excess demand in that market, the aggregate rate of change of wages is a function of aggregate excess labour demand:

$$\dot{W} = f\left(\frac{D-S}{S}\right) \tag{26.2}$$

The next step is to relate aggregate excess labour demand to observable counterparts, the unemployment rate (U) and the job vacancy rate (V):

$$\frac{D-S}{S} = V-U \tag{26.3}$$

This implies a relationship between the rate of wage change and the excess of job vacancies over unemployment:

$$\dot{W} = f(V - U) \tag{26.4}$$

If there is a stable relationship between the unemployment rate and the job vacancy rate, as depicted in Figure 26.1(c), then U alone can be used to measure aggregate excess demand. Substituting the relationship

$$V = g(U) \tag{26.5}$$

Figure 26.1
WAGE CHANGES, EXCESS DEMAND, AND UNEMPLOYMENT

(a) Disequilibrium in Individual Labour Markets

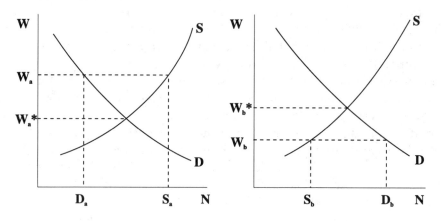

(b) The Relationship between Wage Changes and Excess Demand

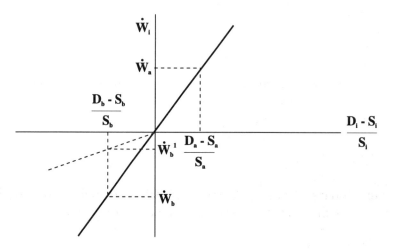

into Equations 26.3 and 26.4 gives:

$$\frac{D-S}{S} = g(U) - U = h(U) \qquad (26.6)$$

which is shown in Figure 26.1(d) and

$$\dot{W} = f(g(U) - U) = F(U) \qquad (26.7)$$

which is the Phillips curve shown in panel (e) $(F'(U) < 0)$.

Figure 26.1 (continued)

(c) The Relationship between Unemployment and Job Vacancies

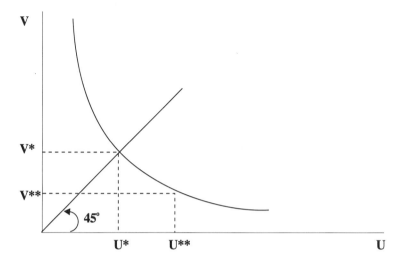

(d) The Relationship between Aggregate Excess Demand and Unemployment

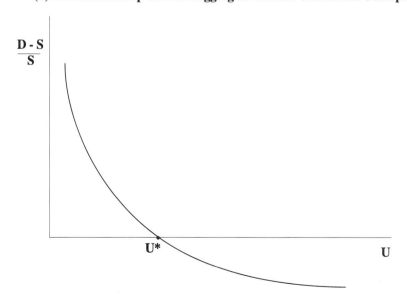

Figure 26.1 (continued)

(e) The Relationship between Wage Changes and Unemployment

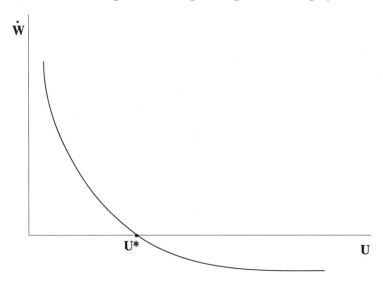

From Equation 26.3 there will exist values of the job vacancy rate and unemployment rate which imply aggregate excess labour demand equals zero. These are shown as V° and U° in panels (c), (d), and (e) of Figure 26.1. At U° the aggregate labour market is in equilibrium in the sense that aggregate demand for labour equals aggregate supply and the aggregate rate of wage change equals zero (i.e., F(U°) = 0). For this reason, U° is called the *equilibrium* or *natural unemployment rate*. Note, however, that individual labour markets need not be in equilibrium. As discussed in Chapters 24 and 25 on unemployment, because of imperfect information, the process of matching workers and jobs takes time. Thus unfilled job vacancies and unemployed workers will coexist at any point in time. This disequilibrium at the micro level is nonetheless consistent with aggregate labour market equilibrium in that the excess demand (unfilled job vacancies) in some markets is offset by an equal amount of excess supply (unemployed job seekers) in other markets. The aggregate rate of wage change equals zero because markets in which wages are rising due to excess demand are offset by markets in which wages are declining.

Under the assumptions illustrated in Figure 26.1, the aggregate labour market is in equilibrium when the unemployment rate equals the job vacancy rate, as depicted in panel (c). However, this outcome need not occur in general. In particular, if the relationship between wage changes and excess demand is kinked at the origin as depicted by the dashed line in panel (b), aggregate labour market equilibrium will require an unemployment rate in excess of the job vacancy rate, such as is illustrated by U°° and V°° in Figure 26.1(c). In these circumstances, because wages fall less rapidly in response to excess supply than they rise in response to excess demand, the aggregate amount of excess supply must exceed the amount of excess demand in order for the aggregate rate of wage change to equal zero. Thus downward wage rigidity — or, more generally, asymmetry in the response of wages to excess demand versus excess supply — results in a higher equilibrium unemployment rate.

The natural unemployment rate depends on numerous factors discussed in Chapters 24 and 25 such as the magnitude and frequency of seasonal, cyclical and other economic disturbances, the job search behaviour of employers and workers and the efficiency of the matching process, the use of layoffs to respond to changes in demand, the amount of labour force turnover, the age-sex composition of the labour force, and labour market policies such as minimum wages and unemployment insurance. Although some of these factors remain approximately constant over time, others change and therefore alter the equilibrium level of unemployment. In these circumstances, the appropriate measure of aggregate excess demand for labour is the difference between the observed and natural unemployment rates, $U - U^*$, with $U > U^*$ implying excess supply of labour and $U < U^*$ implying excess demand. That is, when the equilibrium unemployment rate itself may be changing, the unemployment rate is not a reliable measure of aggregate excess demand and $U-U^*$ should be employed in the wage equation in lieu of U. This requires a modification of Equation 26.7:

$$\dot{W} = F(U-U^*), \ F'\ (\cdot) < 0, \ F(0) = 0 \tag{26.7}'$$

This clearly presents a challenge for empirical research on wage determination in that the natural unemployment rate is not directly observed. We discuss the way empirical studies have dealt with this issue in a subsequent section.

This discussion assumes that there is a unique equilibrium or natural unemployment rate U^* and that U^* is independent of the actual unemployment rate U. However, it is possible that there may be more than one unemployment rate at which the economy is in macroeconomic equilibrium. Similarly, it is also possible that economic shocks which lead to changes in the aggregate unemployment rate may also cause changes to occur in the natural unemployment rate. Both these possibilities have recently received attention from macro and labour economists, and are discussed later in this chapter under the heading "Unemployment Persistence."

An increase in the natural unemployment rate will shift the Phillips curve upward, resulting in larger wage increases at each unemployment rate, because there will be increased excess demand (or reduced excess supply) at each measured unemployment rate. The unemployment-vacancies relationship will also generally shift to the right because increased excess demand for labour at each unemployment rate is usually associated with more job vacancies at each level of measured unemployment. However, in some circumstances an increase in the natural rate is associated with a movement along, rather than a shift in, the U-V relationship. An example is the movement from U^* to U^{**} in Figure 26.1(c), which could have been caused by an increase in the degree of downward wage rigidity in the economy, as illustrated by the shift from the linear to the kinked relationship between wage changes and excess demand in Figure 26.1(b).

In summary, Lipsey's (1960) explanation of the Phillips curve is based on two stable relationships: a positive relation between wage changes and excess demand for labour and an inverse relation between excess demand for labour and the unemployment rate. The latter can be derived by expressing excess labour demand as the difference between the number of unfilled job vacancies and the number of unemployed workers, and assuming a stable (inverse) relation between unemployment and vacancies. The assumption of a stable relationship between the vacancy rate V and the unemployment rate U implies that there is a one-to-one relationship between excess demand and the unemployment rate.

Phillips' original paper indicated that the relationship between wage changes and unemployment in the U.K. had remained stable for almost one hundred years. However, in many countries, including Canada, the temporal stability of this relationship began to be questioned in the late 1960s and early 1970s; the upward drift in wage and price inflation beginning in the mid-1960s resulted in rates of wage increase well above those predicted by estimated Phillips curves. One explanation for this development was that the relationship between the unemployment rate and excess demand for labour had changed; as noted above, such a change could be due to an increase in the natural unemployment rate, perhaps associated with the changing composition of the labour force and revisions to the unemployment insurance system. Evidence relating to this explanation is discussed subsequently. Another leading explanation involved inflationary expectations.

Expected Inflation

Seminal contributions by Friedman (1968) and Phelps (1967, 1968) attacked the theoretical basis for a stable or permanent relationship between wage inflation and unemployment. Friedman and Phelps emphasized the role of inflationary expectations in the wage determination process. If both employers and employees expect prices to increase, they will adjust wage changes upward by the amount of expected inflation in order to achieve the desired change in real wages.

Expected wage or price inflation is a general term which includes several components: changes in the firm's product prices, in wages in similar firms and industries, and in the cost of living. These reflect the fact that what matters to employers and employees are *relative wages* and *real wages*. To workers, what is important are wages relative to those received by comparable workers in other firms and industries (relative wages) and relative to the cost of living (real wages). Thus, expectations about wage changes elsewhere and about changes in the cost of living will influence wage determination, especially when wages are not set or negotiated frequently. To firms, what is important are wages relative to wages elsewhere (which affects the firm's ability to attract and keep workers) and relative to the prices that the firm can charge for its products (which affects the firm's ability to pay). Thus expectations about product price increases should also be a factor in wage determination. In the analysis that follows we will focus on price expectations and thus on issues relating to real wages. The analysis of wage expectations and issues relating to relative wages results in similar conclusions (Phelps, 1968).

The role of expected inflation is illustrated in Figure 26.2. Two individual labour markets, market A with excess supply for labour at the current price level (p_o) and market B with excess demand at p_o, are used for purposes of illustration. During the period for which the wage rate is being determined, the price level is expected to rise from p_o to p_1^e, equivalent to an expected inflation rate of $\dot{p}^e$. Thus both the labour demand and labour supply curves shift up by the expected increase in the price level. Labour demand is a function of the price level because, as indicated in Chapter 9, the firm's labour demand schedule is the marginal revenue product of labour, defined as the marginal physical product times the price at which the product is sold. Hence, an increase in the price of output, other things being equal, will shift up the demand for labour by the amount of the price increase. The labour supply schedule is also a function of the price level, as discussed in Part 1 on Labour Supply, because the worker's desired supply of labour depends on the real wage — the money wage divided by the price level for consumption goods and services. An increase in these prices will thus increase the asking wage for any specific quantity of labour by exactly the increase in the price level. The labour supply curve thus shifts vertically upward by the amount of expected inflation.[1] Even though the current equilibrium wages in markets A and B are W_a^* and W_b^* respectively, the equilibrium wages for the period for which the wage rate is being determined are W_a^{**} and W_b^{**}. Thus the observed wage change will be from W to W^{**} in market A and B. This wage change can be expressed as the sum of two components: (1) from W_i to W_i^*, reflecting the current amount of excess demand or supply in market i, and (2) from W_i^* to W_i^{**}, reflecting the expected increase in the price level. Following the previous derivation of the relationship between wage changes and the current amount of aggregate excess demand for labour as measured by the unemployment rate, the Phillips curve becomes:

$$\dot{W} = F(U-U^*) + \dot{p}^e \tag{26.8}$$

---

1. The relevant price level in the labour supply function is the price level of consumption goods (e.g., the Consumer Price Index) while the relevant price in the labour demand function is the firm's product price. For expositional purposes, these are assumed to increase at the same rate. The analysis can easily be generalized to deal with differential rates of price increase.

Figure 26.2
WAGE CHANGES AND EXPECTED INFLATION

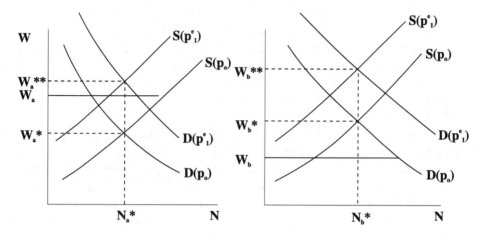

Notes:
1. $W_i^{**} = W_i^* + \dot{p}^e$ $i = a,b$

2. $\dfrac{W_i^*}{p_o} = \dfrac{W_i^{**}}{p^e_1}$ $i = a,b$

This relationship — often referred to as the "expectations-augmented Phillips curve," reflecting the addition of the expected inflation term $\dot{p}^e$ — is illustrated in Figure 26.3. There are a family of Phillips curves, each based on a particular expected inflation rate. Increases in expected inflation shift the Phillips curve vertically upward by the amount of the increase in $\dot{p}^e$; decreases in inflationary expectations shift the relationship downward.

As before, the natural unemployment rate U^* is defined as the level at which aggregate demand for labour equals aggregate supply. Because the aggregate labour market is in equilibrium, real wages are constant at U^*.[2] However, changes in nominal wages may be taking place due to expected inflation, as illustrated by points a, c, and d in Figure 26.3. At each of these points money wages are increasing at the expected inflation rate, with the result that expected real wages are constant. At unemployment rates below U^*, such as the point b in Figure 26.3, aggregate labour demand exceeds aggregate supply and expected real wages are increasing. At unemployment rates above U^*, such as point e, expected real wages are declining, reflecting aggregate excess supply of labour.

---

2. The possibility of real wage growth due to productivity growth is discussed below. For the moment, aggregate productivity growth is assumed to be zero.

Figure 26.3
WAGE CHANGES, UNEMPLOYMENT, AND EXPECTED INFLATION

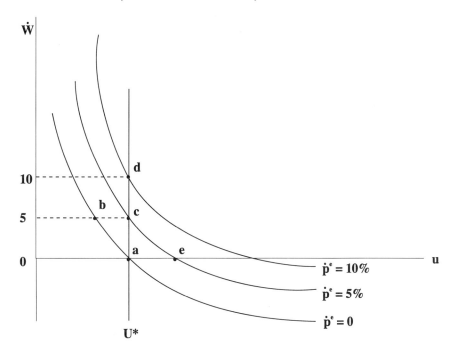

An important aspect of the Friedman-Phelps theory of inflationary expectations is the prediction that both the labour demand and labour supply functions will shift upward by the anticipated increase in the price level. This prediction arises because firms' profits and workers' utility depend on the real wage; increases in the price level thus require an equal increase in the nominal wage at each level of employment to maintain the real wage. As the consequence, the equilibrium level of employment in each individual labour market is not altered by a change in the price level, once nominal wages have adjusted to the higher (expected) price level. This property can be seen in Figure 26.2. In both markets, the labour demand and labour supply functions shift up by the same amount, resulting in the same equilibrium level of employment N_i^* and equilibrium real wage as prevailed before the expected increase in the price level. Because the equilibrium in each market is unchanged, the unemployment rate at which aggregate demand for labour equals aggregate supply is also not altered by changes in the price level, once these changes are anticipated and taken into account by both employers and employees in wage determination. The Phillips curve thus shifts vertically upward by the anticipated increase in the price level, as depicted in Figure 26.3.

In contrast, if workers display "money illusion" by not raising their asking wage for a given quantity of labour by the same proportion as the anticipated increase in prices, then a change in the price level may alter the equilibrium level of employment in each market and the equilibrium unemployment rate for the economy as a whole. In these circumstances, wage changes do not adjust fully by the expected rate of inflation so the expectations-augmented Phillips curve becomes

$$\dot{W} = F(U-U^*) + \lambda \, \dot{p}^e \tag{26.9}$$

where $\lambda < 1$ is a parameter reflecting the degree of money illusion, $\lambda = 1$ implying no money illusion. A number of empirical studies (discussed below) have tested the Friedman-Phelps prediction of a unit coefficient ($\lambda = 1$) in the wage equation.

In summary, Friedman (1968) and Phelps (1967, 1968) emphasized the role of inflationary expectations in the wage determination process. If wages are adjusted simply to eliminate the *current* amount of excess labour demand or supply, then the unemployment rate (or some other measure of excess demand) may be the sole determinant of wage changes. However, because wages are set or negotiated periodically, employers and employees look not only at current conditions but at the conditions *expected to prevail* during the remainder of the wage contract (or period for which the wage is set). This "forward-looking" aspect provides the rationale for including the expected inflation variable.

The role of inflationary expectations in wage determination is widely accepted. It is well grounded in the microeconomic theory of the behaviour of workers and firms; it is capable of explaining much of the upward drift in wage settlements since the 1960s; and it is evident to even the most casual observer of the wage determination process.

The relationship between wage changes, unemployment, and expected inflation summarized by Equation 26.8 has profound implications for macroeconomic policy. In particular, it implies that there is no long-run trade-off between inflation and unemployment even though these two variables may be inversely related in the short run. This important implication is discussed more fully subsequently in this chapter. At this point we continue to examine the main factors that influence wage changes.

Catch-up for Unanticipated Inflation

Wages are typically determined periodically, with the duration of the wage contract usually explicit in the union sector and implicit in the nonunion sector. No matter how much effort employers, employees, and their representatives devote to forecasting inflation and wage increases in other firms and industries, the actual changes in the price level and in wages elsewhere during the contract period may differ from the expected change. Consequently, the real or relative wage at the end of the contract will generally not equal the real or relative wage that was expected to prevail at the time the wage was originally determined. If inflation is greater than anticipated, workers will desire an additional "catch-up" wage increase because of the unanticipated decline in their purchasing power. Similarly, if the increase in wages in related industries and occupations is greater than expected, workers will desire an additional increase to restore wage relativities. In contrast, if prices or other firms' wages increase less rapidly than anticipated, there will be pressure from the firm to adjust wages downward, or to have them increase less quickly in the future.

The effect of unanticipated inflation is illustrated in Figure 26.4. For purposes of illustration, it is assumed that wages are set to yield labour market equilibrium at the end of the contract. At the beginning of the current contract period (end of previous contract), the wage is W_0. Given the anticipated rise in the price level to p_1^e, the appropriate wage increase is from W_0 to W_1^e. However, because the actual rise in the price level (to p_1^a) exceeds the expected rise, the actual equilibrium wage is W_1^a. An additional "catch-up" wage increase from W_1^e to W_1^a is therefore needed to restore equilibrium.

To account for the influence of this catch-up factor, a number of empirical studies have included as an explanatory variable in the wage equation a measure of the difference between the actual rate of inflation over the previous wage contract and the expected rate of inflation at the time the previous contract began (e.g., Turnovsky, 1972; Riddell, 1979; Christofides, Swidinsky and Wilton, 1980a). Note that this variable measures the amount of excess demand or supply due to unanticipated inflation, given by $D_1^a - S_1^a$ in Figure 26.4. Thus this catch-up variable would not be needed (and would not have any explanatory power) if the unemployment rate (or other measures of excess demand) were perfect measures of the excess demand or supply of labour. That is, unanticipated inflation should be reflected in the observed levels of job vacancies and unemployed job seekers. The catch-up variable should therefore be interpreted as a supplementary measure of excess demand or supply in the labour market; it is useful to the extent that movements in the unemployment rate do not fully capture all variations in excess demand for labour.

If inflation or other economic circumstances deviate substantially from the expectations upon which wages were determined, the parties may not wait until the end of the contract period to make an upward or downward adjustment. Such adjustments are more likely to occur in the nonunion sector where the wage rate and the length of the period for which it is in force are not part of a formal contract but rather are implicit understandings about normal behaviour. Nonetheless, even in the union sector, where wage agreements are legal contracts with a fixed duration, a significant change in economic conditions (relative to what was expected) may cause one side to request and the other to agree to renegotiate the terms of the contract. (Indeed, some contracts contain "re-opener provisions" which stipulate the circumstances under which such renego-

Figure 26.4
WAGE CHANGES AND UNANTICIPATED INFLATION

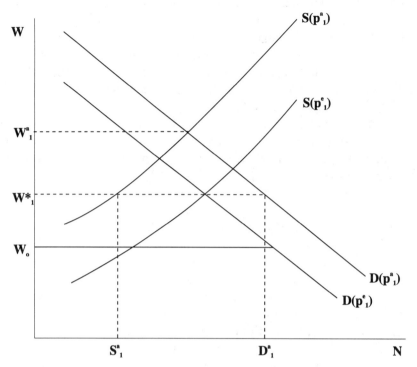

tiation may be necessary.) Examples of such adjustments took place in Canada in the 1970s because of unanticipated increases in inflation and in the 1980s because of an unanticipated decline in inflation and in product market conditions due to the 1981–82 recession.

Productivity Growth

In wage determination, the direct concern of employers and employees may be the nominal or money wage, but their ultimate concern is with the real wage. Thus any factors that influence real wage growth should also be included as explanatory variables in the wage equation. This can be seen by subtracting $\dot{p}^e$ from both sides of Equation 26.8:

$$\dot{W} - \dot{p}^e = F(U - U^*) \tag{26.10}$$

The dependent variable is the (expected) growth in real wages; thus any additional determinants of real wage growth should be included on the right-hand side of Equation 26.10.

The primary determinant of real wage growth is the trend rate of productivity growth. Over the long term, real wages have tended to increase at the rate of growth of productivity (Allen, 1986). This tendency is also evident in the postwar Canadian experience: the sharp decline in productivity growth after 1973 was accompanied by a decline in the rate of growth of real income per capita (Table 26.1) and in the rate of increase in real wages (Table 26.2).

The relationship between productivity and real wage growth can be understood most easily at the level of the economy as a whole. Higher productivity implies that it is possible to produce more output with the same inputs. Thus, other factors being constant (including employment of inputs), an increase in productivity will increase the economy's total output or national income and therefore real income per capita. Wages and salaries make up over 75 percent of national income. Unless there is a massive shift towards profits in the functional distribution of income, an increase in real per capita income will be associated with higher real wages and salaries. The relationship between aggregate output, labour's share of total output or income, productivity and real wages is explained more formally below.

At the level of an individual labour market, the relationship between productivity and real wages is more complex (Allen, 1986). Higher productivity may either increase or decrease labour demand, depending on the magnitude of several offsetting factors. These include a displacement effect — less labour is required per unit of output — and a product demand effect — higher productivity lowers costs and therefore product prices, resulting in an increase in product demand and thus labour demand. As a consequence, productivity gains may result in lower real wages and/or employment in some labour markets.

Technical changes introduced in individual labour markets also have effects throughout the economy. These general equilibrium effects often provide the mechanism through which productivity gains result in real wage increases for the average worker. Increased efficiency in the production of one commodity lowers its price, resulting in higher real incomes for all consumers of that commodity. As a consequence, the demand for all normal goods increases, raising labour demand and therefore real wages throughout much of the economy. Average real wages rise, although there may be lower real wages in some individual labour markets. Productivity improvements tend to reduce costs of production, whereas wage increases raise costs. Thus productivity growth affects the relationship between wage changes and price changes. These aspects are discussed more fully below.

Other Determinants of Money Wage Changes

In addition to productivity growth, expected inflation, catch-up for unanticipated inflation and unemployment, a number of other variables have been suggested as determinants of money wage changes. In many cases, these variables are included (a) to control for certain factors, given the particular nature of the data set employed in the empirical work, or (b) to act as a proxy for other variables for which data was not available. In other cases, they are included in a more or less *ad hoc* fashion, often without a careful theoretical rationale, simply to reflect institutional features that are believed to affect money wage changes. While these factors may truly have an exogenous effect, they may also simply reflect a common set of underlying forces. Hence, it is important to have an appropriate theoretical foundation to sort out cause and effect — the crucial linkage for most policy purposes.

The rate of increase of wages in the United States, for example, may be regarded as a determinant of wage changes in Canada. This could occur through wage spillovers from key groups in the United States, through international unions, multinational corporations, or pressures for wage parity. Again, the extent to which U.S. wages would exert an independent influence, as opposed to Canadian and U.S. wages moving together in response to a common set of forces, would be a difficult issue to disentangle.

Profits have been regarded as a determinant of wage changes largely on the belief that high profits reflect a greater ability to pay or a higher opportunity cost of resisting wage demands and engaging in a strike. They could also reflect short-run demand changes over a business cycle, as well as reverse causation in that wage increases may affect profitability. Again, cause and effect becomes difficult to disentangle without a well-specified theoretical rationale.

Changes in unionization also have been regarded as an independent determinant of money wages. While unions can undoubtedly have an impact on the wages of their members, their impact on the rate of *change* of the *aggregate* wage level is a much more complicated matter. As discussed in Chapter 16, unions affect the wages and employment of both union and nonunion workers, the union-nonunion wage differential reflecting both higher wages for union members and a lower average wage in the nonunion sector (although some nonunion workers receive higher wages due to threat effects). The change in the aggregate wage level depends on the magnitude of these two offsetting effects. The empirical evidence indicates that, although unionization does lower the average wage in the nonunion sector, most of the differential is associated with higher wages in the union sector. Thus unionization may tend to raise the aggregate wage level. However, this constitutes a "one time" increase associated with the creation of a union-nonunion wage differential. In order to have a continuing effect on the aggregate rate of wage change, either an increase in the union-nonunion wage differential or an increase in the extent of unionization must occur. Calculations indicate that these contribute relatively little to the aggregate rate of wage change. For example, the aggregate wage level increased by 44 percent in the U.S. between 1967 and 1973; Ashenfelter (1978) concluded that the combined effect of increases in union density and in the union-nonunion wage differential accounted for only one to two percentage points of this increase. Similar conclusions have been reached in other empirical studies.

Market imperfections also have been discussed, especially in the context of insulating wages from competitive pressures. In the context of dual labour market analysis, for example, aggregate money wage changes could emanate from institutional changes, and from changes in administrative practices, or segmentation of the labour market. This would be facilitated by the ability to pass wage cost increases on to consumers via administrative pricing in noncompetitive product markets. To a certain extent the belief in these market imperfections and noncompetitive forces — most notably monopoly power in the product market and union power in the labour market — were part of the rationale for wage-price controls and guidelines. The exact manner in which these imperfections would affect the wage and inflation process, however, was seldom spelled out clearly. In particular, while imperfections could explain why the price and wage *levels* are higher than they would be in the absence of the imperfections, they do not explain the *changes* in prices and wages, unless one believes that the imperfections themselves change rapidly.

Unusual events — in particular, large wage settlements in specific sectors of the economy — often have been discussed as important determinants of aggregate money wage changes in the economy as a whole. The belief is that unusually high settlements in specific sectors have set off a chain reaction, spilling over into other sectors. In Canada, such key settlements allegedly have included Seaway workers in the late 1960s, construction trades in the 1960s and early 1970s, and the public sector in the 1970s.

While it can be tempting to regard these settlements as setting off inflationary wage settlements elsewhere, there is a danger in such *ad hoc* theorizing about the impact of peculiar events. Certainly they can affect the magnitude of wages in a particular sector, and hence they can affect the wages of that sector relative to other sectors, at least in the short run. Nevertheless, the mechanism whereby this affects aggregate wages in the economy as a whole is not clear. What appears to be a spillover effect may simply be a variety of sectors responding to the same set of economic forces. And even if there are purely institutional spillover effects, it is not clear that they will persist in the long run. There will always be a group that stands out as receiving unusually large settlements in a given short period of time. Some of the gain may reflect a catch-up, some may reflect a short-run demand increase, and some may be a purely transitory gain that will be dissipated over time. Unusually high wage settlements in a particular sector may be a *symptom* of any of a variety of factors; it is another matter, however, to argue that they are the *cause* of general increases in aggregate money wages.

A variety of public policies may also affect aggregate wage changes. In some cases these effects are intended, while in others they are incidental consequences. Incomes policies are designed to restrain wage and price increases. In

Canada these have generally taken the form of guidelines or direct controls on wage and price increases. The economic rationale for such policies and Canada's experience with them are discussed subsequently in this chapter. Changes in taxation may affect wage changes, although such effects are usually unintended consequences. Payroll taxes affect labour demand and income taxes affect labour supply. Thus if tax changes are large enough and apply throughout much of the economy, they may affect the aggregate wage level. As was the case with several other factors discussed in this section, such changes have a one-time impact on the aggregate wage level. However, if income tax rates are not indexed to changes in the price level then taxation may affect wage changes on a continuing basis in an inflationary period, as individuals are pushed into higher tax brackets due to increases in their nominal incomes associated with inflation.

In summary, in addition to the main determinants of wage changes, a number of other factors may affect labour demand or labour supply and therefore the wage level. In many cases these exert a one-time impact; however, some may exert a continuing influence on aggregate wage changes.

EMPIRICAL EVIDENCE

A large number of studies of the determinants of wage changes have been carried out in Canada. Early studies such as Kaliski (1964), Bodkin, Bond, Reuber and Robinson (1966), Vanderkamp (1966, 1972), Swidinsky (1972), and Turnovsky (1972) used as the dependent variable the annual or quarterly rate of change in an aggregate earnings index (such as average hourly earnings). Aggregate earnings indexes cover a substantial proportion of the labour force; however, there are several problems associated with their use in this context. Earnings depend not only on the hourly wage rate but also on factors such as overtime pay and bonuses. Average wage or earnings indexes may also be affected by changes in the composition of employment; for example, if firms lay off mainly junior, low-wage employees in a cyclical downturn and rehire these employees in the upturn, the average wage or earnings index will vary counter-cyclically even if wage rates and earnings for each type of employee do not. For this reason, it is preferable to use indexes based on fixed employment weights.

Another important measurement issue relates to the distinction between current and deferred wage changes. In the postwar period there has been a marked trend toward the use of multi-year contracts in the unionized sector. Multi-year contracts often contain deferred increases. The magnitude of these may be predetermined (i.e., fixed at the time the contract is signed) or indexed to changes in the Consumer Price Index via a cost-of-living-allowance or COLA clause. The observed change in average wages or earnings at any point in time is thus a mixture of current and deferred changes. Yet the determinants of these two types of wage change clearly differ.

Table 26.2 shows that the differences can be large. Column (3) shows the percentage change in base wage rates which occurred during the year; that is, contains both current and deferred increases in all major collective agreements. Column (5) shows the percentage change in base wage rates in new non-COLA settlements reached during the period. These are calculated on a life-of-the-contract basis; that is, they show the compound annual average increase in wages (both current and deferred) to take place over the duration of the agreement. Contracts containing COLA clauses are excluded because the value of these increases is not known at the time the contract is signed. The difference between the two series is at times as much as five percent; at other times the difference is negligible.

Another implication of multi-year contracts is that the fraction of the labour force negotiating a new agreement varies from year to year. In some years as few as 30 percent of major collective agreements are renegotiated while in other years more than 60 percent are renegotiated. Thus, some of the variation in aggregate earnings (or wage) indexes is simply due to the timing of the "bargaining calendar."

In view of these difficulties associated with aggregate wage index data, researchers have increasingly turned to data on individual contracts. This approach treats each negotiated settlement as an observation and to a considerable extent circumvents the two problems discussed above. In particular, deferred increases are not so problematic in that all the increases to take effect during the contract are included in calculating the percentage change in wages, which is then explained in terms of economic conditions prevailing at the time the contract was signed. Similarly, variations in the proportion of the labour force bargaining in each period are evidently less of a problem in that one observes the underlying rate of change of wages. In addition to lessening the measurement difficulties associated with analyzing the determinants of wage changes, the use of contract data also provides considerably more micro detail and therefore permits the testing of some hypotheses which would simply not be feasible with more aggregative data. Canadian studies based on individual wage contracts include Sparks and Wilton (1971), Cousineau and Lacroix (1977), Riddell (1979), Christofides, Swidinsky and Wilton (1980a, 1980b), Christofides and Wilton (1985) and Prescott and Wilton (1992).

Not all of the recent studies treat each individual contract as an observation. Reid (1979) and Riddell and Smith (1982) use more aggregative measures by employing as the dependent variable a weighted (by number of employees) average of the individual wage changes negotiated during the period; that is, a series such as that shown in Column (5) of Table 26.2. The use of these aggregate measures of the rate of change of wages does not raise the problems associated with aggregate measures obtained from wage index data. In addition, some researchers — such as Fortin and Newton (1982) — continue to use aggregate earnings indexes because these cover more of the labour force, including the nonunion sector.

Each of the empirical studies of Canadian wage changes employs the basic expectations-augmented Phillips curve specification discussed above. However, they differ in several ways, including the measurement of excess demand for labour, the measurement of inflationary expectations and in the role of catch-up for unanticipated inflation.

Each of the studies includes a measure of excess demand for labour. Generally this variable is found to be a significant determinant of wage changes; however, in contrast to Phillips' original study, recent analyses find that the aggregate unemployment rate is often insignificant or even perversely signed (e.g., Christofides, Swidinsky and Wilton, 1980a). This outcome is not unexpected. As discussed in Chapters 24 and 25 on unemployment, because of substantial changes in the age-sex composition of the labour force and in social programs such as unemployment insurance, the meaning of (say) a six percent unemployment rate in the 1970s differed considerably from that in the 1950s. In other words, the demographic trends and changes in social policies have raised the equilibrium unemployment rate, and consequently the aggregate unemployment rate has not remained a consistent measure of the "tightness" of the labour market.

There are several ways to deal with this situation. A number of authors have constructed adjusted unemployment-rate series which are intended to provide a consistent measure of excess demand and thus can be used in lieu of the measured unemployment rate to explain wage changes. Fortin and Newton (1982) is an example of this approach in the Canadian setting. These adjusted measures are typically weighted averages of the unemployment rates of different age-sex groups, with groups such as males aged 25–44 being assumed to contribute more to excess labour demand than females and youths. Alternatively, some authors use the job vacancy rate on the assumption that the developments that raised U° did not significantly alter the equilibrium job vacancy rate V° (Reid, 1979; Christofides, Swidinsky and Wilton, 1980a, 1980b).

An alternative procedure, used for example by Riddell and Smith (1982), is to measure aggregate excess demand for labour as the difference between the actual and natural unemployment rates, $U - U^{\circ}$ as in Equation 26.8. A second equation is added to account for changes in U° in terms of the demographic, legislative, and other changes. When this is done, the coefficient on the excess demand variable has the predicted sign and is statistically significant.

The conclusion that excess demand for labour exerts a significant influence on wage changes implies that there is a relationship between wage inflation and unemployment in the short run, holding constant inflationary expectations and the natural unemployment rate. Changes in the natural rate have shifted this short-run Phillips curve upward, resulting in higher wage inflation at each measured unemployment rate. Another important conclusion is that the short-run Phillips curve is fairly flat. Because the relationship is generally found to be nonlinear as depicted in Figure 26.3, the slope depends on the unemployment rate. Evaluated at the average unemployment rate during the sample period, a slope of 0.5 is a typical estimate, implying that a one percentage point increase in the unemployment rate reduces the rate of wage inflation by one-half of one percent. As discussed below, such estimates indicate that the costs of reducing inflation by monetary and fiscal restraint (thereby reducing excess demand in labour and product markets) may be extremely high.

Both expected inflation and catch-up for unanticipated inflation appear to play significant roles in wage determination. Empirical studies indicate that much of the upward shift in the Phillips curve during the 1960s and 1970s was associated with the rise in inflationary expectations. Catch-up forces also contributed to the upward shift in the late 1960s and early 1970s. This factor can work in either direction; more recently, with the decline in inflation in the early 1980s, lower than anticipated inflation has exerted additional downward pressure on wage increases.

A fundamental difficulty is that inflationary expectations are not generally observed. Empirical analysis thus requires constructing a proxy for expected inflation; as a consequence, the results represent a joint test of the theory of the determinants of aggregate wage changes and the theory of expectations formation. This joint hypothesis testing problem is exacerbated by the fact that any errors in measuring expected inflation will also cause errors in the measurement of unanticipated inflation.

Early research was often based on the hypothesis of adaptive expectations, according to which the expected rate of inflation could be expressed as a distributed lag of previous inflation rates. This approach was criticized on the basis that these expectations need not be consistent with the actual process generating inflation. In other words, the adaptive expectations hypothesis implies that those involved in forming expectations may repeatedly over- or under-predict inflation without revising their forecasting behaviour. For this reason, economists have increasingly adopted the hypothesis of rational expectations, according to which individuals' expectations represent an optimal forecast given the available information regarding the process that actually generates inflation. The most controversial aspect of rational expectations is the information set that market participants are assumed to possess and use in forming their expectations. In much theoretical work in macroeconomics, market participants are assumed to understand the structure of the economy, including knowing the parameters of the various relationships among variables. In empirical work on wage determination, much weaker versions of rational expectations are generally assumed. A common assumption is that in forecasting inflation individuals use only the previous inflationary experience as their information set. Rational expectations of inflation can generally be written as a distributed lag of previous inflation rates; however, the distributed lag weights must reflect the actual relationship between current and past inflation rather than being arbitrary as is the case under adaptive expectations.

A central empirical issue is the prediction of Friedman (1968) and Phelps (1967, 1968) that wage changes will adjust fully to reflect changes in expected inflation. In these circumstances there is no long-run relationship between wage inflation and unemployment; the long-run Phillips curve is vertical at the natural unemployment rate as depicted in Figure 26.3. As noted previously, this prediction implies $\lambda = 1$ in Equation 26.9. Tests of this "natural rate hypothesis" have generally confirmed the $\lambda = 1$ prediction. Some apparent exceptions are discussed below.

However expectations of future inflation are formed, they will often turn out to be incorrect. The catch-up variable is intended to capture the effects of over- or under-prediction of changes in the price level. When wage increases fully reflect expected inflation, the catch-up variable can be measured as the difference between the actual rate of inflation since the wage was last determined and the expected rate of inflation at that time, $\dot{p}_{t-1} - \dot{p}^e_{t-1}$. This catch-up specification was originally used by Turnovsky (1972) with aggregate data and subsequently employed by Riddell (1979) with individual contract data. Because the length of wage contracts varies considerably across bargaining units, it is difficult to incorporate the influence of unanticipated inflation adequately with aggregate data. With individual contract data, both expected inflation and catch-up can be precisely related to the actual timing of wage settlements.

If for some reason wage changes do not fully incorporate expected inflation, then the specification $\dot{p}_{t-1} - \dot{p}^e_{t-1}$ will not be appropriate. In this context, Christofides, Swidinsky and Wilton (1980a) have introduced the distinction between unanticipated and uncompensated inflation. The latter is the difference between the actual rate of inflation since the wage was last determined and the amount of expected inflation incorporated in the previous wage change, $\dot{p}_{t-1} - \lambda \dot{p}^e_{t-1}$. When $\lambda = 1$ unanticipated and uncompensated inflation are equal. However, Christofides, Swidinsky and Wilton argue that, because of uncertainty about future inflation, employers and employees may prefer wage settlements that do not fully incorporate expected inflation ex ante, leaving the remainder to be adjusted ex post. Thus they predict $\lambda < 1$ so that ex ante expectations contribute less to the current wage change and ex post adjustments for realized inflation contribute more. Christofides, Swidinsky and Wilton (1980a) find empirical support for this hypothesis. However, although they estimate $\lambda < 1$, the combined ex ante and ex post adjustment for inflation is equal to unity. Thus these empirical results also support the conclusion that there is no long-run relationship between wage inflation and unemployment.

Although the Christofides et al. (1980a) conclusion that λ is systematically less than unity is consistent with the proposition that there is no long-run trade-off between inflation and unemployment, the extent to which wages adjust to expected inflation on an ex ante versus ex post basis is clearly important for the short-run dynamics of unemployment and inflation. Even with full ex ante compensation for expected inflation, the existence of a catch-up adjustment for forecast errors implies that unanticipated disturbances will have effects that will persist for several years. The smaller the ex ante compensation and thus the larger the ex post compensation, the more significant are the delayed responses to economic shocks. Even if expectations are forward-looking, there is therefore an important backward-looking aspect to wage determination. Several Canadian studies suggest that this aspect is quantitatively significant.

The impact of the 1975–78 Anti-Inflation Program on wage and price inflation has been extensively investigated. Studies include Auld, Christofides, Swidinsky and Wilton (1979), Cousineau and Lacroix (1978), Christofides and Wilton (1985), Fortin and Newton (1982), Reid (1979), and Riddell and Smith (1982). As can be seen in Table 26.2, wage increases fell substantially during the 1975–78 period, with new non-COLA settlements declining from over 19 percent prior to the introduction of controls in October 1975 to under 11 percent in the first year of the program (1976) and under 8 percent in the last two years of the program (1977 and 1978). The main objective of the empirical research has been to determine how much of this decline (if any) can be attributed to the wage-price controls program and how much is associated with other factors such as the higher unemployment and slack economic conditions which prevailed during this period.

Empirical studies are unanimous in concluding that the controls program had a significant effect on the rate of wage increase. Most estimates of the impact on new wage settlements are in the range of 3 to 4 percent per year. This is a very large impact. Assuming the slope of the short-run Phillips curve to be approximately 0.5, the same direct effect on wage inflation would have required unemployment rates of 13 to 14 percent throughout the three-year period rather than the 7 to 8 percent unemployment rates actually experienced. Studies by Riddell and Smith (1982) and Christofides and Wilton (1985), which include data from the post-controls period, also conclude that the AIP was not followed by a wage explosion or "post-controls bubble," as has been the case in many other attempts at achieving wage restraint. The policy implications of these findings are discussed below.

PRICE INFLATION AND UNEMPLOYMENT

We have seen how actual and expected changes in the price level affect wages. However, because of their effect on labour costs, wage changes also influence prices. Thus any factor such as unemployment or expected inflation which affects the rate of change of wages will also affect the rate of change in prices as changes in labour costs associated with wage changes feed through into prices. This implies that there will exist a relationship between price inflation and unemployment, holding constant other factors, as well as between actual and expected price inflation. These relationships and their implications are briefly examined in this section. Further discussion is provided in textbooks on macroeconomics.

Because labour costs are a substantial proportion of total costs, a common assumption in empirical work on price determination is that the output price p can be written as a mark-up m times the labour cost per unit of output, or unit labour cost. The mark-up reflects both other costs such as raw materials and a normal rate of profit. Unit labour cost is simply the wage rate (dollars per employee-hour) divided by output Q per employee-hour N or average labour productivity. Thus

$$p = m \cdot \frac{W}{A} \tag{26.11}$$

where A is average labour productivity (Q/N). Totally differentiating (26.11), assuming the mark-up is approximately constant over time (dm = 0), and dividing the left-hand side by p and the right-hand by $m \cdot \frac{W}{A}$ gives:

$$\dot{p} = \dot{W} - \dot{A} \tag{26.12}$$

where $\dot{p} = dp/p$ and similarly for $\dot{W}$ and $\dot{A}$. This relationship between price inflation, wage inflation, and labour productivity growth is derived more formally below under competitive rather than mark-up pricing. Note that Equation (26.12) can be rewritten as

$$\dot{W} - \dot{p} = \dot{A} \tag{26.13}$$

which states that, with a constant mark-up on unit labour costs, real wages will grow at the rate of productivity growth. The historical tendency for real wages to increase at the rate of productivity growth was discussed previously. This analysis provides an explanation for this phenomenon; namely, that mark-ups on unit labour cost are approximately constant in the long run.

Combining Equations 26.12 and 26.9 gives the relationship between price inflation and unemployment:

$$\dot{p} = F(U - U^*) - \dot{A} + \lambda \dot{p}^e \tag{26.14}$$

Holding constant productivity growth and expected inflation, price inflation and unemployment are inversely related in the same way as wage inflation and unemployment. Changes in productivity growth or expected inflation shift this relationship upward or downward. Figure 26.5 illustrates this "price Phillips curve" relationship for the case in which the Friedman-Phelps prediction that $\lambda = 1$ holds. As before, there is a family of price Phillips curves, each for a different expected inflation rate. These are referred to as short-run Phillips curves because they depict the relationship between inflation and unemployment during the period in which inflationary expectations do not adjust to changes in the actual rate of inflation. For example, by employing expansionary monetary and fiscal policy, it may be possible to move the economy from point a in Figure 26.5 to point b. However, the latter outcome is not sustainable in the long run because actual inflation exceeds expected inflation. The set of outcomes that can be maintained on a permanent basis are those points such as a, c, and d in Figure 26.5 at which actual and expected inflation are equal. At any other points, actual and expected inflation will diverge, causing employers and employees to adjust their expectations and consequently their behaviour.

This analysis implies that there is a temporary or short-run tradeoff between inflation and unemployment, other things being equal. However, there is no permanent or long-run tradeoff between these two variables.[3] The long-run relationship between inflation and unemployment is vertical at the natural or equilibrium unemployment rate U*, as depicted in Figure 26.5. Formally, macroeconomic equilibrium requires that

$$\dot{p}^e = \dot{p} \tag{26.15}$$

---

3. If workers display permanent money illusion ($\lambda < 1$ in Equation 26.14), there will be a tradeoff between inflation and unemployment even in the long run. This tradeoff will be steeper than the short-run counterpart. See question 8 at the end of this chapter.

Figure 26.5
THE RELATIONSHIP BETWEEN INFLATION AND UNEMPLOYMENT

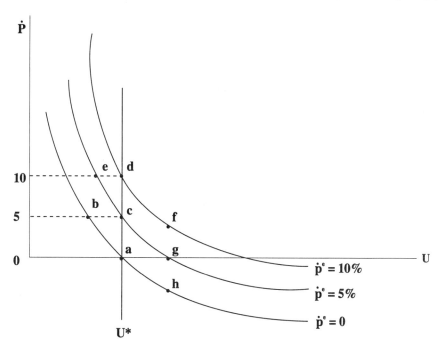

Substituting Equation 26.15 into Equation 26.14 gives the long-run or equilibrium relationship between inflation and unemployment:

$$F(U - U^*) - \dot{A} = 0 \tag{26.16}$$

Because the rate of inflation does not appear in Equation 26.16, Equations 26.15 and 26.16 are consistent with any steady rate of inflation.

The absence of a permanent trade-off can be illustrated by examining the implications of attempting to maintain unemployment below the natural unemployment rate U^* on a continuing basis. For example, beginning in long-run equilibrium at point a in Figure 26.5, the authorities may be able to move the economy to point b via expansionary monetary or fiscal policy. However, once workers and firms adjust to the higher rate of inflation, the short-run Phillips curve will shift up and the economy will return to U^* at c (given the higher level of inflationary expectations). Attempting to maintain the lower level of unemployment will require further stimulus to the economy, resulting in higher inflation such as at point e. Thus a policy of maintaining $U < U^*$ will be associated with steadily increasing inflation. For this reason, U^* is sometimes referred to as the "Non-Accelerating Inflation Rate of Unemployment" or NAIRU. As the NAIRU designation emphasizes, at $U = U^*$ the rate of inflation will tend to remain constant, whereas inflation will tend to accelerate if $U < U^*$ is maintained over an extended period.

This discussion illustrates how overly expansionary aggregate demand policies may have contributed to the upward drift in inflation during the 1960s and early 1970s, as described previously in this chapter. Such policies may have been adopted for several reasons: (1) because of a belief that there existed a permanent trade-off between inflation and unemployment (a view that was prevalent during that period) and that it was preferable to accept a bit more inflation in exchange for lower unemployment; (2) because of a belief that the unemployment rate consistent with steady inflation was in fact lower than the true value of U^*; or (3) possibly with full recognition that the consequences would be increased inflation in the longer term, albeit lower unemployment and a more buoyant economy in the short run.

The reduction of inflation by a policy of demand restraint can also be illustrated with reference to Figure 26.5. Beginning at a point like d with an ongoing inflationary spiral — in which wages are rising because workers and firms expect prices to rise and prices are rising due to higher wages — a sharp reduction in aggregate demand such as occurred in 1981/82 will cause unemployment to rise and inflation to fall, moving the economy to a point such as f in the short run. As expectations adjust to the lower rate of inflation, the Phillips curve shifts down, enabling the authorities to achieve a permanently lower rate of inflation such as at point c. Alternatively, continued restraint would move the economy from f to g to h, eventually resulting in a return to price stability at point a. This process of a gradual reduction in inflationary expectations characterizes the outcome of the anti-inflation policy followed in Canada during the 1980s and early 1990s.

UNEMPLOYMENT PERSISTENCE

During the 1960s the notion of a stable Phillips curve or trade-off between inflation and unemployment became incorporated into macroeconomic thinking, and into the models used by policymakers and economic forecasters. This era began with the important early contributions by Phillips (1958) and Lipsey (1960). Canadian research along these lines included Kaliski (1964) and Bodkin et al. (1966). However, as discussed above, the notion of a stable trade-off between inflation and unemployment was unable to account for observed behaviour during the late 1960s and 1970s, when both higher rates of inflation and unemployment were observed. In addition, this perspective came under increased attack on both theoretical and empirical grounds for its failure to take into account the role of expectations in the inflationary process. Following a flurry of theoretical and empirical research, together with a process of debate and controversy, the notion of a Phillips curve or trade-off between inflation and unemployment was discarded and replaced by the model developed above, characterized by a unique equilibrium unemployment rate and the absence of a trade-off between inflation and unemployment in the long run. By the mid- to late-1970s, the Phelps-Friedman "natural rate" view had become economic orthodoxy.

However, macroeconomic experience during the 1980s — especially that in Europe — has resulted in considerable questioning of the natural rate paradigm. In many European countries, unemployment rates more than doubled between 1970 and 1980 and by 1990 were three to four times their 1970 levels. Yet, in most countries, inflation during this period was approximately stable, or at best declined modestly. To be consistent with the natural rate view, this behaviour requires massive increases in the natural rate. Although some increases in the equilibrium level of unemployment may have occurred because of changes in social and labour market legislation and changes in the structure of the labour force, such changes appear unlikely to be able to explain a rise in the equilibrium unemployment rate by a factor of three to four times. In Canada, many of the factors used to explain the rise in the equilibrium level of unemployment during the 1960s and early 1970s (increased labour force participation of women and youths, higher real minimum wages, more comprehensive and more generous unemployment insurance) either levelled off or were reversed during the latter half of the 1970s and during the 1980s; on these grounds, the natural rate would have been expected to decline (Kaliski, 1985). However, average levels of unemployment in Canada were higher during the 1980s rather than lower.

Economists have accordingly devoted increased attention to the possibility that changes in the actual unemployment rate may themselves result in changes in the natural rate, so that economic shocks which increase unemployment may also raise the equilibrium level of unemployment. This reassessment of the natural rate paradigm is an area of active research and debate. In this section we describe some of the central issues being examined; however, the conclusions at this stage must necessarily be very tentative.

Two related research issues are being addressed. First, are theoretical and empirical models capable of explaining the persistence of high levels of unemployment? As noted, this persistence is most evident in Europe, but has also characterized the Canadian experience, albeit to a lesser extent, during the 1980s, as inspection of Figure 24.1 and Tables 26.1 and 26.2 will confirm. (See Exhibit 24.1 for a summary of the European experience.) Explanations include insider-outsider models of wage setting, the loss of physical and/or human capital during economic downturns, and persistence generated by the features of the unemployment insurance system. The second issue involves the possibility of multiple equilibria, so that the economy could settle down at an equilibrium with a low natural unemployment rate or one with a high rate.

As discussed in Chapter 25, insider-outsider models emphasize the significant role played by currently employed (and perhaps also recently employed) workers in the wage-setting process (the "insiders") and the much less significant role of unemployed workers (the "outsiders") in that process. A shock to aggregate demand which reduces employment may thus reduce the number of insiders and increase the number of individuals who exert little influence on the wage determination process. The interests of workers laid off may be represented for a certain period of time, after which they drift away and become disenfranchised. Thus the increase in unemployment may not produce much downward pressure on wage changes, in contrast to the standard natural rate model where increases in unemployment above the equilibrium level produce downward pressure on wage settlements (relative to expected inflation).

Persistence in unemployment may also be associated with decay in physical or human capital during a recessionary period. Individuals who have been unemployed for a long period may suffer a deterioration in their labour market skills or in the intensity with which they search for work. Such effects would imply that their probability of re-employment falls with the duration of unemployment. It may also be the case that employers are less likely (other things being equal) to employ those who have been unemployed for a long period.

Decay in physical capital can arise due to bankruptcies and plant closures during a slump in economic activity. These events lower labour demand at each wage rate, thus lowering the equilibrium level of employment and raising the natural unemployment rate. If entry by new firms during the subsequent recovery occurs only slowly, one consequence of a recession may be a higher natural unemployment rate for an extended period of time.

These mechanisms imply that short-run shocks can have long-run (although not necessarily permanent) effects. An economic downturn can reduce the number of insiders, with the consequence that wages are subsequently set at the higher level required to employ the smaller number of insiders. This mechanism appears to be consistent with the combination of persistent high unemployment and substantial real wage growth among employed workers observed in many European countries during the 1970s and 1980s (Blanchard and Summers, 1986; Blanchard, 1991).

The deterioration of physical and/or human capital is also an example of a long-run consequence of a short-run cyclical phenomenon. During a recession, the number of unemployed job seekers rises and employment opportunities decline; consequently the probability of re-employment falls and the duration of unemployment rises. In a severe recession, many of the long-term unemployed may suffer a decline in their job-related skills, the intensity of their search effort, or they may become accustomed to public assistance in the form of unemployment insurance or social assistance. In these circumstances, the long-term unemployed may become "outsiders," a group with little labour market attachment and one which exerts little restraining influence on wage settlements. The reduced impact on wage bargaining (for a given level of the unemployment rate) implies an increase in the equilibrium level of unemployment.

The extreme case of "hysteresis" occurs when the unemployment rate may drift upward or downward, without any tendency to return to an equilibrium level. Hysteresis exists when short-run shocks have permanent effects, so that the concept of a natural unemployment rate becomes irrelevant.

Whether or not cyclical downturns have long-lasting "scarring" effects on the labour market — including whether or not the labour market is characterized by hysteresis — is an important policy issue. For example, according to the natural rate hypothesis, reducing inflation involves a trade-off between temporary costs (higher unemployment and lower output during the period of monetary and fiscal restraint) and permanent gains (a permanently lower rate of inflation). The costs are temporary because once the rate of inflation has been brought down the economy can return to normal levels of output and employment. However, if economic downturns have long-run effects then the costs of reducing inflation by demand restraint are larger; in the extreme case of hysteresis the costs as well as the benefits are permanent. Because a substantial reduction in output and employment is required to achieve even small reductions in inflation, the wisdom of reducing inflation by demand restraint is very dependent on the degree to which recessions have long-run adverse effects.

The structure of Canada's unemployment insurance system may also contribute to persistence in unemployment, an aspect investigated by Milbourne, Purvis and Scoones (1991). As discussed in Chapter 25, most research finds a relationship between the generosity and maximum duration of UI benefits and the duration of unemployment. This relationship, by itself, does not make unemployment more persistent, although a more generous UI program has the dual effects of making an economic downturn less painful (i.e., performing its primary role of providing insurance against the risk of unemployment) and making the increase in unemployment during a recession larger than it would otherwise be (because workers laid off take longer than they would in the absence of UI, or with less generous UI, to obtain work). However, since 1978–79, Canada's UI program has had a "regional extended benefit" provision which can contribute to persistence. Specifically, the maximum duration of UI benefits and the minimum number of weeks of benefits needed to qualify for UI depend on the regional unemployment rate. Thus as unemployment rises in an economic downturn, more and more regions have shorter qualifying periods and longer duration periods, thus making unemployment more likely to last longer. Milbourne, Purvis and Scoones (1991) find some evidence that the degree of persistence of Canadian unemployment did increase after 1978–79, a change which could be due to this institutional feature of Canada's UI program.

Related to the issue of hysteresis is the possibility of multiple equilibria. The natural rate hypothesis is based on the assumption that the economy is characterized by a unique equilibrium level of employment and unemployment. Although a unique general equilibrium is predicted by an important class of economic models, many theoretical models which appear reasonable on a priori grounds are characterized by multiple equilibria. In particular, equilibrium models of employer and employee search and matching in the labour market often have multiple rather than unique equilibria, as discussed by Mortensen (1989) and Pissarides (1990). (See also Diamond, 1982; Howitt, 1988; and Howitt and McAffee, 1987.)

For example, an economy may be characterized by two possible equilibria, one in which employers and workers are optimistic, so that firms are searching for workers because they believe that expanding employment will be profitable in the future and unemployed workers are actively seeking work because they believe employers will be hiring in the future. In contrast to this "high level equilibrium" is an equilibrium at a low level of economic activity in which, because of pessimism about the future, most employers are not hiring and many workers are not actively searching. In these circumstances, the economy can become stuck in a low-level equilibrium — a situation (because it is an equilibrium) which the economy does not, on its own, have a tendency to move away from.

International evidence on hysteresis in unemployment is mixed. The evidence is strongest for Europe in the 1970s and especially the 1980s (Blanchard and Summers, 1986, 1988; Dreze and Bean, 1990), but weaker when behaviour over longer periods and more countries is examined. The dramatic growth of long-term unemployment in Europe appears to have played an important role; many of the long-term unemployed seem to have become excluded from the labour market on a continuing basis. Whether this outcome is due to insider-outsider wage setting, loss of job-related skills during unemployment, reduced search intensity and greater discouragement, or because employers are reluctant to hire those who have been unemployed the longest is an important area of research. Layard, Nickell and Jackman (1991) report detailed findings on these questions for the U.K. labour markets.

Canadian studies almost uniformly reject hysteresis (Cozier and Wilkinson, 1991; Fortin, 1991; Poloz and Wilkinson, 1992; see Jones, 1992, for a useful review and extension of Canadian evidence). However, there is evidence of increased persistence in unemployment during the 1980s (McCallum, 1987; Fortin, 1991). Whether this change is due to the changes made to Canada's UI system in the late 1970s, as is suggested by the work of Milbourne, Purvis and Scoones (1991), or perhaps due to long-term effects of the severe 1981–82 recession, is an important unanswered question.

THE WAGE-PRICE-PRODUCTIVITY NEXUS: MATHEMATICAL EXPOSITION*

The relationship between wage changes, price changes, and productivity growth was derived above under the assumption that prices are set according to a constant mark-up on unit labour costs. This relationship can also be derived under competitive pricing behaviour. To illustrate this we assume output Q is related to the inputs of labour N and capital K according to a Cobb-Douglas production function:

$$Q = \alpha N^\beta K^\gamma \tag{26.17}$$

The marginal product of labour can be derived as

$$\partial Q/\partial N = \beta \alpha N^{\beta-1} K^\gamma = \beta(Q/N) = \beta A \tag{26.18}$$

where A is the average product of labour. As indicated in our earlier discussion of the demand for labour, a firm which is competitive in the product and labour markets will employ labour to the point at which the value of the marginal product of labour equals the wage rate. Therefore

$$W = p\beta A \tag{26.19}$$

Totally differentiating Equation 26.19 and dividing the left-hand side by W and the right-hand side by pβA yields:

$$\dot{p} = \dot{W} - \dot{A} - \dot{\beta} \tag{26.20}$$

Interpretation of this equation is aided by noting that ß is labour's share of the value of output. This can be seen by multiplying both sides of Equation 26.19 by N and dividing by pQ to obtain:

---

*Indicates more difficult material

$$\beta = \frac{WN}{pQ} \qquad\qquad (26.21)$$

Thus Equation 26.20 states that under competitive pricing, the rate of change of prices will equal the rate of change of wages minus the rate of productivity growth when labour's share is constant. This result is similar to that derived earlier under mark-up pricing except that the assumption of a constant mark-up is replaced by constancy of labour's share.

Equation 26.20 also indicates that if real wages increase at the rate of growth of labour productivity then labour's share will be constant. That is, if $\dot{W} = \dot{p} + A$ then $\beta = 0$ (labour's share would not increase). This is important because one could easily make the mistake of thinking that if wage increases exceeded inflation by the amount of the productivity increase, then labour is receiving *all* of the gains of the productivity increase. This is not so: labour's share would not increase under such circumstances. In fact all factors of production can have an increase in their real return equal to the productivity increase and there would be no change in factor shares.

ANTI-INFLATION POLICY

The problems of inflation and unemployment have been at the forefront of the policy agenda throughout the past several decades. They remain major challenges today. This section briefly examines the main policies that have been advocated and/or adopted in order to achieve price stability and low levels of unemployment. These policies are discussed in terms of the two major challenges that policymakers face in dealing with inflation and unemployment.

1. how to maintain desirable levels of employment without steadily increasing inflation; and
2. how to stop or reduce in severity an inflationary wage-price spiral should one develop.

Full Employment and Price Stability

According to the natural unemployment rate paradigm described in this chapter, there is a trade-off between inflation and unemployment in the short run but not in the long run. This conclusion, if correct, has significant implications for economic policy. In particular, it implies that attempting to achieve lower levels of unemployment in exchange for higher rates of inflation is not a feasible option on a continuing basis. Using monetary and fiscal policies to maintain the economy at an unemployment rate below the natural unemployment rate or NAIRU will lead to a continually increasing inflation rate. Thus when the objectives of price stability and high levels of employment are considered together, the natural unemployment rate emerges as an appropriate definition of full employment: it is the unemployment rate that is feasible on a sustained basis and is compatible with price stability.

At the same time, the objectives of full employment and price stability are compatible with each other, providing that full employment is defined as the level of employment at which $U = U^*$. If an inflationary wage-price spiral develops, it may be necessary to accept higher unemployment and reduced output *temporarily* in order to restrain inflationary pressures, but it is not necessary to accept unemployment in excess of the natural rate on a *permanent* basis.

The natural unemployment rate or NAIRU is an appropriate definition of full employment in the sense that it is compatible with price stability (or any steady rate of inflation) whereas lower unemployment rates are not. However, U^* may be considered too high in terms of other social objectives such as minimizing the hardship, poverty, and lost output associated with unemployment and providing ample employment opportunities. In these circumstances, policymakers should attempt to lower U^* directly using structural and labour market policies rather than attempting to lower U below U^* using monetary and fiscal policies. Policies to reduce frictional, structural, and seasonal unemployment include mobility and adjustment assistance, provision of labour market information to improve the search and matching process, and training and retraining. Labour market policies that contribute to a higher NAIRU, such as minimum wages and unemployment insurance, could also be examined. In designing these programs, policymakers need to recognize the trade-off that exists between the immediate objectives of the program (such as providing insurance against the risk of unemployment in the case of UI) and the consequences of the program for the amount of frictional, structural, and seasonal unemployment in the economy.

Pursuing a full employment goal that is consistent with price stability (or a low stable rate of inflation) may succeed in preventing a wage-price spiral from developing, as occurred during the 1960s and 1970s. However, should wage and price inflation again climb to unacceptable levels several alternative policies are available for halting an inflationary spiral — ranging from pure reliance on demand restraint to various forms of incomes policies. The theoretical and empirical analysis discussed in this chapter sheds some light on the costs and benefits of these alternative anti-inflationary strategies.

Demand Restraint

There is general agreement among economists that restraint in aggregate demand — in particular in the rate of increase of the money supply — must play a role in halting an inflationary wage-price spiral. There are, however, differences in opinion regarding how costly such a policy will be in terms of high unemployment and reduced output. Those who believe that the costs of reducing inflation are extremely high often advocate additional policies such as wage and price controls.

In the context of the framework developed in this chapter, demand restraint reduces the amount of excess demand for labour (lowers U relative to U*), thus reducing the rates of wage and price increases. As actual inflation falls, expectations of future inflation decline and the short-run Phillips curve shifts down. Demand restraint may also impact directly on expectations. If those involved in wage and price determination form expectations rationally, they will realize that a policy of demand restraint will reduce wage and price inflation. Thus, if they believe the authorities are determined to follow this policy, they will lower their expectations, thus directly reducing wage and price increases.

The problem with this approach is that it may require a prolonged period of high unemployment and weak economic conditions in order to achieve a significant reduction in the rate of inflation. This will especially be the case if the restraint policy has little direct impact on expectations — either because those involved in wage and price determination do not form their expectations in a rational, forward-looking manner or because they do not believe that the announced policy will be followed. In these circumstances, the initial impact of the reduction in aggregate demand (or its rate of increase) will be on output and employment. The inflationary spiral can be broken, but only at a very high cost. Whether the costs are worth incurring depends on the benefits of achieving a (permanent) reduction in inflation. The approximate magnitude of these costs can be calculated from the empirical studies of the determinants of wage and price changes discussed previously in this chapter. If the slope of the short-run Phillips curve is approximately 0.5, each one percent reduction in the rate of wage and price inflation requires an increase in the unemployment rate of two percentage points. This estimate does not allow for any direct impact of demand restraint on expectations or for a variety of other factors which a complete analysis would want to take into account. Nonetheless, inspection of Table 26.2 indicates that this rough estimate predicts the experience of the demand restraint policy implemented in 1981–82 reasonably well. Between 1981 and 1985, price inflation declined from 12.5 to 4.0 percent, a drop of 8.5 percentage points. The decline in wage inflation was similar in magnitude. This reduction in wage and price inflation was accompanied by an increase in unemployment of 3.5 percentage points in 1982 (11.0–7.5), followed by increases of 4.4 in 1983 (11.9–7.5), 3.8 in 1984 and 3.0 in 1985. During this four year period, the cumulative increase in unemployment above its previous level was 14.7 percentage points. The reduction in the rate of inflation was 8 percent, just slightly more than would be predicted by assuming a slope of 0.5 for the short-run Phillips curve and no direct impact on expectations.

The experience of the 1981–82 recession and its aftermath provides a vivid illustration of both the effectiveness of a determined policy of demand restraint in bringing about a substantial reduction in inflation and the extraordinarily high costs associated with this approach. As a consequence of the latter, economists and policymakers have searched for — and in some cases implemented — alternative ways of making the transition to a lower inflation rate with less severe adverse consequences. The following are discussed in turn: enhancing credibility, incomes policies, and increased wage and price flexibility.

Enhancing Credibility

If those involved in wage and price determination form their expectations in a rational, forward-looking manner, they will take into account the policy stance of the monetary and fiscal authorities in predicting future inflation. In these circumstances, if the authorities can make believable their commitment to a policy of demand restraint, the transition to a lower rate of inflation can be made more easily due to the policy's direct impact on expectations. As a consequence, more of the reduction in aggregate demand falls on wages and prices and less on employment and output.

Establishing or enhancing credibility is not a simple matter. Simply announcing that a restraint policy will be followed is unlikely to be effective. The citizenry may be skeptical about the ability or willingness of the authorities to continue with such a potentially costly policy. Over time, credibility can be enhanced by adopting and maintaining a policy stance — such as nonaccommodation of inflation — which can be learned by firms, unions, and workers and taken into account in wage and price determination. For this reason, macroeconomists have increasingly focused on the question of the appropriate policy rules to be followed by the monetary and fiscal authorities (Taylor, 1982).

However, experience indicates that enhanced credibility is not easily achieved in the short run. A dramatic shift to a disinflationary stance was made by the monetary authorities in the United States and the United Kingdom in 1979. Canada followed in 1981. The determination to persist with this policy was clearly and repeatedly stated by the relevant authorities. This experience, particularly that of the United States and the United Kingdom, provides a test of the hypothesis that inflation can be reduced at low cost in terms of lost output and employment by committing to and carrying out a determined policy of demand restraint. A large number of studies of these policy experiments, reviewed in Riddell (1986, Ch. 3), have been carried out. The results do not support the hypothesis. The costs of reducing inflation by this determined policy of demand restraint were extremely high and do not appear to have been significantly reduced by the attempt to make a clear commitment to a disinflationary stance.

Incomes Policies

The term "incomes policy" refers to a wide range of programs that intervene in wage and price setting in order to influence the rate of inflation. These include voluntary mechanisms such as the setting of guidelines or attempting to reach agreement on norms for wage and price increases, mandatory controls on wages, prices or profits, and incentive-based incomes policies that employ taxes, subsidies, or other incentives to restrain wage and price increases.

In the postwar period governments have experimented with a variety of incomes policies. In Canada, these experiments included a number of attempts at voluntary restraint such as that involving the Prices and Incomes Commission during the late 1960s and early 1970s. In addition, mandatory wage and profit controls were a central part of the Anti-Inflation Program of 1975–78. Wage controls were also applied to public sector employees in the federal "6-and-5" program and associated provincial programs introduced in 1982 to complement the demand restraint policy in effect at that time. Similarly, in the United States, incomes policies were employed during the Korean War period (the Wage Stabilization Board of 1950–53), during the Kennedy and Johnson administrations (the guideposts of 1962–66), during the Nixon administration (the wage-price controls of 1971–74), and during the Carter administration (1977–81). In many European countries, periods without incomes policies of some form were less common than periods with such policies. There is thus a rich history of experience with incomes policies from which we may be able to infer their advantages and disadvantages.

In principle, temporary wage and price controls can achieve a lasting reduction in inflation provided they are carefully coordinated with a program of demand restraint. In effect, the direct controls ensure that the reduction in aggregate demand (or in its rate of growth) falls primarily on wages and prices rather than on employment and output. Expectations of future inflation will also decline if the citizenry observes both lower actual inflation and monetary and fiscal policy that is appropriate for maintaining the lower rate of wage and price increase. This rationale for a temporary incomes policy combined with demand restraint is very similar to that for enhancing credibility. Both indicate the possibility of breaking an inflationary wage-price spiral without incurring severe adverse consequences in the form of high unemployment and reduced output.

Experience with wage and price control programs in the United Kingdom and the United States, reviewed in more detail in Riddell (1986, Chapter 4), has been mixed. In several cases the programs appear to have had little impact other than of a temporary nature. In other cases the policy appears to have achieved some moderation in inflation. A common characteristic of the unsuccessful programs is the failure of the authorities to combine the incomes policy with the appropriate demand restraint. In these circumstances the controls program and monetary and fiscal policy are working at cross purposes. Thus even if controls are successful in restraining inflation in the short run, their long-run impact is minimal as wage and price increases move sharply upward in the post-controls period. In some cases the overly expansionary monetary and fiscal policy and the concomitant excess demand in product and labour markets led to the demise of the incomes policy.

The Canadian experience, in particular the Anti-Inflation Program of 1975–78, has been more favourable. As discussed previously, econometric studies generally conclude that the wage and price controls reduced new wage settlements by 3–4 percent in each of the three years of the program, and price inflation by 1–3 percent per year, with the reduction in price inflation concentrated in the second half of the program and continuing well beyond the end of the AIP. The somewhat greater success of Canada's 1975–78 controls program relative to the experience of the United States and the United Kingdom can probably be attributed to two key design features: the coordination of the wage and price controls with monetary and fiscal restraint, and the use of gradually declining norms for wage and price increases. The latter are important in Canada (and in the United States) because wage contracts in the union sector are often two to three years in length. Attempting to achieve an extremely rapid reduction in the rate of wage change will cause larger inequities in the relative wage structure as groups whose contracts come up for renewal early in the controls program are treated substantially differently than those who renegotiated wages prior to the imposition of controls. Gradually declining norms for wage increases facilitate winding down the wage-price spiral without causing severe inequities that may undermine support for the program and impose high penalties on certain groups.

Although both economic theory and empirical evidence indicate that the combination of wage and price guidelines and appropriate restraint in aggregate demand represent an attractive alternative to reliance on demand restraint alone to reduce inflation, it would be a mistake to resort to controls on a regular basis. The interference with individual freedom and the inevitable inequities associated with mandatory limits on wage and price increases result in a loss of goodwill between different groups in society and alter the relationship between the citizens and the state. For this reason and others, controls should be limited to situations in which there is widespread agreement on the need to reduce inflation to more acceptable levels.

Because direct controls are a tool that should be used infrequently, and because the experience of some countries with their use has been disappointing, there has also been considerable interest in alternative policies for restraining inflation. Incentive-based incomes policies are schemes that create incentives for moderating wage and price increases without otherwise interfering with market forces. Tax-based incomes policies have received the most attention (Seidman, 1978, 1981; Layard, 1982). These utilize taxes or subsidies to provide incentives for wage and price restraint. There is little experience with these policies, but they continue to receive attention as an alternative to direct wage and price controls.

Some economists (Lerner, 1978; Lerner and Colander, 1980) have advocated elaborate schemes in which firms would require permits to increase prices (or wages) by a certain amount. The number of permits issued to firms would be restricted in order to achieve the desired overall rate of inflation. Permits would trade in an open market so that firms that wish to raise their prices more than average could purchase additional permits from firms wishing to raise prices less than average. In this way, relative prices can change despite the restraints on the overall rate of price increase.

Encouraging Wage and Price Flexibility

Disinflation via demand restraint is a costly policy to pursue because of the persistence or momentum in the inflationary process. Because of this inertia, most of the initial impact of the reduction in aggregate demand falls on output and employment. Only subsequently, as weak product and labour market conditions take their toll, do wage and price increases moderate. If wages and prices were more responsive to changes in economic conditions, inflation could be reduced without massive increases in unemployment. Enhanced credibility may be an effective means of increasing wage and price responsiveness if the main source of persistence is slowly adjusting expectations. However, if inertia is due to other factors, alternative approaches may be more successful. An approach that has received considerable attention involves altering the institutional features of wage determination so that wages and prices adjust more quickly to changing economic conditions.

Countries differ considerably in their wage-setting arrangements (Sachs, 1979; Gordon, 1982; Riddell, 1983). A distinguishing feature of the Japanese economy is the extensive use of bonus payments. These semi-annual bonuses represent a substantial proportion of Japanese employees' total income (about 25 percent on average); their magnitude in any year depends in part on the economic performance of the firm and industry. Japan is also characterized by short-term wage contracts (one year) and synchronized bargaining (wage negotiations take place at approximately the same time) in the form of the "spring wage offensive." Canada and the United States are unique in the degree to which long-term (often two or three years) overlapping (rather than synchronized) wage contracts are common in the union sector. The extent of wage indexation in the form of cost-of-living-allowance clauses is low; thus real wages are not rigid but nominal wages display considerable inertia. European countries vary. Some (e.g., Italy) employ long-term contracts but these do not overlap. Several (e.g., Sweden, Norway, Austria, Germany) have short duration contracts and synchronized, usually centralized bargaining. The United Kingdom is characterized by one-year contracts (although these do not have a predetermined expiry date as in North America) and nonsynchronized negotiations. Many European countries have wages either explicitly or implicitly indexed to changes in the price level; wages thus display "real wage rigidity."

These countries also differ substantially in their macroeconomic performance during the past two decades. Japan weathered the contractions of the postwar period, including the 1973 and 1979 OPEC oil price shocks, with relatively little increase in unemployment. Some of this is undoubtedly due to other factors, but their wage-setting institutions (synchronized bargaining, annual contracts, substantial bonus payments) probably contributed to this remarkable performance. Canada and the U.S. found it difficult to restrain inflation at low cost, and suffered significant increases in unemployment in response to the two oil price shocks. In both countries, however, unemployment subsequently began to decline. However, two severe recessions (in 1981–82 and 1990–92) were evidently needed to achieve a return to low levels of inflation. Although there are important differences among the European countries, as a group they suffered the largest increases in unemployment in response to the 1973 and 1979 OPEC oil price shocks. Furthermore, in many of these countries unemployment continued to rise during the 1980s, often to levels that are three to four times those experienced in the 1960s.

Research by Taylor (1980, 1983) and others has indicated that the long-term overlapping contracts used in the union sector in the United States are an important source of persistence in wage and price inflation. Even with rational expectations and a fully credible disinflationary policy, wage and price changes respond only slowly to a reduction in aggregate demand. In Canada, the use of long-term overlapping contracts and the extent of observed persistence in wage settlements is broadly similar to that in the U.S. (Riddell, 1983).

Although some economists have proposed imposing shorter wage contracts as a means of increasing wage and price responsiveness and thus reducing the costs of following a disinflationary policy, another proposed reform that has received considerable attention is Weitzman's (1984) suggestion that profit-sharing be adopted on a widespread basis. The bonus payments used in Japan can be viewed as a form of profit-sharing (although bonuses are not explicitly tied to profits), and Weitzman has attributed Japan's remarkable economic performance with respect to inflation and unemployment to the widespread use of these bonuses.

Weitzman (1984) argues that there is a fundamental difference between an economy in which workers are paid a fixed hourly wage and one in which workers are paid a share of revenue or profit. In the wage economy, the labour market is in equilibrium when the demand for labour equals the supply of labour. In the share economy, however, in equilibrium there is unsatisfied demand or unfilled job vacancies at the prevailing negotiated shares. This excess demand for labour will act as a cushion, protecting the economy from significant deviations from full employment. Any reduction in labour demand will be offset by the existence of unfilled vacancies.

The reason for this important difference is that with the share-compensation system the firm can increase profits by expanding employment and output because, with profit-sharing, the marginal cost of labour is a declining function of employment. Thus as the firm expands employment the average cost of labour (and compensation per employee) declines. With the wage system the firm will expand employment only to the point at which the additional revenue generated by the extra employment equals the additional cost which is the hourly wage.

Flexible compensation arrangements such as profit-sharing and gain-sharing, which relate employee remuneration to the economic performance of the firm or industry, are often recommended for their *micro* benefits at the level of the individual enterprise. By giving employees a greater financial stake in the enterprise and an opportunity for benefiting directly from its performance, the adversarial nature of labour-management relations may be replaced by a more cooperative, problem-solving approach. Additional potential benefits include increased levels of employee motivation, leading to higher productivity and organizational effectiveness and thus greater income to be shared by employers and employees. However, it may be the case that the most significant effects of these compensation arrangements are *macro* rather than micro benefits. Indeed, as Weitzman (1984) argues, there may be a rationale for public policy to encourage these compensation arrangements (for example by taxing income in the form of profit shares at a lower rate than income from a fixed wage) because of a "macroeconomic externality." In particular, benefits in the form of greater stability in employment and output accrue to society as a whole and not just to the parties choosing a compensation system.

In summary, the high degree of real wage rigidity displayed by many European countries is believed by many analysts to be a dominant factor accounting for the sharp rise and severe persistence of unemployment in those societies. Canada and the U.S., characterized by nominal wage inertia but more flexible real wages, have suffered large increases in unemployment but the higher levels of unemployment have not persisted as in Europe. Nonetheless, the process of reducing inflation in North America has been very costly and alternative approaches warrant attention. Japan, with the most flexible wages, has not suffered increases in unemployment to as great an extent and has been able to reduce inflation at relatively low cost. The thesis that institutional changes to compensation arrangements should be made in order to increase wage and price flexibility is likely to continue to receive attention in research and policy debates.

QUESTIONS

1. Discuss the theoretical rationale, if any, for the use of the unemployment rate as a determinant of the rate of increase in aggregate money wages.

2. Distinguish between the wage inflation-unemployment trade-off curve and the price inflation-unemployment trade-off curve.

3. In the econometric estimation of aggregate money wage equations, the unemployment rate is often entered in reciprocal form (i.e., 1/U), with an expected positive sign, to capture the nonlinear relationship whereby the Phillips curve is convex to the origin. Illustrate why this is the case, and why a positive sign would reflect a negative relationship between $\dot{W}$ and U.

4. The original Phillips curve literature was often criticized for simply reflecting an unstable empirical relationship without any theoretical rationale for the relationship. Consequently, the policy implications were often wrong, and when the underlying causal determinants changed, the old empirical relationship no longer predicted well. This illustrates the importance of theory for accurate prediction and for policy prescriptions. Discuss.

5. Trace out the wage and price inflation-unemployment curves, both in the short run and the long run, from the adjustment process portrayed in Figure 26.2. What if the asking wage of labour increased by more than the inflation rate of a given year, perhaps because of a catch-up process or because of an anticipation of even higher inflation in the future? Illustrate this in Figure 26.2, and illustrate the new point in the inflation-unemployment trade-offs. What if this were prevented by a wage-price control program that restricted wage increases to the price inflation of that year? (Illustrate on the diagram and on the trade-off curves). Could this provide a rationale for a controls program to move the economy towards its long-run equilibrium without experiencing a phase of high unemployment necessary to moderate wage demands? Does it matter if the original wage demands in excess of actual inflation were based on a catch-up for past inflation, or an expectation of future inflation? What if the controls were not accompanied by the appropriate monetary and fiscal policies to curb inflationary pressures? What if inflation continued in spite of the wage controls, perhaps because the price-markup over labour costs were increased?

6. Illustrate how the adjustment process of Figure 26.2 could arise from a job search process whereby job seekers face a distribution of money wage offers and have an acceptance or reservation wage in money terms. That is, they will continue to remain unemployed and search (sample the distribution of money wage offers) until they receive their reservation wage. Indicate how an increase in aggregate demand and its accompanying increase in the aggregate price level may reduce search unemployment in the short run, but may not reduce it in the long run when the reservation wage adjusts to the inflation.

7. "Any microeconomic theory of behaviour that requires wage rigidities due to such things as unanticipated inflation, wage lags or money illusion in the collective bargaining or job search process, could only explain phenomena in the short run, not in the long run." Discuss.

8. Illustrate the long-run relationship between price inflation and unemployment for different values of λ in Equations 26.9 and 26.14; for example, $\lambda = 0.5$, $\lambda = 0.8$, $\lambda = 0.95$. What happens as λ approaches unity?

9. "If price inflation is 6 percent and productivity growth 2 percent, and if money wages increase by 8 percent, then labour receives *all* of the productivity increase and nothing is left over for other factors of production." Discuss.

10. Discuss the extent to which noncompetitive forces such as monopolies and powerful unions could lead to wage and price inflation.

11. Assume that the following short-run *hypothetical* Phillips curve is estimated econometrically:

$$\dot{W} = 2.5 + .9\,\dot{p}^e - .25U$$

where $\dot{W}$ is the annual rate of change in money wages, $\dot{p}^e$ is the annual expected inflation rate, and U is the unemployment rate. All variables are expressed in percentage terms and their averages over the sample period are 10 for $\dot{W}$, 10 for $\dot{p}^e$, and 6 for U. The estimated coefficients are all significantly different from zero, and the coefficient of $\dot{p}^e$ is not significantly different from one, according to conventional significance tests.

(a) What is the rate of change of money wages that would result if expected inflation were 10 percent and the unemployment rate 6 percent?

(b) What would happen if the unemployment rate were raised to 10 percent?

(c) What would happen if expected inflation were moderated to 5 percent?

(d) What does the coefficient for the inflation-expectations variable imply about the formation of expectations?

(e) What does the coefficient of the unemployment rate variable imply about the shape of the Phillips curve?

(f) Plot a Phillips curve assuming expectations of inflation of 5 percent. Plot a Phillips curve assuming expectations of inflation of 10 percent. What happens to the Phillips curve if expectations of inflation increase from 5 to 10 percent?

(g) Assume that the short-run aggregate price equation was estimated as $\dot{p} = 2 - .25U + 1.1\dot{W}$ and that, in the long run, actual inflation equals expected inflation: that is, $\dot{p} = \dot{p}^e$. Utilize the short-run Phillips curve and aggregate price equations, and the long-run equilibrium condition to solve for the long-run Phillips curve. Compare the shape of the long-run and short-run Phillips curves. Solve for the long-run or natural rate of unemployment as a function of money wages.

REFERENCES AND FURTHER READINGS

A. Phillips Curve and the Wage-Price-Unemployment Relationship

Alkerlof, G. Relative wages and the rate of inflation. *QJE* 83 (August 1969) 353–374.

Allen, R.C. The impact of technical change on employment, wages and the distribution of skills: a historical perspective. *Adapting to Change: Labour Market Adjustment in Canada*, W.C. Riddell (ed.) Toronto: University of Toronto Press, 1986, 77–110.

Alogoskoufis, G.S. and A. Manning. Wage setting and unemployment persistence in Europe, Japan, and the USA. *EER* 32 (1988) 698–706.

Archibald, G.C. The Phillips curve and the distribution of unemployment. *AER* 59 (May 1969) 124–134.

Ashenfelter, O. Union relative wage effects: new evidence and a survey of their implications for wage inflation. *Econometric Contributions to Public Policy*, R. Stone and W. Peterson (eds.). New York: St. Martin's Press, 1978, 31–60.

Ashenfelter, O., G. Johnson and J. Pencavel. Trade unions and the rate of change of money wages in United States manufacturing industry. *R.E. Studies* 39 (January, 1972) 27–54.

Ashenfelter, O., and J. Pencavel. Wage changes and the frequency of wage settlements. *Economica* 42 (May 1975).

Ball, L. Insiders and outsiders: a review essay. *Journal of Monetary Economics* 26 (1990) 459–469.

Bills, M. Wage and employment patterns in long-term contracts when labor is quasi-fixed. *NBER* Macroeconomics Annual (1990) 187–235.

Black, S. and H. Kelejian. The formulation of the dependent variable in the wage equation. *R.E. Studies* 39 (January 1972) 55–59.

Blank, R.M. Why are wages cyclical in the 1970s? *JOLE* 8 (January 1990) 16–47.

Blanchard, O. Wage bargaining and unemployment persistence. *Journal of Money, Credit, and Banking* 23 (August 1991) 277–292.

Blanchard, O. and L.H. Summers. Hysteresis and the European unemployment problem. *NBER Macroeconomic Annual* (1986) 15–78.

Blanchard, O. Hysteresis in unemployment. *EER* 31 (1987) 288–295.

Blanchard, O. and L. Summers. Beyond the natural rate hypothesis. *AER* 78 (1988) 182–187.

Bodkin, R., E. Bond, G. Reuber and T. Robinson. *Price Stability and High Employment*. Special Study No. 5. Ottawa: Economic Council of Canada, 1966.

Bruno, M. and J. Sachs. *Economics of Worldwide Stagflation*. Cambridge, MA: Harvard University Press, 1985.

Christofides, L.N. The interaction between indexation, contract duration and non-contingent wage adjustment. *Economica* 57 (August 1990) 395–410.

Christofides, L., R. Swidinsky and D. Wilton. A microeconometric analysis of the Canadian wage determination process. *Economica* 47 (May 1980a) 165–178.

Christofides, L., R. Swidinsky and D. Wilton. A microeconometric analysis of spillovers within the Canadian wage determination process. *R.E. Stats.* 62 (May 1980b) 213–221.

Christofides, L. and D. Wilton. Wage determination in the aftermath of controls. *Economica* 52 (February 1985) 51–64.

Cousineau, J.M. and R. Lacroix. *Wage Determination in Major Collective Agreements in the Private and Public Sectors.* Ottawa: Economic Council of Canada, 1977.

Crafts, N.F.R. Long-term unemployment and the wage equation in Britain, 1925–1939. *Economica* 56 (May 1989) 247–254.

Cross, R. (ed.). *Unemployment, Hysteresis and the Natural Rate Hypothesis*, Oxford & New York: Basil Blackwell, 1988.

Davis, S.J. Gross job creation and destruction: microeconomic evidence and macroeconomic implications. *NBER Macroeconomics Annual* (1990) 123–186.

Dicks-Mireaux, L. and J. Dow. The determinants of wage inflation: the United Kingdom, 1946–1956. *The Journal of the Royal Statistical Society* Series A, 22 (1959).

Drazen, A. Cyclical determinants of the natural level of economic activity. *IER* 26 (1985) 387–397.

Dreze, J.H. *Underemployment Equilibria: Essays in Theory, Econometrics and Policy*, Cambridge: Cambridge University Press, 1991.

Dreze, J.H. and C.R. Bean. *Europe's Unemployment Problem*, Cambridge, MA: MIT Press, 1990.

Durlauf, S.N. Multiple equilibria and persistence in aggregate fluctuations. *AER* 81 (1991) 70–74.

Ehrenberg, R.G., Danziger, L, and G. San. Cost-of-living adjustment clauses in union contracts: a summary. *JOLE* 1 (July 1983) 215–245.

Ellwood, D. Teenage unemployment: temporary scar or permanent blemish? *The Youth Labor Market: Its Nature, Causes and Consequences*, R. Freeman and D. Wise (eds.). Chicago: University of Chicago Press, 1982.

Frank, J. Monopolistic competition, risk aversion, and equilibrium recessions. *QJE* 105 (1990) 921–938.

Fortin, P. How "Natural" is Canada's high unemployment rate? *EER* 33 (1989) 89–110.

Fortin, P. and K. Newton. Labour market tightness and wage inflation in Canada. *Workers, Jobs and Inflation*, N.M. Baily (ed.). Washington: Brookings Institution, 1982, 243–278.

Friedman, M. The role of monetary policy. *AER* 58 (March 1968) 1–17.

Friedman, M. Nobel lecture: Inflation and unemployment. *JPE* 85 (June 1977) 451–472.

Gordon, R. Wage-price controls and the shifting Phillips curve. *BPEA* (No. 2, 1971) 385–421.

Grubb, D. Topics in the OECD Phillips curve. *EJ* 96 (March 1986) 55–79.

Hall, R. The process of inflation in the labour market. *BPEA* (No. 2, 1974) 343–410.

Hamermesh, D. Wage bargains, threshold effects and the Phillips curve. *QJE* 84 (August 1970) 501–517. Comment by A. Gustman and reply, 86 (May 1972) 332–341.

Howitt, P. Business cycles and search with costly recruiting. *QJE* 103 (1988) 147–165.

Howitt, P. and R.P. McAfee. Costly search and recruiting. *IER* 28 (1987) 89–107.

Jackman, R. and R. Layard. Does long-term unemployment reduce a person's chance of a job? A time series test. *Economica* 58 (February 1991) 93–106.

Jones, S.R.G. *Hysteresis in Canadian Labour Markets.* Mimeo, McMaster University, 1992.

Jovanovic, B. Observable implications of models with multiple equilibria. *Econometrica* 57 (1989) 1431–1438.

Juhn, C., K.M. Murphy, and R.H. Topel. Why has the natural rate of unemployment increased over time? *BPEA* (1991) 75–126.

Kaliski, S.F. The relation between unemployment and the rate of change of money wages in Canada. *IER* 5 (January 1964) 1–33.

Kaliski, S.F. and D.C. Smith. Inflation, unemployment and incomes policy. *CJE* 6 (November 1973) 574–591.

Kotowitz, J. *The Effect of Direct Taxes on Wages.* Anti-Inflation Board Report. Ottawa: Supply and Services, 1979.

Laidler, D. and D. Purdy. *Inflation and Labour Markets.* Toronto: University of Toronto Press, 1974.

Layard, R., S. Nickell and R. Jackman. *Unemployment: Macroeconomic Performance and the Labour Market.* Oxford: Oxford University Press, 1991.

Levi, M. and J. Makin. Inflation uncertainty and the Phillips curve: some empirical evidence. *AER* 70 (December 1980) 1022–1027.

Lipsey, R. The relationship between unemployment and the rate of change of money wage rates in the United Kingdom, 1862–1957: a further analysis. *Economica* 27 (February 1960) 1–31.

Lucas, R. and L. Rapping. Price expectations and the Phillips curve. *AER* 59 (June 1969) 342–350.

Lucas, R. and L. Rapping. Real wages, employment and inflation. *JPE* 77 (September/October 1969) 721–754.

Matsukawa, S. The equilibrium distribution of wage settlements and economic stability. *IER* 27 (June 1986) 415–438.

McCallum, J. Unemployment in Canada and the United States. *CJE* 20 (1987) 802–822.

Mortensen, D. Job search, the duration of unemployment and the Phillips curve. *AER* 60 (December 1970) 847–862. Comment by P. Gayer and R. Goldfarb and reply 62 (September 1972) 714–719.

Mortensen, D. The persistence and indeterminacy of unemployment in search equilibrium. *Scandinavian Journal of Economics* 91 (1989) 347–360.

Perry, G. Determinants of wage inflation around the world. *BPEA* (No. 2, 1975) 403–448.

Phaneuf, L. Wage contracts and the unit root hypothesis. *CJE* 23 (August 1990) 580–592.

Phelps, E. Phillips curves, expectations of inflation and optimal unemployment over time. *Economica* 34 (August 1967) 254–281.

Phelps, E. Money-wage dynamics and labor-market equilibrium. *JPE* 76 (July/August 1968) 678–711.

Phillips, A. The relation between unemployment and the rate of change of money wage rates in the United Kingdom, 1861–1957. *Economica* 25 (November 1958) 283–299. Comment by G. Routh 36 (1959) 299–315.

Pissarides, C.A. Trade unions and the efficiency of the natural rate of unemployment. *JOLE* 4 (October 1986) 582–595.

Pissarides, C.A. Unemployment and macroeconomics. *Economica* 56 (February 1989) 1–14.

Prescott, D. and D. Wilton. The determinants of wage changes in indexed and non-indexed contracts: a switching model. *JOLE* 10 (July 1992) 331–55.

Poloz, S.S. and G. Wilkinson. Is hysteresis a characteristic of the Canadian labour market? A tale of two studies. Bank of Canada, 1992.

Reid, F. Unemployment and inflation: an assessment of Canadian macroeconomic policy. *CPP* 6 (Spring 1980) 283–299.

Reid, F. and N. Meltz. Causes of shifts in the unemployment-vacancy relationship. *R.E. Stats.* 61 (August 1979) 470–475.

Rees, A. The Phillips curve as a menu for policy choice. *Economica* 37 (August 1970) 227–238.

Riddell, W.C. The empirical foundations of the Phillips curve: evidence from Canadian wage contract data. *Econometrica* 47 (January 1979) 1–24.

Riddell, W.C. and P.M. Smith, Expected inflation and wage changes in Canada, 1967–1981. *CJE* 15 (August 1982) 377–394.

Rowley, J. and D. Wilton. Quarterly models of wage determination: some new efficient estimates. *AER* 63 (June 1973) 380–389.

Rowley, J. and D. Wilton, Empirical foundations for the Canadian Phillips curve. *CJE* 7 (May 1974) 240–259.

Rowley, J. and D. Wilton. *The Determination of Wage Change Relationships*. Ottawa: Economic Council of Canada, 1977.

Setterfield, M.A., D.V. Gordon and L. Osberg. Searching for a Will O'the Wisp: an empirical study of the NAIRU in Canada. Dalhousie University, Department of Economics, Working Paper No. 90-04, 1990.

Sparks, G. and D. Wilton. Determinants of negotiated wage increases. *Econometrica* 39 (September 1971) 739–750.

Summers, L.H. *Understanding Unemployment*. Cambridge, MA: The MIT Press, 1990.

Swidinsky, R. Trade unions and the rate of change of money wages in Canada, 1953–1970. *ILRR* 25 (April 1972) 363–375.

Taylor, J. Staggered wage setting in a macro model. *AER* 69 (May 1979) 108–113.

Turnovsky, S. The expectations hypothesis and aggregate wage equation: empirical evidence for Canada. *Economica* 39 (February 1972) 1–17.

Vanderkamp, J. Wage and price level determination: an empirical model for Canada. *Economica* 33 (May 1966) 194–218.

Vanderkamp, J. Wage adjustment, productivity and price change expectations. *R.E. Studies* 39 (January 1972) 61–72.

Weitzman, M.L. *The Share Economy: Conquering Stagflation.* Cambridge, Mass: Harvard University Press, 1984.

Zaidi, M. The determination of money wage rate changes and unemployment-inflation "trade-offs" in Canada. *IER.* 10 (June 1969) 207–219.

B. Anti-inflation Policy

Ashenfelter, O. The withering away of a full employment goal. *CPP* 9 (March 1983) 114–125.

Ashenfelter, O. and R. Layard. Incomes policy and wage differentials. *Economica* 50 (May 1983) 127,196144.

Auld, D., L. Christofides, R. Swidinsky and D. Wilton. The impact of the Anti-Inflation Board on negotiated wage settlements. *CJE* 12 (May 1979) 195–213.

Baily, M. Contract theory and the moderation of inflation by recession and by controls. *BPEA* (No. 3, 1976) 585–634.

Barber, C.L. and J.C.P. McCallum. *Controlling Inflation: Learning from Experience in Canada, Europe and Japan.* Toronto: James Lorimer, 1982.

Brunner, K. and A. Meltzer (eds.). *The Economics of Price and Wage Controls.* New York: North Holland, 1976.

Bruno, M. and J. Sachs. *The Economics of Worldwide Stagflation.* Cambridge, Mass: Harvard University Press, 1984.

Cecchetti, S.G. Indexation and incomes policy: a study of wage adjustment in unionized manufacturing. *JOLE* 5 (July 1987) 391–412.

Cousineau, J.M. and R. Lacroix. L'impact de la politique Canadienne de contrôle des prix et des revenues sur les ententes salariales. *CPP* 4 (Winter 1978) 88–100.

Cozier, B. and G. Wilkinson. Some evidence on hysteresis and the costs of disinflation in Canada. Bank of Canada, Technical Report No. 55, 1991.

Flanagan, R.J., D.W. Soskice and L. Ulman. *Unionism, Economic Stabilization and Incomes Policies: The European Experience.* Washington: Brookings Institution, 1983.

Foot, D.K. and D.J. Poirier. Pubic decision making in Canada: the case of the anti-inflation board. *IER* 21 (June 1980) 489–504.

Fortin, P. The Phillips Curve, macroeconomic policy, and the welfare of Canadians. *CJE* 24 (1991) 774–803.

Fung, K.C. Profit-sharing and European unemployment. *EER* 33 (December 1989) 1787–1798.

Gordon, R.J. Wage-price controls and the shifting Phillips curve. *BPEA* (No. 2, 1972) 385–421.

Gordon, R.J. Why U.S. wage and employment behaviour differs from that in Britain and Japan. *EJ* 92 (March 1982) 13–44.

Gregory, R. Wages policy and unemployment in Australia. *Economica* 53 (1986) 53–74.

Hagens, J.B. and R.R. Russell. Testing for the effectiveness of wage-price controls: an application to the Carter program. *AER* 75 (March 1985) 191–207.

Howitt, P. Indexation and the adjustment to inflation in Canada. *Postwar Macroeconomic Developments*, J. Sargent (ed.). Toronto: University of Toronto Press, 1986, 175–224.

Kahn, G.A. International differences in wage behavior: real, nominal, or exaggerated? *AER* 74 (May 1984) 155–159.

Kaliski, S. and D. Smith. Inflation, unemployment and incomes policy. *CJE* 6 (November 1973) 574–591.

Layard, R. Is incomes policy the answer to unemployment? *Economica* 49 (August 1982) 219–239.

Lerner, A.P. A wage increase permit plan to stop inflation. *BPEA* 2 (1978) 491–505.

Lerner, A.P. and D. Colander. *MAP: A Market Anti-Inflation Plan.* New York: Harcourt Brace Jovanovich, 1980.

Lipsey, R.G. (ed.) *Zero Inflation: The Goal of Price Stability.* Toronto: C.D. Howe Institute, 1990.

Lipsey, R. and J. Parkin. Incomes policy: a reappraisal. *Economica* 37 (May 1970) 115–138.

Lipsey, R.G. The understanding and control of inflation: is there a crisis in macroeconomics? *CJE* 14 (November 1981) 545–576.

Lucas, R.F. The Bank of Canada and zero inflation: a new cross of gold. *CPP* 15 (March 1989) 84–93.

Maslove, A.M. and G. Swimmer. *Wage Controls in Canada 1975–1978.* Montreal: Institute for Research on Public Policy, 1980.

Pencavel, J. The effects of income policies on the frequency and size of wage changes. *Economica* 45 (May 1982) 147–160.

Reid, F. Canadian wage and price controls. *CPP* 2 (Winter 1976) 104–113.

Reid, F. The effect of controls on the rate of wage change in Canada. *CJE* 12 (May 1979) 214–227.

Reid, F. Control and decontrol of wages in the United States: an empirical analysis. *AER* 71 (March 1981) 108–120.

Riddell, W.C. The responsiveness of wage settlements in Canada and economic policy. *CPP* 9 (March 1983) 9–23.

Riddell, W.C. *Dealing with Inflation and Unemployment in Canada.* Toronto: University of Toronto Press, 1986.

Sachs, J.D. Wages, profits and macroeconomic adjustment: a comparative study. *BPEA* (No. 2, 1979) 269–319.

Sachs, J.D. Real wages and unemployment in OECD countries. *BPEA* (No. 1, 1983) 255–289.

Scarfe, B.L. Economic fluctuations and stabilization policy in Canada: the state of the art. *CPP* 13 (March 1987) 75–85.

Schelling, T.C. Establishing credibility: strategic considerations. *AER* 72 (May 1982) 77–80.

Seidman, L.S. Tax-based incomes policies. *BPEA* 2 (1978) 301–348.

Seidman, L.S. Equity and tradeoffs in a tax-based incomes policy. *AER* (May 1981) 295–300.

Smith, D. *Incomes Policies: Some Foreign Experiences and their Relevance for Canada.* Special Study No. 4. Ottawa: Economic Council of Canada, 1966.

Taylor, J.B. Aggregate dynamics and staggered contracts. *JPE* 88 (February 1980) 1–23.

Taylor, J.B. Establishing credibility: a rational expectations viewpoint. *AER* 72 (May 1982) 81–85.

Taylor, J.B. Union wage settlements during a disinflation. *AER* 73 (December 1983) 981–993.

Ulman, L. and R. Flanagan. *Wage Restraint: a Study of Incomes Policies in Western Europe.* Berkeley: University of California Press, 1971.

Wadhwani, S. and M. Wall. The effects of profit-sharing on employment, wages, stock returns and productivity: evidence from micro-data. *EJ* 100 (March 1990) 1–17.

Weitzman, M.L. *The Share Economy: Conquering Stagflation.* Cambridge, Mass.: Harvard University Press, 1984.

York, R.C. (ed.). *Taking Aim: The Debate on Zero Inflation.* Toronto: C.D. Howe Institute, 1990.

INDEX

— — — — — — — — — *cut here* — — — — — — — — —

STUDENT REPLY CARD
LABOUR MARKET ECONOMICS, Third Edition

You can help us to develop better textbooks. Please answer the following questions and return this form via Business Reply Mail. Your opinions matter: thank you in advance for sharing them with us!

Name of your college or university: _____

Major program of study: _____

Course title: _____

Were you required to buy this book? _____ yes _____ no

Did you buy this book new or used? _____ new _____ used ($_____)

Do you plan to keep or sell this book? _____ keep _____ sell

Is the order of topic coverage consistent with what was taught in your course?

Are there chapters or sections of this text that were not assigned for your course? Please specify:

— — — — — — — — — · *fold here* — — — — — — — —

Were there topics covered in your course that are not included in the text? Please specify:

What did you like most about this text?

What did you like least?

Please add any comments or suggestions:

- - - - - - - - - - - - - - - *cut here* - - - - - - - - - - - - - - - - -

- - - - - - - - - - - - - - - - *fold here* - - - - - - - - - - - - - - - - -

Postage will be paid by

MAIL ➤ POSTE

Canada Post Corporation / Société canadienne des postes

Postage paid	**Port payé**
If mailed in Canada	si posté au Canada
Business Reply	**Réponse d'affaires**
0183560299	01

0183560299-L1N9B6-BR01

Attn.: Sponsoring Editor
College Division

MCGRAW-HILL RYERSON LIMITED
300 WATER ST
WHITBY ON   L1N 9Z9

cut here

tape shut